HRW

ALGEBRA

explore

communicate

APPLY

Integrating

Mathematics
Technology
Explorations
Applications
Assessment

HOLT, RINEHART AND WINSTON
Harcourt Brace & Company
Austin • New York • Orlando • Atlanta • San Francisco • Boston • Dallas • Toronto • London

AUTHORS

James E. Schultz

Dr. Schultz is one of the math education community's most renowned mathematics educators and authors. He is especially well regarded for his inventive and skillful integration of mathematics and technology. He helped establish standards for mathematics instruction as a co-author of the NCTM "Curriculum and Evaluation Standards for Mathematics" and "A Core Curriculum, Making Mathematics Count for Everyone." Following over 25 years of successful experience teaching at the high school and college levels, his dynamic vision recently earned him the prestigious Robert L. Morton Mathematics Education Professorship at Ohio University.

Kathleen A. Hollowell

Dr. Hollowell is widely respected for her keen understanding of what takes place in the mathematics classroom. Her impressive credentials feature extensive experience as a high school mathematics and computer science teacher, making her particularly well-versed in the special challenges associated with integrating math and technology. She currently serves as Associate Director of the Secondary Mathematics Inservice Program, Department of Mathematical Sciences, University of Delaware and is a past-president of the Delaware Council of Teachers of Mathematics.

Wade Ellis, Jr.

Professor Ellis has gained tremendous recognition for his reform-minded and visionary math publications. He has made invaluable contributions to teacher inservice training through a continual stream of hands-on workshops, practical tutorials, instructional videotapes, and a host of other insightful presentations, many focusing on how technology should be implemented in the classroom. He has been a member of the National Research Council's Mathematical Sciences Education Board, the MAA Committee on the Mathematical Education of teachers, and is a former Visiting Professor of Mathematics at West Point.

Printed in the United States of America
  2 3 4 5 6 7   041   00 99 98 97 96
ISBN: 0-03-097771-1

CONTRIBUTING AUTHOR

Irene "Sam" Jovell   An award winning teacher at Niskayuna High School, Niskayuna, New York, Ms. Jovell served on the writing team for the New York State Mathematics, Science, and Technology Framework. A popular presenter at state and national conferences, her workshops focus on technology-based innovative education.

PROGRAM CONCEPTUALIZER

Larry Hatfield   Dr. Hatfield is Department Head and Professor of Mathematics Education at the University of Georgia. He is a recipient of the Josiah T. Meigs Award for Excellence in Teaching, his university's highest recognition for teaching. He has served at the National Science Foundation and is Director of the NSF-funded Project LITMUS.

● ●

Editorial Director of Math
Richard Monnard

Executive Editor
Gary Standafer

Senior Editor
Ronald Huenerfauth

Project Editors
Stephen C. Johnson
Paula Jenniges

Design and Photo
Pun Nio
Diane Motz
Julie Ray
Rhonda Holcomb
Lori Male
Anne Wright
Linda Wilbourn
Katie Kwun
Mavournea Hay
Cindy Verheyden
Sam Dudgeon
Victoria Smith

Editorial Staff
Steve Oelenberger
Andrew Roberts
Joel Riemer
Ann Farrar
Editorial Permissions
Jane Gallion,
Desktop Systems Operator
Jill Lawson,
Department Secretary

Production
Donna Lewis
Amber Martin
Shirley Cantrell
Jenine Street
Laura Cuellar

CONTENT CONSULTANT

Lloy Lizcano Dr. Lizcano is a Mathematics Teacher and Department Chair at Bowie High School, Austin, Texas. She is also a Lecturer at the University of Texas at Austin, teaching Secondary Mathematics Methods and supervising student teachers. Her many areas of expertise include the integration of science and technology into the mathematics curriculum.

MULTICULTURAL CONSULTANT

Beatrice Lumpkin A former high school teacher and associate professor of mathematics at Malcolm X College in Chicago, Professor Lumpkin is a consultant for many public schools for the enrichment of mathematics education through its multicultural connections. She served as a principal teacher-writer for the *Chicago Public Schools Algebra Framework* and has served as a contributing author to many other mathematics and science publications that include multicultural curriculum.

REVIEWERS

Toni Antonelli
St. Ignatius High School
Cleveland, Ohio

Judy B. Basara
St. Hubert High School
Philidelphia, Pennsylvania

Robert E. Bauman
Appleton East High School
Appleton, Wisconsin

Blanche Brownley
DC Public Schools
Washington, D.C.

Larry Clifford
Patterson Career Center
Dayton, Ohio

Carol Goehring
Dr. Phillips High School
Orlando, Florida

Michael Moloney
Snowden International High School
Boston, Massachusetts

Lisa Naxera
Westwood High School
Austin, Texas

Curt Perry
Plymouth-Salem High School
Canton, Michigan

Diane Young
Covington Middle School
Austin, Texas

TABLE OF CONTENTS

MATH *Connections*

Coordinate Geometry
80, 85
Geometry   11, 13,
14, 15, 20, 33,
49, 64, 68
Statistics   27, 33, 40,
44, 78

APPLICATIONS

Science
Ecology 24
Physics 18, 20, 66, 69, 84, 94
Physiology 37
Science 63, 94

Social Studies
Demographics 42, 59, 85
Geography 44
Music History 46
Psychological Experiment 83

Language Arts
Communicate 12, 19, 26, 32,

38, 43, 48, 62, 67, 74, 81, 87
Eyewitness Math 76

Business and Economics
Advertising 33
Cost of Living 43
Economics 49
Finance 60, 62, 63, 94
Fund-raising 27, 75
Investments 89

Life Skills
Consumer Economics 27, 89
Consumer Math 68, 82

Health 39
Home Economics 21, 82
Home Improvement 32

Sports and Leisure
Entertainment 27, 54
Music 74
Sports 45, 88

Other
Transportation 75, 89

MATH
Connections

Geometry 106,
115, 116, 117,
127, 129, 130,
142, 156, 161,
164, 166, 172,
179, 184, 191,
194, 202
Maximum/Minimum
193
Number Theory
184
Probability 179
Statistics 122, 123,
126, 129, 148,
150, 179, 181,
184, 202

APPLICATIONS

Science
Physics 105, 130, 163
Temperature 102, 110, 136

Social Studies
Demographics 179
Government 198

Language Arts
Communicate 105, 110, 115,
121, 128, 135, 149, 155, 164,
171, 178, 184, 189, 197
Eyewitness Math 158

Business and Economics
Banking 150, 167

Business 106
Clothing Manufacturing 116
Discounts 176, 179
Finance 142
Fund-raising 122, 125, 172
Inventory 123
Investment 127, 168
Quality Control 192
Sales Tax 129, 177
Small Business 157
Wages 156

Life Skills
Carpentry 160, 165

Consumer Economics 172,
175, 185, 191
Health 197
Home Economics 129
Savings 171
Woodworking 195

Sports and Leisure
Ballet 189, 190
Hobbies 142
Picture Framing 116
Sports 102, 129, 136, 162,
171, 172, 202
Travel 129, 165, 185

MATH
Connections

Coordinate Geometry
244, 261, 284
Geometry 213, 214,
232, 245, 247,
248, 250, 251,
270, 271, 276,
279, 282, 283,
289, 291, 302
Maximum/Minimum
288
Number Theory 283
Numeration 270,
294, 297
Statistics 211, 266

APPLICATIONS

Science
Chemistry 296, 302
Ecology 222
Physical Science 222, 232, 246
Physics 229
Science 209

Social Studies
Demographics 283
Social Studies 215, 239

Language Arts
Communicate 212, 221, 230, 237, 243, 250, 264, 270, 275, 281, 289, 296

Eyewitness Math 216
Language Arts 244

Business and Economics
Accounting 276
Equipment Rental 257
Finance 276, 302
Fund-raising 234, 238, 239, 271, 291
Income 276
Investment 268, 276
Money 293

Life Skills
Budget 244
Budgeting Time 291

Career Options 283
Consumer Mathematics 231

Sports and Leisure
Auto Racing 226
Recreation 213, 240, 245, 271, 296
Sports 214, 226, 266, 291
Travel 210, 256, 277, 294

Other
Age 295
Aviation 265, 270, 297
Mixture 292

MATH Connections

Coordinate Geometry
336, 343, 365,
374, 387
Geometry 321, 330,
335, 342, 349,
370, 381, 384,
386, 387, 392,
394
Statistics 319, 321,
359, 378, 379,
380

APPLICATIONS

Science
Chemistry 343, 387
Meteorology 362, 372, 374
Science 336, 392
Space Exploration 313
Weather 364

Social Studies
Demographics 365
Geography 313, 392

Language Arts
Communicate 312, 318,
328, 334, 341, 357, 364,
371, 379, 386, 391, 397
Eyewitness Math 322, 366
Language 387, 392

Business and Economics
Advertising 370, 373

Business 312, 348, 349
Fund-raising 325, 330
Inventory Control 316, 319,
324, 342, 384
Manufacturing 329
Marketing 404
Pricing 329
Sales Tax 329
Stocks 348
Work Schedule 336

Life Skills
Appointments 363, 394
Consumer Economics 337
Employment 343
Health 393
Job Opportunities 342
Membership Fees 359
Savings 381

Scheduling 382, 387, 399

Sports and Leisure
Contests 399
Games 352, 358, 378, 379,
385, 386, 390, 392, 395,
396, 397, 398
Making Choices 388
Sports 320, 364, 368, 373,
381, 383
Ticket Sales 342
Travel 399

Other
Logic 376
Opinion Polls 380
Route Planning 398
Survey 391
Transportation 308, 313, 392

MATH *Connections*

Coordinate Geometry 462, 482
Geometry 420, 423, 435, 442, 452, 462, 467, 500
Maximum/Minimum 418
Number Theory 473
Probability 417, 430, 447, 462, 472, 491, 494, 500
Statistics 423, 430, 435, 439, 442, 462, 468, 482
Transformations 473

APPLICATIONS

Science
Astronomy 476, 481, 500
Bacteriology 459
Biology 461
Carbon Dating 494
Physics 435, 478
Science 417, 481, 489
Space Travel 477
Temperature 452

Social Studies
Demographics 488, 492, 494, 501

Social Studies 480, 483, 485
Taxes 501

Language Arts
Communicate 415, 421, 428, 433, 440, 446, 460, 466, 471, 480, 487, 493
Eyewitness Math 474

Business and Economics
Business 452, 453
Economics 481
Finance 490

Packaging 447

Life Skills
Investment 491, 494, 501

Sports and Leisure
Art 429
Travel 443, 468

Other
Databases 416
Time 500

MATH *Connections*
Coordinate Geometry
  518, 555
Geometry 509, 512,
  515, 517, 518,
  523, 526, 528,
  529, 533, 534,
  535, 544, 557,
  563, 569, 572,
  580
Maximum/Minimum
  549, 560, 581
Probability 522, 528,
  534, 563, 580
Statistics 534, 573

APPLICATIONS

Science
Biology 522
Optics 584
Physics 528, 551, 557, 562, 563, 573, 585

Language Arts
Communicate 510, 517, 522, 527, 532, 538, 550, 556, 561, 571, 579, 584
Eyewitness Math 564

Business and Economics
Accounting 580
Advertising 518
Discount Sales 518

Sales Tax 539
Small Business 521

Sports and Leisure
Crafts 544
Photography 572
Sports 590

MATH
Connections
Coordinate Geometry
626, 627, 628, 632
Geometry 601, 610,
622, 623, 629,
637, 643, 662,
663, 667, 668,
686, 688, 693, 698
Probability 669, 675
Statistics 693

APPLICATIONS

Science
Biology 668
Ecology 658, 674
Physics 611, 617, 618, 648, 660, 662

Social Studies
Politics 683

Language Arts
Communicate 601, 609, 616, 622, 629, 636, 641, 656, 662, 667, 673, 682, 687, 692

Eyewitness Math 676

Business and Economics
Investment 661, 663

Life Skills
Auto Painting 683
Baking 688
Decorating 688
Fitness 698
House Construction 642
Landscaping 623
Savings 669
Surveying 634

Sports and Leisure
Boating Navigation 621
Model Airplanes 642
Music 663
Recreation 683
Sports 617, 623, 648, 678, 686
Travel 688, 698

Other
Emergency Service 625
Highways 642
Rescue Services 630
Transportation 656, 663, 688

CHAPTER 1

From Patterns to Algebra

Throughout history, people from all parts of the world have been fascinated with patterns. The early native tribes of the American Southwest were aware of the patterns that influenced their lives. They recorded these patterns by weaving them into rugs.

In the technological world of modern society, patterns provide a basis for discoveries in science and engineering. Scientists use patterns to study, understand, and predict nature. Mathematicians look for regular patterns when investigating systems of numbers. Patterns can provide a powerful tool for solving problems.

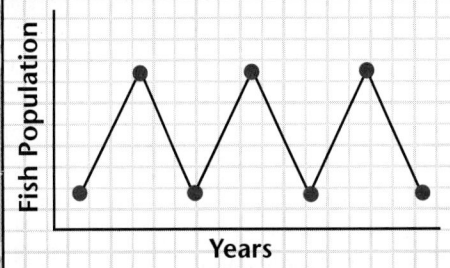

Yearly Changes in Fish Population

Fish Population

Years

In biology, a mathematical pattern can describe the changes in a region's fish population. The graph has a striking resemblance to the pattern shown on the Native American blanket.

Patterns are used in the structure and design of modern cities and the transportation networks that connect them. This student is using a computer to design a trestle bridge.

PORTFOLIO ACTIVITY

1. Construct the fourth and fifth dot patterns for this sequence of peaks. The fourth dot pattern should use 25 dots.

1   5   13

2. How many dots are in the pattern with 7 peaks?

Mathematical Power

Why *As you explore and investigate ideas from Algebra you will see how the power of mathematics is used to solve problems that occur in the everyday world. Put this power to work, and you will be successful in solving many problems.*

As you explore patterns and examine data in tables and graphs, you will learn to solve real problems for practical reasons and sometimes just for fun. Applying mathematics to the real world means interacting with science, economics, statistics, business, sports, entertainment, and music. As you explore mathematical concepts in the context of real world applications, math will seem more relevant and will answer the question, "Why do I have to learn this?"

As you explore, you will have the opportunity to see how technology can help you understand mathematics and process real world data. This technology takes the form of scientific calculators, graphics calculators, interactive dynamic graphics programs, and spreadsheets.

Algebra in Marketing

Statistics and probability are used to determine buying patterns, project future sales, determine packaging size and style, and create new products.

Algebra in Science

Algebra in Medicine

A person's heart rate is a function of many physiological factors such as age, diet, and heredity. A medical professional must understand these relationships to perform his or her job effectively.

Mathematics has been referred to as the language of science. Whether it be in the field of biology, chemistry, physics, earth science, or ecology, scientists cannot investigate or represent their findings without understanding and using mathematics.

Advertising Spending

Category	Total ad spending (in billions)
Automotive	$5.9
Food	$3.5
Entertainment	$3.1
Travel & Hotels	$2.2
Snacks	$1.2

Algebra in Mass Media

Today's information comes to us in many forms—newspapers, television, computer networks. Data is collected and stored in microseconds, making it possible to understand and predict in ways we never thought possible only a few years ago. As you explore mathematical concepts and relationships, you will learn how mathematics helps you make sense of real-world data.

Exploring with Algebra

As you explore, you may use various tools such as algebra tiles, integer squares, calculators, and other manipulative devices. As you use tools, make tables, and draw graphs, you will be given instructions to construct models. These models define and represent mathematical concepts you will use now and in the future. You will build your own mathematical power.

You can begin now . . . .

Exploration Graphing Changes

1 The table shows the distances covered by a mountain-bike racer on a level but rough road. Assuming a constant rate of speed, compute the values missing from the table.

Time	0	0.5	1.0	1.5	2.0	2.5	3.0
Distance	0	10	20	30	?	?	?

2 Write a sentence that tells how to find the distance if you know the time and rate of speed.

3 If the mountain bike-rider maintains the same speed, how far will the rider travel in 5 hours?

4 A graph is another way to show the relationship between time and distance. It can also be used to make predictions. Use the graph to predict how long it will take to cycle 100 miles.

The graph in 4 describes the cyclist's movement traveling at a constant rate. Examine the graphs below.

5 How does each graph describe the changes in the movement of a cyclist during each 5-hour trip?

Representing Number Patterns

Why *Patterns appear everywhere—in the architecture of buildings, in the designs of clothing, and in the advertisement of products. In music, sound patterns influence some people to like classical music and some to like rock and roll. Just as some people are fascinated by the patterns that occur in music, others are fascinated by the patterns formed by shapes and numbers.*

Small portions of the fractal, when magnified, produce images similar to the entire fractal.

Suppose 10 people have just returned to school, and they all want to tell each other about what they did during summer break. Each person has just one conversation with each of the other people. Guess how many conversations will take place.

Using Problem-Solving Strategies

If you are not sure how to solve a problem directly, experiment with various problem-solving strategies. For example, you can draw a picture, think of a simpler problem, make a table, or look for a pattern.

Think of a Simpler Problem Begin with the most basic situation for the problem, then make regular changes to build a pattern. Start with just 1 person. How many conversations can 1 person have if talking to oneself doesn't count? Organize the results, and make a table.

2 people
1 conversation

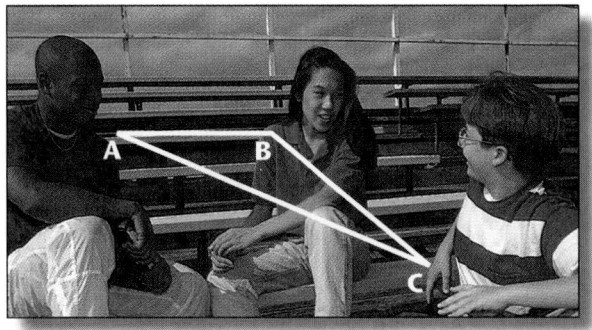

3 people
3 conversations

When you add a fourth person, D, the same three conversations are possible, but there are also three new conversations possible—A with D, B with D, and C with D, for a total of 6.

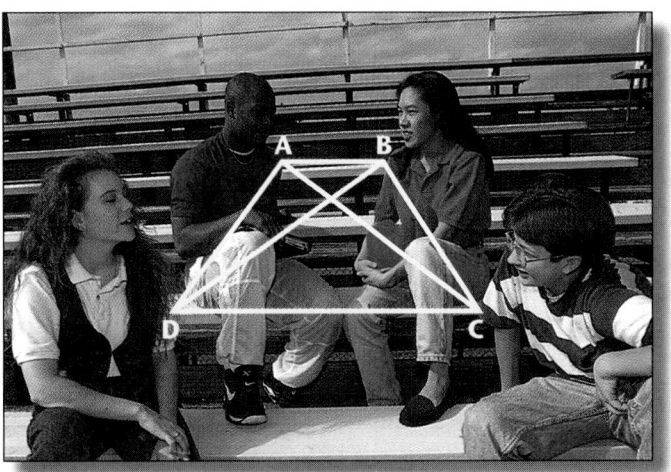

4 people
6 conversations

Look for a Pattern   When a regular number pattern emerges, organize the information in a table. Are there other patterns that might help solve this problem?

People	1	2	3	4	5
Conversations	0	1	3	6	?
Increase		+1	+2	+3	

Once you discover how the pattern in the number of conversations increases, continue the pattern to find the number of conversations possible among 5 people.

Reason logically from the pattern   When the fifth person joins, add 4 new conversations (one with each of the other four people) to the previous 6 conversations. Five people will have **6 + 4**, or **10**, conversations.

People	1	2	3	4	5
Conversations	0	1	3	6	10
Increase		+1	+2	+3	+4

When a sixth person joins, add 5 new conversations to the previous 10, and so on. Continue to build the pattern to 10 people.

People	1	2	3	4	5	6	7	8	9	10
Conversations	0	1	3	6	10	15	21	28	36	45
Increase		+1	+2	+3	+4	+5	+6	+7	+8	+9

Lesson 1.1   Representing Number Patterns   **9**

When mathematicians study number sequences for patterns, they often make *conjectures*. A **conjecture** is a statement about observations that they believe to be true. Mathematicians try to prove the conjecture or find a counterexample to show that the conjecture is not true.

The sum of conversations among any number of students is the sum of the numbers from 1 to one less than the number of students. Test this conjecture for 11, 12, 13, and 14 students.

Which figure completes the pattern?

Pattern

?

Choices

     A   B   C   D

Number Sequences

When a sequence shows a pattern, discover as much as you can about how the numbers relate to each other.

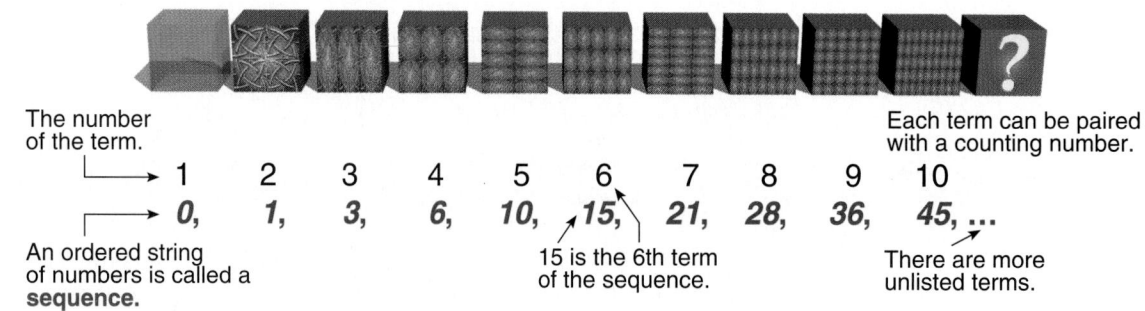

The number of the term.

1	2	3	4	5	6	7	8	9	10
0,	1,	3,	6,	10,	15,	21,	28,	36,	45, ...

Each term can be paired with a counting number.

An ordered string of numbers is called a **sequence.**

15 is the 6th term of the sequence.

There are more unlisted terms.

EXAMPLE 1

Predict the next three terms for each of the following sequences.

A 15, 18, 21, 24, 27, . . .   **B** 2, 6, 12, 20, 30, . . .

Solution▸

A The first sequence begins with 15. Each of the following terms is 3 more than the one before. A calculator can be helpful when you explore sequences. A calculator with a *constant feature* can easily provide more terms of the sequence by repeating an operation.

Calculator

For example, the first sequence can be generated by entering

Continue pressing the *equals key* three more times. The next three terms of the sequence are **30**, **33**, and **36**.

On some *graphics* calculators, enter

15 [ENTER] [+] 3 [ENTER] [ENTER] [ENTER] [ENTER].

Continue pressing the *enter key* three more times to generate **30**, **33**, and **36**.

B

The next three terms are found by adding **12**, **14**, and **16**.

30 + 12 = 42   **42 + 14 = 56**   **56 + 16 = 72** ❖

The sequence in Example 1B appears when counting the number of dots in the pattern below.

The next three terms are **6 · 7 = 42**, **7 · 8 = 56**, and **8 · 9 = 72**.

GEOMETRY
Connection

Why do you think the numbers in the sequence 2, 6, 12, 20, 30, . . . are called **rectangular numbers?**

EXAMPLE 2

Find the sum $1 + 2 + 3 + 4 + 5$ using a geometric pattern.

Solution►

There are many ways to find this sum. This method shows how geometric patterns can be used to find the sum of $1 + 2 + 3 + 4 + 5$.

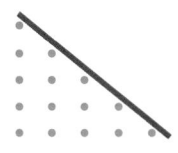

Represent the sum of 1 to 5 by a triangle of dots. Form a rectangle of dots by copying the triangle and rotating it.

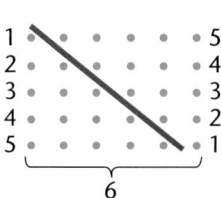

The total number of dots is $5 \cdot 6 = 30$. There are half as many dots in the triangle. $\frac{5 \cdot 6}{2} = 15$.

The sum of the numbers from 1 to 5 is 15. ❖

Spreadsheet

Technology can simplify the exploration of sequences and can provide tools to calculate sums of sequences. Try using a computer *spreadsheet* to add the numbers 1 to 9. A spreadsheet allows you to control lists or tables of data. Familiarize yourself with the layout of this spreadsheet.

Spreadsheet Title

Cell Reference (Shows where the results appear) — C2

Sum of numbers

=SUM (A2:A10) — **Formula Bar**

Column Identifiers

	A	B	C
1	Numbers		Sum
2	1		45
3	2		
4	3		
5	4		
6	5		
7	6		
8	7		
9	8		
10	9		
11			

Active Cell

Row

Cell

Row Identifiers   **Column**   **(Spreadsheet)**

Try This   Find the sum $1 + 2 + 3 + 4 + 5 + 6 + 7 + 8 + 9$ using a geometric pattern. Then use your understanding of the number pattern to find a simple way to calculate the sum $1 + 2 + 3 + 4 + 5 + \cdots + 100$.

EXERCISES & PROBLEMS

Communicate

1. Explain how you would set up a table to solve the following problem. If each of the teams A, B, C, and D play each other, one of the games will be A versus B. Show how you would list all the games.

2. Describe how to find the sum of the numbers from 1 through 20.

Explain how you would find the pattern for the following sequences.

3. 3, 7, 11, 15, 19, 23, . . .   4. 1, 1, 2, 3, 5, 8, 13, . . .

5. Give four problem-solving strategies that are used in Lesson 1. Choose one strategy from your list, and explain how it helps in solving problems.

Practice & Apply

6. If each of the teams A, B, C, D, and E play each other, one of the games will be A versus B. List all of the games, and tell how many there are.

7. In 1993, after Penn State joined the Big Ten athletic conference, the Big Ten had 11 teams. If each of the 11 teams played each of the other teams, how many games would there be? Make a table similar to the conversation problem.

8. If an athletic conference has 12 teams and each of the teams play each of the other teams, how many games will there be?

9. Use the geometric dot-pattern method. Based on the sketch, the red dots represent the numbers being added. After forming the dot rectangle, what is the next step for finding the sum? What is the sum of the numbers being added?

10. Use the geometric dot-pattern method. Based on the sketch, the red dots represent the numbers being added. After forming the dot rectangle, what is the next step for finding the sum? What is the sum of the numbers being added?

11. Use the geometric dot-pattern method to find the sum $1 + 2 + 3 + 4 + 5 + 6 + 7$. Include the sketch.

12. Find the sum $1 + 2 + 3 + 4 + 5 + \cdots + 40$. Think of the sketch without actually drawing it.

13. How can you find $1 + 2 + 3 + 4 + 5 + \cdots + 50$?

Find the next three terms in each sequence.
Then explain the pattern used to find the terms for each.

14. 4, 9, 14, 19, 24, . . .   **15.** 7, 16, 25, 34, 43, . . .   **16.** 9, 19, 29, 39, 49, . . .

17. 2, 4, 8, 16, 32, . . .   **18.** 5, 7, 9, 11, 13, . . .   **19.** 3, 9, 27, 81, 243, . . .

20. 8, 10, 12, 14, 16 . . .   **21.** 16, 8, 4, 2, 1, . . .   **22.** 5, 12, 19, 26, 33, . . .

23. Solve the conversation problem for 10 people by drawing a sketch. Draw 10 dots around a circle, and connect them with segments to represent conversations. How many segments did you draw?

24. **Geometry** Diagonals are segments other than sides that connect the vertices of a polygon. Draw a hexagon and its diagonals. How many diagonals does a regular hexagon have? How does this problem relate to the conversation problem?

Karl Friedrich Gauss (1777–1855) was one of the greatest of all mathematicians.

Cultural Connection: Europe When Gauss was very young, a teacher told him to find the sum of all the whole numbers from 1 to 100, thinking this would keep him busy for a long time. But Gauss very cleverly paired the numbers and observed that $1 + 99 = 100$, $2 + 98 = 100$, $3 + 97 = 100$, and so on.

25. How many pairs could be formed in this way without repeating any numbers?

26. Which whole numbers from 1 to 100 would not appear in any of the pairs?

27. Based on the previous two exercises, what is the sum of the numbers from 1 to 100?

Geometry The numbers in each array are square numbers. The first four square numbers are 1, 4, 9, and 16.

28. Find the tenth square number.

29. Find the twentieth square number.

1   4   9   16

30. Suppose 8 cities are to be linked by phone lines, with 1 phone line between each pair of cities. How many phone lines will there be?

31. Use the geometric dot-pattern method to find the sum $1 + 3 + 5 + 7 + 9$.

Geometry The numbers in each array are triangular numbers. The first four triangular numbers are 1, 3, 6, and 10.

32. Find the tenth triangular number.

33. Find the twentieth triangular number.

1   3   6   10

34. Show visually that the square number 16 is the sum of two triangular numbers. Which two?

35. Show visually that 100 is the sum of two triangular numbers. Which two?

36. Copy and complete the following table.

Number	1	3	5	7	9	11	13	15	17
Sum	1	4	9	16	?	?	?	?	?

37. Examine the pattern in the table in Exercise 36. Guess the sum of the first 100 odd numbers. Explain your method.

38. Copy and complete the following table.

Number	1	2	3	4	5	6	7
Cube of the number	1	8	27	64	?	?	?
Sum of the cubes	1	9	36	100	?	?	?

39. Examine the pattern in the table in Exercise 38. Guess the sum of the first 10 cubes. Explain your method.

Look Back

David and his 3 friends collected 196 aluminum cans.

40. They divided the cans among themselves equally into 4 bags. How many cans are in each bag?

41. The 4 friends take the cans to the recycling center where they can get $0.20 a pound for the cans. The total weight of the cans is 9 pounds. How much money will they get all together?

42. How much will each of the 4 friends get from recycling the cans if they split the money equally?

43. If a teacher collects $2.75 from each of 28 students for the field trip, how much does the teacher collect?

Recycled aluminum cans are compressed before shipping.

44. The citywide concert sold 6702 tickets to students and 3749 tickets to parents. How many tickets were sold all together?

45. **Geometry** What is the area of a square 23 centimeters on a side?

46. What are 3 fractions that are equivalent to $\frac{3}{5}$?

Find the value of the following expressions.

47. $6(9 + 4)$ **48.** $\sqrt{225}$

49. Kim had $4.75 and spent $3.12. How much money does she have left?

Look Beyond

50. The following pattern was first explored in China and Iran and was later called Pascal's Triangle. Extend the pattern by finding rows 6, 7, and 8.

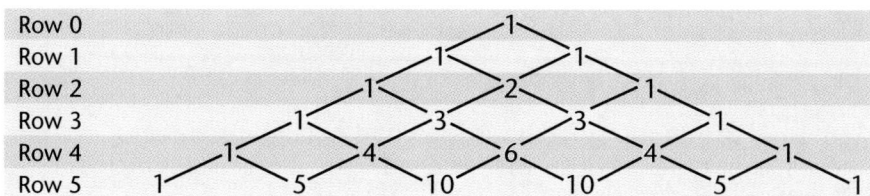

51. Find the sum of the numbers in each row of Pascal's Triangle.

Using Differences to Identify Patterns

A drum barometer keeps a running record of air pressure, one of the many factors important in the prediction of weather patterns.

People are always making predictions to answer questions they have about the future. We use data from the past and regular patterns in the present to make predictions about the future.

Recall from the previous lesson the problem of finding the number of possible conversations among 10 people. Examine the table for 5 people again. Look at the differences between consecutive terms in the sequence of numbers 0, 1, 3, 6, 10, . . .

People	1	2	3	4	5
Conversations	0	1	3	6	10

1   2   3   4   ← First differences

Extend the table. This time find the second differences.

People	1	2	3	4	5
Conversations	0	1	3	6	10

1   2   3   4   ← First differences

1   1   1   ← Second differences constant

You can see from the table that the second differences are *constant*. When the constant differences are greater than 0, the sequence is increasing. When the constant differences are less than 0, the sequence is decreasing.

If you know the differences, you can predict other terms of the sequence.

EXAMPLE 1

Find the next two terms of each sequence.

A 80, 73, 66, 59, 52, . . .   **B** 1, 4, 9, 16, 25, . . .

Solution

A Find the first differences.

80 ⟍ 73 ⟍ 66 ⟍ 59 ⟍ 52
  −7   −7   −7   −7   ⟵ First differences are constant.

Notice that each term decreases by 7. The first difference is represented by − 7. Subtract 7 from the previous term to find each new term.

$$52 - 7 = 45 \qquad 45 - 7 = 38$$

The next two terms are 45 and 38.

B Find the first differences.

1 ⟍ 4 ⟍ 9 ⟍ 16 ⟍ 25
  3   5   7   9   ⟵ First differences are not constant.

Since the first differences are not constant, find the second differences.

1 ⟍ 4 ⟍ 9 ⟍ 16 ⟍ 25
  3   5   7   9
    2   2   2   ⟵ Second differences are constant.

Now work backward to find the first differences. To find each of the next first differences, add 2 to the previous first difference.

$$9 + 2 = 11 \qquad 11 + 2 = 13$$

1 ⟍ 4 ⟍ 9 ⟍ 16 ⟍ 25 ⟍ 11 ⟍ 13   ⟵ First differences
  3   5   7   9   11   13
    2   2   2   2   2   ⟵ Second differences are constant.

Continue working backward to find the next two terms of the sequence. To find each of the next terms, add the first difference to the previous term.

$$25 + 11 = 36 \qquad 36 + 13 = 49$$

1 ⟍ 4 ⟍ 9 ⟍ 16 ⟍ 25 ⟍ 36 ⟍ 49
  3   5   7   9   11   13   ⟵ First differences
    2   2   2   2   2   ⟵ Second differences are constant.

The next two terms are 36 and 49. ❖

CRITICAL
Thinking

The numbers in the sequence 1, 4, 9, 16, . . . are called perfect square numbers. The first term is $1 \cdot 1 = 1^2$, or 1. The second term is $2 \cdot 2 = 2^2$, or 4. Make a conjecture about the term number and its square in the square number sequence.

EXAMPLE 2

Physics   The data shown in the table is from the flight of a small rocket during the first 4 seconds of its flight. The flight of the rocket ends when it hits the ground 14 seconds after takeoff. Use the method of finding differences to find the maximum height during the rocket's flight.

Seconds	0	1	2	3	4	...	14
Height in feet	0	208	384	528	640	...	0

Solution▶

Look at the differences in the table to discover the pattern.

Seconds	0	1	2	3	4
Height in feet	0	208	384	528	640

208   176   144   112   ◀—— First differences
  −32   −32   −32   ◀—— Second differences constant

The second differences are each a constant − 32 feet. This means the differences *decrease* by 32 feet each time. Use the strategy of **working backward** from the second differences to extend the table.

Seconds	0	1	2	3	4	5	6	7	8	9	10
Height in feet	0	208	384	528	640	720	768	784	768	720	640

208   176   144   112   80   48   16   −16   −48   −80
  −32   −32   −32   −32   −32   −32   −32   −32   −32

After 7 seconds, the numbers for the height begin to repeat in reverse order. By continuing the table to 14 seconds, the pattern shows points along the complete path of the rocket.

Seconds	11	12	13	14
Height in feet	528	384	208	0

The table shows the highest point reached is 784 feet. This occurs at 7 seconds. Why does this information seem reasonable? ❖

Try This　Suppose a rocket flight takes 15 seconds from launch to return to the Earth. After how many seconds do you think it will reach its maximum height?

Spreadsheet　Using a spreadsheet can extend your ability to use differences to discover sequence patterns, especially when the differences and calculations become more complicated.

Compute the first difference of 208 in cell C2 by subtracting the value in cell B2 from the value in cell B3. What instructions do you think are needed to compute the values in cells C3 and in D2 of the spreadsheet?

C2	=B3-B2		
	A	**B**	**C**	**D**
1	seconds	height	1st diff	2nd diff
2	0	0	**208**	−32
3	1	208	176	−32
4	2	384	144	−32
5	3	528	112	
6	4	640		
7	5	720		

EXERCISES & PROBLEMS

Communicate

1. Explain how to find the first and second differences for the sequence that the students are discussing.

2. Describe the method for predicting the next two terms in the following sequence. 88, 76, 64, 52, 40, . . .

3. Suppose a rocket took 17 seconds to return to the Earth after launching. Explain how to find the time it takes the rocket to reach its maximum height.

4. **Technology**　Explain how to find the entry in cell C4 on the spreadsheet above.

5. Explain why you can stop calculating additional differences when the differences become constant.

6. Explain how to work backward from a second difference to extend a sequence.

7. Find the first and second differences for the sequence 20, 27, 36, 47, 60.

Find the next two terms of each sequence.

8. 18, 32, 46, 60, 74, . . .

9. 33, 49, 65, 81, 97, . . .

10. 20, 21, 26, 35, 48, . . .

11. 30, 31, 35, 42, 52, . . .

12. 100, 94, 88, 82, 76, . . .

13. 44, 41, 38, 35, 32, . . .

14. 12, 12, 18, 31, 53, 87, . . .

15. 1, 7, 23, 50, 89, . . .

Apply the method of finding differences until you get a constant. Which difference produces a constant? What is it?

16. 1, 2, 3, 4, 5, . . .

17. $1^2, 2^2, 3^2, 4^2, 5^2,$ . . . HINT: This is the same as 1, 4, 9, 16, 25, . . .

18. $1^3, 2^3, 3^3, 4^3, 5^3,$ . . . HINT: This is the same as 1, 8, 27, 64, 125, . . .

19. Tell how many differences you will have to compute before you get a constant for $1^4, 2^4, 3^4, 4^4, 5^4, 6^4$. . .

20. The first three terms of a sequence are 7, 11, and 16. Find the first and second differences. If the second differences are constant, what are the next three terms?

21. The first three terms of a sequence are 2, 6, and 12. The second differences are a constant 2. What are the next three terms of the sequence?

22. The third and fourth terms of a sequence are 15 and 23. If the second differences are a constant 2, what are the first 5 terms of the sequence?

23. If the second differences of a sequence are a constant 3, the first of the first differences is 7, and the first term is 2, find the first 5 terms of the sequence.

24. Physics If a rocket lands after 20 seconds of flight, at what time do you think it reached its maximum height? How can you tell?

25. Suppose a rocket took 24 seconds to reach its maximum height. After how many seconds do you think it hit the ground?

Technology Complete the spreadsheet that follows Example 2. What spreadsheet instruction formula should you place in the following cells?

26. C6

27. D5

28. ▱ **Geometry** Complete this table for the perimeter of a square with the length of a side given in centimeters.

Side	1	2	3	4	5	6	7	8
Perimeter	4	8	12	16	?	?	?	?

29. Apply the method of differences to find the next four terms of 4, 8, 12, 16, . . . Compare the results with the previous exercise. Do you get the same answer?

30. **Home Economics** As the altitude increases, the boiling point of water decreases according to the table. For example, in Las Vegas, Nevada, at an altitude of 2180 feet, the boiling point of water is about 98° Celsius. Estimate the boiling point of water in Colorado Springs, Colorado, which is at an altitude of 6170 feet.

Altitude in thousands of feet	0	1	2	3	4	5
Boiling point C°	100	99	98	97	96	95

31. **Portfolio Activity** Complete the problem in the portfolio activity on page 3.

Look Back

32. If you read a 603-page novel at a rate of 24 pages a day, how many days will it take you to read it?

33. If 675 people came to see the play and the admission was $1.60, how much money was taken in?

34. There were 162 students who volunteered to help 3 organizations. How many students went to each of the organizations if each got the same number?

Perform each of the following operations.

35. $\sqrt{169}$

36. 15^2

37. Find the sum of $1 + 2 + 3 + 4 + \cdots + 18$. **[Lesson 1.1]**

38. Find the next 3 numbers in the following pattern. 43, 49, 55, 61, 67, . . . **[Lesson 1.1]**

Look Beyond

The method of finding differences works only for certain kinds of patterns. Try it on the following patterns and tell what happens. Predict the next three terms of each pattern without using differences. Explain your method.

39. 1, 2, 4, 8, 16, 32, . . .

40. 1, 10, 100, 1000, 10,000, . . .

Exploring Variables and Equations

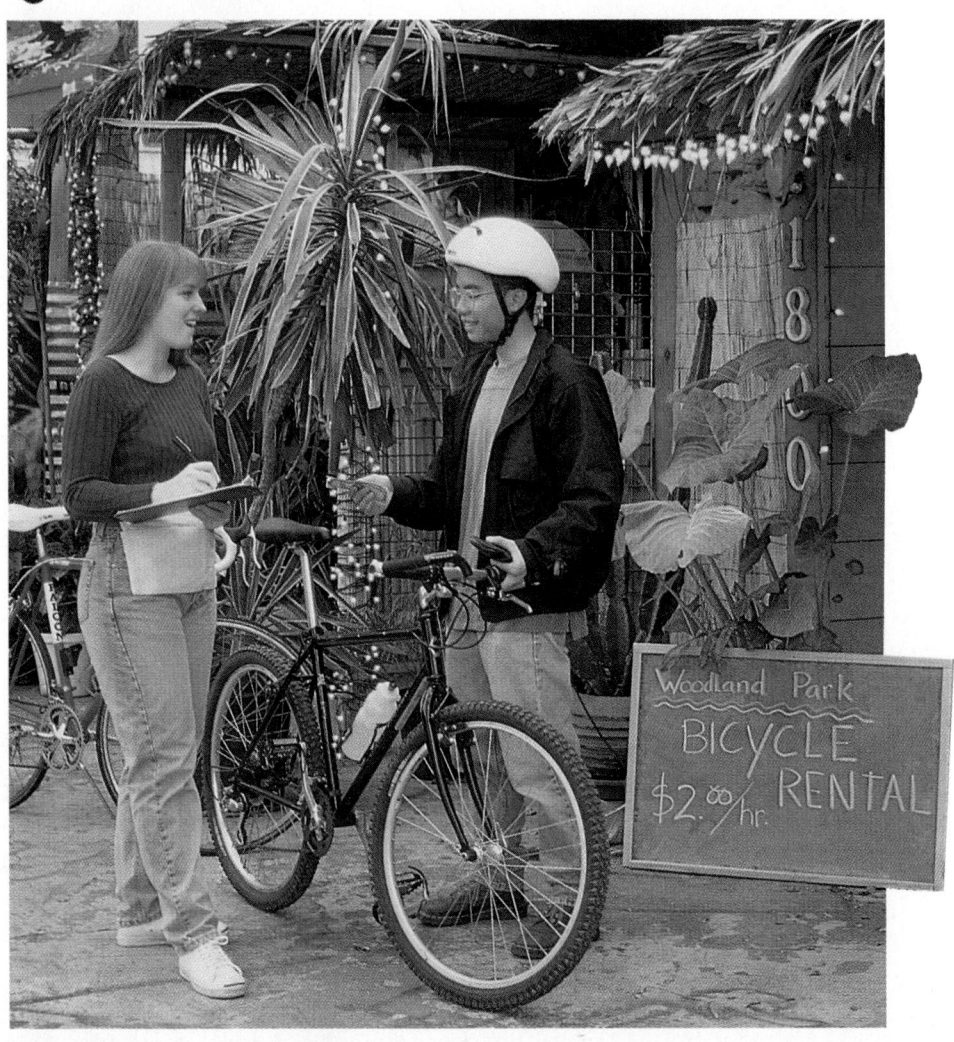

why *You want to rent a bicycle, but you don't know how long you can rent it with the money you have. In mathematics, one technique used to solve this kind of problem begins when you represent the unknown number in the problem with a symbol. You then work with the symbol until the algebra reveals the symbol's value.*

•Exploration 1 Defining a Variable

For how many hours can you rent a bike if you have $15 to spend?

This relatively simple problem provides an example that can be used to explore the basic tools of algebra. Examine the pattern for the bike rental.

Hours	1	2	3	4	5	...	h
Cost in $	2	4	6	?	?	...	?

1 What is the cost for 1 hour of bike rental?

2 What is the cost for 2 hours of bike rental?

3 What is the cost for 4 hours of bike rental?

4 When you rent a bike for 5 hours, what number do you multiply by 5 to get the value for the *cost*?

5 Let *h* stand for the number of hours. Follow the same pattern that you use to generate the values for *cost*. What number would you multiply times *h*?

6 What should the *cost* entry be in this table for *h* hours?

7 When *h* is 5, what is the value of 2*h* (which means 2 times *h*)? ❖

You can find the value of 2*h* when *h* is 6 by substitution. Replace *h* with 6, and simplify.

$$2h = 2 \cdot 6 = 12$$

Extend the table.

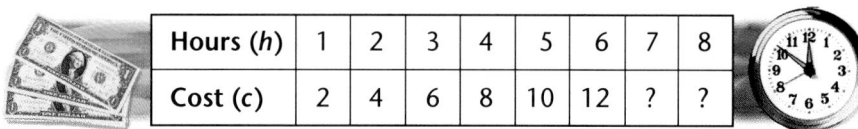

Hours (*h*)	1	2	3	4	5	6	7	8
Cost (*c*)	2	4	6	8	10	12	?	?

What is the cost for bike rental if the hours rented are halfway between 7 and 8 hours, and you can pay by the half-hour? For how many hours can you rent a bike at the park if you have $15?

If the cost for renting a bike is $3 an hour, how many hours can you rent a bicycle for $15? Try other values to see how they affect the hour and cost table.

Another way to represent the cost for renting the bicycle is to use the letter *c*. Since both *c* and 2*h* represent *cost,* they are equivalent. This is written as *c* = 2*h* or 2*h* = *c*.

In algebra, a **variable** is a letter or other symbol that can be replaced by any number or other expression. Variables combine with numbers and operations (addition, subtraction, multiplication, and division, for example) to form **expressions.** Two equivalent expressions, when separated by an equal sign, form an **equation.**

Examples of Variables	Examples of Expressions	Examples of Equations
a, x, y	$3x + 4$, $2h$,	$3x + 4 = 8$,
c, h, d, t	$-16t^2 + 4t$	$c = 2h$, $2h = c$

When you solve an equation, you find the value or values of the variable that make the equation a true statement. The solution for 2*h* = 15 is $7\frac{1}{2}$ because $2 \cdot 7\frac{1}{2} = 15$.

The land area of Olympic National Park is 922,654 acres. How long would it take to lose forest land equal to the area of Olympic National Park if 57,600 acres were destroyed each day?

Ecology

•Exploration 2 A Diminishing Sequence

Calculator

On a calculator with a constant feature, enter the area of Olympic National Park. Subtract 57,600 repeatedly, counting the number of times you subtract.

922654
Ans-57600

922654
865054

922654
Ans-57600

922654
865054
807454

922654
Ans-57600

922654
865054
807454
749854

As you continue to subtract, you can see how quickly the daily destruction diminishes the forest area. Continue repeating the subtraction until you have depleted the entire forest area.

To express this problem as an equation, you need to interpret and relate the information from the problem.

1 How many acres are in Olympic National Park?

2 How many acres are lost each day?

3 How would you express *57,600 fewer acres each day,* using the variable *d* to represent a day?

4 If you let *f* represent the number of days, then how much forest is left when *d* is 10?

5 If you let *f* represent the amount of Olympic National Park forest left, explain what it means when $f = 0$.

6 How many days does it take to completely destroy an area the size of Olympic National Park at the given rate? ❖

In one region there are 15,000 acres of forest. An average of 16 acres are destroyed every day from September 1 to October 30. Project the amount of forest lost after 30 more days. There are, in fact, 13,923 acres of the forest left. What could account for the difference between your projection and the actual amount?

APPLICATION

Danica places 3 photos on the cover of her picture album. On all of the other pages she places 4 photos. She has 178 photos in all. How many pages will she need if she uses all of her photos? This problem can be solved by writing an equation and using a guess-and-check strategy.

Let p equal the number of pages in Danica's picture album. To solve this problem, solve for p in the equation $4p + 3 = 178$. Guess the number of pages, substitute the number you guess for p, and check your guess. Keep track of your attempts in a table. Notice that in algebra, parentheses are used to show multiplication.

Guess 1: Try $p = 10$.
$$4p + 3 = 178$$
$$4(10) + 3 = 43$$

43 is too small. Try a larger number for p.

Guess 2: Try $p = 50$.
$$4p + 3 = 178$$
$$4(50) + 3 = 203$$

203 is too large. Try a smaller number for p.

Guess 3: Try $p = 40$.
$$4p + 3 = 178$$
$$4(40) + 3 = 163$$

163 is too small. You will find the number for p is between 43 and 44.

For 178 photos, 43 pages is not enough. The problem asks for the number of pages Danica will use, so the answer is 44 pages. One page more than 43 is needed for the extra photos even though the page will not be filled. ❖

Always be careful to interpret your answer in a way that makes sense. How many concert tickets can you buy for $15? The answer is 1, not $1\frac{7}{8}$. You cannot buy $\frac{7}{8}$ of a concert ticket, and you do not have enough money to buy 2 tickets.

EXERCISES & PROBLEMS

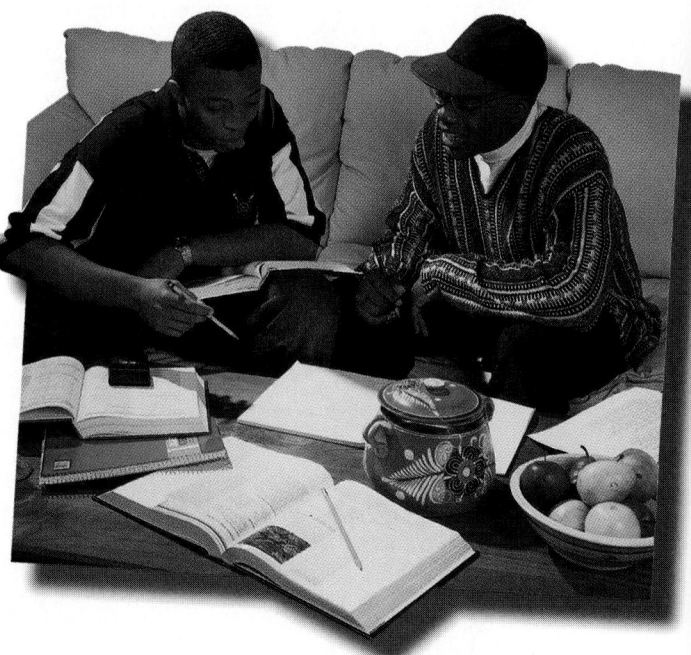

Communicate

1. Define *variable, expression,* and *equation.*

2. Explain how to identify the unknown variable in the following problem: Jeff has 3 more apples than Phil, and together they have 9 apples. How many apples does Phil have?

3. Discuss how you would set up the equation for the following problem: If 12 pencils cost $1.92, find the cost of 1 pencil.

4. Tell how you would set up the equation for the following problem: If tickets for a concert cost $10 each, how many can you buy with $35?

5. Explain how you would use the guess-and-check strategy to solve the equation $10x + 3 = 513$.

Practice & Apply

6. Guess and check to solve the equation $4x + 3 = 49$.

7. Guess and check to solve the equation $49x + 3 = 4$.

8. Is $14y$ an expression or a variable?

If pencils cost 20 cents each, find the cost of

9. 0 pencils.   **10.** 4 pencils.   **11.** 10 pencils.   **12.** p pencils.

Technology   Tell how to generate each sequence on a calculator.

13. 6, 8, 10, 12, 14, . . .   **14.** 15, 25, 35, 45, 55, . . .

15. 100, 90, 80, 70, 60, . . .   **16.** 52, 48, 44, 40, 36, . . .

Make a table to show the substitutions of 1, 2, 3, 4, and 5 for the variables in the expressions below.

17. $4x$ **18.** $5y$ **19.** $7s + 4$ **20.** $3n$

Beatrice solved the equation $7x = 14$ and got 2 for x.

21. Explain what Beatrice did to get this answer.

22. How can you determine if the answer is correct?

23. If tickets for a concert cost $11 each, how many tickets can you buy with $126?

24. If tickets for a concert cost $9 each, how many can you buy with $135?

25. Consumer Economics Write an equation that models the following situation. How many apples can you buy with 99 cents if apples cost 30 cents each?

26. Solve the equation you wrote in Exercise 25. Does the exact solution to the equation answer the problem posed about the apples? Explain.

27. Entertainment How many $5 movie tickets can you buy with $32?

28. Fund–raising How many $5 raffle tickets must you sell to raise $32?

29. If 5 people split the cost of a $32 birthday gift equally, how much should each pay?

 Statistics The average number of people in a family in 1995 was 3.18.

30. How many families had exactly 3.18 people?

31. How large would a typical family be?

 Look Back

Date and Nut Bars
1 Cup pecans or walnuts
1/2 Cup dates
3/4 Cup sifted all–purpose flour
3 eggs
1 1/2 Cups brown sugar, firmly packed
3/4 Teaspoon baking powder
1/4 Teaspoon salt

32. How much of each ingredient should be used if the recipe is tripled?

33. A can of a soft drink contains 355 milliliters of liquid. If 5 people split 2 cans evenly, how many milliliters will each get?

34. If each pizza is cut into 12 pieces, how many pieces will there be in $3\frac{1}{2}$ pizzas?

Cultural Connection: Africa Thousands of years ago Diophantus and Hypatia, who is the first known woman mathematician, investigated the relationship between square numbers and triangular numbers.

35. Find at least three more examples of how a square number can be represented as the sum of two triangular numbers. **[Lesson 1.1]**

36. Are there any square numbers that can't be represented as the sum of two triangular numbers? **[Lesson 1.1]**

37. In the first lesson you found that the sum $1 + 2 + 3 + 4 + 5$ is $\frac{5 \cdot 6}{2}$.
Examine the next two cases, the sum of the first 6 numbers and the sum of the first 7 numbers. Make a conjecture about the sum of the first n numbers. **[Lesson 1.1]**

38. The first differences of a sequence are 7, 9, and 11. The first term of the sequence is 5. What are the first eight terms of the sequence? **[Lesson 1.2]**

Look Beyond

39. If x^2 means x times x, then guess and check to solve for x in the equation $x^2 = 256$.

Order of Operations

Mary plans to buy carpet for two rooms. One room is 12 feet by 15 feet, and the other room is 20 feet by 14 feet. She uses her calculator to find the number of square feet.

$$12 \boxed{\times} 15 \boxed{+} 20 \boxed{\times} 14 \boxed{=} \qquad \boxed{2800}$$

Mary knows that 2800 square feet is way too much. Her entire house is less than 2000 square feet. But calculators don't make mistakes. What is wrong?

CRITICAL *Thinking* What is the correct answer for Mary's computation? What steps did her calculator use to get 2800?

To avoid misunderstandings and errors, mathematicians have agreed on certain rules for computation called the **order of operations.** One rule requires that multiplication be performed before addition. By following this rule, Mary can make the correct calculation.

$$12 \times 15 + 20 \times 14 =$$
$$180 + 280 =$$
$$460 \text{ square feet}$$

The Order of Operations

1. Perform all operations enclosed in symbols of inclusion (parentheses, brackets, braces, and bars) from innermost outward.
2. Perform all operations with exponents.
3. Perform all multiplications and divisions in order from left to right.
4. Perform all additions and subtractions in order from left to right.

Calculators that follow the order of operations use **algebraic logic.** You can use a computation like 2 + 3 · 4 to test for algebraic logic.

Key in 2 [+] 3 [×] 4 [=]. If the answer is 14, the calculator uses algebraic logic. On some calculators you might get 20. These calculators do not use algebraic logic, but you can use parentheses to get the correct answer.

2 [+] [( ] 3 [×] 4 [ )] [=]

You can also enter the multiplication first.

3 [×] 4 [+] 2 [=]

The examples you will see in this book assume that you have a calculator with algebraic logic.

EXAMPLE 1

How do you use a calculator with algebraic logic to work Mary's problem?

Solution▻

Key the numbers and operations just as they appear. 12 [×] 15 [+] 20 [×] 14 [ENTER]. The answer is 460. ❖

12*15+20*14
460

Using Inclusion Symbols

Symbols like parentheses, ( ), brackets, [ ], braces, { }, and the fraction bar are called **symbols of inclusion.** These symbols group numbers and variables. Treat any grouped numbers and variables as a single quantity. Operations should always be done within the innermost symbols of inclusion first. Then work outward.

EXAMPLE 2

Insert inclusion symbols to make $30 + 4 \div 2 - 1 = 16$ true.

Solution▸

Use parentheses to group 30 and 4 before dividing by 2. Begin with the operations in the innermost symbols of inclusion.

$$[(30 + 4) \div 2] - 1 \overset{?}{=} 16$$
$$[34 \div 2] - 1 \overset{?}{=} 16$$
$$17 - 1 \overset{?}{=} 16$$
$$16 = 16 \qquad \text{True} ❖$$

Try This Insert inclusion symbols to make $5 + 30 \div 7 - 4 = 15$ true.

In algebra you are often asked to *evaluate* an expression. To do this, you replace the variables with the numbers that are assigned to those variables. Then proceed with the computation using the order of operations.

EXAMPLE 3

Evaluate $5x^2 + 7y$ when x is 3 and y is 2.

Solution▸

$$5x^2 + 7y$$

Replace x with **3** and y with **2**.	$5 \cdot 3^2 + 7 \cdot 2$
First, square the 3.	$5 \cdot 9 + 7 \cdot 2$
Perform all of the multiplications.	$45 + 14$
Finally, add the results.	$45 + 14 = 59$ ❖

EXAMPLE 4

Show the keystrokes and the answer for $\dfrac{57 + 95}{16} + \dfrac{220}{88 + 104}$.

Solution▸

Since the entire quantity $57 + 95$ is divided by 16, place parentheses around $57 + 95$. Do the same for $88 + 104$.

$$(57 + 95) \div 16 + 220 \div (88 + 104)$$

Scientific Calculator

The keystrokes are shown.

$$\boxed{(}\ 57\ \boxed{+}\ 95\ \boxed{)}\ \boxed{\div}\ 16\ \boxed{+}\ 220\ \boxed{\div}\ \boxed{(}\ 88\ \boxed{+}\ 104\ \boxed{)}\ \boxed{=}$$

The answer is 10.646 to 3 decimal places. Is this reasonable? The first fraction is about $\frac{160}{16}$, or 10. The second fraction is about $\frac{200}{200}$, or 1. Thus, the estimated sum is about $10 + 1$, or 11. The answer 10.646 is reasonable. ❖

EXAMPLE 5

On Monday, John's father borrowed $3 and said that he would repay John double the amount plus $1 on Friday. The next Monday, John's father borrowed the amount he gave John the previous Friday and said that he would repay him double that amount plus $1 on Friday. The next week he did the same. On the third Friday, John figured his father owed him 2{2[2(3) + 1] + 1} + 1 dollars. How much money would John receive that Friday?

Solution

When you simplify an expression that contains several pairs of inclusion symbols begin with the innermost pair and work outward.

$$2\{2[2(3) + 1] + 1\} + 1$$
$$2\{2[7] + 1\} + 1$$
$$2\{15\} + 1$$
$$31$$

If you use a calculator or a computer, the expression will usually contain only parentheses. The example would appear as 2(2(2(3) + 1) + 1) + 1. This does not affect the way you simplify the expression. ❖

If this pattern continues for one more week, how much money will John get on the fourth Friday?

Pattern Exploration and Technology

Calculator

Complete the pattern using a calculator.

$$1 \cdot 1 = 1$$
$$11 \cdot 11 = ?$$
$$111 \cdot 111 = ?$$

Predict the next three numbers in the pattern. Check your prediction with your calculator. What happens when you try to compute 111111 · 111111? Continue developing the pattern. What happens as the number of ones increases? What are the limitations of using a calculator to explore this pattern?

EXERCISES & PROBLEMS

Communicate

1. Explain how $3 + 2 \cdot 4$ can give two different answers.
2. Give two possible answers to the computation $20 \div 2 \cdot 5$.
3. Which answer to $20 \div 2 \cdot 5$ is correct? Why?
4. Describe the order of the steps for simplifying $\{[3(8 - 4)]^2 - 6\} \div (4 - 2)$.
5. Explain why rules called *the order of operations* are necessary for computation.
6. Explain why 3 is a reasonable estimate for $\frac{173 + 223}{151 - 21}$.

Practice & Apply

Place inclusion symbols using the correct order of operations to make each equation true. Tell how you would use a calculator to check your answers.

7. $28 - 2 \cdot 0 = 0$
8. $59 - 4 \cdot 6 - 4 = 51$

Simplify using the correct order of operations. If answers are not exact, round them to the nearest thousandth.

9. $57 \cdot 29 + 89$
10. $7.2(9.8) + 1.2$
11. $89 + 57 \cdot 29$
12. $0.3(1.5) + 9$
13. $43 \cdot 32 + 91 \cdot 67$
14. $4.5(7.5) + 9.0(2.4)$
15. $\frac{28 + 59}{97 - 17}$
16. $\frac{9.6 - 1.7}{7.2 + 0.7}$
17. $\frac{43 \cdot 91}{8 \cdot 25}$
18. $157 - 29 + 23 \cdot 9$
19. $91 \div 7 + 6$
20. $187 - 34 \div 17$

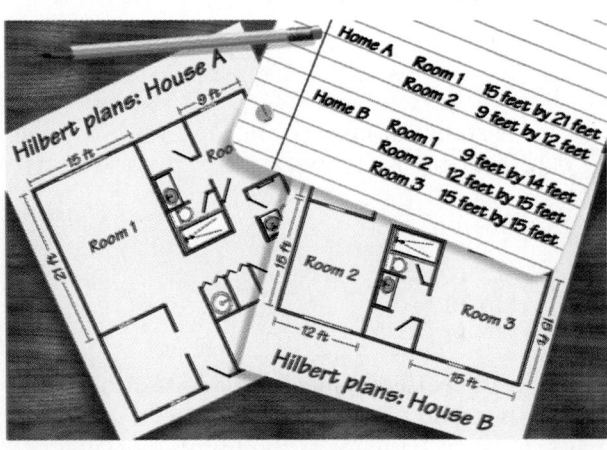

Home Improvement Mr. Hilbert owns two rental properties, Home A and Home B. He wants to carpet two rooms in Home A and three rooms in Home B. He finds a carpet store selling carpet for \$12.99 per square yard. The dimensions of the rooms in his homes are shown on the blueprint. HINT: $1\text{yd}^2 = 9\text{ft}^2$

21. Find the total number of square feet of carpet needed for Home A.
22. Find the total number of square feet of carpet needed for Home B.
23. Find the cost of carpeting the rooms in Home A.
24. Find the cost of carpeting the rooms in Home B.

Evaluate each expression.

25. $2(5 + 4) \div 9$

26. $12 - 7 \cdot 3 + 9^2$

27. $3 - 1 + 24 \div 6$

28. $7 + 6 \div 2 \cdot 10$

Given that *a* is 5, *b* is 3, and *c* is 4, evaluate each expression.

29. $a + b - c$

30. $a^2 + b^2$

31. $a^2 - b^2$

32. $(a + b) \cdot c$

33. $a^2 - b - c$

34. $a^2 - (b + c)$

35. Technology What answer do you get if you enter

32 $\boxed{\times}$ 38 $\boxed{\div}$ 11 $\boxed{\times}$ 19 $\boxed{=}$?

Explain how the calculator gets this answer.

36. 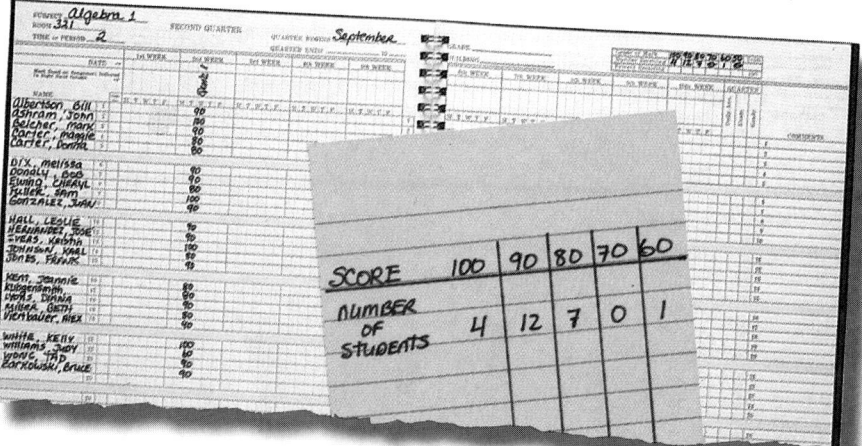 **Statistics**
The teacher's grade
book shows the results
of an algebra quiz. Find
the average for the class.
Show your method.

Look Back

37. Advertising A photographer arranges cans of soup in a large triangle for a supermarket ad. The top row contains 1 can, and each of the rows contains 1 can more than the row above it. If there are 10 rows of cans, how many cans will the photographer need to form the triangle display? **[Lesson 1.1]**

Determine when the first constant difference appears, and find the next 2 terms of each sequence. **[Lesson 1.2]**

38. 2, 5, 8, 11, 14, . . .

39. 1, 6, 13, 22, 33, . . .

40. 2, 9, 20, 35, 54, 77, . . .

41. **Geometry** The perimeter of a rectangle is the distance around the border, or the sum of twice the length and twice the width. The area is the product of the length and width. For what length and width will the perimeter and the area be the same number?

Look Beyond

An exponent indicates repeated multiplication. The 3^2 means $3 \cdot 3$. Indicate whether each equation is true or false.

42. $(3 + 4)^2 = 3 + 4^2$

43. $(3 + 4)^2 = (3 + 4)(3 + 4)$

44. $(3 + 4)^2 = 3^2 + 2(3)(4) + 4^2$

45. $(3 + 4)^2 = 3^2 + 4^2$

Representing Linear Functions by Graphs

Sometimes it is difficult to recognize patterns in a numerical problem. It is often easier to see the pattern from a picture. In algebra, a graph is used to display a picture of a mathematical pattern.

You can describe the location of your hometown on a road map by using the letters and numbers in the margins.

To find Northfield, Minnesota, on a map, for instance, you look up from the letter K on the bottom and across from the number 18 in the margin of a state road map. The region where these strips overlap contains Northfield.

Graphing With Rectangular Coordinates

Cultural Connection: Africa The idea of graphing has been around for over 4500 years. Around 2650 B.C.E., Egyptians sketched curves using pairs of numbers to locate points. This idea gave rise to the use of coordinates.

The set of real numbers can be represented on a number line.

* To construct the number line, mark a point on the line as zero.

* Mark off equal spaces to the right from zero, and number them 1, 2, 3, 4, . . .

* Then mark off corresponding spaces to the left from zero, and number them − 1, − 2, − 3, − 4, . . .

$$-4 \quad -3 \quad -2 \quad -1 \quad 0 \quad 1 \quad 2 \quad 3 \quad 4$$

Where is $\frac{1}{2}$ on this number line? $-\frac{1}{2}$?

Cultural Connection: Europe In the seventeenth century, Rene Descartes, a French mathematician and philosopher, used a horizontal and vertical number line to divide a plane into four regions, called **quadrants**.

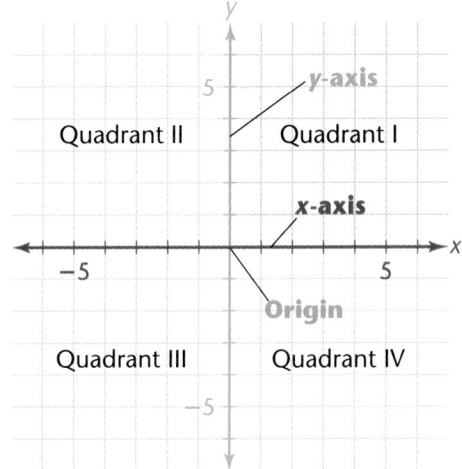

The rectangular, or Cartesian, coordinate system was named for Rene Descartes.

The horizontal line is called the **x-axis**, and points on this axis are called **x-coordinates**.

The vertical line is called the *y-axis*, and points on this axis are called *y-coordinates*.

The *x-* and *y*-axes intersect at the origin.

Letters other than *x* and *y* can represent the axes. For example, you might see the *x*-axis shown as the *t*-axis when *t* refers to time in a problem.

Graphing on the entire coordinate system will appear later in this book. For the rest of this chapter, however, the emphasis is on the first quadrant.

Coordinates give the address of a point. They are written as an **ordered pair**, indicated by two numbers in parentheses, (*x*, *y*). The number from the *x*-axis will appear first in an ordered pair.

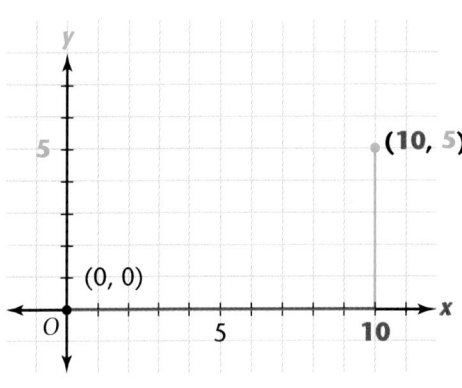

To locate the point shown on the graph, start at the origin, (0, 0). Move to the right 10 units along the *x*-axis, then move up 5 units to the point. The ordered pair is (**10**, **5**).

CRITICAL Thinking Find a way to locate a point on a plane without using the rectangular coordinate system.

Linear Functions

Once you organize the data from a problem, you can form ordered pairs and graph the data.

Sabrina normally rides her bicycle at a constant rate of 0.2 miles per minute.

EXAMPLE 1

On Wednesday, Sabrina left school and rode her bicycle for 15 minutes in a direction away from home. If Sabrina lives 2 miles from school, how far is Sabrina from home?

Solution►

The equation that models Sabrina's distance from home with respect to time spent bicycling is based on the following relationship.

Distance = rate · time + miles from home to school.

Replace *distance* with *d*, *rate* with 0.2, *time* with *t*, and miles from home to school with 2.

The equation is now written as $d = 0.2t + 2$.

To make a table of values, substitute values for *t* in minutes, and find the corresponding values for *d* in miles. The ordered pairs are then plotted and connected by a line.

t	0.2t + 2.0	d	(t, d)
0	0.2(0) + 2.0	2.0	(0, 2.0)
1	0.2(1) + 2.0	2.2	(1, 2.2)
2	0.2(2) + 2.0	2.4	(2, 2.4)
3	0.2(3) + 2.0	2.6	(3, 2.6)
4	0.2(4) + 2.0	2.8	(4, 2.8)
5	0.2(5) + 2.0	3.0	(5, 3.0)
. . .			
15	0.2(15) + 2.0 =	5.0	(15, 5.0)

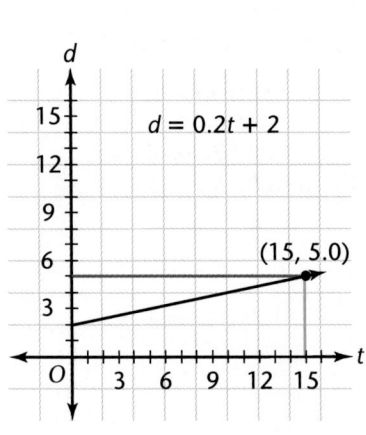

The table and the graph indicate that after 15 minutes, Sabrina is 5 miles from home. ❖

The graph of the equation $d = 0.2t + 2$ is a straight line. For this reason the equation is called a **linear equation.** This equation generates a set of ordered pairs, (t, d). We say that d is a function of t, or that d is dependent on t. Thus, $d = 0.2t + 2$ is a **linear function.** In the ordered pairs, the set of first values, t, is the **domain** of the function. The set of second values, d, is the **range** of the function.

In the function $d = 0.2t + 2$, t is called the independent variable. The d is called the dependent variable because its value depends on the value chosen for t.

Try This   How far would Sabrina be from home if she rides for 60 minutes?

EXAMPLE 2

Physiology   Represent the maximum heart rate by r and the age by a. Write a function to express the maximum heart rate in terms of a. Graph the function for ages 10 to 50.

A reasonable estimate for the maximum heart rate during exercise should be no more than 220 beats per minute minus the person's age.

Solution

Since the rate, r, equals 220 beats per minute minus the age, write the equation $r = 220 - a$. Values for a are elements of the domain, and values for r are elements of the range. When graphing a function, it is customary to represent the domain on the x-axis and the range on the y-axis.

When the values 10, 20, 30, 40, and 50 are in the domain, the table shows how substitution allows you to find the corresponding values in the range.

To graph the function, enter **220 − x** after Y = on a graphics calculator. Set the limits of the domain and range in the range window. Then press GRAPH .

a	$220 - a$	r
10	$220 - 10$	210
20	$220 - 20$	200
30	$220 - 30$	190
40	$220 - 40$	180
50	$220 - 50$	170

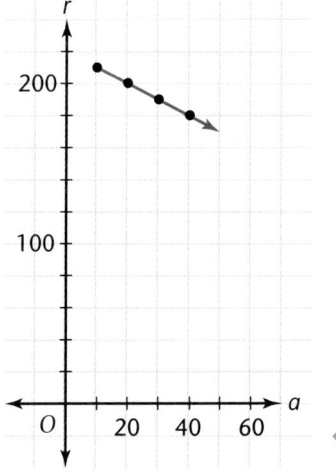

For some people the proper level of heart rate should not exceed 60% of the maximum found in this example. Individuals should consult a doctor if there is any doubt about their capability.

Exercises & Problems

Communicate

1. Explain how you find the coordinates of a given point from a graph.

2. Is (6, 7) the same point as (7, 6)? Explain.

3. Describe the steps for plotting the point with the coordinates (7, 3) on a graph.

4. Describe the relationship between the coordinates of an ordered pair and the domain and range of a function.

5. Compare the advantages of displaying data in a table and on a graph.

6. How would you make a table of the values for x and y for the equation $y = 2x + 5$?

Practice & Apply

Graph each list of ordered pairs. State whether they lie on a straight line.

7. (1, 3), (2, 6), (3, 9)

8. (1, 5), (2, 4), (3, 1)

9. (1, 10), (2, 7), (3, 2)

10. (1, −3), (2, −6), (3, −9)

11. (5, 2), (7, 2), (9, 2)

12. (4, 1), (4, 5), (4, 9)

What are the coordinates of the given points?

13. A 14. B 15. C 16. D

Graph and compare the following two linear functions. What effect does the operation before the x have on the graph of the line?

17. $y = 3 + x$ 18. $y = 3 - x$

Graph each of the following functions on the same set of coordinate axes, and compare the graphs. Explain your conclusions.

19. $y = x + 7$ 20. $y = x - 7$ 21. $y = 7 - x$ 22. $y = -7 - x$

Find the values for y by substituting 1, 2, 3, 4, and 5 for x. Make a table.

23. $y = x + 3$ 24. $y = x + 4$ 25. $y = 2x$ 26. $y = 2x + 5$

Don walks at a rate of 3 miles per hour. You can determine the distance that he walks by multiplying the rate times the number of hours that he walks.

27. Represent hours by h, and write an equation for the distance, d.

28. Make a table to show how far Don walked in 0, 1, 2, and 3 hours.

29. Draw a graph of the function.

30. **Health** Using the equation given in Example 2, what should be the maximum heart rate for a 36-year-old person?

Suppose a T-shirt company charges a $3 handling fee per order.

31. How much does an order of 2 shirts cost?

32. How much does an order of 5 shirts cost?

33. Make a set of ordered pairs from the given information, and plot them as points on a graph. Do the points lie on a straight line?

34. Use your graph to see how many shirts you can buy with $75.

Look Back

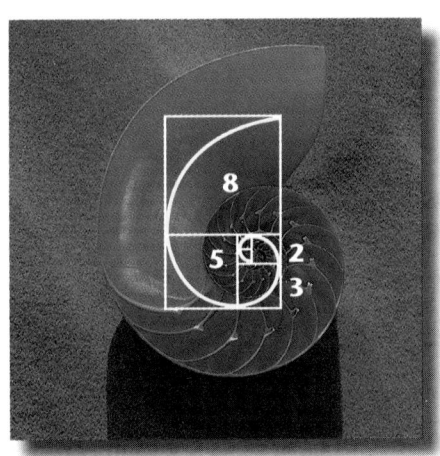

The Fibonacci sequence is found in many examples from nature.

35. What are the next 3 terms for the sequence 1, 1, 4, 10, 19 . . .? **[Lesson 1.1]**

36. How many differences are needed to reach a constant for the sequence 1, 2, 6, 15, 31, . . . ? What is the constant difference for the sequence? **[Lesson 1.2]**

37. The sequence 1, 1, 2, 3, 5, 8, . . . is a famous sequence named for the Italian mathematician Fibonacci. Find the next two terms. Will the method of differences help? What kind of patterns do the differences show? **[Lesson 1.2]**

38. Guess the closest whole number value for x that satisfies the equation $3x + 406 = 421$. Test your guess using substitution. **[Lesson 1.3]**

39. Make a table to solve the equation $7x + 4 = 40$. **[Lesson 1.3]**

Simplify. **[Lesson 1.4]**

40. $2 \cdot 14 \div 2 + 5$

41. $6 + 12 \div 6 - 4$

42. $4[(12 - 3) \cdot 2] \div 11$

43. $[3(4) - 6] - [(15 - 7) \div 4]$

Look Beyond

44. The graph of the path of a rocket is a parabola. The highest point of this curve is called the vertex. Graph each set of points, and determine which point is the vertex.

x	0	1	2	3	4	5	6	7	8	9	10
y	0	9	16	21	24	25	24	21	16	9	0

Exploring Scatter Plots and Correlation

Why *People frequently try to determine relationships between different factors such as math and science aptitude. How can you tell whether these factors are related? Correlations provide a way to measure numerically how well the sets of data are related.*

Scatter Plots

A newspaper article reported that students who do well in math also do well in science.

An effective way to see a relationship from data is to display the information as a **scatter plot**. Before you can use a scatter plot, however, you need to

- have clearly defined variables you can assign to the axes,
- form ordered pairs from the data you have, and
- locate each data point on the scatter plot the way you would graph a point on the coordinate plane.

You can then examine the patterns for clues to see how closely the variables are related.

 Exploration *Find a Correlation*

STATISTICS
Connection

Ten students took a series of aptitude tests. The table shows the scores on two parts of the test. You want to know if students who do well in math also do well in science. Once you create a scatter plot, you can see the pattern for the relationship.

Student	A	B	C	D	E	F	G	H	I	J
Math	61	40	80	21	62	54	20	33	75	51
Science	70	38	92	50	68	41	38	20	73	48

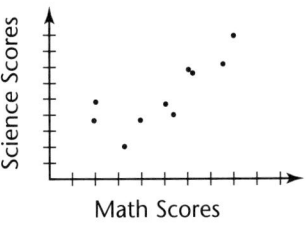

1 Look at the scores of the students who did well in math. How well did they do in science?

2 Look at the scores of the students who did *not* do well in math. How well did they do in science?

3 Do the data points appear broadly scattered or do they seem to cluster along a line?

4 Does the line of the data rise or fall as you go from left to right?

5 What does the data pattern show about the relationship between the variables? ❖

Correlation

Scatter plots show a picture of how the variables relate to each other by displaying how well data points can fit the line.

A strong positive correlation As one variable increases, the other also tends to increase. This shows a *positive* correlation. When the points are nearly in a line, there is a *strong* correlation.

A strong negative correlation As one variable increases, the other tends to decrease. This shows a *negative* correlation. When the points are nearly in a line, there is a *strong* correlation.

Little or no correlation As one variable increases, you cannot tell if the other tends to increase or decrease. There is a weak correlation or none at all.

APPLICATION

Compare the scatter plots with the statements.

1. The variables studied are *age* and *distance* from home to school.

2. The variables studied are *age* and the *time it takes to run a fixed distance*.

3. The variables studied are *age* and *height*.

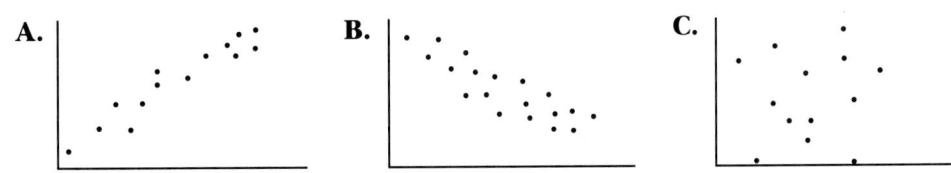

Match the statements 1, 2, and 3 with the appropriate scatter plot.

Scatter plot A

Scatter plot A shows a positive correlation. A positive correlation occurs when both variables increase. The correlation is strong. The data points are close to the shape of a straight line. This indicates that it is easy to predict the behavior of one variable by the behavior of the other. As a child's age increases, his or her height increases. This scatter plot would fit Statement 3.

Scatter plot B

Scatter plot B shows a negative correlation. A negative correlation occurs when one variable increases and the other variable decreases. The data points are close to the shape of a straight line, which is characteristic of a strong correlation. As a child gets older, bigger, and stronger, the time it takes the child to run a fixed distance decreases. This scatter plot would fit Statement 2.

Scatter plot C

Scatter plot C shows little or no correlation. The data points are randomly scattered. You cannot tell whether the correlation is positive or negative. The distance that a student lives from school has no relation to age. This scatter plot would fit Statement 1. ❖

Do not confuse a strong positive or negative correlation with a cause-and-effect relationship. Over the past few years there has been a strong positive correlation between annual consumption of diet soda and the number of traffic accidents reported in one Midwestern state. No one would claim that drinking diet soda would cause accidents, or that the trauma of being in an accident would cause people to drink more soda. Often a third variable can be responsible for the unexpectedly strong correlation between two otherwise unrelated variables. Can you think of what another variable might be in the Midwestern state?

APPLICATION

Demographics Make a scatter plot for the given data, showing the number of people working on farms for various years. Describe the correlation as strong positive, strong negative, or little to none.

Decline in Number of Farm Workers

Year	Number
1940	8995
1950	6858
1960	4132
1970	2881
1980	2818
1990	2864

1. Mark off the vertical axis from 0 to 9000

2. Label the horizontal axis from 1940 to 1990

3. Plot the points

As the years increase, the number of workers decreases, so this is a strong negative correlation. However, you can notice a leveling off over the last 20 years. ❖

CRITICAL
Thinking

Think of two examples of related data. Describe the kind of correlation you can expect from the data.

EXERCISES & PROBLEMS

Communicate

1. Use the scatter plot from the Exploration to explain how you would decide what type of correlation the math and science scores have.

2. How would you label the axes of a scatter plot involving the *cost* of notebooks and the *number of pages* each has? Which axis would you use for the cost? Which axis would you use for the number of pages?

3. Discuss how a correlation provides you with information about the variables.

Describe the correlation for the following scatter plots as strong positive, strong negative, or little to none. Explain the reason for your answer.

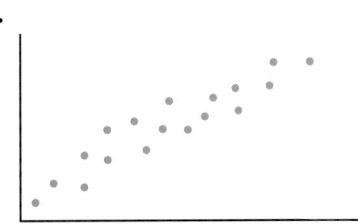

4.

5.

Cost of Living The following table shows the value of food items that cost $1.00 in 1982 has increased in cost over the years.

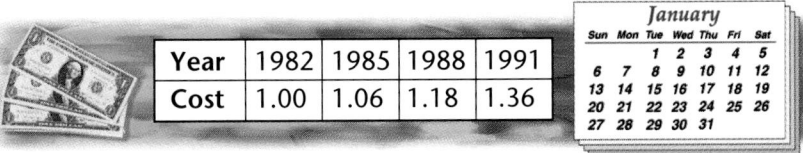

Year	1982	1985	1988	1991
Cost	1.00	1.06	1.18	1.36

6. Tell which axis you would use for the cost of food, and which you would use for the year. Why?

7. Tell how you would plot the points for the data in the table.

Practice & Apply

Describe the correlation as strong positive, strong negative, or little to none.

8.

9.

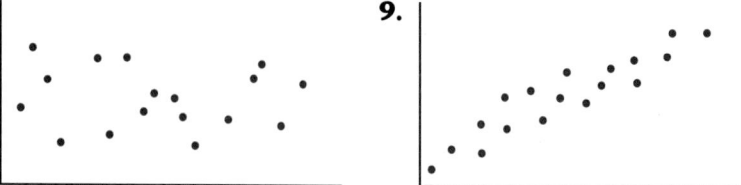

Statistics Use the data in the table for the following exercises.

Student	A	B	C	D	E	F	G	H	I	J
English	50	63	70	48	70	52	43	78	70	65
Math	61	40	80	21	62	54	20	33	75	51
Science	70	38	92	50	68	41	38	20	73	48

[34 ÷ 2] − 1 ? 16
Read
17 − 1 ? 16
Chapters
16 = 16
8 & 9

10. Graph the student scores on math against the scores on English, and explain the correlation.

11. Graph the student scores on science against the scores on English, and explain the correlation.

Geography Use the following data to answer Exercises 12–15.

Look down the column for north latitude, which is arranged in increasing order. As north latitude increases, is there an obvious tendency in what happens to

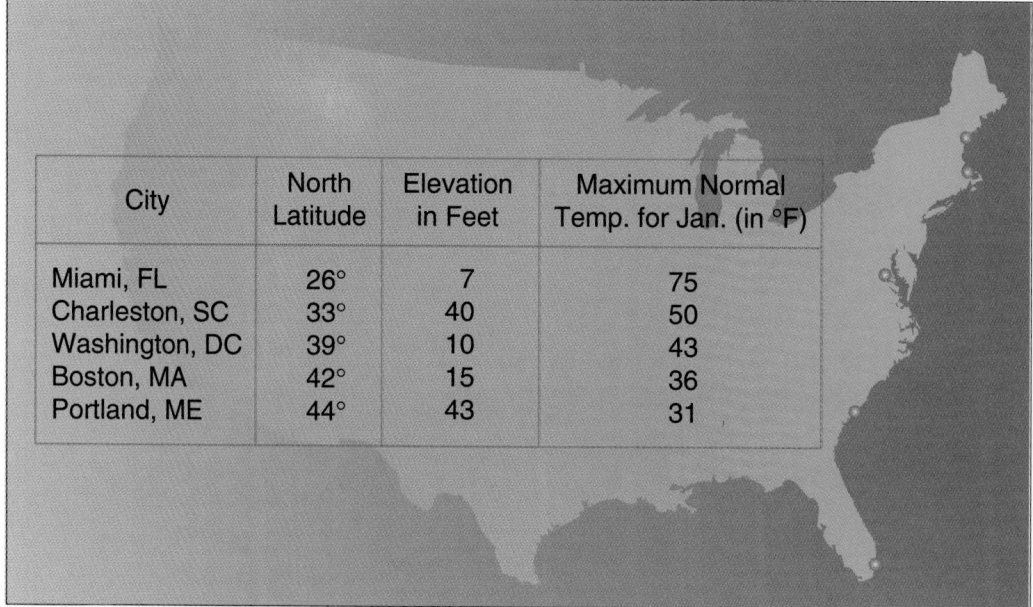

City	North Latitude	Elevation in Feet	Maximum Normal Temp. for Jan. (in °F)
Miami, FL	26°	7	75
Charleston, SC	33°	40	50
Washington, DC	39°	10	43
Boston, MA	42°	15	36
Portland, ME	44°	43	31

12. elevation?

13. temperature?

14. Describe the correlation between latitude and elevation.

15. Describe the correlation between latitude and temperature.

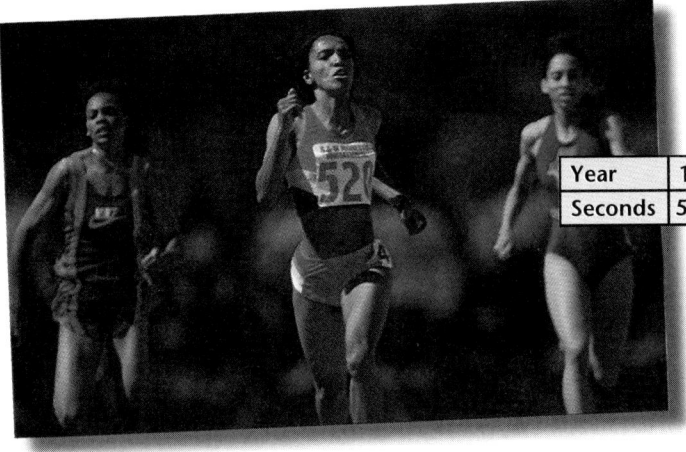

16. Sports Make a scatter plot of the following data for the women's 400-meter run in the Olympics.

Year	1964	1968	1972	1976	1980	1984	1988	1992
Seconds	52.00	52.00	51.08	49.29	48.88	48.83	48.65	48.83

17. Describe the correlation.

18. Draw these two coordinate axes on your graph paper. Plot the same points, (1, 2), (2, 3), and (3, 4), on each set of axes. Connect the points with a line.

a.

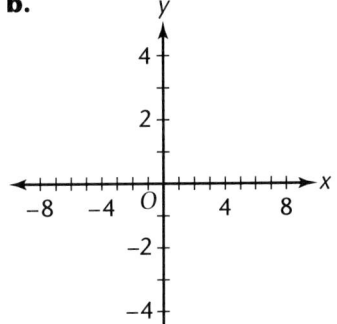

b.

19. Which graph is steeper, a or b?

20. What caused one graph to be steeper, even though the points are the same?

21. How does a change in scale on the axes affect the steepness of the line? *Scale* refers to the distance between units on an axis.

22. Does the scale of the graph affect a correlation? Explain your answer.

Look Back

Simplify.

23. $\sqrt{144}$ **24.** $\sqrt{7+2}$ **25.** $\sqrt{171-2}$ **26.** $2\sqrt{16}$

27. What is the greatest common factor for 30, 60, and 15?

28. Pair the equivalent expressions.
   a. $2 \cdot 2 \cdot 2 \cdot 2$ **b.** $4(2)$ **c.** $2 + 2 + 2 + 2$ **d.** 2^4

Plot each point on the coordinate plane.   [Lesson 1.5]

29. (5, 3)   **30.** (5, 5)   **31.** (3, 4)   **32.** (7, 2)

Look Beyond

33. Make a scatter plot from the data. Use a straightedge to draw a line that best fits the points on the scatter plot.

x	0	3	6	9	12
y	14	20	26	32	38

34. What is the relationship of the points to the line?

Finding Lines of Best Fit

Total Number of Pieces Written by Mozart for Each Age From 14 to 35 Years

(35,684)
(30,544)
(25,388)
(20,279)
(15,144)

Pieces: 700, 600, 500, 400, 300, 200, 100, 0

Mozart's Age: 15, 20, 25, 30, 35

Scatter plots are used to show a trend in the data. This scatter plot presents an overall picture of Mozart's musical output. Since the points lie very nearly on a straight line, it tells you that Mozart wrote at an amazingly steady pace. When a trend is represented by a line that fits the data, you can study the line to see how the data behave in general.

Music History Wolfgang Amadeus Mozart (1756–1791) composed many musical pieces in his lifetime. The scatter plot shows the total number of musical pieces that Mozart had written as he reached different ages. For example, the point (35, 684) shows that by age 35 Mozart had written 684 pieces. The graph starts at (14, 121), indicating that by age 14 he had already written 121 musical pieces.

The Line of Best Fit

The line of best fit represents an approximation of the data on the scatter plot. The following steps should help you find this line.

Use a straightedge, such as a clear ruler, to model the line.

To fit the line to the points, choose a line that best matches the trend.

The line does not necessarily have to pass through any of the points.

If you look at the numbers along the *age*-axis, you see that there is a difference of 21 years between ages 14 and 35. How many pieces of music did Mozart write during that period of time? Form a ratio that compares the difference in the number of pieces to the difference in the number of years. What does this ratio tell you about Mozart's rate of composition?

Correlation and the Line of Best Fit

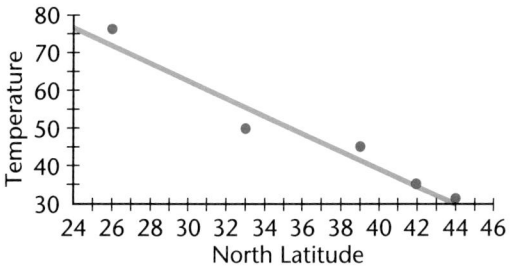

The scatter plot shows the maximum normal temperatures in January plotted for selected cities at different latitudes along the Atlantic coast. It also shows a computer-drawn line that best fits the data. In this first example, the points are close to the line.

The next scatter plot shows elevation in feet for selected cities along the Atlantic coast plotted against different average January temperatures. It shows a computer-drawn line that best fits the data. In this second example, the data points do not fit the line very closely.

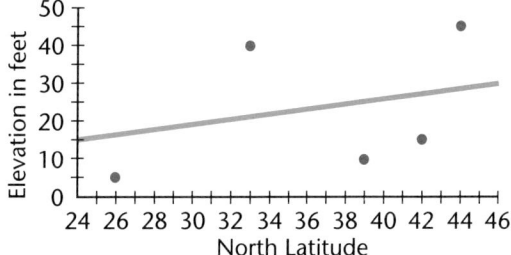

The measure of how closely a set of data points falls along a line is the **correlation coefficient,** r. The correlation coefficient can be computed automatically using a graphics calculator. The possible values for a correlation coefficient range from -1 to 1. The better the data fit the line, the closer the numerical value of the correlation coefficient is to *either* 1 or -1.

In the first scatter plot, the correlation coefficient computed on a calculator is -0.98, which is near -1. This indicates that the data are packed closely to a line that falls from left to right.

In the second scatter plot, the correlation coefficient from a calculator is 0.36, which is not close to 1 or -1. This indicates that the data do not cluster near the best fit line. Because the correlation is positive, the line rises from left to right.

CRITICAL Thinking Think of several pairs of things that correlate well positively and several that correlate negatively. Try to identify the cause of the positive or negative correlation between the pairs.

> ### EXAMPLE

Examine the scatter plots and the lines. Which line best fits the points, Line 1 or Line 2?

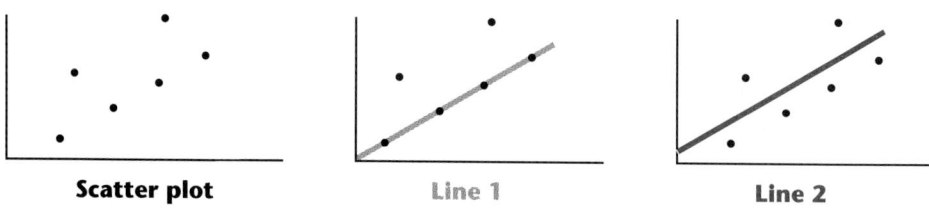

Scatter plot   **Line 1**   **Line 2**

A line of best fit should take all of the points into consideration.

Solution▶

Line 1 goes through four of the points but ignores the others.

Line 2 is closer to the four points than it is to the other two, but it takes all of the points into account. **Line 2** is a better fit, even though it doesn't go through any of the points. ❖

EXERCISES & PROBLEMS

Communicate

1. Explain how to determine the line of best fit between points using a piece of uncooked spaghetti for a line.

2. If the coefficient of correlation is 1 or − 1, what does this tell you about the data points and the line of best fit?

3. What does a correlation coefficient very close to + 1 tell you?

4. What does a correlation coefficient very close to − 1 tell you?

5. What does a correlation coefficient of 0.23 tell you?

Practice & Apply

Examine the airlines timetable spreadsheet.

6. What tends to happen to the minutes as the miles increase?

7. Is the correlation positive or negative?

8. The line of best fit rises and fits the points quite well. What does this tell you about the numerical value of the correlation?

City	miles	minutes
Jacksonville	270	64
St Louis	484	102
Pittsburgh	526	89
New York	765	120
Minneapolis	906	154
Boston	946	145
Denver	1208	190
Phoenix	1587	236
Los Angeles	1946	286
San Francisco	2139	312

Tell whether the given correlation coefficient describes a line of best fit that rises or falls. Also tell whether the line is a good fit or not.

9. 0.09   **10.** -0.92 **11.** 0.89   **12.** -0.45

Tell whether the correlation coefficient for the scatter plot is nearest to -1, 0, or 1.

13.   **14.**

15. Given the scatter plot, which line best fits the points? Explain.

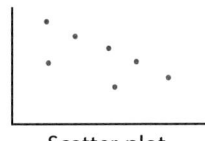

Scatter plot   Line 1   Line 2

Graph the points (1, 1), (1, 2), (2, 1), and (2, 2).

16. What shape do the points suggest?

17. According to one calculator, the line of best fit is a horizontal line $y = 1.5$. What other line fits equally well?

For Exercises 18–20, match the scatter plots to the correlation coefficients.   **A.** -0.9 **B.** 0.2   **C.** 0.9

18. 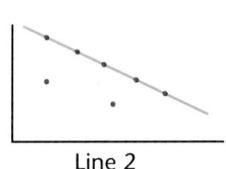   **19.**   **20.**

21. **Economics** The chart shows how food items that cost $1.00 in 1982 have increased in cost over the years. Make a scatter plot, and use a straightedge to estimate the line of best fit.

Year	1982	1985	1988	1991
Cost	1.00	1.06	1.18	1.36

Look Back

What are the next 2 terms of each sequence?   [Lesson 1.1]

22. 1, 4, 7, 10, 13, . . .   **23.** 2, 5, 10, 17, 26, . . .   **24.** 1, 2, 3, 4, 0, 1, 2, . . .

25. Create two sequences, and an expression that generates the terms. **[Lesson 1.3]**

26. Explain how to plot the point (5, 3) on a coordinate axis.   **[Lesson 1.5]**

27. **Geometry** The side of a square is 8 centimeters. What is its area?

Look Beyond

28. What value for x in $2x + 4 = 12$ will make the equality true?

29. What value for n in $32 - 2n = 10$ will make the equality true?

Patterns in numbers are sometimes closely related to patterns in art. The work of a quilt maker, for example, is filled with mathematically intricate designs. This project will give you a method for creating artistic patterns from numbers.

Many patterns can be created by using the remainders from dividing numbers. Divide the whole numbers from 0 to 9 by 4, and examine the remainders.

Number	0	1	2	3	4	5	6	7	8	9
Remainder	0	1	2	3	0	1	2	3	0	1

Repeating Patterns

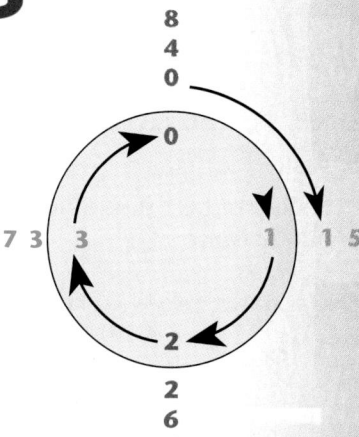

Notice how the remainders repeat. A pattern that repeats is sometimes called a cycle. This is like a clock that cycles the hours from 1 to 12 and back to 1. In this case, the cycle goes from 0 to 3 before repeating. For example, 4 brings you back to 0, 5 brings you to 1, and so on.

Building a Design Table

On the left and top margins on a piece of graph paper, place the whole numbers from 0 to 9. Make a standard multiplication table. On another piece of graph paper, make a table of the remainders when each product is divided by 4.

Standard Multiplication Table

X	0	1	2	3	4	5	6	7	8	9
0	0	0	0	0	0	0	0	0	0	0
1	0	1	2	3	4	5	6	7	8	9
2	0	2	4	6	8	10	12	14	16	18
3	0	3	6	9	12	15	18	21	24	27
4	0	4	8	12	16	20	24	28	32	36
5	0	5	10	15	20	25	30	35	40	45
6	0	6	12	18	24	30	36	42	48	54
7	0	7	14	21	28	35	42	49	56	63
8	0	8	16	24	32	40	48	56	64	72
9	0	9	18	27	36	45	54	63	72	81

Remainder Table

X	0	1	2	3	4	5	6	7	8	9
0	0	0	0	0	0	0	0	0	0	0
1	0	1	2	3	0	1	2	3	0	1
2	0	2	0	2	0	2	0	2	0	2
3	0	3	2	1	0	3	2	1	0	3
4	0	0	0	0	0	0	0	0	0	0
5	0	1	2	3	0	1	2	3	0	1
6	0	2	0	2	0	2	0	2	0	2
7	0	3	2	1	0	3	2	1	0	3
8	0	0	0	0	0	0	0	0	0	0
9	0	1	2	3	0	1	2	3	0	1

Creating the Designs

1. Choose a color for each number in the remainder table.

2. Create an artistic design for each number in the remainder table.

3. On one grid, replace the numbers with the colored squares. On the second grid replace the numbers with the artistic designs.

0   1   2   3

Activity 1

Create a pattern graph using the remainders from division by 5. Be creative in your choice of cell designs to replace the remainders. Extend beyond nine numbers for the left and top margin numbers to enlarge the pattern.

Activity 2

On another piece of graph paper, create a pattern using a similar set of numbers greater than 5. Try an operation other than multiplication. Use addition or some combination of operations. Vary the colors. Be creative.

Activity 3

Compare the patterns. See what conclusions you can draw from comparing the patterns for addition with the patterns for multiplication. How do the patterns compare when you use more numbers? What other interesting information can you find from the patterns?

Chapter 1 Review

Vocabulary

Key Skills & Exercises

Lesson 1.1

➤ **Key Skills**

Generate the terms of a sequence.

Examine the sequence 6, 15, 24, 33, 42, . . . The first term is 6. Each of the terms is 9 more than the one before. To find the next three terms add 9.

$$42 + 9 = 51 \qquad 51 + 9 = 60 \qquad 60 + 9 = 69$$

➤ **Exercises**

Find the next three terms of the sequence, then explain the pattern used to find the terms for each.

1. 1, 4, 7, 10, 13, . . . **2.** 1, 4, 16, 64, 256, . . . **3.** 27, 9, 3, 1, $\frac{1}{3}$, . . .

Lesson 1.2

➤ **Key Skills**

Extend a sequence by working backward from the differences.

Apply the method of differences until you get a constant. Then use the differences to find the next two terms.

$$10, 13, 19, 28, 40, . . .$$

```
       10    13    19    28    40
1st        3     6     9    12
2nd           3     3     3
```

To find the next first differences, add 3 to the previous first differences.

$$12 + 3 = 15 \qquad 15 + 3 = 18$$

To find the next terms, add the first difference to the previous term.

$$40 + 15 = 55 \qquad 55 + 18 = 73$$

The next two terms are 55 and 73.

➤ **Exercises**

Find the next two terms of each sequence.

4. 1, 5, 11, 19, 29, . . . **5.** 1, 1, 6, 16, 31, . . . **6.** 90, 70, 54, 42, 34, . . .

7. The fourth and fifth terms of a sequence are 21 and 32. If the second differences are a constant 3, what are the first 5 terms of the sequence?

Lesson 1.3

➤ *Key Skills*

Evaluate an expression using substitution and use the results to solve an equation.

Make a table to show the substitutions of the values 1, 2, 3, 4, 5 for x in the expression $2x + 3$. Then use the table of values to solve the equation

$$2x + 3 = 9.$$

x	1	2	3	4	5
$2x + 3$	5	7	9	11	13

We can see from the table above that $2x + 3 = 9$ when $x = 3$.

➤ *Exercises*

Make a table to show the substitutions of the values 1, 2, 3, 4, 5 for the variable.

8. Find the values of $10y$ by substituting for y.

9. Find the values of $8t - 2$ by substituting for t.

10. Use the table of values in 9 to solve the equation $8t - 2 = 14$.

Lesson 1.4

➤ *Key Skills*

Perform calculations in the proper order.

To simplify the expression $5(7 - 4) - 6^2 \div 3 + 1$, follow the order of operations: parentheses, then exponents, next multiplication/division left to right, finally addition/subtraction left to right.

$$5(7 - 4) - 6^2 \div 3 + 1$$
$$5(3) - 6^2 \div 3 + 1$$
$$5(3) - 36 \div 3 + 1$$
$$15 - 12 + 1$$
$$4$$

➤ *Exercises*

Simplify.

11. $17 - 4 \cdot 3$

12. $32 - 24 \div 6 - 4$

13. $3 \cdot 4^2 - [24 \div (6 - 4)]$

Lesson 1.5

➤ *Key Skills*

Graph points from the ordered pairs.

Graph the ordered pairs (7, 2), (1, 4), and (3, 5). Tell whether they lie on a straight line.

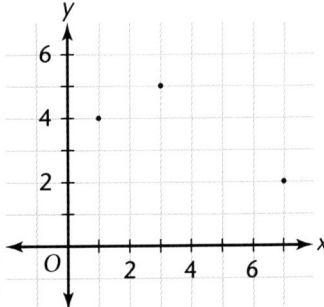

The points do not lie on a straight line.

➤ *Exercises*

Graph each list of ordered pairs. Tell whether they lie on a straight line.

14. (1, 2), (2, 3), (3, 4)

15. (2, 5), (4, 2), (5, 2)

16. (1, 3), (1, 4), (3, 5)

17. (0, 1), (1, 2), (3, 6)

18. Graph the linear function $y = x + 2$.

Lesson 1.6

➤ Key Skills

Identify the fit of data to a line.

Describe the correlation for the scatter plot as strong positive, strong negative, or little or none.

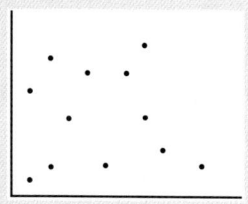

The correlation is little or none since the points do not cluster about a line.

➤ Exercises

Describe the correlation as strong positive, strong negative, or little or none.

19.   **20.**

Lesson 1.7

➤ Key Skills

Understand the relationship of the measure of correlation between the values − 1 and 1 to the closeness of the fit of data to the line.

A correlation coefficient of 0.99 describes a line of best fit which rises. It is a good fit because 0.99 is close to 1.

A correlation coefficient of − 0.98 describes a good fit, which is falling instead of rising.

➤ Exercises

Tell whether the given correlation coefficient describes a line of best fit which rises or falls. Also tell whether it is a good fit or not.

21. 0.02   **22.** − 0.95

23. Tell whether the correlation coefficient for the scatter plot in 19 is nearest to − 1, 0, or 1.

24. Tell whether the correlation coefficient for the scatter plot in 20 is nearest to − 1, 0, or 1.

Applications

25. Entertainment If it costs $13 per person to enter the amusement park, how many people can get in for $98?

Suppose the cost to order tickets to the ballet is $22 per person, plus a $5 handling charge per order (regardless of how many tickets are ordered).

26. How much does an order of 4 tickets cost?
27. How much does an order of 8 tickets cost?
28. Represent the number of tickets by t, and write an equation for the cost of an order of tickets.

Chapter 1 Assessment

Find the next three terms of each sequence.

1. 4, 8, 16, 32, 64, . . .   **2.** 49, 40, 32, 25, 19, . . .

3. If a softball league has 15 teams and each of the teams play each of the other teams, how many games will there be?

4. If the second differences of a sequence are a constant 5, the second of the first differences is 11, and the second term is 12, find the first 5 terms of the sequence.

5. Use guess and check to solve the equation $3x + 4 = 16$. Find the value of x.

If notebooks cost 59 cents, find the cost of the following.

6. 2 notebooks   **7.** 5 notebooks   **8.** 12 notebooks

9. Write an equation to model the situation. How many notebooks can you get for $14.75?

10. Find the values of $9x - 5$ by substituting 1, 2, 3, 4, 5 for x.

11. How much is $3 + 27 \div 3^2 - (7 - 5)$?

What are the coordinates of the given points?

12. A
13. B
14. C
15. D

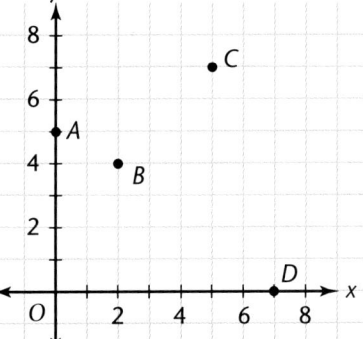

Teri drives at an average rate of 55 miles per hour.

16. Represent hours by h, and write an equation for the distance.

17. Draw a graph of the function. What type of function is shown by the graph?

18. Make a scatter plot of this data for the amount of time spent studying and the score on the test the next day.

Time Studying in minutes	15	12	20	30	25	40	60	50	55	45
Test Score	65	68	75	80	79	85	97	80	88	73

19. Describe the correlation as strong positive, strong negative, or little or none.

20. Tell whether the correlation coefficient for the scatter plot is nearest to -1, 0 or 1.

21. Use a straightedge to estimate the line of best fit for this scatter plot.

CHAPTER 2

LESSONS

A Preview of Functions

Many activities that people take for granted are rich in mathematics. Functions are particularly valuable because they let you see the way numerical relationships behave.

Linear functions appeared in the first chapter. In this chapter, you will have the opportunity to examine other common functions that are part of the mathematician's toolbox. These functions include those that appear on standard scientific and graphics calculators. You will be able to study these functions in more detail in later chapters.

Paula walks into the post office to mail a package to her friend and notices that the cost for postage relates to the package's weight. A function shows this relationship.

Alex changes the length of the vibrating part of a guitar string by moving his fingers. A function shows how the number of vibrations per second relates to the length of the string.

Reciprocal Function Graph

Portfolio Activity

"Walk-and-Tie" Problem

Years ago when people traveled on horseback, they might find themselves away from home without enough money for the return trip. A common solution for a pair of people with two horses was to sell one horse and take turns riding the other one home. Since a horse walks faster than a person, one person rode ahead for a while, tied up the horse, and continued walking toward home.

The second person would reach the point where the horse was tied and begin riding until he or she caught up with the first person. They would repeat this until they got home.

Suppose a horse travels 5 miles per hour and two people walk 3 miles per hour. How long will it take to make a 20-mile trip if the "walk-and-tie" method is used?

Previewing Exponential Functions

Why *What happens each time you double your money? An exponential function is used to determine the total amount at any time.*

> **Prize A**
> Start with $100,
> add $100 each
> day, for 20 days!
>
> **Prize B**
> Start with 1¢, double
> your money each
> day, for 20 days!

Students at Lincoln High have been playing the Math Know-How game. After the game, the winners will have 10 seconds to decide which prize to choose. They get to keep the money from the 20th day.

•Exploration 1 *Evaluating Your Options*

Calculator

To help the winners decide, try this exploration with a partner. One person finds the amount of Prize A and the other the amount of Prize B. If calculators are available, use the constant feature.

Prize A
Start with 100. Repeatedly *add* 100 to the previous day's total.

Prize B
Start with 0.01. Repeatedly *multiply* the previous day's amount by 2.

	Prize A		Prize B
Day	Total Amount	Day	Total Amount
1	$100	1	$0.01
2	$200	2	$0.02
3	$300	3	$0.04
4	$400	4	$0.08

1 After four days, which prize would you pick?

2 Do you think Prize B will ever exceed Prize A?

3 Continue each table through 20 days.

4 Did Prize B ever exceed Prize A? If so, when?

5 Extend each table to 30 days. What is each prize worth? ❖

Introducing Exponential Functions

In Chapter 1 you worked with linear functions. Are all functions linear functions?

Spreadsheet

The population of Parksburg grew from 10,000 to 11,000 in one year. Calculate and graph the population growth for 10 years if

Ⓐ the population grows by a fixed *amount* each year.

Ⓑ the population grows at a fixed *rate* each year.

Solution➤

Demographics

Ⓐ The first year's growth is 1000 people. If the population grows by a fixed amount, it will gain 1000 people each year. You can use a spreadsheet to display this information. Start with 10,000 and repeatedly *add* 1000 to the previous number. The results appear in column B.

	A	B	C
	Year	Fixed amount	Fixed rate
1	0	10,000	10,000
2	1	11,000	11,000
3	2	12,000	12,100
4	3	13,000	13,310
5	4	14,000	14,641
6	5	15,000	16,105
7	6	16,000	17,716
8	7	17,000	19,487
9	8	18,000	21,436
10	9	19,000	23,579
11	10	20,000	25,937

Ⓑ The first year's growth is 1000 people. To figure the *fixed rate* of growth for each year, divide 1000 by 10,000.

$$fixed\ rate = \frac{\text{first year change}}{\text{original population}} = \frac{1000}{10,000} = \frac{10}{100} = 10\%$$

The new population after one year is 100% of the original population plus an additional 10%. Each year, the population is 110% of the previous year's. Enter this on the spreadsheet by starting with 10,000 and repeatedly *multiplying* by 1.10 or 1.1.

Use the data to draw each graph on the same axis.

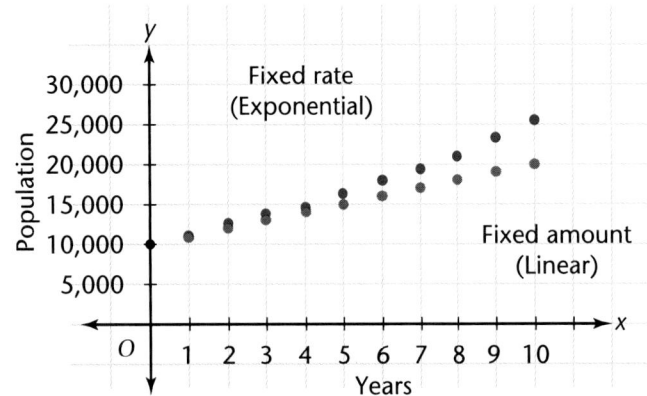

When population grows by a *fixed amount,* the points lie along a line. This relationship is **linear**. When the population grows at a *fixed rate,* the relationship is **exponential**. You can see from the table and the graph that exponential growth is only slightly greater than linear growth for the first few years. To see how quickly the exponential function increases, examine the data for 10 more years.

Comparing Exponential Growth and Decay

When you found the population of Parksburg by repeatedly multiplying by 1.1, the population grew exponentially. The function modeled **exponential growth**.

EXAMPLE 2

Calculator

What would happen to Parksburg's population if you repeatedly multiplied it by 0.9?

Solution▶

If you use a calculator with a constant feature, enter 10,000. Now multiply by 0.9 repeatedly. The output is 9000, 8100, 7290, 6561, . . . The sequence decreases, as does the population. The function models **exponential decay**. ❖

Try This   Start with 100. First multiply repeatedly by 1.2. Write the first five terms of the sequence that results. Then multiply 100 repeatedly by 0.8, and write the first five terms of the sequence that results.

CRITICAL
Thinking

What happens to the growth and decay of 100 as your multiplier gets closer and closer to 1?

Compound Interest

Escalating costs are making it necessary to save for years to cover college expenses. Compound interest accounts earn increasing amounts of interest over long periods of time.

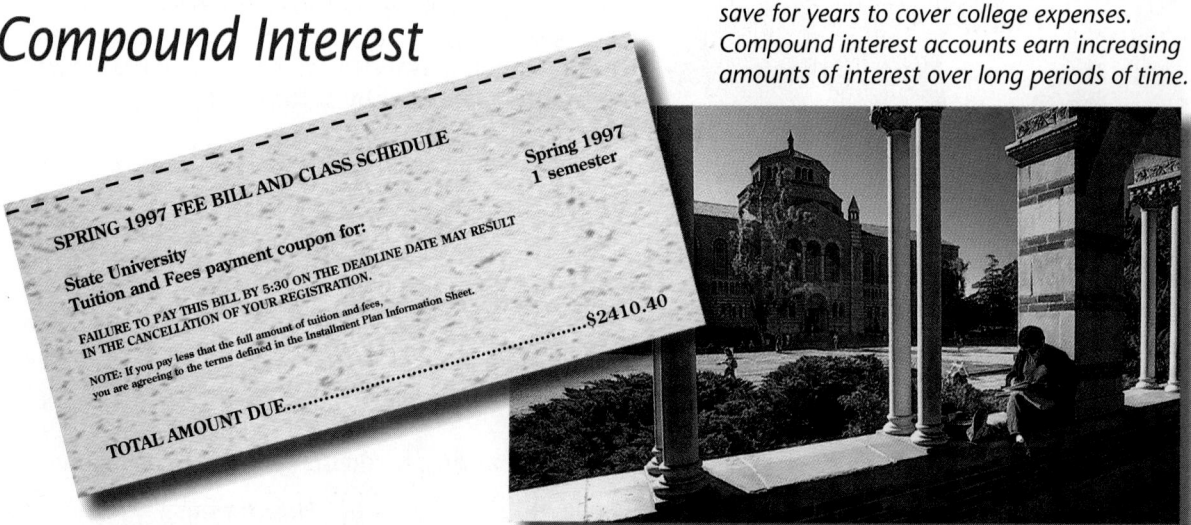

Finance   The money paid or earned on a given amount of money often involves **compound interest**. Compound interest is an example of exponential growth.

EXAMPLE 3

When Jill was eight years old, her parents invested $10,000 in a college tuition fund. The interest rate is 8% compounded once at the end of each year. How much will Jill have for tuition after 10 years?

Solution➤

The interest for the *first year* is 8% of $10,000, which is

$$0.08 \cdot 10{,}000 = \$800.$$

After one year the investment is $10,000 + $800 = $10,800.

The interest for the *second year* is 8% of $10,800, which is

$$0.08 \cdot \$10{,}800 = \$864.$$

After two years the investment is $10,800 + $864 = $11,664.

Since you are adding 8% to 100%, multiply Jill's balance by 1.08.

Spreadsheet

A calculator or spreadsheet will help you construct a table to find what the investment is worth after 10 years. Start with $10,000 and multiply repeatedly by 1.08.

Will Jill have enough tuition for four years at State U? ❖

	A Year	B Cmp. interest
1	Year	Cmp. interest
2	0	10,000.00
3	1	10,800.00
4	2	11,664.00
5	3	12,597.12
6	4	13,604.89
7	5	14,693.28
8	6	15,868.74
9	7	17,138.24
10	8	18,509.30
11	9	19,990.05
12	10	21,589.25

An equation such as $y = 2^x$ is an example of an **exponential function**. The base, 2, is the number that is repeatedly multiplied. The exponent, x, represents the number of times that the base, 2, occurs in the multiplications.

Exploration 2 Growth and Decay

 Copy and complete the table for $y = 2^x$.

x	1	2	3	4	5
y	2	4	8	?	?

 Graph the function $y = 2^x$ for the values in Step 1.

 What do you notice about the graph you made for $y = 2^x$?

 Make a table of x and y values for $y = \left(\frac{1}{2}\right)^x$. Use the same x-values that you used for Step 1. Then graph the function.

 Compare the graphs of $y = 2^x$ and $y = \left(\frac{1}{2}\right)^x$. How are they different? ❖

EXERCISES & PROBLEMS

Communicate

Explain how to find the rule for each of the following sequences.

1. 10, 20, 40, 80, 160, . . .   **2.** 20, 10, 5, 2.5, 1.25, . . .

3. 10, 20, 30, 40, 50, . . .   **4.** 160, 140, 120, 100, 80, . . .

5. How can you tell if any of the sequences in Exercises 1–4 are exponential?

6. Describe the characteristics of a linear sequence. Which of the sequences in Exercises 1–4 are linear?

7. Explain what is meant by *exponential decay*. Explain what is meant by *exponential growth*.

8. How do you determine the number that multiplies the principal when you solve the following problem? How much will Sabrina have to pay if she borrows $1000 at 10% interest, compounded once each year for 4 years?

Practice & Apply

9. Write the first five terms of a sequence that starts with 5 and doubles the previous number.

What are the next three terms of each sequence?

10. 100, 300, 900, 2700, 8100, . . .   **11.** 100, 300, 500, 700, 900, . . .

12. 100, 10, 1, 0.1, 0.01, . . .   **13.** 40, 35, 30, 25, 20, . . .

14. Which of the sequences in Exercises 10–13 show exponential growth?

15. Which of the sequences in Exercises 10–13 are linear?

16. Is the sequence 64, 16, 4, 1, $\frac{1}{4}$, . . . linear or exponential?

17. Generate a table for the function $y = 3^x$ when x is 1, 2, 3, 4, and 5.

18. **Cultural Connection: Africa** An ancient Egyptian papyrus of mathematical problems includes the statement, "Take $\frac{1}{2}$ to infinity." Explore the sequence you get by starting with 0.5 and repeatedly multiplying by 0.5. What happens to each successive product?

19. **Finance** Jacy invests $1000. His investment earns 5% interest per year and is compounded once each year. How long will it take Jacy to double his money?

Science Suppose a ball bounces $\frac{1}{2}$ its previous height on each bounce. Find how high the ball bounces on the

20. second bounce.   **21.** third bounce.

22. fourth bounce.   **23.** fifth bounce.

24. Write the sequence you get if you start with 1 and keep tripling the previous number.

25. Is the sequence 100, 90, 80, 70, 60, . . . linear or exponential?

26. Define *fixed rate*.

27. Define *fixed amount*.

28. Define *compound interest*.

29. Start with the number 8, and double the amount 5 times. What number do you get?

30. **Finance** Consider the terms for a First Money credit card. The company has a promotion where interest is not charged for the first 6 months. What is the balance to the nearest cent after the sixth payment on $1000 if only minimum payments are made?

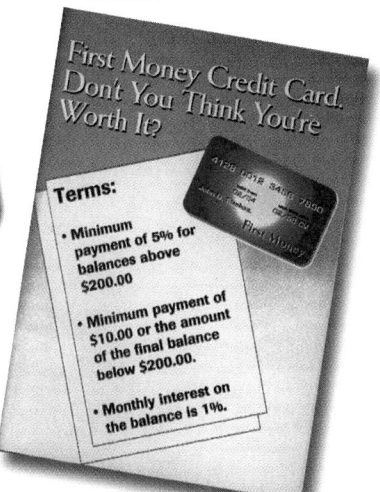

Terms:
- Minimum payment of 5% for balances above $200.00
- Minimum payment of $10.00 or the amount of the final balance below $200.00.
- Monthly interest on the balance is 1%.

Look Back

Write the rule used to find the next term in each pattern. Find the missing terms. [Lesson 1.1]

31. 3, 7, 11, 15, 19, __, __   **32.** 4, 2, 1, $\frac{1}{2}$, $\frac{1}{4}$, __, __   **33.** 33, 30, 27, 24, 21, __, __

Find the next two terms for each sequence. [Lesson 1.1]

34. 10, 21, 34, 49, 66, __, __   **35.** 99, 78, 72, 51, 45, __, __

Plot the following points on the coordinate plane. [Lesson 1.5]

36. $A(3, 5)$ **37.** $B(5, 6)$ **38.** $C(6, 5)$ **39.** $D(4, 5)$

Look Beyond

40. **Technology** Explore the sequence you find when you begin with 100 and repeatedly multiply by -2. Do the numbers get close to any particular number? Some calculators have a special key for "negative," such as ⟨(-)⟩. On other calculators you enter 2 and then ⟨+/-⟩ to get -2.

41. Explore the sequence you find when you begin with 100 and repeatedly multiply by -0.5. Do the numbers get close to any particular number?

LESSON 2.2
Exploring Quadratic Functions

Why *The path of a softball, basketball, or volleyball is often a parabola. This path can be modeled by a quadratic function. In geometry, quadratic functions model problems that relate to area.*

•Exploration 1 *Edges and Surface Area*

GEOMETRY
Connection

edge   face

1 m
1 m
1 m

Suppose you are going to paint all six faces of a cube. Guess how much paint will be needed if you decide to double the length of each edge. You might be surprised to find that it will take more than twice the amount to paint the faces of the larger cube.

Suppose the edge of a cube is 1 meter. Since the area of each face is 1 square meter ($1m^2$), the surface area of the six faces is 6 m^2. Examine how the surface area of the six faces grows as the length of each edge is increased.

2 m
2 m
2 m

1 Compare the surface area of a 1-meter cube with the surface area of a 2-meter cube. What happens to the surface area of the six faces when the length of each edge doubles?

3 m
3 m
3 m

2 Examine a cube whose edge is triple the length of the first cube's edge. What happens to the surface area of the six faces when the length of each edge triples?

3 Examine the surface area patterns for the six faces of the 1-, 2-, and 3-meter cubes. What is the ratio of the surface area of a 2-meter cube to the surface area of a 3-meter cube? How does this ratio relate to the ratio for the lengths of the edges of the two cubes?

4 Extend the following table to include a length of 5 meters for an edge. Notice the pattern.

Length of each edge in meters	1	2	3	4	5	...	e
Area of each face in m²	1	4	9	16	?	...	e^2
Formula for the total surface area of the cube	6(1) 6(1²)	6(4) 6(2²)	6(9) 6(3²)	6(16) 6(4²)	? ?	... ...	?
Total surface area of the cube in m²	6	24	54	96	?	...	

5 Write a formula for the total area of the six faces of a cube based on the pattern you see in the table.

In the exploration the relationship involves the *square* of the length of an edge. Functions that involve squaring, such as $y = 6e^2$ or $y = x^2$, are called **quadratic functions**.

How can you use this pattern to find how much paint you need to paint cubes of different sizes?

These students are painting a portion of the set for a school play.

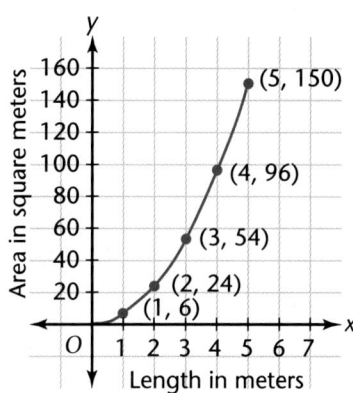

If you want to get a picture of the information in the table on page 65, you can draw a graph. Start by writing five ordered pairs that relate edge length to surface area (edge length, surface area).

These ordered pairs are (1, 6), (2, 24), (3, 54), (4, 96), and (5, 150). You can plot the points and connect them with a smooth curve like the one at the left.

Only part of the quadratic curve appears. Why does the graph show only the positive values of the quadratic function? ❖

Physics   The following spreadsheet shows the data from the rocket problem in Lesson 1.2. The graph of the data is called a **parabola**.

seconds	height (ft)
0	0
1	208
2	384
3	528
4	640
5	720
6	768
7	784
8	768
9	720
10	640
11	528
12	384
13	208
14	0

Rocket Data

Use the data from the spreadsheet to draw a graph. The curve is a parabola that opens downward. You can see that the rocket reaches its maximum height, 784 feet, after 7 seconds.

A parabola may open either **downward** or **upward**. The point where the curve changes direction is the **vertex**. ❖

•Exploration 2 *Graphing Sequences*

Use the method of differences (Lesson 1.2) to discover the next two terms in the sequence 1, 4, 9, 16, 25, . . .

1   Calculate the first differences. Are they constant? If not, what are the second differences?

2   Work backward using the pattern, and predict the missing terms.

3 What do you notice about the second differences?

4 Graph the sequence. Use the term number for x and the value of the term for y. For example, the first three points will be (1, 1), (2, 4), and (3, 9). What kind of curve appears when you connect the points?

5 Find the first differences for the sequence 4, 8, 12, 16, 20. Are the differences constant?

6 Graph the sequence. For example, the first three points will be (1, 4), (2, 8), and (3, 12). What kind of curve appears when you connect the points?

7 How does the graph in Step 4 differ from the graph in Step 6? Which differences were constant for each sequence? ❖

EXERCISES & PROBLEMS

Communicate

1. Explain what happens to the area of a square if you double the length of each side.

2. What clues allow you to recognize a quadratic relationship? List at least two clues.

3. Explain how you would use a pattern to determine the type of relationship between the given pairs.

x	1	2	3	4	5
y	4	0	4	0	4

How would you decide which graph matches which relationship?

a. Linear
b. Exponential
c. Quadratic
d. None of these

4.

5.

6.

7.

8.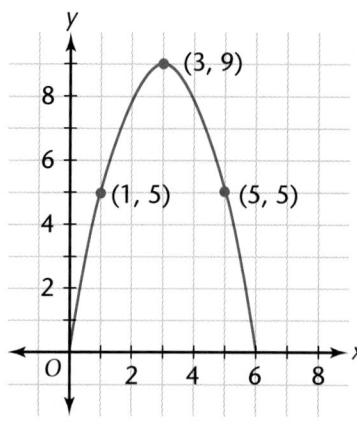

Explain how to determine the vertex in the parabola shown. Explain how to determine the vertex from a table.

9. Discuss how differences are used to determine sequences and how the differences relate to quadratic relationships. What is the importance of constant differences?

Practice & Apply

Match each table with the correct type of relationship.

a. Linear   **b.** Exponential   **c.** Quadratic   **d.** None of these

10.

x	1	2	3	4	5
y	3	6	11	18	27

11.

x	1	2	3	4	5
y	1	2	4	8	16

12.

x	1	2	3	4	5
y	9	8	7	6	5

13.

x	1	2	3	4	5
y	4	6	8	10	12

14.

x	1	2	3	4	5
y	10	17	26	37	50

15.

x	1	2	3	4	5
y	80	40	20	10	5

Geometry What happens to the area of a square if you

16. triple the length of each side?

17. multiply the length of each side by 10?

18. take $\frac{1}{3}$ the length of each side?

19. **Geometry** What is the area of each pizza shown? Use 3.14 for π.

20. Consumer Math Which pizza is the better buy?

$d = 15''$
$9.00

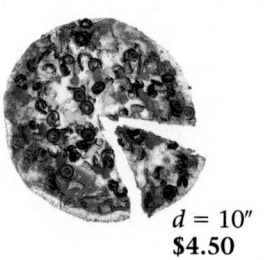

$d = 10''$
$4.50

Match each relationship with its graph.

a. Linear   **b.** Exponential   **c.** Quadratic   **d.** None of these

21.   **22.**   **23.**   **24.**

25. How many times larger is the side of the larger square than the side of the smaller square?

26. A parabola has a vertex at $(8, -2)$ and contains the point $(5, 3)$. Give the coordinates of another point on the parabola.

27. Tell whether the parabola opens upward or downward.

Physics Find the height of each waterfall if the time the float takes to fall is

28. 1 second.
29. 2 seconds.
30. 5 seconds.

31. Graph the points you found for the waterfalls in Exercises 28–30.

32. Use your graph to estimate how long it would take a float to fall from a waterfall with a height of 600 feet.

 Look Back

Write the next term in each pattern. Leave answers in fraction form.   [Lesson 1.1]

33. 11, 33, 99, 297, 891, ___
34. 3125, 625, 125, 25, ___

35. $\frac{4}{5}, \frac{8}{10}, \frac{16}{20}, \frac{32}{40},$ ___

36. If each package of notebook paper costs $1.29, find the cost of 5 packages.   **[Lesson 1.4]**

Evaluate each expression for the values $x = 2$, $y = 1$, and $z = 4$.   [Lesson 1.3]

37. $3xy$ **38.** $4z$ **39.** $21yz$ **40.** xyz

Substitute values for x to find y. Form ordered pairs, and plot the points. [Lesson 1.5]

41. Graph the line for the equation $y = 4x + 6$.
42. Graph the line for the equation $y = 3x + 2$.

Look Beyond

Identify the coordinates for the vertex of each function.

43. $y = (x - 2)^2 + 7$ **44.** $y = x^2 + 2x + 1$

The height, h, (in feet) of a waterfall can be determined by the time, t, (in seconds) that it takes a float in the water to fall from the top to the bottom, according to the formula $h = 16t^2$. For example, if the float takes 3 seconds to fall, the height is $16 \cdot 3^2 = 144$ feet.

LESSON 2.3 Previewing Reciprocal Functions

why *How many cars must the students wash to meet their goal? If they decide to charge $4 per car, how many cars will they have to wash? As the price per car decreases, what happens to the number of cars they have to wash? A reciprocal function models this relationship.*

The carwash was successful. The students met their goal of $500, but the school needs $1000 more to help a needy family. If the students find 10 contributors, how much money will each person need to contribute to meet the goal?

For 10 contributors, divide the goal of $1000 by 10. Each of the 10 contributors will need to pay $100. Look at what happens when a greater number of contributors donate. For 20 contributors, divide $1000 by 20. What do you get? Calculate the amounts donated when there are 50, 100, and 1000 contributors. Compare your results with the data in the table.

Let *n* be the number of contributors	10	20	50	100	1000
Let *d* be the dollars given per person	100	50	20	10	1

Graphing Reciprocal Functions

To graph a function, substitute numbers into the equation to get ordered pairs. Then plot the points on the coordinate plane and connect the points to form a curve. You can save time by using a graphics calculator or computer to construct the graph directly from the function.

Plot the ordered pairs (10, 100), (20, 50), (50, 20), and (100, 10), and connect them with a curve. The point (1000, 1) is omitted because it would make the scale too large to see most of the other points. As the number of contributors increases, what happens to the number of dollars each contributor makes?

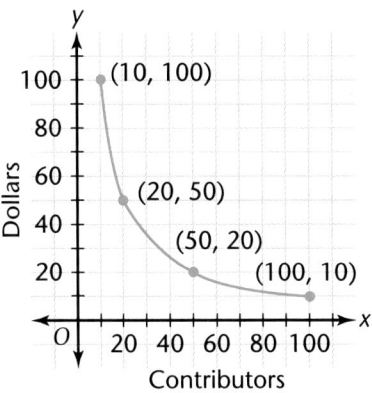

Examine the ordered pairs on the graph. Notice that in each case, the y-value is found by dividing 1000 by x. Functions such as $y = \frac{1000}{x}$ and $y = \frac{1}{x}$ are examples of **reciprocal functions**.

Try This Copy and complete the table which shows the amount, y, each contributor must make to raise $500 when x represents the number of contributors.

x	1	2	5	10	20	25	50	100	250	500
y	500	250	100	50	25	?	?	?	?	?

EXAMPLE 1

Make a table of the values for the reciprocal function $y = \frac{1}{x}$. Then graph the function.

Solution▶

Substitute values for x in the equation $y = \frac{1}{x}$ to make a table of ordered pairs. The y-values represent the reciprocals of the values that you choose for x.

x	1	2	4	5	$\frac{1}{2}$	$\frac{1}{4}$	$\frac{1}{5}$
y	1	$\frac{1}{2}$	$\frac{1}{4}$	$\frac{1}{5}$	2	4	5

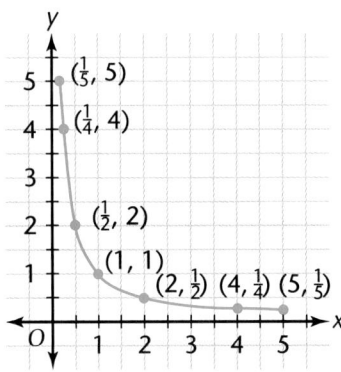

Plot the ordered pairs on the graph, and connect the points. Examine the curve formed from the values of the function. How can you find the reciprocal of $\frac{1}{8}$? ❖

Graphics Calculator

You can use a graphics calculator to display $\frac{1}{x}$. Enter x, and press the $\boxed{x^{-1}}$ key to enter the function after **Y=**. Draw the graph. Use the trace feature to find the reciprocal of $\frac{1}{8}$.

CRITICAL Thinking Examine the graph of $y = \frac{1}{x}$. What happens to the y-value of the function as the x-value decreases from 1 to 0? Why do you think the function is not defined when x is 0?

Reciprocals and Music

The frequency of the open string is 98 vps. Stopping the string halfway produces a frequency of 196 vps.

Cultural Connection: Europe A famous example of a reciprocal relationship occurs in music. Every culture that had stringed instruments probably discovered that the longer the string, the lower the frequency, or pitch, of the note it produced. By the time of the Greek scholar Aristotle, it was known that *doubling the length of a musical string would produce half the frequency* in vibrations per second.

EXAMPLE 2

Suppose a musical string 48 centimeters long produces a frequency of 440 vibrations per second (vps). This frequency produces the musical note A, the usual tuning note for an orchestra. What happens to the frequency when the length of the vibrating string is 96 centimeters?

Solution▶

To determine the frequency for the 96-centimeter string, find the relationship of the new length to the original 48-centimeter string.

$$96 = 2 \cdot 48$$

The string is 2 times as long as the 48-centimeter string. Based on the reciprocal relationship, *doubling the length halves the frequency.* The frequency of the 96-centimeter string is $\frac{1}{2}$ of the 440 vps you get from the 48-centimeter string.

$$\frac{1}{2} \cdot 440 = 220 \text{ vps} \;❖$$

Toccata and Fugue in D minor by Johann Sebastian Bach

Note	A	G	A		D		A	G	A
VPS	880	784	880		588		440	392	440

EXAMPLE 3

Suppose a 15-inch organ pipe produces a frequency of 440 vps. What happens to the frequency when the length of the pipe is

A 5 inches?

B 60 inches?

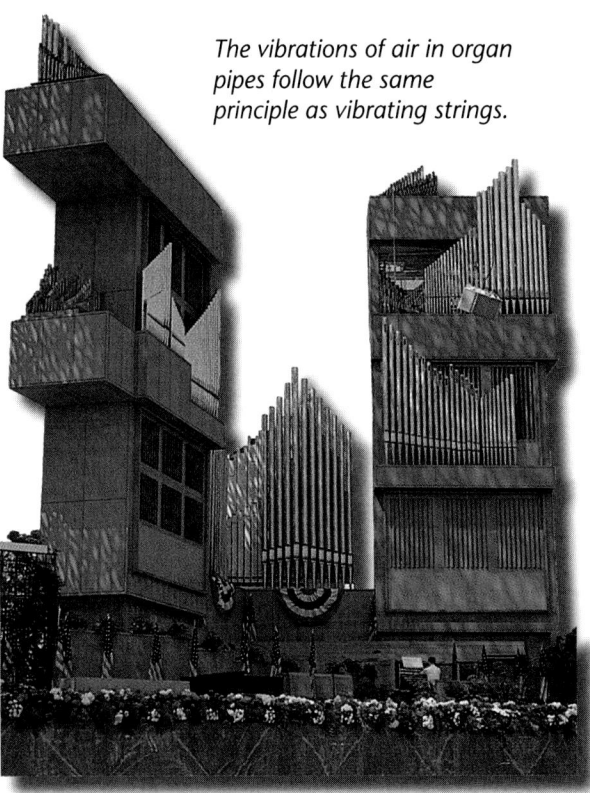

The vibrations of air in organ pipes follow the same principle as vibrating strings.

Solution▶

A To determine the frequency for a 5-inch pipe, find the relationship of the new length to the original 15-inch length.

$$5 = \frac{1}{3} \cdot 15$$

The pipe is $\frac{1}{3}$ as long as the 15-inch pipe.

Based on the reciprocal relation, $\frac{1}{3}$ of the length has 3 times the frequency. The frequency of the 5-inch pipe is 3 times the 440 vps you get from the 15-inch pipe.

$$3 \cdot 440 = 1320 \text{ vps}$$

B To determine the frequency for a 60-inch pipe, find the relationship of the new length to the original 15-inch length.

$$60 = 4 \cdot 15$$

The pipe is 4 times as long as the 15-inch pipe. Based on the reciprocal relation, 4 times the length has $\frac{1}{4}$ of the frequency. The frequency of the 60-inch pipe is $\frac{1}{4}$ of the 440 vps you get from the 15-inch pipe.

$$\frac{1}{4} \cdot 440 = 110 \text{ vps} ❖$$

D
294

A G A
220 196 220

D
147

EXERCISES & PROBLEMS

Communicate

1. Explain a reciprocal function using the terms *increasing* and *decreasing*. Use examples to describe the relationship.

Tell how to find the reciprocal of each number.

2. 5   **3.** 100   **4.** $\frac{1}{4}$ **5.** $\frac{1}{6}$

Discuss how you would find the amount each person would have to contribute to reach a goal of $1000 if there were

6. 4 people.   **7.** 200 people.

8. Describe how you would set up the following problem. Jane's co-workers are chipping in to buy her a birthday present. How much must each contribute if they expect to spend $30 and there are 5 co-workers?

Practice & Apply

x	6	5	3	1	$\frac{1}{3}$	$\frac{1}{5}$	$\frac{1}{6}$
y							

9. Complete the table. Let the y-value for each x be the reciprocal of x. Graph the function.

10. Use variables to set up the following problem. Suppose a trip takes 4 hours if you average 45 miles per hour. How long will it take if you average 60 miles per hour?

Find the reciprocal of each number.

11. 7   **12.** 5   **13.** 25   **14.** $\frac{1}{9}$ **15.** $\frac{1}{10}$ **16.** $\frac{1}{7}$

17. Explain why zero has no reciprocal.

18. Explain why 1 is its own reciprocal.

Suppose a string on a guitar is 60 centimeters long and produces a frequency of 300 vibrations per second.

Music What frequencies will be produced by similar guitar strings with the following lengths?

19. 30 cm   **20.** 20 cm   **21.** 90 cm

Music Given that a string 32 centimeters long produces a frequency of 660 vibrations per second, find the frequencies produced by similar strings with the following lengths.

22. 96 cm   **23.** 16 cm   **24.** 24 cm

Plot several ordered pairs for each of the given rules, and join them using a smooth curve.

25. $y = \frac{12}{x}$ **26.** $y = \frac{20}{x}$ **27.** $y = \frac{24}{x}$

Fund-raising   How much will each person have to contribute to reach a goal of $100 if there are

28. 10 people?   **29.** 20 people?   **30.** 100 people?   **31.** 1000 people?

Transportation   Sixty students and teachers from two algebra classes are going on a field trip to a museum. How many cars are needed if each car holds the number of passengers shown?

32. 3 passengers   **33.** 4 passengers   **34.** 5 passengers

How fast would you have to drive to complete a 240-mile trip in

35. 6 hours?   **36.** 5 hours?   **37.** 4 hours?

38. 3 hours?   **39.** 1 hour?   **40.** $\frac{1}{2}$ hour?

41. Suppose a trip takes 4 hours if you average 50 mph. How long will it take if you average 40 mph?

42. If other factors are held constant, the frequency of a musical note is a constant multiplied by the reciprocal of the length. If a musical string 48 centimeters long produces a frequency of 440 vps, find the constant.

Look Back

43. Find the first differences for the following sequence. Are they constant?   **[Lesson 1.2]**

89, 81, 74, 66, 60

44. How many guitar picks can you buy with $2.25? Write an equation that models this problem. **[Lesson 1.3]**

45. How many $6 movie tickets can you buy with $21?   **[Lesson 1.3]**

46. Make a table to solve the equation $3x + 4y = 12$. **[Lesson 1.5]**

Guitar Picks
25¢

Graph each function on the same coordinate plane, and compare the graphs.   [Lesson 1.5]

47. $y = 10 - x$ **48.** $y = x + 6$ **49.** $x + y = 8$

Plot the following points on a graph.   [Lesson 1.5]

50. (1, 1), (2, 2), (6, 6), (7, 7)

51. (0, 0), (1, 4), (2, 6), (3, 2)

52. (0, 6), (2, 5), (3, 4), (4, 2)

Look Beyond

53. Graph $y = \frac{3}{x}$ for $x = 1, 2, 4, 5, \frac{1}{2}, \frac{1}{4}, \frac{1}{5}, -1, -2, -4, -5, -\frac{1}{2}, -\frac{1}{4}, -\frac{1}{5}$.

Are the DINOSAURS Breeding?

In the novel *Jurassic Park*, dinosaurs created from preserved DNA roam an island theme park. New batches of these cloned creatures are added to the population every six months. During a pre-opening tour, Dr. Malcolm, a mathematician, and Mr. Gennaro, a lawyer, discuss a graph of the heights of one of the dinosaur species on the island.

Height Distribution: Procompsegnathide

"Yes," Malcolm said. "Look here. The basic event that has occurred in Jurassic Park is that the scientists and technicians have tried to make a new, complete biological world. And the scientists in the control room expect to see a natural world. As in the graph they just showed us. Even though a moment's thought reveals that nice, normal distribution is terribly worrisome on this island."

"It is?"

"Yes. Based on what Dr. Wu told us earlier, one should never see a population graph like that."

"Why not?" Gennaro said.

"Because that is a graph for a normal biological population. Which is precisely what Jurassic Park is not. Jurassic Park is not the real world. It is intended to be a controlled world that only imitates the natural world."

Later, Dr. Malcolm discusses the graph with one of the park's scientists, Dr. Wu. They debate whether the supposedly all-female population is somehow breeding.

"Notice anything about it?" Malcolm said.

"It's a Poisson distribution," Wu said. "Normal curve."

"But didn't you say you introduced the compys in three batches? At six-month intervals?"

"Yes . . ."

"Then you should get a graph with peaks for each of the three separate batches that were introduced," Malcolm said, tapping the keyboard. "Like this."

"But you didn't get this graph," Malcolm said.

"The graph you actually got is a graph of a breeding population. Your compys are breeding."

Height Distribution: Procompsegnathide

Wu shook his head. "I don't see how."

"You've got breeding dinosaurs out there, Henry."

"But they're all female," Wu said. "It's impossible."

Cooperative Learning

To help settle the breeding debate, you can model dinosaur populations with random numbers.

1. Model a breeding population by following these steps.
   a. Represent 15 baby dinosaurs with 15 letters A–O as shown.

A	B	C	D	E	F	G	H	I	J	K	L	M	N	O

   b. Use a random number table to generate digits, 0–9.
   Let each digit represent a week's growth for one dinosaur.

A	B	C	D	E	F	G	H	I	J	K	L	M	N	O	
8	1	0	1	7	4	9	0	2	7	7	9	0	3	1	(week 1)
5	0	9	1	2	0	9	3	9	9	2	3	5	0	1	(week 2)
2	2	6	4	2	6	3	0	8	1	0	8	1	9	1	(week 3)

   c. To simulate breeding population, add one new letter (for a new dinosaur) after every 3 weeks.

A	B	C	D	E	F	G	H	I	J	K	L	M	N	O	P	
8	1	0	1	7	4	9	0	2	7	7	9	0	3	1		(week 1)
5	0	9	1	2	0	9	3	9	9	2	3	5	0	1		(week 2)
2	2	6	4	2	6	3	0	8	1	0	8	1	9	1		(week 3)
8	9	4	2	0	6	7	8	0	0	5	5	1	3	7	5	(week 4)

   d. Run the simulation for about 15–20 weeks. Then add the digits for each dinosaur to find its height.

   e. Display the dinosaur heights in a frequency distribution graph. Which of the graphs on page 76 is shaped more like your graph?

2. Now model an artificially controlled population.

   a. Start with 10 dinosaurs labeled A–J, and run the simulation for 8 weeks. To model a population that is not breeding, do not add a new dinosaur every 3 weeks.

   b. After the first 8 weeks, introduce a new batch of 10 dinosaurs by adding letters K–T.

   c. Run the simulation for 8 more weeks. Now you will need 20 digits for each week.

   d. Find the height of each dinosaur. Display the heights on a frequency distribution graph. How is your graph like the bottom one on page 76? How is it different? Why does the difference make sense?

3. Do you agree with Dr. Malcolm or Dr. Wu? Why?

The absolute value function and greatest integer function model special relationships. The graph of each of these functions has a distinctive appearance. One looks like a V, and the other looks like a flight of steps.

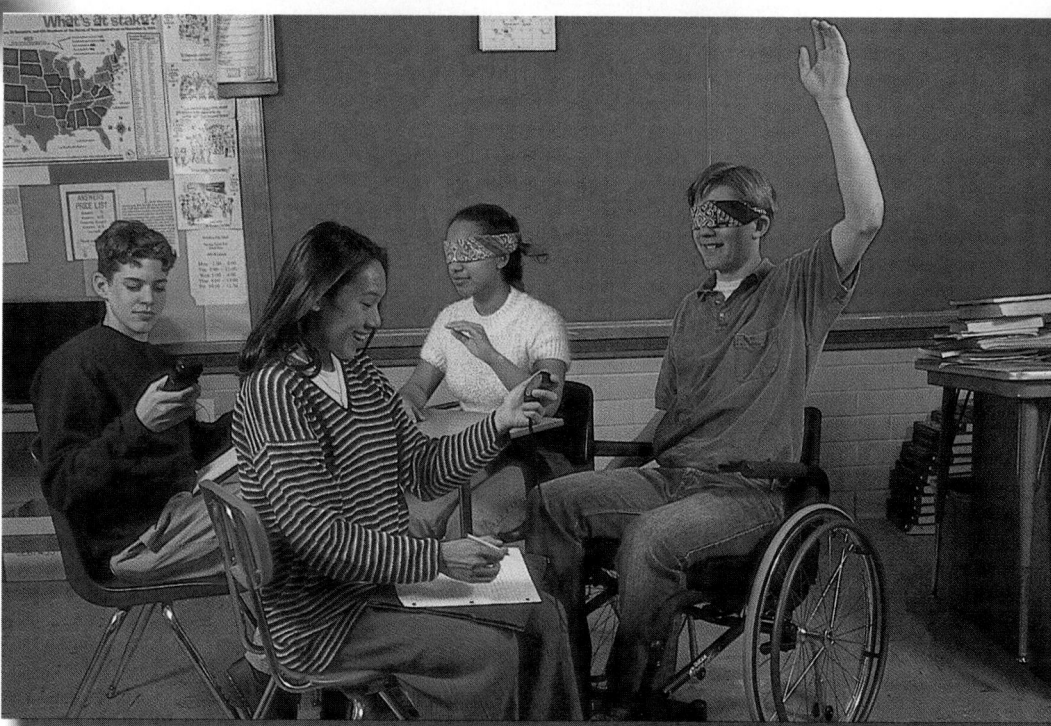

These students are participating in an experiment in which some of the students estimate elapsed time.

STATISTICS
Connection

Students work in pairs. Half of the students close their eyes and raise their hands. They are told to put their hands down when they think a minute has passed. The other half of the students act as timers. Later they switch roles.

Data from the first group is entered in a spreadsheet with the students' names in column A and their estimates of the time elapsed in column B.

Compute the error *in seconds* in column C by subtracting 60 from each student's time.

Spreadsheet In column C, a negative number means the guess was *under* 1 minute, and a positive number means the guess was *over* 1 minute. The absolute error in column D shows only the *amount* of the error.

D2		=ABS(C2)		
	A	B	C	D
1	Student	Time	Error	Abs. error
2	Tricia	49	−11	11
3	Keira	59	−1	1
4	Tom	51	−9	9
5	Louise	65	5	5
6	James	68	8	8
7	Sakeenah	77	17	17
8	Hong	66	6	6
9	Louis	54	−6	6
10	Mary	67	7	7
11	Maria	46	−14	14
12	Marcus	62	2	2
13	Shamar	73	13	13
14	Lois	61	1	1
15	Dianne	53	−7	7
16	Suzanne	64	4	4

For example, row 10 shows that Mary thought 1 minute had elapsed after 67 seconds. Her error was 67 − 60 = 7, indicating that her answer was 7 seconds *over* 1 minute. Row 15 shows that Dianne's time was 53 seconds, so her error was 53 − 60 = −7 or 7 seconds *under* 1 minute.

Look carefully at column D in the spreadsheet. Compare the numbers in column C with column D. What is the sign of each number in column D compared with the sign of the corresponding number in column C?

Absolute Value Function

The **absolute value** of a number can be shown on a number line. The absolute value of a number is its distance from zero.

A function written in the form $y = |x|$ or $y = \text{ABS}(x)$ is an example of an **absolute value function**.

7 is 7 units from 0, so the absolute value of 7 is 7.

−7 is 7 units from 0, so the absolute value of −7 is 7.

Thus, $|7| = 7$.

Thus, $|-7| = 7$.

Exploration *How Far From Zero?*

1 Complete the table for $y = |x|$.

x	−4	−3	−2	−1	0	1	2	3	4
y	4	?	?	?	0	?	?	?	4

2 Graph the function $y = |x|$ for the values in Step 1.

3 Describe the graph for the absolute value function you drew in Step 2. ❖

EXAMPLE 1

Graph the function $y = |x|$ from -3 to 3.

Solution

Make a table of ordered pairs. For every value of x in the table, write the absolute value of x as the y-value. For example, if x is -2, then y is $|-2|$ or 2.

x	-3	-2	-1	0	1	2	3
y	3	2	1	0	1	2	3

COORDINATE GEOMETRY
Connection

Plot the ordered pairs, and connect the points.

The graph of the absolute value function has a distinctive V shape. The tip of the V shape is at the origin, (0, 0). ❖

CRITICAL *Thinking*

Compare these graphs. How are they alike? different?

Integer Function

Graphics Calculator

Another function that is especially important when using technology is the **integer function**, also called the **greatest integer function**. It appears on some calculators as a menu item (INT) or as a key INT . When given a number, the integer function rounds down to the next integer. The following are some examples.

$$\text{INT}(1) = 1 \qquad \text{INT}(2) = 2$$
$$\text{INT}\left(\frac{5}{3}\right) = 1 \qquad \text{INT}(-2.1) = -3$$
$$\text{INT}(1.9) = 1 \qquad \text{INT}(-3.9) = -4$$

It is important to note that the integer function always rounds *down*, and not *up*, to the nearest integer. When you use this function, you will see that INT(1.9) is rounded *down* to 1, even though it is nearer to 2. You can also represent these examples using ordered pairs, such as (1, 1), $\left(\frac{5}{3}, 1\right)$, (1.9, 1), (2, 2), (−2.1, −3), and (−3.9, −4).

EXAMPLE 2

If you have $8.00 and want to buy movie tickets that cost $3.00 each, how many movie tickets can you buy?

Solution

You can buy only 2 tickets, even though you have enough money for $2\frac{2}{3}$ tickets. You cannot buy $\frac{2}{3}$ of a ticket. In this situation, $2\frac{2}{3}$ tickets rounds down to 2 tickets. ❖

Try This How many movie tickets can you buy if you have $7.00? if you have $8.50? if you have $7.75? Remember, the integer function *rounds down* to the *next lower integer.*

EXAMPLE 3

Marilyn is ordering raffle tickets for her 12 friends. The tickets cost $1.50 each. Her friends have $2.00, $2.50, $2.75, $3.50, $3.75, $4.00, $4.50, $5.00, $5.25, $6.00, $6.25, and $6.50, respectively. Draw a graph representing the amount of money each has and the number of tickets that each can purchase.

Solution

Make a table to show the amount of money each has and the number of tickets that amount will purchase. Plot the pairs of numbers on a graph.

You can see that the *y*-value of the function stays level until the *x*-value reaches a certain point. Then the *y*-value jumps. This suggests that the graph is in steps. ❖

Amount	Number of tickets
$2.00	1
$2.50	1
$2.75	1
$3.50	2
$3.75	2
$4.00	2
$4.50	3
$5.00	3
$5.25	3
$6.00	4
$6.25	4
$6.50	4

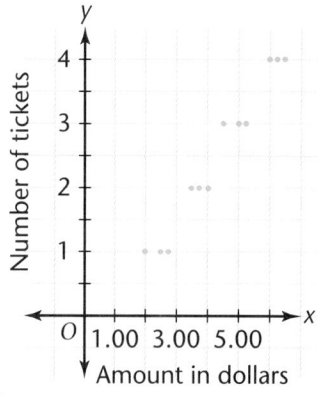

EXERCISES & PROBLEMS

Communicate

Explain how to find the absolute value (ABS) of each.

1. 29　　　**2.** -34　　　**3.** 7.99　　　**4.** -3.44

Explain how to find the greatest integer (INT) of each.

5. 29　　　**6.** $\frac{34}{5}$　　　**7.** 7.99　　　**8.** 3.44

9. Describe the graph of an absolute value function.

10. Describe the graph of an integer function.

Practice & Apply

In the minute-timing experiment, one student's estimate of a minute was 52 seconds.

11. Was the student over or under the correct amount?

12. What was the student's error?

13. What was the absolute value of the student's error?

Find the absolute value (ABS).

14. 17

15. -33

16. 8.67

17. -7.11

Evaluate.

18. $|4.8|$

19. $|-3.2|$

20. INT(5.8)

21. INT(11/2)

22. ABS(5.8)

23. ABS(11/2)

Find the greatest integer (INT).

24. 17

25. $\frac{33}{4}$

26. 8.67

27. 7.11

Evaluate.

28. $|8|$

29. $|-8|$

30. $-|8|$

31. $-|-8|$

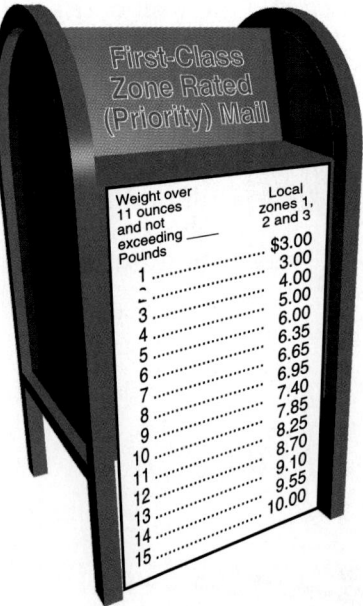

32. Consumer Math Another name for the integer function is the postal function. Paula often mails several packages and needs to know the cost for the various weights. Make a graph of the weight and cost from the table.

Weight over 11 ounces and not exceeding — Pounds	Local zones 1, 2 and 3
	$3.00
1	3.00
2	4.00
3	5.00
4	6.00
5	6.35
6	6.65
7	6.95
8	7.40
9	7.85
10	8.25
11	8.70
12	9.10
13	9.55
14	10.00
15	

Home Economics Melissa is making place mats, each requiring 1 yard of fabric. How many place mats can she make from the following amounts of fabric?

33. $6\frac{3}{4}$ yards

34. $3\frac{1}{2}$ yards

35. $5\frac{1}{4}$ yards

36. $\frac{9}{10}$ yard

Technology After collecting the data shown for the minute-timing experiment, students switched roles. Only this time, the students estimating the time were distracted by having to carry on conversations at the same time. The spreadsheet shows the data for this group.

37. On your own paper, write the errors for column C.

38. Write the absolute errors for column D.

39. Psychological Experiment Perform both parts of the minute-timing experiment in your class. Give the estimate, the error, and the absolute value of the error for each student. Use a spreadsheet to display the results for the two groups.

D3		=ABS(C3)	
	A	B	C	D
1	Student	Time	Error	Abs. error
2	Buster	57	−3	3
3	Tony	74	14	14
4	Charlotte	59		
5	Amy	60		
6	Gerti	71		
7	Nathan	56		
8	Jamie	58		
9	Marni	70		
10	Lee	58		
11	Wei	62		
12	Ned	68		
13	Debbie	74		
14	Wynton	70		
15	Bob	73		
16	Jill	65		

40. **Portfolio Activity** Complete the problem in the portfolio activity on page 57.

 Look Back

Find the next three numbers in each pattern. [Lesson 1.1]

41. 2, 6, 10, 14, 18, __, __, __

42. 27, 9, 3, 1, $\frac{1}{3}$, __, __, __

43. If Miguel has $2, how many bookmarks can he buy if the bookmarks are $0.38 each? **[Lesson 1.3]**

Graph each function on the same coordinate plane, and compare the graphs. [Lesson 1.5]

44. $y = x + 2$

45. $y = x + 9$

46. $y = 5 - x$

47. $y = x + 1$

48. Use this data to make a graph. Which point is the vertex? **[Lesson 2.2]**

x	1	2	3	4	5
y	4	3	4	7	12

Look Beyond

Choose positive and negative values for x, and graph each of the following functions.

49. $y = |x| + 3$ **50.** $y = |x + 3|$

Identifying Types of Functions

why *Each second that an automobile accelerates increases the distance it travels. This relationship is not like a linear function.*

Physics Have you ever heard the phrase "Zero to sixty in eight seconds"? The following table shows how time and distance are related.

Time in seconds	0	1	2	3	4	5	6	7	8
Distance in feet	0	5.5	22	49.5	88	137.5	198		

How far do you think the car will have traveled after 8 seconds? Describe the graph of the data.

The method of differences can be used to find the distance for 7 and 8 seconds.

Time in seconds	0		1		2		3		4		5		6		7		8	
Distance in feet	0		5.5		22		49.5		88		137.5		198		269.5		352	
		5.5		16.5		27.5		38.5		49.5		60.5		71.5		82.5	← First differences
			11		11		11		11		11		11		11		← Second differences

The data indicate that after 8 seconds the car will have traveled 352 feet.

When you graph the ordered pairs, (time, distance), the results show that the relationship is quadratic. What was it about the constant differences that indicated that you should expect a quadratic function?

EXAMPLE 1

The table shows the speed of a car with respect to the time it travels when accelerating at a constant rate of 11 feet per second per second. Find the speed of the car after 8 seconds.

| Time in seconds | 0 | 1 | 2 | 3 | 4 |
| Speed in feet per second | 0 | 11 | 22 | 33 | 44 |

COORDINATE GEOMETRY
Connection

Solution▸

Plot the ordered pairs.

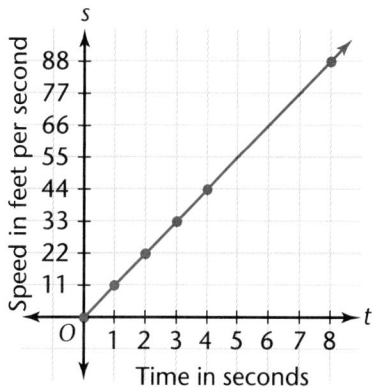

The points lie on a straight line. A linear function appears to be a good model. The graph indicates that after 8 seconds the car will be traveling 88 feet per second.

Notice that $\dfrac{60 \text{ miles}}{1 \text{ hour}} = \dfrac{60 \cdot 5280 \text{ feet}}{60 \cdot 60 \text{ seconds}} = \dfrac{88 \text{ feet}}{1 \text{ second}}$. ❖

CRITICAL *Thinking*

Once the driver reaches a speed of 88 feet per second, the speed limit, he or she stops accelerating, but doesn't decelerate. What happens to the graph in Example 1 after 8 seconds?

EXAMPLE 2

Demographics Draw a graph of the following United States population figures. Describe what pattern appears in the data as shown by the graph.

Year	1800	1820	1840	1860	1880	1900	1920	1940	1960	1980	1990
Population in millions	5	10	17	31	50	76	106	132	179	179	249

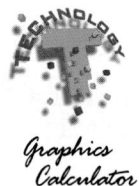
Graphics
Calculator

Solution▶

For changes in population, a good model is often an exponential function. A scatter plot drawn by a graphics calculator shows this model. The graphics calculator draws the exponential graph that best fits the points. According to the graphics calculator, the multiplier is 1.02, which means the growth is about 2% per year.

Piecewise Functions

In some cases, functions that are made up of pieces of other functions provide a model for the situation you are studying. These functions are called **piecewise functions**.

EXAMPLE 3

A car rental company charges $40 per day and allows 200 free miles. There is an additional charge of 10 cents per mile in excess of 200 miles. Graph the function that models this situation.

Solution▶

Make a table of values, and graph the ordered pairs.

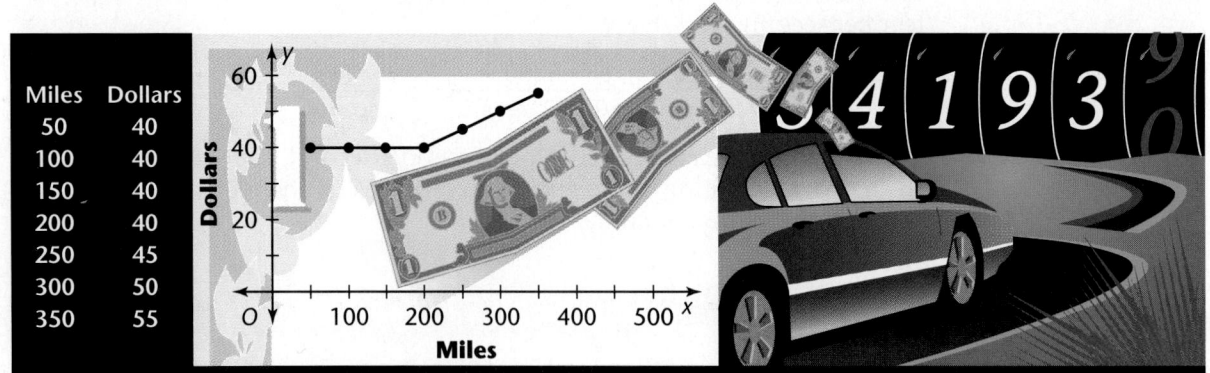

Miles	Dollars
50	40
100	40
150	40
200	40
250	45
300	50
350	55

Notice that the graph is constant until *x* is equal to 200. At that point it becomes a non-constant linear function because the cost increases by a fixed amount. ❖

Try This   What is the cost for 450 miles?

Summary

The following is a summary of the different types of functions you have studied in this chapter. As you continue to study algebra, you will discover and use many properties of these functions.

Type	Growth Pattern	Graph	Typical Application
Linear	Fixed amount	Line ╱	Cost per item
Exponential	Fixed rate	Exp. curve ╯	Population growth
Quadratic	Up and down	Parabola ∩	Gravity
Reciprocal	Inverse	$\frac{1}{x}$ curve ╰	Fund-raising
Absolute value	Linear pieces	V shape ∨	Absolute error
Integer	In steps	Steps ⌐	Postage

EXERCISES & PROBLEMS

Communicate

Match the graphs with the types of functions. Describe the characteristics that distinguish the functions.

  a. Linear   **b.** Exponential   **c.** Quadratic

  d. Reciprocal   **e.** Absolute value   **f.** Integer

1.   **2.**   **3.**

4.   **5.**   **6.**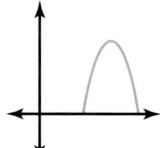

Explain how to use the method of finite differences to determine whether the relationship is linear, quadratic, or neither.

7.

x	1	2	3	4	5
y	12	20	30	42	56

8.

x	1	2	3	4	5
y	3	4	12	24	48

9.

x	1	2	3	4	5
y	8.0	6.8	5.6	4.4	3.2

10.

x	1	2	3	4	5
y	2	4	6	8	10

Practice & Apply

Match the graphs of the data points with the correct type of function.

- **a.** Linear
- **b.** Exponential
- **c.** Quadratic
- **d.** Reciprocal
- **e.** Absolute value
- **f.** Integer

11.

12.

13.

14.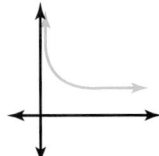

15.

16.

Use the method of differences to determine whether the relationship is linear, quadratic, or neither.

17.

x	1	2	3	4	5
y	2	6	18	54	162

18.

x	1	2	3	4	5
y	44	55	66	77	88

19.

x	1	2	3	4	5
y	8	15	24	35	48

20.

x	1	2	3	4	5
y	1	1	2	3	5

21.

x	1	2	3	4	5
y	1	3	6	10	15

22.

x	1	2	3	4	5
y	45	54	63	72	81

23. Sports On graph paper, graph the data for the number of points scored (*y*-axis) versus

Player	Minutes	Points
Amy	23	14
Ling	18	10
Sarah	14	8
Trina	17	3
Latissa	10	6

the number of minutes played (*x*-axis) for a team's five players in a high school basketball game. Use spaghetti or thread to fit a straight line to the data. Write a sentence that explains what the fit tells you about the team's performance.

24. Describe the absolute value function as a piecewise function.

Consumer Economics

25. Graph the function for the shipping charges.

26. What type of function is it?

A catalog company charges $2 per pound (16 ounces), or fraction of a pound, for shipping its merchandise.

Transportation   A car rental charges $30 per day and allows 100 free miles. There is an additional charge of 5 cents per mile in excess of 100.

27. Graph the function that models the situation.

28. What kind of function is Exercise 27?

Investments   The Rule of 72 is a method of estimating how long it will take to double your money at various interest rates. To make an estimate, divide 72 by the number of hundredths. For example, if the interest rate is 8%, it will take about $\frac{72}{8}$ or 9 years. Use the Rule of 72 to predict how long it will take to double an investment if the interest rate is

29. 4%.   **30.** 6%.   **31.** 9%.

32. Use the Rule of 72 to write an equation for the number of years (t) it will take to double an investment at x% interest.

33. What kind of function did you write for Exercise 32?

 Look Back

34. Find the sum of $0.1 + 0.2 + 0.3 + 0.4 + 0.5 + \cdots + 0.8$.
[Lesson 1.3]

Perform calculations in the proper order.   **[Lesson 1.4]**

35. $12 + 3^2 - 4 \div 2$ **36.** $74 - 10 \cdot 5 \div 2$ **37.** $5(1 \cdot 4 - 2) - 6 \div 3$ **38.** $60 \div \frac{6}{3} + 2^2 \cdot 9$

Cultural Connection: Asia   Babylonian mathematicians 3800 years ago solved problems in compound interest. They asked how many years it would take to double money at 20% interest compounded annually. Today, bankers answer that question by looking it up in a table. The Babylonians did the same thing. Use your calculator to make a table that starts with $1000.00. Find how many years it would take to double the amount if it is compounded annually at 20% interest.   **[Lesson 2.1]**

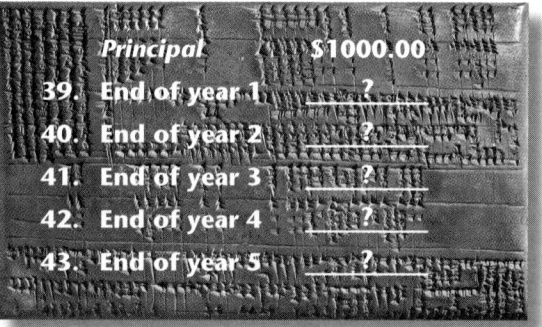

Principal	$1000.00
39. End of year 1	?
40. End of year 2	?
41. End of year 3	?
42. End of year 4	?
43. End of year 5	?

Look Beyond

Graph the following two equations.

44. $y = 2x + 5$ **45.** $y = x + 6$

Cubes and Pyramids

▼ ◆ ■ ◇ ▼ ◆ ■ ◇ ▼ ◆ ■ ■ ▼ ◆ ■ ◇ ▼ ◆ ■

The cube below is made from 8 small unit cubes. All sides have been painted bright red, including the top and bottom. When the cubes are separated you can see that each small cube has red paint on 3 sides.

What happens as the length of each side of the red cube increases by 1?

Activity 1
What is the formula?

Use a set of small unpainted cubes (such as sugar cubes) to build a large cube with 3 units on a side. All six faces of the big cube, including the top and bottom, are painted.

How many small cubes will be needed to make the big cube? How many of the small cubes will have paint on

3 faces?   2 faces?
1 face?   0 faces?

Copy and complete this table for cubes of different lengths. The last row gives the general formula for a cube with sides of length n.

Number of small-cubes on each bottom edge	Total Number of small cubes	Number of small cube faces painted			
		3	2	1	0
2	8	8	0	0	0
3	27	12			
4					
5					
6					
n					

Activity 2

How many painted faces?

Again use a set of small, unpainted cubes. This time, build a pyramid with a square base. Each cube in each of the levels must rest directly over another cube. All of the exposed faces of the cubes that form the pyramid are painted, including those on the bottom of the pyramid.

How many small cubes will be needed to make each pyramid? How many of the small cubes will have paint on

5 faces?   4 faces?   3 faces?
2 faces?   1 face?   0 faces?

Count the painted faces for each of the different-sized pyramids. Fill in the table. Look for a pattern in the column sequences to help you fill in the table. See if you can predict the number of painted faces for a pyramid with 11 sides. Try to predict the number of painted faces for pyramids with longer sides. Explain how you predicted the next number for each column.

Number of small-cubes on each bottom edge	Total Number of small cubes	Number of small cube faces painted					
		5	4	3	2	1	0
3	10	1	4	4	0	1	0
5	35			16			
7							
9							

The pyramid-like structure at the left is I.M.Pei's recent addition to the Louvre in Paris. The building at the right is the Transamerica building in San Francisco, California.

Chapter 2 Review

Vocabulary

absolute value function	79	integer function	80	quadratic function	65
compound interest	60	linear	60	reciprocal function	71
exponential function	60	parabola	66	vertex	66
exponential growth and decay	60	piecewise function	86		

Key Skills & Exercises

Lesson 2.1

➤ Key Skills

Identify linear and exponential functions.

Examine 4, 8, 16, 32, 64, . . . Each term is 2 times the previous term. To find the next three terms, multiply by 2. The next three terms are $64 \cdot 2 = 128$, $128 \cdot 2 = 256$, and $256 \cdot 2 = 512$. Since the sequence grows at the same rate, the sequence is exponential.

Identify an exponential function as showing growth or decay.

The sequence 50, 25, 12.5, 6.25, 3.125, . . . shows exponential decay. The multiplier is 0.5, and the terms decrease. When the multiplier is greater than 1, a sequence shows exponential growth.

➤ Exercises

Find the next three terms of each sequence. Then, identify each sequence, as linear or exponential.

1. 45, 43, 41, 39, 37, . . . **2.** 9000, 900, 90, 9, 0.9, . . . **3.** 3, 9, 27, 81, 243, . . .
4. 7, 12, 17, 22, 27, . . . **5.** 2, 4, 6, 8, 10, . . . **6.** 15625, 3125, 625, 125, 25, . . .

7. If the multiplier for an exponential function is 1.2, does the function show growth or decay?
8. Which of the exponential sequences in 1–6 show growth? decay?

Lesson 2.2

➤ Key Skills

Graph basic quadratic functions.

You can use this data to make a graph and identify its vertex.

x	2	3	4	5	8
y	9	4	1	0	9

The vertex is (5, 0).

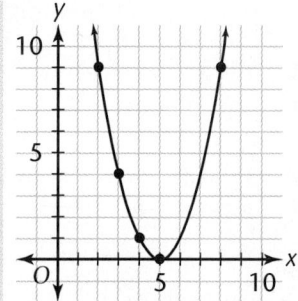

Use the differences method for sequences to identify quadratics.

Examine these sequences.

x	1	2	3	4	5
y	3	5	7	9	11

x	1	2	3	4	5
y	3	5	10	17	26

First differences for the first sequence are constant, so it is linear. Second differences for the second sequence are constant, so it is quadratic.

➤ Exercises

9. Which point is the vertex of this parabola?

10. Use the data to make a graph. Which point is the vertex?

x	5	6	7	8	9
y	1	0	1	4	9

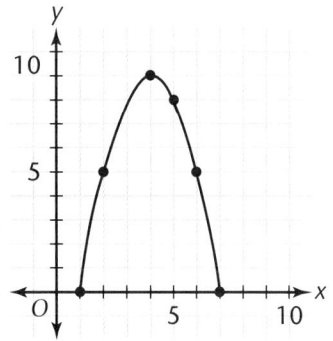

Identify each of the following sequences as quadratic or not quadratic. Explain your answer.

11.

x	1	2	3	4	5
y	2	5	10	17	26

12.

x	1	2	3	4	5
y	6	12	24	48	96

13.

x	1	2	3	4	5
y	3	12	27	48	75

Lesson 2.3

➤ Key Skills

Find the reciprocal of a number.

To find the reciprocal of 4, find a number which, when multiplied by 4, equals 1.

Since $4 \cdot \frac{1}{4} = 1$, $\frac{1}{4}$ is the reciprocal of 4.

Since $\frac{2}{3} \cdot \frac{3}{2} = 1$, $\frac{3}{2}$ is the reciprocal of $\frac{2}{3}$.

➤ Exercises

Find the reciprocal of each number.

14. 6   **15.** $\frac{6}{7}$ **16.** 3.5   **17.** $\frac{1}{15}$

Lesson 2.4

➤ Key Skills

Find the absolute value of a number.

The absolute value of a number is its distance from 0.

$|9.8| = 9.8$ $|-9.8| = 9.8$

9.8 and −9.8 are both 9.8 units from 0.

$\left|6\frac{1}{4}\right| = 6\frac{1}{4}$ $\left|-6\frac{1}{4}\right| = 6\frac{1}{4}$

$6\frac{1}{4}$ and $-6\frac{1}{4}$ are both $6\frac{1}{4}$ units from 0.

Find the greatest integer of a number.

To find INT $\left(3\frac{1}{2}\right)$, round $3\frac{1}{2}$ down to the next integer. So INT $\left(3\frac{1}{2}\right) = 3$.

➤ Exercises

Evaluate the following. $|x|$ means the same as ABS(x), the absolute value of x.

18. $|-3|$ **19.** $|6.7|$ **20.** $|-1.7|$ **21.** ABS(-45)

22. INT(45)   **23.** INT(3.68)   **24.** INT(7.5)   **25.** INT(36)

Lesson 2.5

➤ Key Skills

Identify specific types of functions by their graph.

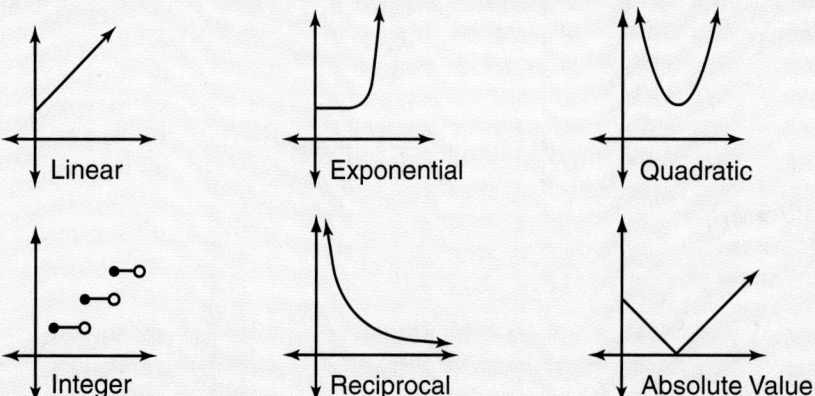

Linear   Exponential   Quadratic

Integer   Reciprocal   Absolute Value

➤ Exercises

What kind of function does each graph model?

26.

27.

28.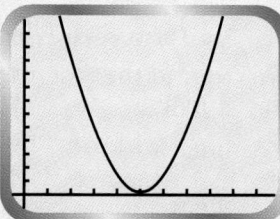

Applications

29. Finance Roy invests $1000. His investment earns 7.9% interest per year, which is compounded once each year. How long will it take Roy to double his money?

30. Physics The distance, d, in meters that a race car travels when accelerating at a rate of 6.2 meters per second per second is determined by the time, t, in seconds that it travels, according to the formula $d = 3.1t^2$. For example, if the car travels for 10 seconds, the distance is $3.1 \cdot 10^2 = 310$ meters. Find the distance the car travels for 5 seconds, 8 seconds, and 20 seconds.

31. Graph the points you found in Item 30. Use your graph to estimate how far the car would travel in 35 seconds.

32. How much would each person need to contribute toward a gift that costs $50 if 20 people are donating equal amounts?

Science One form of uranium has a half-life of 4.5 billion years. This means that after 4.5 billion years half of the atoms in any amount of uranium will have changed into a form of lead. After another 4.5 billion years, half of the remaining uranium atoms will have changed into lead, and so on.

33. What fraction of atoms would remain unchanged after 18 billion years?

34. How long will it be before only $\frac{1}{64}$ of the atoms would be uranium?

Chapter 2 Assessment

Find the next three terms of each sequence. Then identify each sequence as linear or exponential.

1. 10, 20, 40, 80, 160, . . .   **2.** 100, 96, 92, 88, 84, . . .

3. Does the exponential sequence 4000, 400, 40, 4, 0.4, . . . show growth or decay? Explain your answer.

4. Use the following data to make a graph. Which point is the vertex?

x	1	3	4	5	7
y	11	3	2	3	11

5. Mica's gymnastics team is raising money to buy equipment. How much does each team member need to raise toward a goal of $500 if there are 10 team members?

Find the reciprocal of each number.

6. 6   **7.** $\frac{1}{12}$ **8.** 5.5

Evaluate.

9. $|-10|$ **10.** ABS(7.9)   **11.** INT(9.75)

12. Tasha is planting a flower garden. She has $10 to spend. If each pack of flowers costs 49¢, how many packs of flowers can Tasha buy?

Determine whether each relationship is linear, quadratic, or neither.

13.

x	1	2	3	4	5
y	75	70	65	60	55

14.

x	1	2	3	4	5
y	4	7	12	19	28

Match the graphs with the type of function.

a. Linear   **b.** Exponential   **c.** Quadratic
d. Reciprocal   **e.** Absolute value   **f.** Integer

15.

16.

17.

18.

19.

20.

Chapter 1 – 2
Cumulative Assessment

College Entrance Exam Practice

Quantitative Comparison For Questions 1–4, write

A if the quantity in Column A is greater than the quantity in Column B;

B if the quantity in Column B is greater than the quantity in Column A;

C if the two quantities are equal; or

D if the relationship cannot be determined from the information given.

	Column A	Column B	Answers	
1.	$\lvert -4.8 \rvert$	$\lvert 3.2 \rvert$	Ⓐ Ⓑ Ⓒ Ⓓ	**[Lesson 2.4}**
2.	INT(5.8)	INT(3.1)	Ⓐ Ⓑ Ⓒ Ⓓ	**[Lesson 2.4]**
3.	c $c = 4$, $c + 8$ box; Perimeter	Area	Ⓐ Ⓑ Ⓒ Ⓓ	**[Lesson 1.3]**
4.	Play tickets: $7 each for balcony seats, $9 each for main floor seats. 9 balcony tickets	7 main-floor tickets	Ⓐ Ⓑ Ⓒ Ⓓ	**[Lesson 1.3]**

5. What are the next three terms of the sequence 5, 8, 11, 14, 17, . . . ?
[Lesson 1.1]
  a. 20, 23, 26   **b.** 14, 11, 8   **c.** 22, 27, 32   **d.** 19, 21, 23

6. The cost to order CDs by mail is $14 per CD plus a $7 handling charge
per order (regardless of how many CDs are ordered). Which equation
describes this situation? **[Lesson 1.5]**
  a. $c = 7d + 14$ **b.** $c = 14d + 7$ **c.** $c = 14(d + 7)$ **d.** $c = 14d - 7$

7. Which sequence is exponential and shows growth? **[Lesson 2.1]**
  a. 2, 4, 6, 8, 10, . . .   **b.** 20, 10, 5, 2.5, 1.25, . . .
  c. 20, 18, 16, 14, 12, . . .   **d.** 2, 4, 8, 16, 32, . . .

8. Which sequence is linear? **[Lesson 2.1]**
  a. 400, 200, 100, 50, 25, . . .   **b.** 400, 350, 300, 250, 200, . . .
  c. 400, 390, 370, 340, 300, . . .   **d.** 400, 800, 1600, 3200, 6400, . . .

9. Which point is the vertex of this parabola? **[Lesson 2.2]**

 a. (0, 4) **b.** (0, 2) **c.** (2, 0) **d.** (4, 4)

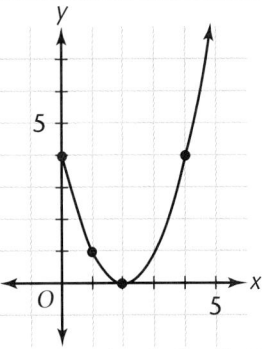

Name the graph shown. [Lesson 2.5]

10.

11.

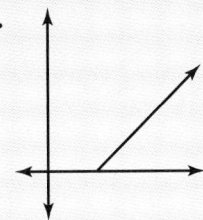

12. Find the values for y by substituting the values 1, 2, 3, 4, and 5 for x in $y = 3x - 2$. **[Lesson 1.4]**

13. Describe the correlation coefficient for the scatter plot at the right as strong positive, strong negative, or little to none. **[Lesson 1.6]**

14. Is the correlation for the scatter plot at the right nearest to -1, 0, or 1? **[Lesson 1.7]**

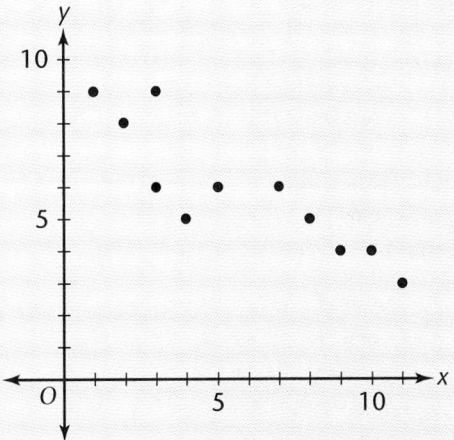

15. Find the next three terms of the sequence 3, 6, 12, 24, 48, . . . Is the sequence linear or exponential? **[Lesson 2.1]**

16. Use the data to make a graph. Which point is the vertex? **[Lesson 2.2]**

x	1	2	3	4	5
y	4	3	4	7	12

17. The first three terms of a sequence are 5, 7, and 13. The second differences are a constant 4. What are the next three terms? **[Lesson 1.2]**

Free-Response Grid The following questions may be answered using a free-response grid commonly used by standardized test services.

18. Misty's team is in a league with 12 other teams. If each of the teams plays each of the other teams once, how many games will be played? **[Lesson 1.1]**

19. Simplify $2(3 + 4) \div 2 + 3^2(4) - 6 \cdot 1$. **[Lesson 1.4]**

20. Sam is making bookshelves that are $1\frac{3}{4}$ feet long. How many shelves can he cut from a board that is 6 feet long? **[Lesson 2.4]**

CHAPTER 3

Addition and Subtraction in Algebra

An equation is a statement that two mathematical expressions are equal. If each of these expressions were placed on a mathematical "scale," they would balance.

In this chapter you will study a number of mathematical properties. These properties maintain the balance in an equation as you add and subtract to find the value of the unknown.

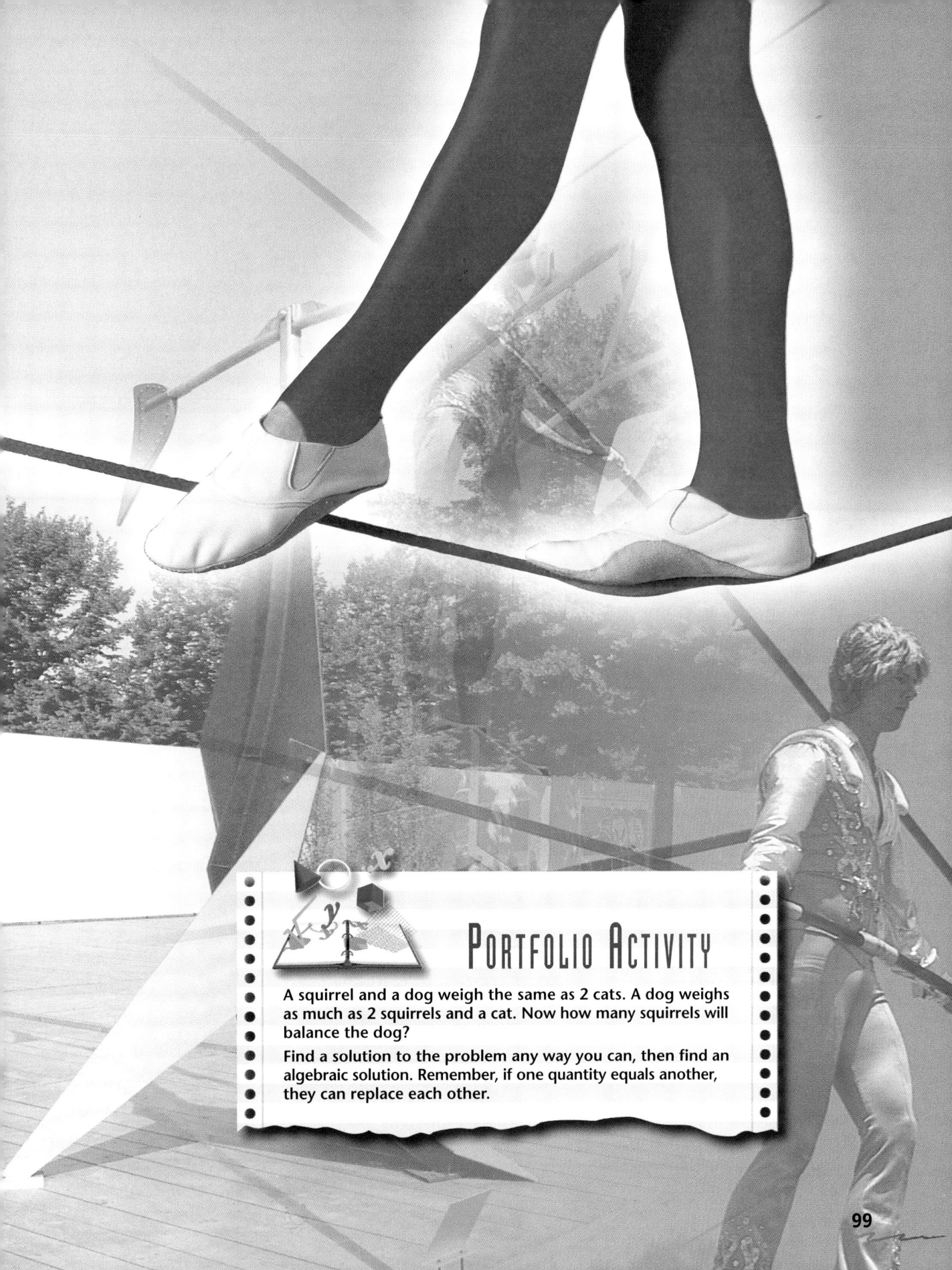

PORTFOLIO ACTIVITY

A squirrel and a dog weigh the same as 2 cats. A dog weighs as much as 2 squirrels and a cat. Now how many squirrels will balance the dog?

Find a solution to the problem any way you can, then find an algebraic solution. Remember, if one quantity equals another, they can replace each other.

LESSON 3.1 Adding Integers

Why *"T minus 2 seconds and counting."* *During the launch of the space shuttle, time is measured before and after the liftoff. Integers are needed to count units that occur on opposite sides of zero.*

Numbers such as +1, +31, and +53 are examples of **positive integers**. However, they are usually written without the positive sign. Thus + 31 and 31 represent the same number. Numbers such as − 1, − 31, and − 53 are **negative integers**. The integers consist of the set of all positive integers, all negative integers, and **zero**. This set can be represented as points on a number line.

Negative integers   Positive integers

$$-5\ -4\ -3\ -2\ -1\quad 0\quad 1\ 2\ 3\ 4\ 5$$

Zero

Two numbers are **opposites** if they are

• on opposite sides of zero, and
• the same distance from zero.

The opposite of 2 shown in symbols is
$$-(2) = -2.$$
The opposite of − 2 shown in symbols is
$$-(-2) = 2.$$
The opposite of zero is zero itself,
$$-(0) = 0.$$

In Lesson 2.4 you were introduced to the absolute value function. The distance from zero to an integer is the absolute value of the integer. The absolute value of an integer less than 0 is the opposite of that integer. For example, $|-5| = -(-5) = 5$.

ABSOLUTE VALUE

For any number x,

if x is greater than or equal to 0, $|x| = x$, and

if x is less than 0, $|x| = -x$.

Using Models to Add Integers

Tiles can be used to model integers.

Represents 1

Represents -1

•Exploration *Patterns in Addition*

Tiles can also be used to model integer addition.

$3 + 2 = 5$ $-2 + -4 = -6$

 Use tiles to model the addition shown, and solve each of the equations.

a. $7 + 3 = ?$ **b.** $4 + 5 = ?$
c. $-7 + (-3) = ?$ **d.** $-4 + (-5) = ?$

 Describe how to add two integers with like signs. ❖

EXAMPLE 1

Sports *After the first lap, Chad was 2 seconds ahead of his competition.*

Chad loses 2 seconds of lead on the second lap. What is his position in the race relative to his competition after the second lap? Represent this problem using tiles.

Solution▶

Let one positive tile represent 1 second gain, and let one negative tile represent 1 second loss, relative to the competition. Then, $2 + (-2)$ represents Chad's position after the second lap.

$2 + (-2) = 0$

Each pair of tiles consists of a positive tile and a negative tile. Since each positive tile is paired with a negative tile, the entire sum is 0. Chad is even with the competition after the second lap. ❖

This example leads to an important property of integers.

PROPERTY OF OPPOSITES
For any number a, $-a$ is its opposite, and
$a + (-a) = 0.$

The numbers a and $-a$ are also called **additive inverses.**

EXAMPLE 2

Temperature Anna recorded temperatures one cold day in January. At noon the temperature was 4°F. By 6 P.M. the temperature had dropped 7°F. What is the 6 P.M. temperature?

Solution A ▶

The sum $4 + (-7)$ represents the temperature in the evening.

Each pair consisting of a positive tile and a negative tile is equal to 0, so three negative tiles remain. This represents $-3°F$.

$$4 + (-7) = -3$$

Solution B ➤

A thermometer is a vertical number line. To use the number line to model $4 + (-7)$, start at 4. Move 7 steps in the negative direction. You will end at -3.

$$4 + (-7) = -3 ❖$$

Cultural Connection: Asia More than 2000 years ago, the Chinese used different colored rods to represent positive and negative numbers. To compute how much remains after a transaction involving a gain, $+23$, and a loss, -54, the Chinese combined sets of red and black rods.

The corresponding red and black rods are eliminated. This leaves three horizontal black rods and one vertical black rod which represent -31.

There are only two conditions to consider when adding integers—either the signs are alike or the signs are different.

RULES FOR ADDING INTEGERS

A. Like signs
   Find the **sum** of the absolute values.
   Use the sign common to both integers.
B. Unlike signs
   Find the **difference** of the absolute values.
   Use the sign of the integer with the greater absolute value.

EXAMPLE 3

Find each sum.   **Ⓐ** $-2 + (-8)$ **Ⓑ** $15 + (-23)$

Solution➤

Ⓐ If the terms have like signs, find the sum of the absolute values, then use the sign that is the same for both integers.

$$|-2| = 2 \qquad |-8| = 8 \qquad -2 + (-8) = -10$$

Ⓑ If the terms have unlike signs, find the difference of the absolute values, then use the sign of the integer with the greater absolute value.

$$|15| = 15 \qquad |-23| = 23 \qquad 15 + (-23) = -8 ❖$$

Try This Find the sums.   **a.** $27 + (-16)$ **b.** $-27 + (-16)$

Adding More Than Two Integers

You will often need to add more than two integers.

EXAMPLE 4

Find each sum. **A** $[-9 + (-2)] + 12$ **B** $-9 + [(-2) + 12]$

Solution

A Add the first pair of integers, and remove the brackets. Then add that sum and the remaining integer.

$$[-9 + (-2)] + 12 = -11 + 12 = 1$$

B The brackets instruct you to add the second pair of integers first. Next, remove the brackets. Then add the first integer to the sum.

$$-9 + [(-2) + 12] = -9 + 10 = 1$$

These examples illustrate the *Associative Property of Addition*. When adding integers, the grouping does not matter.

ASSOCIATIVE PROPERTY OF ADDITION
For all numbers *a*, *b* and *c*,

$$(a + b) + c = a + (b + c).$$

CRITICAL *Thinking*

The Associative Property of Addition applies to adding 3 numbers. Examine several examples, and make a conjecture about adding 4 or more numbers.

Calculator

A calculator will save time when you need to add a series of numbers. Check your calculator to see how to key in a negative number. Try adding $-618 + 587 + (-10)$. Your calculator should show -41 as the answer. Explain how to use your calculator to find the sum of a loss of $26.46 followed by a second loss of $12.27 followed by a gain of $18.25.

EXERCISES & PROBLEMS

Communicate

1. Explain and give an example of what is meant by the opposite of a number.

2. Discuss and give examples of what is meant by the absolute value of a number.

3. Describe how to add two integers that have the same sign.

4. Describe how to add two integers that have unlike signs.

5. How would you add 25 and -13 using the ancient Chinese system? (Hint: Use colored toothpicks as counting sticks.)

6. Explain how to add -43, -9, and 128 in two different ways.

Practice & Apply

Find the opposite of each number.

7. 17   **8.** -17 **9.** 0   **10.** $(12 - 5)$

Use algebra tiles to find the following sums.

11. $-5 + (-2)$ **12.** $-3 + 3$ **13.** $2 + (-6)$ **14.** $-2 + (-6)$

Use a number line to find the following sums.

15. $8 + (-3)$ **16.** $-4 + (-5)$ **17.** $-1 + 5$ **18.** $-1 + (-2)$

Find each sum.

19. $-28 + 50$ **20.** $17 + (-34)$ **21.** $38 + (-72)$

22. $14 + (-29) + (-12)$ **23.** $-43 + 82 + |-19|$ **24.** $-308 + |-24| + (-29)$

Physics Electrons have a charge of -1 and protons have a charge of $+1$. The total charge of an atom is the sum of its electron charges and proton charges. Find the total charge of the following atoms.

25. 16 protons; 18 electrons   **26.** 10 protons; 10 electrons

27. The Lincoln High football team completes two series of downs with the following gains and losses. Find the total gain or loss.

Down	1	2	3	4
Gain/Loss	8	-4	-3	15

Down	1	2	3	4
Gain/Loss	-3	-5	4	-6

28. Business In business, a loss is recorded by placing the amount inside parentheses. What is the yearly profit or loss for the Family Shoe Store?

1st Quarter	2nd Quarter	3rd Quarter	4th Quarter
(389.75)	(794.28)	1796.50	(809.50)

29. During one week Paul Phillips' stock had gains and losses. If the opening stock price was $34\frac{1}{2}$ on Monday, what was the price after the market closed on Friday?

M	T	W	Th	F
$+\frac{1}{2}$	$-\frac{3}{4}$	$+\frac{1}{4}$	0	$-1\frac{1}{2}$

30. The graph starts at -600 feet. It shows the change in depth of a submarine over six 4-hour intervals. What is the depth of the submarine after 24 hours?

Let $a = 2$, $b = -3$, $c = 5$. Evaluate each expression.

31. $(a + b) + c$ **32.** $|a + b| + c$ **33.** $a + |b + c|$ **34.** $a + (b + c)$ **35.** $a + (c + b)$

Look Back

36. Find the sum of $1 + 2 + 3 + \cdots + 40$.   **[Lesson 1.1]**

Find the next two terms in each number sequence.   [Lesson 1.3]

37. 40, 37, 34, 31, ___, ___   **38.** $-5, -1, 3, 7$, ___, ___   **39.** 1, 3, 6, 10, ___, ___

40. **Geometry** What is the perimeter of a square with a side of 3.5 centimeters?   **[Lesson 1.3]**

Calculate.   [Lesson 1.4]

41. $15 - 21 + 3 + 4$ **42.** $[3(4 - 2)^2] + 7$ **43.** $12 + 3^2 + (9 - 6)$

Plot each point in the same coordinate plane.   [Lesson 1.5]

44. (4, 4)   **45.** (5, 2)   **46.** (3, 4)   **47.** (6, 3)

Look Beyond

48. You may recall from arithmetic that adding the same number repeatedly can be represented by multiplication.

$$2 + 2 + 2 + 2 + 2 = 5(2) = 10$$

With this in mind, represent $7(-2)$ as an addition problem, then find the value of $7(-2)$.

49. Recall that -14 is the opposite of 14, and 14 is the opposite of -14. Use this information and the method in Exercise 48 to find the product $-[7(-2)]$.

LESSON 3.2
Exploring Integer Subtraction

why Each time Mark writes a check, the bank subtracts money from his account. When he deposits his paycheck, the bank adds money to his account. You use integers to model subtraction when working with financial transactions.

Suppose Mark starts with $50 in his account and writes a check for $20. This transaction can be represented in two ways.

subtraction: $50 - 20$ or   addition: $50 + (-20)$

In either case, Mark will have $30 left.

$50 - 20 = 30$ or $50 + (-20) = 30$

Using Tiles to Subtract Integers

You can model the subtraction of integers by taking away tiles.

$6 - 4 = 2$ $-4 - (-2) = -2$

In Lesson 3.1 you found the result of adding the same number of positive and negative tiles does not change the *value* of the tiles you have. The result is the same as adding 0. Why do you think the number 0 is called the **additive identity**?

ADDITION PROPERTY OF ZERO

For any number a,

$a + 0 = a = 0 + a$.

The next activity makes use of the Addition Property of Zero. The tile model shows how it is possible to subtract, or *take away*, -4 from 2.

$$2 - (-4) = \underline{\ ?\ }$$

1. Start with 2 positive titles.

2. Since you want to subtract -4, add 4 positive and 4 negative tiles. The total value of the tiles is still 2.

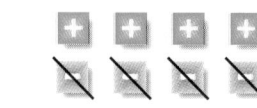

3. Now you can subtract, or *take away*, the 4 negative tiles and rewrite the original problem as $2 + 4$.

Thus, $2 - (-4) = 2 + 4 = 6$.

CRITICAL
Thinking

Describe how you would use algebra tiles to calculate $-2 - 4$. What is the addition expression that is equivalent to $-2 - 4$?

•Exploration *Subtraction With Tiles*

Use tiles to subtract 5 from 3.

1 Begin with 3 positive tiles.

2 The number you want to subtract is 5, so add 5 positive and 5 negative tiles. What value did you add to the expression when you added these tiles?

3 How many and what kind of tiles do you take away?

4 Remove the tiles. How many of each kind are left?

5 What addition expression is equivalent to $3 - 5$? What is the sum?

6 Use this procedure to calculate $-3 - (-4)$.

7 Explain how to subtract an amount with a greater absolute value from an amount with a lesser absolute value. ❖

Relating Addition and Subtraction

When working with integers, you can use addition to calculate a difference. To subtract an integer, add its opposite. The same is true for any numbers.

THE DEFINITION OF SUBTRACTION

For all numbers a and b,

$$a - b = a + (-b).$$

EXTENSION

You will sometimes need to determine the distance between points on the number line. One method uses absolute value.

To find the distance between points A and B, subtract the values, then find the absolute value of the difference.

The distance between -4 and 2 is $|2 - (-4)| = 6$, or $|(-4) - 2| = 6$.

In general, if you let a represent the value for the point A and let b represent the value for the point B, the distance between points on the number line is $|b - a|$ or $|a - b|$. ❖

What is the distance between -3 and 5 on the number line?

Calculator

On all calculators ⎡ – ⎤ is used for the subtraction operation. Recall from the previous lesson that some calculators use ⎡ +/- ⎤ or ⎡ ± ⎤ to change the sign of a number. Other calculators use ⎡ (-) ⎤ to find the opposite of a number.

How does your calculator treat the difference between subtraction and the opposite of a number? What keystrokes on your calculator let you calculate $-35 - 27$?

EXERCISES & PROBLEMS

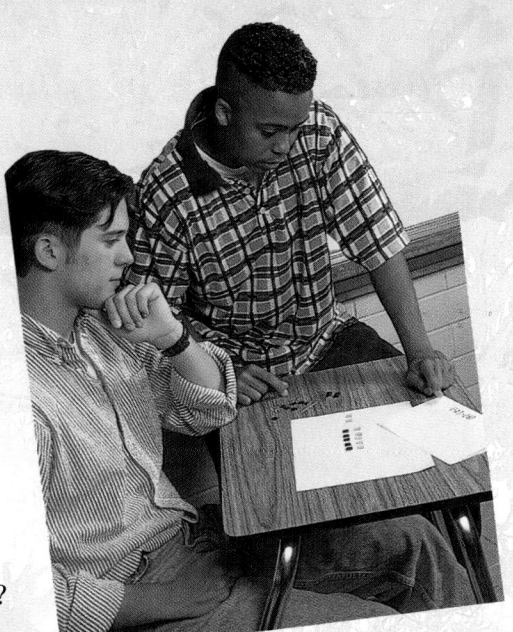

Communicate

1. What is the effect of adding 0 to a number?

2. How does adding pairs of positive and negative tiles help to model a subtraction problem?

3. Mark's account balance is $5. Explain what would happen if Mark were to write a check for $25 dollars.

4. For the problem $-4 - (-7)$, how many positive and negative tiles do you add to solve the problem? Why?

5. What is meant by adding the opposite?

6. Explain how addition can replace subtraction when you use integers.

Practice & Apply

Find the solutions of the following problems.

7. $67 - 3$ 8. $42 - (-9)$ 9. $-10 - (-21)$ 10. $-35 - 17$

11. $33 - (-33)$ 12. $-78 + (-45)$ 13. $990 - (-155)$ 14. $-97 - 88$

15. $-43 + 23 + (-43)$ 16. $-77 - 77 + 5$ 17. $108 + (-18) - 8$ 18. $85 - (-12) - (-9)$

Let $x = 5$, $y = -3$, and $z = -10$. Evaluate each expression.

19. $x - y$ 20. $x + y - z$ 21. $(x - z) - y$ 22. $x - (z - y)$

23. $y - x$ 24. $y - x + z$ 25. $(x + y) - (x - y)$ 26. $y - y - y - y$

Find the distance between the following pairs of points on the number line.

27. $4, 9$ 28. $-6, 15$ 29. $-47, -23$ 30. $-12, 74$

Mike has a balance of $145 in his savings account.

31. If Mike withdraws $37, how much money does he have left in his account?

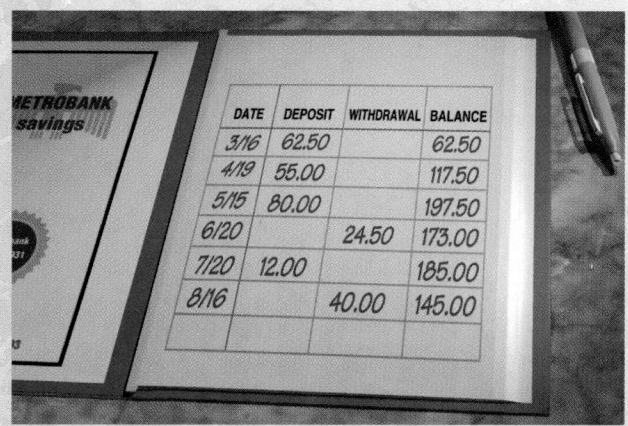

DATE	DEPOSIT	WITHDRAWAL	BALANCE
3/16	62.50		62.50
4/19	55.00		117.50
5/15	80.00		197.50
6/20		24.50	173.00
7/20	12.00		185.00
8/16		40.00	145.00

32. **Temperature** Theresa noticed that the temperature in her freezer was 5°F. She lowered the thermostat by 7°F. Later, she set the thermostat another 2°F lower. What temperature would you expect in the freezer after Theresa finished changing the thermostat?

This wind chill table depends on the air temperature and the speed of the wind.

Wind Speed in Miles per Hour		10	20	30
Air Temperature	20°F	3	–10	–18
Air Temperature	10°F	–9	–24	–33
Air Temperature	0°F	–22	–39	–49
Air Temperature	–10°F	–34	–53	–64

33. The air temperature is 20°F. The wind speed is 20 miles per hour. What is the wind-chill factor?

34. The air temperature is -10°F. How much colder does it feel when the wind speed is 20 miles per hour than when there is no wind?

35. The air temperature is -10°F. The wind speed increases from 10 to 30 miles per hour. How many degrees does the temperature seem to drop?

Look Back

36. Cultural Connection: Africa The earliest woman mathematician whose name we know is Hypatia. She was a professor at the University of Alexandria, Egypt, about 400 C.E. In her commentary on Diophantus, an earlier Egyptian mathematician, Hypatia spoke of pentagonal numbers.

The sequence for pentagonal numbers is 1, 5, 12, 22, 35, . . . Find how many differences it takes to reach a constant. What is the constant? **[Lesson 1.2]**

37. What kind of function will generate the terms of the pentagonal numbers? **[Lesson 1.2]**

Simplify. [Lesson 1.4]

38. $36 - 12 \div 3 - 20$ **39.** $3 \cdot 5 + 7 \div 2$ **40.** $28 \div 2 \cdot 7 + 4$

41. What function is associated with the graph of a parabola? **[Lesson 2.2]**

42. What kind of function will generate the values shown for y in the table? **[Lesson 2.3]**

x	5	6	7	8	9
y	$\frac{1}{5}$	$\frac{1}{6}$	$\frac{1}{7}$	$\frac{1}{8}$	$\frac{1}{9}$

Look Beyond

43. One point on the number line is at 4. The location of the other point is not known. Let the unknown point be at x. If the distance between the points is 7, what is x? (HINT: Draw the number line, and try different values of x.

44. What is x if the distance between the points is 10?

LESSON 3.3 Adding Expressions

why *In Algebra you work with expressions that contain numbers and variables. When you add expressions, you need to write their sum in a simplified form.*

Juan is collecting comic books. He has 3 cartons of comics and 4 loose comics in his room. He has 2 more cartons of comics and 2 loose comics in the trunk of his car.

How can you write an expression that describes the total number of cartons and comics?

In his room	3 cartons of comics and 4 comics	$3x + 4$
In the car	2 cartons of comics and 2 comics	$2x + 2$
All together	5 cartons of comics and 6 comics	$5x + 6$

The quantity, $5x + 6$, is an example of an **algebraic expression.** Both $5x$ and 6 are **terms** of the expression $5x + 6$. The terms $3x$ and $2x$ are **like terms** because each contains the same form of the variable. The number 5 is a **factor** of $5x$. The 5 is also called the **coefficient** of x. The number 6, which represents a fixed amount, is often referred to as a **constant.**

Tiles can be used to model an expression containing a variable such as x.

Represents positive x Represents negative x

EXAMPLE 1

Simplify $(3x + 4) + (2x - 1)$ using tiles.

Solution▸

Represent each of the expressions using the appropriate tiles. Then combine like tiles.

First expression	Second expression	Combination
$3x + 4$	$2x - 1$	$(3x + 4) + (2x - 1)$

Remember, a positive 1-tile and a negative 1-tile equal 0 when combined.

$5x + 3$

When you simplify the expression, the result is $5x + 3$. ❖

How would you model $(2x - 3) + (4x + 5)$? What is the result?

The Distributive Property

Combining like terms such as $3x$ and $2x$ uses an important property of numbers: multiplication can be distributed over addition.

The Distributive Property works in two directions. To multiply $3 \cdot (5 + 2)$, the multiplication is distributed over the addition.

$$3 \cdot (5 + 2) = 3 \cdot 5 + 3 \cdot 2 = 21$$

When used in reverse, a common factor can be removed from the terms.

$$3 \cdot 5 + 3 \cdot 2 = 3 \cdot (5 + 2) = 21$$

This can be shown on a number line.

DISTRIBUTIVE PROPERTY
For all numbers a, b, and c,
$a(b + c) = ab + ac$ and $(b + c)a = ba + ca$.

The Distributive Property can be used to simplify an expression that contains like terms.

$$3x + 2x = (3 + 2)x$$
$$= 5x \text{ ❖}$$

Explain how to add like terms such as $3x$ and $2x$ on the following number line.

$$-5x \quad -4x \quad -3x \quad -2x \quad -1x \quad 0x \quad 1x \quad 2x \quad 3x \quad 4x \quad 5x$$

Rearranging Terms

Changing the order of two terms is called commuting the terms.

> ### COMMUTATIVE PROPERTY FOR ADDITION
> For any numbers a and b,
> $$a + b = b + a.$$

Now the expressions in Example 1 can be added without using tiles for models.

$(3x + 4) + (2x - 1)$	Given
$3x + 4 + 2x + (-1)$	Definition of Subtraction
$3x + (4 + 2x) + (-1)$	**Associative Property**
$3x + (2x + 4) + (-1)$	**Commutative Property**
$(3x + 2x) + (4 + (-1))$	**Associative Property**
$5x + 3$	Combine like terms.

The Associative and Commutative Properties can be used to rearrange terms. For example, $(3B + 4) + (2 + 2B)$ can be rearranged as $(3B + 2B) + (4 + 2)$. Often the middle three steps of the process shown above are combined.

$(3x + 4) + (2x - 1)$	Given
$(3x + 2x) + (4 - 1)$	**Rearrange terms.**
$5x + 3$	Combine like terms.

To check the addition, replace the variable with a number that makes the expression easy to evaluate. For example, replace x with 10, and evaluate each expression.

$$(3x + 4) + (2x - 1) = 5x + 3$$
$$3(10) + 4 + 2(10) - 1 \stackrel{?}{=} 5(10) + 3$$
$$53 = 53 \qquad \text{True}$$

Try This   Add each expression to $5b + 1$.
     a. $2b - 3$ **b.** $-2b + 3$ **c.** $6 - 4b$

EXAMPLE 2

GEOMETRY
Connection

The lengths of the sides of triangle *ABC* are shown in the diagram. What is the perimeter of triangle *ABC*?

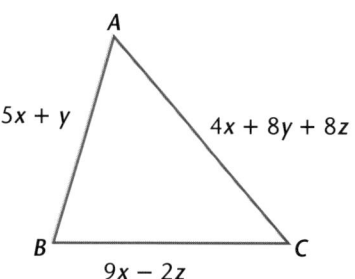

Solution ➤

To find the perimeter, add the lengths of the sides.

$$P = (5x + y) + (4x + 8y + 8z) + (9x - 2z) \quad \text{Definition of Perimeter}$$
$$= (5x + 4x + 9x) + (y + 8y) + (8z - 2z) \quad \text{Rearrange terms.}$$
$$= 18x + 9y + 6z \quad \text{Combine like terms.}$$

The perimeter is $18x + 9y + 6z$. ❖

EXERCISES & PROBLEMS

Communicate

1. Define an *algebraic expression*. Give two examples.
2. Discuss what is meant by a *coefficient*. Give two examples.
3. What is the Commutative Property for Addition? Give an example.
4. Describe the steps in using the Distributive Property to add $3x$ and $8x$.
5. Identify which of the following are like terms. Explain why.

$$7x, 2x, 3z, 5, 7y, 3x, -z, 3y, \text{ and } 23$$

6. Discuss how you would use algebra tiles to model $(5x + 2) + (3x - 4)$.

Practice & Apply

Add.

7. $(5a - 2) + (3a - 6)$

8. $(2x + 3) + (7 - x)$

9. $(4x + 5y) + (x + 9y)$

10. $(1.1a + 1.2b) + (2a - 0.8b)$

11. $\left(\frac{x}{2} + 1\right) + \left(\frac{x}{3} - 1\right)$

12. $\left(\frac{2m}{5} + \frac{1}{2}\right) + \left(\frac{m}{10} + \frac{5}{2}\right)$

13. $(2a + 3b + 5c) + (7a - 3b + 5c)$

14. $(x + y + z) + (2w + 3y + 5)$

15. If 6 cartons and 3 loose tiles are combined with 2 cartons and 1 loose tile, describe the result algebraically.

Picture Framing   A picture is surrounded by a frame which is 3 inches wide. Find the total height of the frame and the picture if the height of the picture is

16. 12 inches.

17. 20 inches.

18. h inches.

3"

A lid 1 inch thick is placed on top of a box. Find the total height if the height of the box is

19. 10 inches.

20. 27.5 inches.

21. x inches.

22. On Tuesday, John bought 2 large boxes of cassette tapes and 3 additional tapes. The next week he bought 1 large box and 7 additional tapes. Represent the boxes and tapes algebraically, and determine the expression for the sum.

23. Each month for 7 months Don buys 2 boxes of cookies. In all, he let his friend have a total of 14 cookies. Let c represent the number of cookies in a box. Represent in an algebraic expression the total number of cookies Don has for himself during the 7 months.

24. **Clothing Manufacturing**   A piece of fabric is cut into 12 equal lengths with 3 inches left over. Define the variable, and write the length of the original piece as an expression.

25. If m is an unknown integer, represent the next two consecutive integers in terms of m.

26. If n is an unknown odd integer, represent the next two odd integers in terms of n.

27. If apples cost 30 cents each, bananas cost 25 cents each, and plums cost 20 cents each, write an expression to find the total cost of a apples, b bananas, and p plums.

 Geometry   Find the perimeter of each figure.

28.

$3m + 2r$

$r - 5m$ $5r - m$

$4r + 2m$

29.

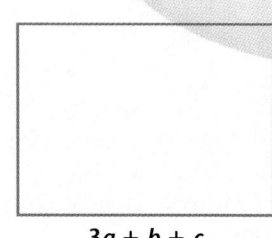

$3a + b + c$

30.

$3s - 3t$

$2p - 5s$

$4b - 2c$

$5t - 4s + p$

31. 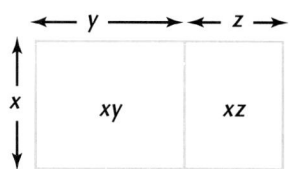 **Geometry** Write an expression for the area of the whole rectangle. Explain how the Distributive Property can be used to find this area.

Use the Distributive Property to write equivalent expressions for the following.

32. $3(5 + 7)$ **33.** $6(x - 3)$ **34.** $(x + y)n$

35. $(4 \cdot 3 + 4 \cdot 5)$ **36.** $4x - 4 \cdot 7$ **37.** $az + bz$

38. Explain how you can use the Distributive Property to find $5 \cdot 12$.

Look Back

39. You use the method of differences on a sequence, and a constant appears in the second difference. What type of function would you expect to find? **[Lesson 1.2]**

40. Find the next 3 terms in the sequence 10, 20, 40, 80, 160, __, __, __. **[Lesson 1.2]**

41. Which ordered pair does not lie on the same line as the others? $(3, 5), (4, 8), (6, 11), (4, 7)$ **[Lesson 1.5]**

Name the following type of function.

42. $y = 10x + 3$ **[Lesson 1.5]** **43.** $y = \frac{1}{x}$ **[Lesson 2.3]**

44. $y = x^2$ **[Lesson 2.2]** **45.** $y = |x|$ **[Lesson 2.4]**

Simplify. **[Lessons 3.1, 3.2]**

46. $-2 + (-3)$ **47.** $-7 + 6$ **48.** $-5 - (-2)$ **49.** $-10 - 5$

Look Beyond

50. Find $(3A + 2B + 4P) - (5A + 3P)$.

51. Draw a diagram to model the area represented by $(x + y)(x + y)$.

52. In the center of the City the blocks are in the shape of a grid. You are at the bank at A, and you want to walk to the post office at B. If all the different ways you take are exactly 7 blocks long, how many different ways are there to get from A to B?

Subtracting Expressions

Why *You can subtract one integer from another by adding its opposite. The same procedure can also be used to subtract expressions.*

Bonnie bought 5 cartons of soda and 7 extra cans of soda for her party. Her guests drank 3 cartons and 4 cans of soda.

How do you write an algebraic expression that represents the total number of cans left after the party?

Before the party there are 5 cartons and 7 cans.

$$5x + 7$$

During the party the guests drank 3 cartons and 4 cans of soda.

$$3x + 4$$

After the party, the amount left is 2 cartons and 3 cans.

$$2x + 3$$

$$(5x + 7) - (3x + 4) = 2x + 3$$

If *x* represents the number of cans in a carton, then the algebraic expression $2x + 3$ represents the total number of cans left after the party.

Modeling Subtraction With Tiles

Subtraction can be modeled with tiles.

EXAMPLE 1

Use algebra tiles to simplify $(2x + 3) - (4x - 1)$.

Solution▸

Start with 2 positive x-tiles and 3 positive 1-tiles to represent $2x + 3$.

2x + 3

Now subtract $(4x - 1)$.

You need to subtract 4 positive x-tiles and 1 negative 1-tile. Add 2 pair of positive and negative x-tiles and 1 pair of positive and negative 1-tiles. Now you can take away 4 positive x-tiles and 1 negative 1-tile.

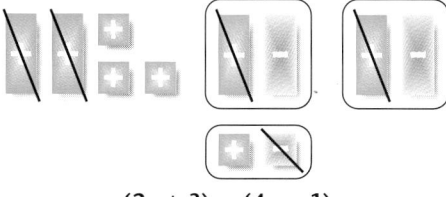

(2x + 3) − (4x − 1)

This leaves 2 negative x-tiles and 4 positive 1-tiles.

$(2x + 3) - (4x - 1) = -2x + 4$ ❖

−2x + 4

A tile pair that has opposite signs is sometimes referred to as a **neutral pair**. Explain how to use neutral pairs to subtract $6x + 2$ from $3x - 5$.

Neutral Pairs

Adding the Opposite of an Expression

You can subtract a number from another by adding its opposite. This is also true for expressions.

DEFINITION OF SUBTRACTION FOR EXPRESSIONS
To subtract an expression, add its opposite.

When you subtract a quantity by adding the opposite, first change the sign of *each term* in the quantity that you are subtracting. Then rearrange to group the like terms, and add.

$(2x + 3) - (4x - 1)$	Given
$(2x + 3) + (-4x + 1)$	Definition of Subtraction
$(2x - 4x) + (3 + 1)$	Rearrange terms.
$-2x + 4$	Combine like terms.

EXAMPLE 2

Subtract.

A $(7m + 2) - (3m + 5)$ **B** $(10d - 3) - (4d + 1)$ **C** $(4x - 2) - (5x - 3)$

Solution▸

A

$(7m + 2) - (3m + 5)$	Given
$(7m + 2) + (-3m - 5)$	Definition of Subtraction
$(7m - 3m) + (2 - 5)$	Rearrange terms.
$4m - 3$	Combine like terms.

B

$(10d - 3) - (4d + 1)$	Given
$(10d - 3) + (-4d - 1)$	Definition of Subtraction
$(10d - 4d) + (-3 - 1)$	Rearrange terms.
$6d - 4$	Combine like terms.

C

$(4x - 2) - (5x - 3)$	Given
$(4x - 2) + (-5x + 3)$	Definition of Subtraction
$(4x - 5x) + (-2 + 3)$	Rearrange terms.
$-x + 1$	Combine like terms. ❖

Explain how to simplify $(3x - 4) - (-2x + 3)$ mentally.

A negative sign in front of a variable does not necessarily mean its *value* is negative.

EXAMPLE 3

Decide whether the expression $-W$ is positive or negative when

A $W = 5$.   **B** $W = -7$.   **C** $W = 0$.   **D** $W = x$.

Solution▸

Substitute each value for W, and examine the value of $-W$.

A $-W = -(5) = -5$ The opposite of a positive 5 is negative 5. The value is negative.

B $-W = -(-7) = 7$ The opposite of a negative 7 is positive 7. The value is positive.

C $-W = -(0) = 0$ The value of 0 is neither positive nor negative.

D $-W = -(x)$ It is impossible to tell whether the expression is positive or negative. Since x is a variable, the value will depend on the value that x represents. ❖

CRITICAL Thinking

Examine the expression $-(5x + 2y)$. The negative sign means to find the opposite of *each term*. Explain why this might be considered a special case of the Distributive Property.

Try This   If a is -4 and b is 3, find $-(a + b)$.

> ### EXAMPLE 4

Simplify $(8x + 4y - z) - (6y + 3z - 5x)$.

Solution▶

$(8x + 4y - z) - (6y + 3z - 5x)$	Given
$8x + 4y - z - 6y - 3z + 5x$	Definition of Subtraction
$(8x + 5x) + (-6y + 4y) + (-z - 3z)$	Rearrange terms.
$13x - 2y - 4z$	Combine like terms. ❖

EXERCISES & PROBLEMS

Communicate

Tell how to represent each subtraction using algebra tiles.

1. $4x - 3x$ **2.** $(3x + 2) - (2x + 1)$ **3.** $(5x + 3) - (2x + 4)$

4. For the following problem, assign a variable to represent the item. Express the information using algebraic expressions. Explain how to subtract the expressions. Then answer the question.

Ms. Green had 5 reams of paper and 100 loose sheets. She gave Mr. Black 2 reams of paper and 50 loose sheets. How much did she have left?

5. Explain what is meant by adding the opposite when subtracting expressions.

6. Explain how to perform the following subtraction.
$$(7y + 9x + 3) - (3y + 4x - 1)$$

7. Describe the set of values for m that make $-6m$ positive.

Practice & Apply

Find the opposite of each expression.

8. 17   **9.** -13 **10.** $2x$ **11.** -6 **12.** $9y + 2w$ **13.** $5a + 3b$
14. $2n - 3m$ **15.** $9c - 5d$ **16.** $-7x + 9$ **17.** $-7r + 4s$ **18.** $-3p - q$ **19.** $-j - k$

Perform the indicated operations.

20. $9x - 3x$ **21.** $8y - 2y$ **22.** $5c - (3 - 2c)$ **23.** $7d - (1 - d)$
24. $(7r + 2s) + (9r + 3s)$ **25.** $(9k + 2k) + (11j - 2j)$
26. $(2a - 1) - (5a - 5)$ **27.** $(9v - 8w) - (8v - 9w)$
28. $(2x + 3) - (4x - 5) + (6x - 7)$ **29.** $(4y + 9) - (8y - 1) + (7 - y)$

Tell whether each statement is true or false. Then explain why.

30. $x - (y + z) = x - y + z$ **31.** $x - (y - z) = x - y - z$

32. $-x + (y - z) = -x - y + z$ **33.** $-x$ is a negative number

 Statistics The graph is called a stacked bar graph. It shows the population of the United States by gender. For example, it shows that in 1970 there were approximately 100 million males and 100 million females in the United States.

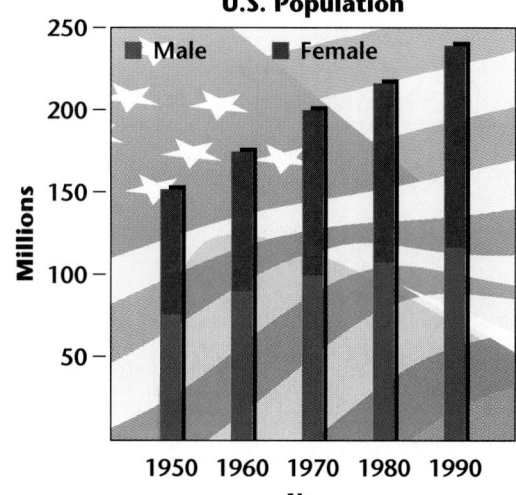

U.S. Population

34. What was the female population in 1950?

35. What was the male population in 1950?

36. How many people were there in 1950?

37. What was the female population in 1980?

38. What was the male population in 1980?

39. How many people were there in 1980?

Tell whether the given statement about the stacked bar graph is true or false.

40. The population rose during each 10-year period.

41. The number of males and females were approximately equal during each 10-year period.

42. The number of females rose much faster than the number of males.

Fund-raising Students at Valley View High School are selling fruitcakes for a fund raiser. They earn $3 for each deluxe fruitcake and $2 for each regular fruitcake. How much would the students earn for selling

43. 89 deluxe fruitcakes and 234 regular fruitcakes?

44. d deluxe fruitcakes and r regular fruitcakes?

Suppose students earn $3.75 for each deluxe fruitcake and $2.50 for each regular fruitcake. How much would they earn for selling

45. 89 deluxe fruitcakes and 234 regular fruitcakes?

46. d deluxe fruitcakes and r regular fruitcakes?

Joni sold 9 deluxe fruitcakes and 23 regular fruitcakes.
Wang sold 7 deluxe fruitcakes and 27 regular fruitcakes.

47. How much did they earn together?

48. Who earned more? How much more?

Fund Raiser Records

Fruitcake	Deluxe	Regular
	$3.75	$2.50
Joni	9	23
Wang	7	27

49. Inventory The school kitchen had 11 cases of juice plus 3 extra cans of juice. After lunch they had 6 cases of juice and no extra cans. How much juice was distributed during lunch?

Cultural Connection: Africa The idea of adding and subtracting numbers has been around for thousands of years. Early Egyptians represented addition by feet walking towards the number and subtracting by feet walking away from the number.

11 plus 2   11 minus 3

50. How would you represent 11 + 5 using the early Egyptian notation if | = 1 and ∩ = 10?

51. Represent 23 − 12 using the early Egyptian notation.

Look Back

52. Write 5 problem-solving strategies.   **[Lesson 1.1]**

53. Create a sequence whose differences reach a constant at the third difference.   **[Lesson 1.2]**

Place parentheses and brackets to make each equality true. [Lesson 1.4]

54. $28 \div 2 - 4 \cdot 1 = 10$

55. $16 \div 5 + 3 \div 2 = 1$

56. $40 \cdot 2 + 10 \cdot 4 = 1680$

57. What is the value of INT(x) for $x = \frac{7}{4}$?   **[Lesson 2.4]**

58. Describe the graph of an integer function.   **[Lesson 2.4]**

59. What is the shape of the graph for the absolute value function? **[Lesson 2.4]**

60. The first reading from a gauge is 100. The changes recorded at 1 hour intervals are $-4, +51, 0, +7, -12, -78, +2, -13, -1$. What is the current reading on the gauge?   **[Lesson 3.1]**

Look Beyond

61. **Statistics** Fred has test scores of 87, 74, and 90. How many points does he need on the next test to have an average of 85?

62. Paul and Dan leave from the bandstand at one end of a lake and sail around the lake in the same direction. Paul sails at 12 miles per hour, while Dan sails at 10 miles per hour. The lake is 2 miles around, and they leave at noon. When will the two first meet again at the bandstand?

Addition and Subtraction Equations

why *Many problems can be modeled algebraically by writing an equation. A sequence of steps that isolates the unknown on one side of the equal sign provides the solution.*

How many degrees must the temperature rise to reach the freezing point, 32°F?

If Riva has $20, how much more money does Riva need to buy the boots?

Although each situation is different, each can be modeled by the same equation $x + 20 = 32$.

Tiles can be used to demonstrate the steps in solving the equation. Suppose the x-box represents an unknown number of tiles. The x-box and the tiles represent the equation $x + 20 = 32$.

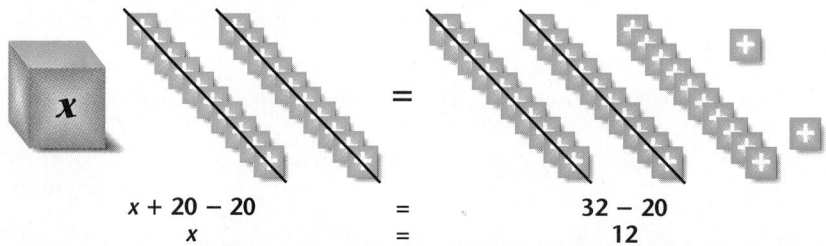

$$x + 20 - 20 \qquad = \qquad 32 - 20$$
$$x \qquad = \qquad 12$$

After you take away 20 tiles from each side, the amounts on each side of the equal sign remain equal. The x-box is now alone on one side of the equal sign. Count the remaining tiles on the other side of the equal sign to find the number that the box represents.

The idea of subtracting the same number of tiles from each side of the equation is called the Subtraction Property of Equality.

CRITICAL *Thinking*

How can you use a model to represent and solve the equation $3x + 4 = 2x + 10$?

SUBTRACTION PROPERTY OF EQUALITY

If equal amounts are subtracted from the expressions on each side of an equation, the expressions remain equal.

EXAMPLE 1

Fund-raising The school band needs new uniforms. Write and solve an equation to find how much more money the band must raise.

BAND UNIFORM FUND RAISER

Goal $5000.00

$2344.10 — Money Raised

0

Solution▶

Organize the information.
Study the problem. Identify the known and unknown information. Then write a sentance using words and operation symbols.

(the amount to be raised) + (the amount they have) = (the goal)

Clearly identify a variable to represent the unknown value you are trying to find.

Let x represent the amount to be raised.

Write the equation.
Replace the words you wrote in the first step with the variable and the known information from the problem.

$$x + 2344.10 = 5000.00$$

Solve the equation.
Use the Subtraction Property of Equality.

$$x + 2344.10 - 2344.10 = 5000.000 - 2344.10$$
$$x = 2655.90$$

The equation is solved when the variable is isolated on one side of the equation.

Check the answer.
Replace the variable in the original equation with the value you found. See if it makes a true statement. Be sure that the solution is reasonable and that it answers the question in the problem.

$$2655.90 + 2344.10 \stackrel{?}{=} 5000.00$$
$$5000.00 = 5000.00 \quad \text{True}$$

The band must raise $2655.90. ❖

Solving Subtraction Equations

Some equations contain subtraction. To solve $x - 2 = 8$, use a procedure similar to that used for equations containing addition.

Suppose an x-box represents an unknown number of tiles. Represent the equation with tiles. To solve the equation, add 2 positive tiles to each side of the equation, and simplify. The unknown value is 10.

Equation   **Solution**

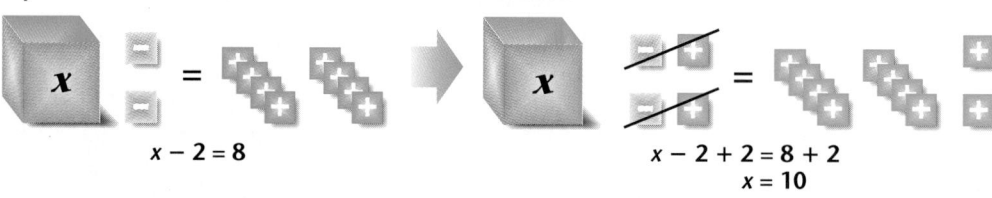

$$x - 2 = 8$$

$$x - 2 + 2 = 8 + 2$$
$$x = 10$$

Try This   Solve with tiles.   **a.** $x - 5 = 4$ **b.** $x - (-3) = 5$

ADDITION PROPERTY OF EQUALITY
If equal amounts are added to the expressions on each side of an equation, the expressions remain equal.

In a dart game, the range of scores is 47 points, and the lowest score is 52.

EXAMPLE 2

Write and solve an equation to find the highest score in the dart game.

Solution

Organize
In statistics, the *range* is the difference between the highest and lowest scores.

range = highest score − lowest score

Let H represent the highest score.

Write
Replace the words in the range formula with the variable, H, and the known values from the problem.

$$47 = H - 52$$

STATISTICS
Connection

Solve

$47 = H - 52$	Given
$47 + 52 = H - 52 + 52$	Addition Property of Equality
$99 = H$	Combine like terms.

Check
Substitute 99 for H.

Since $47 = 99 - 52$, the highest score is 99. ❖

Literal Equations

Scientists often express numerical relationships as **formulas**. For example, $C = \frac{5}{9}(F - 32)$ relates Fahrenheit and Celsius temperatures. Because formulas often contain a number of different letters that represent variables, they are called **literal equations**.

EXAMPLE 3

Investment

The formula $A = P + I$ shows that the total amount of money you receive from an investment equals the principal (the money you started with) and added interest. Write a formula for the interest, I, based on the principal and the amount.

Solution➤

To isolate I, subtract P from *both sides* of the equation.

$A = P + I$	Given
$A - P = P + I - P$	Subtraction Property of Equality
$A - P = I$	Group like terms and simplify.
$I = A - P$	Rewrite with I on the left. ❖

Literal equations that contain subtraction are treated in much the same way as equations that contain addition. In business, the equation $P = s - c$ is a formula for profit, where s represents the selling price and c represents the cost.

EXAMPLE 4

Solve the equation $P = s - c$ for s.

Solution➤

To solve the equation for s, add c to both sides of the equation.

$P = s - c$	Given
$P + c = s - c + c$	Addition Property of Equality
$P + c = s$	Combine like terms.
$s = P + c$	Rewrite with s on the left. ❖

Explain how you would solve $a = x - b$ for x.

Try This

GEOMETRY
Connection

The figure shows the sector of a circle with center O and radius R. The formula for the length h is $h = R - k$. Solve the equation for k.

Technology that solves equations on a computer or a calculator already exists. For example, to solve the equation $23 = C + 14$ using the Maple™ software program, type in solve($23 = C + 14$);. The computer will display 9.

Computer

To solve the equation $5 - 2r = 7 - (3r - 8)$, type in solve($5 - 2 * r = 7 - (3 * r - 8)$);. The computer will display 10. Note that $2 * r$ is entered for 2r. What do you think will result if you type in solve($4 + 3 * t = 6 + 5 * t$);?

EXERCISES & PROBLEMS

Communicate

1. Compare the Addition Property of Equality with the Subtraction Property of Equality. Tell how to decide which one to use.

2. Give a situation that can be modeled by
   a. $x + 20 = 50$.   **b.** $x - 20 = 50$.

3. Tell how to solve each equation using algebra tiles.
   a. $x + 6 = 10$ **b.** $x - 6 = 10$

4. Explain how to solve $s + x = r$ for s.

5. Explain how to solve $m + b = n$ for m.

Practice & Apply

State which property you would use to solve each equation. Then solve.

6. $a - 16 = 15$ **7.** $t + 29 = 11$ **8.** $m + 54 = 36$ **9.** $r - 10 = -80$

10. $l - 27 = 148$ **11.** $b - 109 = 58$ **12.** $y + 37 = -110$ **13.** $396 = z + 256$

14. $x + \frac{3}{4} = \frac{5}{4}$ **15.** $7.4 + t = 5.2$ **16.** $3r + 5.78 = 2r + 7$ **17.** $\frac{3}{8} - x = \frac{3}{4}$

Solve each equation using algebra tiles.

18. $x + 9 = 6$ **19.** $x - 7 = 3$ **20.** $x - 10 = -4$ **21.** $x + 6 = -4$

Solve each equation for a.

22. $a + b = c$ **23.** $a - b = c$ **24.** $a + b = -c$ **25.** $a - b = -c$

26. If $a - \frac{2}{3} = 4$, what is the value of $3a$?

27. If $2.5 + s = 5.3$, what is the value of $2s - 3$?

Exercises 28-30 refer to the photo.

28. Solve equation a.

29. Solve equation b.

30. Compare the steps in solving the 2 equations.

Equation a: $x + 35 = 96$

Equation b: $x - 35 = 96$

For Exercises 31-41, assign a variable and write an equation for each situation. Then solve the equation.

31. Sports The first- and second-string running backs on a football team ran for a total of 94 yards. If the first-string back ran 89 yards, how many yards did the other back run?

32. The first- and second-string running backs on a football team ran for a total of 89 yards. If the first-string back ran 94 yards, how many yards did the other back run?

33. The first-string running back on a football team ran for a total of 94 yards. If the second-string running back on a football team ran for a total of 89 yards, what was their combined yardage?

34. Carl sees that the calendar indicates that his birthday, December 21, is the 355th day of the year and that the current day, October 15, is the 288th day of the year. In how many days is his birthday?

35. Home Economics A recipe for turkey gravy says you should add water to the drippings (with most of the fat removed) to get $1\frac{1}{2}$ cups of liquid. If there is $\frac{7}{8}$ of a cup of drippings, how much water should be added?

36. Travel If the odometer on Sarah's car registered 23,580 when she finished a vacation trip of 149 miles, what did it read when she started?

37. If the odometer on Elizabeth's car registered 23,580 when she started a trip of 149 miles, what did it read when she finished her trip?

38. Sales tax Sam bought two items costing $12.98 and $14.95 plus tax. He got 39¢ change from $30. How much was the tax?

39. Statistics The range of a set of scores was 28. If the highest score was 47, what was the lowest score?

40. Geometry Supplementary angles are pairs of angles whose measures total 180 degrees. Find the measure of an angle supplementary to an angle with a measure of 92 degrees.

41. Two of the angles of a triangle each measure 50 degrees. Find the measure of the third angle. (HINT: The sum of the measures of the angles of a triangle is 180 degrees.)

42. 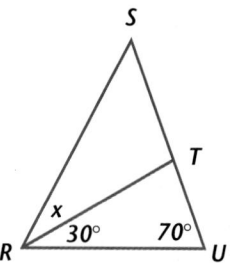 **Geometry** The angles *SRU* and *SUR* of triangle *RSU* measure 70°. Write and solve an equation to find the measure of angle *SRT*.

43. What would the measure of angle *SRT* be in triangle *SRU* if the measure of angle *TRU* were 18°?

 Geometry

44. Find *x* if *S* lies on line *RT*.

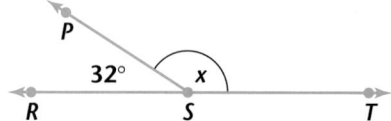

45. Find *y* if angle *UVW* is a right angle.

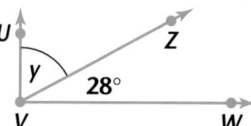

Technology Some computer software packages, such as Maple™ can solve equations automatically, as shown in the following examples. The line after > is what you type in. The next line is what the computer types back when you hit ENTER.

> solve(x + 98765432123456789 = 44444444444444444);

345679012320987655
_____

> solve(A = P + I,I);

A − P

Give Maple™ commands to solve the following equations.

46. x + 12345.6789 = 55555555

47. A + B = C + D (Solve for D.)

48. 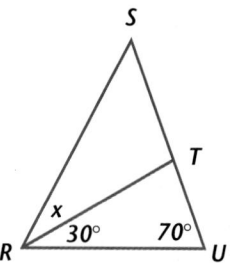 **Portfolio Activity** Complete the problem in the portfolio activity on page 99.

Look Back

49. State the pattern and find the next three numbers for the sequence 6, 0, 12, 6, 18, 12, __, __, __. **[Lesson 1.1]**

50. Find the perimeter of a square with sides 2.5 inches long. **[Lesson 1.3]**

51. If poster board cost 39¢ for each piece, how many pieces of poster board can Shannon buy with $2.50? **[Lesson 1.3]**

52. **Physics** Suppose a rocket took 23 seconds from take off to its return to Earth. Find the time it reached its maximum height. **[Lesson 2.2]**

Simplify. **[Lessons 3.3, 3.4]**

53. $(3x − 7) − (x − 4)$

54. $2(8 − 7y) + 5(2y)$

55. $−4(n − 2) − (−3n + 4)$

Look Beyond

Solve these equations using algebra tiles.

56. $x − 3 = 8$

57. $2x = 12$

58. $2x + 5 = 11$

LESSON 3.6 Solving Related Inequalities

why When the weather report says "The high temperature for today will be 75°," it implies that the temperatures throughout the day will be less than or equal to 75°F. This inequality describes a range of values. Inequalities can also be used to determine the items you can purchase when you have a given amount of money.

Hamburger $1.45
Cheeseburger 2.50
Brownie/
Ice Cream 3.00
Fruit
30¢

Michael can spend at most $3.10 for lunch. He buys a hamburger for $1.45. How much can he spend on dessert and stay within his spending limit?

Let x be the amount Michael spends on dessert. The total amount he spends is then $x + 1.45$. Michael's situation can be modeled by the statement $x + 1.45 \leq 3.10$. It is read "x plus 1.45 *is less than or equal to* 3.10." Michael will stay within his limit as long as $x + 1.45$ remains less than or equal to 3.10.

Suppose Michael buys a dish of fruit. Substitute 0.30 for x. The result is $0.30 + 1.45$, or 1.75. This is within the limit, since 1.75 is less than 3.10.

Suppose Michael buys a brownie with ice cream. Substitute 3.00 for x. The result is $3.00 + 1.45$, or 4.45. This is beyond Michael's limit because 4.45 is *not* less than or equal to 3.10.

For Michael, any amount for x that is less than or equal to 1.65 will make the inequality $x + 1.45 \le 3.10$ a true statement. The *solution of an inequality* is the set of numbers that make the inequality statement true.

There are several inequality statements used in algebra.

STATEMENTS OF INEQUALITY	
a is less than b	$a < b$
a is greater than b	$a > b$
a is less than or equal to b	$a \le b$
a is greater than or equal to b	$a \ge b$
a is not equal to b	$a \ne b$

 Exploration *Solving an Inequality*

1. Tell whether each of the following values makes the inequality $x + 1.25 \le 2.35$ true or false.
   a. 0.50
   b. 1.00
   c. 1.50
   d. 1.75

2. Continue guessing until you find the largest value of x which makes the statement true.

3. Give a method that does not use guessing to determine the largest value of x that makes the inequality statement true. ❖

The inequality $x + 1.45 \le 3.10$ can be thought of as two statements.

a. $x + 1.45 < 3.10$ an inequality
b. $x + 1.45 = 3.10$ an equality

The equation $x + 1.45 = 3.10$ can be solved by using the Subtraction Property of Equality. Thus, $x + 1.45 = 3.10$ is true when $x = 1.65$. The inequality $x + 1.45 < 3.10$ is true when $x < 1.65$. This suggests that there are properties for solving inequalities similar to the ones for solving equations.

ADDITION PROPERTY OF INEQUALITY
If equal amounts are added to the expressions on each side of an inequality, the resulting inequality is still true.

SUBTRACTION PROPERTY OF INEQUALITY
If equal amounts are subtracted from the expressions on each side of an inequality, the resulting inequality is still true.

EXAMPLE 1

Solve the inequality $8m - 8 \geq 7m + 2$.

Solution ➤

One strategy is to arrange the terms with the variable on one side of the inequality sign and the numbers on the other side. Then combine like terms.

$8m - 8 \geq 7m + 2$	Given
$8m - 8 - 7m \geq 7m + 2 - 7m$	Subtraction Property of Inequality
$8m - 7m - 8 \geq 7m - 7m + 2$	Rearrange like terms.
$m - 8 \geq 2$	Combine like terms.
$m - 8 + 8 \geq 2 + 8$	Addition Property of Inequality
$m \geq 10$	Combine like terms.

The solution is $m \geq 10$. Now, substitute values into the original inequality to see if the result is true. For example, see if the values for m are greater than or equal to 10.

Substitute **10** for m.
$$8(\mathbf{10}) - 8 \geq 7(\mathbf{10}) + 2$$
$$80 - 8 \geq 70 + 2$$
$$72 \geq 72$$
True, because 72 equals itself.

Substitute **11** for m.
$$8(\mathbf{11}) - 8 \geq 7(\mathbf{11}) + 2$$
$$88 - 8 \geq 77 + 2$$
$$80 \geq 79$$
True, because 80 is greater than 79.

When m is 10 the inequality is true.   When m is 11, the inequality is true. ❖

Using the Number Line to Represent an Inequality

EXAMPLE 2

Graph the solution to $x - 2 < 8$.

Solution ➤

First, solve the inequality for x.

$x - 2 < 8$	Given
$x - 2 + 2 < 8 + 2$	Addition Property of Inequality
$x < 10$	Combine like terms.

To graph the result on a number line, shade *all* points to the left of 10 on the number line. This includes all points whose coordinates are less than 10. To indicate that 10 *is not included,* put an *open dot* at 10.

$x < 10$

To check, substitute 9 for x.
$$9 - 2 = 7 \quad \text{Since 7 is less than 8, 9 is a solution.}$$
Substitute 10 for x.
$$10 - 2 = 8 \quad \text{Since 8 is not less than 8, 10 is not a solution.} ❖$$

What numbers do you think are represented by the graph on this number line? Write an inequality that describes these points.

Displaying Solutions

The graph of an inequality using integers is not the same as the graph using all numbers. Integers are graphed as dots (●). An inequality using all numbers is graphed as an interval, a ray, or a line.

$$-2 \leq x < 1$$

Using integers

Using all numbers

Remember, included endpoints are shown as dots, while endpoints not included are shown as open dots (○). It is assumed that all numbers are to be used unless otherwise noted.

The solution to an equation or inequality is the set of all numbers that makes the statement true. The solution might be listed, described, or graphed.

EXAMPLE 3

Graph the solution to $-4 \leq x$ and $x < 2$.

Solution ➤

The solution set must be true for both inequalities. This includes -4, since $-4 \leq 2$ and $-4 \leq -4$ are both true. It does not include 2, since $2 < 2$ is false.

The statement **$-4 \leq x$ and $x < 2$** can also be written as **$-4 \leq x < 2$**. This means that x is between -4 and 2 and includes -4. ❖

Try This Graph the solution to $-3 < x \leq 2$.

In everyday conversation, certain inequalities have important meanings. Some common examples are the following.

$x \geq 5$
x is at least 5
$x \leq 5$
x is at most 5

$5 < x < 7$
x is between 5 and 7
$5 \leq x \leq 7$
x is between 5 and 7 inclusive

What is the solution to $5 \geq x \geq 7$? Explain your answer.

EXERCISES & PROBLEMS

Communicate

Explain how you would determine whether the following inequalities are true or false.

1. $5 \geq 2 + 3$ **2.** $4 \leq 2 + 3$

3. $6 < 7 - 3$ **4.** $8 > 10 - 2$

5. Tell how the properties for inequalities are similar to the properties for equality.

6. Tell the steps necessary to solve the inequality $3x - 4 \leq 2x + 1$. Name the property you would use for each step.

7. How do you find what values of the solution to $x - 4 \leq 9$ represent equality and what values represent inequality?

8. Explain how to draw the solution to $x + 3 < 7$ on a number line.

9. How do you write an inequality that describes the points that are shown on the graph?

10. Discuss the word *inclusive* and its meaning in mathematics. List the whole numbers between 5 and 10 inclusive.

Practice & Apply

State whether each inequality is true or false.

11. $8 > 9 - 1$ **12.** $-2 \leq 5 - 7$ **13.** $8 \leq 9 - 1$ **14.** $-2 > 5 - 7$

Solve each inequality.

15. $x + 8 > -1$ **16.** $x - 6 \leq 7$ **17.** $x + \frac{3}{4} < 1$

18. $x + \frac{3}{4} \geq \frac{1}{2}$ **19.** $x + 0.04 > 0.6$ **20.** $x - 0.1 < 8$

21. Describe a real world situation modeled by $x + 10 < 100$.

22. Show the solution to $x - 4 \geq -1$ on a number line.

23. Show the solution to $x + 3 < 2$ on a number line.

Write an inequality that describes the points on the following number lines.

24.

25.

26. List the whole numbers between 57 and 66 inclusive.

Temperature Write an inequality to represent each situation. Use *t* to represent

27. all possible daytime temperatures where you live.

28. a high temperature of 66°F.

29. the low temperature was 54 degrees.

30. 54 degrees.

31. If *t* represents the first reading, then after a rise of 5 degrees the temperature was between 70 and 80 degrees inclusive. What is the inequality?

32. **Sports** A sports stadium holds 15,000 people. Everyone at the game has a seat, but the stadium is not full. Write an inequality that models this situation using *P* as the number of people.

33. The school auditorium can seat 450 people for graduation. Graduates will use 74 seats. Write an inequality to describe the number of others who can be seated in the auditorium.

34. From 30 to 50 people attended a party. From 5 to 10 people left early. Write an inequality to represent the possible range for the number of people who did not leave early.

35. A table is supposed to be 42.3 centimeters long. Write an inequality for *M*, the measure of the table, that allows a possible error of 0.5 centimeters.

36. Students in Ms. Ambrose's algebra class earn an "A" if they average at least 90, and they earn a "B" if they average at least 80 but less than 90. This can be represented using the following inequalities.

$$A \geq 90 \qquad 80 \leq B < 90$$

Translate this inequality into words.

$$70 \leq C < 80$$

37. Represent the following statement using an inequality. Students earn a "B" if they average at least 78, but less than 89.

38. State the pattern and find the next three numbers for the following sequence, 1, 1, 2, 3, 5, ___, ___, ___. **[Lesson 1.1]**

39. If rulers cost 45¢, how many rulers can Mimi buy with $2.35? **[Lesson 1.3]**

40. If the points of a scatter plot lie near a line that rises from left to right, name the type of correlation. **[Lesson 1.6]**

41. If the points of a scatter plot lie near a line that falls from left to right, name the type of correlation. **[Lesson 1.6]**

42. A linear function grows by what kind of amount? **[Lesson 2.4]**

43. If a graph appears to be in steps, describe the type of function. **[Lesson 2.5]**

Find the values. [Lesson 3.1]

44. $|-4.5|$

45. $|-9|$

46. $|-3 + 4|$

Solve for x. [Lesson 3.5]

47. $25x - 47 = 24x + 39$

48. $4x + 110 = 5x + 17$

Look Beyond

Solve each of the following inequalities. Try several values for x, and check the answers.

49. $2x < 8$ **50.** $4x + 5 \leq 16$ **51.** $8x - 3 > 33$

52. How many ways can you make change for a quarter using pennies, nickels, and dimes?

53. Place dots on a circle, and connect the dots with line segments. Count the regions that are formed. Be sure that no three lines meet at the same point unless the point is on the circle. What is the maximum number of regions formed from 6 dots?

| 1 dot | 2 dots | 3 dots | 4 dots |
| 1 region | 2 regions | 4 regions | 8 regions |

Find it FASTER

A database is a tool to organize and manipulate information. A basic example is an address book. A computer database is an electronic version that provides the speed of automatic organization and fast access. Many computer spreadsheets can be used for a database.

Suppose you are planning to continue your education after high school. A database can be constructed and used to help you decide on a college or institute to attend. You can collect information like the college name, region, state, tuition, room and board expenses, and undergraduate student population. You then use a database to store, organize, and manipulate the information.

Familiarize yourself with the terminology of a database. Notice that the first database is arranged alphabetically by school name.

To view one of the other fields in order, select a field like tuition, and use the SORT command. Notice that the records listed for tuition are now sorted in ascending order.

an entry — a field

	A School Name	B Region	C State	D Tuition	E Expenses	F Students
1	Bennett College	Southeast	NC	5600	3095	650
2	Grinnell College	Midwest	IA	15688	4618	133
3	Heidelberg College	East	OH	13500	4400	918
4	Hamline University	Midwest	MN	13252	4194	1362
5	Mills College	West	CA	14100	6000	740
6	St. John's College	Northeast	MD	17430	5720	3932

a record

	A School Name	B Region	C State	D Tuition	E Expenses	F Students
1	Bennett College	Southeast	NC	5600	3095	650
2	Hamline University	Midwest	MN	13252	4194	1362
3	Heidelberg College	East	OH	13500	4400	918
4	Mills College	West	CA	14100	6000	740
5	Grinnell College	Midwest	IA	15688	4618	133
6	St. John's College	Northeast	MD	17430	5720	3932

Different spreadsheets and database software have different features for requesting various combinations of information. A request for information is called a **query**. Some common operations for making a query are comparisons that use AND, OR, and NOT.

A query using OR will retrieve all information from the records that fulfill any specification.

> **Query:** Schools in the **midwest OR** the **west**.
>
> **Output:** Hamline University
> Grinnell College
> Mills College

A query using AND will give only the information from the records that fulfill all the specifications.

> **Query:** Schools with **tuition less than $14,000 AND** schools in the **midwest**.
>
> **Output:** Hamline University

Different databases and spreadsheets treat AND and OR operations in different ways.

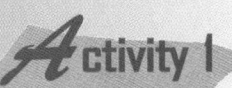 Activity 1

Alternative 1 (For those with a computer and a spreadsheet or database): Make a database from some data that is of interest to you. It should contain regular lists of information.

Alternative 2 (For those without a computer): Create your own database from index cards. As a group project, you can also act out the computer procedure for creating a database. This includes writing, storing, sorting, and retrieving information.

Activity 2

Once you have created your database, devise a strategy for sorting various fields and retrieving a record. Form questions that ask for specific combinations of the data. The search should contain a method for using OR and another using AND.

Chapter 3 Review

Vocabulary

Key Skills & Exercises

Lesson 3.1

➤ **Key Skills**

Add two or more integers.

When the terms have like signs, find the sum of the absolute values, and use the common sign. When the terms have unlike signs, find the difference of the absolute values. Use the sign of the integer with the greater absolute value.

$$-4 + (-7) = -11 \qquad 4 + 7 = 11$$
$$26 + (-9) = 17 \qquad -26 + 9 = -17$$

To add more than two integers, use the Associative Property of Addition.

$$(-5 + 5) + -3 = 0 + (-3) = -3$$
or
$$-5 + (5 + (-3)) = -5 + 2 = -3$$

➤ **Exercises**

Find each sum.

1. $-17 + 6$ **2.** $48 + (-15)$ **3.** $-23 + (-25) + 3$ **4.** $-39 + 68$

5. $33 + (-55)$ **6.** $-214 + 214$ **7.** $6 + (-7) + (-9)$ **8.** $-8 + 8 + (-12)$

Lesson 3.2

➤ **Key Skills**

Subtract integers.

To subtract -8, add the opposite. $-16 - (-8) = -16 + 8 = -8$

To subtract 19, add the opposite. $-38 - 19 = -38 + (-19) = -57$

➤ **Exercises**

Find each difference.

9. $9 - (-15)$ **10.** $48 - (-48)$ **11.** $-13 - 28$ **12.** $39 - (-18)$

13. $-67 - (-42)$ **14.** $-23 - (-72)$ **15.** $-42 - (-42) - 53$ **16.** $8 - 14 - 27$

Lesson 3.3

➤ Key Skills

Simplify expressions with several variables by adding like terms.

To simplify $(3r + 7t) + (4r - 8t)$, use this procedure.

$(3r + 7t) + (4r - 8t)$	Given
$(3r + 7t) + (4r + (-8t))$	Definition of Subtraction
$(3r + 4r) + (7t + (-8t))$	Rearrange terms.
$7r + (-1)t$	Combine like terms.
$7r - t$	Definition of Subtraction

➤ Exercises

Simplify.

17. $(6a - 1) + (5a - 4)$ **18.** $(7 - t) + (3t + 4)$ **19.** $\left(\frac{x}{3} - 2\right) + \left(\frac{x}{2} + 4\right)$

20. $(1.4m - 6.2n) + (2.4m - 5.5n)$ **21.** $(3x + 2y + z) + (6x - 4y - 3z)$

Lesson 3.4

➤ Key Skills

Use adding the opposite to subtract expressions.

To simplify $(7x + 4y - 2z) - (6x - 5y + z)$, use this procedure.

$(7x + 4y - 2z) - (6x - 5y + z)$	Given
$7x + 4y - 2z - 6x + 5y - z$	Definition of Subtraction
$7x - 6x + 4y + 5y - 2z - z$	Rearrange terms.
$x + 9y - 3z$	Combine like terms.

➤ Exercises

Simplify.

22. $3x - 5x$ **23.** $7y - (7 - 5y)$ **24.** $(8m - 4) - (6m - 3)$
25. $(6d + 3) - (4d - 7) + (3d - 5)$ **26.** $(4a - 3b - c) - (6a + 5b - 4c)$

Lesson 3.5

➤ Key Skills

Solve algebraic equations that contain addition and subtraction.

Solve $x + 15 = 11$.

$$x + 15 = 11$$
$$x + 15 - 15 = 11 - 15$$
$$x = -4$$

Check: $-4 + 15 = 11$ True

Solve $-8 = y - 14$.

$$-8 = y - 14$$
$$-8 + 14 = y - 14 + 14$$
$$6 = y$$

Check: $-8 = 6 - 14$ True

➤ Exercises

Solve.

27. $w + 16 = 25$ **28.** $r + 26 = 16$ **29.** $t + 7 = -5$ **30.** $a + 1.5 = 3.6$

31. $m + \frac{1}{2} = \frac{5}{6}$ **32.** $y - 13 = 12$ **33.** $24 = x - 19$ **34.** $-6 = g - 17$

35. $h - \frac{1}{6} = \frac{2}{3}$ **36.** $7k - (6k + 5) = 7$ **37.** $4 - (2 - 3z) = 6z - (4z + 3)$

Lesson 3.6

➤ *Key Skills*

Solve inequalities, and show the solution on the number line.

For inequalities with addition and subtraction, solve as you would solve an equation. Then draw the solution on the number line.

$$5t - 6 \le 4t + 1$$
$$5t - 4t - 6 \le 4t - 4t + 1$$
$$t - 6 \le 1$$
$$t - 6 + 6 \le 1 + 6$$
$$t \le 7$$

0 1 2 3 4 5 6 7 8

➤ *Exercises*

Solve each inequality, and show the solution on a number line.

38. $x + 5 > 10$ **39.** $n - 15 \le -3$ **40.** $y + 0.09 < 3.09$ **41.** $d - \dfrac{2}{3} \ge \dfrac{3}{9}$

Applications

42. Marsha is stocking the art supply cabinet at school. The paint brushes she needs to buy cost $1.75 each, and the paint she needs costs $2.45 per jar. Write an expression to show how much she will spend if she buys b paint brushes and j jars of paint.

43. **Finance**   Lon's checking account is overdrawn by $30. How much does he need to deposit so that he has a balance of $150?

44. Toby bought a car for $9299. He paid $1500 as a down payment. How much money did he have to borrow to pay for the rest of the car?

45. **Hobbies**   Rita has $15 to spend on new baseball cards. She decides to buy a card that costs $3. What is the most she can spend on other baseball cards?

Geometry   The formula for the perimeter, P, of a triangle with sides of length a, b, and c is $P = a + b + c$.

46. Use the formula to find P when $a = 3.5$, $b = 4.7$, and $c = 5.9$.

47. Solve this formula for b.

48. Use the formula to find b when $P = 13.5$, $a = 4.5$, and $c = 4.5$.

49. **Geometry**   Write a formula for the sum of the angles in triangle ABC.

50. Solve the formula for z.

51. Find z when $x = 55$ and $y = 60$.

Chapter 3 Assessment

1. What are the opposites of 3, -5, and 0?

2. Find the absolute values of -27, 82, and 0.

3. Draw algebra tiles to represent $3 + (-5)$, and find the resulting sum.

Find each sum or difference.

4. $-7 + (-5)$

5. $8 - (-10)$

6. $18 + (-65)$

7. $-34 + 34$

8. $-16 - (-58) - 8$

9. $42 + (-78) + |-9|$

10. Kylie had 13 tapes. Then she bought 6 tapes, gave 4 away, and bought 3 more. How many tapes does she have now?

Simplify by performing the indicated operations.

11. $(8n - 2) + (6n + 4)$

12. $5t - (9t - 6)$

13. $(8 - 9z) - (7z + 15)$

There are 5 boxes of books and 5 additional books in the book storage room.

14. If Mr. Weaver puts 2 boxes of books and 2 additional books in the book room, how many boxes and additional books are there now?

15. If Mrs. Thompson *then* removes 3 boxes of books and 1 additional book, how many boxes and additional books are left in the room?

Solve.

16. $c + 18 = 10$

17. $t - 36 = 19$

18. $8 + y = -14$

19. $34 = h - 4$

20. $w + \dfrac{2}{5} = \dfrac{7}{10}$

21. $4x - (2x + 2) = 3 + x$

22. Complementary angles are pairs of angles whose measures total 90°. Find the measure of an angle complementary to an angle with a measure of 48°.

23. The relationship between total cost of an item, I, the cost without sales tax, C, and the amount of sales tax, T, is given by the formula $I = C + T$. Write a formula for C in terms of I and T.

24. Using the formula $I = C + T$, find C when $I = \$25.19$ and $T = \$1.20$.

Solve each inequality.

25. $x + 6 \geq 4$

26. $y - 2.3 < 1.4$

27. $t + \dfrac{1}{2} \geq 2$

28. Show the solution to $x - 8 > -4$ on a number line.

29. Write an inequality that describes the points on the number line.

30. There is room for 40 people in an aerobics class. There are 25 people signed up for the class. Write an inequality to describe the number of other people that can still sign up for the class.

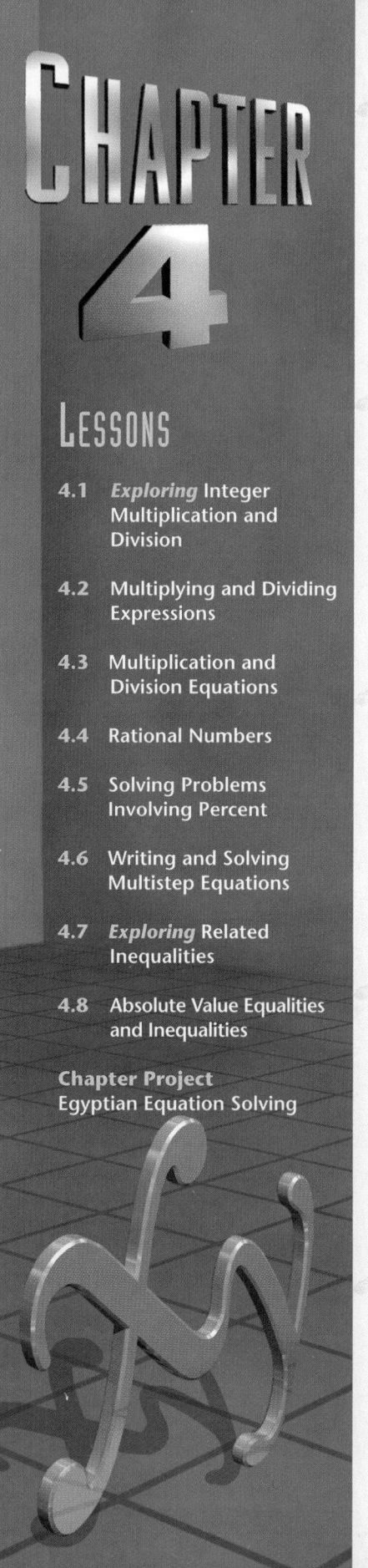

CHAPTER 4

Multiplication and Division in Algebra

Chapter 4 continues the development of solving linear equations. It extends the concept of maintaining balance in an equation when you multiply and divide to find the value of the unknown.

This chapter blends traditional methods with the latest technology. It begins with a very old problem written about Diophantus, who is often called the father of algebra. Applications of linear functions to business are stressed. Although modern applications of mathematics to commerce and business are often rather complex, we still use linear relationships to compute percentages, simple interest, and discounts.

Portfolio Activity

Diophantus lived $\frac{1}{6}$ of his life as a child,

$\frac{1}{12}$ more as a youth,

$\frac{1}{7}$ more before he married, and

5 years more before his son was born.

His son was alive for only $\frac{1}{2}$ Diophantus's life.

Diophantus then found solace in his studies for 4 years more,

Before he too died.

How long did Diophantus live?

145

Exploring Integer Multiplication and Division

why *Measures of gain and loss in financial affairs, time before and after blastoff, and distance above and below sea level are reported using integers. Calculations using this data often involve multiplication and division. When calculating with positive and negative numbers, it is important to know what happens to the sign of the result.*

CHECKS AND OTHER ITEMS ARE RECEIVED FOR DEPOSIT SUBJECT TO THE PROVISIONS OF THE UNIFORM COMMERCIAL CODE OR ANY APPLICABLE COLLECTION AGREEMENT.

Sean & Becky Savage
PH. 512-555-1789
15704 Braker Lane
St. Louis, MO 63114

CASH	CURRENCY	
	COIN	$100
LIST CHECKS SINGLY		$100
		$100

78-460/1229

Date _____ 5/6 _____ 19 95

TOTAL FROM OTHER SIDE	
TOTAL	$300
LESS CASH RECEIVED	
NET DEPOSIT	

AVAILABILITY
OF DEPOSITS
Notice: Deposits
may not be available
for immediate
withdrawal.

SIGN HERE FOR CASH RECEIVED (IF REQUIRED)

Bank of American Nations Association
P.O. BOX 9109 AUSTIN, TX 78731
Payable Thru Town East Bank
Farmers Branch, Tx 2978

USE OTHER SIDE FOR
ADDITIONAL LISTING

⑉⑉111904503⑉617500 10068650⑉

Cliberty Check Printers   Printed with soy based inks on recycled paper containing post-consumer waste.

Examine what happens to an account when transactions are made. Explain how the signs of the integers relate to the transactions.

Transaction	Representation	Result
Add 3 deposits of $100	$(3)(100) = 300$	Increase $300
Add 4 withdrawals of $20	$(4)(-20) = -80$	Decrease $80
Remove 5 deposits of $50	$(-5)(50) = -250$	Decrease $250
Remove 2 withdrawals of $10	$(-2)(-10) = 20$	Increase $20

•Exploration 1 *Multiplying Integers*

1 Complete each pattern.

a.
$(2)(3) = 6$
$(2)(2) = 4$
$(2)(1) = 2$
$(2)(0) = 0$
$(2)(-1) = \underline{\ ?\ }$
$(2)(-2) = \underline{\ ?\ }$
$(2)(-3) = \underline{\ ?\ }$

b.
$(3)(3) = 9$
$(2)(3) = 6$
$(1)(3) = 3$
$(0)(3) = 0$
$(-1)(3) = \underline{\ ?\ }$
$(-2)(3) = \underline{\ ?\ }$
$(-3)(3) = \underline{\ ?\ }$

c.
$(3)(-3) = -9$
$(2)(-3) = -6$
$(1)(-3) = -3$
$(0)(-3) = 0$
$(-1)(-3) = \underline{\ ?\ }$
$(-2)(-3) = \underline{\ ?\ }$
$(-3)(-3) = \underline{\ ?\ }$

Identify the sign for each statement.
a. (positive) • (positive) = ___?___
b. (positive) • (negative) = ___?___
c. (negative) • (positive) = ___?___
d. (negative) • (negative) = ___?___

 Find the products.
a. $(12)(4) = $ _?_
b. $(13)(-3) = $ _?_
c. $(-81)(6) = $ _?_
d. $(-16)(-5) = $ _?_

 To multiply two integers, first multiply their absolute values.
a. What is the sign of the product when two integers have *like signs*?
b. What is the sign of the product when two integers have *unlike signs*? ❖

CRITICAL *Thinking* Is the product positive or negative when an even number of negative numbers are multiplied? Is the product positive or negative when an odd number of negative numbers are multiplied? Explain your answers.

Multiplication and division are related.

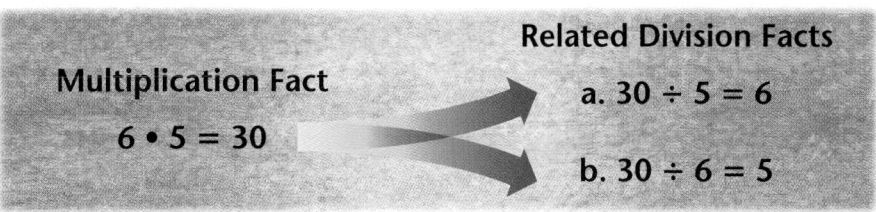

Multiplication Fact

$6 • 5 = 30$

Related Division Facts

a. $30 ÷ 5 = 6$

b. $30 ÷ 6 = 5$

•Exploration 2 *Dividing Integers*

 Complete each multiplication. Examine all the signs.
a. $(8)(7) = $ _?_
b. $(5)(-3) = $ _?_
c. $(-4)(2) = $ _?_
d. $(-8)(-1) = $ _?_

 Write the related division facts for a, b, c, and d. Compare the signs of the completed multiplications with the signs of the related division facts.

 Identify the sign for each statement.
a. (positive) ÷ (positive) = ___?___
b. (positive) ÷ (negative) = ___?___
c. (negative) ÷ (positive) = ___?___
d. (negative) ÷ (negative) = ___?___

 Find the quotients.
a. $(48) ÷ (2) = $ _?_
b. $(18) ÷ (-2) = $ _?_
c. $(-24) ÷ (-3) = $ _?_
d. $(-49) ÷ (7) = $ _?_

 To divide two integers, first divide their absolute values.
a. What is the sign of the quotient when the two integers have *like signs*?
b. What is the sign of the quotient when the two integers have *unlike signs*? ❖

Zero is also an integer. As you know, zero has several special properties.

PROPERTIES OF ZERO

Let *a* represent any number.

1. **The product of any number and zero is zero.**
   $a \cdot 0 = 0$ and $0 \cdot a = 0$

2. **Zero divided by any nonzero number is zero.**
   $\frac{0}{a} = 0, a \neq 0$

3. **A number divided by zero is undefined.**
   That is, NEVER DIVIDE BY ZERO.

APPLICATION

STATISTICS *Connection*

What is the average score for the bowling team?
Is a guess of 133 reasonable?

To see how accurate the guess is, find the differences between each actual score and the guess of 133. Add the differences and divide the total by 5.

Scores	Difference From Guess (133)	Divide by 5.
105	−28	−5 ÷ 5 = −1
117	−16	
133	0	
140	7	
165	32	
	−5	

The average difference is **−1**. Add **−1** to the guess of 133. The average bowling score is 133 + **(−1)** or 132. The guess is reasonable.

You can check the answer by adding all of the scores and dividing by 5. ❖

EXERCISES & PROBLEMS

Communicate

Discuss what would happen to the balance in a savings account if you make the following transactions. Complete Exercises 1-2.

1. Add 4 withdrawals of $25.

2. Subtract 3 withdrawals of $10.

3. Explain how to write a bank transaction modeling deposits or withdrawals from a bank account for the representation $(3)(-5)$.

4. Discuss how to write two division problems and their answers based on the multiplication problem $(3)(-5) = -15$.

5. Explain how to find the average of 95, 119, 110, 130, 141, and 155 using differences. Is a guess of 120 reasonable?

Practice & Apply

Evaluate.

6. $(-12)(-6)$

7. $(-12) - (-6)$

8. $(-8) - (-2)$

9. $(-6.6) \div (3)$

10. $(-0.8)(-2)$

11. $(-8) \div (-2)$

12. $(-8) + (-2)$

13. $(-22) \div (-1)$

14. $(-12) + (-6)$

15. $(-1.2) \div (-6)$

16. $(-5)[6 + (-6)]$

17. $(7)(4)(-6)$

18. $(-9) - [8 + (-3)]$

19. $(-54)(-115)$

20. $(-8)(-2)(-3)$

21. $(-225) \div (-5)$

22. $(-47)(23)$

23. $(-2108) \div (124)$

24. $(-6942) \div (-78)$

25. $(-90) \div (-15)(-3)$

26. $\frac{(7)(-1)}{-7}$

27. $\frac{(-6)(-12)}{3}$

28. $\frac{(-100)(-5)}{-25}$

29. $\frac{(-1)(10)(-80)}{-4}$

THE DUBLIN GROUP
8592 N Mopac Blvd., Ste 409
Austin, Tx 78750

May 16, 1995   No. 0174382319

AMOUNT
$25.00

Pay to the
Order of   JULIE RAMIREZ
17786 Melan Cove
Austin, Tx 78756

Tell whether the following statements are true or false.

30. The sum of two negatives is negative.

31. The difference of two negatives is always negative.

32. The product of two negatives is negative.

33. The quotient of two negatives is negative.

34. The average of a set of negative numbers is negative.

Banking Juan opened a savings account with a $20 deposit. He made a total of 4 more deposits of $20 and withdrawals of $10, $15, and $25.

35. What is the total increase in Juan's account *after* his initial deposit?

36. What is the total decrease in Juan's account *since* his initial deposit?

37. What is the total amount currently in Juan's account?

Statistics A running back in football ran 12 times for the following yardage: − 2, − 2, − 1, 0, 0, 2, 2, 3, 3, 3, 6, 16.

38. What is his total *net* yardage? HINT: To get the net yardage, subtract the total losses from the total gains.

39. What is the *median* (middle value) number of yards? HINT: Order the yardage from smallest to largest to find the middle value.

40. What is the *mode* (most frequent value) for the number of yards?

41. What is the *mean* (average) number of yards?

42. What is his longest run?

43. Find the average of 95, 104, 87, 120, 102, 100, and 99 using differences. Is a guess of 100 reasonable?

44. **Statistics** Five people were asked to estimate the number of beans in a jar. Their errors were −135, −43, −22, 38, and 111. What is the average error? Make a guess and use differences to find the average.

Look Back

Evaluate. [Lessons 3.1, 3.2]

45. $-2 + (-7)$ **46.** $-8 - (-3)$ **47.** $|-7|$ **48.** $|13|$

Solve the following equations.   [Lesson 3.5]

49. $3x - 5 = 2x + 12$ **50.** $4(y - 7) = 3(y + 2)$ **51.** $2(5y + 7) = 11y + 8$

Solve the following inequalities.   [Lesson 3.6]

52. $3p - 9 > 16 + 2p$ **53.** $15 - 3x \le 2x$ **54.** $-3c + 4 > -2c$

Look Beyond

Simplify.

55. $\dfrac{6x + 12}{2}$ **56.** $\dfrac{-5y - 25}{-5}$ **57.** $\dfrac{3w + 24}{-3}$ **58.** $\dfrac{5y - 45}{-5}$

LESSON 4.2 Multiplying and Dividing Expressions

Why *Businesses use spreadsheets to compute employees' earnings. The spreadsheet automatically multiplies and divides any integers needed.*

D2		=6*C2		
	A	**B**	**C**	**D**
1	Month	Day	Hours	Earnings
2	October	1	3	**18**
3	October	2	2	12
4	October	3	4	24
5	October	4	0	0
6	October	5	6	36

Richard has a part-time job making $6 an hour. His supervisor keeps a record of his earnings on a spreadsheet.

Spreadsheet

The formula **= 6*C2** at the top of the spreadsheet computes the earnings for October 1. The value in Cell D2 is 6 times the value in Cell C2.

Just as a spreadsheet uses **6*C2** to mean *6 times the value in Cell C2*, you can write 6 · *h*, or simply 6*h*, to represent 6 times a particular quantity *h*.

Word Expression	Algebraic Expression
six times the number of hours	6*h*

The algebraic expression 6*h* can be used to find how much Richard earned on any day.

To find what Richard earned on October 1, evaluate the expression 6*h* for *h* equal to 3. That is, replace *h* with 3, and then multiply 6 times 3.

$$6h = 6(3) = 18$$

Richard earned $18 on October 1. What expression represents Richard's earnings for 5 hours? What are Richard's earnings for 5 hours? for 5.5 hours?

EXAMPLE 1

Evaluate $5m - 3n$ when m is 6 and n is -2.

Solution

$5m - 3n$ means five times m minus 3 times n. Replace m by **6** and n by -2, and then simplify.

$$\begin{aligned} 5m - 3n &= 5(6) - 3(-2) \\ &= 30 - (-6) \\ &= 30 + 6 \\ &= 36 \end{aligned}$$

Multiplying Expressions

To model $2 \cdot 3$, use 1-tiles. Think of $2 \cdot 3$ as the area of a rectangle with dimensions 2 by 3.

The area of the rectangle is $2 \cdot 3 = 6$.

To model $2 \cdot 3x$, use x-tiles. Think of $2 \cdot 3x$ as the area of a rectangle with dimensions 2 by $3x$.

The area of the rectangle is $2 \cdot 3x = 6x$.

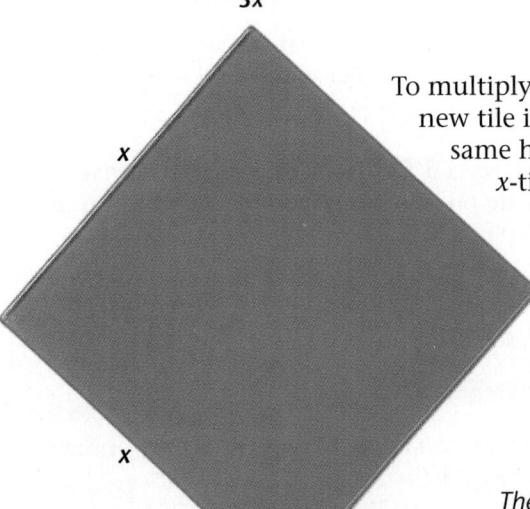

To multiply expressions, each of which contains x, a new tile is needed. Think of a tile that has the same height and length as the longer side of an x-tile. The area of the x^2-tile is x^2.

To model $2x \cdot 3x$, use x^2-tiles. Think of $2x \cdot 3x$ as the area of a rectangle with dimensions $2x$ by $3x$.

The area of the rectangle is $2x \cdot 3x = 6x^2$.

To check the multiplications, substitute 10 for x.

$$2 \cdot 3x = 2 \cdot 3(10) = 60$$
$$6x = 6(10) = 60$$

$$2x \cdot 3x = 2(10) \cdot 3(10) = 600$$
$$6(x^2) = 6(10^2) = 6(100) = 600$$

Thus, $2 \cdot 3x = 6x$ checks.

Thus, $2x \cdot 3x = 6x^2$ checks.

•Exploration• *Multiplying by −1*

Copy the following table.

Scientific Calculator

1. Choose seven values for *n*. Make three positive, three negative, and one 0. Include at least two fractions and two decimals. List your choices in the first row of the table.

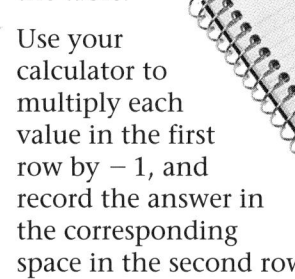

2. Use your calculator to multiply each value in the first row by − 1, and record the answer in the corresponding space in the second row.

3. What do you notice about the entries in the second row?

4. In relation to the original number, what do you call the numbers that have been multiplied by − 1?

To multiply an expression by − 1, find the opposite of the expression.

MULTIPLICATIVE PROPERTY OF − 1

For all numbers *a*,

$$-1(a) = -a.$$

The Distributive Property Over Subtraction

Which expression below is equal to $100 - 2(12 - 9)$?

a. $100 - 2(12) - 9$ **b.** $100 - 2(12) + 9$
c. $100 - 2(12) - 2(9)$ **d.** $100 - 2(12) + 2(9)$

Since $100 - 2(12) + 2(9) = 94$, and $100 - 2(12 - 9) = 94$, **d** is the correct choice. Multiplication can be distributed over subtraction. The sign of the number being distributed must be distributed to every term.

DISTRIBUTIVE PROPERTY OVER SUBTRACTION

For all numbers *a*, *b*, and *c*,

$$a(b - c) = ab - ac, \text{ and } (b - c)a = ba - ca.$$

EXAMPLE 2

Simplify $-2(5a - 4)$.

Solution ➤

An expression is simplified when no more operations can be performed.

$$(-2)(5a - 4) = (-2)(5a) - (-2)(4) \qquad \text{Distributive Property}$$
$$= -10a + 8 \qquad \text{Simplify.}$$

You can distribute a negative sign across the expressions you are subtracting. That is, if a subtraction sign is in front of an expression, change the sign to addition, and then distribute the negative number to every term of the quantity.

EXAMPLE 3

Simplify $(5x + 3y - 7) - 3(2x - y)$.

Solution ➤

Use the definition of subtraction to write an addition problem.
$$(5x + 3y - 7) - 3(2x - y) = (5x + 3y - 7) + (-3)(2x - y)$$

Distribute -3 over $(2x - y)$.
$$(5x + 3y - 7) + (-3)(2x - y) = 5x + 3y - 7 + [(-3)(2x) - (-3)(y)]$$
$$= 5x + 3y - 7 + (-6x + 3y)$$

Complete the addition.
$$= 5x + 3y - 7 - 6x + 3y$$
$$= (5x - 6x) + (3y + 3y) - 7$$
$$= -x + 6y - 7$$

You can check the answer by substituting 2 for x and 3 for y. ❖

Try This Simplify $(2b - 5c) - 4(3b + c)$.

Dividing Expressions

The division problem $\frac{2x + 6}{2}$ can be modeled with tiles.

$= 2x + 6$

$= \frac{2x + 6}{2}$

Since each set contains one x-tile and three 1-tiles, $\frac{2x + 6}{2} = x + 3$.

154   CHAPTER 4

Simplify.   **a.** $\dfrac{16x - 4}{-4}$ **b.** $\dfrac{5x + 7}{5}$

Notice that when you divide an expression by a number, *each term* is divided by that same number.

DIVIDING AN EXPRESSION

For all numbers a, b, and c, $c \neq 0$,

$$\frac{a + b}{c} = \frac{a}{c} + \frac{b}{c}, \text{ and } \frac{a - b}{c} = \frac{a}{c} - \frac{b}{c}.$$

EXAMPLE 4

Simplify $\dfrac{8 - 4x}{-4}$.

Solution ➤

Remember that since the quantity $8 - 4x$ is being divided by -4, each term must be divided by -4.

$$\frac{8 - 4x}{-4} = \frac{8}{-4} - \frac{4x}{-4}$$

$$= \frac{8}{-4} + \frac{-4x}{-4}$$

$$= -2 + x$$

CRITICAL Thinking   Explain why $\dfrac{-4x^2}{-4} = x^2$.

EXERCISES & PROBLEMS

Communicate

Explain how to write an expression that shows Jan's earnings if she earns \$6 an hour and works

1. 4 hours.   **2.** 2.5 hours.   **3.** h hours.

4. Discuss how to evaluate $2j + 3$ when j is 10.

5. Explain how to multiply the expression $3y \cdot 4y$.

6. Discuss how to simplify $-2(4m + 5)$.
What does it mean when an expression is simplified?

7. Describe how to simplify the expression $\dfrac{5p - 15}{-5}$.

Practice & Apply

Evaluate 13r for the following values of r.

8. −4 **9.** 1.5 **10.** 7 **11.** $\frac{1}{2}$

Evaluate 2t + 1 for the following values of t.

12. 10 **13.** 8.5 **14.** −6.2 **15.** $\frac{3}{4}$

Simplify the following expressions.

16. $2 \cdot 6x$ **17.** $-6x \cdot 2$ **18.** $6x \cdot 2x$ **19.** $-66x \div 2$

20. $12x \cdot 3x$ **21.** $-2(6x + 3)$ **22.** $-1.2x \cdot 3x$ **23.** $-12x \div 3$

24. $7x - (3 - x)$ **25.** $-3(7x - 3)$ **26.** $-2(4x - 1)$ **27.** $3x \cdot 5 + 2x \cdot 2$

28. $-2 \cdot 8x$ **29.** $8x \cdot 2x$ **30.** $-8x \div 2$ **31.** $-21x \cdot 3x$

32. $2.1x \cdot 3x$ **33.** $-21x \div 7$ **34.** $8x - (2 - 5x)$ **35.** $8(x + 1) - (2 - 5x)$

36. $(3x + 2y - 9) - 2(x - y)$ **37.** $4(x - 5y) - (2x - 2y)$ **38.** $\frac{5w + 15}{-5}$

39. $8(x - y) + 5(3x - 3y)$ **40.** $(3x + 5y - 9) - 7(x - y)$ **41.** $\frac{8 + 16w}{8}$

42. $\frac{-90x + 2.7}{-9}$ **43.** $-(x - y) - 4(x - y + 9)$ **44.** $9(2x + y) - 3(3x + 3y)$

45. $\frac{11 - 33y}{11}$ **46.** $\frac{-10x + 35}{5}$ **47.** $-6(4x - y) - 4(2x - 3y)$

48. A computer program shows the formula A=−5*B+B.
What is the value of A when B is 3?

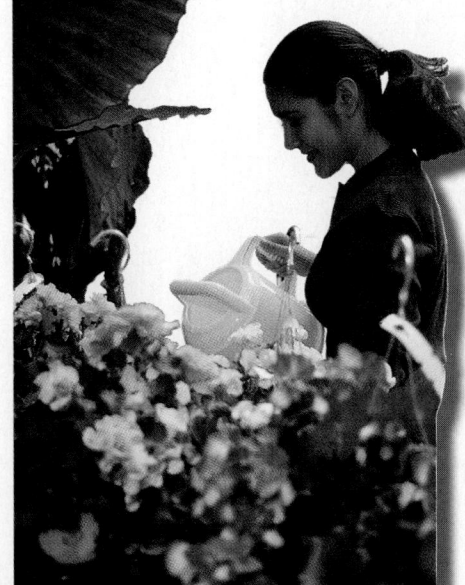

Nicole has a part-time job 3 days after school and on Saturdays.

Wages If Nicole earns $5.25 per hour, find her earnings for the following 4 days.

49. 4 hours on Friday **50.** 6.5 hours on Saturday

51. h hours on Monday **52.** $(h + 1)$ hours on Tuesday

53. Write an expression in simplified form that shows the total number of hours Nicole worked for the 4 days.

54. **Geometry** The formula for area is $A = lw$. Find the area of the rectangle by evaluating the formula for the given length and width.

$l = 6$ cm

$w = 4$ cm

55. **Geometry** Evaluate the volume formula $V = lwh$ of a rectangular prism with the given dimensions.

$l = 8$ in.

$h = 3$ in.

$w = 4$ in.

Charlotte is a plumber who charges $35 an hour, plus a fixed service fee of $20 per job.

Small Business How much does Charlotte charge for a job that takes

56. 1 hour?   **57.** 3 hours?   **58.** h hours?

On Sundays, Charlotte doubles her charges. How much does she charge for a job that takes

59. 1 hour?   **60.** 3 hours?   **61.** h hours?

Look Back

Add the like terms.   **[Lesson 3.3]**

62. $(3x + 2y) + (3x - 2y) - (3x - 2y)$

63. $(x^2 + 2y + 4) - (2x - 3y - 2)$

Solve each inequality.   **[Lesson 3.6]**

64. $x - 4 \leq 3$ **65.** $x + 5 \geq -2$ **66.** $x < -3 + 4$

67. Find the average of 2, 3, 4, 5, and 6.   **[Lesson 4.1]**

Simplify.   **[Lesson 4.2]**

68. $2(-3 + c)$ **69.** $-6(-4 + x)$ **70.** $4(2x + 4)$

Look Beyond

Simplify.

71. $\dfrac{2x^2 + 4x}{4x}$

72. $\dfrac{5y^2 - 15y}{-5y}$

73. In a bowling alley, the pins are set up as shown. Move exactly three of the pins to form a triangle exactly like the original one, but pointing away from you.

74. Sixty-four teams are picked to play in the NCAA tournament. Each team plays until it loses one game. How many games are played in order to have a champion?

Finding Order in Disorder

The science of chaos reveals nature's secrets.

Stock prices lurch and career during weeks of financial mayhem. The number of measles cases inexplicably soars and crashes. A storm unparalleled in more than 40 years slams into England, killing at least 13 people. Such apparently random events have always seemed far beyond the understanding of even the most powerful computers and brilliant researchers. But are they?

A growing and eclectic band of scientists has come to suspect that such chaotic happenings are governed by laws just as orbiting planets and falling apples are. . . .

The premise of chaos theory is the oldest cliché in science: beneath disorder lurks order. "Chaologists" find that, although it may never be possible to precisely predict the weather or the stock market or even the path of a roulette ball, one can foresee the *patterns* of their behavior. These patterns are the order within the chaos. . . .

Chaos's power to explain diverse phenomena has encouraged researchers to seek it everywhere. Cardiologists studying the human heart find that its normal rhythm is subtly chaotic, rather than regular like a metronome. Climatologists pondering the ice ages find that their timing follows no regular pattern.

Order *Turn on a faucet just a little. The flow of water looks as smooth as glass tube. This is order.*

Chaos *Turn the handle a little more. The stream changes. Its outer edges flicker in unpredictable turbulence. This is chaos.*

Some things, like the water stream in the photos, are sometimes orderly and sometimes chaotic. You can use equations to model such dual behavior. To see how this works, explore a biological system that is simpler to model than a stream of water—an imaginary population of insects.

The insects hatch in the spring and die after laying eggs in the fall. The population each year depends on the population of the prior year. The more insects there are, the more eggs there are to hatch next spring. Other factors, such as limited food supply, keep the population from becoming infinitely large.

If you know the size of the insect population one year, you can find the size the next year by using the following equation.

$$P_2 = r \cdot P_1(1 - P_1)$$

P_1 is the size of the population in year 1, P_2 is the size of the population in year 2, and r is some constant.

To find the population in year 3, use $P_3 = r \cdot P_2(1 - P_2)$.

What equation would you use for year 4?

In these equations, all values of P must stay in the range 0 to 1. A value of 0 represents no insects, and a value of 1 represents the final population size.

Now you are ready to see how these equations can model behavior that is sometimes orderly and predictable and sometimes chaotic and unpredictable.

Cooperative Learning

1. Copy and complete the chart for $r = 2$ and $r = 4$.

   a. To start, choose any value for P_1 where $0 < P_1 < 1$. Round each answer to three decimal places.

   b. Do you see a pattern in your results for each value of r? Explain.

   c. Compare your results with the results of groups that started with a different value for P_1.

   d. For which values of r can you make a good prediction about the size of the population in year 6 even if your year 1 data were a little off?

2. Copy and complete the chart for $r = 4$. This time start with a value for P_1 that is 0.001 less or 0.001 greater than the value you originally used.

3. In chaos, a tiny difference at the beginning can make a big difference later on. For which value of r did the equation make chaos? Explain.

$r = 2$	
P_1	
P_2	
P_3	
P_4	
P_5	
P_6	

$r = 4$	
P_1	
P_2	
P_3	
P_4	
P_5	
P_6	
P_7	
P_8	

Multiplication and Division Equations

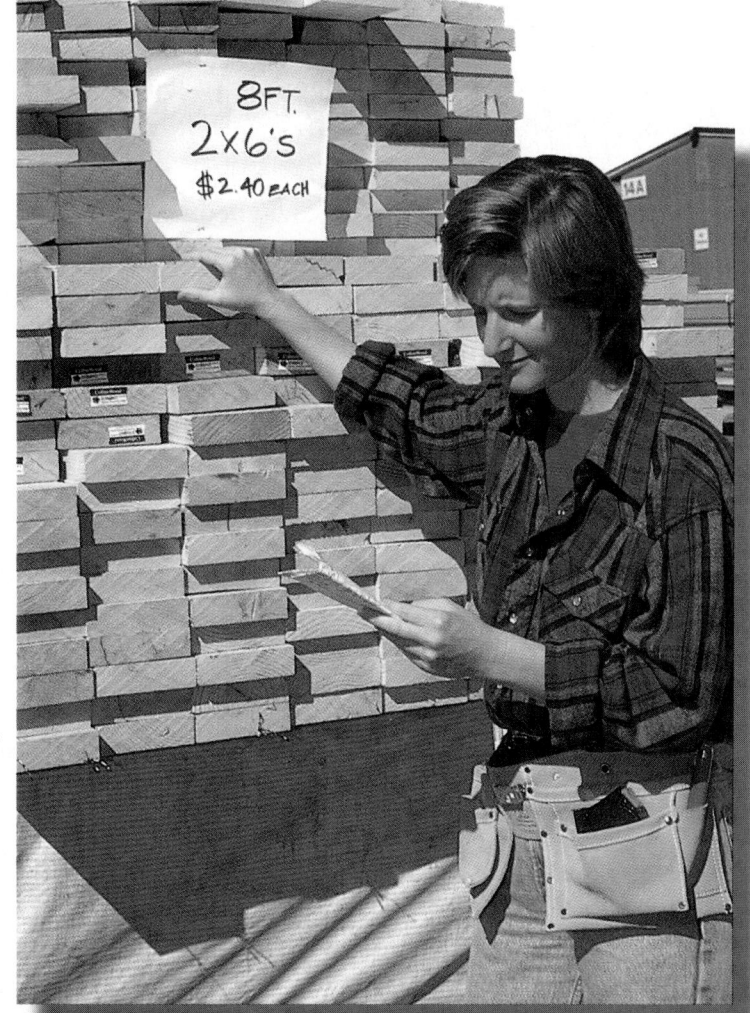

8FT.
2×6's
$2.40 EACH

why *Carpenters use problem-solving strategies, estimation, and mental computation in their work. They write and solve equations to calculate measurements and to compute the cost of materials and labor.*

Alice needs to purchase 20, 8-foot reinforcement beams. What is the cost per foot?

Multiplication Equations

Carpentry To find the cost per foot, you can draw a picture or use an equation.

Draw a picture of a beam divided into 8, 1-foot pieces.

x							8x
1	2	3	4	5	6	7	8
?							$2.40

To find the cost of one piece, divide $2.40 by 8. $\frac{2.40}{8} = 0.30$

The cost per foot is $0.30.

If the cost per foot is represented by x, the equation that models the problem is $8x = 2.40$. To find x, divide both sides of the equation by 8.

$$8x = 2.40 \qquad \text{Given}$$
$$\frac{8x}{8} = \frac{2.40}{8} \qquad \text{Divide both sides by 8.}$$
$$x = 0.30 \qquad \text{Simplify.}$$

Each piece costs $2.40 ÷ 8, or $0.30.

You can solve multiplication equations by using division.

> ### DIVISION PROPERTY OF EQUALITY
> If each side of an equation is divided by equal nonzero numbers,
> the results are equal.

EXAMPLE 1

GEOMETRY
Connection

In a regular pentagon all of the angles have equal
measures. If the sum of the measures is 540°, find
the measure of each angle.

Regular Pentagon

Solution ➤

Organize
Find the information needed to write an equation.
The total number of angles is 5. The sum of the
measures of the angles is 540°.

Write
Let x be the measure of each angle.
  Then $5x$ is the sum of the measures of all 5 angles.
  Thus, $5x = 540$.

Solve
$5x = 540$ Given

$\dfrac{5x}{5} = \dfrac{540}{5}$ Division Property of Equality

$x = 108$ Simplify.

Check
$5(108) \stackrel{?}{=} 540$

  $540 = 540$ True

Each angle has a measure of 108°. ❖

EXAMPLE 2

GEOMETRY
Connection

The formula for the circumference of a circle is
$C = \pi d$. Solve for d.

Solution ➤

Divide both sides of the equation by π.

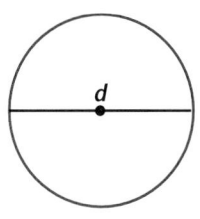

$C = \pi d$

$C = \pi d$ Given

$\dfrac{C}{\pi} = \dfrac{\pi d}{\pi}$ Division Property of Equality

$\dfrac{C}{\pi} = d$ or $d = \dfrac{C}{\pi}$ Simplify. ❖

Try This The area of a rectangle is $A = lw$. Solve for w.

Division Equations

Sports Division is used to find batting averages. The number of hits is divided by the number of times at bat. To find the number of hits, a division equation can be used as a model.

$$\frac{\text{Hits}}{\text{Times at bat}} = \text{Average}$$

Alex batted 150 times during his varsity career. If his final batting average is 0.300, how many hits did he get?

Let H represent the number of hits. The problem can be modeled by the following division equation.

$$\frac{H}{150} = 0.300$$

To solve the division equation, multiply both sides of the equation by the divisor of H.

$$150\left(\frac{H}{150}\right) = 150(0.300)$$
$$H = 45$$

Alex had 45 hits in his varsity career.

The solution to the batting average problem leads to another important property of equality.

MULTIPLICATION PROPERTY OF EQUALITY

If each side of an equation is multiplied by equal numbers, the results are equal.

EXAMPLE 3

Solve the equation $\dfrac{x}{-3} = -6$.

Solution ➤

Use the Multiplication Property of Equality.

$$\frac{x}{-3} = -6 \qquad \text{Given}$$

$$-3\left(\frac{x}{-3}\right) = -3(-6) \qquad \text{Multiplication Property of Equality}$$

$$x = 18 \qquad \text{Simplify.}$$

Check the solution.

$$\frac{18}{-3} \overset{?}{=} -6$$

$$-6 = -6 \qquad \text{True} \; \diamond$$

CRITICAL Thinking

Compare the Division Property of Equality with the Multiplication Property of Equality. How are they alike? different?

You now have tools to solve addition, subtraction, multiplication, or division equations. Make sure you apply the correct property to solve an equation.

EXAMPLE 4

Solve. $y + \frac{1}{3} = 6$ **B** $\frac{x}{0.7} = -56$

Solution ➤

A Since this is an addition equation, subtract $\frac{1}{3}$ from each side.

$$y + \frac{1}{3} = 6$$

$$\left(y + \frac{1}{3}\right) - \frac{1}{3} = 6 - \frac{1}{3}$$

$$y = 5\frac{2}{3}$$

Check $5\frac{2}{3} + \frac{1}{3} \stackrel{?}{=} 6$

$6 = 6$ True

B This is a division equation. Multiply each side by 0.7.

$$\frac{x}{0.7} = -56$$

$$0.7\left(\frac{x}{0.7}\right) = (0.7)(-56)$$

$$x = -39.2$$

Check $\frac{-39.2}{0.7} \stackrel{?}{=} -56$

$-56 = -56$ True ❖

Why is it important to identify what type of equation you are solving before you apply one of the properties of equality?

Try This   Solve.   **a.** $d - \frac{3}{4} = 8$ **b.** $\frac{m}{-1.5} = -6$

EXAMPLE 5

Physics   In science, **density** is the mass of a material divided by its volume. If the formula for density is $\frac{m}{v} = d$, find the mass, m, in terms of the density, d, and the volume, v.

Solution ➤

Multiply each side of the equation by v.

$$\frac{m}{v} = d \qquad \text{Given}$$

$$v\left(\frac{m}{v}\right) = v(d) \qquad \text{Multiplication Property of Equality}$$

$$m = vd \qquad \text{Simplify.} ❖$$

Try This   Solve $m = \frac{F}{a}$ for F.

Rules for solving equations are summarized in the following chart.

SUMMARY FOR SOLVING EQUATIONS	
Type of Equation	**Operation**
Addition	Subtract equal amounts from each side.
Subtraction	Add equal amounts to each side.
Multiplication	Divide each side by equal nonzero amounts.
Division	Multiply each side by equal amounts.

EXERCISES & PROBLEMS

Communicate

1. Make up a real world situation modeled by each equation.
   a. $5x = 100$
   b. $\dfrac{x}{2} = 10$

Explain how to solve for *x*.

2. $592x = 812$

3. $x - 246 = 528$

4. $5x = \dfrac{1}{10}$

5. $x + 10 = 5$

6. **Geometry** From the formula for the circumference of a circle, explain how to find the radius, r, in terms of the circumference, C, and π.

Write a real-world problem modeled by each equation.

7. $4x = 40$

8. $\dfrac{x}{4} = 20$

9. $2.50p = 15$

10. $\dfrac{x}{20} = 4$

Practice & Apply

Give the property needed to solve each equation. Do not solve.

11. $\frac{y}{3} = -13$ **12.** $\frac{x}{27} = -26$ **13.** $x - \frac{1}{3} = 2$ **14.** $7x = 56$

15. $\frac{b}{-9} = 6$ **16.** $-12y = 84$ **17.** $x + \frac{2}{3} = 2$ **18.** $777x = -888$

19. $5.6v = 7$ **20.** $7 = -56w$ **21.** $\frac{x}{-7} = -1.4$ **22.** $3x = 2$

Solve each equation.

23. $888x = 777$ **24.** $y + \frac{3}{4} = 12$ **25.** $x - 888 = 777$ **26.** $-3f = 15$

27. $\frac{y}{7} = -8$ **28.** $\frac{x}{0.5} = 6$ **29.** $w + 0 = -22$ **30.** $-4x = -3228$

31. $4b = -15$ **32.** $\frac{p}{111} = -10$ **33.** $\frac{s}{-1} = -40$ **34.** $c + 7 = 63$

35. $r - \frac{1}{5} = 2$ **36.** $2a = 13$ **37.** $\frac{x}{10} = -1.9$ **38.** $d - 7 = 35$

39. $\frac{w}{-1.2} = 10$ **40.** $7e = -14$ **41.** $\frac{m}{-9} = 0$ **42.** $\frac{b}{15} = 1$

43. $0.55x = 0.55$ **44.** $\frac{p}{-9} = 0.9$ **45.** $-3x = -4215$ **46.** $p + 2300 = 890$

Carpentry Max is apprenticing as a carpenter during his summer vacation. Write and solve an equation to describe each situation in Exercises 47–51.

47. How many rolls of masking tape can Max buy for $6.00?

48. Max finds masking tape in packages of 4 rolls for $4.32. What is the cost of one roll in the package? Is the cost of a single roll of masking tape more than one packaged roll?

49. Max's employer has $19.00 to spend for tape measures for his crew. How many can he buy for that amount?

Masking Tape $1.15
Hammers $7.99
Tape Measure $4.95

50. What is the price of an extension cord if Max can buy a package of 6 for $7.26?

51. Max finds that a package of 4 AA-type batteries costs $2.52. What is the price for each battery?

52. Travel Natalie's family wants to make a 400-mile trip in 10 hours, but they figure that they will stop a total of 2 hours along the way. What speed must they average for the trip?

53. Travel Derrick averaged 50 miles per hour during three hours of driving, but he still has 75 miles to go. How many miles does he drive in all?

54. Maria wants to drive 320 miles in 8 hours. What speed should she average for the trip?

Solve each formula for the variable indicated.

55. $r = \dfrac{d}{t}$ for d **56.** $A = bh$ for b **57.** $V = lwh$ for w

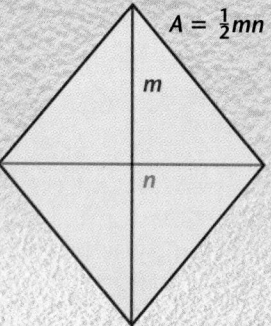

$\square\!\!\!\bigcirc$ **Geometry** In a regular polygon with n sides, the measure, m, of each interior angle is given by the formula $m = \dfrac{180(n-2)}{n}$. Find the measure of an interior angle of the following regular polygons. HINT: Use the number of sides of each polygon for n.

58. equilateral triangle
59. square
60. pentagon
61. octagon

62. What effect does the *number* of angles in each regular polygon have on the *size* of the angles?

63. $\square\!\!\!\bigcirc$ **Geometry** Solve the formula for the area of a rhombus for the variable n.

$A = \tfrac{1}{2}mn$

64. The diagonals of the rhombus are m and n. If the area is 21 square inches and the longer diagonal, m, is 7 inches, what is the length of the other diagonal?

Look Back

65. Write a sequence of 5 terms, starting with 3 and doubling to get the next number. **[Lesson 1.1]**

Evaluate the following if $a = 3$, $b = 2$, and $c = 0$. [Lesson 1.4]

66. $\dfrac{a + b \cdot c}{3}$ **67.** $\dfrac{a \cdot b}{b + c}$ **68.** $\dfrac{a + b}{b + c}$

69. Place parentheses to make $30 \cdot 7 - 4 \cdot 10 \div 2 \div 2 = 200$ true. **[Lesson 1.4]**

Find the absolute value for each number. [Lesson 3.1]

70. $|0|$ **71.** $|-5|$

72. $\left|-\dfrac{1}{6}\right|$ **73.** $|9|$

Look Beyond

Solve each equation.

74. $3y = 10 + 5y$ **75.** $4x + \dfrac{1}{2} = 5x$

76. $x + 4 = 3x - 2$ **77.** $6x - 2 = 14 - 2x$

LESSON 4.4 Rational Numbers

why *In Lesson 4.3 you solved equations involving fractions and decimals. Although fractions and decimals are not integers, the properties for solving equations still apply.*

Geno decides to borrow $25 from his savings account in 2 equal withdrawals.

Banking How can Geno use negative numbers to express the amount of withdrawal?

Since Geno is withdrawing $25, use -25 to represent the total withdrawal.

Let x represent each withdrawal. Then $2x$ represents the 2 equal withdrawals. The equation to model the problem is

$$2x = -25.$$

Apply the Division Property of Equality.

$$\frac{2x}{2} = \frac{-25}{2}$$

$$x = -12\frac{1}{2}$$

The result is $x = -12\frac{1}{2}$ or -12.50. Geno should withdraw $12.50 each time.

Numbers such as -12.5, $-12\frac{1}{2}$, or $\frac{-25}{2}$ are called negative **rational numbers.** Other examples of rational numbers include $\frac{1}{2}$, $-\frac{1}{2}$, 0, and 2. A rational number is a number that can be expressed as the ratio of two integers with 0 excluded from the denominator.

Rational numbers can be graphed on a number line.

How can you show that the integers 0 and -2 are also rational numbers?

EXAMPLE 1

Investment

A share of stock increased by $\frac{1}{4}$ of a point on Monday and decreased by $\frac{5}{8}$ of a point on Tuesday. What is the net effect on the stock?

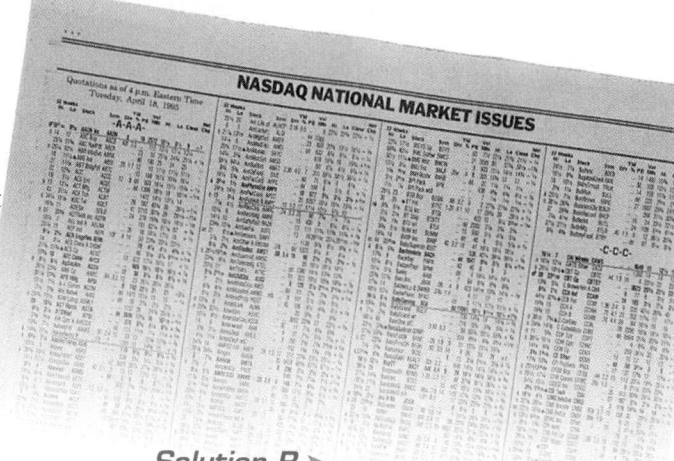

NASDAQ NATIONAL MARKET ISSUES

Solution A ➤

Use a number line to model this problem.

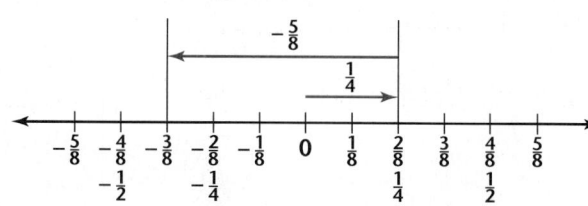

Solution B ➤

Use a common denominator to solve this problem.

$$\frac{1}{4} + \left(-\frac{5}{8}\right) = \frac{2}{8} + \left(-\frac{5}{8}\right)$$

$$= \frac{2 + (-5)}{8}$$

$$= -\frac{3}{8}$$

The net effect is a decrease of $\frac{3}{8}$ of a point. ❖

The set of rational numbers shares all the properties of the set of integers. However, there is one additional rational number property not shared by the set of integers.

RECIPROCAL PROPERTY

For any nonzero number r, there is a number $\frac{1}{r}$ such that

$$r \cdot \frac{1}{r} = 1.$$

In Lesson 2.3 you graphed the reciprocal function $y = \frac{1}{x}$ for positive values of x. Now draw the graph for the negative values of x. How do the two graphs compare?

EXAMPLE 2

Write the reciprocal.

A $\frac{2}{3}$ **B** 7 **C** $-\frac{4}{5}$ **D** $-1\frac{1}{2}$

Solution ➤

A The reciprocal of $\frac{2}{3}$ is $\frac{3}{2}$ $\left(\text{or } 1\frac{1}{2}\right)$ because $\frac{2}{3} \cdot \frac{3}{2} = \frac{6}{6} = 1$.

B The reciprocal of 7 is $\frac{1}{7}$ because $7 \cdot \frac{1}{7} = \frac{7}{7} = 1$.

C The reciprocal of $-\frac{4}{5}$ is $-\frac{5}{4}$ $\left(\text{or } -1\frac{1}{4}\right)$ because $-\frac{4}{5} \cdot -\frac{5}{4} = \frac{20}{20} = 1$.

D The reciprocal of $-1\frac{1}{2}$ $\left(\text{or } -\frac{3}{2}\right)$ is $-\frac{2}{3}$ because $-\frac{3}{2} \cdot -\frac{2}{3} = \frac{6}{6} = 1$. ❖

Why is -1 its own reciprocal?

Calculator

In Lesson 2.3 you used a calculator to find reciprocals. The reciprocal of a number is also called its **multiplicative inverse** and the x^{-1} key is called an inverse key. How can you use the inverse key to find the reciprocal of $-\frac{2}{5}$?

The Reciprocal Property can be used to solve an equation when the coefficient of a variable is a rational number.

EXAMPLE 3

Find the coefficient of the variable for each of the following.

A $\frac{3x}{4}$ **B** $-\frac{5}{6}k$

C $\frac{-y}{8}$ **D** $-\frac{p}{7}$

Solution ➤

A The coefficient of $\frac{3x}{4}$ is $\frac{3}{4}$ because $\frac{3x}{4} = \frac{3}{4}x$.

B The coefficient of $-\frac{5}{6}k$ is $-\frac{5}{6}$.

C The coefficient of $\frac{-y}{8}$ is $-\frac{1}{8}$ because $\frac{-y}{8} = -\frac{1}{8}y$.

D The coefficient of $-\frac{p}{7}$ is $-\frac{1}{7}$ because $-\frac{p}{7} = -\frac{1}{7}p$. ❖

How does the placement of a negative sign affect the coefficient of a variable?

EXAMPLE 4

Solve. (A) $\frac{2}{3}x = 8$ (B) $\frac{-x}{5} = -3$

Solution ➤

Multiply each side of the equation by the reciprocal of the coefficient of x. The coefficient of $\frac{2}{3}x$ is $\frac{2}{3}$. The coefficient of $\frac{-x}{5}$ is $-\frac{1}{5}$.

(A)

$$\frac{2}{3}x = 8$$

$$\frac{3}{2}\left(\frac{2}{3}x\right) = \frac{3}{2}(8)$$

$$x = \frac{24}{2}$$

$$x = 12$$

(B)

$$\frac{-x}{5} = -3$$

$$-\frac{1}{5}x = -3$$

$$-5\left(-\frac{1}{5}x\right) = -5(-3)$$

$$x = 15 \;❖$$

What is the reciprocal for the coefficient of x in the expression $\frac{2x}{-3}$?

Try This Solve.   **a.** $\frac{3}{4}y = -15$ **b.** $\frac{-w}{9} = 8$

Remember, a proportion states that two ratios are equal. Example 5 shows two different methods to solve a proportion.

EXAMPLE 5

Solve the proportion $\frac{x}{-5} = \frac{2}{3}$.

Solution ➤

Method A Multiply both sides by the reciprocal of the coefficient of x.	**Method B** Multiply both sides by the least common denominator.
$$\frac{x}{-5} = \frac{2}{3}$$ $$-5\left(\frac{x}{-5}\right) = -5\left(\frac{2}{3}\right)$$ $$x = \frac{-10}{3}$$ $$x = -3\frac{1}{3}$$	$$\frac{x}{-5} = \frac{2}{3}$$ $$-15\left(\frac{x}{-5}\right) = -15\left(\frac{2}{3}\right)$$ $$3x = -10$$ $$\frac{3x}{3} = \frac{-10}{3}$$ $$x = \frac{-10}{3}$$ $$x = -3\frac{1}{3}$$

❖

CRITICAL Thinking Compare these methods to Example 4. How are they alike? How are they different?

EXERCISES & PROBLEMS

Communicate

1. Define a rational number. Give three examples.
2. Explain why an integer is also a rational number.
3. Describe how to model $\frac{1}{2} + (-3)$ on the number line.
4. How can you find $\frac{1}{2} - \frac{5}{6}$ using the number line?
5. State the Reciprocal Property. Use the number -3 to give an example of the Reciprocal Property.
6. Give two ways to solve the proportion $\frac{y}{8} = \frac{-2}{3}$.

Practice & Apply

Write the reciprocal for each number.

7. -2 8. $\frac{-1}{9}$ 9. $4\frac{1}{5}$ 10. $\frac{10}{3}$ 11. 1

Solve each equation.

12. $\frac{3}{5}x = 3$ 13. $\frac{1}{-7}y = -6$ 14. $\frac{-5}{8}q = 10$ 15. $-\frac{3}{4}x = -9$

16. $\frac{-y}{7} = 11$ 17. $\frac{p}{4} = -1.2$ 18. $\frac{p}{-11} = -5$ 19. $-\frac{t}{4} = -5$

20. $-\frac{w}{8} = 9$ 21. $\frac{-m}{-12} = -10$

Solve each proportion.

22. $\frac{x}{5} = \frac{3}{4}$ 23. $\frac{-x}{3} = \frac{2}{5}$ 24. $\frac{x}{4} = \frac{-3}{6}$ 25. $\frac{p}{-3} = \frac{-4}{7}$

26. $\frac{x}{16} = \frac{-3}{8}$ 27. $\frac{x}{-25} = \frac{3}{-5}$ 28. $\frac{-x}{6} = \frac{5}{3}$ 29. $\frac{w}{-14} = \frac{6}{7}$

30. $\frac{-x}{6} = \frac{3}{7}$ 31. $\frac{-y}{7.5} = \frac{-2}{4}$

32. **Savings**   Miguel plans to mow lawns during summer vacation. He spends $\frac{2}{3}$ of his savings to buy a lawn mower. If Miguel's savings totaled \$210, how much of his savings did he use to buy the lawn mower?

33. **Sports**   If Alicia shoots $\frac{3}{4}$ of her free throws by noon, how many practice free throws has she taken?

Alicia spent Saturday shooting free throws for basketball. Her coach asked her to practice shooting 60 baskets.

A senior class sold $\frac{2}{3}$ of their spirit ribbons on the first day.

Set up a proportion and solve each problem.

34. **Fund-raising** If they had 450 ribbons left, how many ribbons did the senior class sell on the first day?

35. **Sports** Mary can run a 50-yard race in 9 seconds. How many seconds will it take Mary to run 75 yards at the same speed?

36. **Consumer Economics** Miguel went to the store to buy dog food. Six cans of dog food sell for $1.74. He has $5 for his purchase. Does he have enough money to buy 16 cans?

37. **Cultural Connection: Americas** If 9 crowns were exchanged for 7 pesos, how many pesos were exchanged for 63 crowns?

In the days of colonial Mexico, crowns and pesos were forms of money.

38. **Geometry** At a certain time of day a tree casts a shadow 12 meters long. If a meter stick (1 meter long) casts a shadow of 60 centimeters at the same time of day, how tall is the tree?

Look Back

39. Evaluate $6p - 3q + r$ if $p = 3$, $q = 4$, and $r = 10$. **[Lesson 1.4]**

Evaluate. **[Lesson 2.4]**
40. INT(1.6)   41. INT(6.35)   42. INT(9)

Simplify. **[Lesson 3.3]**
43. $(3x - 2) + (4x + 7)$ 44. $2(4q + 3) - (q - 2)$ 45. $(2a + 3b - 1) - 2(a - 2b - 1)$

Solve each inequality. **[Lesson 3.6]**
46. $x + 4 > -12$ 47. $y - 0.05 \leq 10.5$ 48. $w + 1.4 \geq 10.2$

Look Beyond

Find a pair of numbers (x, y) that satisfies both equations.
49. $3x + 2y = 14$ and $6x - 4 = 8$

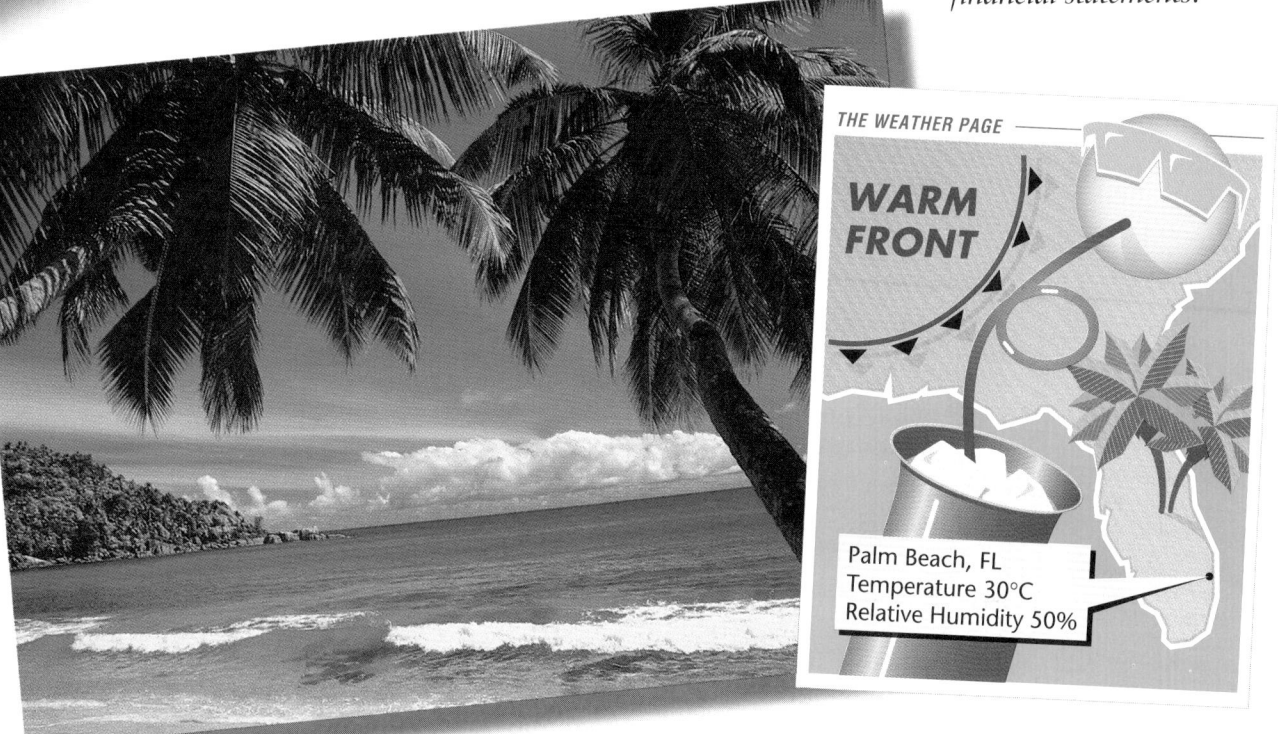

why *The next time you read a newspaper, notice how many times a percent is used. Percents are found in tables, charts, and graphs. They are also in ads, on display signs, and in financial statements.*

The relative humidity is expressed as a percent. For example, at 30°C, if 1 cubic meter of air holds 42 liters of water vapor, the air is saturated, and the relative humidity is 100%. If the air contains 21 liters of water vapor or $\frac{1}{2}$ the saturated amount, the relative humidity is 50%.

Fractions, Decimals, and Percent

A **percent** is a *ratio* that compares a number with 100. 50% is the same as $\frac{50}{100}$ or $\frac{1}{2}$. 100% is the same as $\frac{100}{100}$ or 1. A percent can be expressed in many ways. For example, 25% can be written as a decimal or a fraction. Sometimes you may wish to write the fraction in lowest terms.

$$25\% = \frac{25}{100} = 0.25 \qquad \frac{25}{100} = \frac{1}{4}$$

Suppose the air contains $10\frac{1}{2}$ liters of water vapor. Then the relative humidity is $\frac{10\frac{1}{2}}{42}$ or $\frac{1}{4}$ of 100%. The relative humidity is 25%.

EXAMPLE 1

Write each percent as a decimal and as a fraction.

A 75%   **B** 110%   **C** 4.4%

Solution ➤

A $75\% = \dfrac{75}{100} = 0.75$ $\dfrac{75}{100} = \dfrac{3}{4}$

B $110\% = \dfrac{110}{100} = 1.10$ $\dfrac{110}{100} = \dfrac{11}{10}$

C $4.4\% = \dfrac{4.4}{100} = 0.044$ $\dfrac{4.4}{100} = \dfrac{44}{1000} = \dfrac{11}{250}$ ❖

What percent would you write for the decimal 0.08?

Writing and Solving Percent Problems

In a percent equation such as 25% of 40 = 10, 25% is the percent, 40 is the base, and 10 is the percentage. How can you solve an equation if one of the parts is unknown? Explain how each equation is solved using basic properties.

1. 25% of $40 = x$ $\longrightarrow$ $\dfrac{1}{4} \cdot 40 = x$ $\longrightarrow$ $10 = x$

2. 25% of $x = 10$ $\longrightarrow$ $\dfrac{1}{4} \cdot x = 10$ $\longrightarrow$ $x = 40$

3. $x\%$ of $40 = 10$ $\longrightarrow$ $\dfrac{x}{100} \cdot 40 = 10$

$$x \cdot \dfrac{40}{100} = 10$$

$$x = 10 \cdot \dfrac{100}{40}$$

$$x = 25$$

You can also visualize a percent bar and form a proportion.

1. The percentage is unknown. 25% of $40 = x$

$$\dfrac{x}{25} = \dfrac{40}{100}$$

$$100\left(\dfrac{x}{25}\right) = 100\left(\dfrac{40}{100}\right)$$

$$4x = 40$$

$$x = 10$$

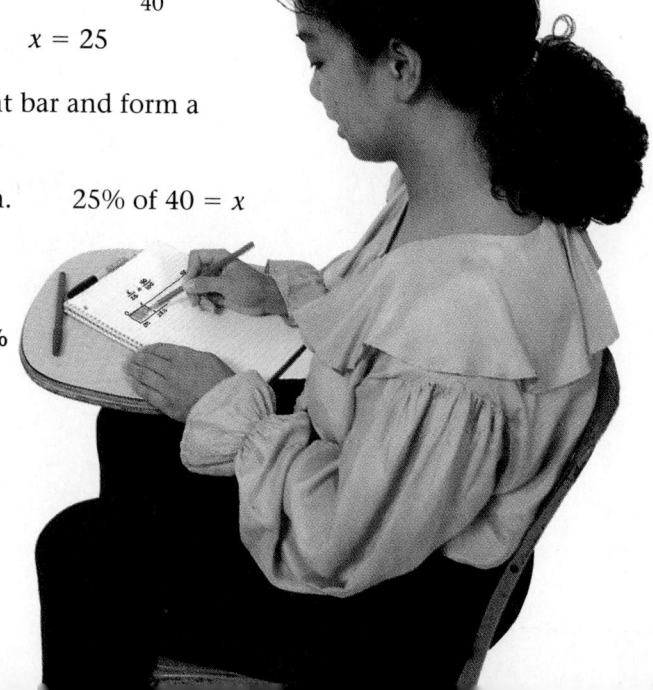

2. The base is unknown.   25% of $x = 10$

0   10 x

0%   25%   100%

$$\frac{10}{25} = \frac{x}{100} \rightarrow 100\left(\frac{10}{25}\right) = 100\left(\frac{x}{100}\right)$$
$$40 = x$$

3. The percent is unknown. x% of $40 = 10$

0   10   40

0% x%   100%

$$\frac{10}{x} = \frac{40}{100} \text{ or } \frac{x}{10} = \frac{100}{40}$$
$$10\left(\frac{x}{10}\right) = 10\left(\frac{100}{40}\right)$$
$$x = 25$$

CRITICAL
Thinking

Explain how to use one step to solve for each variable in the equation a% of $b = c$.

EXAMPLE 2

Consumer Economics

Tom plans to buy a CD player for his mother. He sees that the Royal Appliance Store is having a sale. How much will he save at the sale if he buys a $120 CD player?

Solution ➤

The CD player Tom is buying has an original price of $120. CD players are on sale for 25% off. You want to find the amount of savings. Let x represent the amount of savings.

Write a percent statement for the problem, and then solve the equation.

$$25\% \text{ of } \$120 \text{ is } \underline{\ ?\ }.$$
$$0.25 \cdot 120 = x$$
$$30 = x$$

Tom will save $30 on the CD player. ❖

EXAMPLE 3

Discounts

The freshman class is sponsoring a trip to see a play. If a group of 20 people buy tickets, there is a 30% discount. How much will each ticket cost at the discounted price?

Students $8.00
Group discounts available

The Rialto Theater
Matinee Wednesday 3:00pm
Daily 8:30pm

Solution A ➤

Proportion method You can visualize the problem using a percent bar. The regular cost, $8, is the length of the bar. Since the tickets will be 30% off, the discounted price will be 70% of the original price. Mark off 70% of the bar, and use x to represent the discounted price.

The figure indicates that the price is between $4 and $8. Write the proportion $\frac{x}{70} = \frac{8}{100}$ from the percent bar. Apply the Multiplication Property of Equality, and simplify.

$$70\left(\frac{x}{70}\right) = 70\left(\frac{8}{100}\right)$$

$$x = \frac{560}{100}$$

$$x = 5.60$$

The discounted price is $5.60, which agrees with the estimate.

Solution B ➤

Equation method Use x to represent the discounted price.

The discounted price, x, is 70% of the full price, $8. Model this statement by an equation.

$$x = \frac{70}{100} \cdot 8$$

$$x = 5.60$$

The discounted price is $5.60. ❖

Try This Solve the following percent statements.

a. _?_% of 80 is 15.   **b.** 115% of 200 is _?_.   **c.** 35% of _?_ is 45.

A VCR is on sale for $239.40.

EXAMPLE 4

What was the original price of the VCR before the sale?

Solution ➤

Use the equation method. Let *w* represent the original price, and note that $239.40 is 60% of the original price. Model this statement by an equation.

$$239.40 = 0.60\, w$$

Apply the Division Property of Equality, and simplify.

$$\frac{239.40}{0.60} = \frac{0.60w}{0.60}$$

$$399 = w$$

The original price was $399. ❖

EXAMPLE 5

Sales Tax Some states have a sales tax for certain items. Linda paid $47.74 for a $45.25 item. Find the percent of tax or the tax *rate*.

Solution ➤

Use the proportion method. First, find the amount of tax. Subtract the amount of the item from the total amount paid.

$$\$47.74 - \$45.25 = \$2.49$$

The tax was $2.49. Use the percent bar to show the proportion.

Let *p* represent percent.

Use the Multiplication Property of Equality to solve the proportion.

0 2.49 45.25

0%   *p*%   100%

$$\frac{p}{2.49} = \frac{100}{45.25}$$

$$2.49 \left(\frac{p}{2.49}\right) = 2.49 \left(\frac{100}{45.25}\right)$$

$$p = \frac{249}{45.25} \approx 5.5\% ❖$$

What is a percent statement for Example 5?

Exercises & Problems

Communicate

1. Explain the procedure for changing a percent to a fraction.

Explain how to draw a percent bar to model each problem.

2. 40% of 50
3. 200% of 50
4. 30 is 50% of what number?
5. What percent of 80 is 60?

Estimate each answer as more or less than 50. Explain how you made your estimate.

6. 40% of 50
7. 200% of 50
8. 30 is 50% of what number?
9. What percent of 80 is 60?

Describe how to set up the equation for the following problem.

10. How many ounces of pretzels would you have to eat to get 100% of the U.S. RDA of iron?

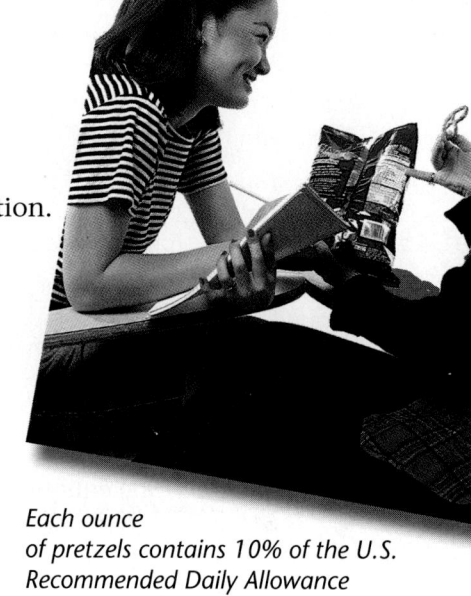

Each ounce of pretzels contains 10% of the U.S. Recommended Daily Allowance (U.S. RDA) of iron.

Practice & Apply

Write the following percents as decimals.

11. 55%
12. 1.2%
13. 8%
14. 145%
15. 0.5%

Write the following decimals as percents.

16. 0.47
17. 0.019
18. 8.11
19. 0.001
20. 9.00

Draw a percent bar to model each problem.

21. 35% of 80
22. 5 is what percent of 25?
23. What number is 10% of 8?
24. 150% of 40
25. What percent of 90 is 40?
26. 18 is 20% of what number?

Estimate each answer as more or less than 50 or 50%.

27. 5% of 80
28. 18 is 20% of what number?
29. 150% of 40
30. 2.5% of 100
31. What percent of 90 is 40?
32. What percent of 60 is 120?

Find each answer.

33. 40% of 50
34. 125% of what number is 45?
35. 8 is 20% of what number?
36. 200% of 50
37. 30 is 60% of what number?
38. What percent of 80 is 10?
39. 35% of 80
40. What number is 3.5% of 120?
41. 3 is what percent of 3000?
42. 75% of 900
43. 72 is 9% of what number?
44. What percent of 60 is 24?

Probability According to the National Highway Safety Administration, a 0.1% blood alcohol concentration increases the odds of a car accident 7 times, and even one-half of that amount impairs reflex time and depth perception.

45. Write the percent that is one-half of 0.1%.

46. Write a percent statement using "of" and "is" for Exercise 45.

47. **Statistics** In a sample of 50 seniors, 52% opposed the school's new driving policy. How many students in the survey opposed the policy?

48. **Discounts** Troy buys a tennis racket for 30% off the original price of $66. What is the sale price of the racket?

Statistics

49. What percent of the popular vote did each candidate get?

50. What percent of the electoral vote did each candidate get?

51. Make a pie graph of the popular vote for all three candidates.

52. Make a pie graph of the electoral vote for all three candidates.

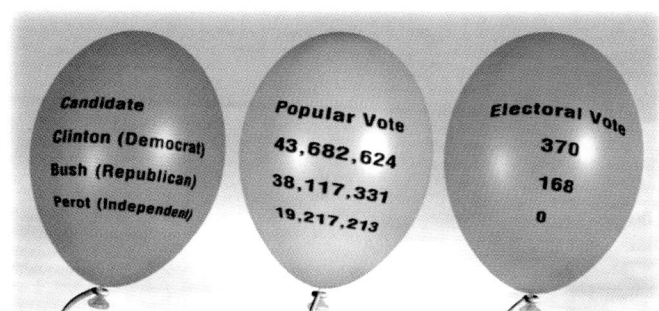

Candidate	Popular Vote	Electoral Vote
Clinton (Democrat)	43,682,624	370
Bush (Republican)	38,117,331	168
Perot (Independent)	19,217,213	0

53. **Demographics** The newspaper says that only 42% of the registered voters voted to determine a school bond election. If 11,960 people voted, how many registered voters are there?

54. **Geometry** The dimensions of a rectangle are increased by 10%. By what percent is the perimeter increased?

55. Mr. Arnes assigns 100 points out of 600 points for a group project. What percent is this?

Look Back

Plot the following points on a coordinate plane. [Lesson 1.5]

56. $A(7, 8)$ **57.** $B(5, 2)$ **58.** $C(3, 3)$ **59.** $D(5, 7)$

60. If x is an even integer, represent the next two even integers in terms of x. [**Lesson 3.3**]

Simplify. [**Lesson 4.2**]

61. $-2 \cdot 3x$ **62.** $-8x \div 4$ **63.** $\dfrac{10x + 25}{-5}$ **64.** $\dfrac{22 - 2y}{2}$

Look Beyond

65. Solve $4x + 6 = 2x - 2$.   **66.** Solve $4x + \dfrac{1}{2} = 5x - \dfrac{3}{4}$.

Writing and Solving Multistep Equations

Why *It usually takes more than one step to solve a problem about money. To find the sales tax, you first add the item prices, then multiply by the tax rate. Solving multistep problems involves solving multistep equations.*

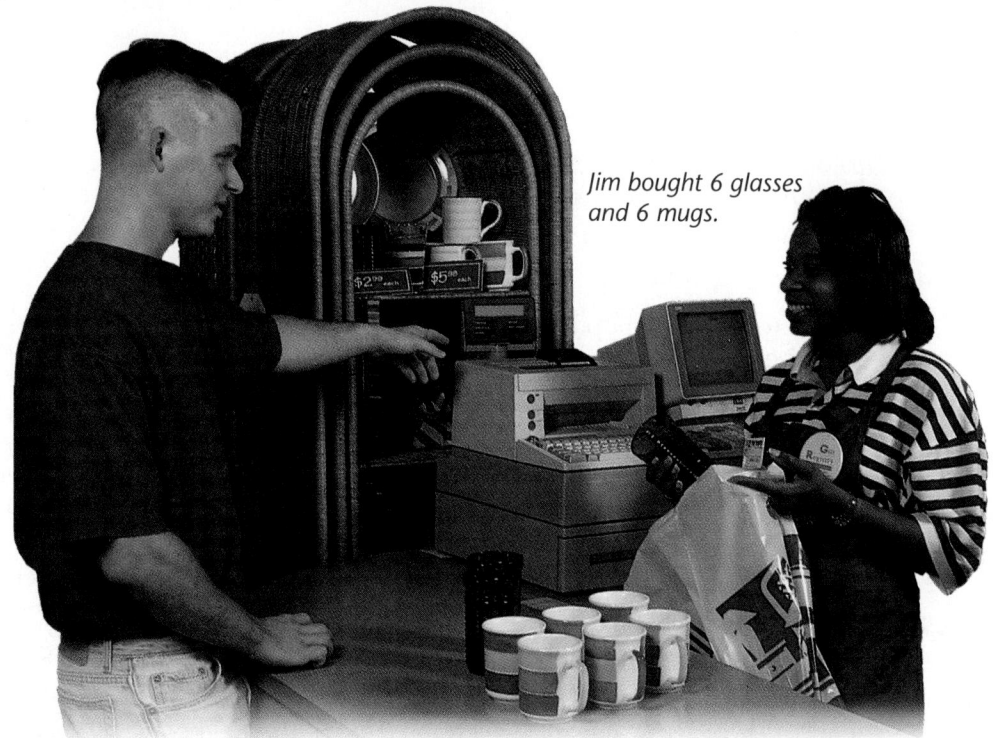

Jim bought 6 glasses and 6 mugs.

Jim knew that one mug and one glass would cost between $9 and $10. Six of each should cost between $54 and $60, so he knew that the amount the salesperson rang up was too low. Jim believed that the salesperson had entered the correct amount for the mugs and wanted to find the number of glasses he had been charged for.

The cost of the mugs is $5.98 · 6 or $35.88. Let *g* represent the number of glasses. The cost of the glasses is $2.99*g*. The total cost for mugs and glasses is $47.84.

Write the equation that models the problem.

$$\text{Cost of mugs} + \text{Cost of glasses} = \text{Total cost}$$
$$35.88 \quad + \quad 2.99g \quad = 47.84$$

Solve for *g*.

$35.88 + 2.99g = 47.84$	Given
$35.88 + 2.99g - 35.88 = 47.84 - 35.88$	Subtraction Property of Equality
$2.99g = 11.96$	Simplify.
$\dfrac{2.99g}{2.99} = \dfrac{11.96}{2.99}$	Division Property of Equality
$g = 4$	Simplify.

Jim thought the salesperson accidentally entered 4 glasses instead of 6.

When the salesperson returned with the register receipt and the purchase, the receipt indicated that the salesperson had indeed miscalculated the bill, but she had charged Jim for 6 glasses and only 5 mugs. How was this possible?

Computer

There are computer programs that solve equations like the one in the previous problem. For example, if you type in

solve(35.94+3.49*n=46.41); ENTER

using a certain Maple™ software program, the computer will immediately display the answer, 3. As equations become more complicated, you can use technology to solve them.

Why is it important to know how to write and solve equations without relying on technology?

EXAMPLE 1

Heather is waiting to take her third test.

What score does Heather need so that her average will be 80?

STATISTICS
Connection

Solution ➤

Let s represent the score Heather needs on her third test. The average after 3 tests can be modeled by $\frac{72 + 79 + s}{3}$.
She wants her average to equal 80.

$$\frac{72 + 79 + s}{3} = 80 \qquad \text{Given}$$

$$\frac{151 + s}{3} = 80 \qquad \text{Simplify.}$$

$$3\left(\frac{151 + s}{3}\right) = (3)(80) \qquad \text{Multiplication Property of Equality}$$

$$151 + s = 240 \qquad \text{Simplify.}$$

$$151 + s - 151 = 240 - 151 \qquad \text{Subtraction Property of Equality}$$

$$s = 89 \qquad \text{Simplify.}$$

Heather needs 89 on her third test.

Check $\qquad \frac{72 + 79 + (89)}{3} \stackrel{?}{=} 80$

$$80 = 80 \qquad \text{True} \; ❖$$

How do you determine the number to divide by to find an average?

CRITICAL
Thinking

Do you think Heather will receive an 80 for her average if she scores 88 on her third test? Why or why not?

EXAMPLE 2

Solve $5x + 6 = 2x + 18$.

Solution ➤

Isolate x on one side of the equation. Begin by subtracting $2x$ from both sides.

$5x + 6 = 2x + 18$	Given
$5x + 6 - 2x = 2x + 18 - 2x$	Subtraction Property of Equality
$3x + 6 = 18$	Simplify.
$3x + 6 - 6 = 18 - 6$	Subtraction Property of Equality
$3x = 12$	Simplify.
$\dfrac{3x}{3} = \dfrac{12}{3}$	Division Property of Equality
$x = 4$	Simplify.

Check the solution.

$$5(4) + 6 \stackrel{?}{=} 2(4) + 18$$
$$26 = 26 \qquad \text{True} \; ❖$$

Try This   Solve $4y - 3 = 6y - 2$.

The next example shows how technology can be used to solve an equation graphically.

EXAMPLE 3

Graphics Calculator

Solve $5x + 6 = 2x + 18$ using a graphics calculator.

Solution ➤

On a graphics calculator, let $Y_1 = 5x + 6$, and let $Y_2 = 2x + 18$. Graph the functions.

Start *with the standard window.*

Zoom out. *Trace to find the intersection.*

Zoom in. *Trace to find the coordinates.*

The value of the x-coordinate is 4. ❖

X=3.8947368 Y=25.473684

X=4   Y=26

Sometimes, there are many steps needed to find a solution. In the next example all four properties of equality are used.

EXAMPLE 4

Solve $x - \frac{1}{4} = \frac{x}{3} + 7\frac{3}{4}$.

Solution ➤

Step 1

Clear the equation of fractions by multiplying both sides by the least common denominator, 12.

$$x - \frac{1}{4} = \frac{x}{3} + 7\frac{3}{4}$$ Given

$$12\left(x - \frac{1}{4}\right) = 12\left(\frac{x}{3} + 7\frac{3}{4}\right)$$ Multiplication Property of Equality

$$12x - 12\left(\frac{1}{4}\right) = 12\left(\frac{x}{3}\right) + 12\left(7\frac{3}{4}\right)$$ Distributive Property

$$12x - 3 = 4x + 93$$ Simplify.

Step 2

Isolate x on one side of the equation.

$$12x - 3 - 4x = 4x + 93 - 4x$$ Subtraction Property of Equality

$$8x - 3 = 93$$ Simplify.

Step 3

Collect the constant terms on the other side of the equation, and divide both sides by 8 to find x.

$$8x - 3 + 3 = 93 + 3$$ Addition Property of Equality

$$8x = 96$$ Simplify.

$$\frac{8x}{8} = \frac{96}{8}$$ Division Property of Equality

$$x = 12$$ Simplify.

Step 4

Check your result by substituting 12 for x in the original equation.

$$12 - \frac{1}{4} \stackrel{?}{=} \frac{12}{3} + 7\frac{3}{4}$$

$$11\frac{3}{4} = 11\frac{3}{4}$$ True ❖

EXERCISES & PROBLEMS

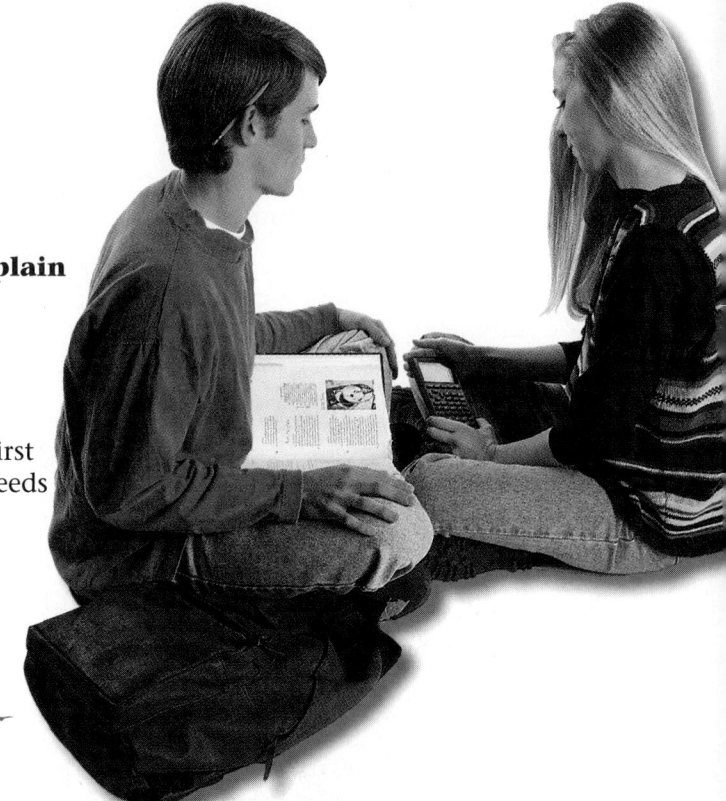

Communicate

Without actually solving the equation, explain the steps needed to solve the problem.

1. $23p + 57 = p + 984$ **2.** $\frac{x}{24} + 33 = 84$

3. $5z - 5 = -2z + 19$

4. **Statistics** Brittany scored 85 on her first 2 tests. Explain how to find what score she needs on her third test to average 90 for all 3 tests.

5. Explain how to solve $2x + 3 = 8 - 2x$ using a graphics calculator.

Practice & Apply

Solve each equation.

6. $2m + 5 = 17$ **7.** $9p + 11 = -7$

8. $5x + 9 = 39$ **9.** $3 + 2x = 21$

10. $6 - 8d = -42$ **11.** $2s + 9 = -13$

12. $2(x + 3) = 14$ **13.** $4(x - 20) = 16$ **14.** $2(5 - x) = -25$

15. $\frac{z}{5} = 22 - 20$ **16.** $\frac{v}{4} + 8 = 11$ **17.** $\frac{x}{10} - 1 = -31$

18. $5x - 7 = 2x + 2$ **19.** $4x + 1 = 12 - 8x$ **20.** $3x - 8 = 5x - 20$

21. $1 - 3x = 2x + 8$ **22.** $15 - 4x = 12 - 8x$ **23.** $2(x - 3) = 3(x - 4)$

24. $3v - 8 = \frac{v}{2} + 2$ **25.** $\frac{w}{2} + 7 = \frac{w}{3} + 9$ **26.** $6 - \frac{t}{4} = 8 + \frac{t}{2}$

Solve each equation for the indicated variable.

27. $3t = r$ for t **28.** $53 + 9s = P$ for s **29.** $ma = q$ for a

30. $al + r = 7$ for l **31.** $2a + 2b = c$ for b **32.** $y = mx + b$ for x

33. **Number Theory** The sum of three consecutive whole numbers is 48. Write an equation, and solve to find the numbers.

34. Lynn has scores of 95, 91, and 88 on three tests. Write an equation and solve to find a fourth score to average 90 on all four tests.

35. **Geometry** If the perimeter of the rectangle is 180 feet, write and solve an equation to find the length and width.

x

$x + 60$

36. Consumer Economics If the total cost, C, is \$98.65, how many compact discs did Kara and a friend buy?

Travel You can estimate the time T (in hours) it takes to fly non-stop a distance D (in miles) by the equation $T = \frac{D}{500} + \frac{1}{2}$. For example, a flight of 1000 miles would take about $2\frac{1}{2}$ hours.

37. Use the formula to estimate the time it takes to fly 1300 miles.

38. Solve the formula for D.

39. Use the new formula to find how many miles you can travel in 4 hours.

40. Cultural Connection: Africa Find the answer to this problem from ancient Egypt about 3800 years ago: Fill a large basket $1\frac{1}{2}$ times. Then add 4 *hekats* (a *hekat* is about half a bushel). The total is 10 *hekats*. How many *hekats* does the basket hold?

Look Back

41. Find the area of a square with a side of 6 inches. **[Lesson 2.2]**

Write the opposite for each expression. [Lesson 3.4]

42. $6c - 3d$ **43.** $-s + t$ **44.** $-a - c$

Solve. [Lesson 3.6]

45. $x + 4 = 10$ **46.** $x - 6 = 23$ **47.** $x - 13 = 10$

48. $5 - x = -12$ **49.** $2.1 + x = -8.3$ **50.** $-\frac{2}{3} + x = -\frac{1}{6}$

Look Beyond

Solve each equation for x.

51. $x^2 = 4$ **52.** $x^2 = 169$ **53.** $x^2 = 24$

54. Paul, Pete, and Pam each took a different type of fruit to school for lunch. Paul does not like bananas and Pete does not like to peel fruit. If the fruits are an apple, an orange, and a banana, who brought the apple?

Exploring Related Inequalities

why *Many real-life situations are not solved by finding a single answer, but by finding a minimum, maximum, or range of answers that satisfy the conditions. For problems of this type, you often need to solve an inequality.*

SECTION R

The Arts

The HOMETOWN GAZETTE • Sunday

Music 6
TV 9
Theater 11

On Display

"Cubism Examined" now at the Midtown Gallery. Museum curator Laura Chang provides a retrospective of the Cubist art movement, as well as some insights on the impact it has had on subsequent artistic styles. Now through November 12th.

Record Season Ticket Sales Predict Promising Year

The Ballet

"The Community Dancers" begin their 8th season next week. Season ticket sales are up nearly 10 percent over this same period last year. Director Manolo Clemente stated, "Everything is progressing more quickly than we had expected. There has never been this much talent in our company at the same time. This looks to be an historic season to be sure."

The critics agree. "This isn't the same old troupe" says Gene Egbert Theater/Dance critic for Evening News, "The public is in for a pleasant surprise." Board members hope this newly energized season will open ballet to a wider audience in the community. "These performances are for everyone", stated Manolo, "We're the Community Dancers... We dance for everyone".

Hours of rehearsal pay off as skilled dancers elevate this years Ballet to an unprecedented level of performance.

In Concert

Puppetry Center opens Today!

The Kids

This ballet company must maintain a certain minimum average attendance to stay in business. Fire codes and health regulations establish a certain maximum number of people allowed in the theater at one time.

Exploration 1 *Multiplying and Dividing Inequalities*

Numbers graphed on the number line are in order from left to right.

$-4 < 2$ $2 > -4$

Complete the pattern by filling in <, =, or >.

 a. $-4 < 2$, so $-4 \cdot 2 < 2 \cdot 2$
b. $-4 < 2$, so $-4 \cdot 1 < 2 \cdot 1$
c. $-4 < 2$, so $-4 \cdot 0 \underline{\ ?\ } 2 \cdot 0$
d. $-4 < 2$, so $-4 \cdot -1 \underline{\ ?\ } 2 \cdot -1$
e. $-4 < 2$, so $-4 \cdot -2 \underline{\ ?\ } 2 \cdot -2$

 a. $-4 < 2$, so $-4 \div 2 < 2 \div 2$
b. $-4 < 2$, so $-4 \div 1 < 2 \div 1$
c. $-4 < 2$, so $-4 \div -1 \underline{\ ?\ } 2 \div -1$
d. $-4 < 2$, so $-4 \div -2 \underline{\ ?\ } 2 \div -2$

 What happens to an inequality when each side is multiplied or divided by the same positive number?

 What happens to an inequality when each side is multiplied or divided by 0?

 What happens to an inequality when each side is multiplied or divided by the same negative number?

 a. Write a rule to explain what you must do to the inequality when you multiply or divide each side by the same positive number.
b. Write a rule to explain what you must do to the inequality when you multiply or divide each side by the same negative number. ❖

•Exploration 2 *Multiplying Inequalities*

To visualize what happens when you multiply each side of an equality by the same number, complete this exploration.

 Start with a set of 3 positive tiles and a set of 5 positive tiles. The < sign is used

$3 < 5$

2️⃣ To multiply both sides of the inequality by 2, think of doubling each set of tiles.

Fill in <, =, or >.

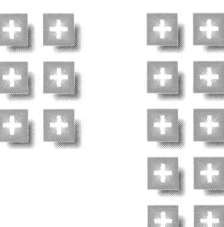

$2 \cdot 3 \underline{\ ?\ } 2 \cdot 5$

3️⃣ To multiply both sides of the inequality by -2, think of doubling each set *and* changing the sign of each tile.

Fill in <, =, or >.

$-2 \cdot 3 \underline{\ ?\ } -2 \cdot 5$

 Draw a picture to show what happens to $-3 > -4$ when each side of the inequality is multiplied by **a.** 2. **b.** -2. ❖

Exploration 3 *Dividing Inequalities*

To visualize what happens when you divide each side by the same number, complete this exploration.

1 Start with a set of 4 positive tiles and a set of 8 positive tiles. The < sign is used for the inequality.

$$4 < 8$$

2 To divide both sides of the inequality by 4, think of dividing each set of tiles by 4.

Fill in <, =, or >.

$$4 \div 4\ \underline{?}\ 8 \div 4$$

3 To divide both sides of the inequality by −4, think of dividing each set by 4 *and* changing the sign of each tile.

Fill in <, =, or >.

$$4 \div -4\ \underline{?}\ 8 \div -4$$

4 Draw a picture to show what happens when each side of the inequality −4 < 2 is divided by **a.** 2. **b.** −2. ❖

The properties for multiplying and dividing inequalities express the results of multiplying or dividing an inequality by a positive or negative number.

MULTIPLICATION AND DIVISION PROPERTY OF INEQUALITIES

If each side of an inequality is multiplied or divided by the same positive number, the resulting inequality has the same solution.

If each side of an inequality is multiplied or divided by the same negative number and the inequality sign is reversed, the resulting inequality has the same solution.

CRITICAL Thinking

A famous inequality states that $|a| + |b| \geq |a + b|$. Explain why this inequality is true for all numbers a and b.

APPLICATION

Special women's ballet slippers cost about $50 a pair and wear out very quickly.

Ballet A dancer rehearsing and performing the role of Clara during one run of *The Nutcracker* can typically use 15 to 25 pairs of these ballet slippers. If a ballet company has at most $1000 to spend, how many pairs of these special ballet slippers can the company buy?

Use an inequality to model the problem.

Let x equal the number of ballet slippers the company can buy. Then, $50x$ equals the amount they can spend.

The amount must be less than or equal to 1000. Thus, $50x \leq 1000$.

Divide both sides of the inequality by 50.

$$\frac{50x}{50} \leq \frac{1000}{50}$$

$$x \leq 20$$

The company can buy at most 20 pairs of special ballet slippers. ❖

EXERCISES & PROBLEMS

Communicate

Tell whether each statement is true or false. Explain your reason.

1. $7 < 8$ **2.** $7 < 7$ **3.** $7 \leq 7$ **4.** $7 \neq 7$

Describe the steps needed to solve each inequality.

5. $x + 1 > 4$ **6.** $x - 3 \leq 13$ **7.** $-3p < 12$ **8.** $4x - 2 \geq 2x + 3$

Practice & Apply

Write an inequality that corresponds to each statement.

9. L is greater than W.

10. r is greater than or equal to 4.

11. V is between 3.1 and 3.2 inclusive.

12. x cannot equal 0.

13. m is positive.

14. y is non-negative.

Tell whether each statement is true or false.

15. $4.2 \geq 4.2$

16. $9.22 \leq 9.22$

17. $3.1 < 3.01$

18. $8.55 > 8.505$

19. $\frac{1}{7} \geq \frac{1}{6}$

20. $\frac{3}{4} \leq \frac{4}{5}$

21. $-8 < -4$

22. $0 \geq -3$

Solve each inequality.

23. $x + 8 \geq 11$

24. $x - 11 < -20$

25. $G - 6 \leq 9$

26. $8 - H > 9$

27. $6 - x > -1$

28. $5 - y \leq 2$

29. $\frac{x}{8} < 1$

30. $\frac{u}{-3} \geq 21$

31. $5b > 3$

32. $9c \geq -21$

33. $2d + 1 < 5$

34. $5x - 2 > 2x + 9$

35. $\frac{x}{3} + 4 < 10$

36. $\frac{-x}{5} - 1 < 3$

37. $8x - 3 \leq 9$

38. $15 + \frac{y}{4} > 10$

A man uses about 15 pairs of ballet shoes a year, and the price ranges from $17 to $20 per pair.

39. Ballet Express the cost, C, of shoes used in a year (for one man) as an inequality.

Cultural Connection: Europe A Russian mathematician named Chebyshev (1821–1894) proved that the next prime after p is less than $2p$. For example, 7 is a prime, so according to Chebyshev, the next prime after 7 is less than 14, which is true since 11 is the next prime and $11 < 14$.

Substitute the following numbers for p to show that Chebyshev's statement is true.

40. 2 **41.** 3 **42.** 5 **43.** 89

Geometry The length, l, of a rectangle is to be at least 5 centimeters more than the width, w.

44. Express the statement as an inequality.

45. Write an inequality for the length if the width is 20 centimeters.

Consumer Economics Robin has $50 to order videos on sale from a catalog.

46. Write *two* inequalities that express the possible number of videos she can buy.
  a. the maximum number
  b. the minimum number

47. List all the possibilities for the number of videos Robin can buy.

48. How many videos can Robin and two friends buy if they have $80 to spend?

Look Back

Evaluate. [Lesson 3.2]

49. $89 - (-14)$ **50.** $400 - (-111)$ **51.** $-16 - (-3)$ **52.** $-674 - 9(-900)$

State if the product is negative or positive. [Lesson 4.1]

53. $(-3)(-3)(-1)(-1)$ **54.** $(-1)(1)(-1)(-1)$ **55.** $\frac{-22}{2}$ **56.** $\frac{-16}{-4}$

Use the Distributive Property to simplify the following expressions. [Lesson 4.2]

57. $(3x - 2y + 1) - 3(x + 2y - 1)$ **58.** $3(a + b) - 2(a - b)$

Look Beyond

Sometimes inequalities are used to indicate values that make sense in an equation, such as in these examples.

If $y = \frac{1}{x}$, then $x \neq 0$.   If $y = \sqrt{x}$, then $x \geq 0$.

Complete each of the following.

59. If $y = \frac{1}{x - 2}$, then $x \neq \underline{?}$.   **60.** If $y = \sqrt{x + 3}$, then $x \geq \underline{?}$.

Absolute Value Equalities and Inequalities

3.50 cm   maximum allowable tolerance of ± 0.01

40508 a

4050

Gear No. 40508 a & b – CNC Spec. 1 / Department R

why *Measurement is important in science, engineering, economics, industry, medicine, and many other fields. Measurements and the errors they contain can be expressed using absolute value.*

Error and Absolute Value

A company manufactures a small gear for a car. If the gear is made too large, it will not fit. If it is made too small, the car will not run properly. How accurate is close enough?

Quality Control   A gear is designed with a specification of 3.50 centimeters for the diameter. It will work if it is within ± 0.01 centimeter of the specified measurement. The **absolute error** is the absolute value of the difference between the actual measure, x, and the specified measure 3.50 cm. This is written $|x - 3.50|$.

If the maximum error permitted is 0.01 centimeters, the acceptable diameters can be shown with an absolute value inequality.

$$|x - 3.50| \leq 0.01$$

To find the maximum and minimum diameters, solve the equation part of the inequality.

Recall the definition for absolute value.

$$|x| = x \text{ if } x \geq 0$$
$$|x| = -x \text{ if } x < 0$$

To solve absolute value equations you must consider two cases.

Case 1
Consider the quantity within the absolute value sign to be positive.

$$|x - 3.50| = 0.01$$
$$x - 3.50 = 0.01$$
$$x = 3.50 + 0.01$$
$$x = 3.51$$

Case 2
Consider the quantity within the absolute value sign to be negative.

$$|x - 3.50| = 0.01$$
$$-(x - 3.50) = 0.01$$
$$-x = -3.50 + 0.01$$
$$x = 3.49$$

The maximum and minimum allowable lengths for the gear's diameter are 3.51 centimeters and 3.49 centimeters.

EXAMPLE 1

Solve $|3x - 2| = 10$.

Solution ➤

Case 1
Consider the quantity within the absolute value sign to be positive.

$$|3x - 2| = 10$$
$$3x - 2 = 10$$
$$3x = 2 + 10$$
$$3x = 12$$
$$x = 4$$

Case 2
Consider the quantity within the absolute value sign to be negative.

$$|3x - 2| = 10$$
$$-(3x - 2) = 10$$
$$-3x + 2 = 10$$
$$-3x = -2 + 10$$
$$-3x = 8$$
$$x = -\frac{8}{3}$$

Check

$$|3x - 2| \overset{?}{=} 10$$
$$|3(4) - 2| \overset{?}{=} 10$$
$$|12 - 2| \overset{?}{=} 10$$
$$|10| = 10 \quad \text{True}$$

Check

$$|3x - 2| \overset{?}{=} 10$$
$$\left|3\left(-\frac{8}{3}\right) - 2\right| \overset{?}{=} 10$$
$$|-8 - 2| \overset{?}{=} 10$$
$$|-10| = 10 \quad \text{True} \quad \diamond$$

Try This   Solve $|2x - 4| = 8$.

Distance and Absolute Value

GEOMETRY
Connection

Absolute value can be used to describe the distance between any two points on a number line. The number line and geometry can help you visualize the meaning of the absolute value equation and its solution.

What values of x are 2 units from 5?

Geometric solution On a number line locate the center of a circle at the point representing 5. Draw an arc 2 units from the center on each side. The arc intersects the number line at 3 and 7.

Algebraic solution Express this relationship as an absolute value equation. The expression $|x - 5|$ represents the distance between 5 and each of the two points marked on the number line.

$$|x - 5| = 2$$

Center   Radius

Then solve the absolute value equation $|x - 5| = 2$.

Case 1
The quantity within the absolute value sign is positive.

$$|x - 5| = 2$$
$$x - 5 = 2$$
$$x = 7$$

Case 2
The quantity within the absolute value sign is negative.

$$|x - 5| = 2$$
$$-(x - 5) = 2$$
$$-x + 5 = 2$$
$$-x = -3$$
$$x = 3$$

These are the points where the arcs of the circle cross the number line. They are each two units from 5.

CRITICAL *Thinking*

Explain why absolute value is always a non-negative number. Why is absolute value a good method for finding distance?

Absolute Values and Inequalities

An artist must fit a square piece of metal in a 48.00 centimeter by 48.00 centimeter space in a wood block.

EXAMPLE 2

Woodworking
The metal square will fit properly in the block if the amount of error in the length of the sides is within 0.05 centimeter. Use absolute value to express the amount of error allowed for the length of the metal square's side.

Solution ➤

The absolute error can be expressed as $|x - 48.00|$. The allowable error is ± 0.05 centimeter. Since the allowable error includes all the lengths within 0.05 centimeters of the specified measure, use an inequality.

Solve $|x - 48.00| < 0.05$.

Case 1
Consider the quantity within the absolute value sign to be positive.

$$|x - 48.00| < 0.05$$
$$x - 48.00 < 0.05$$
$$x < 48.00 + 0.05$$
$$x < 48.05$$

Case 2
Consider the quantity within the absolute value sign to be negative.

$$|x - 48.00| < 0.05$$
$$-(x - 48.00) < 0.05$$
$$-x + 48.00 < 0.05$$
$$-x < -48.00 + 0.05$$
$$x > 47.95$$

In case 2, recall that the inequality reverses when you multiply both sides of the inequality by a negative value.

The allowable lengths for the sides of the stained-glass square are all the measures between 48.05 centimeters and 47.95 centimeters. Write $47.95 < x < 48.05$. ❖

Lesson 4.8   Absolute Value Equalities and Inequalities   **195**

EXAMPLE 3

What values of x are less than or equal to 2 units from 5? Solve the inequality $|x - 5| \leq 2$.

Solution ➤

Case 1
The quantity within the absolute value sign is positive.

$$|x - 5| \leq 2$$
$$x - 5 \leq 2$$
$$x \leq 7$$

Case 2
The quantity within the absolute value sign is negative.

$$|x - 5| \leq 2$$
$$-(x - 5) \leq 2$$
$$-x + 5 \leq 2$$
$$-x \leq -3$$
$$x \geq 3$$

The inequality is true when x is less than or equal to 7 and greater than or equal to 3. We write $x \leq 7$ *and* $x \geq 3$, or $3 \leq x \leq 7$.

This can be represented on the number line.

$$3 \leq x \leq 7$$

EXAMPLE 4

Solve $|x - 6| > 2$.

Solution ➤

Consider two cases.

Case 1
The quantity $x - 6$ is positive.

$$|x - 6| > 2$$
$$x - 6 > 2$$
$$x > 8$$

Case 2
The quantity $x - 6$ is negative.

$$|x - 6| > 2$$
$$-(x - 6) > 2$$
$$-x + 6 > 2$$
$$-x > -4$$
$$x < 4$$

The inequality is true when $x > 8$. The inequality is also true when $x < 4$. We write $x > 8$ *or* $x < 4$.

Check by testing numbers from the solution in the original inequality. ❖

CRITICAL *Thinking*

What problem occurs when you try to solve $|x - 5| < -1$? Test some values to see. Why does this happen?

196   CHAPTER 4

Exercises & Problems

Communicate ~~~~

1. What is the meaning of the specification 45 ± 0.001 centimeters?

2. Explain how to write an absolute value equation that can be used to express 45 ± 0.001 centimeters.

3. Why must you consider two cases when you solve absolute value equations and inequalities?

4. Describe how to use an absolute value inequality to represent all the values on the number line that are within 3 units of the number -7.

5. Explain how to check the values of an absolute value inequality to see if they are inside or outside the boundary.

Practice & Apply ~~~~

Find the values of x that solve each absolute value equation. Check your answer.

6. $|x - 5| = 3$ **7.** $|x - 1| = 6$ **8.** $|x - 2| = 4$ **9.** $|x - 8| = 5$

10. $|5x - 1| = 4$ **11.** $|2x + 4| = 7$ **12.** $|4x + 5| = 1$ **13.** $|-1 + x| = 3$

Find the values of x that solve each absolute value inequality. Graph the answer on a number line. Check your answer.

14. $|x - 3| < 7$ **15.** $|x + 4| > 8$ **16.** $|x - 8| \leq 4$ **17.** $|x - 5| \geq 2$

18. $|x - 2| > 6$ **19.** $|x - 2| \leq 10$ **20.** $|x + 1| < 5$ **21.** $|x - 4| > 2$

Health According to a height and weight chart, Margo's ideal weight is 118 pounds. She will be satisfied if she stays within 5 pounds of her ideal weight.

22. Draw a diagram. Identify boundary values, acceptable weights, and unacceptable weights on your diagram.

23. Write an absolute value equation to describe the boundary values. What is the solution of your equation?

24. Write an absolute value inequality to describe the acceptable weights. What is the solution of your inequality?

The distance between x and 2 is 7.

25. Draw a number line diagram to illustrate the given sentence.

26. Translate the given sentence into an absolute value equation.

Let $|x - 3| < 4$.

27. Write the absolute value inequality as a sentence that begins, "The distance between . . ."

28. Draw a number line solution for the inequality.

29. Describe the solution in words.

30. Give 5 specific numbers that satisfy the inequality.

Let $|x - 3| > 4$.

31. Write the absolute value inequality as a sentence that begins, "The distance between . . ."

32. Draw a number line solution for the inequality.

33. Describe the solution in words.

34. Give 5 specific numbers that satisfy the inequality.

Government In a recent voter preference poll between two candidates, respondents gave Candidate A 50% of the vote. The polling technique used gives results accurate to within 3 percentage points.

35. What are the upper and lower boundaries for the actual percent of voters who favor Candidate A?

36. Describe the upper and lower boundaries using an absolute value equation.

37. Is it possible that Candidate A will lose the election? Explain.

Technology **Use technology to compare each set of two graphs.**

38. $y = 2x - 6$
$y = |2x - 6|$

39. $y = -x + 5$
$y = |-x + 5|$

40. **Portfolio Activity** Complete the problem in the portfolio activity on page 145.

 Look Back

Evaluate the following if $p = 4$, $q = 1$, and $r = 2$. **[Lesson 1.4]**

41. $pqr - q$

42. $\frac{pq}{r}$

43. $\frac{pqr}{q} + pqr$

Simplify. **[Lesson 3.5]**

44. $-5(8c + 3)$

45. $9(7b + 2)$

46. $-4(-5k + 8)$

Solve the following inequalities. **[Lesson 3.6]**

47. $4x + 5 < 25$

48. $6y - 10 > 5$

49. $9m - 8 < 4 + 8m$

50. What percent of 480 is 60? **[Lesson 4.5]**

Look Beyond

51. Subtract $x^2 + 4$ from $x^2 - x + 5$.

52. Subtract $x^2 + 2x + 3$ from $x^2 + 10$.

EGYPTIAN

Equation Solving

About 4000 years ago, Egyptian mathematicians wrote texts to prepare students for jobs in commerce. In 1858, a Scotsman named Rhind bought one of these texts which had been copied by a scribe named Ahmes about 1650 B.C.E.

One problem of the Ahmes Papyrus asks, "What is the amount of meal in each loaf of bread if 2520 *ro* of meal is made into 100 loaves?" In the form of an equation: $100m = 2520$

The Egyptians solved the equation using the methods of make-a-table and guess-and-check.

They first substituted key values, like different powers of ten, for the variable to find the value of the left side of the equation.

m	$100m$
1	100
10	1000
20	2000
5	500
$\frac{1}{5}$	20

They then used multiples of those values to see how close they could get to 2520. Since $2520 = 2000 + 500 + 20$, they asked what value of m would produce 20 for $100m$.

For the equation $100m = 2520$, they needed to compute $100(20) + 100(5) + 100\left(\frac{1}{5}\right)$ on the left side to get 2520 on the right side.

So, m was $20 + 5 + \frac{1}{5}$ or $25\frac{1}{5}$ *ro*.

Activity

Solve these equations as the Egyptians might have.

a. $40m = 4800$ c. $60j = 930$

b. $24p = 252$ d. $10w = 2432$

Make up and solve some equations of your own.

HINT: For problem **a** complete this table.

m	$40m$
1	40
100	—
20	—

Chapter 4 Review

Vocabulary

Key Skills & Exercises

Lesson 4.1

➤ **Key Skills**

Use the rules for multiplying and dividing integers to evaluate the following expressions.

a. $(-14)(-3) = 42$ **b.** $(-72) \div (24) = -3$ **c.** $(-8)[(-7) + 7] = (-8)(0) = 0$

➤ **Exercises**

Evaluate.

1. $(-12)(-5)$ **2.** $(-0.9)(3)$ **3.** $(54) \div (-18)$

4. $(-121) \div (-11)$ **5.** $(-2)(-2)(-2)(-2)$ **6.** $(-6)(-3) \div (-9)$

7. $45[8 + (-8)]$ **8.** $(-4)(4) \div (-1)$ **9.** $(-5)(-5)(-1)(1)$

Lesson 4.2

➤ **Key Skills**

Use the rules for multiplying and dividing expressions.

a. $-4(5x - 6) = (-4)(5x) - (-4)(6)$
$$= -20x + 24$$

b. $(2a - 3b + 1) - 4(a + 6b - 3) = 2a - 3b + 1 - 4a - 24b + 12$
$$= (2a - 4a) + (-3b - 24b) + 13$$
$$= -2a - 27b + 13$$

c. $\frac{12 - 8y}{-4} = \frac{12}{-4} + \left(\frac{-8y}{-4}\right) = -3 + 2y$

➤ **Exercises**

Multiply and divide expressions.

10. $3 \cdot 9x$ **11.** $-33x \div 3$ **12.** $-2(7y - 2)$

13. $-2.4x \cdot 2x$ **14.** $\frac{-30v + 3.6}{-3}$ **15.** $9r^2 - 8(4 - 3r^2)$

Lesson 4.3

➤ Key Skills

Solve multiplication equations.

Solve $-9k = 108$.

$-9k = 108$	Given
$\dfrac{-9k}{-9} = \dfrac{108}{-9}$	Division Property of Equality
$k = -12$	Simplify.

Solve division equations.

Solve $\dfrac{w}{-5} = -2.2$.

$\dfrac{w}{-5} = -2.2$	Given
$-5\left(\dfrac{w}{-5}\right) = -5(-2.2)$	Multiplication Property of Equality
$w = 11$	Simplify.

➤ Exercises

Solve each equation.

16. $17x = -85$ **17.** $-4g = -56$ **18.** $-2.2h = 33$

19. $24f = 150$ **20.** $\dfrac{w}{-8} = 0.5$ **21.** $\dfrac{y}{-2.4} = -10$

Lesson 4.4

➤ Key Skills

Solve an equation using the Reciprocal Property.

Solve $\dfrac{-2}{5}t = 14$.

One way to solve this equation is to multiply both sides by $\dfrac{5}{-2}$.

$$\dfrac{-2}{5}t = 14$$

$$\dfrac{5}{-2}\left(\dfrac{-2}{5}t\right) = \dfrac{5}{-2}(14)$$

$$t = -35$$

➤ Exercises

Solve each equation.

22. $\dfrac{4}{7}x = 4$ **23.** $\dfrac{1}{-9}y = 2.5$ **24.** $\dfrac{-j}{7} = -8$ **25.** $\dfrac{w}{-8} = \dfrac{3}{-4}$ **26.** $-\dfrac{3}{5}m = \dfrac{8}{15}$

Lesson 4.5

➤ Key Skills

Solve percent problems.

a. Find 30% of 15. You can use the equation method.

$$0.30 \cdot 15 = x$$
$$4.5 = x$$

b. What percent of 50 is 75? You can use a percent bar or an equation.

$$\dfrac{50}{100} = \dfrac{75}{x} \quad \text{or} \quad \dfrac{100}{50} = \dfrac{x}{75}$$

$$x\,75\left(\dfrac{100}{50}\right) = 75\left(\dfrac{x}{75}\right)$$

$$150 = x$$

The solution is 150%.

➤ Exercises

Find each answer.

27. 55% of 60   **28.** 28 is 70% of what number?   **29.** 200% of 40

30. 35% of 140   **31.** What percent of 90 is 40.5?   **32.** 4 is what percent of 50?

Lessons 4.6–4.7

➤ Key Skills

Solve multistep equations.

Solve $8c - 9 = 15$.

$8c - 9 = 15$	Given
$8c - 9 + 9 = 15 + 9$	Addition Property of Equality
$8c = 24$	Simplify.
$\dfrac{8c}{8} = \dfrac{24}{8}$	Division Property of Equality
$c = 3$	Simplify.

Solve inequalities.

Solve $14 - x \le -6$.

$14 - x \le -6$	Given
$14 - 14 - x \le -6 - 14$	Subtraction Property of Inequality
$-x \le -20$	Simplify.
$-1(-x) \ge -1(-20)$	Multiplication Property of Inequality
$x \ge 20$	Simplify.

➤ Exercises

Solve.

33. $3a + 7 = 31$

34. $-2x + 10 = -4$

35. $n - 6 = 2n - 14$

36. $\dfrac{y}{5} + 4 = 4.05$

37. $7k - 2(k + 6) = -2$

38. $2q + 5 = \dfrac{q}{5} - 4$

39. $x + 4 < 6$

40. $8 - y \ge 7$

41. $5r > -60$

42. $\dfrac{-p}{8} \le -3$

43. $x + 3 \ge 9 - x$

44. $t + \dfrac{1}{2} < \dfrac{t}{4} + 2$

Lesson 4.8

➤ Key Skills

Solve absolute value equations and inequalities.

Solve $|x - 2| = 3$.

Case 1
The quantity $x - 2$ is positive.

$$|x - 2| = 3$$
$$x - 2 = 3$$
$$x = 5$$

Case 2
The quantity $x - 2$ is negative.

$$|x - 2| = 3$$
$$-(x - 2) = 3$$
$$-x + 2 = 3$$
$$-x = 1$$
$$x = -1$$

Solve $|x - 2| < 3$.

Case 1
The quantity $x - 2$ is positive.

$$|x - 2| < 3$$
$$x - 2 < 3$$
$$x < 5$$

Case 2
The quantity $x - 2$ is negative.

$$|x - 2| < 3$$
$$-(x - 2) < 3$$
$$-x + 2 < 3$$
$$-x < 1$$
$$x > -1$$

$$-1 < x < 5$$

➤ Exercises

Solve each equation or inequality. Graph the inequalities on a number line.

45. $|x - 4| = 8$

46. $|3x + 2| = 6$

47. $|x - 3| \le 1$

48. $|x - 6| < 2$

Applications

Solve.

49. Sports Mark's basketball team scored 66, 74, 70, 82, and 58 points in their last 5 games. What is the average score for the team? Is a guess of 72 reasonable?

50. Geometry The area of a triangle with a base of 6 centimeters is 21 square centimeters. What is the height of the triangle? (HINT: $A = \dfrac{1}{2} bh$)

51. Statistics Suppose that of all of the potatoes grown in the United States, 22% are made into French fries. Out of 50 pounds of potatoes, how many pounds are made into French fries?

Chapter 4 Assessment

Evaluate.

1. $(-25)(-4)$

2. $(-3.6) \div (18)$

3. $(-6)[(-3) + 3]$

4. $(-80) \div (-4)(-5)$

5. Explain how to write a bank transaction modeling deposits or withdrawals from a bank account for the representation $(-6)(-30)$.

6. Are all of the following expressions equivalent? If not, which are different?

    a. $(8x - 32) \div 4$
    b. $(8x - 32)(0.25)$
    c. $\frac{32 - 8x}{4}$
    d. $\frac{-32 + 8x}{4}$

Multiply or divide the following expressions.

7. $-5x \cdot 7$

8. $-3(4 - 11y)$

9. $\frac{-28q}{-14}$

10. $\frac{-10t + 35}{-5}$

Solve each equation.

11. $|x - 4| = 11$

12. $\frac{r}{5} = -22$

13. $-10f = 0.55$

14. $\frac{z}{-3.5} = -7$

15. Farah needs to make a 360 mile trip in 9 hours. What speed should she average for the trip?

16. Are all rational numbers integers? Explain your answer.

17. What is the only rational number that does not have a reciprocal? Explain why it does not have a reciprocal.

Solve each equation or proportion.

18. $\frac{2}{3}y = -8$

19. $\frac{-5}{11}p = -15$

20. $\frac{-c}{14} = -\frac{1}{7}$

21. $\frac{x}{3} = \frac{-2}{5}$

22. $\frac{-a}{6} = \frac{4}{-9}$

23. $\frac{r}{2.8} = -\frac{5}{7}$

24. Paula made $350 in 4 weeks. At this rate, how much will she make in 10 weeks?

Find each answer.

25. 35% of 90

26. 3 is what percent of 15?

27. 150% of 28

28. Jamie made 80% of his free throws in his game last night. If he attempted 15 free throws, how many free throws did he make?

Solve each equation.

29. $3w + 5 = -16$

30. $\frac{m}{2} - 1 = 11$

31. $4x + 9 = 2x + 6$

32. The formula for the area of a triangle is $A = \frac{1}{2}bh$. If the area of a triangle is 336 square centimeters, and the base is 24 centimeters, what is the height of the triangle?

Solve each inequality.

33. $|x - 8| \geq 14$

34. $-9n < 42$

35. $\frac{j}{4} + 3 \leq -10$

36. $|x - 7| < 9$

Chapters 1 – 4
Cumulative Assessment

College Entrance Exam Practice

Quantitative Comparison For Questions 1–5, write

A if the quantity in Column A is greater than the quantity in Column B;
B if the quantity in Column B is greater than the quantity in Column A;
C if the two quantities are equal; or
D if the relationship cannot be determined from the information given.

	Column A	Column B	Answers
1.	$2x + 1$ $\quad x = 6$	$4x - 3$	(A) (B) (C) (D) **[Lesson 1.3]**
2.	The number of $14 CDs you can buy for $86.	The number of $12 CDs you can buy for $75.	(A) (B) (C) (D) **[Lesson 3.5]**
3.	$\lvert -12.4 \rvert$	$\lvert 12.4 \rvert$	(A) (B) (C) (D) **[Lesson 2.4]**
4.	$\text{INT}(4.75)$	$\text{INT}(5.2)$	(A) (B) (C) (D) **[Lesson 2.4]**
5.	$(3y + 2) + (2y - 5)$	$(2y + 6) - (y + 7)$	(A) (B) (C) (D) **[Lesson 3.4]**

6. Which expression is equal to 19? **[Lesson 1.4]**
  a. $15 \cdot (6 + 4) - 57$ **b.** $2^3 - 6 \cdot (3 + 2)$ **c.** $4^2 - 5^2 \div 5 + 2(0)$ **d.** $9 \div 3 \cdot 2^3 - 5$

7. Which decimal number is the equivalent of 143%? **[Lesson 4.5]**
  a. 0.143   **b.** 1.43   **c.** 143.0   **d.** 14.3

8. Which percent is equivalent to 0.5? **[Lesson 4.5]**
  a. 0.5%   **b.** 50%   **c.** 5%   **d.** 0.05%

9. The correlation between two variables is described as strongly positive. What is the relationship between the two variables? **[Lesson 1.6]**
  a. As one variable increases, the other also tends to increase.
  b. As one variable increases, you cannot tell if the other tends to increase or decrease.
  c. As one variable increases, the other tends to decrease.

10. Which sequence is linear? **[Lesson 2.1]**
  a. 60, 30, 15, 7.5, 3.75, . . .   **b.** 6, 12, 24, 48, 96, . . .
  c. 6, 9, 13.5, 20.25, 30.375, . . .   **d.** none

11. What is the value of $\frac{5^2 + 11}{9 \cdot 3 + 9}$? **[Lesson 1.4]**

    a. 36   **b.** 0   **c.** 1   **d.** $\frac{4}{9}$

12. Which expression is *not* equivalent to the others? **[Lesson 4.2]**

    a. $3x - 2(x - 3)$ **b.** $3x + 2(3 - x)$ **c.** $3x - 2(x + 3)$ **d.** $3x - 2(-3 + x)$

13. What is the solution to the equation $\frac{q}{139.2} = -58$? **[Lesson 4.3]**

    a. -2.4 **b.** -8073.6 **c.** -0.417 **d.** -7551.6

14. 60 is 75% of what number? **[Lesson 4.5]**

    a. 0.45   **b.** 0.8   **c.** 45   **d.** 80

15. Which is the value of x for the absolute value equation $|x - 6| = 10$? **[Lesson 4.8]**

    a. -16 **b.** 4   **c.** 10   **d.** -4

16. Is $18t$ a variable or an expression? Explain your answer. **[Lesson 1.3]**

What are the coordinates of the given points? **[Lesson 1.5]**

17. X **18.** Y **19.** Z

20. Solve the absolute value inequality $|x - 4| > 11$. **[Lesson 4.8]**

21. Determine whether the relationship is linear, quadratic, or neither. **[Lesson 2.1]**

x	1	2	3	4	5
y	4	6	10	16	24

Simplify the following expressions. **[Lesson 4.2]**

22. $(9w - 5) - (3w - 4)$ **23.** $(-54s) \div (6s)$ **24.** $4x^2 - 4(3x^2 + 7)$

25. Solve the inequality $y + 4.5 \geq 6$. **[Lesson 3.6]**

26. Jeff is on the 14th floor of a medical building. He rides the elevator up 3 floors, down 5 floors, and then up 4 floors. What floor is Jeff on now? **[Lesson 1.1]**

Evaluate. **[Lesson 3.1]**

27. $[-16 + (-4)] + 6$ **28.** $-8 + [(-23) + 23]$ **29.** $-48 + 8 - 3 + 2$

30. Solve the proportion $\frac{g}{12} = \frac{-3}{20}$. **[Lesson 4.4]**

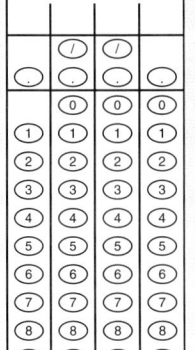

Free-Response Grid The following questions may be answered using a free-response grid commonly used by standardized test services.

31. What is the solution to the equation $3r + 2(r - 4) = 12$? **[Lesson 4.6]**

32. Hannah got 84% of the questions correct on a test. If there were 50 questions on the test, how many did she answer correctly? **[Lesson 4.5]**

33. Maurice has scores of 88, 90, and 80 on three tests. What score must he make on his next test to have a test average of 85? **[Lesson 4.1]**

CHAPTER 5

Linear Functions

Pyramids are part of the heritage of people who trace their roots to Africa or to Central America. The great pyramid of Gizeh in Egypt was built over a 30-year period around 2900 B.C.E. The builders used blocks averaging 2.5 tons, but some weighed as much as 54 tons. The blocks were transported from a quarry 600 miles away. The pyramid has a square base, and the error in constructing the right angles of the base was only one part in 14,000.

In order to build the pyramids, people needed to solve problems about slope. One problem asks to find the *seked* (a form of slope) of a pyramid with a base of 360 cubits (618 feet) on a side and a height of 250 cubits (429 feet).

The Rhind Papyrus

PORTFOLIO ACTIVITY

How steep are the sides of the pyramids of Egypt? The use of mathematics allows us to measure steepness. One measure of steepness is the slope.

1. Use cubit measures to find the slope of the triangular face of the pyramid.

2. Repeat the process using measurement in feet.

3. What do you notice about the slope measured in these two ways?

You may wish to include your calculations and explanation in your portfolio.

429 feet
250 cubits

360 cubits
618 feet

Defining Slope

why *Kara and Ricardo hike up the steep slope of a mountain trail. They remember how easy it was walking along the nearly flat streets back home. If you want to compare the steepness of a mountain trail with a level street, you can compare their slopes. You can also measure the steepness of lines on a coordinate plane the same way.*

Finding the Slope

Slope is the ratio of the vertical rise to the horizontal run. The slope of the hill is the same for both Kara and Ricardo.

EXAMPLE 1

What is the slope of the hill?

Solution ➤

To calculate the slope, find the ratio of rise to run for each hiker. The horizontal length of Ricardo's step (the run) is 35 centimeters, and the vertical length (the rise) is 14 centimeters. The run for Kara's step is 30 centimeters and the rise is 12 centimeters.

<table>
<tr><td align="center">**Ricardo**</td><td align="center">**Kara**</td></tr>
<tr><td align="center">$slope = \dfrac{rise}{run} = \dfrac{14}{35} = \dfrac{2}{5}$</td><td align="center">$slope = \dfrac{rise}{run} = \dfrac{12}{30} = \dfrac{2}{5}$</td></tr>
</table>

The slope of the hill is $\dfrac{2}{5}$. ❖

Try This Find the slope of a hill if the rise is 12 and the run is 24.

> **GEOMETRIC OR GRAPHIC INTERPRETATION OF SLOPE**
> **Slope** measures the steepness of a line by the formula
> $$\text{slope} = \frac{\text{rise}}{\text{run}}.$$

Sign of the Slope

The slope of the mountain trail was found from an upward rise left to right and calculated to be positive $\frac{2}{5}$. In Example 2, you will investigate a situation in which a line slopes downward from left to right.

EXAMPLE 2

Science Maya and her class are studying the effects of friction. To model the concept, they lean a meter stick against a wall. The object is to see how steep an incline is necessary to keep the ruler from slipping. They find that the friction prevents the ruler from slipping when the top of the ruler is 80 centimeters from the table top and the base is 60 centimeters from the wall. The students record the results on a graph. The wall is the *y*-axis, and the table top is the *x*-axis. What is the slope of the line formed by the meter stick?

Solution ➤

To measure the slope, start at the point where the line touches the *y*-axis. Measure the run, then the rise, and end at the point where the line touches the *x*-axis.

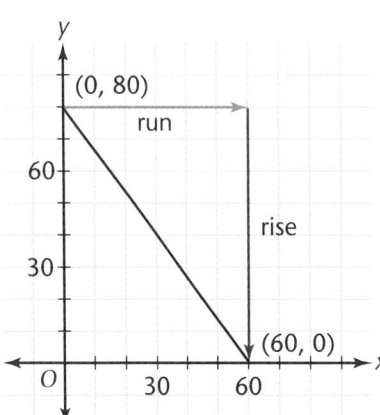

The distance toward the *right* is positive. Thus, the run is 60. The distance moving *down* is negative. Thus, the rise is -80.

$$\text{slope} = \frac{\text{rise}}{\text{run}} = \frac{-80}{60} \text{ or } -\frac{4}{3} \quad \diamondsuit$$

What makes the slope of a line negative?

EXAMPLE 3

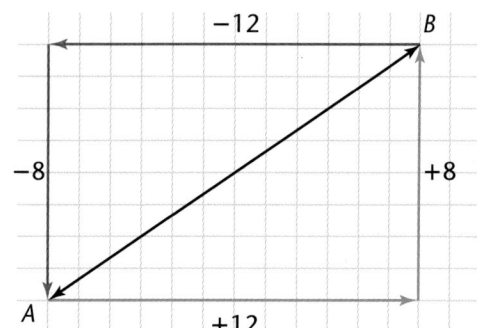

What is the slope of line *AB*?

Solution ➤

The run from *A* to *B* is 12, and the rise is 8.

$$\text{slope} = \frac{\text{rise}}{\text{run}} = \frac{8}{12} = \frac{2}{3}$$

The run from *B* to *A* is −12, and the rise is −8.

$$\text{slope} = \frac{\text{rise}}{\text{run}} = \frac{-8}{-12} = \frac{2}{3}$$ ❖

What combinations of positive and negative values for rise and run produce positive slope? negative slope?

EXAMPLE 4

Travel Jim drives from Kansas City, Kansas, to Denver, Colorado, along Interstate 70. Although the land seems flat, the drive is uphill much of the time. The table shows the elevation for several cities along the way, and how many miles the cities are from Kansas City. What is the average slope of the highway between Kansas City and Denver?

CITY	MILES from Kansas City	ELEVATION in feet
Kansas City, KS	0	744
Lawrence, KS	29	850
Topeka, KS	61	951
Salina, KS	171	1220
Hays,KS	264	1997
Goodland, KS	407	3683
Burlington, CO	437	4160
Denver, CO	600	5280

Solution ➤

To determine the slope, the average increase in elevation, the miles listed in the table are assumed to be horizontal.

STATISTICS
Connection

Use the data in the table to make a scatter plot. Draw a line that best fits the data points. Choose two convenient points on the line, the points (100, 1200) and (500, 4375), for example.

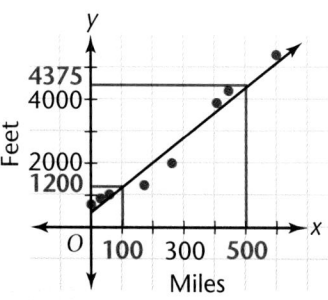

The *x*-coordinate changes from 100 to 500, an increase of 400 miles. The *y*-coordinate changes from 1200 to 4375, an increase of 3175 feet.

$$\text{slope} = \frac{\text{rise}}{\text{run}} = \frac{4375 - 1200}{500 - 100} = \frac{3175}{400} \approx 7.9$$

On the average, the elevation increases by about 7.9 feet per mile. ❖

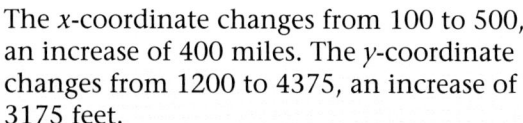

EXAMPLE 5

Burlington, Colorado, elevation 4160 feet, is on Interstate 70 near the border of Kansas and Colorado. The horizontal distance from Denver to Burlington is 163 miles. Is the average slope of the highway steeper in Kansas or in eastern Colorado?

Solution ➤

Use the information from the table in Example 4.

Find the slope from Kansas City to Burlington.

$$\text{slope} = \frac{\text{rise}}{\text{run}} = \frac{4160 - 744}{437 - 0} = \frac{3416}{437} \approx 7.8$$

Find the slope from Burlington to Denver.

$$\text{slope} = \frac{\text{rise}}{\text{run}} = \frac{5280 - 4160}{600 - 437} = \frac{1120}{163} \approx 6.9$$

The slope is steeper in Kansas than in eastern Colorado. ❖

In the last two examples slope was determined by finding the differences in the *y*-coordinates and *x*-coordinates. This gives another interpretation of slope.

SLOPE AS A DIFFERENCE RATIO

If the coordinates of two points on a line are given, the slope, *m*, is

the ratio $\frac{\text{difference in } y\text{-coordinates}}{\text{difference in } x\text{-coordinates}}$.

EXAMPLE 6

Find the slope of the line containing points $M(\mathbf{3}, \mathbf{5})$ and $N(-\mathbf{2}, -\mathbf{6})$.

Solution ➤

$$\text{slope} = \frac{\text{rise}}{\text{run}} = \frac{\textbf{difference in } y\textbf{-coordinates}}{\textbf{difference in } x\textbf{-coordinates}}$$

You must remember to subtract the x-coordinates in the same order as the corresponding y-coordinates.

$$\text{slope} = \frac{5 - (-6)}{3 - (-2)} = \frac{11}{5} \quad \text{or} \quad \text{slope} = \frac{(-6) - 5}{(-2) - 3} = \frac{-11}{-5} = \frac{11}{5} \; \diamond$$

What is the effect on the slope if the coordinates are not entered in the same order?

EXERCISES & PROBLEMS

Communicate

1. Explain how to draw a line with a rise of 4 and a run of 3.
2. Describe the slope of a line with negative rise and positive run.
3. Explain how to find the slope of line AC.
4. Explain why the slope of line AC is the same as the slope of line AB.
5. What is the slope of a line with rise s and run t? Compare this line with a line that has a rise $-s$ and run t. Which is steeper?
6. Point $R(5, -3)$ and point $N(9, -4)$ are on line k. Explain how to find the slope of line k.

Practice & Apply

Examine the graphs below. Which lines have positive slope? Which have negative slope? Which have neither?

7.

8.

9.

10.

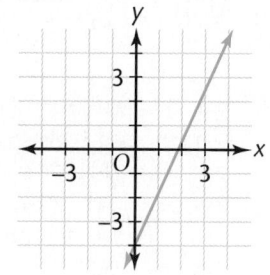

11. **Recreation** Suppose the slopes of two hills are negative. On graph paper draw a line to represent a beginner's ski hill and another to represent an advanced ski hill. Indicate each slope.

Find the slope for each of the given lines.

12. rise 6, run 2

13. rise 1, run 7

14. rise − 1, run 7

15. rise 0, run 5

16. rise 5 − 3, run − 3 − 1

17. rise − 7 + 2, run 3 − 1

18. Draw coordinate axes on graph paper. Place a point somewhere in the upper left of your graph. From that point, move 6 spaces to the right, then down 4. Place a point there. Move again 6 spaces right, then down 4. Repeat this at least four times. What shape appears when you connect the points? What is the slope of the line connecting the points? Try this exercise with patterns of your own.

Each pair of points is on a line. What is the slope of the line?

19. $M(9, 6), N(1, 4)$

20. $M(1, 3), N(1, 5)$

21. $M(-3, 1), N(2, 6)$

22. $M(0, 2), N(3, 0)$

23. $M(-3, 5), N(7, 6)$

24. $M(2, -4), N(5, 9)$

25. $M(3.2, 8.9), N(9.1, 7.2)$

26. $M(8, 10), N(8, 7)$

27. $M(10, 4), N(7, 4)$

28. $M(-3.2, -9.0), N(1.8, 8.2)$

29. $M(0, 7), N(0, 0)$

30. $M(-1, 1), N(-3, -3)$

31. Use the information in the table to find the average amount you climb per mile from Kansas City to Denver.

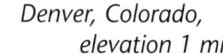
Denver, Colorado, elevation 1 mile

CITY	MILES from Kansas City	ELEVATION in feet
Kansas City, KS	0	744
Denver, CO	600	5280

32. Denver's elevation is 1 mile. The elevation of Loveland Pass, 70 miles west of Denver near Interstate 70, is 12,992 feet. Find the average slope of the land from Denver to Loveland Pass.

33. Find the difference for several pairs of *x*-coordinates and the difference for corresponding pairs of *y*-coordinates. Next, plot the values on a graph, and find the slope. Explain the relationships that you find.

x	1	2	3	4	8
y	3	6	9	12	24

34. **Geometry** Find at least 4 pictures of houses that have roofs with different slopes. Use a protractor to measure the angle for the slope of each roof. Use a ruler to find the slopes. Make a table to match the measure of each angle to each slope.

There are 91 steps in each of the 4 stairways on the Mayan/Toltec pyramid at Chichen Itza, in the Yucatan.

Cultural Connection: Americas
Mexican pyramids are somewhat different in concept from African pyramids. They have stairways on each side.

35. If the height (rise) of the stairway is 27 meters and the length (run) is 25 meters, what is the slope of the stairway of the pyramid?

36. The rise of the average step is 29.7 centimeters, and the run is 27.5 centimeters. Find the slope of a step, and compare it with the slope of the pyramid's stairway.

37. **Geometry** Use a string to measure several round objects. Record their diameters and circumferences in a table. Then graph the data with diameters on the *x*-axis, and fit it with a straight line. How well does the line fit the points? Find the slope of the line. What conjectures can you make from this information?

38. **Sports** On graph paper, graph the following data for the minutes played (*x*-axis) versus the number of points scored (*y*-axis) for the last game of the 1993 NBA playoffs. Use a clear ruler to fit a straight line to the data. Estimate the slope, giving the units. Write a sentence that explains what the slope tells you about the players' performance.

Player	Minutes	Points
Pippen	43	23
Armstrong	41	18
Jordan	44	33
Paxson	22	8
S. Williams	22	5
Tucker	7	9
King	2	0

39. Social Studies Graph the data in the table for the populations and number of representatives in the U.S. Congress for all the states starting with M. Fit the data to a line and find its slope. Explain what information the slope gives us about the 1990 data.

The number of representatives to Congress depends on the population of the state.

State	Representatives	Population
Maine	2	1,227,928
Maryland	8	4,781,468
Massachusetts	10	6,016,425
Michigan	16	9,295,297
Minnesota	8	4,375,099
Mississippi	5	2,573,216
Missouri	9	5,117,073
Montana	1	799,065

40. A line passes through the origin $O(0, 0)$ and point $P(c, d)$. What is the slope of the line?

41. A line passes through point $M(a, b)$ and point $N(c, d)$. What is the slope of the line?

Look Back

Name the type of correlation indicated by the given conditions. As one variable increases, **[Lesson 1.6]**

42. the other variable tends to decrease.

43. you cannot tell if the other variable tends to increase or decrease.

44. the other variable tends to increase.

Evaluate. **[Lesson 3.1]**

45. $|-7|$ **46.** $|50|$ **47.** $-|-9|$ **48.** $-|99|$

49. If there are 8 people contributing to a charity, how much would each person have to contribute to raise $98? **[Lesson 4.3]**

Solve for x. **[Lesson 4.6]**

50. $3x + 16 = 19$ **51.** $28 = -4 + 4x$

Look Beyond

52. Graph $y = 4x - 2$, $y = 4x$, and $y = 4x + 3$ on the same axes. Find the slope of each function. What do you notice about the lines?

53. On graph paper, plot the point $(0, 0)$. From the point, move right 5 and up 3 to plot the second point. Then move right 5 and up 6 to plot the third. Next move right 5 and up 9 for the fourth point. Continue until you have plotted 6 points. Connect the points with a smooth curve. Have you seen a curve like this before? If so, where?

Watch Your Step!

Atlanta Architect Steps Up Quest for Safe Staircases

Numerous Accidents, Deaths Push Researcher into Career

By Lauran Neergaard
THE ASSOCIATED PRESS

ATLANTA – ...Every year, 1 million Americans seek medical treatment for falls on staircases. About 50,000 are hospitalized and 4,000 die.

Templer's interest began as a student at Columbia University, when someone asked him why people were always falling down the steps outside Lincoln Center. Templer discovered there was no research on stair safety. Nobody even counted falls.

Curious, he visited Lincoln Center with his family, and his sister-in-law tripped on the steps.

Templer called his mentor to propose the topic as a thesis and learned that the man had just broken his leg falling down a stairwell of a subway station.

A career was born.

"All stairs are dangerous, it's a matter of degree," Templer said. "There are ways to mitigate that danger, if we could get that message to people."

Stairs evolved from ladders and first were used for defense. Narrow winding staircases, for instance, hampered intruders. Europeans wrought stairs into works of art, building grand palace staircases that gradually steepened, forcing visitors into a slow, stately pace as they approached royalty.

Most stairs now have 9-inch treads and $8\frac{1}{2}$-inch risers, a size determined around 1850, Templer said. But people today have bigger feet that hang over the edges of stairs, throwing them off balance, he said.

Stairs also are too high, Templer concluded after experiments in which he forced volunteers to trip on collapsible stairs. They were harnessed so they didn't tumble all the way to the floor, but Templer used videos to simulate how they would have landed.

He wants building codes revised for stairs with 11-inch treads and 7-inch risers. His proposal prompted a lobbying blitz from the National Association of Home Builders, which contends that larger stairs would add at least 150 square feet and $1,500 in costs to a house.

NAHB's Richard Meyer dismissed Templer's work, saying people fall when stairs are improperly lighted, have loose carpeting or have objects placed in the way.

That's true, too, Templer said. But he said his experiments, funded by the National Science Foundation, prove stair shape is a large problem.

Cutaway Diagram of Steps of Stairway

Tread Width (run)

TREAD

Riser Height (rise)

RISER

Count Your Way to Stair Success

By Karol V. Menzie and Randy Johnson

Baltimore Sun

The riser height (rise) and the tread width (run) determine how comfortable the stairs will be to use. If the rise and run are too great, the stairs will strain your legs and be hard to climb. If the rise and run are too small, you may whack your toe on the back of each step.

Over the years, carpenters have determined that tread width times riser height should equal somewhere between 72 to 75 when the measurements are in inches.

On the main stair, the maximum rise should be no more than $8\frac{1}{4}$ inches and the minimum run should be no less than nine inches.

To determine how many steps, or treads, you need, measure from the top of the finished floor on the lower level to the top of the finished floor on the upper level.

To figure the rise and run in a house with eight-foot ceilings, for instance, start by figuring the total vertical rise. By the time you add floor joists, subfloor and finish floor, the total is about 105 inches.

A standard number of treads in a stair between first and second floors is 14. One hundred five divided by 14 equals $7\frac{1}{2}$. That means the distance from the top of each step to the top of the next step will be $7\frac{1}{2}$ inches.

With a riser height of $7\frac{1}{2}$ inches, tread width (run) should be at least nine inches. Ten inches is a more comfortable run; when you multiply $7\frac{1}{2}$ inches by 10 inches, you get 75–within the conventional *ratio* of 72 to 75. With fourteen 10-inch treads, the total run of the stairs will be 140 inches. In other words, the entire stair will be 105 inches tall and 140 inches deep.

You can alter the rise and run to some extent. If you used 15 risers instead of 14, for instance, the rise would be 7 inches, and the tread width would be $10\frac{1}{2}$ inches (7 times $10\frac{1}{2}$ equals 73.5, within the rise and run guidelines).

Cooperative Learning

1. In the news article above, is the underlined term *ratio* used correctly? Explain.

2. Find the angle of each of the following staircases. Draw the triangle and measure the angle.

   a. the 105-inch by 140-inch staircase described in the article above

   b. the same staircase changed as described in the article to include 15 risers instead of 14

   c. the same staircase changed to meet Templer's preference for 7-inch risers and 11-inch treads

3. Which of the three staircases in activity 2 would take up the most room? Explain.

4. Suppose you did not want the building code for stairs to change. Write an argument (with numbers and a diagram) to show that larger stairs would add at least 150 square feet to a typical house. Make up your own estimates of the size of a typical house.

Project

Measure the rise and run on staircases in your school or home. Do they match the information in the two news articles about standard stairs?

Exploring Graphs of Linear Functions

Why *A linear function can be used to model the descent of an airplane. When you model situations using the graph of a linear function, you need to know the slope and the point at which the line crosses the y-axis.*

The captain announces the beginning of the descent from 24,000 feet for a landing in Detroit in 30 minutes.

The variable t represents the time in minutes after the airplane begins its descent. The starting condition is 24,000 feet. The constant rate of descent is -800 feet per minute because the plane must descend 24,000 feet in 30 minutes. The function $y = 24,000 - 800t$ models the descent.

Recall from Lesson 1.5 that the equation $y = 24,000 - 800t$ is a linear function because its graph is a line. In fact, any equation of the form $y = mx + b$ is a linear function. When $b = 0$, $y = mx$ is a line passing through the origin.

Graphics
Calculator

Graph the point $A(3, 6)$ on a coordinate plane. Start with the equation $y = \mathbf{m}x$. Guess and check numbers for $\mathbf{m}$ by graphing the equations until you find a line that intersects point A.

If you use a graphics calculator, enter point $A(3, 6)$, and let your first guess for m be 4. Enter **Y = 4X**, then graph the function. See if the line passes through $A(3, 6)$.

If you use graph paper, substitute values for x in the equation $y = 4x$ to locate points like $(0, 0)$ and $(2, 8)$. Then draw the line that connects the points. See if the line passes through $A(3, 6)$.

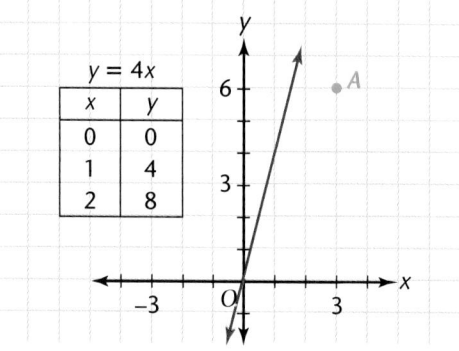

When m is 4, the line representing $y = 4x$ is too steep. It does not go through point A.

Repeat the process with different values for m until you find an equation for the line that intersects point A.

•Exploration 1 *Fitting the Line to a Point*

Repeat the process above for points B through F.

$B(2, 8)$ $C(6, 3)$ $D(3, -6)$ $E(2, -8)$ $F(4, 7)$

What is the connection between the coordinates of the given target point and the value of m? ❖

•Exploration 2 *Changing the Slope*

Draw the graph of each equation on the same axes. Compare the graphs you get from varying the values of m.

$y = x$ $y = 5x$ $y = -2x$ $y = \dfrac{1}{2}x$ $y = -\dfrac{1}{3}x$

1 What stays the same when the lines are drawn?

2 What is different?

3 Guess what the graph of $y = 3x$ looks like.

4 Check your guess by drawing the graph.

5 Make a conjecture about how the coefficient of x affects the graph of $y = mx$. ❖

•Exploration 3 Introducing a Constant

Draw the graphs of the next set of equations on the same axes.

$$y = x + 3 \quad y = 2x + 3 \quad\quad y = 5x + 3 \quad\quad y = -2x + 3$$

1 What stays the same when the lines are drawn?

2 What is different?

3 Guess what the graph of $y = 4x + 3$ looks like.

4 Check your guess by graphing.

5 Make a conjecture about how adding 3 affects the graph of $y = mx$.

Draw the graphs of the next set of equations on the same axes.

$$y = 2x \quad\quad y = 2x + 3 \quad\quad y = 2x + 5 \quad\quad y = 2x - 3$$

6 What stays the same when the lines are drawn?

7 What is different?

8 Guess what the graph of $y = 2x - 4$ looks like.

9 Check your guess by graphing.

10 Make a conjecture about how the value of b affects the graph of $y = 2x + b$. ❖

Picture the graph of each equation in your mind. Check the accuracy of your pictures by graphing each equation.

$$y = -3x \quad\quad y = 5x - 6$$
$$y = 5 + 6x \quad\quad y = \frac{1}{2}x + 3$$

How do the values of m and b affect the graph of $y = mx + b$? Discuss the results of what you have discovered from the explorations.

APPLICATION

An orange carton weighs 3 pounds when empty, and each orange you place in it weighs an average of half a pound. If you know the number of oranges, what is the total weight of the carton and the oranges? Let n equal the number of oranges. Write an equation for the total weight, w, of a carton containing n oranges. Graph the equation with n on the x-axis and w on the y-axis.

total weight = weight of oranges + weight of carton
$$w \quad = \quad \frac{1}{2}n \quad + \quad 3$$

The weight and the number of oranges cannot be negative. Thus, the graph will include only points where n and w are greater than or equal to 0.

Try $n = 0$. When there are 0 oranges, the weight is only that of the carton, or 3 pounds. Plot the first point $(0, 3)$. Each time you add an orange, the weight increases by $\frac{1}{2}$ pound. This is the **rate of change of the function** $w = \frac{1}{2}n + 3$. How does the rate of change compare with the slope of a line?

You can add new points to the graph by moving to the right 1 unit and up $\frac{1}{2}$ unit from each point. If you draw a line through the points, the line will have a slope of $\frac{1}{2}$.

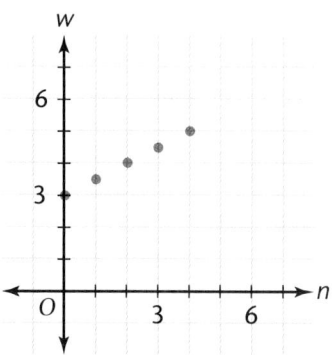

The individual points represent the weight of whole oranges.

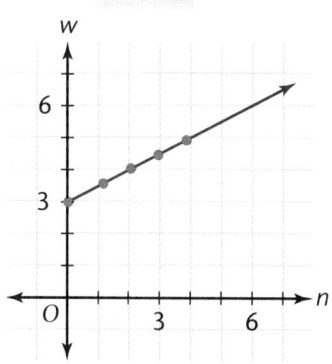

The solid line also includes the weight of partial oranges. ❖

CRITICAL
Thinking

What happens to the function that models the airplane descent if the starting altitude or rate of descent changes? Find the new equation if the starting condition is 30,000 feet and the rate of change is -1000 feet per minute.

EXERCISES & PROBLEMS

Communicate

1. Summarize your discoveries from each of the explorations.

Guess the equation of the line through the origin that intercepts each of the following points. Check by drawing the graph.

2. $G(8, 4)$ **3.** $H(4, 8)$ **4.** $I(4, 9)$ **5.** $J(6, -4)$

6. Line l passes through the origin $(0, 0)$ and $A(3, 6)$. Explain how to write the equation for line l.

7. If the slope of a line through the origin and through a given point is a, what is the equation of the line?

Practice & Apply

Draw each *pair* of graphs on the same axes. In each case, tell what is the same and what is different.

8. $y = 6x$; $y = -6x$

9. $y = 8x$; $y = -8x$

10. $y = \frac{1}{2}x$; $y = -\frac{1}{2}x$

11. $y = \frac{4}{3}x$; $y = -\frac{4}{3}x$

12. $y = 4x - 7$; $y = -4x - 7$

13. $y = 8x - 7$; $y = -8x - 7$

14. What effect does changing $y = ax$ to $y = -ax$ have on the line that is graphed?

15. Physical Science From the information shown, write an equation for the total mass, T, of a beaker containing m milliliters of water. Substitute values for m, and graph the equation.

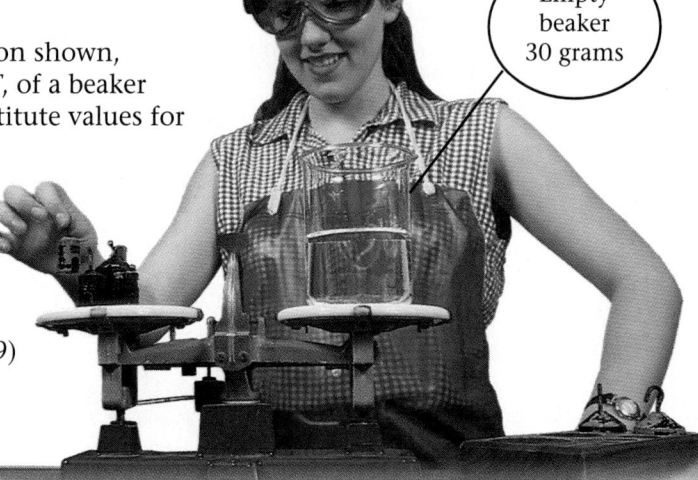

Empty beaker 30 grams

Find the equation of a line that passes through the origin and the following points.

16. (2, 5)

17. (5, 8)

18. (1, 9)

19. (4, 2)

20. (7, 3)

Draw each *pair* of graphs on the same axes. In each case, tell what is the same and what is different.

1 milliliter of water = 1 gram

21. $y = \frac{1}{4}x$ and $y = \frac{1}{4}x + 6$

22. $y = \frac{1}{4}x + 8$ and $y = \frac{1}{4}x - 2$

23. Describe the effect on the lines that are graphed if only the values of b change in the equation $y = ax + b$.

The normal water level of the river is 32 feet.

24. Ecology The water level of a river is rising 0.8 foot per day. Write an equation for the water level after d days. Graph the equation.

25. Suppose the water level of the river is 34 feet and is receding 0.5 foot per day. Write an equation for the water level after d days. Graph the equation. In how many days will the water level be 26 feet?

Guess what each line will look like when it is graphed. Then check your guess by graphing.

26. $y = -5x$ **27.** $y = -6x + 3$ **28.** $y = 7$ **29.** $y = -6$

30. $y = -5x - 1$ **31.** $y = -2$ **32.** $y = -x - 3$ **33.** $y = 2x + 3$

34. $y = 3x + 7$ **35.** $y = 7 - x$ **36.** $y = -3 + \frac{1}{2}x$ **37.** $y = x + 4$

38. **Portfolio Activity** Complete the problem in the portfolio activity on page 207.

Look Back

39. If Calvin earns $8 an hour, find his earnings for 33 hours. Write an equation and find the solution. **[Lesson 4.1]**

40. Dan earns 15% of the price of each $20 box of candles he sells. How many boxes will he have to sell to earn at least $100? **[Lesson 4.5]**

41. Solve the equation $3x + 4y = 24$ for y. **[Lesson 4.6]**

42. Suppose that in the orange carton example on page 220 the total weight is 51 pounds. How many oranges are in the carton? **[Lesson 4.6]**

Is the slope of the line negative, positive, or neither? **[Lesson 5.1]**

43.

44.

45.

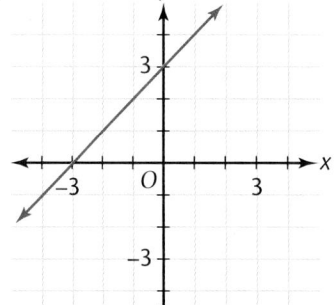

46. **Cultural Connection: Africa** The Ahmes papyrus that describes the mathematics of building the ancient pyramids of Egypt refers to a *seked*. Consider a pyramid with a base of 360 cubits and a height of 250 cubits. Scribes took half the base as the run and the height as the rise. They calculated the *seked* to be $\frac{180}{250}$. Compare the calculation of a *seked* with the calculation of the slope for the same pyramid. Explain the relationship between *seked* and slope. **[Lesson 5.1]**

Look Beyond

47. If you graph the lines $y = 0.5x + 3$ and $y = \frac{3}{6}x + 4$, will the graphs of the lines ever cross?

The Slope-Intercept Formula

why *Once you understand how to interpret the components of a linear function y = mx + b, you can sketch the graph. When you can identify the slope and the y-intercept of a line, you can write its equation.*

Skaters practice several hours a week. Kim pays $50 for club fees and $3 an hour to practice.

Kim can use a graph to see the pattern of her skating expenses. How would you draw this graph?

Examine the data. Let *x* be the number of hours Kim practices. Let *y* be the total cost of practice time. Calculate the first differences to determine the slope from a table of *x* and *y* values.

Skating Expenses

Hours	x	0	1	2	3	4	5	6	7	8	9	10
Cost	y	50	53	56	59	62	65	68	71	74	77	80

first differences → $3 $3 $3 $3 $3 $3 $3 $3 $3 $3

Since the first differences are 3, the table represents a linear function in the form $y = mx + b$. How can you find the slope of the line that represents the function? The table shows that the change in the *y*-variable is 3. Since the *x*-values are consecutive, the change in the *x*-variable is 1.

$$\text{slope} = m = \frac{\text{change in } y}{\text{change in } x} = \frac{3}{1} = 3$$

The slope indicates that Kim pays an additional $3 for every hour she practices.

The table also shows that when x is 0, y is 50. Thus, (0, 50) is on the graph. The y-value, 50, of the point where the line crosses the y-axis, is the **y-intercept**. It means that Kim pays a $50 fixed fee in addition to the hourly rate.

Recall from Lesson 5.2 that the graph of an equation in the form $y = mx + b$ has slope m and crosses the y-axis at b. If you substitute 3 for m and 50 for b, the equation for Kim's practice costs becomes $y = 3x + 50$.

SLOPE-INTERCEPT FORM

The slope-intercept formula or form for a line with slope **m** and y-intercept **b** is **$y = mx + b$**.

EXAMPLE 1

The equation of the line that represents Kim's skating costs is $y = 3x + 50$. How can you use the slope-intercept formula to construct the graph?

Solution ➤

The formula is $y = \boldsymbol{m}x + \boldsymbol{b}$. Since $\boldsymbol{b}$ is 50, measure 50 units up from the origin (0, 0) on the y-axis and graph a point.

From that point, measure the run and then the rise of the slope to locate a second point.

The slope, $\boldsymbol{m}$, is 3. You can use any equivalent of $\frac{3}{1}$ to measure the run and rise. The axes are in intervals of 10 units, so use $\frac{30}{10}$. Move right 10, then up 30.
Graph the second point. Draw a line through the points. ❖

CRITICAL
Thinking

Why do you think the line in Kim's graph does not extend past the y-axis? How far do you think the line would realistically extend in the positive direction?

From Two Points to an Equation

When you know 2 points on a line, you can write the equation for that line. First, calculate the slope, m, using the slope formula. Then calculate b from the slope-intercept formula and one of the points.

> **SLOPE FORMULA**
>
> Given two points with coordinates (x_1, y_1) and (x_2, y_2), the formula for the slope is
>
> $$m = \frac{\text{change in } y}{\text{change in } x} = \frac{\text{difference in } y}{\text{difference in } x} = \frac{y_2 - y_1}{x_2 - x_1}.$$

EXAMPLE 2

Sports   Kim's notes show that after 7 hours of practice the total cost for her skating is $71. Earlier in her notes it shows that after 5 hours of practice the total cost was $65. How can you write an equation for a line knowing only this information?

Solution ➤

You can represent the data as points by writing the hours and cost as ordered pairs, (5, 65) and (7, 71). To write the equation of a line from these two points, substitute the values into the slope formula.

$$m = \frac{\text{difference in } y}{\text{difference in } x}$$

$$m = \frac{71 - 65}{7 - 5} = \frac{6}{2} = 3$$

Substitute 3 for *m* in $y = mx + b$.

$$y = 3x + b$$

Next, choose either point, and substitute the coordinates for *x* and *y* into the equation. If you use the point (5, 65), substitute 5 for *x* and 65 for *y*. Then, solve for *b*.

$$y = 3x + b$$
$$65 = 3(5) + b$$
$$65 = 15 + b$$
$$50 = b$$

Now substitute 3 for *m* and 50 for *b* in $y = mx + b$. The equation for the line is $y = 3x + 50$. ❖

Why is the equation in Example 2 the same as the equation in Example 1?

Try This   Write an equation for a line passing through points (3, 3) and (5, 7).

EXAMPLE 3

Auto Racing   For decades the Indianapolis 500 automobile race has attracted the attention of millions of enthusiasts. Since 1911 the average speed for the race has increased from 74.59 miles per hour to 185.987 in 1990. How steadily has this average increased? How can you find an equation for the line of best fit that shows the trend in average speed? Use the data given for a sample of speed averages from 1915 to 1975 in 5-year intervals. Consider 1900 as year 0.

Year	Average Speed in MPH	Year	Average Speed in MPH	Year	Average Speed in MPH	Year	Average Speed in MPH
1915	89.84	1935	106.240	1955	128.209	1975	149.213
1920	88.62	1940	114.277	1960	138.767	1980	?
1925	101.13	1945	none	1965	150.686	1985	?
1930	100.448	1950	124.002	1970	155.749	1990	?

Solution ▷

One way to find the equation for the line of best fit is to plot the points on a graph and estimate the location of the line using a clear ruler.

Find the slope and y-intercept. Then substitute the numbers into the formula for a line $y = mx + b$.

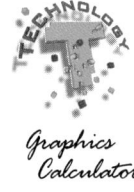

Graphics Calculator

The developments in calculator technology have made graphing a scatter plot much easier. On a graphics calculator, the line of best fit is called the **regression line**. The calculator will automatically calculate the slope, the y-intercept, and the equation, based on the line of best fit.

L1	L2	L3
15.00	89.84	-------
20.00	88.62	
25.00	101.13	
30.00	100.45	
35.00	106.24	
40.00	114.28	
50.00	124.00	

L2={89.84,88.62...

```
LinReg
 y=ax+b
 a=1.16
 b=68.24
 r=.98
```

*Select the **Stat** key of your calculator. Place the information from the table into a 2-variable (x, y) data table.*

*Locate and select the **Reg** or **LinReg** for linear regression. Here, the slope for the line of best fit is a, and the y-intercept is b. The r is the correlation coefficient.*

*The calculator will graph the scatter plot and draw the line of best fit when you enter the equation for the regression line after **Y =**.*

The rate of increase in average speed is the slope, about 1.16 miles per hour per year. The line of best fit (regression line) has the approximate equation $y = 1.16x + 68.24$. ❖

Estimate the average speeds that would account for the missing data in 1980, 1985, 1990, and 1995. Compare your estimates with the actual speeds achieved in those years. What might explain the variation?

Direct Variation

Springs are used in science to measure mass. Some springs are thin, and you can stretch them easily. Others are much stiffer, and you cannot seem to stretch them at all. The relationship between the amount of stretch and the load on the spring is defined by a law of physics called Hook's law.

Hook's law is an example of direct variation. It states that the distance a spring stretches varies directly as the force applied.

EXAMPLE 4

Jill works as a cable technician and charges by the hour. Her records show the hours and pay she receives for various jobs.

Hours on the job	3	5	7	10	15
Pay for the job	$28.80	$48	$67.20	$96	$144

Write an equation that will calculate the hourly wage that Jill earns.

Solution ➤

In looking at Jill's records, as the number of hours she works increases, the pay she earns also increases. This is an example of *direct variation*. The pay that Jill earns varies directly as the number of hours she works. The hourly wage that Jill earns is determined by

$$\frac{\text{Pay for the job}}{\text{Hours on the job}} = \text{Hourly wage}, \quad \text{so } \frac{48}{5} = \$9.60.$$

The hourly wage, $9.60, is a constant.

Let p = the pay for the job. Let h = the hours on the job.

$$\frac{p}{h} = 9.6$$

If you solve for p, you find that $p = 9.6h$. ❖

The equation $\frac{p}{h} = 9.6$ models direct variation. The expression $\frac{p}{h}$ is a ratio, and 9.6 is called the constant of variation.

DIRECT VARIATION

If y varies directly as x, then $\frac{y}{x} = k$ or $y = kx$. The k is called the **constant of variation.**

The direct variation equation written in the form $y = kx$ is a linear function. To find the constant of variation, find the ratio $\frac{y}{x}$. How does this compare with finding the slope of a line?

EXAMPLE 5

Physics   A force of 5 pounds stretches a spring a distance of 17.5 inches. How far will a spring stretch if an 8-pound weight is applied?

Solution ➤

The distance of stretch varies directly as the force applied.

Let d = the distance the spring stretches.

Let f = the force of the weight applied.

17.5 in.

$\frac{d}{f} = k$	From the definition of direct variation, identify the value of k, the constant of variation.
$\frac{17.5}{5} = k$	The ratio of 17.5 inches to a 5-pound weight simplifies to the constant $k = 3.5$. The constant k is the constant of variation.
$\frac{d}{f} = 3.5$	Once you know the constant of variation, you can find the distance of a given stretch, d.

$d = 3.5f$	Solve for d to form a linear function in terms of f.
$d = 3.5(8)$	Substitute 8 for the weight.

$d = 3.5(8)$	The distance the spring stretches is 28 inches. ❖
$\quad = 28$ inches	

If you graph a direct variation equation, you will find that direct variation represents a linear function in the form $y = mx + b$. The y-intercept is 0 and the slope is k, the constant of variation.

EXERCISES & PROBLEMS

Communicate

1. Explain what effect a change in b represents on the graph of the equation $y = mx + b$.

2. Explain what effect a change in m represents on the graph of the equation $y = mx + b$.

3. Describe how to find the slope, m, of a line passing through the points $(3, 5)$ and $(7, 2)$.

4. How do you determine the y-intercept for the equation $y = 5x - 1$?

5. How do you determine the slope for the equation $y = 10 - x$?

6. Tell how to graph the equation $y = 2x - 3$ without plotting points.

7. Explain how to write the equation of a line passing through points $A(-2, 4)$ and $B(1, 5)$.

Practice & Apply

Make a table from the given data points, and find the first differences of the x- and y-values.

$(0, 4), (2, 10), (4, 16), (6, 22), (8, 28), (10, 34)$

8. According to the differences, what is the slope?

9. At what point does the line cross the y-axis? How do you know?

10. Write the equation, and graph the line.

Give the coordinates where the line for each equation crosses the y-axis.

11. $y = 4x + 5$ 12. $y = 8x - 1$

13. $y = -3x + 7$ 14. $y = -5x - 9$

Examine the lines on the graph.

15. Which line has a y-intercept of -2?

16. Which line has a positive slope?

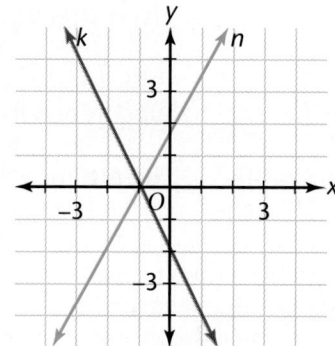

Find the slope of the lines passing through the following points.

17. $(0, 6), (5, 0)$ **18.** $(3, 4), (-1, -2)$ **19.** $(1, 7), (5, 3)$

20. $(7, 2), (-4, -2)$ **21.** $(-5, 0), (8, -4)$ **22.** $(7, -7), (-4, -3)$

23. $(-4, -3), (-2, -6)$ **24.** $(6, 6), (-2, -2)$ **25.** $(-1, 1), (5, -7)$

Write the equation for a line

26. with slope -1 and with y-intercept 0.   **27.** through $(0, -4)$ and with slope -4.

28. with slope 11 and with y-intercept 15.   **29.** with slope -5 and y-intercept 7.

30. through $(0, 5)$ and with slope 1.   **31.** with slope -3 and y-intercept -1.

32. through $(0,3)$ and with slope 3.   **33.** with slope $\frac{2}{3}$ and y-intercept 2.

Write the equation for a line passing through the following points.

34. $(7, 9), (3, -2)$ **35.** $(-1, 0), (0, 3)$ **36.** $(4, 5), (-1, -2)$

37. $(3, 3), (-2, -6)$ **38.** $(-8, -3), (6, -2)$ **39.** $(-1, -4), (-3, -5)$

40. **Consumer Mathematics** A catalog company charges a fixed charge of $2 per order plus $0.50 per pound for shipping. Write an equation for the total shipping cost, c, in terms of the number of pounds, p. What is the form of your equation?

Write equations for the graphs of the following lines.

41.

42.

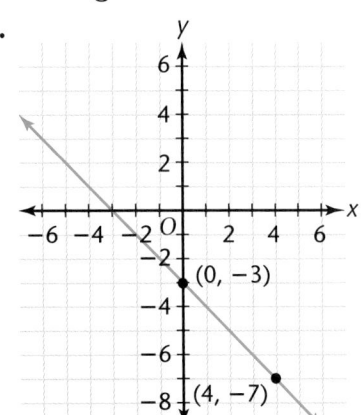

Josh sold 20 animal figures. The cost of materials was $43.80.

43. Write an equation for the amount of money Josh will have after each sale.

44. How many figures must Josh sell before the cost of materials is less than the income from sales?

45. Graph the equation for Josh's business.

46. The distance a car travels at a constant rate of 65 miles per hour varies directly as the time in hours. Write an equation to express this direct variation. Then find the distance if the time elapsed is 3 hours.

$ 6.50 each

47. Physical Science Find a sturdy rubber band or spring and suspend it so that you can measure its length. Measure the length without any weight on it. Find several weights that the rubber band or spring will hold without breaking. Hang the weights on the rubber band or spring and measure the length of the stretch. Make a table of the weights and the stretch lengths of the rubber band or spring. Make a scatter plot and find the line of best fit. Use a graphics calculator if possible to find the equation of the regression line. Write a description of your findings.

48. Geometry If the base area of a container is kept constant, the volume of the container varies directly as the height. The table shows the volumes and heights. Find the constant of variation. Then find the volume if the height is 12 centimeters.

V	27	36	45	54
h	3	4	5	6

Look Back

49. Technology The population of Cooperstown is 12,345 and is growing by 678 people per year. Make a table or use the repeating operations feature of a calculator to determine when the population will reach 20,000. **[Lesson 1.3]**

50. Write an equation for the population, p, in terms of y, the number of years the population increases at this rate. **[Lesson 1.3]**

51. Technology Plot a graph (try using a graphics calculator) to find when the population will reach 20,000. **[Lesson 1.3]**

Perform the indicated operations.

52. $(3a - 2) + (2a - 2)$ **[Lesson 3.3]**   **53.** $(4x + 3y) - (3x - 2y)$ **[Lesson 3.4]**

Solve each inequality.   **[Lesson 3.6]**

54. $x + 7 \leq -2$ **55.** $x - 4 \geq 6$ **56.** $x - 5 < 21$

Look Beyond

57. If a line is horizontal, what is the rise when you compute the slope?

58. If a line is vertical, what is the run when you compute the slope? What does this indicate about the slope?

LESSON 5.4

Other Forms for Equations of Lines

Wynton Marsalis studied at the Julliard School of Music and played trumpet with the New Orleans Philharmonic Orchestra before gaining fame as a jazz musician. He was born in New Orleans on October 16, 1961.

why *The equation of a line can be written in different forms. If you know which form to use, you can often save yourself time and effort.*

You can write the birth date of Wynton Marsalis many ways. Each way has its advantages. The long form is easy to read. The short form is quick and easy to write. The computer form is easy to sort by year, month, and day.

October 16, 1961	10-16-61	611016
long form	**short form**	**computer form**

You have been using the slope-intercept form of a line, $y = mx + b$, to solve problems. As with the dates, there are other forms for the equation of a line. Each form has its advantages. The standard form provides a simple method for graphing the equation of a line.

Standard Form

Consider an equation written in the form $Ax + By = C$. After values for A, B, and C are determined, a solution to this equation is any ordered pair of numbers (x, y) that makes the equation true. When these ordered pairs are graphed, they form a straight line.

STANDARD FORM

An equation in the form $Ax + By = C$ is in standard form when
- A, B, and C are integers,
- A and B are not both zero, and
- A is not negative.

EXAMPLE 1

Fund-raising Jackie is in charge of selling tickets for the school jazz concert. She hopes the total ticket sales will be about $588. This will cover expenses and make a modest profit.

A Write an equation in standard form that models this situation.

B Graph the equation.

Solution ➤

A The equation must show that $4 times the number of adult tickets added to $2 times the number of student tickets is $588.

Let x represent the number of adult tickets sold. Let y represent the number of student tickets sold. The equation that models the problem is $4x + 2y = 588$. Notice that the equation is in standard form $Ax + By = C$.

B Graph the adult tickets, x, on the horizontal axis and student tickets, y, on the vertical axis. The points (x, y) represent (adult tickets, student tickets).

To graph an equation in standard form, substitute 0 for x, and solve for y. Then substitute 0 for y, and solve for x. This gives the intercepts $(0, y)$ and $(x, 0)$, which are the points where the line crosses each axis. Draw the line connecting these 2 points.

Start with the equation $4x + 2y = 588$.

First let $x = 0$.
$$4x + 2y = 588$$
$$0 + 2y = 588$$
$$y = 294$$

$(0, 294)$ is on the graph.

Then let $y = 0$.
$$4x + 2y = 588$$
$$4x + 0 = 588$$
$$x = 147$$

$(147, 0)$ is on the graph

Draw the line connecting the points. The coordinates of any point on the line will solve the equation. ❖

To solve the jazz concert ticket problem, however, the values for x and y must be whole numbers.

What does the point (97,100) represent with respect to the problem?

Adult tickets

EXAMPLE 2

Use a graphics calculator to graph the equation $4x + 2y = 588$.

Graphics Calculator

Solution ➤

To enter an equation in a graphics calculator, first rewrite the equation in slope-intercept form $y = mx + b$.

$$4x + 2y = 588$$
$$2y = 588 - 4x$$
$$y = 294 - 2x$$
$$y = -2x + 294$$

The equation can now be entered into the calculator.

Y = − 2X + 294 ❖

CRITICAL Thinking

If the equation of a line is written in standard form $Ax + By = C$, the slope is $-\frac{A}{B}$ and the y-intercept is $\frac{C}{B}$. Explain why.

Point-Slope Form

If you know the slope of a line and the coordinates of one of its points, you can write a third form of an equation for a line.

Remember that the value for the slope **m** can be calculated from any two points (x_1, y_1) *and* (x_2, y_2), and the formula $m = \frac{y_2 - y_1}{x_2 - x_1}$.

Multiply both sides of the slope formula by $x_2 - x_1$, and simplify.

If you replace the specific point (x_2, y_2) with a general point on the line, represented by (x, y), the result is $y - y_1 = m(x - x_1)$.

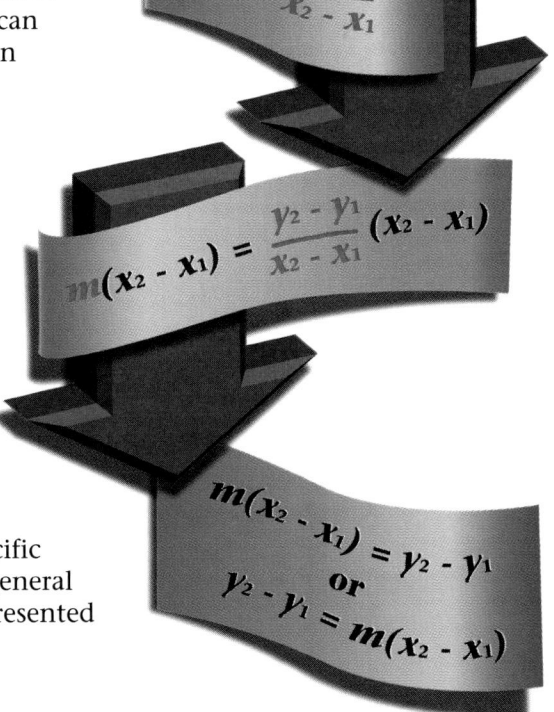

$$m = \frac{y_2 - y_1}{x_2 - x_1}$$

$$m(x_2 - x_1) = \frac{y_2 - y_1}{x_2 - x_1}(x_2 - x_1)$$

$$m(x_2 - x_1) = y_2 - y_1$$
or
$$y_2 - y_1 = m(x_2 - x_1)$$

POINT-SLOPE FORM

The form $y - y_1 = m(x - x_1)$ is the **point-slope form** for the equation of a line. The coordinates x_1 and y_1 are taken from a given point (x_1, y_1), and the slope is **m**.

EXAMPLE 3

A line with slope 3 passes through point (2, 7). Write the equation of the line in point-slope form.

Solution ➤

Let $m = 3$, $x_1 = 2$, and $y_1 = 7$. Substitute the given values into the point-slope equation.

$$y - y_1 = m(x - x_1)$$
$$y - 7 = 3(x - 2)$$ ❖

EXAMPLE 4

Janet's class is ordering T-shirts. Write the point-slope equation that models the information on the order form.

T-Shirts inc.
506 Front Street

T-Shirts $9.00 ea.
* Plus Shipping and Handling

COLOR	S	M	L	TOTAL
Red	2			2
Blue	1	4	5	10
White		1	3	4
TOTAL				16
TOTAL COST				$149.00

Solution ➤

Each time a shirt is ordered, the cost increases by $9, so 9 is the rate of change or slope. Since 16 shirts cost $149, (16, 149) represents a point on the graph. Substitute 9 for m, 16 for x_1, and 149 for y_1 into $y - y_1 = m(x - x_1)$. The result is $y - 149 = 9(x - 16)$. ❖

EXAMPLE 5

Change $y - 149 = 9(x - 16)$ to an equation

A in slope-intercept form.　　**B** in standard form.

Solution ➤

A Begin with $y - 149 = 9(x - 16)$　　Given
　　　　　　　　$y - 149 = 9x - 144$　　Distributive Property
　　　　　　　　　　　$y = 9x + 5$　　Addition Property of Equality

The equation is now in slope-intercept form. Note that the slope represents $9, the cost per shirt. The y-intercept represents $5, the shipping and handling charge.

B Next, change the slope-intercept form to standard form, $Ax + By = C$.

　　　　　　$y = 9x + 5$　　Given
　　　　$y - 9x = 5$　　Subtraction Property of Equality
　　　$9x - y = -5$　　Multiply each side by -1 and
　　　　　　　　　　　rearrange terms. ❖

EXAMPLE 6

Find an equation for the graph of a line that passes through the points $(-1, 10)$ and $(5, 8)$.

Solution ➤

The point-slope form simplifies the way this problem is solved. Let $(x_1, y_1) = (-1, 10)$ and $(x_2, y_2) = (5, 8)$. Find the slope.

$$\text{Slope} = \frac{y_2 - y_1}{x_2 - x_1} = \frac{8 - 10}{5 - (-1)} = \frac{-2}{6} = -\frac{1}{3}$$

Use the point $(-1, 10)$ for (x_1, y_1). Substitute the slope and the coordinates of the point into the equation $y - y_1 = m(x - x_1)$. Then $y - 10 = -\frac{1}{3}(x - (-1))$ or $y - 10 = -\frac{1}{3}(x + 1)$. ❖

Try This Use the point-slope form to find the equation of the line from the points $(5, 65)$ and $(7, 71)$.

SUMMARY OF THE FORMS OF A STRAIGHT LINE

Name	Form	Example
Slope-intercept	$y = mx + b$	$y = 3x + 5$
Standard	$Ax + By = C$	$3x - y = -5$
Point-slope	$y - y_1 = m(x - x_1)$	$y - 11 = 3(x - 2)$

EXERCISES & PROBLEMS

Communicate

1. Explain how to write the equation $5y - 2 = -3x$ in standard form.

2. Tell how you would find the intercepts for the equation $3x + 6y = 18$.

3. Describe how to graph $2x + 3y = 12$ by finding the intercepts.

4. Explain how to change $x - 3y = 9$ into slope-intercept form.

5. How would you use the point-slope formula to write the equation of a line through points $(-2, 4)$ and $(4, -8)$?

6. The equation of a given line is $5x + 2y = 40$. Tell how you would find the slope of the line.

Practice & Apply

Write the following equation in standard form.

7. $4x = -3y + 24$ **8.** $7y = -5x - 35$ **9.** $6x + 4y + 12 = 0$ **10.** $2x = 4y$

11. $6x - 8 = 2y + 6$ **12.** $x = \frac{2}{3}y + 6$ **13.** $2 + 7x + 14y = 3x - 10$ **14.** $5 = y - x$

Find the intercepts for the following equations.

15. $x + y = 10$ **16.** $3x - 2y = 12$ **17.** $5x + 4y = 20$ **18.** $x = 2y$

19. Draw a graph of a line that intercepts the axes at (3, 0) and (0, 7).

20. Graph the equation $2x + 6y = 18$ by finding the intercepts.

21. The equation of a given line is $6x + 2y = 40$. What is the slope of the line?

22. Compare the graphs of $4a + 2s = 588$ and $2a + s = 294$. What do you find?

23. Fund-raising   Suppose that in Example 1 on page 234, 50 is selected for y in the equation $4x + 2y = 588$. What is x? What is the maximum number of adult tickets you can buy for $588? What is the maximum number of student tickets you can buy for $588?

Write the equation of a line

24. through (5, 2) and with slope -8.

25. that crosses the x-axis at $x = 7$ and the y-axis at $y = 2$.

26. through (5, 2) and with slope 8.

27. through (5, 2) and with slope 0.

28. that crosses the x-axis at $x = 1$ and the y-axis at $y = \pi$.

29. through (2, 3) and (8, -3).

30. through (5, 9) and (10, 9).

31. Technology   On some graphing calculators, equations must be in function form, $y = f(x)$, to draw graphs. The symbol $f(x)$ indicates a function. Which of the three forms for the equation of a line given in the summary at the end of the lesson is/are in $y = f(x)$ form?

Rewrite each of the following in the function form $y = f(x)$, where the equation is solved for y.

32. $2x + 3y = 12$ **33.** $y - 2 = 3(x - 7)$ **34.** $2x + 2y + 5 = 0$

35. $3x + 4y = 6 + 5y$ **36.** $\frac{x}{2} + \frac{y}{3} = 18$ **37.** $\frac{5x - y}{7} = 14$

38. Write an equation that shows the value of n nickels and d dimes is $5. HINT: Be careful about the units.

Fund-raising Write an expression for the sales in dollars of

39. 40 adult tickets.

40. 20 student tickets and 37 adult tickets.

41. *s* student tickets and *a* adult tickets.

42. Write an equation that says the total sale of *s* student tickets and *a* adult tickets is $700. Which of the given forms did you choose for your equation?

Solve the equation in Exercise 42 when

43. 90 student tickets are sold.

44. 80 adult tickets are sold.

Social Studies Complete this table showing various forms of the same date.

	Long form	Short form	Computer form	Importance
45.	—	—	081001	1st Model T Ford
46.	January 31, 1958	—	—	1st U.S. satellite
47.	—	7-20-69	—	1st Moon walk
48.	Why do you think the computer form uses 07 for July instead of 7?			

 Look Back

Evaluate the following if *x* = 1, *y* = 1, and *z* = 2,   [Lesson 1.3]

49. $x^2 + y + z^2$ **50.** $x - y + z$ **51.** $x + y - z$ **52.** $-(x + y + z)$

53. Describe the scatter plot of a strong negative correlation.   **[Lesson 1.6]**

Solve the following equations.   [Lesson 4.3]

54. $-5y = 30$ **55.** $3x = 420$ **56.** $\frac{y}{9} = 36$ **57.** $\frac{x}{2} = 108$

Find the slope of the line containing the origin (0, 0) and the given point. Then give the equation of the line passing through the origin and the given point.   [Lesson 5.1]

58. $A(3, 6)$ **59.** $B(2, 8)$ **60.** $C(6, 3)$ **61.** $D(-5, -7)$

Look Beyond

62. If two lines intersect, how many points do they have in common?

63. If two lines are parallel, how many points do they have in common?

64. Michael has a 5-liter and a 3-liter bottle. Neither have markings. He has a supply of water. How can he measure 1 liter exactly?

Vertical and Horizontal Lines

Why *A line along the diving tower, seen from the side is horizontal. A line down the tower is vertical. The slopes and equations for such horizontal or vertical lines have a special significance in algebra.*

A pool offers two options. Residents can either pay $2 per visit or $25 for a season pass. How many times do you have to use the pool to make the season pass the better option?

Recreation The data points showing the number of visits to the pool, x, and the total cost, y, can be modeled for both options. The graphs show only the first quadrant because the number of visits and the amount of money spent will always be 0 or greater.

Per-visit option For 0 visits the cost is $0. Thus, the y-intercept is 0, and (0, 0) is one point on the graph. Since the rate of increase is $2 per visit, the slope is 2.

The first graph is a line drawn through **(0, 0)** with slope 2. The line represents a typical equation in slope-intercept form, $y = 2x + 0$. You can check this by seeing that it contains the point **(10, 20)**. Ten visits cost $20.

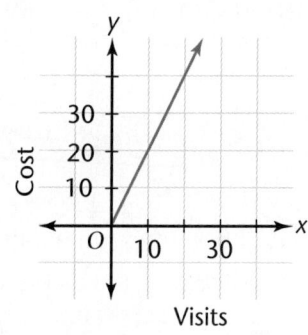

Season-pass option The cost is a constant $25. If you go to the pool twice, the cost is $25. Thus, **(2, 25)** is a point on the graph. If you go to the pool 40 times, the cost is still $25. The point **(40, 25)** is on the graph.

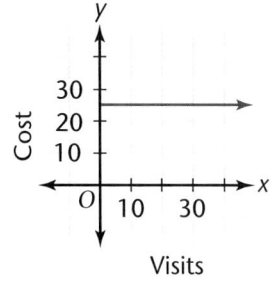

The second graph is a horizontal line consisting of all y-values equal to 25. The line drawn through the two points is a special equation in slope-intercept form, $y = 0x + 25$ or $y = 25$.

Now draw both graphs on one coordinate plane.

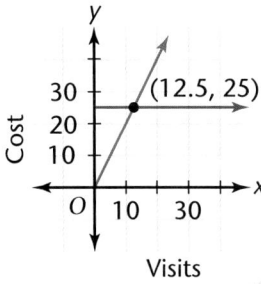

The graph shows that for $12\frac{1}{2}$ visits the cost is $25. Since you cannot make $\frac{1}{2}$ visit, the season-pass option is better for 13 or more visits.

Notice that the graph of $y = 25$ is a horizontal line. The graph of $y = b$ is a horizontal line that crosses the y-axis at the point $(0, b)$. The equation $y = b$ is a **constant function** where b is the **constant**. It can be written as $y = mx + b$ where the slope is zero. In standard form the equation is written $0x + 1y = b$.

EXAMPLE 1

Compare the graphs of $y = 4$ and $x = 4$.

Solution ➤

The graph $y = 4$ consists of all points with a y-coordinate of 4. The graph $x = 4$ consists of all points with an x-coordinate of 4.

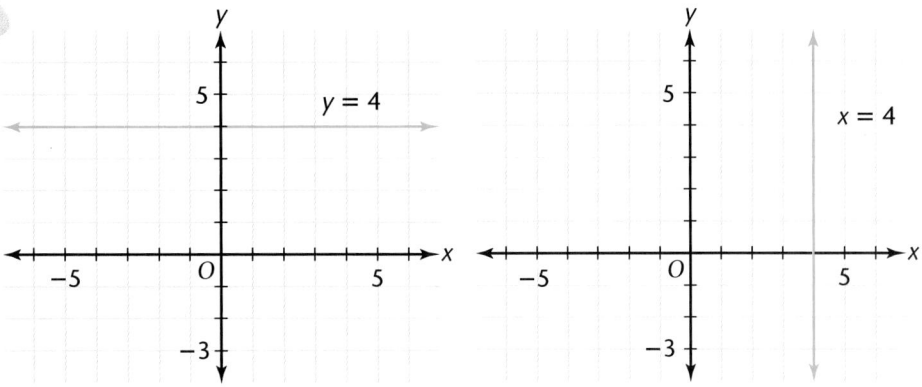

The equation y = 4 is a horizontal line 4 units above the origin.

The equation x = 4 is a vertical line 4 units to the right of the origin. ❖

Try This   Compare the graphs of $y = -3$ and $x = -3$.

EXAMPLE 2

Find the slope of **(A)** $y = 5$. **(B)** $x = -3$.

Solution ➤

(A) The graph of $y = 5$ has the same value of y for any value of x. Use any two points with a y-coordinate of 5 to find the slope. The points $(3, 5)$ and $(7, 5)$ will work.

$$\text{slope} = \frac{y_2 - y_1}{x_2 - x_1} = \frac{5 - 5}{7 - 3} = \frac{0}{4} = 0$$

(B) The graph of $x = -3$ has the same value of x for any value of y. Use any two points with an x-coordinate of -3 to find the slope. The points $(-3, 3)$ and $(-3, 7)$ will work.

$$\text{slope} = \frac{y_2 - y_1}{x_2 - x_1} = \frac{7 - 3}{(-3) - (-3)} = \frac{4}{0}$$

The run is 0. Since dividing by 0 is impossible, the *slope* is *undefined*. ❖

The graph of an equation in the form $x = a$ is a vertical line that crosses the x-axis where x equals a. The equation cannot be written in slope-intercept form, but it can be written in standard form.

$$1x + 0y = a$$

HORIZONTAL AND VERTICAL LINES

The equation for a horizontal line is written in the form $y = b$.
The slope is 0.

The equation for a vertical line is written in the form $x = a$.
The slope is undefined.

EXAMPLE 3

Compare the graphs of $y = 3x$ and $x = 3y$.

Solution ➤

Write the equations in slope-intercept form. Graph the lines.

The equation $y = 3x$ is in slope-intercept form, and its graph is a line with slope 3 and y-intercept 0.

Change $x = 3y$ to slope-intercept form by solving for y. Since $x = 3y$ is equivalent to $3y = x$,

$$3y = x \rightarrow y = \frac{x}{3} \rightarrow y = \frac{1}{3}x + 0.$$

The equation $x = 3y$ is written $y = \frac{1}{3}x + 0$ in slope-intercept form. Its graph is a line with slope $\frac{1}{3}$ and y-intercept 0.

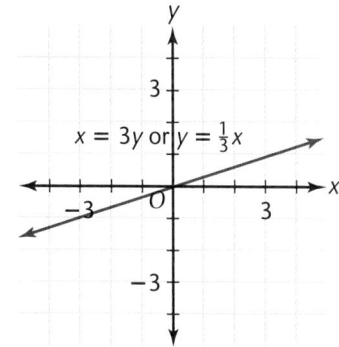

Recall that numbers are reciprocals when their product equals 1. Since $3 \cdot \frac{1}{3} = 1$, the slopes of $y = 3x$ and $y = \frac{1}{3}x$ are reciprocals.

The graphs of the original equations $y = 3x$ and $x = 3y$ are lines with reciprocal slopes and a common point at the origin, (0, 0). ❖

Draw the graph of $y = x$. On the same set of axes, draw the graphs of $y = 2x$ and $x = 2y$. Try other pairs of lines in the form $y = mx$ and $x = my$ with different m values. What relationships do you see?

Communicate ~~~~

1. Describe how to find the slope of $y = 3$. Is the line vertical or horizontal?

2. Explain how to find the slope of $x = 1$. Is the line vertical or horizontal?

3. Discuss how to find the slope of $y = 5x$. Choose your own points to graph the equation. At what point does the line cross the y-axis?

4. Tell how to find the slope of $x = 5y$. Choose your own points to graph the equation. At what point does the line cross the y-axis?

5. Explain why the graph of $y = 15$ is a constant function.

6. Tell why the slope for a vertical line is unusual.

7. Tell why the slope for a horizontal line is 0.

Practice & Apply

Find the slope for each equation. Plot points, graph a line for each equation, and indicate if the line is vertical or horizontal.

8. $y = 5$ **9.** $x = 7$ **10.** $y = 9x$ **11.** $x = 5y$ **12.** $y = \dfrac{1}{3}x$

13. Language Arts Where do you think the word *horizontal* comes from?

Match each equation with the appropriate description.

14. $x + y = 9$ **A.** a horizontal line 9 units above the origin

15. $xy = 9$ **B.** a vertical line 9 units to the right of the origin

16. $x = 9$ **C.** a line through the origin with slope 9

17. $y = 9$ **D.** a line through the origin with slope $\dfrac{1}{9}$

18. $y = 9x$ **E.** a line with slope -1 and y-intercept 9

19. $x = 9y$ **F.** something other than a straight line

Coordinate Geometry The equation $x = 4$ cannot be written in slope-intercept form because the slope is undefined. It can, however, be written in standard form as $1x + 0y = 4$. Copy and complete the table. Write an equivalent form of each equation, writing "undefined slope" when appropriate.

	Given	Slope-intercept Form	Standard Form
20.	$x = 1$	?	?
21.	$y = 4$	?	?
22.	$x + y = 5$	?	?
23.	$y = 4x$	?	?
24.	$x = 4y$	?	?

Determine whether the following lines are vertical or horizontal.

25. $y = 8$ **26.** $x = -2$ **27.** $y = -4$ **28.** $x = 9$

29. Technology Find how to graph a vertical line on a graphics calculator.

30. Find the slope for the equations $y = 6x$ and $x = 6y$. What point do these lines have in common?

31. Write the reciprocal of 0.25.

32. Write the reciprocal of $-4\dfrac{1}{2}$.

33. Budget Saul's father is thinking of buying his son a six-month movie pass for $40. If matinees are $3.00 each, how many times must Saul attend before it would benefit his father to buy the pass?

In the swimming pool example from the beginning of the lesson, which option is cheaper if you visit the pool

34. 8 times?   **35.** 29 times?

36. Recreation Repeat the swimming pool problem, but change the pool costs to those shown at the right. After how many visits does it pay to buy a season pass?

37. Technically, should the graphs of the problem consist of points or lines? For example, does it make sense to visit the pool $4\frac{2}{3}$ times? For ease in making and reading graphs, often lines are used instead of points. What is another example where this might happen?

~~~~ Look Back

Use the order of operations to simplify the following expressions. [Lesson 1.4]

38. $30 \cdot 2 \div 1 + 10$ **39.** $[2(105 - 5)] \div 2$ **40.** $\dfrac{(3 + 9)^2 - 4}{2^2 \cdot 15 - 10}$

Simplify.   [Lesson 3.3]

41. $(6a + 3) + (2a + 6)$ **42.** $(x + y + z) + (3x + y + 4z)$ **43.** $-7 + 3p - 9 - 7p$

44. Mary needs $1\frac{1}{3}$ yards of fabric to make a skirt. A yard of fabric costs $4.58. How many skirts can she make from 5 yards of fabric? **[Lesson 4.3]**

Lunch cost $15.50. There is a 7% sales tax, and the customary tip is 15%.   [Lesson 4.5]

45. How much is the tax on the lunch?

46. How much is the tip?

47. Given points $(3, 5)$ and $(-4, -2)$, write the slope-intercept equation for the line.   **[Lesson 5.3]**

Look Beyond ~~~~

48. **Geometry** What kind of angle is formed when a horizontal and a vertical line intersect?

49. Will two distinct vertical lines ever intersect?

50. Will two distinct horizontal lines ever intersect?

Parallel and Perpendicular Lines

Why *The graphs of parallel and perpendicular lines are related by their slopes. Algebra can help you solve problems in geometry.*

Parallel lines often look like they meet, but they never do.

Physical Science Examine the three different thermometer readings. It is possible to derive a linear equation for changing degrees in one scale to degrees in another. Notice how slope is used in each solution.

EXAMPLE 1

Use the data in the table to find the formula that changes

Ⓐ degrees Celsius (°C) to degrees Fahrenheit (°F).

Ⓑ kelvins to degrees Fahrenheit.

Reference	°F	°C	K
Water boils	212	100	373
Water freezes	32	0	273
Absolute 0	−460	−273	0

Solution ➤

Ⓐ The relationship between any two of the temperature scales is linear. Compare the Fahrenheit and Celsius scales. In the equation $y = mx + b$, let the y-values be Fahrenheit and the x-values be Celsius. Use the data from the table to calculate the slope and y-intercept.

$$\text{slope} = \frac{\text{change in Fahrenheit}}{\text{change in Celsius}} \quad \rightarrow \quad m = \frac{212 - 32}{100 - 0} = \frac{180}{100} = \frac{9}{5}$$

Notice that when C is 0, F is 32. The y-intercept, b, is 32. Find the equation that changes Celsius to Fahrenheit by substituting $\frac{9}{5}$ for m and 32 for b in $y = mx + b$. The equation $y = \frac{9}{5}x + 32$ gives the formula $F = \frac{9}{5}C + 32$.

Ⓑ You can change from kelvins to degrees Fahrenheit in much the same way. The equation for this change is $y = \frac{9}{5}x - 460$. The formula is $F = \frac{9}{5}K - 460$. ❖

Parallel Lines

Graphics Calculator

Graph the equations $F = \frac{9}{5}C + 32$ and $F = \frac{9}{5}K - 460$. Let y represent degrees in Fahrenheit and x represent either degrees in Celsius or kelvins. On a graphics calculator, enter the Celsius formula as the function **Y = (9/5)X + 32**. Then enter the Kelvin formula as the function **Y = (9/5)X − 460**. Place both graphs on the same axes.

GEOMETRY
Connection

If you use graph paper, use the y-intercept to locate the first point for one equation. Then use the run and rise of the slope for the second point. Once you have determined 2 or 3 points for the equation, use a straightedge to connect the points. Repeat the process for the second equation.

How are the slopes of the two lines related?

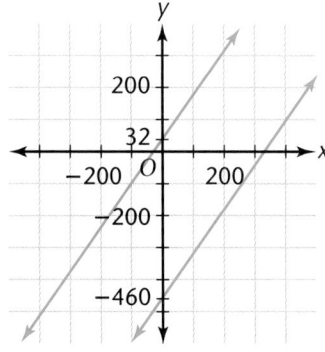

> **PARALLEL LINES**
> If two different lines have the same slope, the lines are **parallel**.
>
> If two non-vertical lines are parallel, they have the same slope. Two parallel, vertical lines have undefined slopes.

Perpendicular Lines

GEOMETRY
Connection

In geometry, **perpendicular lines** form right angles. In the next exploration, you can discover the algebraic relationship between two lines that are perpendicular. The symbol for perpendicular is ⊥.

$AB \perp PQ$

$SR \perp RT$

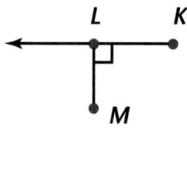

$LM \perp KL$

Exploration 1 *Slopes and Perpendicular Lines*

On a piece of graph paper, draw a line that has a run of 5 squares and a rise of 3 squares, as shown on the grid. Make a copy on clear plastic or tracing paper. Rotate the copy of the graph 90 degrees clockwise, as shown. When one graph is placed on the other, you can see the relationship of the two lines.

The lines are **perpendicular,** because the line on the second graph was rotated 90 degrees. Where the lines on the two graphs intersect, they will form right angles.

 What is the slope of the original line?

 Find the slope of the rotated line. Count the squares to find the rise over the run. Notice, for example, that the new line shows a run of 3. Consider the direction of the line.

 What is the sign for the slope of the perpendicular line?

 What is the relationship of the slope of the perpendicular line to the original line?

 What happens when this exploration is repeated with a negative slope? ❖

Exploration 2 *Perpendicular Slope on a Calculator*

Graphics Calculator

You can use a graphics calculator to find the equation of a line perpendicular to a given line. Be sure to select a *square window* when you set the range on your graphics calculator. Then the right angles will have the correct appearance on the screen.

1 Enter the graph for the line **Y = 2X**.

2 On the same axes, graph new lines in the form $y = mx$ by trying different values for m until the new line looks perpendicular to the original line. Keep a record of your guesses.

3 Decide on your best guess for the slope. Write an equation for a line perpendicular to $y = 2x$ from the slope you find. ❖

On the same set of axes graph $y = 2x$ and the line perpendicular to $y = 2x$ that passes through (0, 0). Find the slope of each line. Compare the signs. Compare the absolute values. Multiply the slopes. What is the result?

 CRITICAL *Thinking*

Suppose a given line has positive slope m. Make a conjecture about the product of m and the slope of a line perpendicular to the given line.

PERPENDICULAR LINES

If the slopes of two lines are m and $-\dfrac{1}{m}$, the lines are perpendicular.

If the slope of a line is m, then the slope of a line perpendicular to it is $-\dfrac{1}{m}$.

EXAMPLE 2

Find an equation for the line that contains point (4, 5) and is

Ⓐ parallel to the line $2x + 3y = 7$.

Ⓑ perpendicular to the line $2x + 3y = 7$.

Solution ➤

Ⓐ First, write the equation of the given line in slope-intercept form.

$$2x + 3y = 7$$
$$3y = -2x + 7$$
$$y = -\frac{2}{3}x + \frac{7}{3}$$

The slope of the line is $-\dfrac{2}{3}$. Any line parallel to the original line must also have slope $-\dfrac{2}{3}$. The coordinates of the given point are (4, 5), and the slope is $-\dfrac{2}{3}$. Now use the *point-slope form* to write an equation for the parallel line through the given point. Substitute $-\dfrac{2}{3}$ for m and (4, 5) for (x, y) into the point-slope form of the equation, $y - y_1 = m(x - x_1)$.

$$y - 5 = -\frac{2}{3}(x - 4)$$

B When a line has slope $-\frac{2}{3}$, any line perpendicular to that line has slope $\frac{3}{2}$. Since the line also contains the point (4, 5), you can substitute the coordinates into $y - y_1 = m(x - x_1)$.

$$y - 5 = \frac{3}{2}(x - 4)$$

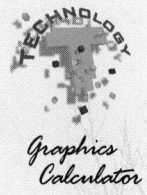

Graphics Calculator

As a check, change the equations $y - 5 = -\frac{2}{3}(x - 4)$ and $y - 5 = \frac{3}{2}(x - 4)$ to slope-intercept form. Graph the lines using a square window. The lines will appear perpendicular. Both lines will also contain the point (4, 5) where they intersect. ❖

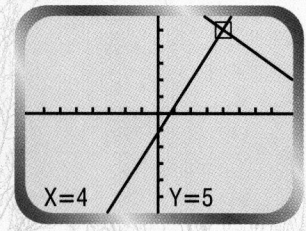

X=4   Y=5

Try This Find an equation for a line containing the point (4, 1) and perpendicular to $y = 2x - 7$.

GEOMETRY
Connection

If line l has slope $\frac{2}{5}$, and line m has slope $-\frac{5}{2}$, then $l \perp m$.

If $r \perp s$ and r has slope $\frac{2}{5}$, then s has slope $-\frac{5}{2}$.

EXERCISES & PROBLEMS

Communicate

1. Explain how to write an equation for a line parallel to $y = 4x + 3$.

2. The slope of a line is $\frac{3}{2}$. Explain how to find the slope of a line that is perpendicular to that line.

3. How would you find the slope of a line perpendicular to the line $y = \frac{1}{3}x + 2$?

4. Describe how to find an equation for a line perpendicular to $y = 4x + 3$.

5. Explain how to write the standard form of an equation for a line with y-intercept -4 that is parallel to a line with slope 3.

6. Tell how to write an equation in standard form for a line perpendicular to the line with $m = -6$ and $b = 12$.

7. Discuss how to write the equation for a line that passes through the point (0, 0) and is perpendicular to $x - 5y = 15$.

Practice & Apply

Graph the line $y = 5x$.

8. Sketch a graph that is parallel to $y = 5x$. Write its equation.

9. Sketch a graph that is perpendicular to $y = 5x$. Write its equation.

10. What relationship is there between the graph of $y = 3x + 2$ and $y = -3x + 2$? Are these lines perpendicular?

What is the slope of a line

11. parallel to a horizontal line?　　**12.** perpendicular to a horizontal line?

13. parallel to a vertical line?　　**14.** perpendicular to a vertical line?

Write an equation for a line according to the instructions.

	Contains	Is parallel to		Contains	Is perpendicular to
15.	$(3, -5)$	$5x - 2y = 10$	**16.**	$(3, -5)$	$5x - 2y = 10$
17.	$(-2, 7)$	$y = 3x - 4$	**18.**	$(-2, 7)$	$y = 3x - 4$
19.	$(2, 4)$	$y = 7$	**20.**	$(2, 4)$	$y = 7$

Write the slope of a line that is parallel to the following lines.

21. $y = 3x + 14$　　**22.** $2x + y = 6$　　**23.** $8 = -4x + 2y$　　**24.** $-2x + \frac{1}{2}y = 16$

Write the slope of a line that is perpendicular to the following lines.

25. $y = -\frac{1}{3}x + 10$　　**26.** $-\frac{1}{2}x - y = 20$　　**27.** $13 = -x + y$　　**28.** $3x + 12y = 12$

Geometry Write equations of 4 lines that meet to form a square whose sides are

29. *parallel* to the axes.　　**30.** are *not parallel* to the axes.

Look Back

31. Place parentheses to make $2 \cdot 7 + 35 \div 7 - 10 = 2$ true.　**[Lesson 1.4]**

Evaluate the following.　**[Lesson 3.1]**

32. $-4 + (-3) + 1$　　**33.** $-2 + 3 + (-7) + 3$　　**34.** $-12 + 4 - (-4)$

Add the like terms.　**[Lesson 3.3]**

35. $2x^2 + 3y + 4y + 3x^2$　　**36.** $3x + 2 + 4y + 2 + 3y$　　**37.** $2x + 3xy + 5x^2 + 7xy$

Look Beyond

How many ordered pairs will satisfy both equations simultaneously, if two linear equations have graphs

38. that are parallel?　　**39.** that are perpendicular?

Diophantine
EQUATIONS

Donna sees only the legs of horses and the legs of the riders walking the horses. She counts 22 legs in all. How many horses and how many riders does she see?

This is a version of a classic algebra problem. The numbers of legs must be whole numbers, so the number of solutions is restricted.

Equations that require the condition that *the solutions must be integers* are called **Diophantine equations.** The equations are named for Diophantus, who lived about 2000 years ago in the city of Alexandria in Egypt.

Using a Table Solution

Find all the solutions to the horse-and-rider problem.

Let h be the number of horses and r be the number of riders. Since each horse has 4 legs and each rider has two legs, $4h + 2r = 22$. Notice that all the coefficients are even numbers, so you can divide both sides of $4h + 2r = 22$ by 2 to get $2h + r = 11$.

Make a table to find the solutions. Remember, all solutions must be whole numbers. The first solution can be found by letting $h = 0$. When you substitute, you will find that $r = 11$. To find other solutions, note that each time there is one more horse, there are 4 more legs. That means there are 2 fewer riders. The table shows the solutions.

Horses	0	1	2	3	4	5
Riders	11	9	7	5	3	1

It is impossible to have more than 5 horses, because there would then be more than 22 legs.

Using a Graph Solution

Find all of the whole-number solutions to the equation $4h + 2r = 22$.

If $r = 0$, the equation becomes $4h = 22$, which has no whole-number solution. If $r = 1$, the equation becomes $4h + 2 = 22$, or $4h = 20$, which does have a solution, $h = 5$. So $(1, 5)$ is one solution to the Diophantine equation. Now look at the Diophantine equation in slope-intercept form. Solve for h.

$4h + 2r = 22$ becomes $h = -\frac{2}{4} r + \frac{22}{4}$.

The slope of the equation in lowest terms is $-\frac{1}{2}$.

Start with $(1, 5)$. Use a run of 2 and a rise of -1 to find any other solutions. The second solution is $(3, 4)$. After $(11, 0)$, another run of 2 and rise of -1 would place you at $(13, -1)$. But -1 is not a whole number.

How can you be sure that no other whole-number solutions exist between $(1, 5)$ and $(11, 0)$?

Try whole-number values for r between 1 and 11, to be sure.

Notice that if you had used $-\frac{2}{4}$ for the slope from $(1, 5)$, you would have missed the point $(3, 4)$. Why must the slope be in lowest terms?

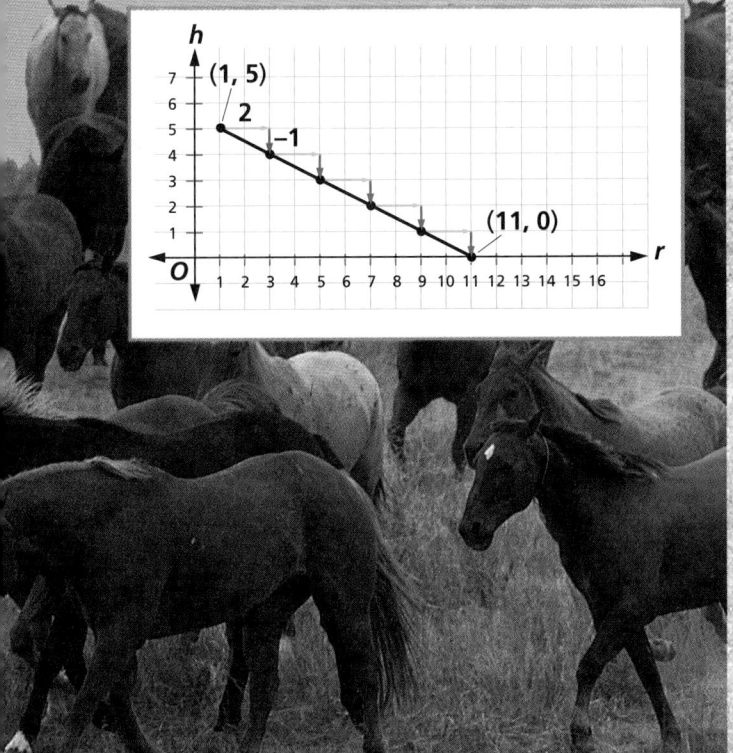

Activity 1

If flour comes in 2-pound bags and in 5-pound bags, list all the ways to buy exactly 18 pounds of flour. Draw a graph, and locate the solutions on the graph. What can you say about the solutions that appear on the graph?

Activity 2

Ms. Smiley has written a math exam with two parts. Part I has 9 questions, and part II has 8 questions. She wants to know how to assign points to each part so that the total will be exactly 100 points. Find a Diophantine equation that represents this situation and solves Ms. Smiley's problem.

Make a table to represent tests that have 9 questions in part I, 8 questions in part II, and 200 points in all.

Make tables to represent tests that have different numbers of questions but have 100 total points.

Activity 3

How can you tell that $4x + 6y = 125$ has no whole-number solution?

How can you tell that $5x + 10y = 112$ has no whole-number solution?

How can you tell whether $6x + 9y = 100$ has whole-number solutions?

Chapter 5 Review

Vocabulary

constant	241	perpendicular lines	248	run	209
constant function	241	point-slope form	235	slope	209
constant of variation	228	rate of change	221	slope-intercept form	225
direct variation	228	regression line	227	standard form	233
parallel lines	247	rise	209	y-intercept	225

Key Skills & Exercises

Lesson 5.1

➤ **Key Skills**

Find the slope of a line passing through two points.

Find the slope of a line passing through the points $A(-2, 4)$ and $B(3, 5)$.

$$m = \text{slope} = \frac{\text{difference in } y\text{-values}}{\text{difference in } x\text{-values}}$$
$$= \frac{5 - 4}{3 - (-2)} = \frac{1}{5}$$

➤ **Exercises**

If each pair of points is on a separate line, what is the slope of each line?

1. $A(-3, 2), B(2, 3)$ **2.** $A(2, 5), B(4, 1)$

3. $A(-5, 4), B(1, 4)$ **4.** $A(-3, -1), B(-3, 3)$

Lesson 5.2

➤ **Key Skills**

Find the equation of a line passing through the origin and a given point.

Find the equation of a line passing through the origin and the point $(2, 6)$.

Since the slope is $\frac{6}{2}$ or 3, the equation is $y = 3x$.

➤ **Exercises**

Find the equation of a line passing through the origin and the following points.

5. $(-3, 2)$ **6.** $(2, 3)$

7. $(-2, -5)$ **8.** $(-5, 4)$

Lesson 5.3

➤ Key Skills

Find the equation of a line in slope-intercept form.

Write the equation for a line

a. with slope 3 and y-intercept 5.
b. passing through $(-2, 3)$ and $(-1, 5)$.

a. Substitute 3 for m and 5 for b.
$y = mx + b$ $y = 3x + 5$
b. Slope $= m = \dfrac{5 - 3}{-1 - (-2)} = \dfrac{2}{1} = 2$

Substitute 2 for m, -2 for x, and 3 for y in the equation $y = mx + b$. Then solve for b.

$$3 = 2(-2) + b$$
$$3 = -4 + b$$
$$7 = b$$

Substitute 2 for m and 7 for b.
$y = mx + b$ $y = 2x + 7$

Graph an equation in slope-intercept form.

Graph the equation $y = \dfrac{1}{2}x + 5$.

The slope is $\dfrac{1}{2}$ and the y-intercept is 5.

Locate $(0, 5)$. From that point move 2 to the right and up 1 to $(2, 6)$. Draw the line.

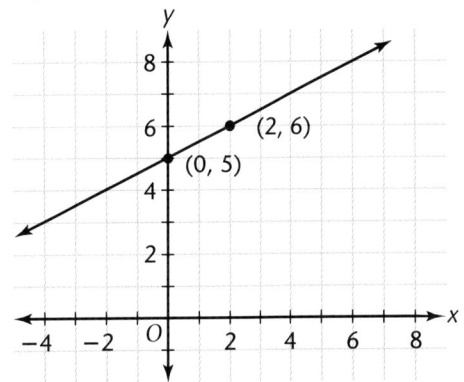

➤ Exercises

Write the equation in slope-intercept form

9. with slope 2 and y-intercept 1.

10. with slope $-\dfrac{1}{2}$ and y-intercept 5.

11. through $(0, -4)$ and with slope 8.

12. through $(4, 2)$ and $(5, -4)$.

13. through $(2, -1)$ and $(-3, -1)$.

14. through $(-6, 0)$ and $(0, -2)$.

Graph each equation.

15. $y = -\dfrac{1}{2}x - 2$ **16.** $y = -x + 6$

17. $y = -\dfrac{3}{4}x + 4$ **18.** $y = 4x - 1$

Lesson 5.4

➤ Key Skills

Write equations in standard form.

Write $y + 10 = -8x - 22$ in standard form.

$$y + 10 = -8x - 22$$
$$8x + y + 10 = -22$$
$$8x + y = -32$$

Write equations in point-slope form.

Write an equation for the line passing through the point $(4, 5)$ and $(2, 1)$.

$$m = \frac{1 - 5}{2 - 4} = \frac{-4}{-2} = 2$$

Use the point $(4, 5)$ for (x_1, y_1).

$$y - 5 = 2(x - 4)$$

➤ Exercises

Write the following equations in standard form.

19. $y + 9 = 4x - 8$ **20.** $y - 4 = -x + 1$ **21.** $y - 13 = 2x + 4$ **22.** $3x + y + 6 = 9$

Write the equations in point-slope form

23. with slope of 2 and y-intercept of -1.
24. through $(0, 4)$ and $(1, 2)$.
25. through $(3, 5)$ and with a slope of 4.
26. that crosses the x-axis at $x = -2$ and the y-axis at $y = 1$.

Lesson 5.5

➤ *Key Skills*

Recognize and find the slope of vertical and horizontal lines.

Identify each of the following lines as vertical or horizontal. State the slope for each.

  a. $y = -8$ **b.** $x = 7$

a. The line is horizontal and the slope is 0.

b. The line is vertical and the slope is undefined.

➤ *Exercises*

Identify the following lines as vertical or horizontal, and state the slope.

27. $y = -6$ **28.** $x = 2$ **29.** $y = \frac{1}{2}$ **30.** $x = -\frac{1}{3}$

Lesson 5.6

➤ *Key Skills*

Recognize and write equations of lines parallel and perpendicular to given lines.

Write equations in slope-intercept form containing the point (0, 4) that are parallel and perpendicular to the line $2x + 3y = 5$.

$2x + 3y = 5$ in slope-intercept form is $y = -\frac{2}{3}x + \frac{5}{3}$. The slope is $-\frac{2}{3}$. Since (0, 4) tells us the y-intercept is 4, the equation of the parallel line is $y = -\frac{2}{3}x + 4$.

The slope of a perpendicular line is the negative reciprocal of $-\frac{2}{3}$, or $\frac{3}{2}$. Its y-intercept is also 4. The equation of the perpendicular line is $y = \frac{3}{2}x + 4$.

➤ *Exercises*

Write the slope of one line parallel to and one line perpendicular to each of the following lines.

31. $y = \frac{2}{3}x + 4$ **32.** $y = -7x - 1$

Write an equation for a line that contains the point (1, 5) and is

33. parallel to $2x + y = -1$.   **34.** perpendicular to $2x + y = -1$.

Applications

Travel   Marie had driven 110 miles when she stopped at a gas station. After she left the station, she recorded the hours that she drove. If she averaged 50 miles per hour for the rest of the trip, write an equation for the total distance she drove after

35. 1.6 hours.   **36.** 2.9 hours.   **37.** 3.3 hours.

38. Suppose that during a flood the water level, L, of a river is 38 feet on Monday and is receding by 2 feet a day. Write and solve an equation of the water level in d days. Graph the equation. In how many days will the water level be 26 feet?

Chapter 5 Assessment

1. Copy and complete the table for $y = 2x + 3$.

x	-2	-1	0	1	2
y					

Write the following equations in standard form.
2. $y + 3 = -5x + 6$ **3.** $4x + 2y + 8 = 10$

Write the following equations in slope-intercept form.
4. $2x + y = 10 - 3$ **5.** $3x - y = 2y - 5$

Find the slope and the y-intercept for the following equations.
6. $3x + 6y = 12$ **7.** $2x + 3y = -3$

8. On a sheet of graph paper, sketch the graph $y = 3x + 1$.

9. Find the slope of a line passing through the points $(5, 3)$ and $(8, -2)$.

Write the slope-intercept form of an equation for a line that contains the point (2, 4) and is

10. parallel to the graph of $y = -3x - 7$.

11. perpendicular to the graph of $2x + y = 13$.

12. On a sheet of graph paper draw a graph of a line that has slope -4 and y-intercept 1.

13. Write the equation in slope-intercept form for a line with slope $-\frac{2}{3}$ and y-intercept 5.

14. Write the equation in point-slope form for a line that has a slope of -3 and passes through the point $(2, -3)$.

15. Write the equation in slope-intercept form for a line that passes through the points $(-5, -1)$ and $(1, 5)$.

16. What is the slope-intercept form of the equation for the graph at the right?

17. Give the slope of a line drawn through the points $(1, 8)$ and $(-2, 8)$.

18. Equipment Rental Steve can rent a small lawn mower for a daily rate of $15, or he can rent a mower for 10 days at a flat rate of $100. How many days must Steve use the mower before it would benefit him to pay the flat rate?

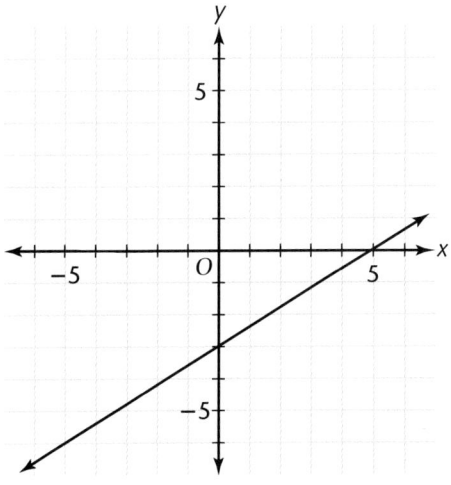

CHAPTER 6

Systems of Equations and Inequalities

The atmosphere is electric! It's minutes before your favorite rock star begins her huge benefit concert. You can't believe you're here. You had to work hard to make the money and time to attend, but you love her music, and the money goes to charity. Thousands of people like you crowd the hall, eager for the concert to start. A hush passes over the crowd. The curtain begins to rise. The crowd cheers. It has begun!

PORTFOLIO ACTIVITY

You are so excited about the concert that you ask all of your friends to join you. The tickets are sold at two different prices, with the center section costing $10 more than the balcony seats. You buy 7 center seats and 5 balcony seats for a total of $310. What is the cost of each type of seat?

Use *c* as the cost of a center section ticket and *b* as the cost of a balcony ticket. Write two equations using *c* and *b* that show the relationships in this problem. Save them. As you progress through the chapter, you will be asked to solve this problem using various methods. You may wish to use these exercises as part of your portfolio.

BALCONY

STAGE

Graphing Systems of Equations

why *One of the best ways to present mathematical data is to use a graph. For business decisions and customer choices, graphs can represent costs and show comparisons. When you graph two linear equations on the same coordinate system, you can see where the lines intersect.*

If the Smiths decide to buy the VCR, they plan to spend $20 per month to rent 8 tapes. Should the Smiths buy the VCR or install cable TV?

VCRs on sale NOW
Starting at
$185.00
12:00
12:00

One way to help the Smiths decide on the right purchase is to compare the cost of each purchase over several months. An equation can be written to model each purchase.

First, define two variables.

Let c = the total costs for a system.
Let m = the number of months.

Next, write an equation in words to represent the costs. Then substitute variables and numbers for the words.

Installation
$45.00
Movies
Sports
Only
$34.00
a month!
News
EAST COAST CABLE

Cable TV costs = installation + cost for m months
$$c = 45 + 34m \quad \text{(or } c = 34m + 45\text{)}$$

VCR and tape costs = machine + tape rental for m months
$$c = 185 + 20m \quad \text{(or } c = 20m + 185\text{)}$$

**COORDINATE
GEOMETRY**
Connection

Create a table of values from the equations to compare the data. Then graph each equation on one coordinate plane.

	Table of Costs	
	Cable	VCR
m	$34m + 45$	$20m + 185$
1	79	205
2	113	225
3	147	245
4	181	265
5	215	285
6	249	305
7	283	325
8	317	345
9	351	365
10	385	385

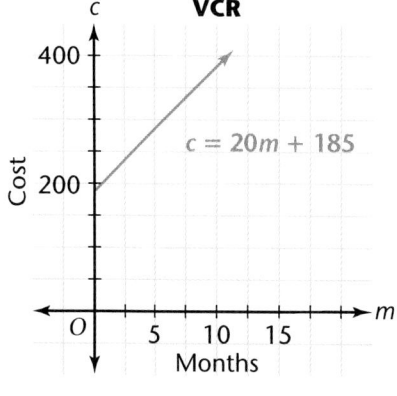

The table and the graphs show that each cost continues to increase, but one increases at a faster rate than the other. At what point do the graphs intersect?

At 10 months the value for both equations is $385. When the value is the same for both equations, the equations have a **common solution**.

CRITICAL
Thinking

Use the information shown in the graph to explain which decision is best for the Smith family.

A System of Equations

Two or more equations in two or more variables is called a **system of equations**. When a system of two equations in two variables has a single solution, the solution is the ordered pair, (x, y), that satisfies both equations. The solution to the system used in the Smith family problem is $(10, 385)$.

EXAMPLE 1

*Graphics
Calculator*

Use a graphics calculator or graph paper to graph the equations $3x + y = 11$ and $x - 2y = 6$. Find the common solution by examining the graph.

Solution ➤

Change each equation to slope-intercept form, $y = mx + b$. Graph the two equations to find a common solution.

$$3x + y = 11 \qquad\qquad x - 2y = 6$$
$$y = -3x + 11 \qquad\qquad -2y = -x + 6$$
$$y = \frac{1}{2}x - 3$$

The graphs show that the lines intersect at exactly one point. To find the common solution, read the coordinates for the point of intersection from the x- and y-axes. It appears from the graph that the point of intersection is $(4, -1)$.

To check the solution, substitute 4 for x and -1 for y in each equation.

$$3x + y = 11 \qquad\qquad\qquad x - 2y = 6$$
$$3(4) + (-1) \stackrel{?}{=} 11 \qquad\qquad 4 - 2(-1) \stackrel{?}{=} 6$$
$$11 = 11 \quad \text{True} \qquad\qquad\qquad 6 = 6 \quad \text{True}$$

Since the substitution is true for each equation, $(4, -1)$ is the common solution. ❖

Try This Graph the equations $y = 2x + 2$ and $y = -x. - 1$. Find a common solution by examining the graph. Check your answer.

EXAMPLE 2

Ms. Alyward gives a science midterm with 200 possible points. There are a total of 38 questions on the test. How many of each type of question are on the test?

Solution ➤

First define two variables. Then write equations to model the problem.

Organize

Let x = the number of 4-point multiple-choice questions.
Let y = the number of 20-point essay questions.

Write

number of 4-point items + number of 20-point items = 38 items

$$x \qquad + \qquad y \qquad = \quad 38$$

points for multiple choice + points for essay = 200 points

$$4x \qquad + \qquad 20y \quad = \quad 200$$

Solve

To find how many of each type question appear on the test, solve the system of equations for (x, y). Make a table, or graph the two equations to find the point of intersection for a common solution. Change the equations to slope-intercept form to find values for the variables.

$$x + y = 38 \qquad\qquad 4x + 20y = 200$$
$$\quad y = -x + 38 \qquad\qquad 20y = -4x + 200$$
$$\qquad\qquad\qquad\qquad\qquad y = -\frac{1}{5}x + 10$$

Use the table or the graph to find the point of intersection.

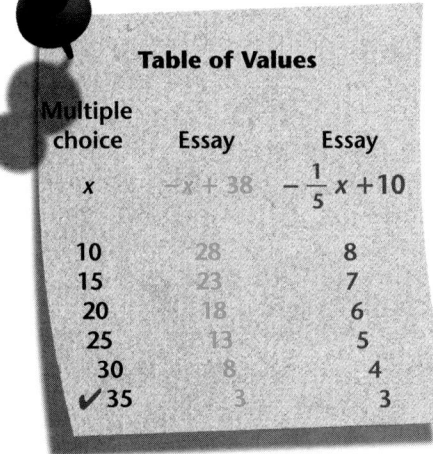

Table of Values

Multiple choice	Essay	Essay
x	$-x + 38$	$-\frac{1}{5}x + 10$
10	28	8
15	23	7
20	18	6
25	13	5
30	8	4
✔ 35	3	3

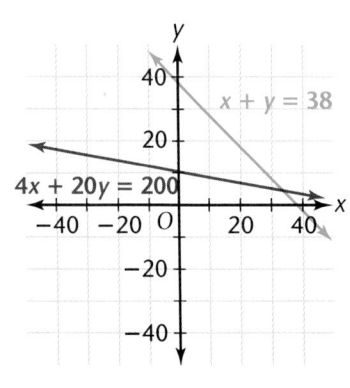

The solution is (35, 3).

There will be 35 multiple-choice questions and 3 essay questions on Ms. Alyward's science midterm.

Check

$$35 + 3 \overset{?}{=} 38 \qquad\qquad 4(35) + 20(3) \overset{?}{=} 200$$
$$38 = 38 \quad \text{True} \qquad\qquad 140 + 60 \overset{?}{=} 200$$
$$\qquad\qquad\qquad\qquad\qquad 200 = 200 \quad \text{True} \ ❖$$

Why were the equations written in slope-intercept form before they were graphed?

Approximate Solution

It is sometimes difficult to find an exact solution from a graph. A reasonable estimate for a point of intersection is an **approximate solution** for a system of equations.

EXAMPLE 3

Solve by graphing. $\begin{cases} 4x + 3y = 6 \\ 2y = x + 2 \end{cases}$

The brace { is used to indicate that the equations form a system.

Solution

Write the equations in slope-intercept form.

Graphics Calculator

$$4x + 3y = 6$$
$$3y = -4x + 6$$
$$y = -\frac{4}{3}x + 2$$

$$2y = x + 2$$
$$y = \frac{1}{2}x + 1$$

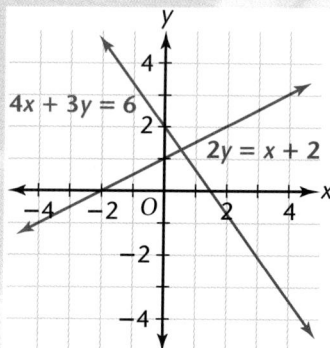

From the graph, locate the point of intersection. Use a graphics calculator, or make a reasonable estimate by inspecting the graph to approximate a common solution. An approximate solution is (0.55, 1.27).

Check the solution by substituting 0.55 for x and 1.27 for y in the original equations.

$$4x + 3y = 6$$
$$4(0.55) + 3(1.27) \stackrel{?}{=} 6$$
$$2.2 + 3.81 \stackrel{?}{=} 6$$
$$6.01 \approx 6$$

$$2y = x + 2$$
$$2(1.27) \stackrel{?}{=} 0.55 + 2$$
$$2.54 \approx 2.55$$

A check shows the approximate solution is reasonable. ❖

EXERCISES & PROBLEMS

Communicate

Exercises 1–3 refer to the following system. $\begin{cases} 2x - 3y = 4 \\ x + 4y = -9 \end{cases}$

1. How do you write each equation in slope-intercept form? Explain why this form is used to find values for the variables.

2. Explain how to create a table of values for the system of equations. Discuss how to find the point of intersection from a table.

3. Discuss how to graph the system of equations from the table in Exercise 2. Tell how to find a common solution from the graph.

4. Discuss how to graph the system at the right. Tell how to make a reasonable estimate of the solution by inspecting the graph.

$$\begin{cases} x + y = 3 \\ x - y = 4 \end{cases}$$

5. Why is it important to check your estimate?

Practice & Apply

Solve by graphing. Round approximate solutions to the nearest tenth. Check algebraically.

6. $\begin{cases} 2x + 10y = -5 \\ 6x + 4y = 2 \end{cases}$ **7.** $\begin{cases} x - 2y = 60 \\ 3y = 30 - 2x \end{cases}$ **8.** $\begin{cases} 3y = 4x - 1 \\ 2x + 3y = 2 \end{cases}$ **9.** $\begin{cases} x + 3y = 6 \\ -6y = 2x + 12 \end{cases}$

10. $\begin{cases} 5x + 6y = 14 \\ 3x + 5y = 7 \end{cases}$ **11.** $\begin{cases} x = 400 - 2y \\ x - 100 = y \end{cases}$ **12.** $\begin{cases} 3x + 5y = 12 \\ 7x - 5y = 8 \end{cases}$ **13.** $\begin{cases} x = 2 \\ 2y = 4x + 2 \end{cases}$

Algebraically determine whether the point (2, 10) is a solution for each pair of equations.

14. $\begin{cases} y = 2x - 4 \\ y = x + 8 \end{cases}$ **15.** $\begin{cases} y = -x + 12 \\ y = -3x + 16 \end{cases}$ **16.** $\begin{cases} y = x + 8 \\ y = -3x + 16 \end{cases}$ **17.** $\begin{cases} x + 3y = 6 \\ -6y = 2x + 12 \end{cases}$

18. Graph the equations for Exercises 14-17 to check the results.

19. Find the point of intersection of $y = -x + 8$ and $y = -3x + 16$.

20. Use a graph to find the approximate point of intersection of $y = -x + 12$ and $y = 2x - 4$. Estimate the exact fractional value of these decimals. Algebraically check your estimate.

Write a system of equations. Then graph the equations to determine the solution.

21. Aviation The plane at 3,800 feet is descending at a rate of 120 feet per minute, and the other plane is climbing at a rate of 40 feet per minute. In how many minutes will they be at the same altitude?

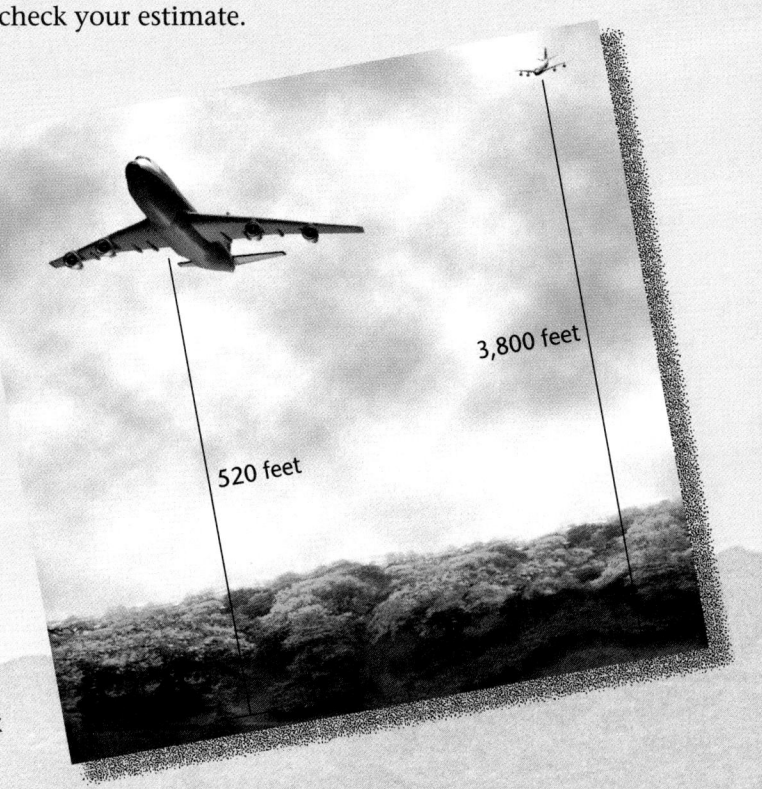

3,800 feet

520 feet

Refer to points A(3, 5), B(4, −1), and C(9, 3). Two lines meet at point A.

22. Write an equation for line *AB*.

23. Write an equation for line *AC*.

24. Check your equations by graphing.

25. Use the point of intersection to check your equations algebraically.

Statistics Use the following data to solve Exercises 26–30.

26. **Technology** Use a graphics calculator to plot the data as ordered pairs (year, degrees). Use the statistical functions of your calculator to show a scatter plot of the data.

27. Fit lines to this data.

28. Find the approximate point of intersection of the two models. About how many years beyond the given data will it be before the number of education degrees and the engineering degrees conferred are the same?

29. If you had to make a career choice based on this data, which career would you choose and why?

30. Estimate the point of intersection for $y = -\frac{1}{2}x + 5$ and $y = \frac{1}{4}x - 5$.

Degrees Conferred for Five Consecutive Years

Year	1	2	3	4	5
Education degrees	7473	7151	7110	6909	6544
Engineering degrees	2981	3230	3410	3820	4191

Sports From 1980 to 1990, there was a shift in salaries for basketball players and baseball players. The graph reflects the average salaries for the players.

31. Estimate the year in which their salaries became equal.

32. Estimate the salary.

33. 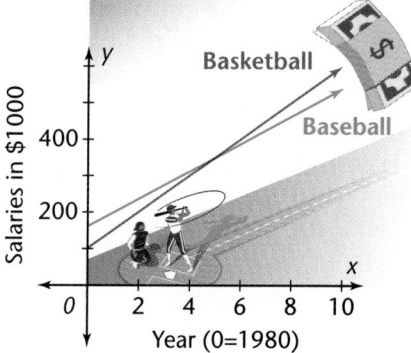 **Portfolio Activity** Complete the problem in the portfolio activity on page 259 using graphs.

Look Back

34. Place parentheses and brackets to make $21 \div 3 - 4 \cdot 0 + 6\,(-3) = -39$ true. **[Lesson 1.4]**

35. If data points are randomly scattered, what type of correlation exists? **[Lesson 1.6]**

36. Graph $-3 < x$ or $x < 5$ on the number line. State the solution set clearly. **[Lesson 3.6]**

37. Solve $3(5 - 2x) - (8 - 6x) = -9 + 2(3x + 4) - 10$. **[Lesson 4.6]**

Find the slope of the following lines. **[Lesson 5.3]**

38. $3x + 2y = 12$ 39. $8y = -6x + 12$ 40. $2y = \frac{1}{2}x + 6$ 41. $y = -3x - 18$

Look Beyond

Substitute each value of x into the equation $-2x + 4y = 12$, and solve for y. EXAMPLE: Let $x = 2$. $-2(2) + 4y = 12$ $y = 4$

42. Let $x = -6$.   43. Let $x = 8y$.   44. Let $x = y - 3$.   45. Let $x = 3y + 1$.

LESSON 6.2

Exploring Substitution Methods

why *Sometimes graphing can be used to find the solution to a system of equations. However, when an exact solution cannot be determined from a graph, other methods are needed to find the solution.*

Sue paid a total of $27 for 3 shares of Allied Sports and 4 shares of Best Design.

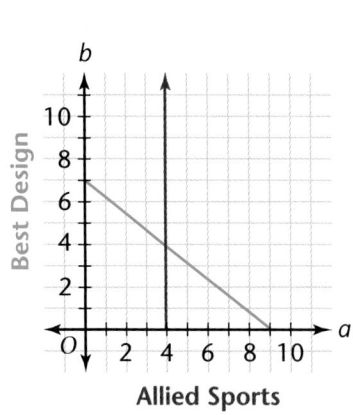

•Exploration 1 What Is the Cost?

This problem can be solved by graphing.

Let a = price of one share of Allied Sports.
Let b = price of one share of Best Design.

Then, $\begin{cases} 3a + 4b = 27 \\ a = 4 \end{cases}$ is the system

that models the problem.

1 Examine the graph. Guess the value of b at the point of intersection.

2 Explain why your guess is an approximation.

3 Substitute 4 for a in $3a + 4b = 27$, and solve the resulting equation for b.

4 How can you find the solution to a system if you know the value of one variable?

5 Give two methods to solve the system. $\begin{cases} 8x + 2y = 19 \\ x = 3 \end{cases}$

6 Which method results in an exact solution? Use what you have discovered to solve the system in Step 5. ❖

(graph shown at right, axis b vs a, labeled "Best Design" on vertical axis and "Allied Sports" on horizontal axis)

•Exploration 2 *Substituting Expressions*

Examine the system. $\begin{cases} 3x + 2y = 4 \\ x + y = 1 \end{cases}$

 Solve $x + y = 1$ for x.
  a. Substitute this expression for x in $3x + 2y = 4$.
  b. What values of x and y solve the system?

 Solve $x + y = 1$ for y.
  a. Substitute this expression for y in $3x + 2y = 4$.
  b. What values for x and y solve the system?

 How do the methods of Step 1 and Step 2 compare?

What substitutions can be made to solve a system if each coefficient in one equation is equal to 1? ❖

 •Exploration 3 *Finding a Value*

Examine the system. $\begin{cases} 15x - 5y = 30 \\ -2x + y = -6 \end{cases}$

 Solve $-2x + y = -6$ for y.

 Substitute this expression for y in the first equation, and then solve the system for x.

Explain how to use substitution to solve a system if at least one variable has a coefficient of 1. ❖

 CRITICAL *Thinking*

How are Explorations 2 and 3 alike? Summarize what you have learned.

APPLICATION

Investment

Each share of New Sport Times stock costs $2 more than a share of Modern Design stock. Find the cost for one share of each stock.

A student in a social studies class purchased 2 shares of Modern Design and 3 shares of New Sport Times for $96.

Define the variables.
  Let m = the cost of a share of Modern Design stock.
  Let n = the cost of a share of New Sport Times Stock.

Write the equations as a system.
$\begin{cases} 2m + 3n = 96 \\ n = m + 2 \end{cases}$

Use the substitution method to solve the system of linear equations. Substitute $m + 2$ for n in the equation $2m + 3n = 96$.

$$2m + 3n = 96$$
$$2m + 3(m + 2) = 96$$
$$2m + 3m + 6 = 96$$
$$5m + 6 = 96$$
$$5m = 90$$
$$m = 18$$

Now substitute 18 for m in the equation $n = m + 2$.

$$n = m + 2$$
$$n = 18 + 2$$
$$n = 20$$

The solution is (18, 20).

Check your answer to see that the values satisfy the original equations. The cost of each share of Modern Design is $18, and the cost of each share of New Sport Times is $20. ❖

APPLICATION

One share of Disposable Inc. and 2 shares of Software Today are worth $65. What is the cost for one share of each stock?

A student in a social studies class purchased 2 shares of Disposable Inc. and 3 shares of Software Today for $105.

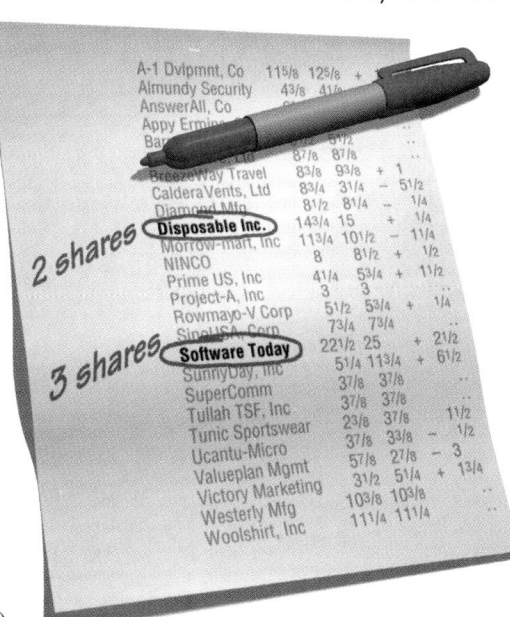

Define the variables.

Let d = the cost of each share of Disposable Inc.

Let s = the cost of each share of Software Today.

Write the system. $\begin{cases} 2d + 3s = 105 \\ d + 2s = 65 \end{cases}$

To use the substitution method, first solve $d + 2s = 65$ for d.

$$d = -2s + 65$$

Now substitute $-2s + 65$ for d in the equation $2d + 3s = 105$.

1. Solve the equation $2(-2s + 65) + 3s = 105$ for s.

2. Use your value of s to find the value of d in the equation $d = -2s + 65$.

3. Is (15, 25) your solution?

The cost of one share of Disposable Inc. is $15. The cost of one share of Software Today is $25. ❖

EXERCISES & PROBLEMS

Communicate

1. If you know $y = 42$, explain how to use substitution to solve $y = 2x + 8$.

2. Given the equations $-4x + y = 2$ and $2x + 3y = 34$, tell how to find the expression to substitute and tell how to find the common solution.

3. Explain how to use substitution to solve the system.

$$\begin{cases} x - 2y = 8 \\ 2x + 3y = 23 \end{cases}$$

Practice & Apply

Solve and check by the substitution method.

4. $\begin{cases} 2x + 8y = 1 \\ x = 2y \end{cases}$
5. $\begin{cases} x + y = 7 \\ 2x + y = 5 \end{cases}$
6. $\begin{cases} 3x + y = 5 \\ 2x - y = 10 \end{cases}$
7. $\begin{cases} y = 5 - x \\ 1 = 4x + 3y \end{cases}$

8. $\begin{cases} 2x + y = -92 \\ 2x + 2y = -98 \end{cases}$
9. $\begin{cases} 4x + 3y = 13 \\ x + y = 4 \end{cases}$
10. $\begin{cases} 6y = x + 18 \\ 2y - x = 6 \end{cases}$
11. $\begin{cases} 2x + y = 1 \\ 10x - 4y = 2 \end{cases}$

First graph each system and estimate the solution. Then use the substitution method to get an exact solution.

12. $\begin{cases} 5x - y = 1 \\ 3x + y = 1 \end{cases}$
13. $\begin{cases} 2x + y = 1 \\ 10x = 4y + 2 \end{cases}$
14. $\begin{cases} 5x = 3y + 12 \\ x = y \end{cases}$
15. $\begin{cases} 3x - 2y = 2 \\ y = 2x + 8 \end{cases}$

16. $\begin{cases} 2x + 3y = 7 \\ x + 4y = 9 \end{cases}$
17. $\begin{cases} 4x - y = 15 \\ -2x + 3y = 12 \end{cases}$
18. $\begin{cases} 2y + x = 4 \\ y - x = -7 \end{cases}$
19. $\begin{cases} 4y - x = 4 \\ y + x = 6 \end{cases}$

754 ft

Write a system of equations for each problem, and solve for an exact solution. Show checks.

20. **Aviation** A hot-air balloon is rising at the rate of 4 feet every second. A small aircraft in the vicinity at 7452 feet is losing altitude at a rate of 30 feet every second. In how many seconds will both be at the same altitude?

21. **Numeration** The sum of two numbers is 27. The larger is 3 more than the smaller. Find the two numbers.

22. **Numeration** One number is 4 less than 3 times a second number. If 3 more than twice the first number is decreased by twice the second, the result is 11. Find both numbers.

23. 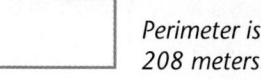 **Geometry** Find the dimensions of this rectangle if its length is twice its width.

Perimeter is 208 meters.

Fund-raising At the Boy Scout "all you can eat" Spaghetti dinner, the troop served 210 people and raised $935. **Complete Exercises 24–27.**

24. Write an equation for the *total amount raised* from adult and child dinners served.

25. Write an equation for the *total number* of adult and child dinners served.

26. Solve the system of equations from Exercises 24 and 25. How many adult and child dinners were served?

27. Name two methods of solving systems of equations that can be used to solve Exercise 26. Why would you choose one over another?

28. **Geometry** The sum of the measures of angle *A* and angle *B* is 90°. If angle *A* is 30° less than twice angle *B*, then find the degree measures for each angle.

29. Cultural Connection: Asia A Chinese problem states that several people pooled their money to buy a farm tool to share. If each person paid 8 coins, 3 coins were left over. If each paid 7 coins, they were 4 coins short. How many people were there, and what was the price of the farm tool?

30. **Portfolio Activity** Complete the problem in the portfolio activity on page 259 using substitution.

Look Back

31. Recreation In a foot race Sam finished 20 feet in front of Joe. Joe was 5 feet behind Mark, and Mark was 10 feet behind Rob. Tom was 15 feet ahead of Rob. In what order did they finish? **[Lesson 3.1]**

Solve each equation for *x*. [Lesson 4.4]

32. $\frac{x}{15} = 3$
33. $\frac{3}{x} = 15$
34. $\frac{15}{x} = 3$
35. $\frac{x}{3} = 15$

36. The number 12.6 is 42% of what number? **[Lesson 4.5]**

Tell if the following equations represent lines that are parallel, perpendicular, or neither. [Lesson 5.6]

37. $3y = 2x - 15$
$3x + 2y = 24$

38. $2y = x - 12$
$2y - x = 12$

39. $y = x - 1$
$y = -x + 3$

Look Beyond

Examine the following equations. Use substitution to solve for *x* and *y*.

40. $\begin{cases} x + 2y + 3z = 8 \\ \quad\ y + 2z = 3 \\ \quad\quad\ \ z = 2 \end{cases}$

41. $\begin{cases} 2x + 3y + 5z = 44 \\ \quad\ 2y - 6z = 4 \\ \quad\quad\ \ z = 4 \end{cases}$

42. $\begin{cases} x + y = a \\ \ \ 2y = 6b \end{cases}$

43. $\begin{cases} x - y - d = 0 \\ 6x - 4y = 2d \end{cases}$

LESSON 6.3 The Elimination Method

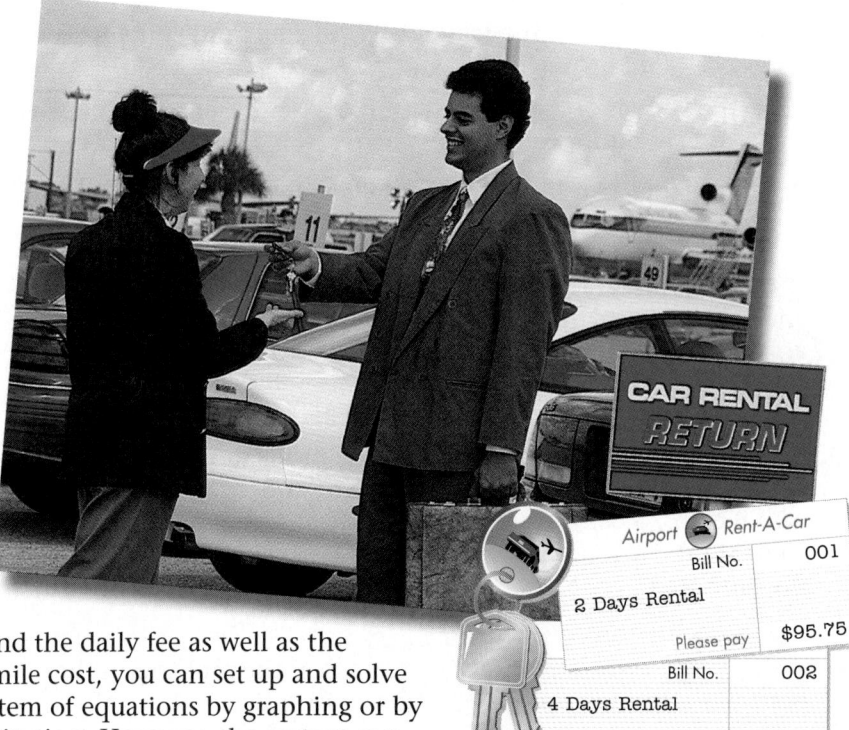

The graphing method and the substitution method both provide a way to solve systems of equations. Another method used to solve systems is the elimination method.

To find the daily fee as well as the per-mile cost, you can set up and solve a system of equations by graphing or by substitution. However, the system can also be solved by another method called *elimination*. The **elimination method** for solving equations eliminates one variable by adding or subtracting opposites.

Airport Rent-A-Car

Bill No.	001
2 Days Rental	
Please pay	$95.75
Bill No.	002
4 Days Rental	
Please pay	$226.50

Jason drove 125 miles on the 2-day trip. He drove 350 miles on the 4-day trip.

•Exploration• Using Opposites

 What is true about the sum of any expression and its opposite?

 Consider the system of equations. $\begin{cases} 3x + 2y = 7 \\ 5x - 2y = 9 \end{cases}$

   a. Which terms of the equations are opposites?
   b. Use the Addition Property of Equality to add the terms on each side of the equal signs.
   c. Solve the resulting equation for x, and then solve for y.
   d. Check the solution in each equation of the system.

 Consider the system of equations. $\begin{cases} 2a + b = 9 & \text{Equation 1} \\ 3a - 4b = 8 & \text{Equation 2} \end{cases}$

   a. Multiply both sides of Equation 1 by 4.
   b. Use the Addition Property of Equality. Solve for a, and then solve for b.
   c. Check the solution in each equation of the system.

 How are opposites used to solve a system of equations? ❖

The Exploration used the method of elimination to solve a system of equations. This method used opposites to eliminate one of the variables. Examine how this method is used to solve the rental car problem.

Define the variables.

Let x = the cost per day. Let y = the cost per mile.

Write the equations in standard form.

$$\begin{cases} 2x + 125y = 95.75 \\ 4x + 350y = 226.50 \end{cases} \begin{array}{l} \leftarrow \text{ Bill No. 001} \\ \leftarrow \text{ Bill No. 002} \end{array}$$

Multiply each term of the first equation by -2.

$$-2(2x) + (-2)(125)y = -2(95.75)$$
$$4x \quad + \quad 350y \quad = \quad 226.50$$

Explain why x rather than y was chosen to be eliminated.

Simplify by using the Addition Property of Equality.

$$\begin{array}{r} -4x + (-250)y = -191.50 \\ \underline{4x + \quad 350y = \quad 226.50} \\ 100y = \quad 35 \\ y = \quad 0.35 \end{array}$$

Substitute (0.35) in the first equation to find x.

$$2x + 125(0.35) = 95.75$$
$$2x + \quad 43.75 \quad = 95.75$$
$$2x = 52$$
$$x = 26$$

The solution is (26, 0.35). A check proves the solution correct. Jason rented the car for $26 a day, and the cost per mile was $0.35.

EXAMPLE 1

Solve. $\begin{cases} 2x + 3y = 1 & \textbf{Equation 1} \\ 5x + 7y = 3 & \textbf{Equation 2} \end{cases}$

Solution ➤

Multiply **Equation 1** by 5 and **Equation 2** by -2.

$$\begin{array}{ll} (5)2x + (5)3y = (5)1 & \rightarrow \quad 10x + 15y = 5 \\ (-2)5x + (-2)7y = (-2)3 & \rightarrow \quad -10x - 14y = -6 \end{array}$$

Use the Addition Property of Equality.
$$15y + (-14y) = 5 + (-6)$$
$$y = -1$$

Then solve for x by substituting.
$$10x + 15y = 5$$
$$10x + 15(-1) = 5$$
$$10x = 20$$
$$x = 2$$

Check each value by substituting the solutions in the original equations. ❖

Try This Solve by elimination. $\begin{cases} 3a - 2b = 6 \\ 5a + 7b = 41 \end{cases}$

You now have several methods for solving a system of equations. Graphing is used as a visual model for a problem involving a system of equations. You can use graphing to check the reliability of a calculated solution or to approximate a solution. The substitution method and the elimination method are used to find an exact solution. Substitution may be the more efficient method if either of the given equations has at least one coefficient of 1 or -1.

EXAMPLE 2

Choose a method for solving the following systems of equations. Explain why you chose each method.

System ① $\begin{cases} y = 10 - 6x \\ y = 3x - 6 \end{cases}$ **System ②** $\begin{cases} 2x - 5y = -20 \\ 4x + 5y = 14 \end{cases}$

System ③ $\begin{cases} 9a - 2b = -11 \\ 8a - 7b = 25 \end{cases}$ **System ④** $\begin{cases} 324p + 456t = 225 \\ 178p - 245t = 150 \end{cases}$

Solution ➤

There may be more than one correct answer for each set of equations.

① Use the substitution method. Because y has a coefficient of 1, the equivalent expression can be substituted.

② Use the elimination method. $5y$ and $-5y$ are opposites.

③ Use the elimination method. No term has coefficient 1 or -1.

④ Use technology. Because the coefficients are such large numbers, ordinary algebra or manual graphing techniques would be cumbersome. ❖

Try This Choose any method to solve the following system of equations. Explain why you chose that method.

$$\begin{cases} y + 2x = 12 \\ 3y = 5x - 19 \end{cases}$$

CRITICAL *Thinking* Which word, **AND** or **OR**, connects the equations in a system? How does this help to explain why a solution to a system of equations must be checked in *both* equations?

EXERCISES & PROBLEMS

Communicate ~~~

In the following systems, which terms are opposites? How would you solve each system?

1. $\begin{cases} x + y = 13 \\ x - y = 5 \end{cases}$ **2.** $\begin{cases} 2x - 3y = 8 \\ 5x + 3y = 20 \end{cases}$ **3.** $\begin{cases} 2a + b = 6 \\ -2a - 3b = 8 \end{cases}$

How would you use opposites to solve each system?

4. $\begin{cases} 4x + 3y = 8 \\ x - 2y = 13 \end{cases}$ **5.** $\begin{cases} a - 2b = 7 \\ 4b + 2a = 15 \end{cases}$ **6.** $\begin{cases} 3m - 5n = 11 \\ 2m - 3n = 1 \end{cases}$

Explain how you would use elimination to solve each system.

7. $\begin{cases} 2x + 3y = 9 \\ 3x + 6y = 7 \end{cases}$ **8.** $\begin{cases} 2x - 5y = 1 \\ 3x - 4y = -2 \end{cases}$ **9.** $\begin{cases} 9a + 2b = 2 \\ 21a + 6b = 4 \end{cases}$

10. The sum of two numbers is 156. Their difference is 6. Write a system to model the problem. Tell what method you would choose to solve the system of equations. Tell why you chose that method.

Practice & Apply ~~~

Solve each system of equations by elimination and check.

11. $\begin{cases} -x + 2y = 12 \\ x + 6y = 20 \end{cases}$ **12.** $\begin{cases} 2x + 3y = 18 \\ 5x - y = 11 \end{cases}$ **13.** $\begin{cases} -4x + 3y = -1 \\ 8x + 6y = 10 \end{cases}$

14. $\begin{cases} 2x - 3y = 5 \\ 5x - 3y = 11 \end{cases}$ **15.** $\begin{cases} 6x - 5y = 3 \\ -12x + 8y = 5 \end{cases}$ **16.** $\begin{cases} 4x + 3y = 6 \\ -2x + 6y = 7 \end{cases}$

17. Mrs. Jones is celebrating Gauss' birthday by having a pizza party for her two algebra classes. She orders 3 pizzas and 3 bottles of soda for her 2nd period class and 4 pizzas and 6 bottles of soda for her 7th period class. How much does each pizza and bottle of soda cost?

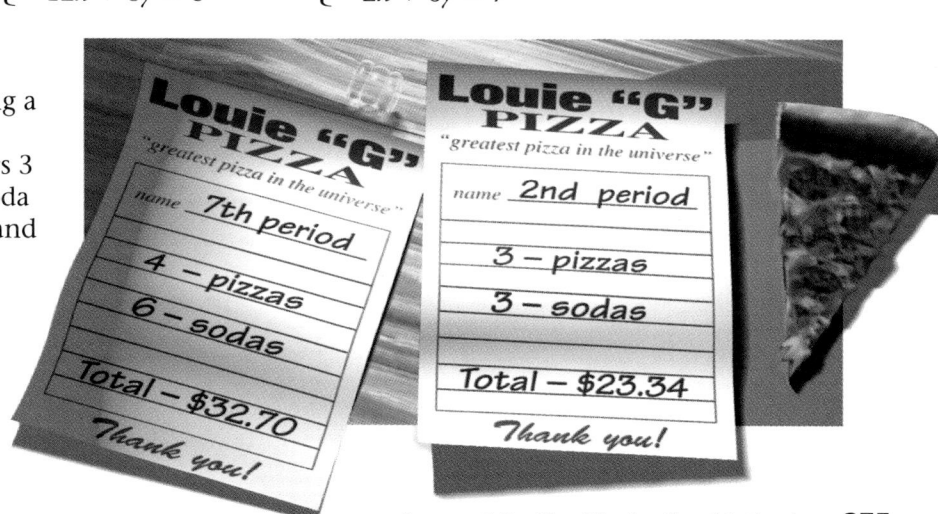

Solve each system of equations using any method.

18. $\begin{cases} 2m = 2 - 9n \\ 21n = 4 - 6m \end{cases}$

19. $\begin{cases} x - 7 = 3y \\ 6y = 2x - 14 \end{cases}$

20. $\begin{cases} 2x = 3 - y \\ y = 3x - 12 \end{cases}$

21. $\begin{cases} 3b = -6a - 3 \\ b = 2a - 1 \end{cases}$

22. $\begin{cases} y = 1.5x + 4 \\ 0.5x + y = -2 \end{cases}$

23. $\begin{cases} 2x = 3y - 12 \\ \frac{1}{3}x = 4y + 5 \end{cases}$

Write a system of equations for each problem, and choose the best method to solve the system. Solve and show checks. **Complete Exercises 24–30.**

At Highview Towers, a one-time deposit is required with the first month's rent.

24. Finance Roberto paid $900 the first month and a total of $6950 during the first year. Find the monthly rent and the deposit.

25. Income As a parking attendant at the Saratoga Raceway, Juan earns a fixed salary for the first 15 hours worked each week and then additional pay for overtime. During the first week, Juan worked 25 hours and earned $240, while the second week he worked 22.5 hours and earned $213.75. What is Juan's weekly salary for 15 hours and his overtime pay per hour?

26. Shopping at Super Sale Days, Martha buys her children 3 shirts and 2 pairs of pants for $85.50. She returns during the sale and buys 4 more shirts and 3 more pairs of pants for $123. What is the sale price of the shirts and pants?

Bill's grandfather sells his tractor for $6000.

27. Investment With the money from the sale of his tractor, Bill's grandfather makes two investments, one at a bank paying 5% per year and the rest in stocks yielding 9% per year. If he makes a total of $380 in interest in the first year, how much is invested at each rate?

28. Accounting At a local music store, single-recording tapes sell for $6.99 and concert tapes sell for $10.99. The total number of these tapes sold on Monday was 25. If the sales total for Monday is $230.75 for these two items, find the number of each type sold.

29. 🗨️ **Geometry** The perimeter of a rectangle is 24 centimeters, and its length is 3 times its width. Find the length and width of this rectangle.

30. Travel At the rates shown, what is the daily room rate and cost of a meal at the Shamrock Inn?

Use the equations $y = x - 2$ and $y = 2x$ for Exercises 31–33.

31. Graphically find the common solution.

32. Algebraically solve the system.

33. What method did you use to solve Exercise 32?

34. 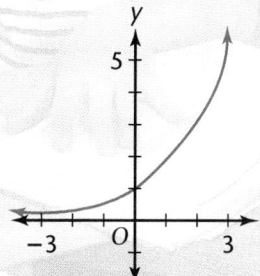 **Portfolio Activity** Complete the problem in the portfolio activity on page 259 using the elimination method.

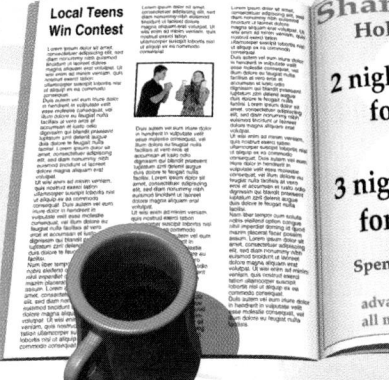

The Shamrock Inn offers two special holiday weekends.

Look Back

35. Find the pattern and give the next 3 terms of the sequence. **[Lesson 1.1]**

5, 8, 11, 14, 17, ___, ___, ___

Evaluate the following. [Lesson 2.4]

36. INT $\left(31\frac{4}{5}\right)$ **37.** INT (31.4)   **38.** INT (6.91)

39.

This graph represents a function that would be best described as

a. absolute value.

b. quadratic.

c. exponential.

d. linear. **[Lesson 2.5]**

40. Cultural Connection: Africa The following problem is in the Rhind papyrus by Ahmes. A bag contains equal weights of gold, silver, and lead purchased for 84 *sha'ty*. What is the weight (in *deben*) of each metal if the price for 1 *deben* weight is 6 *sha'ty* for silver, 12 *sha'ty* for gold, and 3 *sha'ty* for lead? **[Lesson 4.6]**

Find the slope for the following lines. [Lesson 5.3]

41. $7x - 3y = 22$ **42.** $3x + 2y = 6$ **43.** $y = 2x$

44. Write the equation of the line through $(-3, 8)$ and parallel to $y = 0.8x - 7$. **[Lesson 5.6]**

Look Beyond

Technology Graph the system of equations on the same set of axes, and describe your graph. Use a graphics calculator if you have one.

45. $\begin{cases} 2x - 3y = 6 \\ 4x - 6y = 18 \end{cases}$

Non-Unique Solutions

20.0 21.9	20.8 22.7	21.8 23.6	23.0 24.8	23.9 25.8
1970	**1975**	**1980**	**1985**	**1990**

Average Ages for First Marriages

Sara's brother is getting married in a few weeks. While reading a magazine, the information in the chart catches her eye.

why *You have studied systems of equations that have exactly one solution. There are also systems that have an infinite number of solutions or no solution at all. A comparison of systems can be made by studying the graphs of their lines and their solutions.*

In the chart you can see that the average age for first marriages is increasing for both men and women. If the ages continue to increase at the same rate for both men and women, will the average age ever become the same?

Inconsistent Systems

The data in the chart can be compared by making a graph.

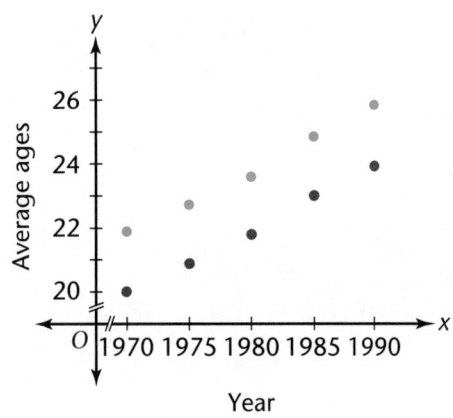

Let x = the year in 5-year intervals.
Let y = the average age for first marriages.

On the graph the years begin with 1970 and the ages begin with 20. What pattern do you see? A line of best fit can be used to model each data set.

The data for men can be modeled by $y = 0.2x - 368$.
The data for women can be modeled by $y = 0.2x - 374$.

Since the slopes of the equations are equal, the lines are parallel. If the trend continues, the lines never intersect, and the mean ages for first marriages will never be equal.

Now consider the system formed by the lines of best fit. $\begin{cases} y = 0.2x - 374 \\ y = 0.2x - 368 \end{cases}$

Solve the system by substituting $0.2x - 368$ for y in the first equation.

$$0.2x - 368 = 0.2x - 374$$
$$-368 = -374 \qquad \text{False}$$

Notice that both variables have been eliminated and 2 unequal numbers remain. This means there is no possible solution for this system. Since the graphs of the two equations are parallel, this system is called an *inconsistent system*.

An **inconsistent system** has no ordered pair that satisfies both of the original equations. Inconsistent systems have no common solutions. Their graphs are parallel and have the same slope but different y-intercepts.

CRITICAL Thinking

The graph for mean ages for marriages between 1970 and 1990 is a good example of an inconsistent system. Give another real life problem that might be modeled by parallel lines.

Consistent Systems

When a system has one or more solutions, the system is **consistent**.

•Exploration I *Dependent Systems*

GEOMETRY *Connection*

Line AB passes through $A(3, 2)$ and $B(-2, -8)$.
Line CD passes through $C(1, -2)$ and $D(7, 10)$.

1 Draw line AB and line CD on the same coordinate plane. What do you notice about the geometric relationship of the lines?

2 Use the point-slope formula, $y - y_1 = m(x - x_1)$, to write the equation for line AB and line CD. Compare your equations. What do you notice about the slopes of the two equations?

3 Solve the two equations as a system.

4 What happens to the variables? Is the statement of equality true? Do the two lines have any points in common? Is more than one solution possible? How many? ❖

Exploration 2 *Other Systems*

Let line 1 have slope m_1, y-intercept b_1, and be represented by equation 1.
Let line 2 have slope m_2, y-intercept b_2, and be represented by equation 2.

Let Equation 1 and Equation 2 form the following system.

$$\begin{cases} \text{Equation 1} \\ \text{Equation 2} \end{cases}$$

 Suppose line 1 and line 2 intersect in exactly 1 point. What can you say about
a. m_1 and m_2?
b. b_1 and b_2?
c. the solution to the system?

 Suppose line 1 and line 2 are parallel. What can you say about
a. m_1 and m_2?
b. b_1 and b_2?
c. the solution to the system?

 Suppose line 1 and line 2 are the same line. What can you say about
a. m_1 and m_2?
b. b_1 and b_2?
c. the solution to the system? ❖

Study the systems shown below.

Independent system
$$\begin{cases} y = x - 3 \\ y = -2x + 3 \end{cases}$$

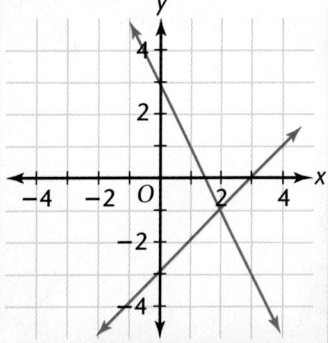

different slope
intersecting lines
one solution
same or different
y-intercepts
consistent

Dependent system
$$\begin{cases} y = x - 3 \\ 2y = 2x - 6 \end{cases}$$

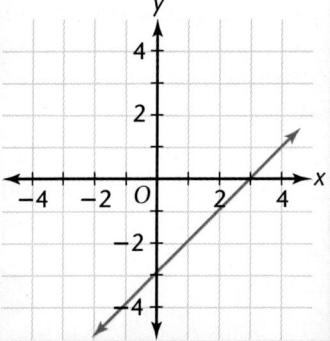

same slope
same line
infinite solutions
same
y-intercepts
consistent

Inconsistent system
$$\begin{cases} y = x + 3 \\ y = x - 3 \end{cases}$$

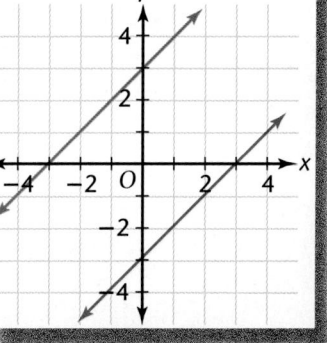

same slope
parallel lines
no solution
different
y-intercepts
inconsistent

What characteristics are the same or different for each pair of systems?

Cultural Connection: Asia In B.C.E. 213, a Chinese ruler ordered all books to be burned. All of the written knowledge of mathematics was lost. Four hundred fifty years later, Liu Hui came to the rescue. He was able to redo the lost mathematics because people still used rod numerals. The rod numerals were used to represent a decimal system with place value, much like the one we use today. Recall from Chapter 3 that red rods were used for positive numbers and black rods for negative numbers. Vertical and horizontal numerals were used alternately to keep the place values separate.

6,510,230 is

−6,510,230 is

First equation   Second equation

To represent a system of equations, the counting rods were arranged in such a way that one column was assigned to each equation of the system.

The figure at the left represents the following system.

$$\begin{cases} 4x + 3y = 8 \\ x - 2y = 13 \end{cases}$$

One row was assigned to the coefficients of each unknown in the equations. The elements of the last row consisted of the entries on the right-hand side of each equation. Red rods were used to represent positive *(cheng)* coefficients and black rods for negative *(fu)* coefficients.

EXERCISES & PROBLEMS

Communicate

Discuss how to identify the following systems of equations as inconsistent, dependent, or independent. Give reasons.

1. $\begin{cases} y = 2x - 3 \\ 3y = 6x - 9 \end{cases}$ **2.** $\begin{cases} y = -3x + 2 \\ y = 2x + 2 \end{cases}$ **3.** $\begin{cases} x + y = 4 \\ x + y = 5 \end{cases}$

4. Explain how to write an equation that would form a dependent system with the equation $y = 2x + 3$. Name what parts of the original equation you would use to write a new equation.

5. If two lines are parallel, discuss how you would write another equation that would form an inconsistent system with the equation $y = -3x + 4$.

6. If two equations in a system intersect at only one point, describe their slopes and y-intercepts. Tell how you determined your answer.

Practice & Apply

Algebraically solve each system. State your solution set clearly.

7. $\begin{cases} x + y = 7 \\ x + y = -5 \end{cases}$

8. $\begin{cases} 3y = 3x - 6 \\ y = x - 2 \end{cases}$

9. $\begin{cases} 3y = 2x - 24 \\ 4y = 3x - 3 \end{cases}$

10. $\begin{cases} y = -2x - 4 \\ 2x + y = 6 \end{cases}$

11. $\begin{cases} 2x + 3y = 11 \\ x - y = -7 \end{cases}$

12. $\begin{cases} 4x = y + 5 \\ 6x + 4y = -9 \end{cases}$

13. $\begin{cases} 4x + y = 8 \\ y = 4 - 2x \end{cases}$

14. $\begin{cases} y = \frac{3}{2}x + 4 \\ 2y - 8 = 3x \end{cases}$

15. $\begin{cases} y = \frac{1}{2}x + 9 \\ 2y - x = 1 \end{cases}$

Determine which of the following systems are dependent, independent, or inconsistent.

16. $\begin{cases} 2y + x = 8 \\ y = 2x + 4 \end{cases}$

17. $\begin{cases} 3x + y = 6 \\ y = 3x + 9 \end{cases}$

18. $\begin{cases} y + 2 = 5x \\ y = -3x - 2 \end{cases}$

19. $\begin{cases} y + 6x = 8 \\ y = -6x + 8 \end{cases}$

20. $\begin{cases} x - 5y = 10 \\ -5y = -x + 6 \end{cases}$

21. $\begin{cases} y + 4x = 12 \\ y = 5x + 12 \end{cases}$

Use the graph to answer Exercises 22–27.

22. Write the equations of the lines AB and CD.

23. What kind of system do lines AB and CD represent? Write a complete sentence that states at least two reasons for your answer.

24. Write the equation of a line that would form an independent system with line AB and that contains the point A. Graph your line.

25. Write the equation of a line that would form a dependent system with line CD and that contains the point D. Graph your line.

26. What is the point of intersection of $y = 6$ with line AB? with line CD?

27. Write the equation of a line that would form an inconsistent system with $y = 6$.

28. Geometry Triangle ABC has a perimeter of 18 centimeters while rectangle $PQRS$ has a perimeter of 24 centimeters. Write a system of equations for the figures. How many solutions are possible?

29. The equations for the lines AB and BC form a dependent system. If the slope of the line BC is -3 and the points $A(-3, p)$ and $B(5, 2 - p)$ are given, find the value of p.

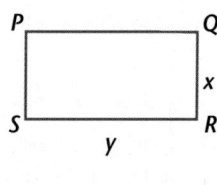

30. Career Options After you work for 10 years, you are offered the choice of two different jobs. The first has a starting salary of $2500 per month with a $750 per month expense account and an annual $800 bonus. The other job pays $3250 per month and has two semi-annual bonuses of $350 each. Write a short paragraph that explains which of the two job offers you would take and why.

31. 🖳 **Geometry** The sum of two angles is 180°. What is the measure of each angle if one angle is 30° more than twice the other?

32. Number Theory One number is 24 more than another number. If the sum of the numbers is 260, what are the numbers?

Demography The chart shows the populations of Center City and Bay City from 1970 to 1990. Complete Exercises 33–34.

33. Write a system of equations that can be used to predict the population (y) in years (x) for each city.

34. In how many years will the population of Center City equal that of Bay City?

	Center City	Bay City
1970	40,070	43,750
1975	42,570	46,250
1980	45,070	48,750
1985	47,570	51,250
1990	50,070	53,750

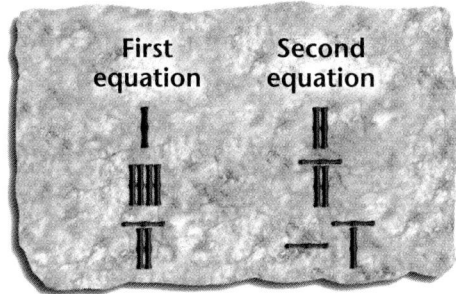

First equation   Second equation

35. Cultural Connection: Asia Change these equations from the Chinese counting-rod system to standard algebraic equations. Then solve the equations.

Look Back

36. Make a table of x and y values, and graph the equation $2x + 5y = 7$.   **[Lesson 1.5]**

37. At what point does a parabola change direction?   **[Lesson 2.2]**

38. Make a table of ordered pairs for the reciprocal function $y = \frac{1}{x}$, when $x = 1, 3, 6, 7, \frac{1}{3}, \frac{1}{6}, \frac{1}{7}$.   **[Lesson 2.3]**

39. What is the value of the integer function, $\text{INT}\left(7\frac{2}{3}\right)$?   **[Lesson 2.4]**

40. Write the y values for ordered pairs for the absolute value function of x for integers from -5 to 5.   **[Lesson 2.4]**

Solve for x.   [Lesson 4.6]

41. $2x - 5 = 15$ **42.** $-3x + 4 = -14$ **43.** $4x + 7 = -5$ **44.** $ax + b = c$

Look Beyond

45. To get to work, Charley travels down Interstate 87 for 15 miles at 60 miles per hour and then 12 miles down Central Avenue at 40 miles per hour. How many minutes should it take Charley to get to work?

why *A tape cassette player costs more than you have saved. Your grade on the last test was less than you expected. These conditions create inequalities. Sometimes unequal conditions can be reversed. Two inequalities form a system that can be solved by graphing.*

Linear Inequalities

Graphing on a coordinate plane allows you to find solutions to linear inequalities that contain more than one variable.

•Exploration• *Graphing Linear Inequalities*

1 Graph the line $y = x$ on a coordinate grid or graphics calculator.

2 Locate and label a point P above the line $y = x$. What is the relationship between the x-coordinate and y-coordinate of P?

3 Locate and label a point Q below the line $y = x$. What is the relationship between the x-coordinate and the y-coordinate of Q? ❖

COORDINATE GEOMETRY *Connection*

The graph of a linear equation such as $y = x$ divides a coordinate plane into two half-planes. The line graphed for $y = x$ is a **boundary line** for the two half-planes. A solution for the linear inequality $y \geq x$ includes *all ordered pairs* that make the inequality true. The shaded area, plus the line $y = x$ contains all possible solutions for $y \geq x$. Test the coordinates for points R and S. Which point has coordinates that satisfy the inequality $y \geq x$?

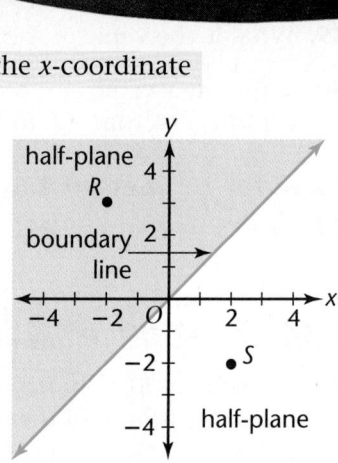

When the inequality symbol is ≤ or ≥, the points on the boundary line *are* included. In this case a solid line is used to represent the boundary. When the inequality symbol is < or >, the points on the boundary line *are not* included. In this case a dashed line is used to represent the boundary.

EXAMPLE 1

Graph the following inequality to find the solution.

$$x - 2y < 4$$

Solution ➤

Graph the boundary line $x - 2y = 4$.

Let $x = 0$.   Let $y = 0$.
$0 - 2y = 4$ $x - 2(0) = 4$
$-2y = 4$ $x - 0 = 4$
$y = -2$ $x = 4$

The points $(0, -2)$ and $(4, 0)$ are on the boundary line. Since the inequality symbol does not include the equal sign, use a dashed line to connect the points.

Try a point such as $(0, 0)$ to see if it satisfies the inequality. Substitute 0 for x and 0 for y.

$$x - 2y < 4$$
$$0 - 2(0) < 4$$
$$0 < 4 \quad \text{True}$$

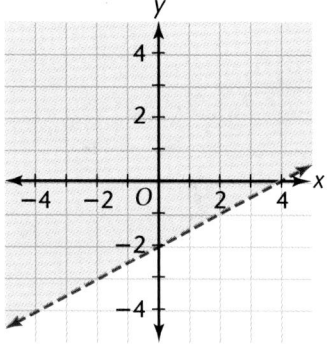

Since the substitution makes the inequality true, $(0, 0)$ is a point in the solution. Shade the half-plane that contains that point. ❖

Systems of Linear Inequalities

For systems of linear inequalities, the solution for the system is the **intersection** of the solution for each inequality. The intersection on the graph is that portion of the plane where the solutions for each inequality overlap. Every point in the intersection satisfies the system.

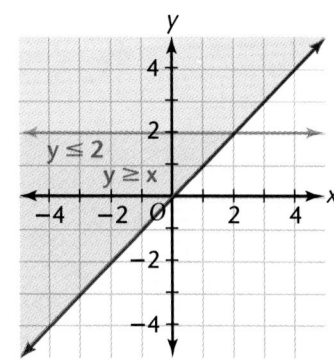

EXAMPLE 2

Solve by graphing. $\begin{cases} x + y \geq -1 \\ -2x + y < 3 \end{cases}$

Solution ➤

Graph the boundary line
$x + y = -1$.

Let $x = 0$.   Let $y = 0$.
$0 + y = -1$ $x + 0 = -1$
     $y = -1$ $x = -1$

The points $(0, -1)$ and $(-1, 0)$ are
on the boundary line. Since the
inequality symbol includes the
equal sign, draw a solid line to
connect the points.

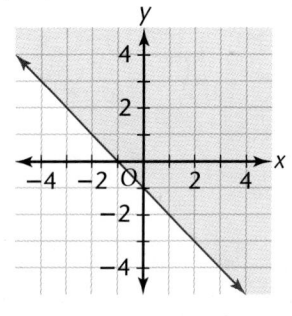

Try a point $(0, 0)$ to see if it satisfies
the inequality.

$$0 + 0 \geq -1$$
$$0 \geq -1 \quad \text{True}$$

Since the substitution is true, shade
the half-plane that contains $(0, 0)$.

Graph the boundary line
$-2x + y = 3$.

Let $x = 0$.   Let $y = 0$.
$-2(0) + y = 3$ $-2x + 0 = 3$
      $y = 3$ $x = -\dfrac{3}{2}$

The points $(0, 3)$ and $(-\frac{3}{2}, 0)$ are
on the boundary line. Since the
inequality symbol does not include
the equal sign, draw a dashed line
to connect the points.

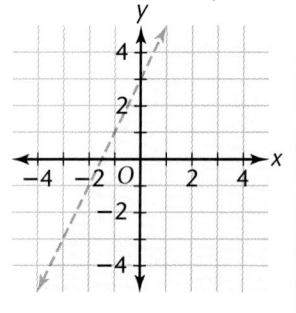

Try a point $(0, 0)$ to see if it satisfies
the inequality.

$$-2(0) + 0 < 3$$
$$0 < 3 \quad \text{True}$$

Since the substitution is true, shade
the half-plane that contains $(0, 0)$.

The graph shows the solution for the two
inequalities. Try a point such as $(1, 1)$
from the intersecting region to test
both inequalities.

$$x + y \geq -1$$
$$1 + 1 \geq -1$$
$$2 \geq -1 \quad \text{True}$$

$$-2x + y < 3$$
$$-2 + 1 < 3$$
$$-1 < 3 \quad \text{True} ❖$$

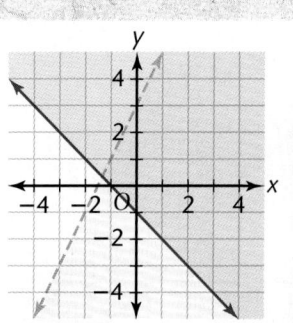

Why is $(1, 1)$ a good point to use as a check for both inequalities?

Try This Solve by graphing. $\begin{cases} y < -2x + 4 \\ y > 3x - 4 \end{cases}$

Kwame gets one job mowing lawns at $4.50 an hour and a second job at the library for $6.00 an hour.

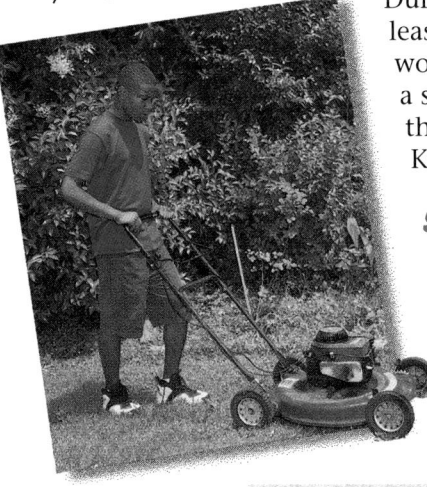

EXAMPLE 3

During the summer, Kwame wants to make at least $90 a week working part-time. He can work no more than 30 hours per week. Write a system of inequalities that represents all of the combinations of hours and jobs that Kwame can work. Graph this solution.

Solution ➤

Define the variables.

Let m = the number of hours mowing.
Let l = the number of hours at the library.

Write the inequalities that satisfy the conditions in the problem. Use the correct inequality symbols.

Kwame can work *no more* than 30 hours a week.	Kwame wants to make *at least* $90 a week.
$m + l \leq 30$	$4.5m + 6l \geq 90$
Graph $m + l = 30$.	Graph $4.5m + 6l = 90$.

Let $l = 0$.	Let $m = 0$.	Let $l = 0$.	Let $m = 0$.
$m + 0 = 30$	$0 + l = 30$	$4.5m + 0 = 90$	$4.5(0) + 6l = 90$
$m = 30$	$l = 30$	$4.5m = 90$	$0 + 6l = 90$
		$m = 20$	$l = 15$

The points (30, 0) and (0, 30) are on the boundary line. Since the inequality symbol includes the equal sign, use a solid line to connect the points.

The points (20, 0) and (0, 15) are on the boundary line. Since the inequality symbol includes the equal sign, use a solid line to connect the points.

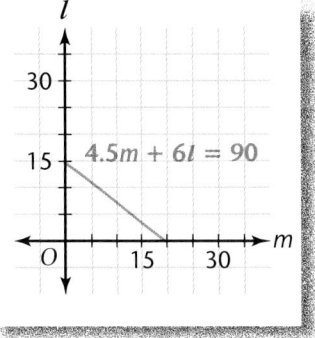

Why do we exclude the points to the left of the *l*-axis and below the *m*-axis?

The point (0, 0) makes the inequality true, so shade below.

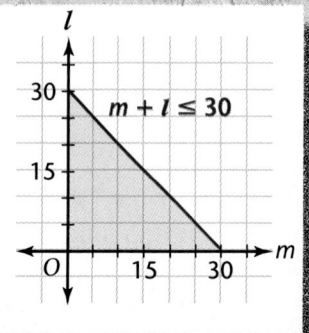

The shaded area shows the greatest number of hours Kwame can work at each job.

The point (15, 5) makes the inequality true, so shade above.

The shaded area shows the least number of hours Kwame works to make at least $90.

The following graph represents the solution for the two inequalities.

Any of the points in the intersecting region will satisfy the conditions for both the inequalities.

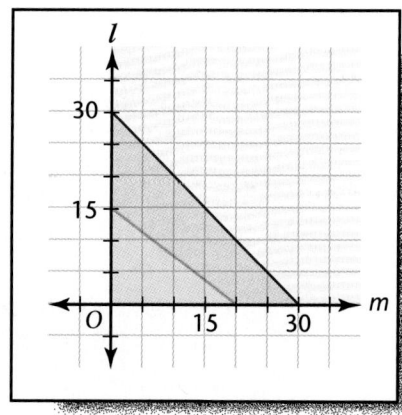

The total hours for each set of points is less than or equal to 30. The total number of dollars Kwame earns for each set of points is $90 or more. ❖

MAXIMUM MINIMUM *Connection*

Explain how you would determine how many hours Kwame can work at each job.

To check, try two points from the intersecting region. For example (15, 5) and (14, 10) show that:

• Kwame can work 15 hours mowing and 5 hours at the library.
• Kwame can work 14 hours mowing and 10 hours at the library.

CRITICAL *Thinking*

What is the least number of hours Kwame can work to earn at least $90? What is the greatest amount of money he can make and still work 30 hours or less?

EXERCISES & PROBLEMS

Communicate

1. Explain how you would decide whether to draw a solid or a dashed line for an inequality.

2. Tell how to choose the area to shade for the inequality $x + 2y < 2$. What is a good point to test in the shaded area?

Discuss how to graph the intersection for the following inequalities. Describe the process for determining the shaded region that satisfies both inequalities in each pair.

3. $\begin{cases} y < 1 \\ x < 1 \end{cases}$
   4. $\begin{cases} 8x + 4y > 12 \\ y < 3 \end{cases}$
   5. $\begin{cases} x \le 3 \\ x - 2y \ge 2 \end{cases}$
   6. $\begin{cases} x + y \le -2 \\ x + y > -2 \end{cases}$

Explain how to write a system of inequalities defined by the problem. Describe how to graph the solution for values that make sense. Show your graph.

7. Each spring Marta and her family add flowers to their garden. They plan on buying no more than 20 new perennials at $5.00 per pot and a number of annuals at $1.50 per pot. Marta knows they have budgeted at least $30 for the plants. What combination of each type of plant can they buy?

Practice & Apply

Graph the common solution of each system. Choose a point from the solution. Check both inequalities.

8. $\begin{cases} 2x - 3y > 6 \\ 5x + 4y < 12 \end{cases}$
   9. $\begin{cases} x - 4y \le 12 \\ 4y + x \le 12 \end{cases}$
   10. $\begin{cases} y < x - 5 \\ y \le 3 \end{cases}$

11. $\begin{cases} 2x + y \le 4 \\ 2y + x \ge 8 \end{cases}$
   12. $\begin{cases} 5y < 2x - 5 \\ 4x + 3y \le 9 \end{cases}$
   13. $\begin{cases} 4y < 3x + 8 \\ y \le 1 \end{cases}$

14. $\begin{cases} x \ge 4 \\ 0 < y \end{cases}$
   15. $\begin{cases} x - 2y < 8 \\ x + y \ge 5 \end{cases}$
   16. $\begin{cases} 3x - y > -2 \\ x - y > -1 \end{cases}$

17. **Geometry** Write the system of inequalities defined by each perimeter, and graph the common solution. Make sure that the solution represents real life possibilities.

Rectangle ABCD's perimeter is at most 30 centimeters, while rectangle PQRS's perimeter is at least 12 centimeters.

You and a friend are going to have lunch at Al's Burgers.

18. Between the two of you, you have $8.00 and each of you wants at least one Alburger and an order of fries. Can either of you have seconds?

Match the set of inequalities to the graph that represents the solution.

19. $y < -2x + 6$ and $y \le 3x - 4$

20. $y \le -2x + 6$ and $y > 3x - 4$

a.

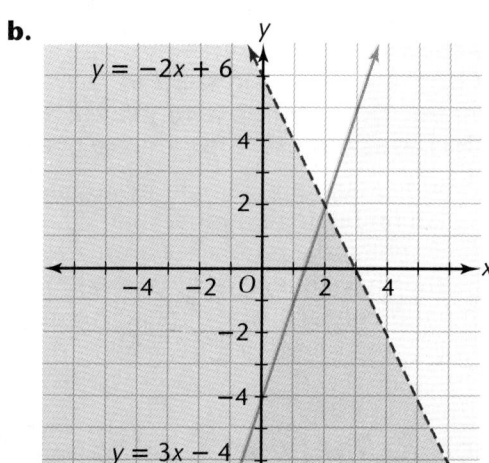

b.

21. Write the ordered pairs for the points A, B, C, D.

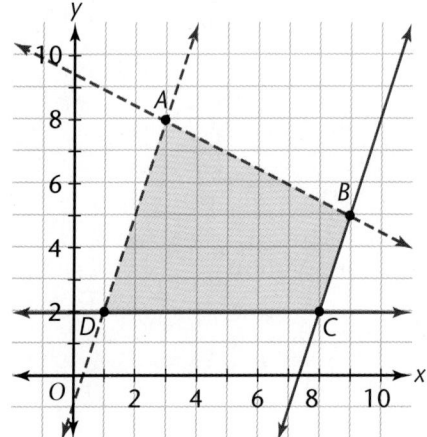

22. Write the system of inequalities that represents the shaded region. Use the points shown to find equations for the boundary lines.

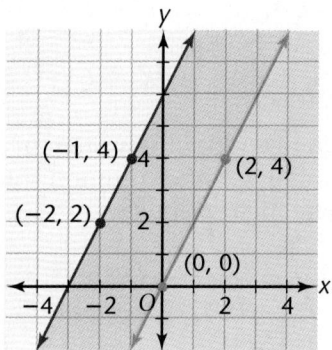

23. Sports You usually average no more than 16 points in a game. Find the combinations of baskets and free throws you could have made.

24. Budgeting Time On most days Abel spends no more than 2 hours on his math and science homework. If math always takes about twice as long as science, what is the maximum time that Abel usually spends on science?

Geometry The length of the first of two rectangles is at least 5 more than its width, while the length of the second is no more than 5 times its width.

25. Represent the length of the first rectangle as an inequality.

26. Represent the length of the second rectangle as an inequality.

27. Draw figures to represent the two rectangles, and label the lengths and widths as inequalities.

Look Back

28. Find the first differences for the following sequence.
   [Lesson 1.2]
      5 17 37 65 101, . . .

29. Find the second differences for the following sequence.
   [Lesson 1.2]
      5 17 37 65 101, . . .

30. Simplify $-8 - 2\{3 + 6\left(5 - \frac{4}{2}\right) + 7\}$.   **[Lesson 1.4]**

31. Fund-raising How much would each person contribute toward a goal of $1200 if there were 20 people contributing equally?
   [Lesson 2.3]

Add the following like terms.   **[Lesson 3.3]**

32. $2x^2 + 4 + 3y - x^2 + 2$ **33.** $3a + a + z + 4a + 2b^2 - b$

Use the slopes to tell whether the following lines are parallel, perpendicular, or neither.   **[Lesson 5.6]**

34. $x + y = 7$ **35.** $-2x + y = -5$ **36.** $-x + 2y = 6$ **37.** $x - 2y = 4$
   $x + 3y = 9$ $-2x - y = 5$ $-x + 2y = -3$ $2x + y = 1$

Look Beyond

For $f(x) = 2x + 4$, evaluate the following.

38. $f(3)$ **39.** $f(-2)$ **40.** $f\left(-\frac{1}{2}\right)$

In the Lincoln high basketball game, you made only 2-point shots and 1-point free throws.

LESSON 6.6 Classic Applications

why *Have you ever seen a classic car? Classic cars usually represent a model that was revolutionary for its time because of either style or technology. Mathematics has classic problems involving mixtures, money, digits, rates, and age.*

Many people collect, restore, and display classic cars that were revolutionary for their day. Only time will tell whether new models will become classics.

Each classic problem from algebra can usually be solved by setting up the equations in special ways.

EXAMPLE 1

Mixture This classic application uses percent to express how much of a chemical is in a solution.

How many ounces of a 15% acid solution should be mixed with a 40% acid solution to produce 60 ounces of a 25% acid solution?

Solution ➤

Draw a diagram and make a table to help solve this problem.

Define the variables.

Let x = the number of ounces of 15% acid solution.

Let y = the number of ounces of 40% acid solution.

Place the information in a chart to help organize the facts.

	% acid	Amount of solution	Amount of acid
First solution	15%	x	$0.15x$
Second solution	40%	y	$0.40y$
Final solution	25%	60	$0.25(60)$

From the table, write two equations.

$$\begin{cases} x + y = 60 \\ 0.15x + 0.40y = 0.25(60) \end{cases}$$

Solving the system gives the solution (36, 24).

To get the required 60 ounces of the 25% acid solution, mix 36 ounces of the 15% acid solution with 24 ounces of the 40% acid solution. ❖

EXAMPLE 2

Money In classic coin problems you can usually set up two equations—one involving the number of coins and the other involving the value of the coins.

In a coin bank there are 250 dimes and quarters worth a total of $39.25. Find how many of each kind of coin are in the bank.

Solution ➤

Create a chart to organize the information.

Coin type	Number	Coin value	Value in cents
Quarters	q	25¢	$25q$
Dimes	d	10¢	$10d$
Total	250		3925

From the table, write two equations based on the totals.

$$\begin{cases} q + d = 250 \\ 25q + 10d = 3925 \end{cases}$$

Solving the system, you find that q is 95 and p is 155.

The coin bank has 95 quarters and 155 dimes. A check indicates that the solution is correct. ❖

EXAMPLE 3

Numeration

Other examples of classic problems are digit problems. In most digit problems the trick is to write the value of a number in expanded form. You can write 52 as $5(10) + 2$. If you reverse the digits in 52, you get 25, and can write it as $2(10) + 5$.

The sum of the digits of a two-digit number is 7. The original two-digit number is 3 less than 4 times the number with its digits reversed. Find the original two-digit number.

Solution ➤

Define the variables.

Let t = the ten's digit of the original number.
Let u = the unit's digit of the original number.

The two-digit number expressed in expanded notation is
$$10t + u.$$

The number with its digits reversed in expanded notation is
$$10u + t.$$

The problem says the sum of the digits is 7.
$$t + u = 7$$

The original 2-digit number is 3 less than 4 times the number with its digits reversed.
$$10t + u = -3 + 4[10u + t]$$

Solving the system $\begin{cases} t + u = 7 \\ 10t + u = -3 + 4[10u + t] \end{cases}$ gives $u = 1$ and $t = 6$.

If t is 6 and u is 1, then the original number is 61. The number 61 is a two-digit number which is 3 less than 4 times itself with its digits reversed. A check proves the solution is correct. ❖

Try This A two-digit number whose ten's digit is 2 more than the units digit is 3 more than 6 times the sum of its digits. Find the original number.

EXAMPLE 4

Travel Classic motion problems have some basic elements. Travel by plane, train, or automobile involves speed, distance, and time. Create charts, draw diagrams, and organize the information to help solve the following problem.

A plane leaves New York City and heads for Chicago, which is 750 miles away. The plane, flying against the wind, takes 2.5 hours to reach Chicago. After refueling the plane returns to New York City, traveling with the wind, in 2 hours. Find the rate of the wind and the rate of the plane with no wind.

Solution ➤

Define the variables. Chart the information.

Let x = the rate of the plane without any wind.
Let y = the rate of the wind.

Rate against wind: $x - y$ Time: 2.5 hr Distance: 750 mi
NYC to Chicago   mph

Rate with wind: $x + y$ Time: 2 hr   Distance: 750 mi
Chicago to NYC   mph

CHICAGO   NEW YORK

To write the system of equations, recall the relationship between rate, time, and distance.

	Rate	× Time	=	Distance
For Chicago to NYC:	$(x - y)$	(2.5)	=	750
For NYC to Chicago:	$(x + y)$	(2)	=	750

To solve, divide to get simpler equations.

$$(x - y)(2.5) = 750 \longrightarrow x - y = 300$$
$$(x + y)(2) = 750 \longrightarrow x + y = 375$$

Solving the system gives $x = 337.5$ and $y = 37.5$.

The rate of the plane is 337.5 miles per hour and the rate of the wind is 37.5 miles per hour. A check proves the solution is correct. ❖

CRITICAL *Thinking*

How would the rate of the plane be affected if there were no wind?

EXAMPLE 5

Age Age problems are classic problems that can usually be solved by writing the two related ages at two points in time. In 4 years the father will be 5 times as old as his son. How old is each now?

A father is 32 years older than his son.

Solution ➤

Write two equations about the age of the father and son at two different times. Define the variables.

Let f = the father's age.   Let s = the son's age.

The father is 32 years older than his son.
$$f = s + 32$$

In 4 years the father will be 5 times the son's age. Add 4 years to each age.
$$f + 4 = 5(s + 4)$$

Solving the system gives $s = 4$ and $f = 36$.

At the present time the father is 36 years old, and his son is 4 years old. A check proves the solution is correct. ❖

EXERCISES & PROBLEMS

Communicate

Discuss the problem-solving strategy you would use to solve the following problems, and explain how to *set up* the equations.

1. When Jim cleaned out the reflecting pool at the library, he found 20 nickels and quarters. The mixture of nickels and quarters totaled $2.60. How many quarters did Jim find?

2. Chemistry You are running a chemistry experiment that requires you mix a 25% alcohol solution with a 5% alcohol solution. This mixture produces 20 liters of a solution that contains 2.6 liters of pure alcohol. How many liters of the 25% alcohol solution must you use?

A bird can fly with the wind 3 times as fast as it can fly against the wind. How can you represent

3. the rate of the bird in still air (no wind) and the rate of the wind.

4. the rate of the bird flying with the wind and against the wind.

5. Discuss how the rate of the bird with the wind compares with the rate of the bird against the wind.

Describe how to set up Exercise 6, and tell how to check your answer after discussing how to solve the problem.

6. The sum of the digits of a two-digit number is 8. If 16 is added to the original number, the result is 3 times the original number with its digits reversed. Find the original number.

Practice & Apply

7. Recreation Ramona turns her canoe around and returns downstream to her cabin in exactly 1 hour. What is the rate of the creek's current and the rate of Ramona's paddling in still water?

Ramona is staying at a vacation resort called Stony Cabins. She rents a canoe and takes 1.5 hours to paddle her canoe upstream.

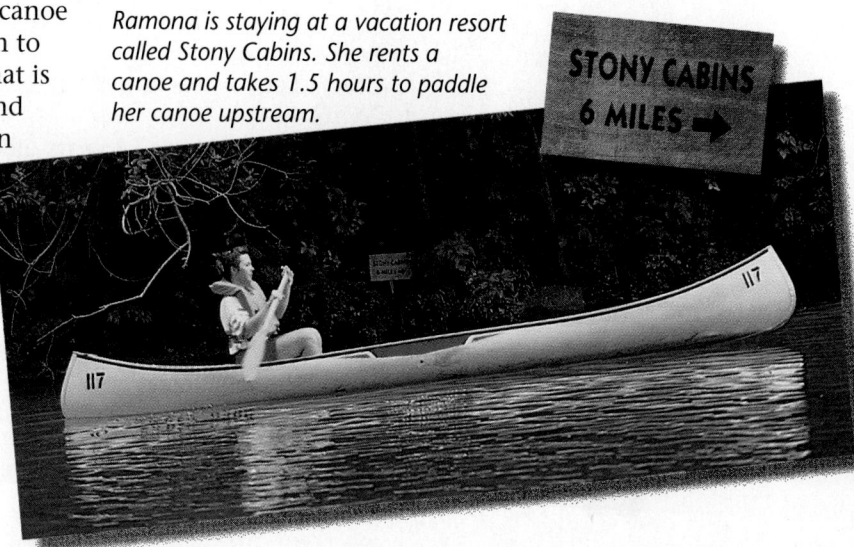

8. The coin box of the hospital's vending machine contains 6 times as many quarters as dimes. If the total amount of money at the end of the day was $28.80, how many of each coin was in the box?

9. Numeration Find the two-digit number whose ten's digit is 4 less than the unit's digit and the original number is 2 more than 3 times the sum of the digits.

With a tailwind, a jet flew 2000 miles in 8 hours, but the return trip against the same wind required 10 hours.

10. Aviation Find the jet's speed and the wind speed.

11. In 15 years, Maya will be twice as old as David is now. In 15 years, David will be as old as Maya will be 10 years from now. How old are they now?

12. A 4% salt solution is mixed with a 16% salt solution. How many milliliters of each solution are needed to obtain 600 milliliters of a 10% solution?

13. Wymon is 20 years older than Sabrina. In 8 years, Wymon will be twice as old as Sabrina. How old is each today?

14. Cultural Connection: Asia Apply the problem-solving strategy of working backwards (inversion) to a problem supplied by Aryabhata, circa 500 C.E.

"O beautiful maiden with beaming eyes, tell me, since you understand the method of inversion:

What number multiplied by 3, then increased by $\frac{3}{4}$ of the product, then divided by 7, then diminished by $\frac{1}{3}$ of the result, then multiplied by itself, then diminished by 52, whose square root is then extracted before 8 is added, and then divided by 10, gives the final result of 2?"

Look Back

Evaluate the following expressions if $x = 2$, $y = 1$, and $z = 1$. [Lesson 1.3]

15. $2x + 4y - 3z$ **16.** $\frac{2x - z}{5}$ **17.** $\frac{x + y}{z}$ **18.** $3xyz$

19. Simplify $(29 + 1) \div 3 \cdot 2 - 4$.   **[Lesson 1.4]**

Multiply. [Lesson 4.2]

20. $-[2(a + b)]$ **21.** $-[-11(x - y)]$ **22.** $2(x - y) + 3[-2(x + y)]$

Find the slope of the following equations. [Lesson 5.4]

23. $3x + 4y = 12$ **24.** $5y = 10x - 20$

Look Beyond

Find each square root.

25. $\sqrt{\dfrac{25}{625}}$ **26.** $\sqrt{a^2}$

Minimum Cost Maximum Profit

Finding the Right Mixture

For a hike you want to make a mixture of nuts and raisins to provide all of the needed nutrients but at the lowest cost. The chart shows the nutritional information and cost per ounce.

You have a weight limit for your pack and can allow no more than 18 ounces for the mix. The snack should supply at least 1800 calories with no more than 117 grams of fat. How many ounces each of raisins and nuts should you buy to minimize the costs?

	Calories	Fat	Price/ounce
Nuts	150	13g	$0.40
Raisins	90	0g	$0.10

1 Define the variables for the unknowns.
   Let *n* = the number of ounces of nuts.

   Let *r* = the number of ounces of raisins.

2 Write a linear equation, called the optimization equation, to express the relationship between the highest and lowest cost.

   Thus, $0.40n$ equals the cost of nuts.

   Thus, $0.10r$ equals the cost of raisins.

   Total cost $0.40n + $0.10r$.

13n ≤ 117

A

n + r ≤ 18

B

150n + 90r ≥ 1800   C

 Write a set of inequalities that define the conditions for the problem.

The snack should supply at least 1800 calories.

$150n + 90r \geq 1800$

The snack should supply no more than 117 grams of fat. $13n \leq 117$

You can carry no more than 18 ounces of the mix $n + r \leq 18$

The conditions defined by the inequalities that must be satisfied are called the **constraints**.

 Sketch or use graphing technology to graph each of the three inequalities. Shade the region that represents the solution set.

The shaded region, **ABC**, represents the solution set. This region, determined by the linear inequalities that contains the possible solutions, is called the **feasibility region**.

 Find the coordinates of the vertices of ABC. They are A(3, 15), B(9, 9), and C(9, 5).

Use each vertex of the shaded region to find values for the optimization equation. Then select the equation with the lowest cost.

Substitute A(3, 15).

Total cost = $0.40n + $0.10r

Total cost = $0.40(3) + $0.10(15)

Total cost = $1.20 + $1.50 = $2.70

Substitute B(9, 9).

Total cost = $3.60 + $0.90 = $4.70

Substitute C(9, 5).

Total cost = $3.60 + $0.50 = $4.10

Three ounces of nuts and 15 ounces of raisins gives a minimum cost of $2.70.

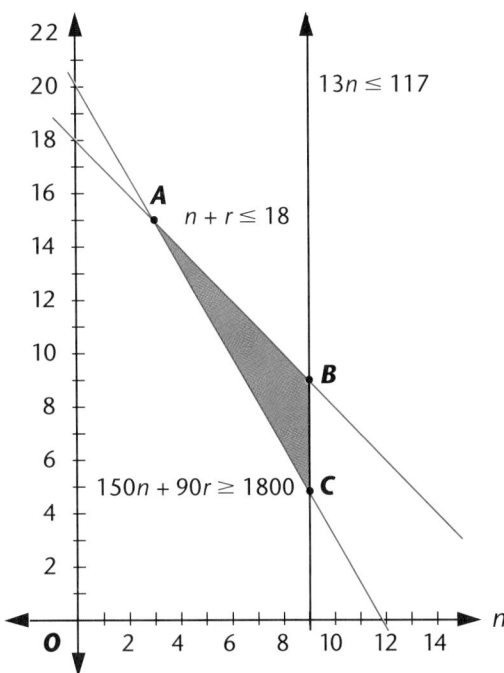

Activity
Solve Using the Method of Linear Programming

The maker of recreational motorcycles lists two models, the Mountain Climber and the Dune Crawler. The chart shows some of the assembly information.

	Labor Hours	Maximum produced	Profit
Mountain Crawler	150	60	$120
Dune Crawler	200	45	$180

If the company has no more than 12,000 hours of labor available for production of motorcycles, find the number of each model that should be built to give the maximum profit.

Chapter 6 Review

Vocabulary

Key Skills & Exercises

Lesson 6.1

➤ **Key Skills**

Find a solution to a system of equations by graphing the equations and finding the point of intersection.

Solve by graphing. $\begin{cases} x + 3y = 12 \\ x - y = 4 \end{cases}$

Write the equations in slope-intercept form.

$$x + 3y = 12 \qquad\qquad x - y = 4$$
$$y = -\frac{x}{3} + 4 \qquad\qquad y = x - 4$$

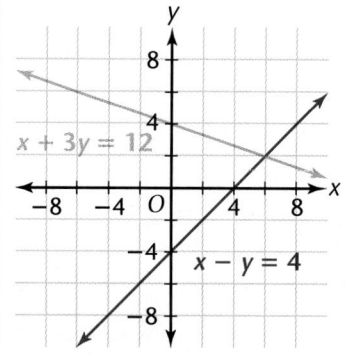

Graph the lines on graph paper or with a graphics calculator. The solution is $(6, 2)$. Check by substituting.

➤ **Exercises**

Solve each system by graphing.

1. $\begin{cases} x - y = 6 \\ x + 2y = -9 \end{cases}$
2. $\begin{cases} y = -x \\ y = 2x \end{cases}$
3. $\begin{cases} x + 6y = 3 \\ 3x + y = -8 \end{cases}$
4. $\begin{cases} 3x + y = 6 \\ y + 2 = x \end{cases}$

Lesson 6.2

➤ **Key Skills**

Find a solution to a system of equations by the substitution method.

Solve. $\begin{cases} c + 4d = 1 \\ 2c + 7d = 6 \end{cases}$

Solve $c + 4d = 1$ for c to get $c = 1 - 4d$.
Then substitute in $2c + 7d = 6$.

$$2(1 - 4d) + 7d = 6$$
$$d = -4$$

Now substitute -4 for d in the equation $c = 1 - 4d$.

$$c = 1 - 4(-4)$$
$$c = 17$$

The solution is $(17, -4)$.

➤ **Exercises**

Solve each system by the substitution method.

5. $\begin{cases} 2x + y = 1 \\ x + y = 2 \end{cases}$ **6.** $\begin{cases} x - y = 6 \\ 2x - 4y = 28 \end{cases}$ **7.** $\begin{cases} x + 2y = 1 \\ 2x - 8y = -1 \end{cases}$ **8.** $\begin{cases} 4x = 3y + 44 \\ x + y = -3 \end{cases}$

Lesson 6.3

➤ **Key Skills**

Find a solution to a system of equations by the elimination method.

Solve by the elimination method.

$$\begin{cases} x + y = -3 \\ 2x = 3y - 11 \end{cases}$$

Write the equations in standard form, and compare the coefficients.

$$\begin{array}{ll} x + y = -3 & \text{Equation 1} \\ 2x - 3y = -11 & \text{Equation 2} \end{array}$$

Choose a variable to eliminate. Multiply both sides of one of the equations by the same number, and solve for the remaining variable.

$$\begin{array}{ll} 3(x + y) = 3(-3) \rightarrow & 3x + 3y = -9 \\ 2x - 3y = -11 \rightarrow & \underline{2x - 3y = -11} \\ & 5x \quad\quad = -20 \\ & x = -4 \end{array}$$

Substitute this value of x in $x + y = -3$ to find the value of y.

$$-4 + y = -3$$
$$y = 1$$

The solution is $(-4, 1)$.

➤ **Exercises**

Solve each system by the elimination method.

9. $\begin{cases} 4x + 5y = 3 \\ 2x + 5y = -11 \end{cases}$ **10.** $\begin{cases} 2x + 3y = -6 \\ -5x - 9y = 14 \end{cases}$ **11.** $\begin{cases} 0.5x + y = 0 \\ 0.9x - 0.2y = -2 \end{cases}$ **12.** $\begin{cases} -2x + 4y = 12 \\ 3x - 2y = -10 \end{cases}$

Lesson 6.4

➤ **Key Skills**

Identify a system of equations as independent, dependent, or inconsistent by its solution.

Solve the system algebraically.

$$\begin{cases} 2x + y = 4 \\ 2x + y = 1 \end{cases}$$

Use the elimination method.

$$\begin{array}{ll} 2x + y = 4 & 2x + y = 4 \\ -1(2x + y) = -1(1) & \underline{-2x - y = -1} \\ & 0 \neq 3 \end{array}$$

All variables are eliminated, and you are left with two unequal numbers. There is no possible solution for this system of equations. The system is inconsistent.

➤ **Exercises**

Solve each system. State your solution set clearly, and identify the system as independent, dependent, or inconsistent.

13. $\begin{cases} 3x - 2y = 7 \\ 4y = -14 + 6x \end{cases}$ **14.** $\begin{cases} 4y = 2x + 20 \\ 3x - y = -20 \end{cases}$ **15.** $\begin{cases} 2x + 5y = 7 \\ 3y = 2x + 17 \end{cases}$ **16.** $\begin{cases} x + y = 7 \\ 28 - 2y = 2x \end{cases}$

Lesson 6.5

➤ *Key Skills*

Graph the solution for a system of linear inequalities.

Graph the solution set for the system.

$$\begin{cases} y < -x - 1 \\ y \le x + 1 \end{cases}$$

Graph $y < -x - 1$ as a dashed line and $y \le x + 1$ as a solid line. Use substitution to determine which half-plane to shade for each inequality. The intersection of the solution sets for the two inequalities is the solution set for the system.

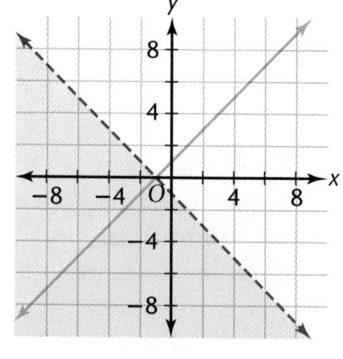

➤ *Exercises*

Graph the common solution. Choose a point to check.

17. $\begin{cases} y \ge 4 \\ y \le x + 1 \end{cases}$

18. $\begin{cases} 4x - y > 1 \\ 2x + y > -2 \end{cases}$

19. $\begin{cases} y - 1 \le x \\ 4y - 2x \ge 2 \end{cases}$

20. $\begin{cases} 2x - y < 3 \\ 2y - 2 \le 6x \end{cases}$

Lesson 6.6

➤ *Key Skills*

Use a system of equations to solve classic problems.

Finance Mr. Carver invested $5000, part at 5% interest and the rest at 3.9% interest. If he earned $233.50 in interest the first year, how much did he invest at each rate?

Make a table to organize the information.

Amount	Rate	Interest
x	5%	$0.05x$
y	3.9%	$0.039y$
Total 5000		233.50

From the table, write two equations.

$$x + y = 5000$$
$$0.05x + 0.039y = 233.50$$

Solving the system gives $x = 3500$ and $y = 1500$.

Mr. Carver invested $3500 at 5% and $1500 at 3.9%.

➤ *Exercises*

Use a system of equations to solve classic problems.

21. Samantha has 43 quarters and nickels in her wallet. If her change is worth $7.75, how many of each kind of coin does Samantha have?

22. Finance Tina invested $600 of her bonus at 4.3% and the rest at 9% interest. If she earned $97.80 in interest in one year, how much was her bonus?

Applications

23. Chemistry How many ounces of a 25% acid solution should be mixed with a 50% acid solution to produce 100 ounces of a 40% acid solution?

24. Geometry The sum of the measures of angles x, y, and z of triangle XYZ is 180°. Angle x has a measure of 60°. The measure of angle y is 15 more than twice the measure of the angle z. What are the measures for angles y and z?

Chapter 6 Assessment

Solve each system by graphing.

1. $\begin{cases} 2x = y + 1 \\ y = 3x + 2 \end{cases}$

2. $\begin{cases} 4y - 2x = -2 \\ 4y - x = 0 \end{cases}$

3. $\begin{cases} 3y - 12x = 18 \\ 3x + y = -1 \end{cases}$

Solve each system by the substitution method.

4. $\begin{cases} y = x + 3 \\ 2x - 4y = -12 \end{cases}$

5. $\begin{cases} 2x - 2y = -6 \\ x + 2y = -9 \end{cases}$

6. $\begin{cases} 2x - 3y = -2 \\ 4x + y = 1 \end{cases}$

7. One number is 4 greater than another number. Twice the larger number is 2 less than 3 times the smaller number. Find both numbers.

Solve each system by the elimination method.

8. $\begin{cases} 2x + 3y = 3 \\ 4x - 3y = 3 \end{cases}$

9. $\begin{cases} 3x + 4y = -2 \\ x - 2y = 6 \end{cases}$

10. $\begin{cases} 5y = 3x - 18 \\ 2x = 3y + 12 \end{cases}$

11. The length of a rectangle is 3 inches more than its width. Find the length and width of this rectangle.

Perimeter is 22 inches.

12. Marla bought 12 books at a garage sale. Some of them were hardback and the rest were paperback. She paid 50¢ for each paperback book and 75¢ for each hardback book. If she spent $6.75, how many of each type of book did she buy?

Solve each system algebraically. State your solution clearly.

13. $\begin{cases} y + 5 = x \\ y = x + 2 \end{cases}$

14. $\begin{cases} 5y - 2x = -15 \\ 4x = 10y + 30 \end{cases}$

15. $\begin{cases} 4x + 5y = 14 \\ 5y = 8x + 2 \end{cases}$

Graph the common solution of each system. Choose a point from the solution, and check both inequalities.

16. $\begin{cases} y > x \\ y - 2x \le 2 \end{cases}$

17. $\begin{cases} y \ge 3x - 1 \\ y < 5 \end{cases}$

18. $\begin{cases} 2y + 10x > -12 \\ y + 5 < 4x \end{cases}$

19. Yvette has $16.30 in dimes and quarters. If she has a total of 88 coins, how many of each type of coin does she have?

20. Henry divides Hannah's current age by 2 and then adds that result to her current age. The result is Henry's current age. Henry notices that in 24 years, his age will be 3 times Hannah's current age. How old are Henry and Hannah?

21. Tamara drove 810 miles in 14 hours over 2 days. If she drove at an average rate of 55 miles per hour the first day and an average rate of 60 miles per hour the second day, how far and how long did she drive each day?

Chapters 1-6
Cumulative Assessment

College Entrance Exam Practice

Quantitative Comparison For Questions 1–5, write
A if the quantity in Column A is greater than the quantity in Column B;
B if the quantity in Column B is greater than the quantity in Column A;
C if the two quantities are equal; or
D if the relationship cannot be determined from the information given.

	Column A	Column B	Answers
1.	$$\begin{cases} 3x + 2y = 6 \\ x + y = 0 \end{cases}$$ x	y	(A) (B) (C) (D) **[Lesson 6.2]**
2.	0.37	$\dfrac{37}{10}$	(A) (B) (C) (D) **[Lesson 4.5]**
3.	The eighth term of the sequence 2, 5, 8, 11, 14, . . .	The eighth term of the sequence 32, 30, 28, 26, 24, . . .	(A) (B) (C) (D) **[Lesson 1.1]**
4.	The solution to $x + 3 \geq 4$	The solution to $x - 7 < 10$	(A) (B) (C) (D) **[Lesson 4.7]**
5.	The slope of the line passing through the points (1, 2) and (0, 3).	The slope of the line $4x + 4y = 12$.	(A) (B) (C) (D) **[Lesson 5.1]**

6. Which expression is the solution to the equation $-4g = -64$?   **[Lesson 4.2]**

    a. $\dfrac{-64}{4}$ **b.** $-64 \cdot (-4)$ **c.** $\dfrac{-64}{-4}$ **d.** -16

7. What is the reciprocal of $-\dfrac{1}{6}$?   **[Lesson 2.3]**

    a. 1.6   **b.** -1.6 **c.** 6   **d.** -6

8. What are the first three terms of a sequence if the fourth and fifth terms are 27 and 39 and the second differences are a constant 2?   **[Lesson 1.2]**

    a. 21, 23, 25   **b.** 3, 9, 17   **c.** 6, 8, 10   **d.** 13, 15, 17

9. If a basketball league has 8 teams and each of the teams plays each of the other teams, how many games will be played?   **[Lesson 1.1]**

    a. 28   **b.** 36   **c.** 72   **d.** 56

10. What is ABS(8.6)?   **[Lesson 2.4]**

    a. 8   **b.** -8 **c.** 8.6   **d.** -8.6

11. Which point is the vertex of the graph of the data in the table? **[Lesson 2.2]**

x	1	2	3	4	5
y	1	0	1	4	9

    a. $(2, 0)$ **b.** $(1, 1)$ **c.** $(3, 1)$ **d.** $(4, 4)$

12. Which number line shows the solution set to the inequality $x - 6 > -4$? **[Lesson 4.8]**

a.
```
←++++++++++++●++++++++→
 -5 -4 -3 -2 -1 0 1 2 3 4 5
```
b.
```
←++++++++++++++○+++++++→
 -5 -4 -3 -2 -1 0 1 2 3 4 5
```

c.
```
←++++++++++++●++++++++→
 -5 -4 -3 -2 -1 0 1 2 3 4 5
```
d.
```
←++++++++++++++○+++++++→
 -5 -4 -3 -2 -1 0 1 2 3 4 5
```

13. What is the solution to $\frac{3}{5}p = -3$? **[Lesson 4.2]**

    a. $-\frac{9}{5}$ **b.** $-\frac{5}{9}$ **c.** -5 **d.** $-\frac{1}{5}$

14. What is 65% of 120? **[Lesson 4.5]**

    a. 78   **b.** 185   **c.** 7800   **d.** 780

15. Which of the following lines is vertical? **[Lesson 5.5]**

    a. $y = -5$ **b.** $x = 3$ **c.** $y = -3x + 1$ **d.** $y = 2x - 2$

Evaluate the following. **[Lesson 2.4]**

16. $|-2.2|$ **17.** ABS(7)   **18.** INT(9.6)

19. Does the exponential sequence 4, 8, 16, 32, 64 show growth or decay? Explain. **[Lesson 2.2]**

Find each sum. **[Lesson 3.1]**

20. $-25 + (-4)$ **21.** $-36 + 6 + |-6|$ **22.** $452 + (-452)$

Solve each equation. **[Lesson 4.7]**

23. $t + \frac{1}{6} = \frac{1}{3}$ **24.** $2a - 5 = 11$ **25.** $\frac{r}{4} + 6 = -4$

26. What is the y-intercept of a line that is parallel to the line $2x + 3y = 4$ and contains the point $(3, -1)$? **[Lesson 5.6]**

Solve each system. **[Lessons 6.1, 6.2, 6.3]**

27. $\begin{cases} y = 7x \\ 2x - y = -10 \end{cases}$ **28.** $\begin{cases} 5x + 2y = 8 \\ x + y = 1 \end{cases}$

Free-Response Grid The following questions may be answered using a free-response grid commonly used by standardized test services.

29. The sum of the digits of a 2-digit number is 7. The tens digit is 1 less than the units digit. Find the number.

30. What is the solution to the proportion $\frac{c}{16} = \frac{3}{8}$? **[Lesson 4.5]**

31. A suit is marked down from an original price of $75 to $63.75. By what percent of the original price has the sweater been marked down? **[Lesson 4.5]**

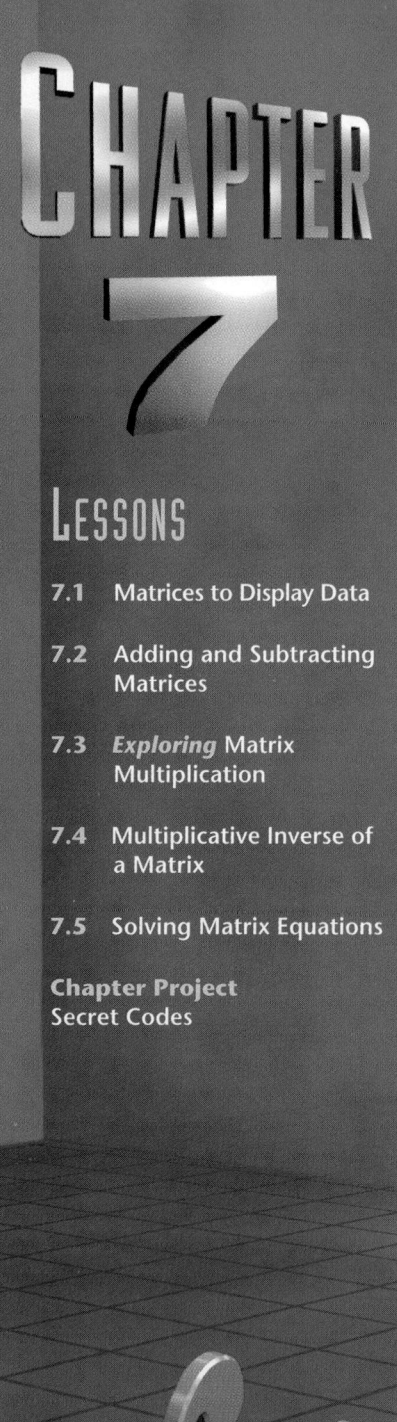

CHAPTER 7

Matrices

Ancient Nubians of Sudan left spectacular pyramids and tombs containing precious artifacts. They left us a puzzle as well. In 1960, the building of the Aswan High Dam on the Nile River meant that the tombs would soon be flooded. Archaeologists from all countries rushed in to excavate the sites and save the artifacts.

Millions of artifacts were recovered, representing civilizations from the Paleolithic to the Middle Ages. But what story do they tell? Researchers used a matrix to organize and classify the treasures at the Fadrus site. The rows and columns of the matrix quickly filled with numbers. These numbers represented the amount of artifacts with similar characteristics found at burial sites. Was it a heart scarab from Fadrus or spiral earrings from Kerma? It was all recorded.

There was a problem with the size of the matrix, 512 rows and 171 columns. The matrix did an excellent job of organizing the data, but in the 1960s, no computer was powerful enough to do the mathematical analysis. For years the artifacts remained unstudied in museum basements. It was not until the 1990s that a computer was available that could handle the analysis and begin to reveal the secrets of these ancient civilizations.

PORTFOLIO ACTIVITY

The many sites on the Aswan High Dam flood plain were divided into three different culture groups: C-Group, Pan-grave, and Transitional. Excitement mounted when excavators realized that some early C-Group sites were royal burial sites. The kings buried at these sites may have been the first pharaohs. Thousands of artifacts needed to be classified. The initial categories were tomb materials, pottery, and nonceramic pieces.

The typical C-Group site might contain 5 pieces of tomb materials, 8 pieces of decorated pottery, and 15 non-ceramic artifacts. These nonceramic pieces were often decorative hair clasps, the most identifying feature of the C-Group sites. The average Pan-grave site would contain 9 tomb materials, 5 pottery pieces, and 12 nonceramic artifacts. The Pan-grave sites were the only places where nonceramic Kohl pots were found. At the Transitional sites, an average of 18 tomb materials, 3 bits of pottery, and 3 nonceramic artifacts were found. The Transitional sites seem to be characterized by burial rituals.

At the end of this phase of digging, the total number of artifacts in each of the categories was 175 tomb materials, 104 pottery pieces, and 195 nonceramic artifacts.

Organize this information into matrix form, and find how many of each type of site were included in this part of the study.

Matrices to Display Data

Why *Do you know exactly where all your papers and projects are? You can easily find what you are looking for when you are organized. You have already seen how tables of values, graphs, and spreadsheets are used in problem solving. Now you will use a closely related and very important organizational tool called a matrix.*

Organizing Data

Imagine you are writing a paper on the effect that automobiles have had on society. You want to include the following data as evidence.

Transportation

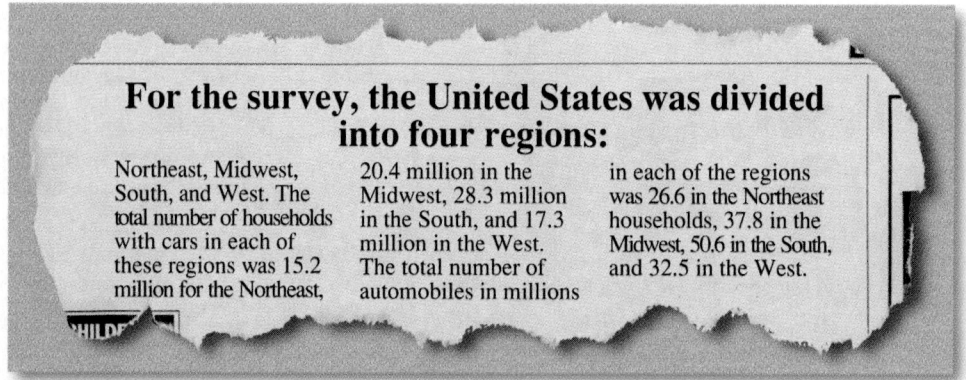

For the survey, the United States was divided into four regions:

Northeast, Midwest, South, and West. The total number of households with cars in each of these regions was 15.2 million for the Northeast, 20.4 million in the Midwest, 28.3 million in the South, and 17.3 million in the West. The total number of automobiles in millions in each of the regions was 26.6 in the Northeast households, 37.8 in the Midwest, 50.6 in the South, and 32.5 in the West.

The amount of information in the paragraph is difficult to interpret. However, when organized in a table the same information is easier to read and understand.

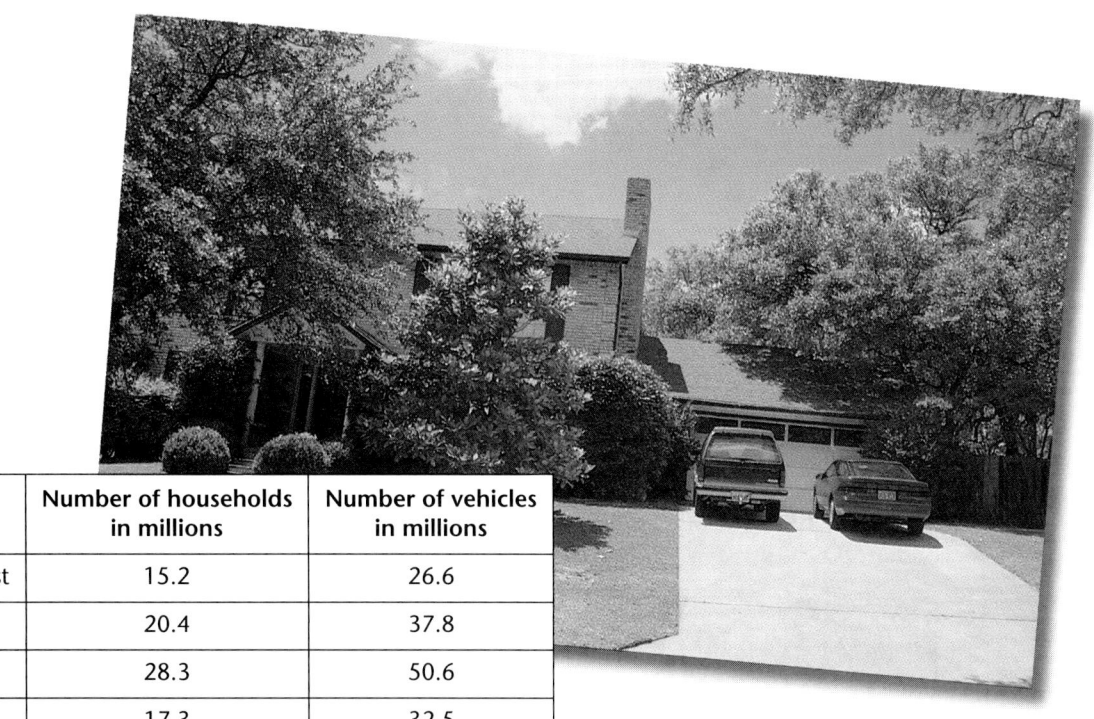

Region	Number of households in millions	Number of vehicles in millions
Northeast	15.2	26.6
Midwest	20.4	37.8
South	28.3	50.6
West	17.3	32.5

$$C = \begin{bmatrix} 15.2 & 26.6 \\ 20.4 & 37.8 \\ 28.3 & 50.6 \\ 17.3 & 32.5 \end{bmatrix}$$

A matrix is named with a capital letter.

When data are arranged in a table of rows and columns and enclosed by brackets, [ ], the structure is a **matrix** (plural, matrices).

Labels describe the information in the rows and columns, but are not part of the matrix.

Matrix C has 4 rows and 2 columns. The number of rows by the number of columns describes the *row by column* dimensions of the matrix. This is sometimes written 4×2 or $C_{4 \times 2}$.

The information that appears in each position of the matrix is called an **entry**. Each entry in the matrix can be located by its matrix **address**.

	Number of households in millions	Number of vehicles in millions	
Northeast	15.2	26.6	
Midwest	20.4	37.8	= C
South	28.3	50.6	
West	17.3	32.5	

Addresses for matrix C

	Column 1	Column 2
Row 1	c_{11}	c_{12}
Row 2	c_{21}	c_{22}
Row 3	c_{31}	c_{32}
Row 4	c_{41}	c_{42}

4 rows by 2 columns
The dimension is 4×2.

The address c_{32} represents the entry in row 3 and column 2 of matrix C. You will find 50.6 at this address. The labels show that c_{32} represents the fact that in the *South* there are 50.6 *million vehicles*.

What is the address in matrix C for the entry 37.8?

EXAMPLE 1

Information in an atlas is in table form. This is easily changed to the format of a matrix. Answer the following questions about the location of the information in the matrix.

Airline Distances in Kilometers

	London	Mexico City	New York	Paris	San Francisco	Tokyo
London	0	8944	5583	344	8637	9590
Mexico City	8944	0	3363	9213	3037	11,321
New York	5583	3363	0	5851	4139	10,874
Paris	344	9213	5851	0	8975	9741
San Francisco	8637	3037	4139	8975	0	8288
Tokyo	9590	11,321	10,874	9741	8288	0

$= D$

A What do the entries in matrix D represent?

B What are the row and column dimensions of D?

C How many kilometers is it from Paris to Mexico City? What is the matrix address of this entry?

D What does the entry d_{36} represent?

E What does the entry d_{22} represent?

Solution ➤

A Each entry in the matrix represents the distance measured in kilometers between the cities whose names appear in the row and column headings.

B D has dimensions 6 by 6. A matrix is called a **square matrix** if its row and column dimensions are equal.

C It is 9213 kilometers between Paris and Mexico City. This information is located at d_{42} and d_{24}.

D The entry at d_{36} is 10,874. It represents the number of kilometers between New York and Tokyo.

E The entry at d_{22} is 0, and it shows that there are 0 kilometers between Mexico City and Mexico City. ❖

Try This What is the entry at d_{35}? What are the addresses for the entry 3037?

CRITICAL Thinking Identify at least 3 reasons why matrices are a good way to organize data.

Matrix Equality

Two matrices are equal when their dimensions are the same and their corresponding entries are equal.

$$A = \begin{bmatrix} 1 & 2 & 3 \\ 4 & 5 & 6 \end{bmatrix} \qquad B = \begin{bmatrix} 1 & 3 & 5 \\ 2 & 4 & 6 \end{bmatrix} \qquad C = \begin{bmatrix} 1 & 4 \\ 2 & 5 \\ 3 & 6 \end{bmatrix}$$

Matrix A and matrix B have the same dimension, but unequal entries, so $A \neq B$. Matrix B and matrix C do not have the same dimensions, so $B \neq C$.

EXAMPLE 2

Does $S = T$? Explain why or why not.

$$S = \begin{bmatrix} 2^2 & (12 - 3) \\ \sqrt{49} & 11 \\ 3(-9) & \dfrac{24}{3} \end{bmatrix} \qquad T = \begin{bmatrix} 2(2) & \sqrt{81} \\ 7 & \dfrac{44}{4} \\ -(30 - 3) & 2^3 \end{bmatrix}$$

Solution ➤

Recall the characteristics that make two matrices equal.

a. The matrices have the same dimensions. In this example, both matrices have dimensions 3 by 2.

b. The matrices have equal entries. Each matrix simplifies to the matrix at the right.

$$\begin{bmatrix} 4 & 9 \\ 7 & 11 \\ -27 & 8 \end{bmatrix}$$

Since the dimensions and corresponding entries of each matrix are equal, $S = T$. ❖

EXAMPLE 3

Matrices U and V are equal. Find the values for x, y, and z.

$$U = \begin{bmatrix} 3x + 2 & x + 2 \\ 2xy & 3y \end{bmatrix} \qquad V = \begin{bmatrix} 14 & 2y \\ 3z & -4y + 21 \end{bmatrix}$$

Solution ➤

Since $U = V$, the corresponding entries are equal. You can form individual equations from the equivalent entries.

$$3x + 2 = 14 \qquad x + 2 = 2y$$
$$2xy = 3z \qquad 3y = -4y + 21$$

When you solve $3x + 2 = 14$, you find that $x = 4$. You can then substitute 4 for x in any of the other equations.

If you substitute 4 for x in the equation $x + 2 = 2y$ and solve, you get $4 + 2 = 2y$, or $y = 3$. You can then use $3y = -4y + 21$ to check the y-value. ❖

How do you find the value for z?

Exercises & Problems

Communicate

1. The paragraph contains information that would be more efficiently represented in matrix form. Explain how to create two matrices to organize this data. Place descriptive labels in the left and top margins to identify the rows and columns.

2. Discuss how to determine the dimensions of each matrix. Select an entry of your own choice from each matrix, and give its address.

> The retail business relies on data to predict the buying habits of the public. During the years 1985, 1989, and 1993, the Retailers' News records show that the sales of men's suits was 14.6, 10.8, and 11.5 units respectively. For all apparel, a sales unit represents 1000 items. For the same years, the sales of women's suits were 17.4, 12.3, and 8.6 units. Shirt sales for women were 25.6, 21.3, and 16.2 units and jeans sales for women were 98.2, 90.1, and 80.3 units. Men's jeans sales were 242.7, 210.5, and 186.9 units. Shirt sales of 16.7, 16.2, and 17.3 units were posted for men.
> ...na knowing last years numbers. *ext season.*

$$P = \begin{bmatrix} x & y \\ -11 & 2 \\ -10 & 6 \\ -9 & 4 \\ -8 & 6 \\ -7 & 2 \end{bmatrix} \quad Q = \begin{bmatrix} x & y \\ -5 & 2 \\ -3 & 6 \\ -1 & 2 \\ -2 & 4 \\ -4 & 4 \end{bmatrix} \quad R = \begin{bmatrix} x & y \\ 3 & 2 \\ 3 & 6 \\ 1 & 6 \\ 5 & 6 \end{bmatrix} \quad T = \begin{bmatrix} x & y \\ 7 & 2 \\ 7 & 6 \\ 7 & 4 \\ 9 & 4 \\ 9 & 6 \\ 9 & 2 \end{bmatrix}$$

3. Discuss how you would tell which matrix or matrices are equal to matrix G.

$$G = \begin{bmatrix} |-5| & \sqrt{4} \\ -(5)^2 & (-2)^3 \end{bmatrix}$$

$$A = \begin{bmatrix} 5 & 2 \\ -25 & -8 \end{bmatrix} \quad B = \begin{bmatrix} -5 & 2 \\ 25 & -8 \end{bmatrix} \quad C = \begin{bmatrix} 8-3 & (\sqrt{2})^2 \\ 20-45 & -20+12 \end{bmatrix} \quad D = \begin{bmatrix} |5| & -2 \\ -\frac{100}{4} & -|8| \end{bmatrix}$$

Practice & Apply

4. Matrices M and N are equal. Find the values of $a, b, c, d, g,$ and k.

$$M = \begin{bmatrix} 2(a+4) & 77 & \frac{1}{3}c \\ -5d-1 & 30 & -\frac{1}{2}k \end{bmatrix} \quad N = \begin{bmatrix} -12 & 11b & 5 \\ -(3-d) & 0.4g & \frac{3}{4}k-3 \end{bmatrix}$$

Business The Ace Auto Repair shop keeps a record of employee absences. Use this record for Exercises 5–8.

	M	T	W	Th	F	
Managers	2	1	0	1	4	
Mechanics	2	0	2	1	0	= A
Secretaries	3	2	1	0	1	

5. What is the entry in the a_{24} address of the auto repair shop matrix? What does this entry represent?

6. On which day were the most employees absent? What reason would you give for this high number of absences?

7. Which group of employees has the least number of absences this week?

8. If there is a total of 30 employees at Ace Auto Repair, what percent of the employees were absent on Friday?

9. **Transportation** Four cities, A, B, C, and D, and the air routes between them are represented by the diagram. The arrows show the only directions of travel allowed. From the diagram you can see there are 2 routes from A to B. In the matrix there is the number 2 for the entry AB. The same number is shown for BC. Complete the given matrix to show the number of allowable air routes between the cities. Use 0 if there is no direct path between cities.

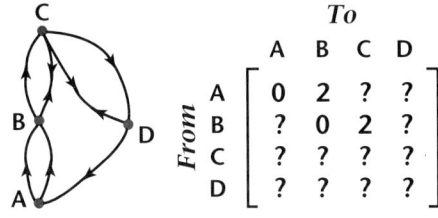

$$\begin{array}{c} \\ \\ From \end{array} \begin{array}{c} \\ A \\ B \\ C \\ D \end{array} \begin{bmatrix} A & B & C & D \\ 0 & 2 & ? & ? \\ ? & 0 & 2 & ? \\ ? & ? & ? & ? \\ ? & ? & ? & ? \end{bmatrix}$$

To

Date	Shuttle	Date	Shuttle
9/88	Discovery	10/90	Discovery
12/88	Atlantis	11/90	Atlantis
3/89	Discovery	12/90	Columbia
5/89	Atlantis	3/91	Discovery
8/89	Columbia	4/91	Atlantis
10/89	Atlantis	5/91	Columbia
11/89	Discovery	7/91	Discovery
1/90	Columbia	8/91	Atlantis
2/90	Atlantis	11/91	Discovery
4/90	Discovery	12/91	Atlantis

Space Exploration The table shows the space shuttle names and launch dates for the years 1988–1991.

10. Form matrix L to display the number of shuttle launches in a given year. Begin with 1988, and label the rows by year. Label the columns with the shuttle names in alphabetical order.

11. What are the dimensions of your matrix?

12. What does the matrix address l_{42} represent in matrix L?

13. Create another column to represent total launches for the year. In which year did the most launches occur?

14. Over the 4-year period, which year had the least number of launches? What historical event resulted in so few launches that year?

Geography The population of Chula Vista, California and Abilene, Texas for the years 1980 and 1990 is represented in the matrix.

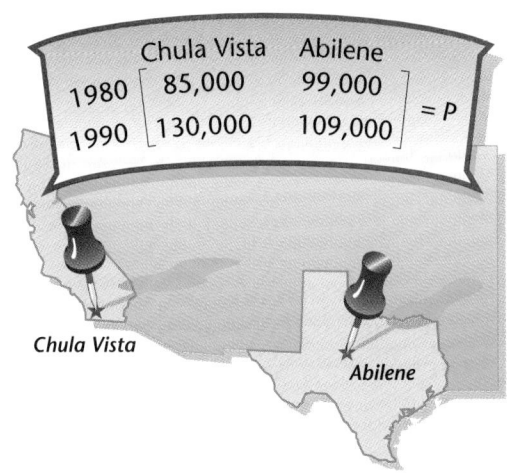

$$\begin{array}{c} \\ 1980 \\ 1990 \end{array} \begin{bmatrix} \text{Chula Vista} & \text{Abilene} \\ 85{,}000 & 99{,}000 \\ 130{,}000 & 109{,}000 \end{bmatrix} = P$$

Chula Vista

Abilene

15. If the rate of growth was assumed to be constant for both cities over this 10-year period, what was each city's population increase per year?

16. Using the assumed population increase for each city, create another matrix that shows each city's population for **each** year from 1980 to 1990.

17. In what year were the populations equal?

18. Write equations that could be used to predict the population for each city. Graphically or algebraically predict the cities' populations in 2000.

Look Back

19. Is the sequence 55, 65, 75, 85, . . . linear or exponential?   **[Lesson 2.1]**

Match the name of the property to the example that best illustrates it. **[Lessons 3.1–3.3]**

Examples	Properties
20. $5 + 0 = 5$	**a.** Property of Opposites
21. $-8 + 8 = 0$	**b.** Associative for Addition
22. $4 + 9 = 9 + 4$	**c.** Addition Property of Zero
23. $12 + 18 = 6(2 + 3)$	**d.** Commutative for Addition
24. $7 + (11 + 4) = (7 + 11) + 4$	**e.** Distributive Property

25. At the end of the season sale at Well-Dressed Women, Ms. Sanchez bought a suit for 40% off the ticket price. She paid $96. What was the original price?   **[Lesson 4.5]**

26. Cultural Connection: Americas The earliest math texts in the Americas were in Spanish. Solve this problem from Guatemala written by D. Juan Joseph de Padilla in 1732.   **[Lesson 4.6]**

Find 3 numbers whose sum is 120. The second number is twice the first. The third number is 3 times the sum of the other two.

27. Solve and graph the solution set on the number line.   **[Lesson 4.7]**
$$3x - (5 - 2x) \leq x - 21 \quad \text{and} \quad -7x - 12 \leq 16$$

28. Graph the equation $3x + 4y = 12$.   **[Lesson 5.4]**

Look Beyond

Technology   Learn how to enter a matrix into a graphics calculator or computer. Enter and display the following matrices. Refer to the calculator manual.

29. $M = \begin{bmatrix} 1 & -5 & 7 \\ 0.5 & 0 & -12 \end{bmatrix}$ **30.** $N = \begin{bmatrix} 4 & 0 \\ 0 & -1 \end{bmatrix}$

The matrix displays the number of people that helped in an environmental clean-up project for 3 areas of a city.

31. Which area had the greatest number of helpers in week 1?

32. How many helpers were in Areas 2 and 3 in week 3?

33. What was the total number of helpers in week 2?

	Area 1	Area 2	Area 3
Week 1	25	35	50
Week 2	15	40	32
Week 3	18	42	45
Week 4	21	53	46

LESSON 7.2 Adding and Subtracting Matrices

why *A matrix is an effective tool for storing data in a way that makes sense visually. What if the data changes? If you keep getting new data tables, you will need a way to combine the information in the old and new matrices without disturbing the organization. You need to know how to add and subtract matrices.*

Your last look at the stock inventory for the Super Sight and Sound Shop was on April 1. Since that time, business has been good, and much of the stock has been sold. The matrices below represent the *stock* available on April 1 and the *sales* for April.

Stock [April 1]:

Videos	DATs
Games	CDs
Laser disks	Tapes

$$A = \begin{bmatrix} 3000 & 1000 \\ 800 & 3000 \\ 1500 & 1000 \end{bmatrix}$$

Sales [April]:

Videos	DATs
Games	CDs
Laser disks	Tapes

$$B = \begin{bmatrix} 2254 & 952 \\ 675 & 1325 \\ 1187 & 548 \end{bmatrix}$$

EXAMPLE 1

Inventory Control   The sales matrix shows how many of each item were sold in April. Since the original stock has decreased, how would you determine the available stock for May?

Solution ➤

Subtract each entry in B from the corresponding entry in A. The result will be a new matrix, C, that shows the stock available May 1. For example, the video stock available **May 1** is $3000 - 2254$ or **746**. Place **746** in address c_{11}.

$$
\begin{array}{ccccc}
A & - & B & = & C \\
\textit{Stock [April 1]} & - & \textit{Sales [April]} & = & \textit{Stock [May 1]}
\end{array}
$$

$$
\begin{bmatrix} \mathbf{3000} & 1000 \\ 800 & 3000 \\ 1500 & 1000 \end{bmatrix} - \begin{bmatrix} \mathbf{2254} & 952 \\ 675 & 1325 \\ 1187 & 548 \end{bmatrix} = \begin{bmatrix} \mathbf{746} & \\ & \\ & \end{bmatrix}
$$

Matrix C is completed by subtracting corresponding entries.

Stock [May 1]:

Videos	DATs
Games	CDs
Laser disks	Tapes

$$C = \begin{bmatrix} \mathbf{746} & 48 \\ 125 & 1675 \\ 313 & 452 \end{bmatrix}$$ ❖

EXAMPLE 2

The next week, the May delivery arrives with more stock. The shipment includes an invoice matrix D to let the store know what is being delivered.

Delivery [May]:

Videos	DATs
Games	CDs
Laser disks	Tapes

$$D = \begin{bmatrix} 2500 & 1500 \\ 625 & 775 \\ 1500 & 400 \end{bmatrix}$$

How would you update your stock matrix to show the increase in stock?

Solution ➤

If you add each entry in C to the corresponding entry in D, you will create a new stock matrix, E, for the available stock for May.

$$
\begin{array}{ccccc}
C & + & D & = & E \\
\textit{Stock [May 1]} & + & \textit{Delivery [May]} & = & \textit{Stock [May]}
\end{array}
$$

For example, video stock in the new matrix **Stock [May]** is $746 + 2500$ or 3246 at address e_{11}.

Stock [May]:

Videos	DATs
Games	CDs
Laser disks	Tapes

$$E = \begin{bmatrix} 3246 & 1548 \\ 750 & 2450 \\ 1813 & 852 \end{bmatrix}$$ ❖

Try This Let $X = \begin{bmatrix} 5 & -3 \\ 2 & 0 \end{bmatrix}$ and $Y = \begin{bmatrix} -1 & 7 \\ 0 & 3 \end{bmatrix}$. Find $X + Y$.

CRITICAL
Thinking

In a large business, how often does this process of updating the stock matrices take place? If there are thousands of different items in the stock of a large business, how does a store manage its inventory and manipulate its matrices?

•Exploration Addition and Subtraction Properties

Use the following matrices to explore and develop conjectures about the algebra of adding and subtracting matrices. This is a good opportunity to use a calculator or computer.

$$A = \begin{bmatrix} 6 & -3 \\ -4 & \frac{9}{2} \end{bmatrix} \qquad B = \begin{bmatrix} -4.2 & 1 \\ 0 & -6.6 \end{bmatrix} \qquad C = \begin{bmatrix} \frac{3}{4} & -10 \\ 9 & -2 \end{bmatrix} \qquad D = \begin{bmatrix} -2 & 11 & 7 \\ 5 & 9 & -12 \end{bmatrix}$$

 Add.   **a.** $A + C$ **b.** $C + A$ **c.** $B + D$
      d. $A + (B + C)$ **e.** $(A + B) + C$ **f.** $C + D$

Describe the procedure for adding matrices.

 Compare and contrast these additions. Are there any restrictions on whether the matrices can be added? Write a conjecture.

 Subtract.   **a.** $C - A$ **b.** $B - B$ **c.** $B - A$
       d. $A - C$ **e.** $B - C$ **f.** $C - D$

Describe the procedure for subtracting matrices.

Compare and contrast these subtractions. Are there any restrictions on whether the matrices can be subtracted? Write a conjecture. ❖

The **identity matrix** for addition is a matrix filled with zeros. It is also called the **zero matrix.** The zero matrix for a 2 by 2 matrix is shown here.

$$\begin{bmatrix} 0 & 0 \\ 0 & 0 \end{bmatrix} = \mathbf{0}_{2 \times 2}$$

Adding the zero matrix to another matrix does not change the original matrix.

$$\begin{bmatrix} 0 & 0 \\ 0 & 0 \end{bmatrix} + \begin{bmatrix} -1 & 5 \\ 7 & 0 \end{bmatrix} = \begin{bmatrix} -1 & 5 \\ 7 & 0 \end{bmatrix}$$
$$\begin{bmatrix} -1 & 5 \\ 7 & 0 \end{bmatrix} + \begin{bmatrix} 0 & 0 \\ 0 & 0 \end{bmatrix} = \begin{bmatrix} -1 & 5 \\ 7 & 0 \end{bmatrix}$$

Consider matrix C and matrix D. Form the zero matrix that can be added to C. Then form the zero matrix that can be added to D.

$$? + C = \begin{bmatrix} \frac{3}{4} & -10 \\ 9 & -2 \end{bmatrix} = C + ? \qquad ? + D = \begin{bmatrix} -2 & 11 & 7 \\ 5 & 9 & -12 \end{bmatrix} = D + ?$$

How do the zero matrices for C and D compare?

$L = \begin{bmatrix} 2 & 3 \\ 6 & -1 \end{bmatrix}$

When you add numbers, you know that certain properties are true.

$M = \begin{bmatrix} 4 & 5 \\ -2 & 0 \end{bmatrix}$

The *Commutative Property for Addition* states that for numbers a and b, $a + b = b + a$.

$N = \begin{bmatrix} -1 & 4 \\ 7 & -5 \end{bmatrix}$

The *Associative Property of Addition* states that for numbers a, b, and c, $a + (b + c) = (a + b) + c$.

Use the given matrices L, M, and N to show that matrix addition is also commutative and associative.

EXERCISES & PROBLEMS

Communicate

Explain how to perform each of the matrix operations. If a solution is not possible, explain why.

1. $\begin{bmatrix} 10 & 6 \\ 4 & -5 \end{bmatrix} + \begin{bmatrix} -1 & 6 \\ 9 & -7 \end{bmatrix}$

2. $\begin{bmatrix} -11 & 6 \\ 13 & 8 \\ 17 & -9 \end{bmatrix} - \begin{bmatrix} -2 & 10 \\ -16 & 12 \\ 4 & -3 \end{bmatrix}$

3. $\begin{bmatrix} \frac{5}{6} \\ \frac{7}{12} \end{bmatrix} + \begin{bmatrix} \frac{-5}{4} & \frac{2}{3} \end{bmatrix}$

4. $\begin{bmatrix} \frac{1}{2} & \frac{1}{4} & \frac{-1}{2} \\ \frac{2}{3} & \frac{4}{5} & \frac{1}{4} \end{bmatrix} + \begin{bmatrix} \frac{1}{2} & \frac{1}{4} & \frac{-1}{2} \\ \frac{2}{3} & \frac{4}{5} & \frac{1}{4} \end{bmatrix}$

5. Is matrix subtraction commutative? Explain.

6. Is matrix subtraction associative? Explain.

7. Is the zero matrix always a square matrix? Explain why or why not.

Practice & Apply

Perform each of the matrix operations. If an operation is not possible, explain why.

8. $\begin{bmatrix} 23 & 46 \\ 4 & -35 \end{bmatrix} + \begin{bmatrix} -51 & 16 \\ 29 & -7 \end{bmatrix}$

9. $\begin{bmatrix} 4 & 2 \\ 6 & -7 \\ 3 & 9 \end{bmatrix} + \begin{bmatrix} -5 & 7 \\ -3 & 9 \\ 3 & -6 \end{bmatrix} - \begin{bmatrix} -11 & 3 \\ 8 & -15 \\ -7 & -2 \end{bmatrix}$

10. $\begin{bmatrix} -6 & 1 \\ -1 & 7 \end{bmatrix} - \begin{bmatrix} -3 & 0 \\ 9 & -2 \end{bmatrix}$

11. $\begin{bmatrix} -1.4 & 4.3 & -9.6 \\ 15.8 & 1 & -3.5 \end{bmatrix} - \begin{bmatrix} 6.2 & 3.2 \\ -9.6 & 7.1 \\ 2.6 & -8.5 \end{bmatrix}$

12. $\begin{bmatrix} 3.2 & \frac{3}{5} \\ \frac{-8}{5} & -4.2 \end{bmatrix} - \begin{bmatrix} \frac{-4}{5} & 5.6 \\ -6.7 & \frac{3}{5} \end{bmatrix}$

13. $\begin{bmatrix} 2.3 & -9.3 & 1.8 \\ -8.2 & -4.1 & 10 \\ 2.1 & 6 & -0.16 \end{bmatrix} + \begin{bmatrix} -1.2 & -2.1 & 5.3 \\ 0.2 & -4 & -2.3 \\ -3.4 & 2.4 & 6 \end{bmatrix}$

14. $\begin{bmatrix} \frac{-2}{3} & \frac{4}{5} \\ \frac{1}{15} & \frac{7}{30} \end{bmatrix} - \begin{bmatrix} \frac{9}{15} & \frac{-4}{3} \\ \frac{-2}{5} & \frac{-4}{15} \end{bmatrix}$

15. $\begin{bmatrix} 0.4 & -1.5 & 0.9 \\ 2.6 & 6.9 & 3.7 \end{bmatrix} + \begin{bmatrix} -4.7 & 2.6 & 6.9 \\ -7.3 & 9.8 & -5.5 \end{bmatrix}$

16. $\begin{bmatrix} -3 & 4 \\ -4 & 0 \\ 2 & -5 \end{bmatrix} + \begin{bmatrix} 0 & -1 \\ -6 & 9 \\ -7 & 5 \end{bmatrix}$

17. $\begin{bmatrix} -25 & 32 & 14 \\ 36 & -42 & -45 \\ -71 & 65 & 29 \end{bmatrix} - \begin{bmatrix} 16 & -34 & -55 \\ 21 & 11 & 22 \\ -43 & -67 & -44 \end{bmatrix} + \begin{bmatrix} 57 & 79 & 64 \\ -38 & -22 & -48 \\ -56 & 88 & 26 \end{bmatrix}$

18. Inventory Control At the Super Sight and Sound Shop, the May sales matrix and June delivery invoice have just appeared on your desk. Make a new matrix to update the May stock matrix to reflect the May sales and June delivery. Label your answer matrix **Stock [June]**.

Stock [May 1]

Videos	DATs
Games	CDs
Laser disks	Tapes

$\begin{bmatrix} 3246 & 1548 \\ 750 & 2450 \\ 1813 & 852 \end{bmatrix}$

Sales [May]

Videos	DATs
Games	CDs
Laser disks	Tapes

$\begin{bmatrix} 2748 & 1081 \\ 702 & 1456 \\ 1679 & 622 \end{bmatrix}$

Delivery [June]

Videos	DATs
Games	CDs
Laser disks	Tapes

$\begin{bmatrix} 3000 & 1500 \\ 750 & 1000 \\ 1800 & 500 \end{bmatrix}$

Statistics In the Woodlake public school system, there are two junior high schools, Glenn and Kelly. The enrollment for the electives music (Mu), art (Ar), technology (Te), and health (He) in each of the schools appears in the matrices below.

Glenn

	Mu	Ar	Te	He
Boys	447	199	514	389
Girls	498	352	432	399

Kelly

	Mu	Ar	Te	He
Boys	387	276	489	367
Girls	505	392	387	437

19. How many girls are enrolled in art? How many boys are taking health?

20. Create a matrix to show the total enrollment in each elective.

21. How many students will take technology while in junior high school?

Sports You keep all of the statistics for the girls' basketball team in a matrix, a portion of which is shown below. You record the number of 3-point baskets, 2-point baskets, 1-point free throws (FTs), and the total points. The matrices show the statistics for the end of the regular season and the total points scored after the playoff series. Use these matrices to complete Exercises 22–24.

Points scored at end of regular season

	3pts.	2pts.	FTs	Total
Waters	3	15	12	51
Riley	1	20	16	59
Sharp	3	17	13	56
Evans	4	6	9	33
Jones	0	14	18	46

Points scored at end of playoff series

	3pts.	2pts.	FTs	Total
Waters	5	25	15	80
Riley	2	22	20	70
Sharp	5	20	15	70
Evans	5	12	12	51
Jones	0	17	22	56

22. Create a matrix to show how many points were scored by each player in the playoff series.

23. Which player should get the Most Valuable Player award for the playoff series based on points?

24. What is the total number of free throws made during the playoffs?

25. Find the values of w, x, y, and z by solving the matrix equation $A + B = C$.

$$A = \begin{bmatrix} -5w + 2 & 2x - 4 \\ 8y - 1 & \frac{1}{5}z \end{bmatrix} \quad B = \begin{bmatrix} -13 & 5 - x \\ -8 & z + 9 \end{bmatrix} \quad C = \begin{bmatrix} 2w + 3 & 7 \\ 3y - 4 & -9 \end{bmatrix}$$

26. Use the values you found for w, x, y, and z to evaluate matrices A, B, and C. Solve the matrix equation $A + B = C$ numerically to check your solutions.

The symbol that represents the additive inverse of a matrix is $-A$. An additive inverse occurs if $-A + A = 0$ (where 0 is the identity matrix for addition).

27. Is there one matrix that can be used as the identity matrix for A, B, and C in Exercise 25? Why?

28. Describe the additive identity matrix that will fit the matrix below.

$$\begin{bmatrix} 1 & 4 & -3 \\ 7 & 12 & 0 \end{bmatrix}$$

29. Form the matrices $-A$, $-B$, and $-C$ for matrices A, B, C in Exercise 25.

Look Back

Statistics Students collect data by measuring the number of inches around the thickest part of their thumb and then around their wrist. The data for 5 students appear in the matrix below.

	Thumb	Wrist
Samir	3.7	7.3
Betty	2.8	4.7
Ruth	3.1	6.2
Todd	3.6	7.1
Han	3.3	6.6

30. Plot the data and describe the correlation. **[Lesson 1.7]**

31. Graph $y = |x|$ and $y = 9$. Find their common solutions. **[Lesson 2.4]**

Write the opposite of each expression. [Lesson 3.4]

32. $-2x + 3y$ **33.** $-4a - 2b$

34. $-x + 3$ **35.** $4(-4x - 7)$

36. Graph $-3 < x$ or $x \leq -5$ on the number line. State the solution set clearly. **[Lesson 3.6]**

37. **Geometry** You have both a square and a rectangle. The rectangle is drawn so that its length is 3 times the side of the square and its width is twice the side of the square. Algebraically express the perimeter and area of the rectangle. **[Lesson 4.6]**

38. Write the equation $5x = 2y - 12$ in standard form, slope-intercept form, and point-slope form. **[Lessons 5.3, 5.4]**

39. Solve the system by elimination and check. $\begin{cases} 3x = 12 + 4y \\ 2y - x = -5 \end{cases}$ **[Lesson 6.3]**

Look Beyond

40. **Geometry** The matrix T represents the coordinates of the vertices of triangle ABC graphed on a coordinate plane. Matrix M represents a translation matrix.

$$T = \begin{bmatrix} x & y \\ 4 & 3 \\ 7 & -2 \\ 10 & 6 \end{bmatrix} \quad M = \begin{bmatrix} x & y \\ -7 & 2 \\ -7 & 2 \\ -7 & 2 \end{bmatrix}$$

Write the matrix equal to $T + M$. Plot the original points and the translated points on a graph.

Barely Enough GRIZZLIES?

Counting Big Bears

Government Wants Them Off the Endangered List; How Many is Enough?

By Marj Charlier

Staff Reporter of The Wall Street Journal

KALISPELL, Mont.—The solitary monarch of the wild, the grizzly rules the rugged high country. Scientists call it *Ursus arctus horribilus,* and the name suggests

the terror it inspires. Fearless and unpredictable, the bear can kill with a single rake of a claw.

Man is the grizzly's only natural enemy, but the clash almost proved fatal to the species. Unrestricted hunting and development of the animal's habitat decimated the bear population over the last century. By 1975 when the grizzly was placed on the federal government's Endangered Species List, only 1,000 were thought to be alive.

Now, federal officials say the grizzly population has rebounded, though they concede there is no hard evidence to support that conclusion, and they are proposing to take the bear off the list.

A Death Sentence?

Many environmentalists, however, believe that even if the number of bears has increased, the grizzly population is still too small to strip the animal of federal protection. Such a plan, they say, would be a death sentence for the species. "If the grizzly bear is delisted, it will be strangled by

development and go extinct quite promptly," says Lance Olsen, president of the nonprofit Great Bear Foundation.

Population Unknown

Complicating the issue is the difficulty of determining how many grizzlies roam the Rockies and their foothills. Everyone concedes that the 1,000 figure for 1975 was only a guess. Attempts to take a bear census in the years since then have proved difficult, and no one knows how many grizzlies there are now.

Richard Mace, a wildlife biologist for the state of Montana, knows how tough it is to count grizzlies. Displaying a battered steel box with a jagged rip down one side, he explains that the box once held a tree-mounted camera that was rigged to snap photos when it sensed body heat from a large animal. Biologists had placed some 43 cameras in a section of Montana's Swan Mountains, hoping to get a more accurate bear count.

Apparently, Mr. Mace says, when the camera flashed, the grizzly slashed. Its two-inch long claws ripped through the steel casing, which is a bit thicker than a coffee can. The bear dug out the camera and chewed it to pieces.

Cooperative Learning

Counting animals in the wild is not easy. How do you know if you have counted some more than once? How can you possibly find them all? Wildlife managers sometimes use a technique called *tag and recapture*. To see how the *tag and recapture* method works, you can try it with objects instead of animals.

You will need:

- a batch of several hundred objects (macaroni, beans, crumpled pieces of paper) in a paper bag or other container
- a marker (or some other way to *tag*, or mark, the objects)

1. Follow steps a–e to estimate the number of objects in your batch.

   a. Remove a handful of objects from the batch and mark them. Record how many you have marked.

   b. Return the marked objects to the batch. Mix thoroughly.

   c. Take 30 objects from the batch at random. Count the number of marked objects in your sample. Copy the chart and fill in the first row. Repeat for each group member.

Sample	Total Number of Objects in Sample (*Ns*)	Number of Marked Objects in Sample (*Ms*)
1	30	
2	30	

   d. How do you think the fraction of marked objects in your samples compares with the fraction of marked objects in the whole batch?

   e. Use the proportion to estimate the size of the whole batch. (Average your samples for M_s and N_s.)

   $$\frac{M_p}{N_p} = \frac{M_s}{N_s}$$

   M_p is the number of <u>marked objects</u> in the batch.
   N_p is the <u>total number of objects</u> in the batch.
   M_s is the number of <u>marked objects</u> in the sample.
   N_s is the <u>total number of objects</u> in the sample.

   f. Check your estimate by counting. Within what percent of the actual total was your estimate?

2. Describe how the *tag and recapture* method could be used to estimate an animal population.

3. Do you think the *tag and recapture* method would work for a grizzly bear population? Why or why not?

Exploring Matrix Multiplication

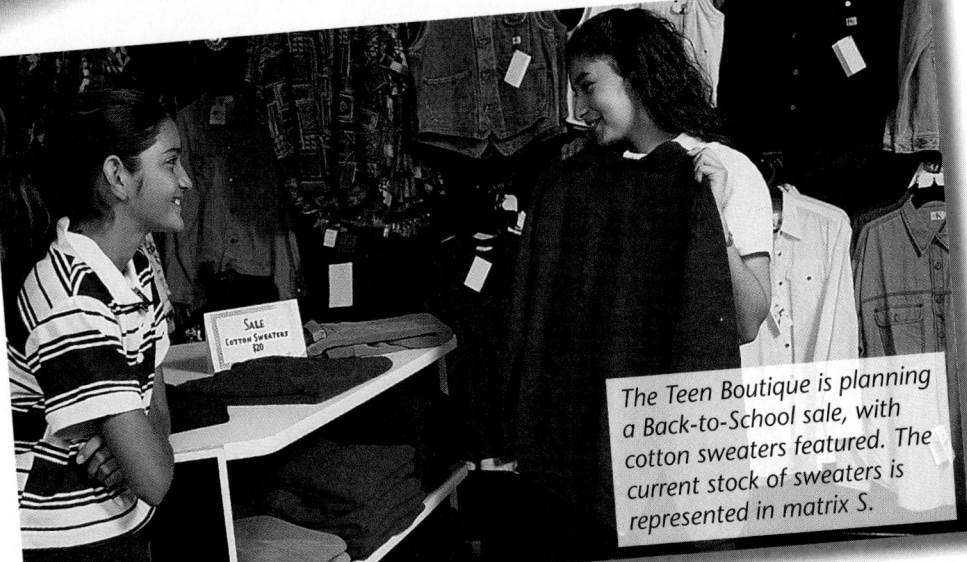

The Teen Boutique is planning a Back-to-School sale, with cotton sweaters featured. The current stock of sweaters is represented in matrix S.

why *If the entries in a matrix are doubled or tripled, you need a way to multiply the data in the matrix by a constant. This is scalar multiplication, and it is related to addition. Another operation called matrix multiplication will enable you to find the product of 2 matrices.*

$$
\text{Stock:} \quad
\begin{array}{c}
\\
\text{Small} \\
\text{Medium} \\
\text{Large}
\end{array}
\begin{array}{ccc}
\text{Blue} & \text{Red} & \text{Black} \\
\left[\begin{array}{ccc}
16 & 18 & 15 \\
14 & 22 & 17 \\
21 & 20 & 19
\end{array}\right] = S
\end{array}
$$

Inventory Control In anticipation of the sale, the boutique orders three times the present stock. How many of each item are ordered? Can you represent the order in matrix form?

If you want three times the stock, find the sum of 3 identical matrices.

$$
\begin{array}{ccccccc}
S & + & S & + & S & = & T
\end{array}
$$

$$
\begin{bmatrix} 16 & 18 & 15 \\ 14 & 22 & 17 \\ 21 & 20 & 19 \end{bmatrix} +
\begin{bmatrix} 16 & 18 & 15 \\ 14 & 22 & 17 \\ 21 & 20 & 19 \end{bmatrix} +
\begin{bmatrix} 16 & 18 & 15 \\ 14 & 22 & 17 \\ 21 & 20 & 19 \end{bmatrix} =
\begin{bmatrix} 48 & 54 & 45 \\ 42 & 66 & 51 \\ 63 & 60 & 57 \end{bmatrix}
$$

An easier way to do this is to multiply each entry by 3.

$$
\begin{array}{ccc}
3S & = & T
\end{array}
$$

$$
3\begin{bmatrix} 16 & 18 & 15 \\ 14 & 22 & 17 \\ 21 & 20 & 19 \end{bmatrix} =
\begin{bmatrix} 48 & 54 & 45 \\ 42 & 66 & 51 \\ 63 & 60 & 57 \end{bmatrix}
$$

This type of multiplication is called **scalar multiplication**. The **scalar** is the number that multiplies each entry in the matrix. In this case, the scalar is 3.

Multiplying two matrices creates a **product matrix**. This matrix combines part of the information from each of the original matrices. Pay close attention to the procedure for multiplying matrices. It is different from scalar multiplication.

Exploration 1 *Understanding the Multiplication of Matrices*

Fund-raising

Each year Sam and Kim make toy cars and trucks for the service club fair. The following matrix examples show how many cars and trucks they plan to make and the number of wheels and nails they will need to construct each of them.

Before you can multiply the two matrices, the *left matrix column labels* (*Cars and Trucks*) must match the *right matrix row labels* (*Cars and Trucks*).

$$
\begin{array}{c}
\text{Sam} \\
\text{Kim}
\end{array}
\begin{array}{cc}
\text{Cars} & \text{Trucks} \\
\begin{bmatrix} 8 & 15 \\ 12 & 10 \end{bmatrix}
\end{array}
\qquad
\begin{array}{c}
\text{Cars} \\
\text{Trucks}
\end{array}
\begin{array}{cc}
\text{Wheels} & \text{Nails} \\
\begin{bmatrix} 4 & 7 \\ 6 & 9 \end{bmatrix}
\end{array}
$$

Sam builds 8 cars with 4 wheels each. He will need 8(4), or 32 wheels. He is also building 15 trucks with 6 wheels each. He will need 15(6), or 90 more wheels. The total number of wheels will be 8(4) + 15(6), or 122.

Place this information in the Sam/Wheels address for the product matrix.

	Wheels	**Nails**
Sam	8(4) + 15(6) 32 + 90 122	
Kim		

The matching column and row labels disappear after you multiply. The product matrix will no longer show labels for (*Cars and Trucks*). *Sam and Kim* and (*Wheels and Nails*) become the labels of the product matrix.

Answer the questions to complete the table.

1. How many wheels will Kim need to complete her work?

2. How many nails will Sam need to complete his work?

3. How many nails will Kim need to complete her work?

4. What matrix represents the product? ❖

The Procedure for Multiplying Matrices

From Exploration 1, it is possible to create a procedure for multiplying matrices. First, check to see if the number of left matrix columns matches the number of right matrix rows. Then follow the pattern to see how to multiply the matrices.

Let $A = \begin{bmatrix} 3 & 5 \\ 4 & -1 \end{bmatrix}$ and $B = \begin{bmatrix} -2 & -7 \\ 6 & 5 \end{bmatrix}$. Find the product, $P = AB$.

First, find the products of each entry in row 1 of matrix A and each corresponding entry in column 1 of matrix B. Then place the sum of the products at the address p_{11} in the product matrix, P.

Matrix A   **Matrix B**

$\begin{bmatrix} 3 & 5 \\ 4 & -1 \end{bmatrix}$ times $\begin{bmatrix} -2 & -7 \\ 6 & 5 \end{bmatrix}$ $\longrightarrow$ $3(-2) + 5(6)$ $\longrightarrow P = \begin{bmatrix} 24 & \\ & \end{bmatrix}$

The entry in p_{11} is the sum of [**row 1** entries times **column 1** entries].

Continue this pattern to complete the remaining entries.

1 What is the entry for address p_{12}? What row and column are multiplied to find p_{12}?

Matrix A   **Matrix B**

$\begin{bmatrix} 3 & 5 \\ 4 & -1 \end{bmatrix}$ times $\begin{bmatrix} -2 & -7 \\ 6 & 5 \end{bmatrix}$ $\longrightarrow$ $3(?) + 5(?)$ $\longrightarrow P = \begin{bmatrix} 24 & ? \\ & \end{bmatrix}$

The entry in p_{12} is the sum of [**row _?_** entries times **column _?_** entries].

2 What is the entry for address p_{21}? What row and column are multiplied to find p_{21}?

Matrix A   **Matrix B**

$\begin{bmatrix} 3 & 5 \\ 4 & -1 \end{bmatrix}$ times $\begin{bmatrix} -2 & -7 \\ 6 & 5 \end{bmatrix}$ $\longrightarrow$ $?(-2) + ?(6)$ $\longrightarrow P = \begin{bmatrix} 24 & p_{12} \\ ? & \end{bmatrix}$

The entry in p_{21} is the sum of [**row _?_** entries times **column _?_** entries].

3 What is the last entry for matrix P? What row and column are multiplied to find the last entry? What is the address for the last entry in matrix P?

Matrix A   **Matrix B**

$\begin{bmatrix} 3 & 5 \\ 4 & -1 \end{bmatrix}$ times $\begin{bmatrix} -2 & -7 \\ 6 & 5 \end{bmatrix}$ $\longrightarrow$ $?(?) + ?(?)$ $\longrightarrow P = \begin{bmatrix} 24 & p_{12} \\ p_{21} & ? \end{bmatrix}$

The entry in p_{22} is the sum of [**row _?_** entries times **column _?_** entries].

4 Summarize in your own words the procedure for multiplying matrices. ❖

Mary places two orders for new tableware. Mary's first order is 1 cup, 4 saucers, and 7 plates. Her second order consists of 3 cups, 6 saucers, and 8 plates. What is the cost of each of the two orders?

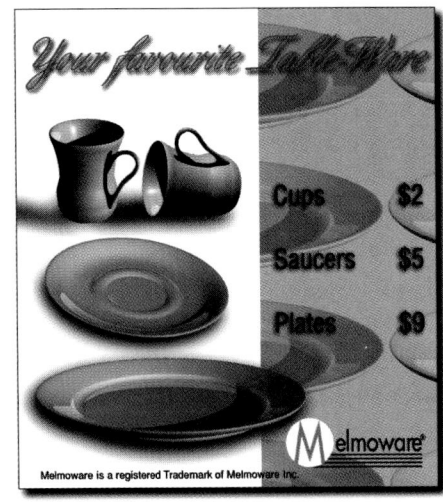

Let matrix C represent the cost for each saucer, cup, and plate. Let matrix D represent the number of pieces in her first order and the number of pieces in her second order.

$$\begin{array}{c} \text{Cost} \end{array} \begin{array}{ccc} \text{Cups} & \text{Saucers} & \text{Plates} \\ [2 & 5 & 9] \end{array} = C$$

$$\begin{array}{c} \\ \text{Cups} \\ \text{Saucers} \\ \text{Plates} \end{array} \begin{array}{c} \text{First} \quad \text{Second} \\ \text{order} \quad \text{order} \\ \begin{bmatrix} 1 & 3 \\ 4 & 6 \\ 7 & 8 \end{bmatrix} \end{array} = D$$

Check the row and column dimensions of each given matrix to see if multiplication is possible. Then find the product matrix.

Since the number of columns in C matches the number of rows in D, multiplication is possible.

C (1 by 3) times D (3 by 2) $\rightarrow P$ (1 by 2)

Matrix C $\cdot$ Matrix D

$$[2 \quad 5 \quad 9] \begin{bmatrix} 1 & 3 \\ 4 & 6 \\ 7 & 8 \end{bmatrix} = P$$

$CD = [2(1) + 5(4) + 9(7) \qquad 2(3) + 5(6) + 9(8)] = [85 \quad 108] = P$

The cost for the first order is $85, and the cost for the second order is $108. ❖

•Exploration 3 *The Multiplicative Identity Matrix*

1. Let $L = \begin{bmatrix} 3 & -7 \\ 16 & 9 \end{bmatrix}$, $M = \begin{bmatrix} 1 & 0 \\ 0 & 1 \end{bmatrix}$, and $N = \begin{bmatrix} 12 & 9 \\ -7 & -4 \end{bmatrix}$. Multiply.

    a. $\begin{bmatrix} 1 & 0 \\ 0 & 1 \end{bmatrix}\begin{bmatrix} 3 & -7 \\ 16 & 9 \end{bmatrix}$ **b.** $\begin{bmatrix} 12 & 9 \\ -7 & -4 \end{bmatrix}\begin{bmatrix} 1 & 0 \\ 0 & 1 \end{bmatrix}$ **c.** $\begin{bmatrix} 1 & 0 \\ 0 & 1 \end{bmatrix}\begin{bmatrix} 12 & 9 \\ -7 & -4 \end{bmatrix}$

2. What do you notice about the products?

3. Notice that $N \cdot M = M \cdot N$. Why is M called an identity matrix?

4. Find $L \cdot N$ and $N \cdot L$. Make a conjecture about the commutative property and matrix multiplication.

5. Explain why a multiplication identity matrix must be a square matrix. ❖

EXERCISES & PROBLEMS

Communicate

Discuss how to identify the row and column dimensions of the following matrices.

1. $\begin{bmatrix} 1 \\ -5 \\ 3 \end{bmatrix}$

2. $\begin{bmatrix} 1 & 3 \\ 0 & 5 \\ 12 & -4 \end{bmatrix}$

3. $\begin{bmatrix} 1 & 4 & 3 & 108 \end{bmatrix}$

Discuss how to recognize if two matrices can be multiplied. Explain how to perform the multiplication with the given matrices.

4. $3\begin{bmatrix} -6 & 1 \\ 4 & -1 \end{bmatrix}$

5. $\begin{bmatrix} 2 & -4 \\ 1 & 0 \end{bmatrix}\begin{bmatrix} 5 & -1 \\ 6 & 5 \end{bmatrix}$

6. $\begin{bmatrix} 1 & 2 \\ 1 & -3 \\ 3 & 5 \end{bmatrix}\begin{bmatrix} 4 & 1 & 3 \\ 2 & 1 & 6 \end{bmatrix}$

Practice & Apply

If you have technology available to assist you with the matrix operations, you should make use of it as directed by your teacher.

The dimensions of two matrices are given. If the two matrices can be multiplied, indicate the dimensions of the product matrix. If they cannot, write *Not Possible*.

	A	B	AB
7.	2 by 4	3 by 5	___
8.	1 by 3	3 by 5	___
9.	4 by 2	2 by 1	___
10.	3 by 3	3 by 2	___
11.	1 by 2	5 by 2	___

Use the given matrices to evaluate the following operations. Write *Not Possible* when appropriate.

$A = \begin{bmatrix} 2 & 0 \\ -1 & 5 \end{bmatrix}$ $B = \begin{bmatrix} 1 & 3 \\ -2 & -3 \\ 4 & 0 \end{bmatrix}$ $C = \begin{bmatrix} 6 & -4 \\ -1 & 12 \\ -2 & 5 \end{bmatrix}$ $D = \begin{bmatrix} -3 & 1 & 4 \\ -2 & 2 & 6 \end{bmatrix}$

12. $B + C$

13. $-5D$

14. $3B - 4C$

15. AB

16. BD

17. DA

18. DC

19. $-BC$

20. CA

Find the missing element(s) in each product matrix.

21. $-3\begin{bmatrix} -1 & 8 \\ 0 & 3 \end{bmatrix} = \begin{bmatrix} 3 & ? \\ 0 & -9 \end{bmatrix}$

22. $\begin{bmatrix} -1 & 8 \\ 0 & 3 \end{bmatrix}\begin{bmatrix} -5 & 6 \\ -9 & 8 \end{bmatrix} = \begin{bmatrix} ? & 58 \\ ? & 24 \end{bmatrix}$

23. $\begin{bmatrix} 0 & 8 \\ -5 & 1 \\ -2 & 3 \end{bmatrix} \begin{bmatrix} 0 & 2 & -1 \\ 4 & 6 & -3 \end{bmatrix} = \begin{bmatrix} 32 & 48 & ? \\ 4 & ? & 2 \\ ? & 14 & -7 \end{bmatrix}$ **24.** $5.7 \begin{bmatrix} 0.03 & -2.3 \\ 1.8 & 0.4 \end{bmatrix} = \begin{bmatrix} 0.171 & ? \\ 10.26 & ? \end{bmatrix}$

Use the given matrices for the following questions about the properties of matrix multiplication. Complete Exercises 25–29.

$A = \begin{bmatrix} 0 & 4 \\ -5 & 1 \\ -2 & 3 \end{bmatrix}$ $B = \begin{bmatrix} 2 & -1 \\ 4 & -3 \end{bmatrix}$ $C = \begin{bmatrix} -1 \\ 2 \end{bmatrix}$ $D = \begin{bmatrix} 1 & -2 \\ 3 & -4 \end{bmatrix}$

25. What are the dimensions of the matrices A, B, and C? Are the dimensions compatible for multiplication?

26. Does $A(BC) = (AB)C$? If they are equal, what is the product?

27. What is your conjecture about the Associative Property for Multiplication? Find 3 more matrices that confirm your conjecture, or find 3 more matrices that show a counterexample.

28. What are the row and column dimensions of matrices B and D? Are these dimensions compatible for multiplication?

29. Does $BD = DB$? Make a conjecture about the Commutative Property for multiplying matrices. Is this true for all matrices?

30. What characteristics of matrices are necessary to make multiplication possible?

31. If $X = \begin{bmatrix} 1 & 2 & 3 \end{bmatrix}$ and $Y = \begin{bmatrix} 3 \\ 2 \\ 5 \end{bmatrix}$, what are the products XY and YX?

32. Manufacturing At the fair, toy cars and trucks must be sold at a price that will generate a profit. Use the matrices below to calculate how much it will cost to make a car and how much it will cost to make a truck. Express your answer as a labeled matrix.

	Supplies					Cost in cents
	Wheels	Nails	Blocks			
Type: Cars	4	7	2	**Supplies:** Wheels	25
Trucks	6	9	3		Nails	3
					Blocks	45

Wooden Toy Craftsman
Denis Grant
764 Harmony Lane
Maplewood, WI 54321
Phone: (123) 456-7891

Cost of supplies:
Wheels $0.25
Nails $0.03
Blocks $0.45

33. Pricing Use the given *Stock by Type* matrix to determine how much it will cost Sam and Kim to produce all of the toy cars and trucks. Express your answer as a labeled matrix in dollars.

	Type	
	cars	trucks
Stock: Sam	8	15
Kim	12	10

34. Sam and Kim decide to charge $5 per car and $6.50 per truck. Create your selling price matrix, and determine how much money Sam and Kim will collect at the fair if all of the cars and trucks are sold.

35. Sales Tax If you make a purchase, the total price you pay is the retail price plus sales tax. If sales tax is 6% in Sam and Kim's town, create the matrix that represents the amount of sales tax for cars and trucks sold at the fair.

36. How much profit will Sam and Kim each make after paying cost and sales tax?

Fund-raising The first matrix shows the number of flags and buttons sold by each class. The second gives the purchase cost and selling price of the flags and buttons.

	Flags	Buttons
Freshmen	125	200
Sophomore	100	200
Junior	50	150
Senior	200	250

	Purchase cost	Selling price
Flags	0.75	2.00
Buttons	0.35	1.00

37. Create the matrix that represents the money both spent and made by each of the classes.

38. Add a column to the matrix to represent the profit made by each class.

39. Why do you think the number of flags and buttons sold decreases each year after the freshman year, but increases again during senior year?

At Central City High School, each class raises funds by selling class flags and buttons.

 Look Back

40. Find the sum of $1 + 2 + 3 + \cdots + 15$. **[Lesson 1.1]**

41. Simplify $5 - 2y\{3 - 2[x + 6y - (3y + x) + 8]\}$. **[Lesson 1.4]**

Find the value of y if x is equal to 5. **[Lessons 3.5, 3.6]**

42. $2x + y = 70$ **43.** $2x - y = 61$ **44.** $2x + 2y = 100$

45. 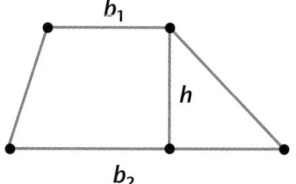 **Geometry** The area of trapezoid is defined by the formula $A = \frac{1}{2}h(b_1 + b_2)$ where h is the height and b_1 and b_2 are the bases. Solve the equation for h. **[Lesson 4.3]**

46. Solve $3x + 5(x - 4) = 10x - 25$ for x. **[Lesson 4.6]**

47. Graph the equation of the line that is perpendicular to $-3x + 9y = 27$ and contains the point $(-6, 7)$. **[Lesson 5.6]**

48. Solve by the elimination method. $\begin{cases} 2x - 5y = 20 \\ 3x - 4y = 37 \end{cases}$ **[Lesson 6.3]**

49. You and a friend can paddle your canoe 6 miles upstream in 3 hours. It takes the two of you 2 hours to return the same distance downstream. Find the rate of the current in the stream. **[Lesson 6.6]**

Look Beyond

Use matrices A and B to complete Exercises 50–51.

$$A = \begin{bmatrix} 2 & -1 \\ 4 & -3 \end{bmatrix} \qquad B = \begin{bmatrix} 1.5 & -0.5 \\ 2 & -1 \end{bmatrix}$$

50. Find AB. Label the resulting matrix C.

51. Why is matrix B called the inverse of matrix A?

LESSON 7.4 Multiplicative Inverse of a Matrix

Why *You have used the Reciprocal Property to solve multiplication equations. When you solve a matrix equation, you use the inverse of a matrix in much the same way.*

You know that $r \cdot \dfrac{1}{r} = 1$, when $r \neq 0$. Since their product is the multiplicative identity, r and $\dfrac{1}{r}$ are called reciprocals or multiplicative inverses. A similar situation occurs with matrices. If the product of two matrices equals an identity matrix, the matrices are inverses. The symbol A^{-1} is used to represent the inverse of matrix A.

Examine the matrix multiplication.

$$\underset{A}{\begin{bmatrix} 3 & -1 \\ 4 & 2 \end{bmatrix}} \cdot \underset{B}{\begin{bmatrix} 0.2 & 0.1 \\ -0.4 & 0.3 \end{bmatrix}} = \underset{I}{\begin{bmatrix} 1 & 0 \\ 0 & 1 \end{bmatrix}} = \underset{B}{\begin{bmatrix} 0.2 & 0.1 \\ -0.4 & 0.3 \end{bmatrix}} \cdot \underset{A}{\begin{bmatrix} 3 & -1 \\ 4 & 2 \end{bmatrix}}$$

The matrices $\begin{bmatrix} 3 & -1 \\ 4 & 2 \end{bmatrix}$ and $\begin{bmatrix} 0.2 & 0.1 \\ -0.4 & 0.3 \end{bmatrix}$ are **inverses.** The product of A and B is the **identity matrix**, $\begin{bmatrix} 1 & 0 \\ 0 & 1 \end{bmatrix}$. Thus, $B = A^{-1}$.

MULTIPLICATIVE INVERSE MATRIX
The matrix A has an inverse matrix, A^{-1}, if
$$AA^{-1} = I = A^{-1}A,$$
where I is the identity matrix.

Finding the Inverse Matrix

EXAMPLE 1

If matrix $A = \begin{bmatrix} 3 & 5 \\ 4 & 6 \end{bmatrix}$, find A^{-1}.

Solution ▶

You know from the definition that $AA^{-1} = I = A^{-1}A$. Since A is a 2 by 2 matrix, A^{-1} must be a 2 by 2 matrix. Let $\begin{bmatrix} a & b \\ c & d \end{bmatrix}$ represent A^{-1}, the inverse matrix. Then substitute the matrices in $AA^{-1} = I$.

$$A \quad \cdot \quad A^{-1} \quad = \quad I$$

$$\begin{bmatrix} 3 & 5 \\ 4 & 6 \end{bmatrix} \begin{bmatrix} a & b \\ c & d \end{bmatrix} = \begin{bmatrix} 1 & 0 \\ 0 & 1 \end{bmatrix}$$

Multiply.

$$AA^{-1} \qquad = \quad I$$

$$\begin{bmatrix} 3a + 5c & 3b + 5d \\ 4a + 6c & 4b + 6d \end{bmatrix} = \begin{bmatrix} 1 & 0 \\ 0 & 1 \end{bmatrix}$$

Equal matrices have equal, corresponding matrix entries. Write the equation for each of the corresponding entries. Pair the equations that have the same variables. Then solve the systems of equations.

$$\begin{cases} 3a + 5c = 1 \\ 4a + 6c = 0 \end{cases} \qquad \begin{cases} 3b + 5d = 0 \\ 4b + 6d = 1 \end{cases}$$

Use elimination to solve for a and b.

$$
\begin{aligned}
18a + 30c &= 6 \\
-20a - 30c &= \\
\hline
-2a \qquad\quad &= 6 \\
a &= -3
\end{aligned}
\qquad
\begin{aligned}
18b + 30d &= 0 \\
-20b - 30d &= -5 \\
\hline
-2b \qquad\quad &= -5 \\
b &= 2.5
\end{aligned}
$$

Substitute and solve for c and d.

$$
\begin{aligned}
3(-3) + 5c &= 1 \\
-9 + 5c &= 1 \\
5c &= 10 \\
c &= 2
\end{aligned}
\qquad
\begin{aligned}
3(2.5) + 5d &= 0 \\
7.5 + 5d &= 0 \\
5d &= -7.5 \\
d &= -1.5
\end{aligned}
$$

Replace the variables in $\begin{bmatrix} a & b \\ c & d \end{bmatrix}$ with the values you found. The inverse matrix A^{-1} is $\begin{bmatrix} -3 & 2.5 \\ 2 & -1.5 \end{bmatrix}$.

$$A \quad \cdot \quad A^{-1} \quad = \quad I \quad = \quad A^{-1} \quad \cdot \quad A$$

Check $\begin{bmatrix} 3 & 5 \\ 4 & 6 \end{bmatrix}\begin{bmatrix} -3 & 2.5 \\ 2 & -1.5 \end{bmatrix} = \begin{bmatrix} 1 & 0 \\ 0 & 1 \end{bmatrix} = \begin{bmatrix} -3 & 2.5 \\ 2 & -1.5 \end{bmatrix}\begin{bmatrix} 3 & 5 \\ 4 & 6 \end{bmatrix}$. ❖

If a matrix has an inverse, what are the requirements for row and column dimensions of the inverse matrix? Why is this the case?

EXAMPLE 2

Find the inverse of $B = \begin{bmatrix} 4 & 6 \\ 2 & 3 \end{bmatrix}$.

Solution ➤

Let $\begin{bmatrix} a & b \\ c & d \end{bmatrix}$ represent the inverse matrix B^{-1}.

Since $BB^{-1} = I$, write the matrix equation.

$$B \quad \cdot \quad B^{-1} \quad = \quad I$$

$$\begin{bmatrix} 4 & 6 \\ 2 & 3 \end{bmatrix} \begin{bmatrix} a & b \\ c & d \end{bmatrix} = \begin{bmatrix} 1 & 0 \\ 0 & 1 \end{bmatrix}$$

Show the matrix multiplication.

$$BB^{-1} \qquad = \quad I$$

$$\begin{bmatrix} 4a + 6c & 4b + 6d \\ 2a + 3c & 2b + 3d \end{bmatrix} = \begin{bmatrix} 1 & 0 \\ 0 & 1 \end{bmatrix}$$

Write the equations from corresponding matrix entries.

$$\begin{cases} 4a + 6c = 1 \\ 2a + 3c = 0 \end{cases} \qquad \begin{cases} 4b + 6d = 0 \\ 2b + 3d = 1 \end{cases}$$

You can use any of the methods you learned in the previous chapter to solve the systems of equations. If you use the elimination method, the result is the following.

$$\begin{array}{r} 4a + 6c = 1 \\ -4a - 6c = 0 \\ \hline 0 = 1 \text{ Not true} \end{array} \qquad \begin{array}{r} 4b + 6d = 0 \\ -4b - 6d = -2 \\ \hline 0 = -2 \text{ Not true} \end{array}$$

In this case, you find 0 equal to 1 and 0 equal to -2, which cannot be true. There is *no solution* for either system. The matrix $\begin{bmatrix} 4 & 6 \\ 2 & 3 \end{bmatrix}$ *has no inverse.* ❖

Carefully study the entries in the rows of the matrix that has no inverse. How is one row related to the other? Make a conjecture about a matrix that has no inverse.

Try This Find the inverse of $\begin{bmatrix} 0 & 1 \\ -6 & 2 \end{bmatrix}$.

Matrix Technology

Working with matrices can be tedious because of the large number of arithmetic calculations. As you continue studying matrices, you will soon encounter larger matrices. The formation of an inverse matrix with dimensions beyond 2 by 2 takes many calculations and a lot of time unless you use technology. Computers and calculators perform these calculations instantaneously, freeing you to explore the more interesting aspects of matrices.

When you use technology, however, you should be aware of unusual situations that cause calculators and computers to display error messages. For example, you can expect an error message if a matrix does not have an inverse or if you try to divide by 0.

Another difficulty could appear when you encounter rounding errors. You might see unusual expressions such as $2E - 13$ for the number 0.

EXERCISES & PROBLEMS

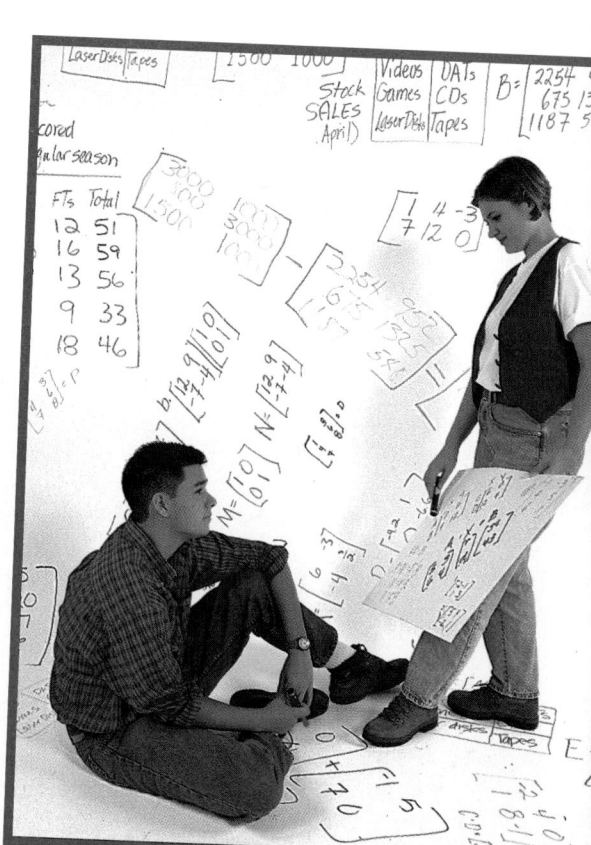

Communicate

Explain how to find the inverse of the following matrices, if possible.

1. $\begin{bmatrix} -1 & 6 \\ 0 & 3 \end{bmatrix}$ **2.** $\begin{bmatrix} 1 & 2 \\ 2 & 4 \end{bmatrix}$ **3.** $[2 \quad -5 \quad 7]$

4. What conditions would cause a matrix not to have an inverse?

5. How is the inverse of a matrix used?

6. What occurs when a number is multiplied by its inverse?

Practice & Apply

Use the definition of inverse matrices to find the inverse matrix for each of the given matrices.

7. $\begin{bmatrix} 5 & -7 \\ -2 & 3 \end{bmatrix}$ **8.** $\begin{bmatrix} 3 & 4 \\ 5 & 5 \end{bmatrix}$ **9.** $\begin{bmatrix} -1 & 0.04 \\ -0.6 & 1 \end{bmatrix}$

10. $\begin{bmatrix} 1 & 0 \\ 0 & -1 \end{bmatrix}$ **11.** $\begin{bmatrix} -1 & -2 \\ 2 & 0 \end{bmatrix}$ **12.** $\begin{bmatrix} -\dfrac{4}{5} & 1 \\ \dfrac{9}{5} & -2 \end{bmatrix}$

Does the matrix have an inverse?

13. $\begin{bmatrix} 5 & -3 \\ 2 & -4 \end{bmatrix}$

14. $\begin{bmatrix} 6 & 8 \\ 3 & 4 \end{bmatrix}$

15. $\begin{bmatrix} -1 & 2 & -3 \\ -4 & 5 & -6 \end{bmatrix}$

16. $\begin{bmatrix} \frac{1}{2} & -1 \\ 1 & -2 \end{bmatrix}$

17. $\begin{bmatrix} -0.2 & 1 \\ 0.3 & -1.5 \end{bmatrix}$

18. $\begin{bmatrix} 5 \\ -6 \\ 7 \end{bmatrix}$

Technology Use technology to find the inverse if it exists.

19. $\begin{bmatrix} 2 & -4 & 6 \\ 5 & 0 & -8 \\ 3 & -6 & 9 \end{bmatrix}$

20. $\begin{bmatrix} 1 & 5 & 10 \\ 0 & -5 & -8 \\ 1 & 5 & -2 \end{bmatrix}$

21. $\begin{bmatrix} 6 & 1 & 1 \\ 1 & 6 & 1 \\ 6 & 6 & 1 \end{bmatrix}$

22. A general 2 by 2 matrix can be written $\begin{bmatrix} a & b \\ c & d \end{bmatrix}$. Calculate the products ad and bc for each of the matrices $\begin{bmatrix} 3 & 5 \\ 4 & 6 \end{bmatrix}$ and $\begin{bmatrix} 4 & 6 \\ 2 & 3 \end{bmatrix}$. Do you find anything significant about these products?

23. Choose at least three of the 2 by 2 matrices that *do not have inverses* from the exercises. Again, for each of these matrices, find the value of the products *ad* and *bc*. Do you find anything significant about these products?

For each matrix below find the products *ad* and *bc*. Write and test a conjecture about the matrices. Use technology to check your conjecture.

24. $\begin{bmatrix} -6 & 60 \\ 0.5 & -5 \end{bmatrix}$

25. $\begin{bmatrix} 7 & -4 \\ 14 & -7 \end{bmatrix}$

26. $\begin{bmatrix} -8 & 3 \\ 7 & 2 \end{bmatrix}$

Geometry The matrices represent the endpoints of two line segments *AB* and *CD*.

$$
\begin{array}{cc}
 & \begin{matrix} x & y \end{matrix} \\
\begin{matrix} A \\ B \end{matrix} & \begin{bmatrix} 0 & 6 \\ 4 & 12 \end{bmatrix}
\end{array}
\qquad
\begin{array}{cc}
 & \begin{matrix} x & y \end{matrix} \\
\begin{matrix} C \\ D \end{matrix} & \begin{bmatrix} 6 & 0 \\ 12 & 4 \end{bmatrix}
\end{array}
$$

27. Graph both line segments. Describe how the line segment slopes are related to each other.

28. If $\begin{bmatrix} 0 & 6 \\ 4 & 12 \end{bmatrix} \cdot T = \begin{bmatrix} 6 & 0 \\ 12 & 4 \end{bmatrix}$, find the matrix T.

29. Draw a line segment of your own choosing, and write a matrix to represent the endpoints. Multiply your matrix by T, and graph the solution. Compare your results to segments *AB* and *CD*.

30. **Technology** Use technology to find the inverse of matrix A.

$$A = \begin{bmatrix} 1 & 2 & 3 \\ 2 & 4 & 5 \\ 3 & 5 & 6 \end{bmatrix}$$

31. Science You want to analyze an experiment in which you have collected data that describes the height in centimeters and body temperature in degrees Fahrenheit for 20 people. The data is collected, and a scatter plot is drawn. Describe the function that would best fit this data. **[Lesson 1.7]**

Make a table to evaluate each of the following functions when x is −3, −2, −1, 0, 1, 2, and 3. Are any values undefined? **[Lessons 2.1, 2.3, 2.4]**

32. $h(x) = x^2 − 2x + 3$ **33.** $f(x) = \dfrac{2}{x − 2}$ **34.** $g(x) = |x|$

Identify the property of equality that you would use to solve each of the following equations. Then solve each equation. **[Lessons 3.6, 4.3]**

35. $y − 11 = 67$ **36.** $\dfrac{2}{3}x = 16$ **37.** $\dfrac{z}{−3} = 20$

38. Coordinate Geometry A line has slope $\dfrac{3}{4}$ and passes through the point $P(2, −4)$. Write an equation in standard form for a line that is perpendicular to the original line and that passes through the common point, P. **[Lesson 5.6]**

39. Work Schedule Rick works 2 jobs. He keeps track of his hours using matrices for each day of a five-day workweek. Matrices A and B represent 2 weeks of work. Create a matrix, C, that shows the total hours worked each day and at each job for the 2 weeks he worked. **[Lesson 7.2]**

Hours Worked

Week 1: Job 1, Job 2
$\begin{bmatrix} M & T & WTh & F \\ 3 & 4 & 2 & 3 & 3 \\ 1 & 6 & 5 & 5 & 4 \end{bmatrix}$

Week 2: Job 1, Job 2
$\begin{bmatrix} M & T & WTh & F \\ 5 & 0 & 2 & 5 & 3 \\ 3 & 6 & 3 & 5 & 1 \end{bmatrix}$

Look Beyond ~~~

The given matrix equation represents a system of equations.

$$\begin{bmatrix} −2 & 5 \\ 4 & −3 \end{bmatrix}\begin{bmatrix} x \\ y \end{bmatrix} = \begin{bmatrix} 10 \\ 1 \end{bmatrix}$$

40. Using the algebra of matrices, write the system of equations in standard form.

41. Solve the system and use matrix multiplication to check your solution.

LESSON 7.5 Solving Matrix Equations

Why *Once you know how to multiply matrices and find inverses, you can solve matrix equations. Matrices can be used to solve systems of equations, even those with many unknowns. As the size of the matrices increases, technology becomes more necessary as a computational tool.*

The Vintage Movie House shows only classic films. Each Friday is "Kids' Night," when children can buy discounted tickets.

NOW SHOWING!

A CARTOON FESTIVAL!

FRIDAY NIGHT IS KID'S NIGHT!

Vintage Movie House
CLASSIC CINEMA

Consumer Economics
One Friday, a group of 3 parents and 5 children went to see *Gone with the Wind,* and the cost was $34. A week later, the group that went to *A Cartoon Festival* consisted of 4 parents and 6 children. The cost was $43. Find the cost of the regular and discounted tickets.

Let r = the regular cost of a parent's ticket.
Let d = the discounted cost of a child's ticket.
Write a system of equations.

3 parents and 5 children cost $34. $\begin{cases} 3r + 5d = 34 \\ 4r + 6d = 43 \end{cases}$
4 parents and 6 children cost $43.

This system can be solved by graphing, elimination, and substitution. It can also be solved by using matrices.

The Matrix Solution for Systems of Equations

The system of equations $\begin{cases} 3r + 5d = 34 \\ 4r + 6d = 43 \end{cases}$ can be written in **matrix equation** form using the following matrices.

Coefficient matrix	Variable matrix	Constant matrix
$A = \begin{bmatrix} 3 & 5 \\ 4 & 6 \end{bmatrix}$	$X = \begin{bmatrix} r \\ d \end{bmatrix}$	$B = \begin{bmatrix} 34 \\ 43 \end{bmatrix}$

Notice that together these three matrices form the matrix equation $AX = B$.

$$\begin{array}{ccc} A & \cdot X & = & B \\ \begin{bmatrix} 3 & 5 \\ 4 & 6 \end{bmatrix} & \begin{bmatrix} r \\ d \end{bmatrix} & = & \begin{bmatrix} 34 \\ 43 \end{bmatrix} \end{array}$$

The product matrix, AX, is $\begin{bmatrix} 3r + 5d \\ 4r + 6d \end{bmatrix}$. When AX is set equal to the constant matrix, B, the result is the original system of equations.

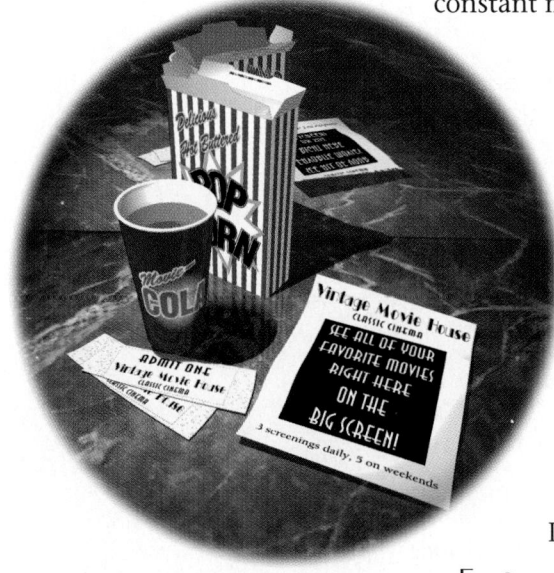

To solve the matrix equation, $AX = B$, multiply **both sides** of the matrix equation by the inverse A^{-1}. Be careful to multiply both sides of the matrix equation with A^{-1} on the left. Matrix multiplication is not generally commutative. Recall that $A^{-1}A = I$ and $IX = X$.

$$\begin{aligned} AX &= B \\ A^{-1}AX &= A^{-1}B \\ IX &= A^{-1}B \\ X &= A^{-1}B \end{aligned}$$

Use this idea to solve the equation $\begin{array}{ccc} A & \cdot X & = & B \\ \begin{bmatrix} 3 & 5 \\ 4 & 6 \end{bmatrix} & \begin{bmatrix} r \\ d \end{bmatrix} & = & \begin{bmatrix} 34 \\ 43 \end{bmatrix} \end{array}$.

In Lesson 7.4, you found that the inverse of $\begin{bmatrix} 3 & 5 \\ 4 & 6 \end{bmatrix}$ is $\begin{bmatrix} -3 & 2.5 \\ 2 & -1.5 \end{bmatrix}$, or A^{-1}. Multiply both sides of $AX = B$ by A^{-1}.

Make sure both multiplications are on the left of A and B.

$$\begin{array}{ccccc} A^{-1} & \cdot & A & \cdot X & = & A^{-1} & \cdot & B \\ \begin{bmatrix} -3 & 2.5 \\ 2 & -1.5 \end{bmatrix} & & \begin{bmatrix} 3 & 5 \\ 4 & 6 \end{bmatrix} & \begin{bmatrix} r \\ d \end{bmatrix} & = & \begin{bmatrix} -3 & 2.5 \\ 2 & -1.5 \end{bmatrix} & & \begin{bmatrix} 34 \\ 43 \end{bmatrix} \end{array}$$

$$\begin{array}{ccc} I & \cdot X & = & A^{-1}B \\ \begin{bmatrix} 1 & 0 \\ 0 & 1 \end{bmatrix} & \begin{bmatrix} r \\ d \end{bmatrix} & = & \begin{bmatrix} -3(34) + 2.5(43) \\ 2(34) - 1.5(43) \end{bmatrix} \end{array}$$

$$\begin{array}{cc} X = A^{-1}B & \\ \begin{bmatrix} r \\ d \end{bmatrix} = \begin{bmatrix} 5.50 \\ 3.50 \end{bmatrix} & \text{so} \quad \begin{array}{l} r = 5.50 \\ d = 3.50 \end{array} \end{array}$$

The cost of the regular ticket, r, for the parents is $5.50, and the cost of the discounted ticket, d, for the children is $3.50.

CRITICAL
Thinking

Why must the inverse matrix be multiplied to the left of both the coefficient and constant matrices? How does this procedure compare to solving the linear equation $ax = b$?

Solving Matrix Equations With Technology

Once you know the inverse of matrix A, you can solve the matrix equation $AX = B$. In Lesson 7.4 you used a system of equations to find A^{-1}. If you have a calculator or computer that works with matrices, you can find A^{-1} directly and solve the equation easily.

EXAMPLE 1

Use matrices to solve. $\begin{cases} 6x - y = 3 \\ 2x - 7y = 1 \end{cases}$

Solution ➤

Write the matrices in the following form.

Coefficient matrix	Variable matrix	Constant matrix
$A = \begin{bmatrix} 6 & -1 \\ 2 & -7 \end{bmatrix}$	$X = \begin{bmatrix} x \\ y \end{bmatrix}$	$B = \begin{bmatrix} 3 \\ 1 \end{bmatrix}$

Substitute the matrices into $AX = B$.
$$\overset{A}{\begin{bmatrix} 6 & -1 \\ 2 & -7 \end{bmatrix}} \cdot \overset{X}{\begin{bmatrix} x \\ y \end{bmatrix}} = \overset{B}{\begin{bmatrix} 3 \\ 1 \end{bmatrix}}$$

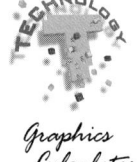

Graphics Calculator

Find the inverse for the coefficient matrix. This step can be simplified by using technology. Enter the original matrix into the calculator, and name it A. Place A on the screen, and press the inverse key $\boxed{x^{-1}}$ to get the inverse matrix. You can now solve the matrix equation. Multiply both sides of the matrix equation by the inverse matrix.

$$\overset{A^{-1}}{\begin{bmatrix} 0.175 & -0.025 \\ 0.05 & -0.15 \end{bmatrix}} \cdot \overset{A}{\begin{bmatrix} 6 & -1 \\ 2 & -7 \end{bmatrix}} \cdot \overset{X}{\begin{bmatrix} x \\ y \end{bmatrix}} = \overset{A^{-1}}{\begin{bmatrix} 0.175 & -0.025 \\ 0.05 & -0.15 \end{bmatrix}} \cdot \overset{B}{\begin{bmatrix} 3 \\ 1 \end{bmatrix}}$$

```
MATRIX[A] 2 ×2
[6      -1         ]
[2      -7         ]
```

$$\overset{I}{\begin{bmatrix} 1 & 0 \\ 0 & 1 \end{bmatrix}} \cdot \overset{X}{\begin{bmatrix} x \\ y \end{bmatrix}} = \overset{A^{-1}B}{\begin{bmatrix} 0.5 \\ 0 \end{bmatrix}}$$

```
[A]⁻¹
     [ [.175 −.025]
       [.050 −.150] ]
```

$$X = A^{-1}B$$
$$\begin{bmatrix} x \\ y \end{bmatrix} = \begin{bmatrix} 0.5 \\ 0 \end{bmatrix}$$

The solution to the system is (0.5, 0); x is 0.5, and y is 0. You can check the results by substituting the values for x and y into the original equation. ❖

Try This   Use matrices to solve this system. $\begin{cases} -x + y = 1 \\ -2x + 3y = 0 \end{cases}$

Lesson 7.5   Solving Matrix Equations   **339**

Once you obtain the matrix equation, you can solve the system of equations quickly and easily with technology.

EXAMPLE 2

Two families buy food at the same store. They decide to have cookouts on consecutive weekends. The next week with prices still the same, the Tates had a cookout and served 10 hamburgers and 7 hot dogs. It cost the Tates $33.00. What was the price for each hamburger and each hot dog?

On the first weekend, the Woodwards served 14 hamburgers and 12 hot dogs. They figured the total cost was $49.50.

Solution ➤

Organize the information. Define the variables. Let h = the cost of each hamburger. Let d = the cost of each hot dog.

Write the equations.

$$14h + 12d = 49.50$$
$$10h + 7d = 33.00$$

Solve the system of equations. Form the matrices for $AX = B$.

$$A = \begin{bmatrix} 14 & 12 \\ 10 & 7 \end{bmatrix} \quad X = \begin{bmatrix} h \\ d \end{bmatrix} \quad B = \begin{bmatrix} 49.50 \\ 33 \end{bmatrix}$$

$$\begin{array}{ccc} A & \cdot \ X \ = & B \end{array}$$
$$\begin{bmatrix} 14 & 12 \\ 10 & 7 \end{bmatrix} \begin{bmatrix} h \\ d \end{bmatrix} = \begin{bmatrix} 49.50 \\ 33 \end{bmatrix}$$

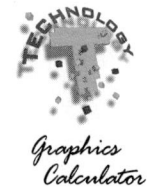

Graphics Calculator

The numbers in the coefficient matrix make it difficult to find an inverse. A graphics calculator saves valuable time.

Enter the coefficient matrix, $\begin{bmatrix} 14 & 12 \\ 10 & 7 \end{bmatrix}$, for A.

Enter the constant matrix, $\begin{bmatrix} 49.50 \\ 33 \end{bmatrix}$, for B.

Select A to recall the coefficient matrix, and take the inverse to get A^{-1}. Then multiply A^{-1} by B. This will give you $\begin{bmatrix} 2.25 \\ 1.50 \end{bmatrix}$.

Each hamburger was $2.25, and each hot dog was $1.50.

Check the answer.

$$14(2.25) + 12(1.50) = 49.50$$
$$10(2.25) + 7(1.50) = 33.00 \ \diamond$$

Cultural Connection: Asia   Over 2000 years ago, the Chinese developed a way to solve systems of linear equations written in matrix form.

Given the equations: $\begin{cases} 4x + 5y = 22 \\ 2x + 3y = 13 \end{cases}$ The Chinese would form the matrices: $\begin{bmatrix} 4 & 5 & 22 \\ 2 & 3 & 13 \end{bmatrix}$

Apply scalar multiplication, $\dfrac{1}{2}\begin{bmatrix} 4 & 5 & 22 \\ 2 & 3 & 13 \end{bmatrix}$, to produce $\begin{bmatrix} 4 & 5 & 22 \\ 4 & 6 & 26 \end{bmatrix}$.

Subtract row 1 from row 2. $\begin{bmatrix} 0 & 1 & 4 \end{bmatrix} \longrightarrow y = 4$

Finally, substitute into the first equation to get $x = \dfrac{1}{2}$.

Explain how to use the Chinese method to solve the system.

$$\begin{cases} 3x + y = 0 \\ 2x + 3y = 7 \end{cases}$$

EXERCISES & PROBLEMS

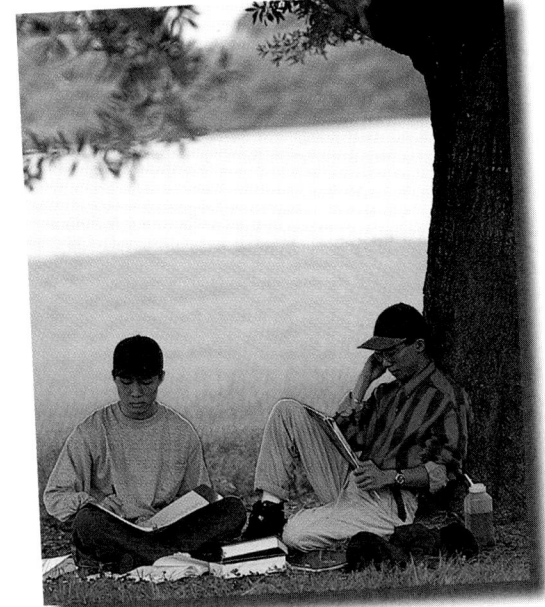

Communicate

Explain how to represent each system in matrix equation form $AX = B$.

1. $\begin{cases} 3a - 6b = 9 \\ 5a + 2b = 9 \end{cases}$ **2.** $\begin{cases} 12x + 5y = 30 \\ 4x - y = 6 \end{cases}$

Explain the method for solving the matrix equation for x and y.

3. $\begin{bmatrix} -2 & -1 \\ 1 & 0 \end{bmatrix}\begin{bmatrix} x \\ y \end{bmatrix} = \begin{bmatrix} -9 \\ -9 \end{bmatrix}$ **4.** $\begin{bmatrix} 1 & 1 \\ 1 & -1 \end{bmatrix}\begin{bmatrix} x \\ y \end{bmatrix} = \begin{bmatrix} 6 \\ 2 \end{bmatrix}$

Practice & Apply

Represent the systems of equations in matrix equation form, $AX = B$.

5. $\begin{cases} 4y = 6x - 12 \\ 5x = 2y + 10 \end{cases}$ **6.** $\begin{cases} 6a - b - 3c = 2 \\ -3a + b - 3c = 1 \\ -2a + 3b + c = -6 \end{cases}$ **7.** $\begin{cases} 2x + 3y + 4z = 8 \\ 6x + 12y + 16z = 31 \\ 4x + 9y + 8z = 17 \end{cases}$

Solve the systems of equations using matrices if possible.

8. $\begin{cases} 2x + y = 4 \\ x + y = 3 \end{cases}$ **9.** $\begin{cases} 2x = 4 \\ 3x + y = 6 \end{cases}$ **10.** $\begin{cases} 2x - 3y = 4 \\ -4x + 6y = 1 \end{cases}$

11. Which of the following ordered pairs represents the solution to $\begin{bmatrix} 3 & \frac{1}{2} \\ -2 & \frac{1}{3} \end{bmatrix}\begin{bmatrix} x \\ y \end{bmatrix} = \begin{bmatrix} -6 \\ 12 \end{bmatrix}$?

   a. $(-4, 6)$ **b.** $(4, 12)$ **c.** $(-4, 12)$ **d.** $(4, -6)$

Job Opportunities The potential for growth in your chosen career is an important consideration. The matrix shows the job growth in two related fields.

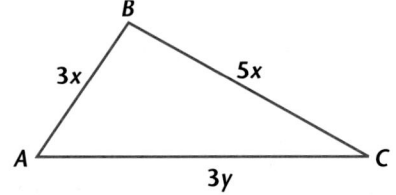

Occupation	U.S. Jobs (in 1000s)	
	1990	2000
Electrical engineers	440	615
Electrical assemblers	235	135

12. What is the projected average increase per year in the engineer's jobs?

13. What is the projected average decline per year in the assembler's jobs?

14. Write the linear equations that best fit this data.

15. In what year were the numbers of these two jobs equal? Use matrices to solve, but check by graphing.

16. **Geometry** The perimeter of rectangle *PQRS* is 30. The perimeter of triangle *ABC* is 40. Find the value of *x* and *y*.

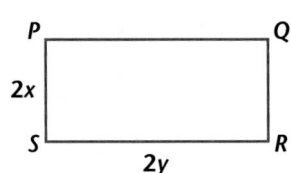

17. At Sid's Burger Barn, after the Friday concert, Kisha and her friend Maya ordered 3 Sid burgers and 2 super shakes. They were joined by Jason and Steve, who ordered 4 Sid burgers and 4 super shakes. When the checks arrived, the girls' check was $14, and the boys' check was $21. What was the cost of a Sid burger, and what was the cost of a super shake?

Inventory Control The matrices below indicate the sweater sales for three days at two branches of the Stylish Woman Sportswear Center.

Store #6 sales

	Cashmere	Cotton	Wool
Day 1	4	8	3
Day 2	3	10	5
Day 3	5	6	8

Store #8 sales

	Cashmere	Cotton	Wool
Day 1	2	5	3
Day 2	4	7	6
Day 3	3	8	8

The total sales for both stores for each day is

Day 1	$465
Day 2	$650
Day 3	$730

18. What was the total number of each type of sweater sold each day?

19. What was the selling price of each type of sweater?

Friday (Concert Ticket Sales)
Adult Tickets 150
Student Tickets 500
Total Tickets 650
Total $1125

Saturday (Football Game Sales)
Adult Tickets 225
Student Tickets 675
Total Tickets 900
Total $1575

20. **Ticket Sales** During Central City High School's spirit weekend, there was a concert on Friday night and an alumni football game on Saturday. For each event, adult and student tickets were sold. What was the price for an adult ticket and for a student ticket?

Employment Kevin and Doug have part time jobs at the local library. One week Kevin worked 14 hours, and Doug worked 16 hours. Together they took home $149.50. The next week they each worked 12 hours and had a combined take-home pay of $120.

21. How much does each boy earn per hour?

22. How much did Kevin earn during the two weeks?

23. Cultural Connection: Asia Solve the system using the Chinese method from the end of the lesson. $\begin{cases} 2x + 5y = -22 \\ 4x - 3y = 8 \end{cases}$

24. What method of solving linear equations does the Chinese method most resemble?

25. **Portfolio Activity**
Complete the problem in the portfolio activity on page 307.

Look Back

26. A ball is tossed between two players. If x represents the distance from the player tossing the ball, y represents the height of the ball. Would a third player between 6 and 7 feet from the player tossing the ball be able to intercept the toss? Explain your answer. **[Lesson 2.2]**

$y = -0.15x^2 + 2x + 2$

27. The given matrix represents 5 ordered pairs of a function. Use differences to show that the function represented by these points is quadratic. **[Lesson 2.2]**

Ordered Pairs

	1	2	3	4	5
x	3	4	5	6	7
y	0	2	6	12	20

28. If the blue jeans you just bought cost $26.70, including the 7% sales tax, what was the ticket price? **[Lesson 4.5]**

Coordinate Geometry Write the equation of the line through the point $(-2, 5)$ and

29. parallel to the x-axis.

30. perpendicular to the x-axis. **[Lesson 5.6]**

31. Chemistry How many ounces of water should be added to a 75% acid solution to produce 30 ounces of 55% acid solution? **[Lesson 6.6]**

Look Beyond

Technology On a calculator, graph $y = \left| \frac{4}{3} x + 8 \right|$. Find the points where this graph intersects the graphs of the following functions.

32. $y = x$ **33.** $y = x + 5$ **34.** $y = 4$

SECRET CODES

Navajo code talkers, one of the best kept secrets of World War II.

Cryptography, the science of coding and decoding information, is an interesting application of the matrix techniques that you have learned.

So, XRXEPENZ!

A	B	C	D	E	F	G	H	I	J	K	L	M	N
1	2	3	4	5	6	7	8	9	10	11	12	13	14

O	P	Q	R	S	T	U	V	W	X	Y	Z	*	
15	16	17	18	19	20	21	22	23	24	25	26	27	

Use asterisk * for space.

1 Use this table to switch to and from the number code .

The coding matrix that you will use is $\begin{bmatrix} 1 & -1 \\ -3 & 4 \end{bmatrix}$

The secret message you want to send is ADIOS AMIGO. Change the letters to numbers.

A D I O S * A M I G O * → 1 4 9 15 19 27 1 13 9 7 15 27

Form a matrix with these numbers. It must be compatible with the 2 by 2 coding matrix so that they can be multiplied. This message matrix is a compatible 2 by 6 matrix with the last position filled with a space, 27.

(Message Matrix) $\begin{bmatrix} 1 & 4 & 9 & 15 & 19 & 27 \\ 1 & 13 & 9 & 7 & 15 & 27 \end{bmatrix}$

2 Multiply the message matrix by the coding matrix.

$$\begin{bmatrix} 1 & -1 \\ -3 & 4 \end{bmatrix}\begin{bmatrix} 1 & 4 & 9 & 15 & 19 & 27 \\ 1 & 13 & 9 & 7 & 15 & 27 \end{bmatrix} = \begin{bmatrix} 0 & -9 & 0 & 8 & 4 & 0 \\ 1 & 40 & 9 & -17 & 3 & 27 \end{bmatrix}$$

3 Some of the numbers in the coded message matrix are not in the alphabet code table because they are either greater than 27 or less than 1. If a number is less than 1, keep adding 27 until the number corresponds to a value on the table. If the number is greater than 27, keep subtracting 27 until the result corresponds to a value in the table.

0	−9	0	8	4	0	1	40	9	−17	3	27
27	18	27	8	4	27	1	13	9	10	3	27

4 Use the alphabet-code table to change the numbers into letters and symbols.

27 18 27 8 4 27 1 13 9 10 3 27 → * R * H D * A M I J C *

5 Send the coded message.

6 Receiver gets **coded message**, and uses the alphabet code table to form the coded-message matrix.

* R * H D * A M I J C * → 27 18 27 8 4 27 1 13 9 10 3 27

$$\begin{bmatrix} 27 & 18 & 27 & 8 & 4 & 27 \\ 1 & 13 & 9 & 10 & 3 & 27 \end{bmatrix}$$

7 Multiply the secret matrix by a **decoding** matrix.

If C (coding) is $\begin{bmatrix} 1 & -1 \\ -3 & 4 \end{bmatrix}$, then C^{-1} is $\begin{bmatrix} 4 & 1 \\ 3 & 1 \end{bmatrix}$.

Use C^{-1} to decode the coded message matrix.

$$\begin{bmatrix} 4 & 1 \\ 3 & 1 \end{bmatrix}\begin{bmatrix} 27 & 18 & 27 & 8 & 4 & 27 \\ 1 & 13 & 9 & 10 & 3 & 27 \end{bmatrix} = \begin{bmatrix} 109 & 85 & 117 & 42 & 19 & 135 \\ 82 & 67 & 90 & 34 & 15 & 108 \end{bmatrix}$$

8 Add or subtract 27 as explained in step 3 to find the numbers that fit the table.

109	85	117	42	19	135	82	67	90	34	15	108
1	4	9	15	19	27	1	13	9	7	15	27

9 Translate from numbers to letters using the alphabet-code table.

A D I O S * A M I G O S *

Activity 1

The message in the introduction of this lesson, XRXEPENZ, was coded with the matrix $\begin{bmatrix} 3 & 7 \\ 2 & 5 \end{bmatrix}$. Decode the message.

Activity 2

Use available technology to code, send, and decode the message BEWARE FEARLESS LEADER. Use the coding matrix $\begin{bmatrix} -1 & 3 \\ -2 & 4 \end{bmatrix}$.

Activity 3

$\begin{bmatrix} 2 & 3 \\ 1 & 2 \end{bmatrix}$ Write your own short message of 15 characters or less. Use the given coding matrix to form the secret message. Give the secret message and the coding matrix to a classmate to decode.

Chapter 7 Review

Vocabulary

Key Skills & Exercises

Lesson 7.1

➤ **Key Skills**

Determine the dimensions and addresses of a matrix.

a. What are the dimensions of each matrix?
b. Find the entry at m_{32} for matrix M.
c. Find the entry at n_{21} for matrix N.
d. Can the matrices be equal? Why or why not?

Let $M = \begin{bmatrix} -5 & 6 \\ 7 & -2 \\ 1 & 0 \end{bmatrix}$ and $N = \begin{bmatrix} -5 & 6 & 7 \\ -2 & 1 & 0 \end{bmatrix}$.

a. Matrix M has dimensions 3×2. Matrix N has dimensions 2×3.
b. 0
c. -2
d. No, the dimensions are unequal.

➤ **Exercises**

Let $R = \begin{bmatrix} -4 & 1 \\ 2 & -3 \\ 2 & 0 \\ 8 & -5 \end{bmatrix}$.

1. What are the dimensions of matrix R?
2. Find the entry at r_{32}.
3. Find the entry at r_{41}.
4. What is the address of -5 in the matrix?
5. Give a matrix that is equal to matrix R.

Lesson 7.2

➤ **Key Skills**

Add matrices.

Evaluate the matrix operation.

$$\begin{bmatrix} -2.1 & 3.5 & 6.7 \\ 8.1 & -0.4 & -9.9 \\ 2.1 & 5.4 & -4.7 \end{bmatrix} + \begin{bmatrix} -6.4 & 3.2 & 3.5 \\ 6.5 & 7.6 & 1.2 \\ -9.3 & -4.2 & 0.5 \end{bmatrix}$$

Add entries with corresponding addresses.

$$\begin{bmatrix} -2.1 + (-6.4) & 3.5 + 3.2 & 6.7 + 3.5 \\ 8.1 + 6.5 & -0.4 + 7.6 & -9.9 + 1.2 \\ 2.1 + (-9.3) & 5.4 + (-4.2) & -4.7 + 0.5 \end{bmatrix}$$

$$= \begin{bmatrix} -8.5 & 6.7 & 10.2 \\ 14.6 & 7.2 & -8.7 \\ -7.2 & 1.2 & -4.2 \end{bmatrix}$$

Subtract matrices.

Evaluate the matrix operation.

$$\begin{bmatrix} -2.1 & 3.5 & 6.7 \\ 8.1 & -0.4 & -9.9 \\ 2.1 & 5.4 & -4.7 \end{bmatrix} - \begin{bmatrix} -6.4 & 3.2 & 3.5 \\ 6.5 & 7.6 & 1.2 \\ -9.3 & -4.2 & 0.5 \end{bmatrix}$$

Subtract entries with corresponding addresses.

$$\begin{bmatrix} -2.1 - (-6.4) & 3.5 - 3.2 & 6.7 - 3.5 \\ 8.1 - 6.5 & -0.4 - 7.6 & -9.9 - 1.2 \\ 2.1 - (-9.3) & 5.4 - (-4.2) & -4.7 - 0.5 \end{bmatrix}$$

$$= \begin{bmatrix} 4.3 & 0.3 & 3.2 \\ 1.6 & -8 & -11.1 \\ 11.4 & 9.6 & -5.2 \end{bmatrix}$$

► **Exercises**

Evaluate each matrix operation.

6. $\begin{bmatrix} 4 & 3 & 5 \\ 2 & 1 & 2 \\ 5 & 3 & 1 \end{bmatrix} + \begin{bmatrix} -3 & 1 & 3 \\ -4 & -3 & 1 \\ -1 & 5 & 9 \end{bmatrix}$ 7. $\begin{bmatrix} 5.6 & 6.5 \\ 4.3 & 4.6 \end{bmatrix} + \begin{bmatrix} 7.1 & 3.4 \\ -6.7 & 1.3 \end{bmatrix}$

8. $\begin{bmatrix} -6 & 1 \\ -1 & 2 \\ -2 & 7 \end{bmatrix} - \begin{bmatrix} 5 & -7 \\ -5 & 1 \\ -2 & 4 \end{bmatrix}$ 9. $\begin{bmatrix} -9.9 & 8.8 \\ -2.3 & 1.2 \end{bmatrix} - \begin{bmatrix} 8.3 & 7.3 \\ -9.6 & -5.5 \end{bmatrix}$

Lesson 7.3

► **Key Skills**

Multiply a matrix by a scalar.

Evaluate the matrix operation.

$$4\begin{bmatrix} -4 & 8 \\ 9 & -5 \end{bmatrix}$$

Multiply each entry of the matrix by 4.

$$\begin{bmatrix} 4 \cdot -4 & 4 \cdot 8 \\ 4 \cdot 9 & 4 \cdot -5 \end{bmatrix} = \begin{bmatrix} -16 & 32 \\ 36 & -20 \end{bmatrix}$$

Multiply two matrices to form a product matrix.

Evaluate the matrix operation.

$$[4 \quad 1 \quad -6]\begin{bmatrix} -1 & 4 \\ 2 & 7 \\ -5 & 3 \end{bmatrix}$$

The product can be found either by the matrix multiplication procedure or by technology. The product is $[28 \quad 5]$.

► **Exercises**

Use the given matrices to evaluate the following operations. Write *Not Possible* when appropriate.

$$A = \begin{bmatrix} -2 & -1 \\ 1 & 3 \end{bmatrix} \quad B = \begin{bmatrix} 1 & -3 & -4 \\ 3 & 5 & -1 \end{bmatrix} \quad C = \begin{bmatrix} 5 & 4 & 7 \\ 2 & 1 & 7 \\ 3 & 2 & 6 \end{bmatrix} \quad D = \begin{bmatrix} -1 & -3 \\ 4 & 5 \\ 6 & 1 \end{bmatrix}$$

10. $3A$ **11.** $-4D$ **12.** AB **13.** BC **14.** BA **15.** $-DA$

Lesson 7.4

► **Key Skills**

Determine the inverse matrix for multiplication.

Find the inverse of $V = \begin{bmatrix} 4 & 3 \\ -1 & -2 \end{bmatrix}$.

By definition, $AA^{-1} = I$.

$$\begin{bmatrix} 4 & 3 \\ -1 & -2 \end{bmatrix}\begin{bmatrix} a & b \\ c & d \end{bmatrix} = \begin{bmatrix} 1 & 0 \\ 0 & 1 \end{bmatrix}$$

The inverse can be found either by the matrix multiplication procedure or by technology.

The inverse is $\begin{bmatrix} 0.4 & 0.6 \\ -0.2 & -0.8 \end{bmatrix}$.

► **Exercises**

Find the inverse matrix if it exists for the given matrices.

16. $\begin{bmatrix} -4 & 1 \\ 3 & -3 \end{bmatrix}$ **17.** $\begin{bmatrix} 0 & -1 \\ -1 & -1 \end{bmatrix}$ **18.** $\begin{bmatrix} 0.1 & 0.2 \\ 1 & 2 \end{bmatrix}$

Lesson 7.5

➤ *Key Skills*

Solve a system of equations using matrices.

Use matrices to solve the system.

$$\begin{cases} 3x + 2y = 16 \\ 2x + 3y = 19 \end{cases}$$

Prepare the matrices so that the matrix equation form is $AX = B$.

$$\begin{bmatrix} 3 & 2 \\ 2 & 3 \end{bmatrix} \begin{bmatrix} x \\ y \end{bmatrix} = \begin{bmatrix} 16 \\ 19 \end{bmatrix}$$

To solve the matrix equation, multiply both sides by the inverse of A which is

$$\begin{bmatrix} 0.6 & -0.4 \\ -0.4 & 0.6 \end{bmatrix}.$$

Since $A^{-1}AX = A^{-1}B$ simplifies to $X = A^{-1}B$, find $A^{-1}B$ to solve the system.

$A^{-1}B = \begin{bmatrix} 2 \\ 5 \end{bmatrix}$, so the solution is (2, 5).

➤ *Exercises*

Solve the systems of equations using matrices if possible.

19. $\begin{cases} 2x + 5y = 12 \\ x + 3y = 7 \end{cases}$

20. $\begin{cases} -3x + y = -5 \\ x + 2y = 10 \end{cases}$

21. $\begin{cases} 4x + 3y = 6 \\ x - 2y = -15 \end{cases}$

22. $\begin{cases} x - 7y = 5 \\ -4x + 5y = -20 \end{cases}$

23. $\begin{cases} -2x - 6y = 3 \\ 4x + 12y = -6 \end{cases}$

Applications

24. **Business**   Mr. Tunney is giving four of his employees an 8% pay raise. Their current hourly wages are $5.50, $4.95, $6.35, and $5.75. Use scalar-matrix multiplication to find the new hourly wage for each employee.

25. **Stocks**   The matrix shows the number of shares of three different

$$\begin{array}{c} \\ \text{Stock:} \\ \end{array} \begin{array}{c} \\ 1 \\ 2 \\ 3 \end{array} \overset{\text{Julie \quad Kira \quad Tony}}{\begin{bmatrix} 14 & 12 & 10 \\ 9 & 4 & 8 \\ 17 & 20 & 2 \end{bmatrix}}$$

Last Friday, stock 1 gained $5.50 a share, stock 2 lost $3.00 a share, and stock 3 lost $4.50 a share. How much did each person gain or lose last Friday?

26. At Frank's Gym, a basic membership costs $250 a year, a silver membership costs $350 a year, and a gold membership costs $500 a year. During the first week of a 4-week sales promotion, 4 basic, 3 silver, and 2 gold memberships were sold. During the second week, 3 basic, 4 silver, and 3 gold memberships were sold. During the third week, 5 basic, 1 silver, and 2 gold memberships were sold. During the fourth week, 1 basic, 1 silver, and 3 gold memberships were sold. What was the total number of memberships sold for each week of the sales promotion?

Chapter 7 Assessment

Use the given matrices to answer items 1–21. Write *Not Possible* when appropriate.

$$M = \begin{bmatrix} -2 & -1 \\ 1 & 3 \end{bmatrix} \quad N = \begin{bmatrix} 1 & -3 & -4 \\ 3 & 5 & -1 \end{bmatrix} \quad P = \begin{bmatrix} 5 & 4 & 7 \\ 2 & 1 & 7 \end{bmatrix} \quad Q = \begin{bmatrix} -1 & -3 \\ 4 & 5 \\ 6 & 1 \end{bmatrix}$$

1. Give the dimensions of each matrix.

2. Find the entry at n_{23} for matrix N.

3. Find the entry at q_{12} for matrix Q.

4. What is the address of 4 in matrix P?

5. What is the address of 3 in matrix M?

6. Discuss how you would determine whether a matrix is equal to matrix N.

7. Discuss how to recognize whether two matrices can be multiplied.

8. $N + P$ **9.** $Q - P$ **10.** $P - N$

11. $2M$ **12.** $-4N$ **13.** MN

14. QP **15.** $-PQ$ **16.** NP

17. M^{-1} **18.** MM^{-1} **19.** P^{-1}

20. Let I be the identity matrix for addition. Evaluate $N + I$.

21. Let I be the identity matrix for multiplication. Evaluate MI.

22. If it is possible to multiply matrix A times matrix B, is it also possible to multiply matrix B times matrix A? Explain your answer.

23. Business Kyle and Kodie both work at the Ice Cream Shop and the Burger Palace. Kyle makes $4.50 per hour at the Ice Cream Shop and $5.00 per hour at the Burger Palace. Kodie makes $4.75 per hour at the Ice Cream Shop and $4.50 per hour at the Burger Palace. Last week, they each worked 15 hours at the Ice Cream Shop and 12 hours at the Burger Palace. How much did each earn last week?

Solve the systems of equations using matrices if possible.

24. $\begin{cases} -2x - 3y = 5 \\ x + 2y = -2 \end{cases}$ **25.** $\begin{cases} -5x + 2y = -3 \\ 10x - 4y = 6 \end{cases}$

26. $\begin{cases} -2x + 7y = 0 \\ x + 3y = 1 \end{cases}$ **27.** $\begin{cases} 4x - 2y = -2 \\ 2x - 4y = 10 \end{cases}$

28. Geometry Recall that two supplementary angles have measures whose sum is 180 and that two complementary angles have measures whose sum is 90. If two supplementary angles have measures of $3x$ and $3y$ and two complementary angles have measures of $4x$ and y, what are the measures of the four angles?

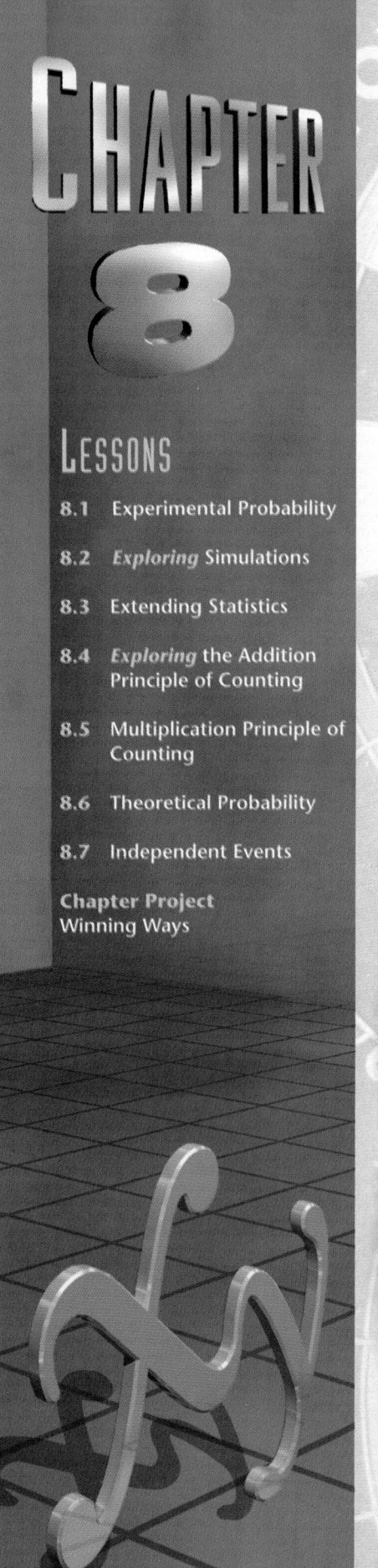

CHAPTER 8

Probability and Statistics

Probability touches your life in many ways. Games, insurance, matters of health and diet, and countless other situations involve an element of chance. Statistics makes extensive use of probability to study, predict, and draw conclusions from data. Sometimes probabilities are found experimentally by looking at data. Other times, probabilities are found theoretically by considering all the possibilities. In this chapter, you can learn about both kinds of probability and explore examples of probability in action.

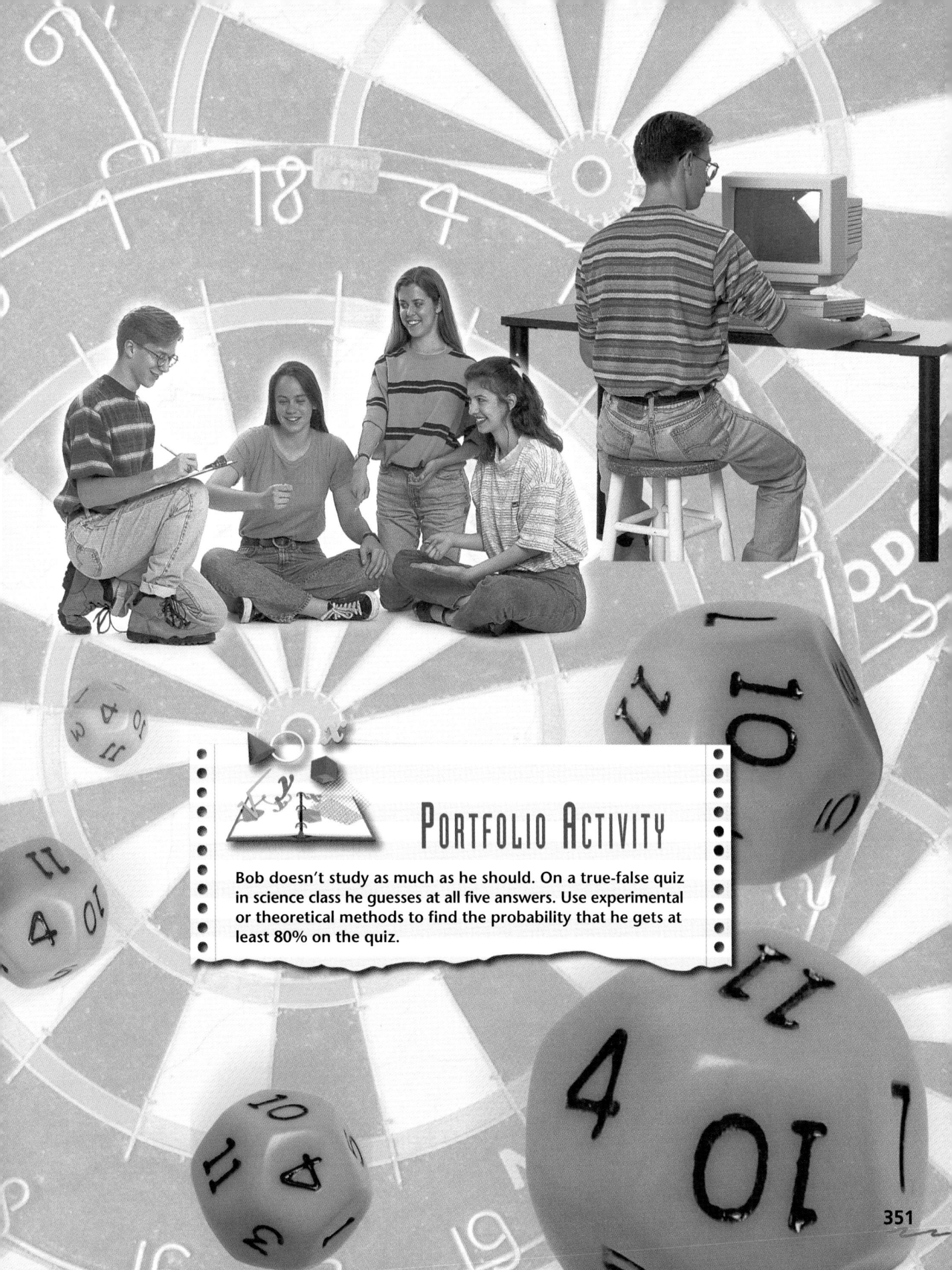

PORTFOLIO ACTIVITY

Bob doesn't study as much as he should. On a true-false quiz in science class he guesses at all five answers. Use experimental or theoretical methods to find the probability that he gets at least 80% on the quiz.

LESSON 8.1 Experimental Probability

why? *People in all cultures seem to enjoy games. Many games involve an element of luck or chance. Chance is often introduced by number cubes or spinners. You can use experimental probability to determine a number that represents the chance that an event will happen.*

Discovering Experimental Probability

Cultural Connection: Americas One American Indian game, *shaymahkewuybinegunug,* involves tossing sticks. It is played by members of the Ojibwa, or Chippewa, tribe. This game uses five flat sticks that have carved pictures of snakes on one side and are plain on the other. Players take turns tossing the sticks to earn points.

•Exploration 1 *Probability and the Coin Game*

Games To play an adaptation of the Ojibwa game, use three coins in place of the sticks. Each person tosses the three coins 10 times. Each time the three coins are the same, that person gets a point. After the person completes 10 tosses, the person with the most points is the winner.

or

Play the game. How many times did all 3 of your coins come up alike?

 If you make 20 tosses, how many times do you think all 3 coins will come up alike?

 Perform an experiment to check your guess. Toss 3 coins 20 times, and record how often the 3 coins come up alike.

Experimental probability is represented as a fraction usually written in lowest terms. The numerator is the number of times all 3 coins come up alike. The denominator is the total number of tosses, 20. Probabilities are sometimes expressed as decimals or percents.

 Give the experimental probability of your result in Step 3. How does your result compare with results from other groups? Notice that there can be many different experimental probabilities.

 Pool the data for the class. What is the experimental probability for the class? Is your result close to the result for the class?

 In this *experiment,* each toss of 3 coins is a *trial.* Let t represent the number of trials. When all 3 coins come up alike it is considered a *successful event.* Let f represent the number of times a successful event occurs. Express the experimental probability, P, in a formula using f and t. ❖

 ## Probability From a Number Cube

 The faces on an ordinary number cube are numbered 1 to 6. If the cube is rolled 10 times, guess how many times a 5 will appear on the top of the cube.

 Roll one cube 10 times. Count how many times you get a 5.

 Define an *event* and a *trial* in this experiment.

 What is the experimental probability of getting a 5 in 10 trials in this experiment? ❖

 Tell how to find the experimental probability for each of the following.

- 3 on one roll of a number cube
- *heads* on a toss of a coin

Experimental probability varies when an experiment is conducted several times.

EXPERIMENTAL PROBABILITY

Let *t* be the number of trials in the experiment. Let *f* be the number of times a successful event occurs.

The **experimental probability**, *P*, of the event is given by

$$P = \frac{f}{t}.$$

•Exploration 3 *A Fair Game With Two Number Cubes*

1 A two-player game involves rolling two number cubes and finding the sum. If the sum is 5, 6, 7, 8, or 9, Player A gets a point. If the sum is 2, 3, 4, 10, 11, or 12, Player B gets a point. Guess which player has a better chance of winning. Explain your answer.

2 When a game is fair, each player has the same chance of winning. Design and conduct a fair game using two number cubes. ❖

Cultural Connection: Asia
Even in prehistoric times people played games of chance and strategy. They must have wondered what the probability of winning would be if they followed a certain strategy. But one of the earliest records of the collection of statistical data had geographical implications.

About 2100 B.C.E., a record was engraved on a clay tablet calculating the surface area of a terrain at Umma, Mesopotamia (Iraq).

Random Numbers

A key ingredient in experimental probability is the use of random numbers. In games, you may have used simple random number generators such as spinners. For making selections, you may have tossed a coin or drawn a number from a hat. There are also books that contain tables of random numbers. Some calculators and computers are useful because they can generate random numbers that can be adapted to fit specific conditions.

Spreadsheet

Spreadsheet software and calculators typically use a command like RAND() or RAND to generate random numbers. The basic random numbers are decimals from 0 to 1, including 0 but not including 1. On some graphics calculators, as you continue to press [ENTER] you get more random numbers.

A6	=RAND()	
	A	**B**
1	0.30261695	
2	0.08310659	
3	0.77245826	
4	0.70603577	
5	0.34317643	
6	**0.66527653**	

Spreadsheet

Rand
.0078387869
.9351587791
.1080114624
.0062633066
.5489861799
.8555803143

Graphics calculator

A6	=RAND()*5	
	A	**B**
1	1.40817388	
2	2.17844473	
3	3.12062745	
4	3.38549538	
5	0.02076914	
6	**4.06690849**	

Rand*5
4.885421232
1.391544132
1.376071474
.6089506775
.2629490302
3.611896579

If each random number is multiplied by 5, the result is six new random numbers. The new random numbers range in value from 0 to 5. Can these new random numbers ever include 5? Explain.

The integer value function INT applied to the random number generator produces random integers from 0 through 4.

A6	=INT(RAND()*5)	
	A	**B**
1	1	
2	0	
3	2	
4	3	
5	4	
6	**0**	

int (rand*5)
3
1
4
2
1
2

Thus the spreadsheet command INT(RAND()*5) gives random integers from 0 through 4. Most software uses a symbol like = or @ to signal a formula.

You can adapt the output of a random number generator to your needs by changing the command.

GENERATING RANDOM INTEGERS

The command INT(RAND() * K) + A generates random integers from a list of K consecutive integers beginning with A.

EXAMPLE 1

Determine the possible numbers that can appear at random when you use the spreadsheet command.

INT(RAND()*2) + 3

Spreadsheet

Solution ➤

Examine the function in steps from the inside out.

Ⓐ **RAND()** results in a decimal value from 0 to 1 excluding 1.

Ⓑ **RAND()*2** results in a decimal value from 0 to 2 excluding 2.

Ⓒ **INT(RAND()*2)** results in the integer 0 or 1.

Ⓓ **INT(RAND()*2) + 3** results in the integer 3 or 4. ❖

Try This Generate sets of random numbers from 3 different commands. Describe the numbers that can possibly appear for each.

EXAMPLE 2

Suppose you are given two random numbers from 1 to 100.
Find the experimental probability that at least one of them is less than or equal to 40.

Solution ➤

Generate 10 pairs of random numbers from 1 to 100. Count how many times at least one of each pair is less than or equal to 40.

Spreadsheet

This spreadsheet shows 10 pairs of random integers from 1 to 100 generated by the command INT(RAND()*100)+1.

In 7 of the 10 trials, at least one of the two numbers is less than or equal to 40. So the experimental probability is $\frac{7}{10}$. ❖

C11	=INT(RAND()*100)+1	
	A	B	C
1	Trial	1st Number	2nd Number
2	1	98	68
3	2	33	82
4	3	21	17
5	4	94	14
6	5	87	36
7	6	73	83
8	7	56	73
9	8	65	12
10	9	18	58
11	10	99	18

CRITICAL *Thinking*

Will you get the same experimental probability if you use a greater number of trials? How does the number of trials affect the probability?

EXERCISES & PROBLEMS

Communicate

1. Explain what is meant by experimental probability.

2. Describe an experiment to find the experimental probability of getting at least 3 *heads* on a toss of 4 coins.

3. Is it possible for two separate groups to conduct the same experiment to determine a probability and get different results? Explain or give an example.

4. Is it possible for someone to conduct the same experiment twice to determine a probability and get different results? Explain or give an example.

5. If RAND generates a number from 0 to 1 (including 0, but not 1), describe the numbers you get from INT(RAND*7) + 1.

6. Look at the spreadsheet data for Example 2. Tell how to find the experimental probability that both numbers are less than 50.

Practice & Apply

7. Describe an experiment to find the experimental probability that if 4 coins are flipped, there will be either 4 *heads* or 4 *tails*.

8. Describe an experiment to find the experimental probability that if 2 number cubes are rolled, at least one of them will show a 6.

9. To determine an experimental probability, Fred conducted 15 trials and Ted conducted 16 trials. Is it possible that they arrived at the same experimental probability? Explain.

Two coins were flipped 20 times with the following results.

Trial	1	2	3	4	5	6	7	8	9	10	11	12	13	14	15	16	17	18	19	20
Coin 1	H	H	H	H	T	H	T	T	H	H	H	T	T	H	T	T	T	H	H	H
Coin 2	T	H	H	T	T	T	T	T	H	T	T	T	H	H	H	H	H	H	T	H

According to the data, find the following experimental probabilities.

10. Both coins are alike.

11. Both coins are *heads*.

12. At least one coin is *heads*.

13. Neither coin is *heads*.

Look at the spreadsheet data for Example 2. Find the experimental probability that

14. both numbers are less than 80.

15. the first number is greater than the second number.

Games Design and conduct experiments to determine the following experimental probabilities. Describe each experiment carefully and give your results for

16. getting *tails* when a coin is flipped.

17. getting 2 *tails* when a coin is flipped twice.

18. getting at least 3 when a number cube is rolled.

19. getting a multiple of 3 when a number cube is rolled.

20. getting an odd sum when two number cubes are rolled.

21. getting "doubles" (the same number on both cubes) when two number cubes are rolled.

22. getting the same result at least 3 times in a row when a coin is tossed 5 times.

23. not getting the same result twice in a row when a coin is tossed 5 times.

Technology The command RAND generates a decimal value from 0 to 1, including 0, but not 1. Describe the output of

24. RAND*2.   **25.** INT(RAND*2).

26. INT(RAND*2)+1.   **27.** 100*(INT(RAND*2)+1).

Before graphics calculators were used, engineers and scientists used books of tables to find data such as random numbers. A table of random digits can be found on page 709.

Look at the command INT(RAND*5)+10, where RAND generates a number from 0 to 1, including 0, but not 1.

28. How many different numbers are there?

29. What is the least number?

30. What is the greatest number?

Technology Write commands to generate random numbers from each of the following lists, where RAND generates a number from 0 to 1, including 0, but not 1. If you can, check your results using technology. Adapt the command to suit the computer or calculator you are using.

31. 0, 1, 2   **32.** 0, 1, 2, 3, 4, 5, 6, 7, 8, 9

33. 1, 2, 3, 4, 5, 6   **34.** 1, 2, 3, 4, 5, 6, 7, 8, 9, 10

35. **Statistics** Use a random number generator to make two sets of 10 numbers. Pair the numbers from the two sets, and make a scatter plot. What kind of correlation does the set of random number pairs show? **[Lesson 1.6]**

36. Write the 6 values of the function $y = 5x + 2$ for x-values of 0, 1, 2, 3, 4, and 5. Examine the differences. Explain what you notice about the differences, and name the function. **[Lesson 2.1]**

Technology Find the values for the following functions. **[Lesson 2.4]**

37. INT(32.573) **38.** INT(34.747) **39.** ABS(-33.725)

Solve each equation for t. [Lesson 4.6]

40. $7(t - r) = 2r - 3(4r + t)$ **41.** $\dfrac{3(4t - 2)}{2} = \dfrac{3}{2} - (5 - t)$

42. Use the data in the table to make a scatter plot. Find the slope of the line that best fits the data. **[Lesson 5.1]**

x	2	5	6	10	12	15
y	4.750	5.875	6.250	7.750	8.500	9.625

43. **Membership Fees** Jenni pays a membership fee of $47, plus $4 per hour at a tennis club. Kari pays a flat fee of $275. For how many hours can Jenni play tennis before her cost equals Kari's? **[Lesson 6.2]**

44. Solve the system of equations. **[Lesson 6.4]**
$$\begin{cases} 3x + 5y = 2 \\ x - 3y = -4 \end{cases}$$

45. Write the system of equations in Exercise 44 as a matrix equation, and solve using matrix methods. **[Lesson 7.5]**

Look Beyond

46. How many pairs of numbers are possible from one red cube and one green cube if each has faces numbered 1 through 6?

47. How many of the pairs in Exercise 46 form a sum of 3?

48. What is the calculator command that will generate the numbers from tossing one number cube?

Complete Exercises 49–54.

49. $1^2 - 0^2$ **50.** $2^2 - 1^2$ **51.** $3^2 - 2^2$

52. $4^2 - 3^2$ **53.** $5^2 - 4^2$ **54.** $6^2 - 5^2$

Use the pattern from Exercises 49–54 to solve.

55. $100^2 - 99^2$ **56.** $a^2 - (a - 1)^2$

Exploring Simulations

Amy makes 50% of her field-goal attempts.

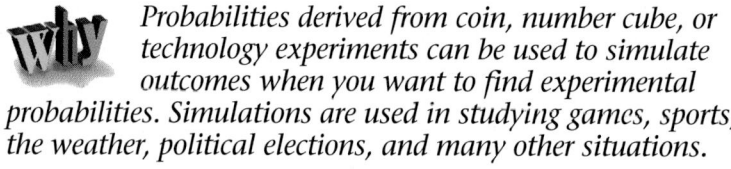

Probabilities derived from coin, number cube, or technology experiments can be used to simulate outcomes when you want to find experimental probabilities. Simulations are used in studying games, sports, the weather, political elections, and many other situations.

Against Johnson High School, Amy made four consecutive shots. How often do you think Amy makes four in a row? If you had the actual shot-by-shot records of Amy's performance, you could see how often this happens. Even without the records, you can find the experimental probability using a *simulation*. A **simulation** is an experiment with mathematical characteristics that are similar to the actual event.

Suppose Amy makes 50% of the shots she takes in a game. If Amy takes 20 shots, a simulation can be designed to find the experimental probability that she will make four in a row.

A spreadsheet can be used to generate random numbers.
1. Let 1 represent each shot made, and let 0 represent each shot missed.
2. Each row represents a trial. Generate 20 random numbers for each row. Thus, each row will have 20 columns.
3. As many rows as needed can be generated. For this simulation examine the first 10 rows or trials as shown. Count the number of trials in which a sequence of at least four consecutive 1s appear.

	A	B	C	D	E
1	Trial	Results			
2	1	1	1	1	0

In the first trial, the beginning of the spreadsheet shows that Amy made her first three shots and missed the next one.

At the end of the first trial, Amy makes 4 in a row.

R	S	T	U
1	1	1	1

U11 =INT(RAND()*2)

	A	B	C	D	E	F	G	H	I	J	K	L	M	N	O	P	Q	R	S	T	U
1	Trial	Results																			
2	1	1	1	1	0	1	1	1	0	1	0	0	0	0	1	1	0	1	1	1	1
3	2	1	0	0	0	1	0	0	0	0	0	0	0	0	1	1	1	0	1	1	0
4	3	0	0	1	1	1	1	1	0	1	1	1	1	0	1	0	0	0	0	0	1
5	4	1	1	1	0	1	0	0	1	0	1	1	0	0	0	1	1	1	0	0	0
6	5	1	1	1	0	1	1	0	1	1	0	1	1	0	0	0	1	0	1	1	1
7	6	0	0	1	0	1	0	0	1	0	0	1	1	1	1	0	1	1	1	1	0
8	7	1	0	0	0	1	0	0	0	1	1	1	1	0	1	0	0	0	0	0	0
9	8	1	0	1	0	1	0	1	1	1	0	1	1	0	0	1	0	1	0	1	0
10	9	1	0	1	0	1	1	1	1	1	0	0	1	0	0	1	0	0	0	0	0
11	10	1	1	1	1	1	1	0	1	1	1	0	1	1	0	0	0	1	0	1	1

Of the 10 trials, there were at least four consecutive 1s recorded in six of the trials: 1, 3, 6, 7, 9, and 10. The experimental probability that Amy will make 4 shots in a row is $\frac{6}{10}$. ❖

•Exploration 1 *Basketball Simulation*

Toss a coin to simulate a 50% chance of making a shot.

 Toss a coin 20 times, and record the sequence of heads and tails.

 Does the sequence have at least four consecutive heads?

 Pool the results of the class.

 Give the experimental probability based on the class experiment. ❖

DESIGNING A SIMULATION

1. Choose a random number generator such as a coin toss, number-cube roll, calculator, computer, or other method that will be used to simulate a situation. Describe what each result represents.
2. Plan how to perform the experiment to simulate one trial.
3. Perform a large number of trials, and record the results.

•Exploration 2 *Weather Simulation*

Spreadsheet

Meteorology Perform a simulation to find the experimental probability that it rains at least one of the two days.

The weather report predicts that the probability of rain is 40% on Saturday and 40% on Sunday.

1 Simulate a 40% chance of rain. If you use technology, generate a sequence of random integers from 1 to 100. A number less than or equal to 40 means it will rain. A number greater than 40 means it will not rain.

You can also use slips of paper or chips. Make 4 slips that show *rain* and 6 that show *no rain*. Then draw at random from a bag.

2 To simulate the weather for two days, generate two numbers. This represents one trial.

3 Perform a large number of trials, and record the results.

The spreadsheet shows the simulation for 10 trials.

C11	=INT(RAND()*100)+1		At least one
A	**B**	**C**	number less than
1 Trial	1st Number	2nd Number	or equal to 40?
2 1	98	68	no
3 2	33	82	yes
4 3	21	17	yes
5 4	94	14	yes
6 5	87	36	yes
7 6	73	83	no
8 7	56	73	no
9 8	65	12	yes
10 9	18	58	yes
11 10	99	18	yes

4 What do the numbers 98 and 68 in Trial 1 indicate about rain on those days?

5 What do the numbers 33 and 82 in Trial 2 indicate about rain on those days?

6 According to the data, how many weekends had rain at least one day in the 10 trials?

7 What is the experimental probability of having rain on at least one of the two days? ❖

APPLICATION

*The probability that Allessandro shows up on time is 60%.
The probability that Zita shows up on time is 80%.*

Appointments   Allessandro and Zita are absent-minded students
who plan to meet for lunch. A simulation can
be designed to find the experimental
probability that both Allessandro and
Zita show up on time.

Design 1: Using number cubes

1. Roll two number cubes, a red one
   and a green one, for example.
   Let a 1, 2, or 3 on the red cube
   mean Allessandro is on time, and
   a 4 or 5 mean that he is late.
   Roll again if you get a 6. Note
   that 60% of the rolls that
   count means he is on time and
   40% means he is late. Let a 1,
   2, 3, or 4 on the green cube
   mean Zita is on time and a 5
   mean she is late. Roll again if you get a 6.

2. Roll the two number cubes to represent one trial, which represents one
   lunch together.

3. Roll the cubes 10 times. Disregard any rolls that include a 6. Determine
   how many times, out of the 10 attempts, they meet on time for lunch.
   How many times does the red cube show a 1, 2, or 3 and the green one
   show a 1, 2, 3, or 4?

4. Express the experimental probability as a fraction with *the number of
   times the friends are on time* over *the total number of trials*. This represents
   the experimental probability that both friends are on time for lunch.

Spreadsheet

Design 2: Using technology

1. Use a calculator or computer to generate a pair of random integers
   from 1 to 100. If the first number is less than or equal to 60, it means
   Allessandro is on time. If the second number is less than or equal to 80,
   it means Zita is on time.

2. Each pair of numbers represents one trial, or meeting.

3. Generate 10 pairs. Determine how many times out of 10 that they
   meet on time.

4. Write the fraction with the number of successful trials over the total
   number of trials. This represents the experimental probability that both
   friends are on time for lunch. ❖

CRITICAL *Thinking*   Make a list of real life situations that can be simulated with a coin flip and
with a toss of a number cube. Compare the characteristics of the different
types of situations.

EXERCISES & PROBLEMS

Communicate

1. What are the three steps for designing a simulation?

2. Assume that the chance of making a shot is 50%. Give three different ways to simulate making or missing a shot.

3. Suppose you want to simulate that it rains when there is a 30% chance of rain. Tell how to do this using random numbers from 1 to 10.

4. Name three things that could be simulated by flipping a coin once.

5. Tell how you might simulate guessing correctly a multiple-choice item with 5 possible responses.

6. Suppose 90% of the flights for an airline are on time. Explain how to design a simulation to find the experimental probability that three consecutive flights are on time.

Practice & Apply

Sports   Look at Amy's basketball game at the beginning of the lesson.

7. How many shots did Amy make in the first trial?

8. In how many trials did Amy make more than 50% of her shots?

9. Give the value in cell Q4. What formula was used to generate the value?

10. What does the value in cell Q4 tell you?

Weather   Suppose you want to simulate two days during which there is an 80% chance of rain and it rains at least one of the two days. Design the simulation and explain how you would perform it using

11. random numbers from 1 to 10.   12. a six-sided number cube.

Describe one way to simulate selecting at random

13. a day of the week.   14. a day of the year.

15. 1 student out of 24 students.   16. which of 8 team captains gets to choose first.

Describe a simulation for Exercises 17–22.

17. You are given two random numbers from 1 to 100. Find the experimental probability that both numbers are less than or equal to 20.

18. There are 10 ways to make a choice. There are 3 possible winning choices. Find the experimental probability that a person will win 2 out of 5 choices.

19. Demographics Find the experimental probability that a family with 4 children includes 2 boys and 2 girls. Assume that births of boys and girls are equally likely.

20. Find the experimental probability that a student correctly guesses all 3 questions on a true-false quiz.

21. Find the experimental probability of guessing the correct answer at least 70% of the time on a 4-question multiple choice quiz, if each item has 5 possible responses.

22. Repeat the previous problem for a 6-question quiz.

Look Back

23. If the variable n represents the number of the term, write an expression based on n that will generate any term of the following sequence.
4, 8, 12, 16, 20, 24, . . . **[Lesson 1.3]**

24. Draw examples, and explain the difference in the graphs of a quadratic and an exponential function. **[Lessons 2.1, 2.2]**

25. Write an expression to represent the length of a piece of fabric that is 74 centimeters longer than the width, w. **[Lesson 3.3]**

26. Solve the previous problem for the width, w, if the length is 86 centimeters. **[Lesson 3.5]**

27. Represent the inequality $-7 < n < 8$ on a number line. **[Lesson 3.6]**

28. Solve $ac = p$ for c. **[Lesson 4.6]**

29. **Coordinate Geometry** A line passes through the origin and the point $A(3, -2)$. Write the equation for the line that passes through point A and is perpendicular to the original line. **[Lesson 5.6]**

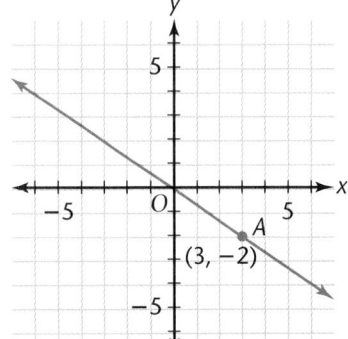

30. Multiply the two pairs of matrices. Are the results the same for each pair? **[Lesson 7.3]**

$$\begin{bmatrix} -1 & 3 \\ 5 & 2 \end{bmatrix}\begin{bmatrix} 2 & -4 \\ 1 & -3 \end{bmatrix} \quad \text{and} \quad \begin{bmatrix} 2 & -4 \\ 1 & -3 \end{bmatrix}\begin{bmatrix} -1 & 3 \\ 5 & 2 \end{bmatrix}$$

31. Multiply the two pairs of matrices. Are the results the same for each pair? **[Lesson 7.3]**

$$\begin{bmatrix} -1 & 3 \\ 3 & -1 \end{bmatrix}\begin{bmatrix} 2 & 1 \\ 1 & 2 \end{bmatrix} \quad \text{and} \quad \begin{bmatrix} 2 & 1 \\ 1 & 2 \end{bmatrix}\begin{bmatrix} -1 & 3 \\ 3 & -1 \end{bmatrix}$$

32. What do Exercises 30-31 show about the commutativity of matrix multiplication?

Look Beyond

33. Find the *theoretical* probability that a student correctly guesses all 3 questions on a true-false quiz.

Hot Hand or Hoop-la?

'Hot Hands' Phenomenon: A Myth?

The gulf between science and sports may never loom wider than in the case of the hot hands.

Those who play, coach or otherwise follow basketball believe almost universally that a player who has successfully made his last shot or last few shots—a player with hot hands—is more likely to make his next shot. An exhaustively statistical analysis led by a Stanford University psychologist, examining thousands of shots in actual games found otherwise: the probability of a successful shot depends not at all on the shots that came before.

To the psychologist, Amos Tversky, the discrepancy between reality and belief highlights the extraordinary differences between events that are random and events that people perceive as random. When events come in clusters and streaks, people look for explanations; they refuse to believe they are random, even though clusters and streaks do occur in random data.

To test the theory, researchers got the records of every shot taken from the field by the Philadelphia 76ers over a full season and a half. When they looked at every sequence of two shots by the same player—hit–hit, hit–miss, miss–hit or miss–miss, they found that a hit followed by a miss was actually a tiny bit likelier than a hit followed by a hit.

They also looked at sequences of more than two shots. Again, the number of long streaks was no greater than would have been expected in a random set of data, with every event independent of its predecessor.

Researchers Survey of 100 Basketball Fans

No 9%

Yes 91%

Does a player have a better chance of making a shot after having just made his last two or three shots than he does after having just missed his last two or three shots?

No 16%

Yes 84%

Is it important to pass the ball to someone who has just made several (2, 3, or 4) shots in a row?

Discuss Do you agree with the majority on the two survey questions? Why or why not?

Cooperative Learning

To form your own opinion about the hot hand study, it helps to know how the researchers tabulated their data. For simplicity, you will use a portion of their data, which shows field goal attempts by two players for several games. Both players usually make about half of their shots from the field.

A—11100010 01110111010010 000011101 0100000101 010101001110 1001111 01001010001 011110110

B—10101000100 11100000110011 0101010111001111 111001001101 0001101010 10010001111

(1 = hit   0 = miss   Space = new game)

1. First, look at what happens after a shot is made.

   a. Copy and complete Table 1, using the data for either Player A or Player B.

   b. What percentage of the times was a hit followed by another hit? What percentage of the times was a hit followed by a miss?

   c. What would you expect your results in **b** would be if there no such thing as a *hot hand*, that is, if the shots were hit or missed just by chance? Explain. (Assume the player makes 50% of his shots on average.)

2. Now look at what happens after a shot is missed.

   a. Copy and complete Table 2. Use the same player you used for Table 1.

   b. What percentage of the times was a miss followed by a hit? What percentage of the times was a miss followed by another miss?

   c. What would you expect your results in **b** would be if the shots were hit or missed just by chance?

TABLE 1	
After a Hit	
Number of times next shot is made	Number of times next shot is missed

TABLE 2	
After a Miss	
Number of times next shot is missed	Number of times next shot is made

3. Do you think there is such a thing as a hot hand in basketball? Why or why not? How does your belief affect the strategy you would use if you were coaching a basketball team?

Extending Statistics

*You already may be
familiar with several topics from
statistics, such as mean, median,
mode, range, scatter plots, correlation,
and lines of best fit. Sometimes the
same statistics can be presented to
show opposite sides of an argument.
To be well informed, you need to
understand when statistics are used
properly and when they are used deceptively.*

National League			
Year	Name	Team	Avg.
1989	Tony Gwynn	Pittsburgh	.336
1990	Willie McGee	St. Louis	.335
1991	Terry Pendleton	Atlanta	.319
1992	Gary Sheffield	San Diego	.330
1993	Andres Galarraga	Colorado	.370

American League			
Year	Name	Team	Avg.
1989	Kirby Puckett	Minnesota	.339
1990	George Brett	Kansas City	.329
1991	Julio Franco	Texas	.342
1992	Edgar Martinez	Seattle	.343
1993	John Olerud	Toronto	.363

Descriptive Statistics

There are many ways
statistics can help you
describe data. For example,
mean, median, and *mode* indicate
the central tendency of data. The
range provides information
about the spread of the data.

EXAMPLE 1

Give the mean, median, mode, and range for the batting averages (avg.) in
each league. Use these statistics to write a summary statement.

Solution ➤

Sports   To find the *mean,* add the batting averages for each player in each league.
Then divide each sum by 5. The means are .338 and .3432, respectively.
The .3432 is customarily rounded to .343.

To find the *median,* arrange the batting averages in order, then choose the middle number. If the number of batting averages is even, the median is the mean of the two middle numbers. The National League batting averages in order from least to greatest are .319, .330, .335, .336, and .370. The median is .335. The median for the American League is .342.

To find the *mode,* choose the batting average that occurs the most. Since no number appears more than once in either set of data, the mode is not reported.

To find the *range,* determine the difference between the highest and lowest batting averages. The range of the National League is .370 − .319 = .051. The range of the American League is .363 − .329 = .034.

In summary, the mean and median batting averages were higher in the American League. The batting averages in the National League had a greater spread, or range. ❖

EXAMPLE 2

Make a scatter plot of the batting averages for each year for the National League. Use a calculator or computer to find the correlation and the line of best fit.

Solution ➤

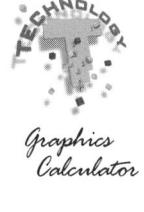

Graphics Calculator

A graphics calculator will help locate the data points for a scatter plot and find the line of best fit.

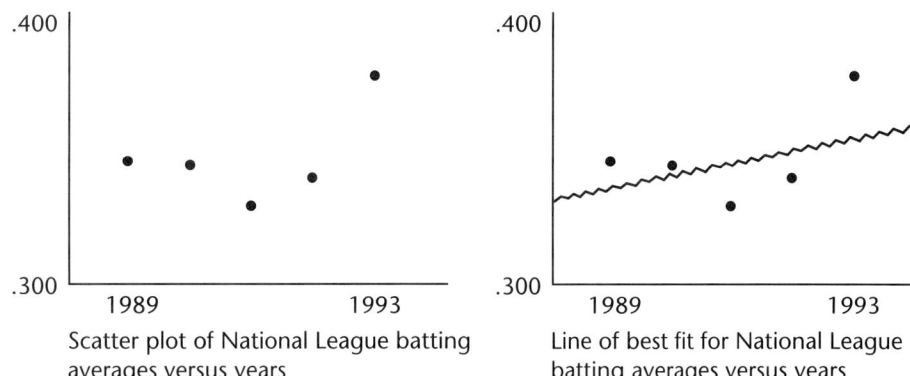

Scatter plot of National League batting averages versus years

Line of best fit for National League batting averages versus years

The variables show a moderate relationship. The correlation is about 0.52, and the line of best fit has a slope of about 0.006. What does this trend suggest about the batting averages over the five-year period? ❖

Try This

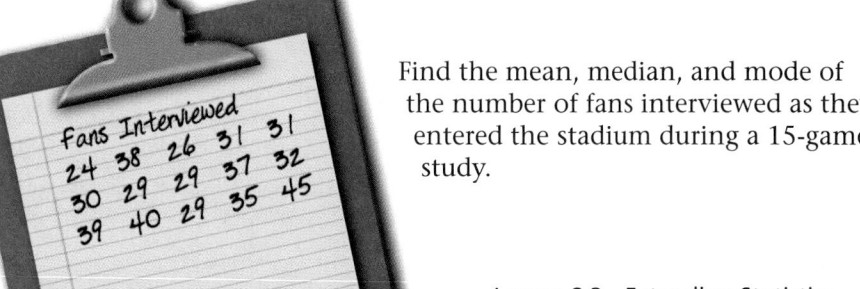

Fans Interviewed
24 38 26 31 31
30 29 29 37 32
39 40 29 35 45

Find the mean, median, and mode of the number of fans interviewed as they entered the stadium during a 15-game study.

Deceptive Statistics

Advertising While statistics are very powerful and useful, they are sometimes used deceptively. Some graphs can be deliberately misleading because of the way the scale is selected.

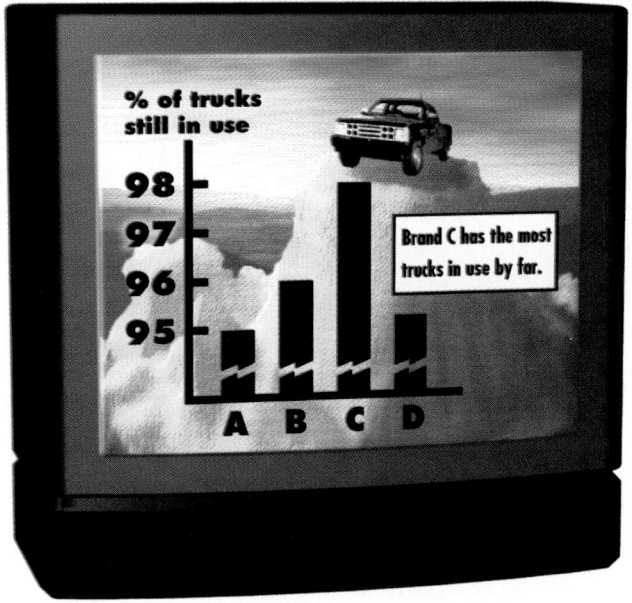

It appears at a glance that Model C, the commercial's sponsor, has far more trucks still in use than the other models. It looks like Model C has about twice as many trucks still in use as Model D. According to the scale, however, Model C differs from Model D by only a small amount, 3%.

GEOMETRY
Connection

Graphs can also be deliberately distorted by using area or volume to compare amounts. The fact that twice as many doctors recommended Brand B can be represented by a 3-dimensional box. A box twice as high with the same width and depth would fairly represent the difference. But a box that has twice the height, twice the width and twice the depth makes Brand B appear much greater.

CRITICAL
Thinking

If box B has twice the dimensions of box A, how many times as great is its volume?

Surveys are used to sample people's opinions about everything from politics to new cosmetics. Major decisions rest on how a sample of the population responds. There is often a question of how certain anyone can be with the results. A phrase used when discussing the results of a survey is *margin of error*.

•Exploration •Margin of Error

Each trial represents the number of heads thrown in 10 tosses of a coin. The table shows the results after 20 trials.

 How many trials show 1 head? 2 heads? 3 heads? How many trials show 5 heads?

Trial	1	2	3	4	5	6	7	8	9	10	11	12	13	14	15	16	17	18	19	20
Heads	3	5	8	3	6	4	5	5	7	3	3	7	6	7	6	5	3	8	5	1

You can see the distribution if you make a tally of the trials for the number of heads thrown.

 What is the mean number of heads thrown for this set of 20 trials?

 What percent of the 20 trials show from 2 to 8 heads?

 How many trials are contained within 3 heads of the mean?

 In general, what do you expect the mean to be for flipping a coin?

 According to the data you can be 95% certain that the results are within 3 heads on either side of 5 for the 20 trials. The 3 heads on either side of the mean represent the *margin of error.*

What percent of all the trials show only 1 head? Is this inside or outside the margin of error? ❖

EXERCISES & PROBLEMS

Communicate

1. Explain the difference between the mean, median, and mode.

2. In Example 1, tell how the median was found for the American League.

3. Bob thought the median for the data 10, 20, 8, 5, and 17 was 8. Explain what he did wrong.

4. Tell how to find the median for a set of six pieces of data.

5. In an ad to show that its lawn products were preferred 2 to 1, a company showed a lawn that its product covered, twice as long and twice as wide as another lawn. This resulted in an area how many times as large? Explain.

6. In the margin-of-error exploration, how confident can you be that the true mean is within 2 of the mean for the total number trials?

7. Describe two ways statistics can be used deceptively.

Practice & Apply

Given the data 24, 26, 19, 20, and 33, find

8. the mean.   **9.** the median.   **10.** the mode.   **11.** the range.

Given the data 12.4, 14.1, 14.1, 14.6, 15.0, and 15.3, find

12. the mean   **13.** the median.   **14.** the mode.   **15.** the range.

16. Tell how to find the median for a set of 9 pieces of data.

17. Tell how to find the median for a set of 10 pieces of data.

18. If the lowest score is 62, the median score is 76, and the highest score is 98, find the range.

19. If the range of scores is 32, the median score is 80, and the highest score is 100, find the lowest score.

20. Construct a set of five pieces of data with mode 18, median 20, and mean 21, or explain why this is impossible.

21. Construct a set of five pieces of data with mode 18, median 20, and mean 19, or explain why this is impossible.

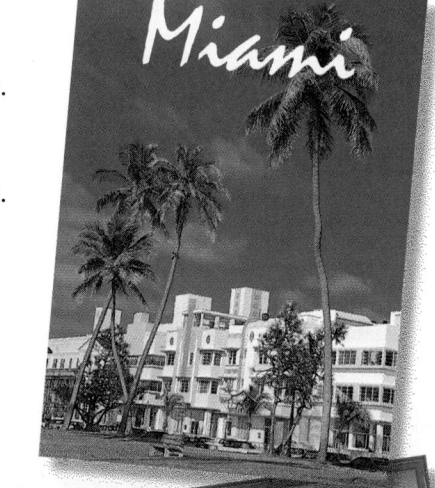

Meteorology The monthly normal mean temperatures for three cities are shown below.

CITY	J	F	M	A	M	J	J	A	S	O	N	D	Avg.
San Francisco, CA	48.7	52.2	53.3	55.6	58.1	61.5	62.7	63.7	64.5	61.0	54.8	49.4	57.1
Washington, DC	34.6	37.5	47.2	56.5	66.4	75.6	80.0	78.5	71.3	59.7	49.8	39.4	58.0
Miami, FL	67.2	68.5	71.1	75.2	78.7	81.4	82.6	82.8	81.9	78.3	73.6	69.1	75.9

22. Compute the range of temperatures for each city.

23. Which city stands out because of its average temperature?

24. Which city stands out because of its temperature range?

25. Write several sentences comparing the temperatures of the cities without using any numbers.

National League			
Year	Name	Team	Avg.
1989	Tony Gwynn	Pittsburgh	.336
1990	Willie McGee	St. Louis	.335
1991	Terry Pendleton	Atlanta	.319
1992	Gary Sheffield	San Diego	.330
1993	Andres Galarraga	Colorado	.370

American League			
Year	Name	Team	Avg.
1989	Kirby Puckett	Minnesota	.339
1990	George Brett	Kansas City	.329
1991	Julio Franco	Texas	.342
1992	Edgar Martinez	Seattle	.343
1993	John Olerud	Toronto	.363

26. Sports Make a scatter plot of the batting averages versus years for the American League.

27. Describe the trend for the American League. Are the batting averages getting better or worse? How much is the change per year?

28. Technology Find the correlation for the data in Exercise 26.

29. Compare the correlation in Exercise 28 with the correlation you found for the National League in Example 2. Is the relationship stronger or weaker?

30. Use an almanac or another source to find data for more recent National League batting champions. Analyze the data, and compare the results with the data given in Example 1.

31. Repeat Exercise 30 for the American League.

32. Advertising Suppose a lawn-care company used an ad to show that its products were preferred 3 to 1 by showing that its product covered a lawn three times as long and three times as wide as a lawn of its competition. This would result in an area how many times as large?

33. Suppose a cereal manufacturer used an ad to show that 20% more people preferred its products by showing a cereal box 20% longer, 20% wider, and 20% taller than its competition. This would result in a box with a volume how many times as large as the competition?

Advertising Examine the graph on pain relievers. Based on the graph, tell whether each of the following is true or false.

34. About 20 doctors recommended Brand A.

35. About twice as many doctors recommended Brand X as Brand A.

36. The majority of doctors recommended Brand X.

37. What percent of doctors recommended Brand C?

38. If there were 200 doctors in the survey, how many recommended Brand X?

39. If there were 200 doctors in the survey, how many more recommended Brand X than Brand C?

Find the margin of error for each situation.

40. The mean is 78, and 95% of the scores are between 75 and 81, inclusive.

41. The mean is 8.4, and 95% of the scores are between 7.8 and 9.0 inclusive.

Look Back

42. The sequence 1, 6, 13, 22, 33, 46, ... is represented by what kind of function? How do you know? **[Lesson 2.2]**

43. Meteorology The lowest temperature on record for each of several cities is shown in the table. What are the differences in temperature? **[Lesson 3.1]**

CITY	RECORD LOW
Juneau, AK	−22
Duluth, MN	−39
Houston, TX	−7

44. A radio is marked $63. What is the actual price of the radio if you also pay a sales tax of 6%? **[Lesson 4.5]**

45. An automatic coin counter totals Denise's collection of dimes and nickels at $36.45. Earlier, she counted 231 nickels. Write and solve an equation to find the number of dimes. **[Lesson 4.6]**

46. Coordinate Geometry What figure is shown when you connect the points $A(-1, -2)$, $B(1, -1)$, $C(3, 0)$, $D(1, 2)$, $E(0, 0)$, and then A again? What is the equation for a line containing points A and D? **[Lesson 5.4]**

47. In Exercise 46 what is the slope of a line parallel to a line containing points D and C? **[Lesson 5.6]**

48. Solve the matrix equation. Then write the matrix equation as a system of linear equations, and solve the system. Are your answers the same? **[Lesson 7.5]**

$$\begin{bmatrix} 9 & -4 \\ -3 & 2 \end{bmatrix} \begin{bmatrix} x \\ y \end{bmatrix} = \begin{bmatrix} 61 \\ -23 \end{bmatrix}$$

Look Beyond

49. Find the mean of the numbers .336, .335, .319, .330, and .370.

50. Find the mean of the whole numbers 336, 335, 319, 330, and 370.

51. If you multiply each piece of data by 1000, by what value does this multiply the original mean?

Exploring the Addition Principle of Counting

Why

The process of counting entries in lists such as databases can often be tedious. Once you learn the shortcuts to counting, you save yourself time and effort. You can also discover many useful and interesting mathematical properties.

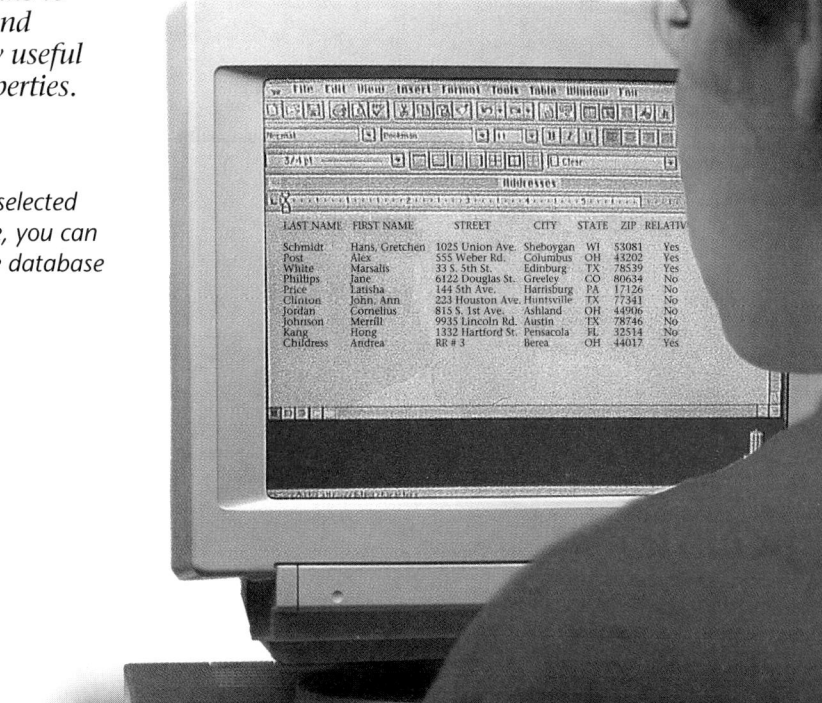

Databases display records that have a selected entry in one or more fields. For example, you can print the records of all the people in the database that live in Texas.

White	Marsalis	TX	Yes
Clinton	John and Ann	TX	No
Johnson	Merrill	TX	No

Similarly, if you print the records of all the people who are relatives, you will get the following.

Schmidt	Hans, Gretchn	WI	Yes
Post	Alex	OH	Yes
White	Marsalis	TX	Yes
Childress	Andrea	OH	Yes

Meg keeps a data base of addresses of her friends and relatives. Each line, like the information about Alex Post, is a *record*. Each category, like State is a *field*.

Databases can also combine information from two different fields. When you request combined information, you will need to use the instructions AND and OR. The words **AND** and **OR** have special meaning when working with databases. For example, if you print the records of all the people who are living in Texas OR who are relatives, you will get the following.

Schmidt	Hans, Gretchn	WI	Yes
Post	Alex	OH	Yes
White	Marsalis	TX	Yes
Clinton	John, Ann	TX	No
Johnson	Merrill	TX	No
Childress	Andrea	OH	Yes

If you print the records of all the people who are living in Texas AND who are relatives, you will get the following.

White	Marsalis	TX	Yes

Logic The relationship between AND and OR can be illustrated with a *Venn diagram.*

Name and Address Database

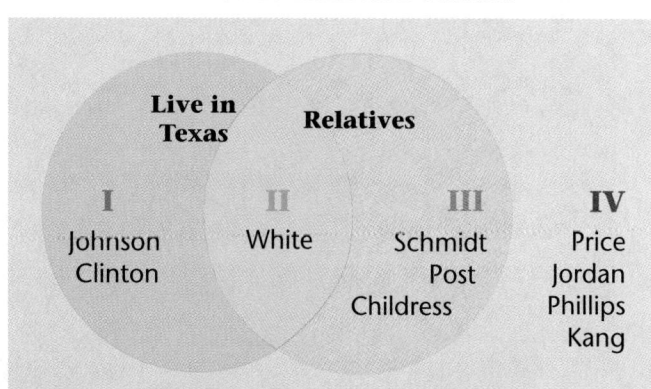

Region I *contains people who live in Texas and who are not relatives.*

Region II *contains people who live in Texas and are relatives.*

Region III *contains people who are relatives and who do not live in Texas.*

Region IV *contains people who do not live in Texas and who are not relatives.*

Region II represents the overlapping regions, or **intersection**, of the people who are in the first set **AND** the second set. Regions I, II, and III together represent the combined regions or **union** of the two sets of people. These people are in the first set **OR** the second set.

When two sets have no members in common, the sets are called **disjoint**. For example, the set of people who live in Ohio and the set of people who live in Texas are disjoint.

intersection

union

disjoint

What is the relationship between set *A* and set *B* if *A* is contained in *B* and *B* is contained in *A*?

Original Database

Last name	First name	Street	City	State	Zip	Relative?
Schmidt	Hans, Gretchen	1025 Union Ave.	Sheboygan	WI	53081	Yes
Post	Alex	555 Weber Rd.	Columbus	OH	43202	Yes
White	Marsalis	33 S. 5th St.	Edinburg	TX	78539	Yes
Phillips	Jane	6122 Douglas St.	Greeley	CO	80634	No
Price	Latisha	144 5th Ave.	Harrisburg	PA	17126	No
Clinton	John, Ann	223 Houston Ave.	Huntsville	TX	77341	No
Jordan	Cornelius	815 S. 1st Ave.	Ashland	OH	44906	No
Johnson	Merrill	9935 Lincoln Rd.	Austin	TX	78746	No
Kang	Hong	1332 Hartford St.	Pensacola	FL	32514	No
Childress	Andrea	RR # 3	Berea	OH	44017	Yes

Exploration 1   *Interpreting a Venn Diagram*

Use the original database that shows all 10 records.

1. Draw a Venn diagram using the last names of people living in Ohio and people living in Texas as your sets.

2. How many of the people live in Ohio?

3. How many of the people live in Texas?

4. How many of the people live in either Ohio OR Texas?

5. How many of the people live in both Ohio AND Texas?

6. Recall the number of people that answers the question in Step 4. Then find the sum of the numbers that answers Steps 2 and 3. What do you notice when you compare these two amounts? ❖

Exploration 2   *Counting from a Venn Diagram*

1. Draw a Venn diagram that represents the people who live in Ohio and who are relatives.

2. How many of the people live in Ohio?

3. How many of the people are relatives?

4. How many of the people either live in Ohio OR are relatives?

5. How many of the people live in Ohio AND are relatives?

6. Does your answer to Step 4 equal the sum of the answers to Steps 2 and 3? Explain.

7. Show how to produce the answer to Step 4 based on the numbers from Steps 2, 3, and 5. ❖

The next application illustrates the use of a counting technique in statistics.

	Favor rule	Oppose rule	Total
Boys	4	9	13
Girls	7	10	17
Total	11	19	30

Mark took a class survey to get student opinions about a new rule concerning students driving to school.

STATISTICS *Connection*

After reviewing the data, the following are some of the questions that were asked. Notice that you need to know what is meant by AND and OR.

Q: How many students sampled are boys?
A: The total for the first row is 13, the total number of boys.

Q: How many students favor the rule?
A: The total for the first column is 11, the total number who favor the rule.

Q: How many students are boys AND favor the rule?
A: If you look under *Favor rule* and across from *Boys,* the intersection shows the number 4. This is the number of students who favor the rule and are boys.

Q: How many students are boys OR favor the rule?
A: The answer is the number of boys plus the number favoring the rule, except that some have been counted twice. Subtract the number who are boys AND favor the rule. If you use the information from the responses to the previous questions, the answer is $13 + 11 - 4 = 20$. ❖

The response to these questions leads to an important principle used in counting.

ADDITION PRINCIPLE OF COUNTING

Suppose there are m ways to make a first choice, n ways to make a second choice, and t ways that have been counted twice. Then there are $m + n - t$ ways to make the first choice OR the second choice.

APPLICATION

Games In an old card game called *Crazy Eights,* if the queen of hearts is showing, you can play either a queen, a heart, or an eight. How many different kinds of cards can be played? Keep in mind that an ordinary deck of cards has 4 queens, 13 hearts, and 4 eights.

You do not count the queen of hearts, since it has already been played. There are 3 other queens, 12 other hearts, and 4 eights, for a total of 19 cards. The eight of hearts has been counted twice, so subtract 1. Thus, $3 + 12 + 4 - 1$ totals 18 cards that can be played. ❖

EXERCISES & PROBLEMS

Communicate

1. Explain the difference between a field and a record in a database.

2. In the original data base for this lesson, name another entry beside the Schmidts that is in the intersection of the set of relatives and the set of people who are not from Texas.

3. In the game of *Crazy Eights* application, explain how the 8 of hearts is counted twice.

4. What do you know about the sets if, in the union of the sets, you do not have to subtract?

Statistics A survey of participation in school music programs has the results shown in the table. Explain how to find each of the following.

5. How many students are girls?
6. How many students participated?
7. How many students are girls AND participated?
8. How many students are girls OR participated?

	Participated	Did not participate
Boys	43	59
Girls	49	57

Practice & Apply

List the integers from 1 to 10, inclusive, which are

9. even.

10. multiples of 3.

11. even AND multiples of 3.

12. even OR multiples of 3.

Games Tell how many of each of the following are in an ordinary deck of 52 cards.

13. aces

14. red cards

15. cards that are red OR aces

In the first application on page 378, how many students

16. are girls?

17. oppose the rule?

18. are girls AND oppose the rule?

19. are girls OR oppose the rule?

20. When does the number of members in the union of two sets equal the sum of the number of members in each set?

List the integers from 1 to 20, inclusive, that are multiples

21. of 5. **22.** of 3.

23. of 5 AND multiples of 3. **24.** of 5 OR multiples of 3.

25. People waiting to hear a concert were given numbers. At one point they let people enter with tickets numbered 100 to 150. How many people entered?

26. If it takes 15 minutes to saw a log into 3 pieces, how long does it take to saw it into 4 pieces? HINT: The answer is *not* 20 minutes.

27. A spreadsheet has 25 rows of data beginning in row 5. Which row has the last row of data?

Statistics In a class of 28 students, 17 have brown eyes and 13 have black hair. This includes 10 who have both.

28. Show this in a Venn diagram. Include all 28 students.

29. How many students have either brown eyes OR black hair?

30. In a poll, 13 students said they had taken music lessons and 5 said they had taken dance lessons. How many students had taken either music lessons OR dance lessons? What is the most? What is the least? HINT: There is more than one answer possible.

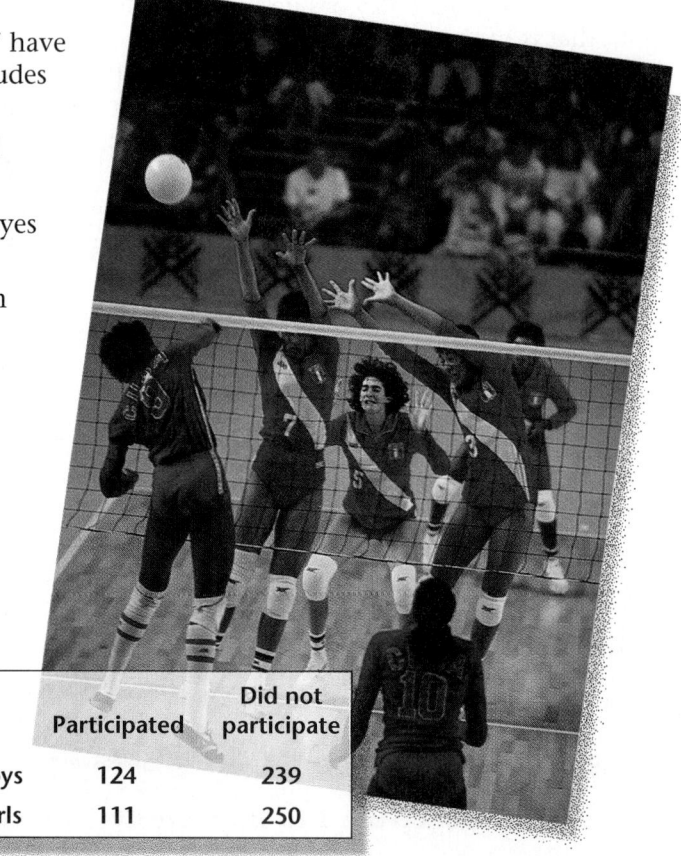

A survey of participation in school team sports had the results shown in the table.

31. How many students are girls?

32. How many students participated?

33. How many students are girls AND participated?

34. How many students are girls OR participated?

	Participated	Did not participate
Boys	124	239
Girls	111	250

Opinion Polls The survey in the first application on page 378 had the following results for the entire school.

Grade	Favor rule	Oppose rule	Total
9	2	97	99
10	34	61	95
11	57	40	97
12	81	7	88

35. How many 9th graders opposed the rule?

36. How many students were either in 9th or 10th grade AND opposed the rule?

37. How many students were either in 9th or 10th grade OR opposed the rule?

38. How many non-seniors favored the rule?

39. How many non-seniors opposed the rule?

Look Back

40. Savings Chou deposits $150 in a bank that pays 5% interest per year compounded annually. How much money will she have in this account after 5 years? **[Lesson 2.1]**

Simplify.

41. $-(-3 + (-7)) + 23 - 2(7 - 5)$ **[Lesson 3.3]**

42. $4(a + 2b) - 5(b - 3a)$ **[Lesson 3.4]**

43. **Geometry** The volume of a cone is $V = \frac{1}{3}\pi r^2 h$. Solve the equation for h.
[Lesson 4.3]

$V = \frac{1}{3}\pi r^2 h$

Solve.

44. $3(x + 5) - 23 = 2x - 47$
[Lesson 4.6]

45. $\frac{x}{3} = -3x + 5$ **[Lesson 4.6]**

46. Sports A major league baseball player has a .300 batting average. Use a simulation to estimate the probability that the batter gets at least two hits in his next four at bats. Describe the 3 steps.
[Lesson 8.2]

Look Beyond

Of the 28 equally-likely ways of choosing a student, there are 17 ways of choosing a student with brown eyes.

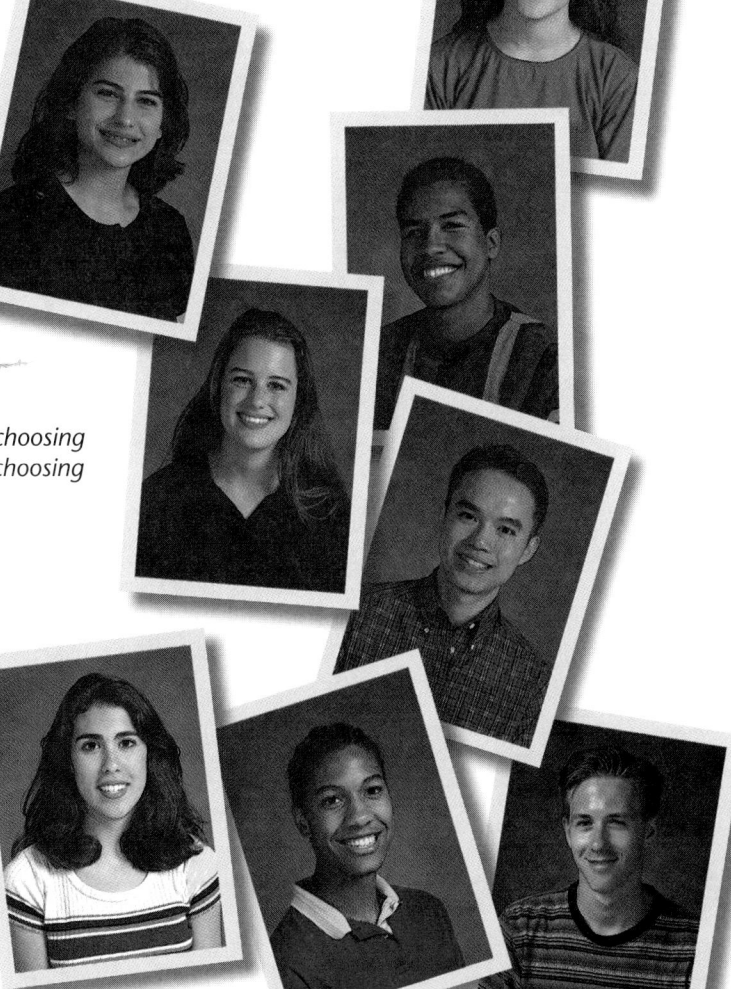

In an earlier exercise there was a class of 28 students. Of those, 17 had brown eyes and 13 had black hair. There were 10 who had both. If a student is chosen at random, the (theoretical) probability that the student has brown eyes is $\frac{17}{28}$. Find the experimental probability that a student chosen at random has

47. brown eyes AND black hair.

48. black hair.

49. brown eyes OR black hair.

50. A book has pages numbered from 1 to 203. How many of the page numbers contain at least one 7?

LESSON 8.5 Multiplication Principle of Counting

why *In the previous lesson you learned how to count in situations in which you made one choice OR another. It is also important to know how to handle situations where you make one choice AND another.*

Jordanna is scheduling her classes for next year. After she registers for her required courses, Jordanna may choose one elective from cluster 1 and one from cluster 2.

Scheduling One way to find all the possible selections is to make a **tree diagram.**

The tree diagram shows that for each of the 3 ways to choose a music elective there are 2 ways to choose a cluster 2 elective.

You can see by following along the branches of the tree diagram that there are 3 · 2, or 6 possible choices.

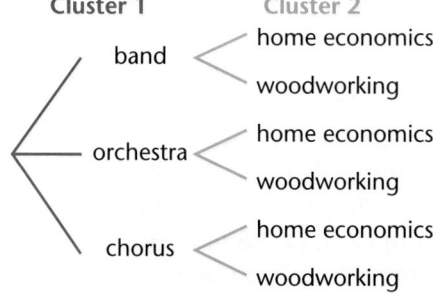

Cluster 1   **Cluster 2**

band
    home economics
    woodworking

orchestra
    home economics
    woodworking

chorus
    home economics
    woodworking

- band and home economics
- orchestra and home economics
- chorus and home economics

- band and woodworking
- orchestra and woodworking
- chorus and woodworking

Jordanna has six possible selections.

EXAMPLE 1

Sports There are 2 volleyball teams, each with 6 members. One player is to be selected at random. Flip a coin to select the team, then roll a number cube to select a player from that team. How many pairings of outcomes are possible?

Solution ➤

Make a tree diagram.

For each of the 2 outcomes from the coin, there are 6 outcomes from the number cube. That is, $2 \cdot 6 = 12$ pairings. ❖

What if there were 3 teams instead of 2? How would you find how many outcomes would be possible?

These examples lead to another useful principle used in counting.

coin   number cube

```
            1
            2
   heads    3
            4
            5
            6

            1
            2
   tails    3
            4
            5
            6
```

MULTIPLICATION PRINCIPLE OF COUNTING

If there are *m* ways to make a first choice and *n* ways to make a second choice, then there are *m* · *n* ways to make a first choice AND a second choice.

The Multiplication Principle of Counting is often used instead of a tree diagram, especially when the answers to counting problems are large. Even when the original numbers are small, some multiplications can produce numbers too inconvenient for a tree diagram.

EXAMPLE 2

Inventory

A shoe store wants to stock 20 styles of shoes in various sizes and colors. How many pairs of shoes must the store carry to have one pair of each combination?

Solution ➤

For each of the 20 styles there are 8 sizes. This means there are 20 · 8, or 160 possible combinations of styles and sizes. For each of these 160 combinations there are 3 colors, so there are 160 · 3, or 480 combinations in all. You can also perform the multiplications together.

$$20 \cdot 8 \cdot 3 = 480$$

As you can see, the Multiplication Principle of Counting can be extended beyond two choices. ❖

Try This Tammy is selecting colored pens. The first characteristic is the kind—felt tip or ballpoint. The second characteristic is the color—red, blue, green, or black. The third characteristic is the brand—Able or Blakely. Use a tree diagram to show the possible selections there are for a pen. Explain how this relates to the Multiplication Principle of Counting.

The next example shows how counting problems occur in geometry.

EXAMPLE 3

GEOMETRY
Connection

Points *A*, *B*, and *C* lie on the same line. How many ways are there to name the line using two of the letters?

Solution A ➤

Make a tree diagram.

There are 6 ways to name the line.

AB, AC, BA, BC, CA, and *CB*

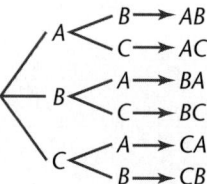

384 CHAPTER 8

Solution B ➤

Use the Multiplication Principle of Counting. There are 3 ways to choose the first letter. Once this is done, there are 2 ways remaining to choose the second letter. There are 3 · 2 = 6 ways. ❖

What are the advantages of using a tree diagram, and what are the advantages of using the Multiplication Principle of Counting?

The addition principle and the multiplication principle are very useful methods to simplify counting. Nevertheless, you must be careful to understand when and where to use each principle.

EXAMPLE 4

Games   A regular deck of 52 playing cards contains 26 red cards and 26 black cards. The deck contains 4 aces, 2 black and 2 red.

Ⓐ One card is drawn. How many ways are there to draw a black ace?

Ⓑ One card is drawn. How many ways are there to draw a black card or an ace?

Ⓒ Two cards are drawn. The first card is replaced before drawing the second card. How many ways are there to draw a black card followed by an ace?

Solution ➤

Ⓐ Two aces are black, the ace of spades and the ace of clubs. Thus, there are 2 ways to draw a black ace.

Ⓑ There are 26 black cards and 4 aces, but the 2 black aces are counted twice. According to the addition principle, there are 26 + 4 − 2 = 28 ways to draw a black card OR an ace.

Ⓒ There are 26 ways to draw a black card. No matter which card is drawn, there are 4 ways to draw an ace as the second card. Using the multiplication principle, there are 26 · 4 = 104 ways to draw a black card AND an ace in two draws. ❖

EXERCISES & PROBLEMS

Communicate

1. Explain the difference between these two sentences and explain the special meaning of AND and OR.
"I am going to the mall AND I am going to study."
"I am going to the mall OR I am going to study."

Suppose in the opening example on page 382, there were four cluster 1 electives instead of three for Jordanna. Explain how to determine how many choices there are if Jordanna were to select

2. one cluster 1 elective OR one cluster 2 elective.

3. one cluster 1 elective AND one cluster 2 elective.

4. Suppose there are 2 volleyball teams, and each team had 8 players. One player is to be selected at random. How can you find the number of branches the corresponding tree diagram has?

5. Two cards are drawn. The first card is replaced before drawing the second card. How can you find the number of ways there are to draw a three followed by a red card?

6. **Geometry** To name a ray, always start with the endpoint. In this case it is *A*. How many ways are there to use two of the letters to name the ray?

Practice & Apply

7. The Raholls are ordering a new car. There are 9 exterior colors from which to choose. For each of these there are 4 interior colors. How many color combinations are possible?

8. **Games** In chess, at the start of the game White moves first and has 20 possible moves. For each of these possibilities, Black has 20 possible countermoves. How many ways are there for the game to start with each player making one move?

9. Tricia bought 2 CDs and 3 tapes. How many recordings did she buy?

10. Ed is trying to choose 1 CD out of 2 that he likes and 1 tape out of 3 that he likes. In how many ways can he do this?

A menu contains 4 appetizers, 3 salads, 5 main courses, and 6 desserts.

11. How many ways are there to choose one of each?

12. How many ways include a chocolate sundae, one of the 6 desserts?

Language   A word which is obtained by rearranging the letters of another word is called an *anagram*. For example, *art* is an anagram of *rat*.

13. Find another anagram of *rat*.

14. How many ways are there to arrange the letters of the word *rat*, using each letter exactly once?

15. How many ways are there to arrange the five letters of the word *stake?* HINT: There are 5 ways to choose which letter is first. For each of these 5 ways, there are 4 ways to choose which letter is second.

16. Write two or three anagrams of *stake*.

17. How many ways are there to arrange the six letters of the word *foster?*

18. Find two anagrams for *foster*.

19. What anagrams can you find within the sentence, "Ed cares that no races on acres in the country scare the birds"?

20. Make up at least 3 anagrams of your own.

21. Geometry   How many ways are there to name the given triangle using 3 letters?

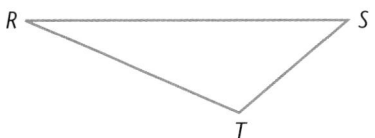

Look Back

22. Hexagonal numbers are in a sequence 1, 6, 15, 28, 45, 66, 91, . . . Find the next two hexagonal numbers.   **[Lesson 1.2]**

Solve.   **[Lesson 4.4]**

23. $\frac{3}{2}x = 6$ **24.** $\frac{x}{-4} = -\frac{5}{8}$ **25.** $\frac{x}{5} = 10$

26. Coordinate Geometry   The points $P(7, -4)$ and $Q(-2, 5)$ are on a line. What is the slope of that line?   **[Lesson 5.1]**

27. If lines on a graph are parallel, how are their slopes related? **[Lesson 5.6]**

28. Chemistry   How many milliliters of a 90% acid solution are needed to mix with 450 milliliters of a 16% acid solution to produce a 50% acid solution?   **[Lesson 6.6]**

Look Beyond

Scheduling   In the opening to this lesson, the probability that Jordanna chooses band and home economics is $\frac{1}{6}$ because making that choice is 1 of 6 equally likely possibilities. Find the following probabilities.

29. She chooses band for her cluster 1 elective.

30. She chooses woodworking for her cluster 2 elective.

31. She chooses either band or woodworking (or both).

32. She chooses band and woodworking.

Theoretical Probability

In the first lesson of this chapter you had the chance to explore experimental probability. Another way to find the numerical measure of chance is to use theoretical probability. Many probabilities can be found more quickly and exactly using theoretical probability.

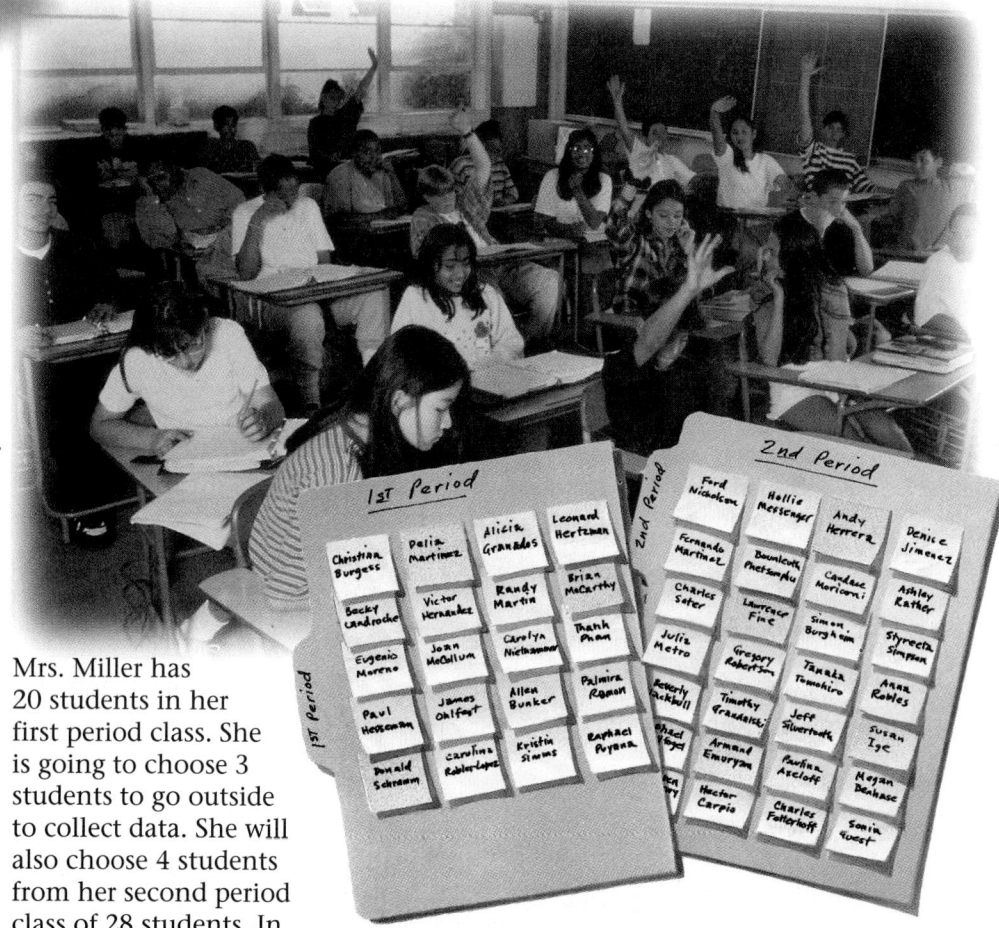

Making Choices

Mrs. Miller has 20 students in her first period class. She is going to choose 3 students to go outside to collect data. She will also choose 4 students from her second period class of 28 students. In which class is the chance of being selected greater?

More students will be chosen from Mrs. Miller's second period class, but does this mean that the chances of being selected are greater? There are also more students from which to choose. To make it easier to compare, find the fraction that represents the number to be chosen out of the total number for each class.

First period   (3 out of 20)   The chances are $\frac{3}{20}$, or 15%.

Second period   (4 out of 28)   The chances are $\frac{4}{28}$, or about 14%.

The chances are slightly greater that a certain person will be chosen in the first period class even though fewer students will be chosen from that class.

The *event* is *to be chosen to go outside*. Since each of the students has an equal chance of being chosen, the number of possible outcomes is the number of students in the class. The number of successful outcomes in the event is the number of students to be chosen.

THEORETICAL PROBABILITY

Let *n* be the number of *equally-likely* outcomes in the event. Let *s* be the number of successful outcomes in the event. Then, the **theoretical probability** that an event will occur is $P = \frac{s}{n}$.

The condition that the outcomes are *equally likely* is important. It means that each outcome has the same probability of happening.

Consider a simple example. A 5 on a toss of an ordinary six-sided number cube is considered a success. You might think that the probability is $\frac{1}{2}$ since only two things can happen—either you will roll a 5 or you will not. Experimentally, however, 5 does not usually come up half the time, so you know this is not correct.

The problem is that the two possibilities are *not equally likely*. The correct probability must take into account six *equally-likely outcomes*. Notice that you can roll either a 1, 2, 3, 4, 5, or 6, but only one of these is a 5. Each of the numbers has 1 chance in 6 that it will appear. The probability of rolling a 5 is $\frac{1}{6}$, not $\frac{1}{2}$.

Always be sure to consider *equally-likely outcomes* when determining the theoretical probability. Note that when you are asked to find probability, it will be the theoretical probability unless otherwise specified.

EXAMPLE 1

Find the probability that an address selected at random from this data base is from Ohio (OH).

ADDRESSES

Schmidt		
Post	Sheboygan	WI
White	Columbus	OH
Phillips	Edinburg	TX
Price	Greeley	CO
Clinton	Harrisburg	PA
Jordan	Huntsville	TX
Johnson	Ashland	OH
Kang	Austin	TX
Childress	Pensacola	FL
	Berea	OH

Solution ➤

There are 10 addresses in all, so there are 10 possible outcomes. There are 3 addresses from Ohio, so the probability of selecting an address from Ohio is $P = \frac{s}{n}$ or $\frac{3}{10}$. This can also be expressed as a decimal, 0.3, or as a percent, 30%. ❖

EXAMPLE 2

Games   Roll a number cube once. What is the probability that the result is less than 3?

Solution ➤

The equally-likely outcomes are 1, 2, 3, 4, 5, or 6. Two of these outcomes (the event 1 appears or the event 2 appears) are successes, so the probability is $\frac{2}{6}$, or $\frac{1}{3}$. ❖

EXAMPLE 3

Roll two number cubes. What is the probability that the sum is 5?

Solution ➤

There are 6 equally-likely outcomes for the first number cube. For each of these, there are 6 equally-likely outcomes for the second number cube. By the Multiplication Principle of Counting, there are 6 · 6, or 36, ways to get a number on the first cube AND a number on the second cube. These are summarized in the table.

Second number cube

	1	2	3	4	5	6
1	2	3	4	5	6	7
2	3	4	5	6	7	8
3	4	5	6	7	8	9
4	5	6	7	8	9	10
5	6	7	8	9	10	11
6	7	8	9	10	11	12

First number cube

Look at the table. How many sums are there in all? How many sums of 5 are there? The probability of rolling a sum of 5 is $\frac{4}{36}$, or $\frac{1}{9}$. ❖

Try This   Find the probability that a sum of 9 is tossed on one roll of two dice.

Notice that there are 11 possible sums: 2, 3, 4, 5, 6, 7, 8, 9, 10, 11, and 12. But the probability of getting a sum of 7 is not $\frac{1}{11}$. These sums are *not* equally likely. For example, a sum of 7 can happen 6 ways, while a sum of 12 can happen only 1 way. The probability of a sum of 7 is $\frac{6}{36}$, or $\frac{1}{6}$. The possibility of a sum of 12 is $\frac{1}{36}$. A sum of 7 is more likely to occur than a sum of 12.

CRITICAL Thinking   Describe an event that results in a sum with a probability of 0. Then describe an event with a probability of 1. What is the range of values for the probability of an event occuring?

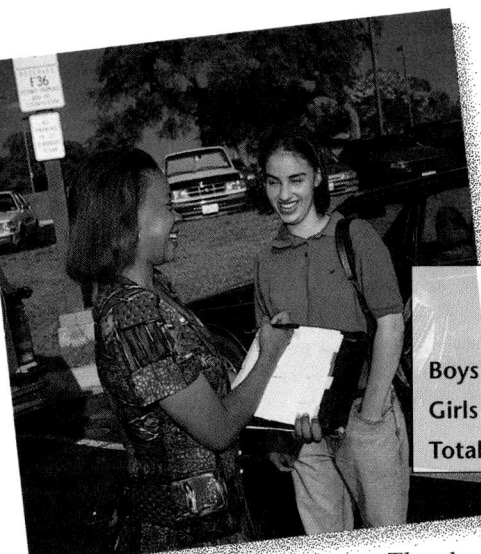

EXAMPLE 4

Statistical Survey   The table shows the results of a class survey about a new rule concerning students driving to school. A student who participated in the survey is chosen at random. What is the probability that the student

	Favor rule	Oppose rule	Total
Boys	4	9	13
Girls	7	10	17
Total	11	19	30

A is a boy?

B favors the rule?

C is a boy AND favors the rule?

D is a boy OR favors the rule?

Solution ➤

The denominator of the fraction is the number of *equally-likely* possibilities. In this case, it is the number of students in the survey, 30. You can use counting principles to find the values for the numerator.

A 13 students are boys. The probability is $\frac{13}{30}$.

B 11 students favor the rule. The probability is $\frac{11}{30}$.

C 4 students are boys who favor the rule. The probability is $\frac{4}{30}$, or $\frac{2}{15}$.

D 13 students are boys and 11 favor the rule, but this counts 4 students twice. By the Additional Principle of Counting there are $13 + 11 - 4$, or 20, students who are boys OR who favor the rule. The probability is $\frac{20}{30}$, or $\frac{2}{3}$. ❖

EXERCISES & PROBLEMS

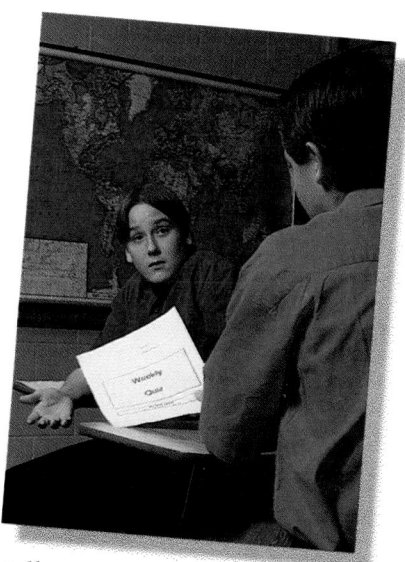

Communicate

1. Explain the difference between experimental probability and theoretical probability.

2. John said he had a 50-50 chance of getting all 10 questions right on a quiz because either he would or he would not get each one right. Is he correct? Explain.

3. Based on his record so far this season, the probability that Nat will make a free throw is 40%. Is this theoretical probability or experimental probability? Explain how you can tell.

4. In a two-player game, four coins are flipped. The first player wins if the coins come up all heads or all tails. Otherwise the second player wins. Is this game fair? Explain.

5. When finding theoretical probabilities, why must the possibilities be equally likely?

Practice & Apply

Games   An integer from 1 to 25 is drawn at random. Find the probability it is

6. even.   **7.** odd.   **8.** a multiple of 3.

9. a prime.   **10.** even AND a multiple of 3.   **11.** even OR a multiple of 3.

Language   Select a letter of the alphabet at random. Find the probability it is

12. r.   **13.** a vowel (a, e, i, o, u, or y).   **14.** a consonant.

15. one of the letters of the word *mathematics*.

16. Geography   Find the probability that the name of a state chosen at random begins with the letter M. HINT: The states starting with M are Maine, Maryland, Massachusetts, Michigan, Minnesota, Mississippi, Missouri, and Montana.

17. Science   Find the probability that a planet in our solar system chosen at random is closer to the sun than Earth is. HINT: Of the 9 planets, only Mercury and Venus are closer.

Transportation   According to the National Transportation Safety Board there were about 0.243 fatal crashes for each 100,000 departures of commuter planes in 1992.

18. Express this ratio as a fraction with a numerator of 1.

19. Is this experimental probability or theoretical probability?

20. For larger planes (carrying over 30 passengers), the number of fatal crashes was 0.05 per 100,000. Express this ratio as a fraction with a numerator of 1.

21. According to these figures, a commuter plane is how many times more likely to have a fatal crash than a larger plane?

22. Language   If the letters of the word *tap* are rearranged, find the probability that an anagram is formed.

23. John is trying to solve the equation $2x - 1 = 15$ on a quiz in algebra. He doesn't remember how to do it, but he hopes that the answer is a positive one-digit number and makes a guess. Find the probability that he guesses the correct solution.

24. Repeat Exercise 23 if the equation is $1 - 2x = 15$.

25.   **Geometry**   A dart is thrown at a circular target. If it is equally likely to land anywhere in the large circle, what is the probability that it hits the bull's-eye, the dark region?

Look Back

26. Technology Make a scatter plot for the points (2, 3), (3, 4), (5, 3), (5, 5), (7, 5), and (8, 7), and find a line of best fit. **[Lesson 1.7]**

27. Draw the graph of the function $y = \dfrac{1}{x}$ for values $0 < x \le 4$. What do you notice about the graph? **[Lesson 2.3]**

28. If $-6 \le n - 4 < 10$, show the solution set on a number line when the values of n are integers. **[Lesson 3.6]**

29. Write an explanation of why you cannot divide by zero. **[Lesson 4.1]**

30. Health A study released at the 1991 annual meeting of the American Heart Association found that taller people have less risk of heart attack. It was found that every inch of height reduced the chance of heart attack by about 3%. Compare the risk of someone 5 feet tall with that of someone 6 feet tall. **[Lesson 4.5]**

31. Carlos bought $53 worth of merchandise and paid $56.18 including the sales tax. Demonstrate the proportion method for finding the rate of the sales tax. **[Lesson 4.5]**

32. Solve the equation $y = mx + b$ for m. **[Lesson 4.6]**

33. A coin is improperly balanced and shows the probability of *heads* to be 0.64. If Suzanne flips the coin 50 times, how many times would you expect *heads* to appear? **[Lesson 8.1]**

Karl's typing rates on 5 tests were 19, 21, 26, 21, and 27 words per minute.

34. What is the mode for Karl's typing rates? **[Lesson 8.3]**

35. What is the median for the typing rates? **[Lesson 8.3]**

Look Beyond

The symbol 3! is 3 *factorial* and means the product of the whole numbers from 3 to 1, inclusive. Thus, 3! = 3 · 2 · 1 = 6.

Find the following values.

36. 4!   **37.** 5!   **38.** 6!   **39.** $\dfrac{3!}{2!}$ **40.** $\dfrac{7!}{5!}$ **41.** $\dfrac{10!}{6!}$

LESSON 8.7 Independent Events

why *Many situations involve more than one event. Sometimes the occurrence of a previous event will affect the probability of the events that follow. As probability becomes more difficult to determine, geometry can provide assistance.*

In Lesson 8.5, Jordanna is selecting one elective from cluster 1, band, orchestra, or chorus, and one elective from cluster 2, home economics or woodworking. If all of the choices are equally likely, find the probability that she chooses band and woodworking.

Cluster 2 — Woodworking — Home ec

GEOMETRY *Connection*

Use 2 vertical lines to divide a square into three equal regions, one for each cluster 1 elective. Then divide it again with a horizontal line, so that there is one region for each cluster 2 elective. The six regions of the square represent the six equally-likely possibilities or outcomes. One region represents the pair, *band* and *woodworking*. The probability for this choice is $\frac{1}{6}$.

EXAMPLE 1

Appointments

Recall that Allessandro and Zita are absent-minded students who plan to meet for lunch. Use an area model to find the probability that both Allessandro and Zita show up on time.

100%
90%
80%
70%
60%
50%
40%
30%
20%
10%
0%

Allessandro

Zita

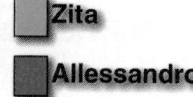

Zita

Allessandro

Solution >

On graph paper, divide a 10 by 10 square into two parts with a vertical line as shown in the first figure. This shows a probability of 60% that Allessandro is on time. Divide the square into two parts with a horizontal line as shown in the second figure. This shows a probability of 80% that Zita is on time.

When the squares for Allessandro and Zita are combined, they divide the square into four regions. The part marked **BOTH** represents the probability that both are on time. It contains $8 \cdot 6$ or 48 out of 100 possible squares. The probability is $\frac{48}{100} = 0.48$, or 48%. ❖

Try This   Use the combined figure to determine the probability that Allessandro is on time and Zita is not.

Sometimes you can solve a problem without using a model.

EXAMPLE 2

Games   Sarah draws a card, then draws another from a regular 52-card deck. Find the probability that the 2 cards she holds are black. In a regular deck, half of the cards are black.

Solution >

There are 52 ways of drawing the first card. After Sarah draws the first card, there are 51 ways of drawing the second card. Since she draws a first card and a second card, apply the Multiplication Principle of Counting. There are $52 \cdot 51 = 2652$ equally-likely ways of drawing the two cards when the first card is not replaced.

A success occurs if Sarah draws two black cards. There are 26 ways to draw the first black card. After the first is drawn, there are 25 ways to draw the second card. Apply the Multiplication Principle of Counting. There are $26 \cdot 25 = 650$ ways to draw two black cards.

The probability of drawing two black cards is then $\frac{650}{2652}$, or $\frac{25}{102}$. ❖

Probability of Independent Events

Independence has a special significance in probability. In Jordanna's problem she first selects an elective from cluster 1 and then selects an elective from cluster 2. The outcome of her first selection has no effect on her second selection. The two events are independent.

> ## INDEPENDENT EVENTS
> Two events are **independent** if the occurrence of the first event does *not* affect the probability of the second event occurring.

EXAMPLE 3

What is the probability that the red cube shows an odd number and the green cube shows a number greater than or equal to 5? Are the events independent?

Todd rolls a red number cube and records the number shown. He then rolls a green number cube, and records the number.

Solution ➤

Apply the Multiplication Principle of Counting. There are 6 · 6, or 36, equally-likely ways that two numbers can appear in two rolls. On the first roll, 3 of the 6 numbers are odd. On the second roll 2 of the 6 numbers are greater than or equal to 5. There are 3 · 2, or 6 ways of rolling a success on both rolls.

The probability of success is $\frac{6}{36}$, or $\frac{1}{6}$. Since the result of the first roll does not affect the second roll, the events are independent. ❖

Examine the probabilities of success on the first roll, the second roll, and both rolls.

The probability of rolling an odd number on the red cube is $\frac{1}{2}$.

The probability of rolling 5 or more on the green cube is $\frac{1}{3}$.

The probability of rolling both an odd number and 5 or more is $\frac{1}{2} \cdot \frac{1}{3}$, or $\frac{1}{6}$.

What pattern do you see?

Events that are not independent are called **dependent**. Give an example of a dependent event.

Finding Probabilities With the Complement

In Lesson 8.6, you determined theoretical probability in three steps.

1. Find the denominator, the number of equally-likely possibilities.

2. Find the numerator, the number of successes.

3. Express the probability as a fraction.

Another useful technique in solving probability problems is to count the ways something *does not* happen. In some cases this provides an easier way to solve the problem.

EXAMPLE 4

Games   Two different cards are drawn from an ordinary deck of 52 cards. Find the probability that at least one of them is red.

Solution ➤

The event *at least one card is red* includes every case that can happen when two cards are drawn *except* getting two black cards. Thus, out of the 2652 ways of drawing two cards, the event includes all except the 650 ways of getting two black cards. This amounts to $2652 - 650$, or 2002, ways. The probability of getting at least one red card is $\frac{2002}{2652}$, or $\frac{1001}{1326}$.

When you find the number of possible ways that an event occurs by considering the number of ways that it does not occur, you are using the **complement**.

Try This   Use the complement to determine the following theoretical probability. When two number cubes are tossed, what is the probability that the sum is at least 3?

CRITICAL Thinking   A box contains 7 quarters, 6 dimes, 5 nickels, and 8 pennies. One coin is drawn from the box. Explain how to find the probability that the coin is worth at least ten cents.

EXERCISES & PROBLEMS

Communicate

1. Name two ways to represent independent events using a diagram.

2. Explain what it means for two events to be independent.

3. When you find probabilities of independent events, are you usually using the Addition Principle of Counting or the Multiplication Principle of Counting?

4. How many paths are there through this tree diagram?

5. Explain the method of finding a probability by finding the complement.

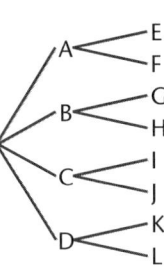

In this area model, the small squares to the left of the vertical red line represent that event A occurs; the small squares above the horizontal blue line represent that event C occurs; etc. Tell how to find the following probabilities.

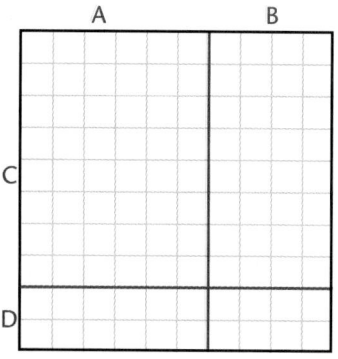

6. A occurs.

7. C occurs.

8. A occurs and C occurs.

9. A occurs or C occurs.

Practice & Apply

10. Find the probability that when two number cubes are thrown, the sum is less than 12.

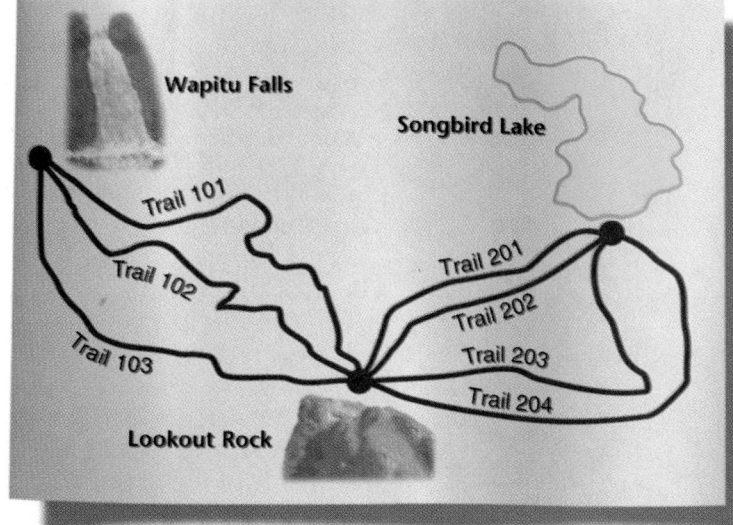

Route Planning John and Jim are each hiking from Wapitu Falls to Songbird Lake using the trails shown on the map.

11. Find the probability that John takes Trail 101 from Wapitu Falls to Lookout Rock.

12. Find the probability that John takes Trail 201 from Lookout Rock to Songbird Lake.

13. Find the probability that John takes Trail 101 from Wapitu Falls to Lookout Rock AND he takes Trail 201 from Lookout Rock to Songbird Lake.

14. Find the probability that John takes Trail 101 from Wapitu Falls to Lookout Rock OR he takes Trail 201 from Lookout Rock to Songbird Lake.

Games Five chips numbered 1 through 5 are in a bag. A chip is drawn and replaced. Then a second chip is drawn. Find the probability that

15. both are even.

16. both are odd.

17. the first is even and the second is odd.

18. one is even and the other is odd.

Games A fair coin is tossed, the result is recorded, and then it is tossed again. Is the event dependent or is it independent? Explain why.

19. Second toss is heads.

20. Second toss is tails.

21. Two heads occur in a row.

Games One number is selected from the list 1, 3, 5, 7. Another is selected from the list 5, 6, 7. Find the probability that

22. both are even.   **23.** they are the same.

24. their sum is even.

25. the number from the second list is greater than the number from the first list.

Use the grid to find the probability that

26. A occurs.

27. C occurs.

28. A occurs AND C occurs.

29. A occurs OR C occurs.

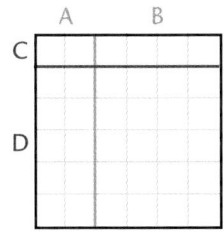

30. **Portfolio Activity** Complete the problem in the portfolio activity on page 351.

Look Back

Find the *y*-values of the following functions for *x* = − 2, − 1, 0, 1, and 2. Then identify the function. [Lesson 2.5]

31. $y = \dfrac{4}{x}$ **32.** $y = 4x^2$

33. $y = |4x|$ **34.** INT(4x)

35. Find the product. $-6(2)\left(\dfrac{5}{-3}\right)$ **[Lesson 4.1]**

36. Tim leaves home on his bike and rides at a rate of 12 miles per hour. Two hours after he leaves, his mother remembers he has a doctor's appointment. She sets out after him in her car at 40 miles per hour. How long does it take her to catch up? **[Lesson 6.6]**

37. **Scheduling** At one school, 48 students signed up to take Spanish, and 23 signed up to take French. If there were 12 that signed up for both, how many signed up for these foreign languages? **[Lesson 8.4]**

38. If Rich has 5 books, how many ways can he arrange them on a shelf of his bookcase? **[Lesson 8.5]**

39. **Travel** There are 3 highways from Birchville to Pine Springs. From Pine Springs to Clearwater there are 5 roads. How many possible ways are there to travel from Birchville to Clearwater? **[Lesson 8.5]**

Look Beyond

40. **Contest** Two radio stations are having a contest. At the first station, the prize is $1000, and your probability of winning is $\dfrac{1}{100,000}$. At the second station, the prize is $25, and your probability of winning is $\dfrac{1}{2000}$. Which provides a better return for the listener's time? Compare your chances of winning the two contests. HINT: What could you expect to happen if you played each contest 100,000 times?

WINNING
W · A · Y · S

Only four out of ten students will be selected to compete against an academic team from another school. If the order in which the students are chosen to appear matters, how many ways can the four students be selected?

If you use the Multiplication Principle of Counting, there are 10 ways to make the first choice. For each of these choices, there are 9 ways to make the second choice.

First choice	Second Choice	Third Choice	Fourth Choice
10	**9**		

That means that there are 10 • 9 ways to make the first two choices. If you continue in the same way, there are 10 • 9 • 8 • 7 or 5040 ways to make the 4 choices.

First choice	Second Choice	Third Choice	Fourth Choice
10	**9**	**8**	**7**

When r objects are chosen and the order matters, the result is called a permutation on n things taken r at a time. The number of possible **permutations** is written $_nP_r$. The example, $_{10}P_4 = 10 \cdot 9 \cdot 8 \cdot 7 = 5040$.

Other examples are:
a. The permutations of 5 letters taken 2 at a time is $_5P_2 = 5 \cdot 4 = 20$.
b. The permutation of 7 digits taken 3 at a time is $_7P_3 = 7 \cdot 6 \cdot 5 = 210$.

Of the 5040 permutations, some of the choices contain the same students. For example, the choice Andy, Bob, Cardi, and Donna is the same as Cardi, Bob, Donna, and Andy when order does not matter. There are $4 \cdot 3 \cdot 2 \cdot 1$ or 24 permutations for each group of 4 students. Thus, if the total number of permutations, 5040, is divided by 24, the result is 210.

Each of these 210 sets of 4 students is called a **combination**. So, while there are 5040 permutations of 10 objects taken 4 at a time, there are only 210 combinations of 10 objects taken 4 at a time.

Thus, when r objects are chosen from n objects and the order does not matter, it is called the **combination** of n things taken r at a time. This is written as $_nC_r$. In the example, $_{10}C_4 = \dfrac{10 \cdot 9 \cdot 8 \cdot 7}{4 \cdot 3 \cdot 2 \cdot 1} = \dfrac{5040}{24} = 210$.

Other examples are:
a. The combination of 5 letters taken 2 at a time is $_5C_2 = \dfrac{5 \cdot 4}{2 \cdot 1} = \dfrac{20}{2} = 10$.

b. The combination of 7 digits taken 3 at a time is $_7C_3 = \dfrac{7 \cdot 6 \cdot 5}{3 \cdot 2 \cdot 1} = \dfrac{210}{6} = 35$.

Activity 1

Use a tree diagram to show the number of permutations of the letters A, B, C, and D taken 2 at a time. Explain how the result can be used to find the number of combinations of 4 objects taken 2 at a time.

Activity 2

Find the number of arrangements consisting of 1, 2, 3, 4, and 5 letters that can be made from the letters in the word *point*. What is the sum of all possible arrangements?

Show all the 3 letter arrangements, and determine which ones form English words. How many English words did you find? What is the probability that a 3-letter word can be formed from the letters in the word *point*?

Chapter 8 Review

Vocabulary

Key Skills & Exercises

Lesson 8.1

➤ Key Skills

Calculate experimental probability.

Two coins are flipped 5 times with the following results.

Trial	1	2	3	4	5
Coin 1	H	T	T	T	H
Coin 2	T	T	H	H	H

The experimental probability that both coins are alike is $\frac{2}{5}$.

Use technology to generate random numbers.

If RAND generates a number from 0 to 1 (including 0, but not 1), then the output of the command INT(RAND( ) * 4) is 0, 1, 2, 3, and the output of the command INT(RAND( ) * 4) + 2 is 2, 3, 4, 5.

In general, the function INT(RAND( ) * K) + A generates random integers from a list of K consecutive integers beginning with A.

➤ Exercises

Use the above data to find the following experimental probabilities.

1. At least one coin is heads.

2. Both coins are tails.

Write commands to generate random numbers from the following lists.

3. 4, 5, 6, 7, 8

4. 0, 1, 2, 3, 4, 5, 6, 7, 8, 9, 10

Lesson 8.2

➤ Key Skills

Design a simulation to determine experimental probability.

Rolling a number cube can be used to simulate choosing at random the correct response to a multiple-choice question. Suppose there are 5 possible choices. Let 1, 2, 3, 4, or 5 represent one of the possible responses, with a 1 representing a correct response. Roll again if you get a 6. One roll represents one trial. The results are as follows.

Trial	1	2	3	4	5	6	7	8	9	10
Roll	1	4	2	1	3	6	1	4	3	5

➤ Exercises

Suppose you want to simulate answering all of the questions on a 10-question quiz correctly. Design a simulation using

5. random numbers.   **6.** coins.

Lesson 8.3

➤ Key Skills

Identify deceptive uses of statistics.

A graph advertising the amount of iron in Cereal X is shown at the right. It appears at a glance that Cereal Z has about twice as much iron as Cereals X and Y. According to the scale, however, Cereal Z differs from Cereals X and Y by only 10%.

➤ Exercises

Refer to the graph for Cereals W, X, Y, and Z above.

7. How much more iron does there appear to be in Cereal Z than in Cereal W?

8. By what percent does Cereal Z actually differ from Cereal W?

Lesson 8.4

➤ Key Skills

Use the Addition Principle of Counting to determine the number of elements in the union of sets.

In a music class, 11 students play the clarinet and 15 play the flute, including 8 who play both. You can use the Addition Principle of Counting to determine the number of students in the music class. There are 11 students who play the clarinet, 15 who play the flute, and 8 who have been counted twice. So, there are $11 + 15 - 8 = 18$ students.

➤ Exercises

Consider an ordinary deck of 52 playing cards.

9. Draw a Venn diagram showing the cards that are black OR face cards.

10. How many cards are black OR face cards?

11. How many cards are red OR numbered cards?

Lesson 8.5

➤ Key Skills

Use the Multiplication Principle of Counting to find the number of possible ways to make multiple choices.

A craft store stocks 5 styles of T-shirts in 4 sizes and in 10 different colors. The craft store stocks $5 \cdot 4 \cdot 10 = 200$ T-shirts.

➤ Exercises

12. How many ways are there to arrange 4 different letters?

13. A dress shop stocks 6 styles of blouses. Each of these comes in 7 sizes, and each in 5 colors. How many different selections are possible?

Lesson 8.6

➤ Key Skills

Find the theoretical probability.

Suppose you roll a number cube once. The probability that you will roll a number greater than 4 is $\frac{2}{6}$, or $\frac{1}{3}$.

➤ Exercises

An integer from −10 to 10 is drawn from a bag containing numbered chips. Find the probability that it is

14. negative.   **15.** zero.   **16.** even.

Lesson 8.7

➤ Key Skills

Find the probability of independent events.

There are 2 red marbles, 4 blue marbles, and 3 white marbles in a bag. If you select a marble, replace it, and then select another marble, the probability of selecting a red marble *and* a blue marble is $\frac{2}{9} \cdot \frac{4}{9} = \frac{8}{81}$.

➤ Exercises

Ten cards numbered 1 through 10 are placed in a box. One card is drawn and replaced. Then another card is drawn. Find the probability that both are

17. less than 5.   **18.** multiples of 3.   **19.** the same.

Applications

Marketing The Hamburger Hut is giving away free food during their grand opening. They are distributing 1,000 cards, of which 100 are for a free hamburger, 150 are for free fries, and 200 are for a free soft drink. What is the probability of winning

20. a hamburger?   **21.** fries?   **22.** a soft drink?

23. The Hamburger Hut advertises 150 different ways to order their hamburgers, 2 different types of fries, and 8 different flavors of soft drinks. How many ways are there to choose one of each?

24. The probability that a household in the United States will have some kind of pet is about $\frac{6}{10}$. The probability that a household will have at least one child is $\frac{1}{3}$. What is the probability of a household in the United States having a pet and a child?

25. Tim, Mary, and Paul have applied for a loan to buy a used car. The probability that Tim's application is approved is 75%. The probability that Mary's application is approved is 80%. The probability that Paul's application is approved is 62.5%. What is the probability that all three loans will be approved?

26. In Item 25, what is the probability that only Mary's application will be approved?

Chapter 8 Assessment

If RAND( ) generates a random number from 0 to 1 (including 0, but not 1), then describe the output of the following commands.

1. RAND( ) * 5   **2.** INT(RAND( ) * 5)   **3.** INT(RAND( ) * 5) + 4

4. Write a command using RAND( ) to generate random numbers from the list 1, 2, 3, 4, 5, 6, 7, 8.

5. Show how to simulate randomly picking a phone number from a list of 6 phone numbers.

6. If the range for a set of scores is 45, the mean is 75, and the highest score is 98, find the lowest score.

Suppose the manager of a health club used an ad to show that people preferred her club 4 to 1 by showing a locker four times as long and four times as wide as the competition.

7. This would result in an area how many times as large?

8. Would this accurately show the comparison of the preference of health clubs? Explain.

In an ordinary deck of 52 cards tell how many of the cards are

9. numbered.   **10.** red.

11. red AND numbered.   **12.** red OR numbered.

13. There are 16 girls on a softball team and 11 girls on a track team. If there are 5 girls on both teams, how many girls are on the softball team and the track team?

14. The Tiberi's are ordering a new couch. There are 5 styles from which to choose. For each of these there are 15 fabrics from which to choose. How many combinations are possible?

15. What does it mean to say that outcomes are equally likely? Why is this important?

Two number cubes are rolled. Find the probability that the sum is

16. 10.   **17.** less than 5.

18. a multiple of 2 AND 3.   **19.** a multiple of 2 OR 3.

An integer from 1 to 10 is selected. Then a letter from the alphabet is selected. Find the probability that

20. the integer is odd AND the letter is a vowel (a, e, i, o, u, or y).

21. the integer is a multiple of 5 AND the letter is a consonant.

22. the first letter of the name of the integer chosen is t AND the letter chosen is t.

A box contains 3 blue stickers, 4 red stickers, and 11 yellow stickers. Maria chooses one sticker and then another. Find the probability that she chooses

23. a blue sticker, puts it back, and then chooses a yellow sticker.

24. a blue sticker, does *not* put it back, and then chooses a yellow sticker.

25. two red stickers without putting the first one back.

26. Explain the difference between independent and dependent events.

Chapters 1 – 8
Cumulative Assessment

College Entrance Exam Practice

Quantitative Comparison   For Questions 1–4, write
A if the quantity in Column A is greater than the quantity in Column B;
B if the quantity in Column B is greater than the quantity in Column A;
C if the two quantities are equal; or
D if the relationship cannot be determined from the information given.

	Column A	Column B	Answers
1.	reciprocal of -2	reciprocal of -3	(A) (B) (C) (D) **[Lesson 2.3]**
2.	$-18 + 4$	$-10 + (-4)$	(A) (B) (C) (D) **[Lesson 3.1]**
3.	$\|-5.2\|$	$\|4.9\|$	(A) (B) (C) (D) **[Lesson 2.4]**
4.	45% of 16	60% of 12	(A) (B) (C) (D) **[Lesson 4.5]**

5. What is INT(4.5)?   **[Lesson 2.4]**
   a. 4   **b.** 4.5   **c.** -4.5 **d.** -5

6. What are the next three terms of the sequence 6, 12, 24, 48, . . .?
   [Lesson 1.1]
   a. 72, 96, 120   **b.** 96, 192, 384   **c.** 86, 162, 240   **d.** 50, 52, 54

7. What is the solution to $16y = -120$?   **[Lesson 4.3]**
   a. 1920   **b.** -1920 **c.** 7.5   **d.** -7.5

8. What is the equation of a line passing through $(-1, 2)$ and parallel to
   $y = 2x - 1$?   **[Lesson 5.6]**
   a. $y = -\frac{x}{2} + 4$ **b.** $y = -\frac{x}{2} + 2$ **c.** $y = 2x + 4$ **d.** $y = -x - 1$

9. Find a solution to the system. $\begin{cases} 2x + 3y = 9 \\ x - 4y = -23 \end{cases}$
   [Lessons 6.1, 6.2, 6.3]

   a. $(3, 5)$ **b.** $(-3, 5)$ **c.** $(3, -5)$ **d.** $(-3, -5)$

10. A television is marked $198. What is the actual cost of the television if
   the sales tax is 7.5%?   **[Lesson 4.5]**
   a. $212.85   **b.** $14.85   **c.** $183.15   **d.** $205.50

11. Graph the list of ordered pairs. Tell whether they lie on a straight line. $(-2, 2), (1, 4), (-4, 0)$ **[Lesson 1.5]**

12. Use this data to make a graph. Which point is the vertex? **[Lesson 2.2]**

x	-2	-1	0	1	2
y	3	0	-1	0	3

13. Is the graph of the equation $x = -4$ vertical or horizontal? What is the slope of the line? **[Lesson 5.5]**

14. Solve $4 - 3t \geq 19$. Show the solution set on a number line. **[Lesson 3.6]**

Let $X = \begin{bmatrix} 2 & -1 & 5 \\ 1 & 0 & -2 \end{bmatrix}$ **for Items 15–17.**

15. What are the dimensions of matrix X? **[Lesson 7.1]**

16. Find the entry at x_{23}. **[Lesson 7.1]**

17. Find $3X$. **[Lesson 7.3]**

18. A jacket has been marked down from an original price of $85 to $59.50. What percent from the original price has the jacket been marked down? **[Lesson 4.5]**

19. Mark has 3 nickels, 5 dimes, and 7 quarters in his pocket. He chooses one coin and then another. If he does not replace the first coin, what is the probability that he chose a dime first and then a quarter? **[Lesson 8.1]**

20. The perimeter of a rectangle is 52 and its width is 6 less than its length. Find the length and width of this rectangle. **[Lesson 3.3]**

Free-Response Grid The following questions may be answered using a free-response grid commonly used by standardized test services.

21. Simplify $4^2 \div 8 + 5(8 - 2) \cdot 2$. **[Lesson 1.4]**

22. How much would each of 16 people need to contribute toward a gift that costs $136? **[Lesson 2.3]**

23. Solve $7(p + 4) = 49$. **[Lesson 4.6]**

24. Find the slope of the line that passes through the origin and $(4, 2)$. **[Lesson 5.3]**

25. What is the value of $54 - 68 + |-80|$? **[Lesson 3.1]**

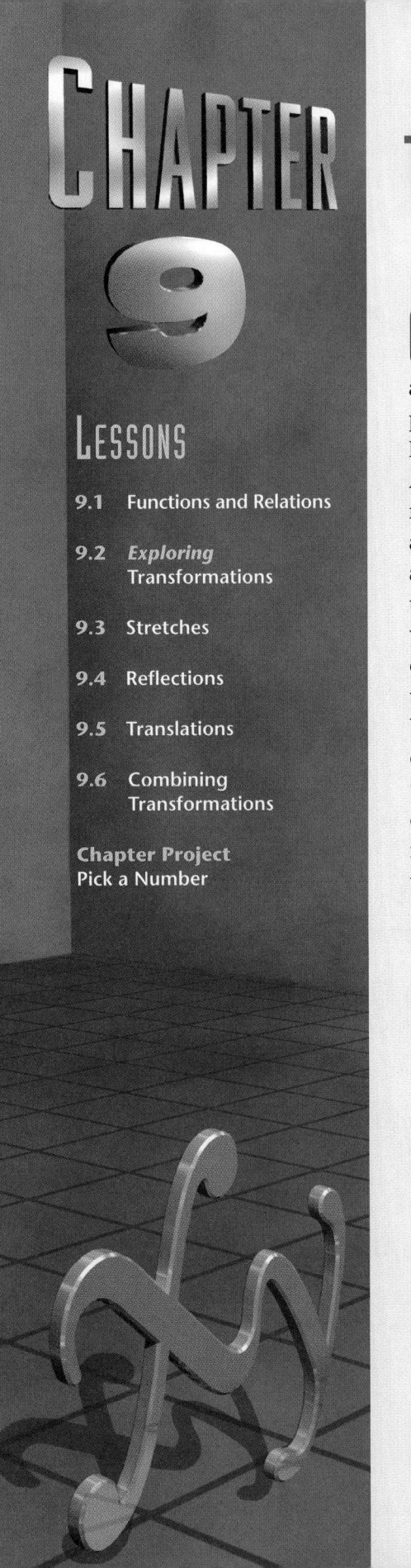

Transformations

Once a basic function and its graph are understood, it is possible to create many adaptations that help fit the function to specific needs. Adaptations, known as transformations in mathematics, appear in such fields as science, business, advertising, and artistic design. In mathematics, the use of technology lets you explore the effects of a broad range of transformations. With technology, even complex transformations of functions are now much easier to perform.

PORTFOLIO ACTIVITY

Draw coordinate axes on a piece of graph paper. Leave most of the space to the right of the y-axis. Then design a figure in the first quadrant like the one shown in the example. Reflect the figure through the x-axis. Translate the reflection to the right far enough so the translation does not overlap the reflection. Then reflect the translation through the x-axis. Continue the procedure to create a repeating pattern.

Create a new figure, and perform several different reflections and translations to create other repeating patterns.

Identify one point on your original design by its coordinates. List the coordinates of that point after each successive transformation.

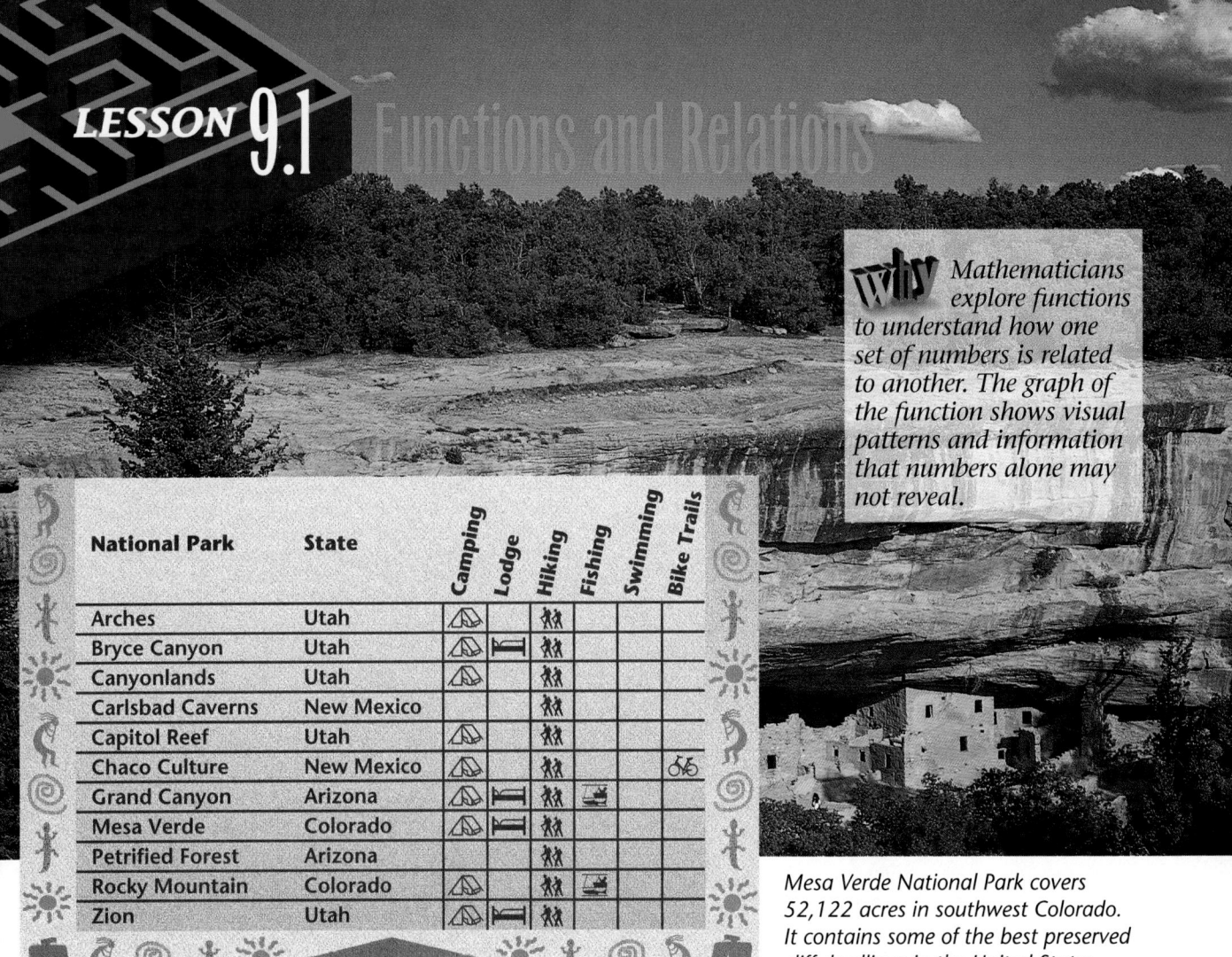

Mathematicians explore functions to understand how one set of numbers is related to another. The graph of the function shows visual patterns and information that numbers alone may not reveal.

National Park	State	Camping	Lodge	Hiking	Fishing	Swimming	Bike Trails
Arches	Utah	🏕		🚶			
Bryce Canyon	Utah	🏕	🛏	🚶			
Canyonlands	Utah	🏕		🚶			
Carlsbad Caverns	New Mexico			🚶			
Capitol Reef	Utah	🏕		🚶			
Chaco Culture	New Mexico	🏕		🚶			🚲
Grand Canyon	Arizona	🏕	🛏	🚶	🎣		
Mesa Verde	Colorado	🏕	🛏	🚶			
Petrified Forest	Arizona			🚶			
Rocky Mountain	Colorado	🏕		🚶	🎣		
Zion	Utah	🏕	🛏	🚶			

Mesa Verde National Park covers 52,122 acres in southwest Colorado. It contains some of the best preserved cliff dwellings in the United States.

The Alvarez family is deciding which national parks to visit on their summer vacation. In their road atlas they have a chart that gives the camping features at each park.

Defining Relation and Function

Two different pairings appear in the table. The first pairs a park with the features available at that park. For example, Mesa Verde is paired with camping, a lodge, and hiking trails. This pairing is an example of a **relation**.

Mesa Verde — Camping / Lodge / Hiking Trails

A **relation** pairs elements from one set with elements of another. In other words, a relation is a set of ordered pairs.

The second pairs each park with *exactly one* state. For example, Mesa Verde is paired only with Colorado. This pairing is an example of a **function**. A function is a special relation.

Mesa Verde — Colorado

FUNCTION

A **function** is a set of ordered pairs for which there is exactly one second coordinate for each first coordinate.

You have already been introduced to several elementary functions and have examined their graphs by plotting ordered pairs. Remember, to graph a function use the coordinates of ordered pairs such as (x, y) to locate a set of points on a coordinate plane.

The graph shows speed as a function of time. The points on the graph form a set of ordered pairs (time, speed). There is exactly one speed for any moment in time.

The graph shows the relation between

- the speed in miles per hour at which the Alvarez's car travels and

- each moment of time during the first 25 minutes of their trip.

CRITICAL Thinking Use the graph to determine what might have happened during the first 25 minutes of the trip to cause the changes in the graph.

Testing for a Function

Once a relation is graphed, a **vertical line test** will show whether the relation is a function. A relation is a function *if any vertical line intersects the graph of the relation no more than once.*

EXAMPLE 1

Which of the following graphs represent functions?

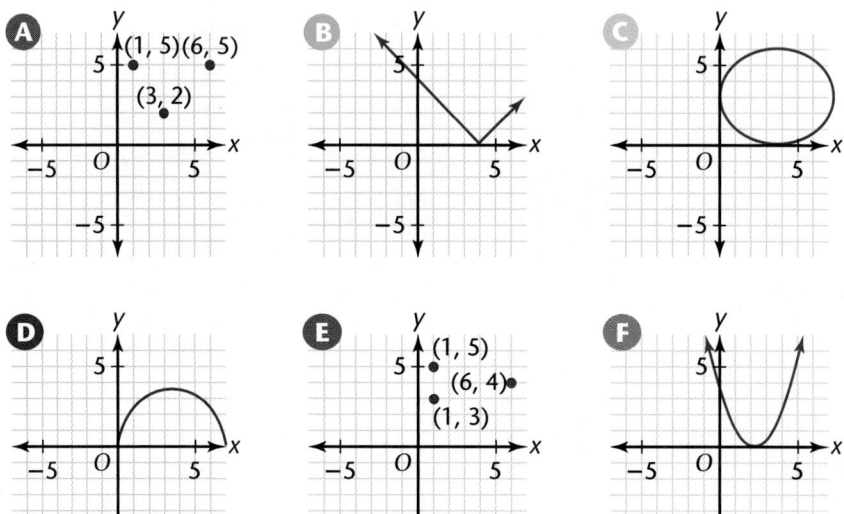

Solution ➤

Any vertical line will intersect graphs A, B, D, and F *at most once,* so these graphs represent functions. The vertical lines on graphs C and E intersect the graph of the relation twice. These graphs do *not* represent functions.

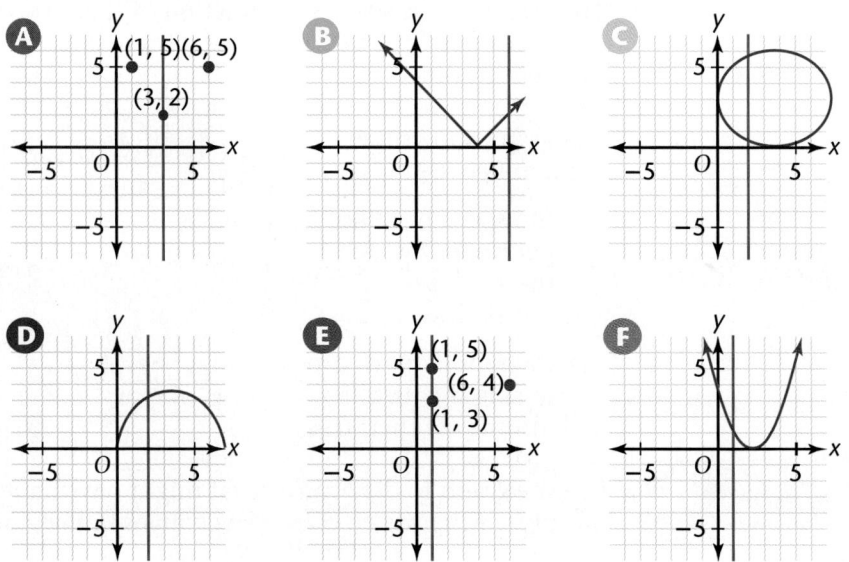

For any function, the set of x-values is the **domain** and the set of y-values is the **range**.

The domain contains only the numbers that can be used in the function. The range contains only the numbers that the function can produce. For example, division by 0 is not permitted, so 0 cannot be in the domain of $y = \frac{1}{x}$. How do you know that there are no negative numbers in the range of the function, $y = |x|$?

Examine the graphs in Example 1. Can you identify the domain and range of the functions A, B, D, and F?

Example	Domain (x-values)	Range (y-values)
A	1, 3, and 6	2 and 5
B	all numbers	all numbers greater than or equal to zero
D	$0 \leq x \leq 7$	$0 \leq y \leq 3.5$
F	all numbers	all numbers greater than or equal to zero

Function Notation

You have used the equation $y = x^2$ to represent a quadratic function. Another way to express $y = x^2$ is to use *function notation*. Function notation uses a symbol such as $f(x)$ to write $y = x^2$ as $f(x) = x^2$.

This is read "*f* of *x* equals *x* squared." The variable in parentheses is the *replacement* variable. The expression on the right side of the equal sign is the *function rule*.

To evaluate a function, replace the variable *x* with a value from the domain. Then perform the operations given in the rule.

For example, to evaluate the function $f(x) = x^2$ for $x = 3$, substitute 3 for *x* and simplify.

$$f(3) = 3^2 = 9$$

In algebra it is customary to use *f*, *g*, and *h* to represent functions. The domain of a function is sometimes called the input values. The range is sometimes called the output values.

input value $\longrightarrow$ function rule $\longrightarrow$ output value

12

$f(x) = x^2$

144

EXAMPLE 2

Suppose h is a function and $h(x) = x^2 - 2x + 1$. The domain of the function is *all of the integers*. Evaluate $h(4)$.

input

function box

output

4

$h(x) = x^2 - 2x + 1$

9

Solution ➤

The notation $h(4)$ indicates that you substitute 4 for each x in the rule $x^2 - 2x + 1$. Since 4 is in the domain, simplify to find the value of the function.

$$h(x) = x^2 - 2x + 1$$
$$h(4) = 4^2 - 2(4) + 1$$
$$h(4) = 9 ❖$$

The h identifies or names the function. The input values for x in Example 2 are all integers. The *function rule* is the expression $x^2 - 2x + 1$. The values you get after substituting each input value and simplifying the expression are the output values.

The values shown in $h(4) = 9$ represent the ordered pair $(4, 9)$. Notice that 4 is an element of the domain, and 9 is an element of the range. If you graph the rule, you can see that it satisfies the vertical line test for a function.

Try This For the function $d(t) = 10t^2$ evaluate $d(-3)$.

The distance between two points on a number line is always a nonnegative number. This makes the absolute value function a good representation for distance on the number line. For example, the absolute value function can represent the distance between a given number, such as 5, and some other number, x.

EXAMPLE 3

Determine a function for the distance between any number and the fixed point, 5. Find the distance between -4 and 5 and between 0 and 5.

Solution ➤

The function $g(x) = |x - 5|$, where the domain is all numbers, gives the distance between any number and 5. To find the distance between -4 and 5, evaluate $g(-4)$.

The distance between x and 5 when $x = -4$ is $g(-4) = |-4 - 5| = 9$.

To find the distance between 0 and 5, evaluate $g(0)$.

The distance between x and 5 when $x = 0$ is $g(0) = |0 - 5| = 5$.

EXERCISES & PROBLEMS

Communicate

1. What is the difference between a function and a relation?

2. Examine the table of national parks at the beginning of the lesson. If national parks and states are paired, they form a function. Explain why this is a function.

3. Describe the relationship between the definition of a function and what the vertical line test shows.

4. Explain the procedure for evaluating the function $f(x) = 2x^2 - 5x$ if the domain consists of integers from -2 to 2.

5. You are given an ordered pair $(6, -46)$ from a function $h(x)$. How do you know which element is from the domain and which is from the range?

6. Explain how to change the information $h(-24) = 16$ to an ordered pair so that it can be graphed.

Practice & Apply

Which of the following are functions? Explain.

7. {(3, 4), (4, 4), (5, 4)}   **8.** {(1, 1)}   **9.**

10.

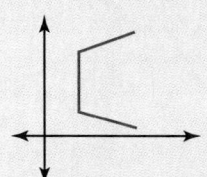

Find the domain and range of each of the following functions.

11. {(3, 4), (4, 4), (5, 4)}   **12.**

13.

14. Given $y = \dfrac{1}{x}$, find the second coordinate of an ordered pair if the first coordinate is 3.

15. If $h(x) = x + 3$, find $h(7)$.

Answer these questions about the national parks trip.

16. Which parks have fishing?   **17.** What are the features at Canyonlands?

18. Which parks have bike trails?   **19.** Which have swimming?

20. Which have a lodge or fishing?   **21.** Which have a lodge and fishing?

Databases Enter the national parks information in a data base. Sort the database to list the parks

22. in Utah.   **23.** with camping.

24. that do *not* have a lodge.   **25.** that have a lodge and fishing.

Which of the following are functions?

26. {(9, 5), (9, −5)}   **28.**

29.

27. {(5, 9), (−5, 9)}

Find the domain and range of each of the following functions.

30. $\left\{ \left(1, \frac{1}{2}\right), \left(2, \frac{1}{2}\right), \left(3, \frac{1}{3}\right), \left(4, \frac{1}{4}\right) \right\}$ **32.**

33.

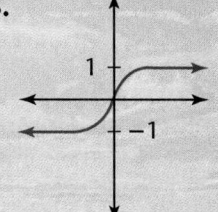

31. {(0.1, 1), (0.2, 2), (0.3, 3)}

If $f(x) = 5x$, find the following.

34. $f(3)$ **35.** $f(0)$ **36.** $f(-2)$ **37.** $f(-6)$

Evaluate the functions for $x = 3$.

38. $g(x) = x^2$ **39.** $f(x) = 2^x$ **40.** $h(x) = |x|$ **41.** $k(x) = \dfrac{1}{x}$

42. Technology Tell how to graph $|y| = x$ on a graphics calculator. HINT: Graph the top half using $\mathbf{Y_1} =$ and the bottom half using $\mathbf{Y_2} =$.

Which of these points satisfy $|x| + |y| = 8$?

43. $(5, 3)$ **44.** $(5, -3)$ **45.** $(-5, 3)$ **46.** $(-5, -3)$ **47.** $(7, 1)$

48. Graph $|x| + |y| = 8$. Is the graph a function? Explain.

49. Science If you know the Celsius temperature, can you necessarily know the Fahrenheit temperature? Is Fahrenheit temperature a function of Celsius temperature?

50. If you know the Fahrenheit temperature, can you necessarily know the Celsius temperature? Is Celsius temperature a function of Fahrenheit temperature?

Look Back

Solve the following equations. [Lesson 4.6]

51. $3r + 4 = -2 + 6r$ **52.** $8(x - 7) = 3x + 4$

53. Solve $A = \dfrac{h}{2}(B + b)$ for B. **[Lesson 4.6]**

54. Graph the linear equation $4x - 2y = 24$. **[Lesson 5.4]**

55. A line crosses the y-axis at 5 and is perpendicular to $y = -\dfrac{3}{5}x$. Write the equation in standard form for that line. **[Lesson 5.6]**

56. Find the product of $\begin{bmatrix} 1 & 3 \\ -5 & 7 \end{bmatrix}\begin{bmatrix} 6 & 0 \\ 1 & -2 \end{bmatrix}$. **[Lesson 7.3]**

57. Find the inverse of $A = \begin{bmatrix} 3 & -5 \\ -1 & 2 \end{bmatrix}$. **[Lesson 7.4]**

58. There are 5 friends that play tennis. How many games must occur for each player to play all the others once? **[Lesson 8.5]**

Probability There are 8 red, 4 green, 9 yellow, and 3 white tickets in a jar. The tickets are thoroughly mixed. **[Lesson 8.6]**

59. Jacob draws one ticket from the jar without looking. What is the probability that he will draw a red ticket?

60. Jacob puts the ticket back and draws another without looking. What is the probability that he will now draw a yellow or a green ticket?

Look Beyond

Graph each on a graphics calculator. HINT: **Graph the top half using $\mathbf{Y_1} =$ and the bottom half using $\mathbf{Y_2} =$.**

61. $x = y^2$ **62.** $x^2 + y^2 = 25$

Exploring Transformations

why *Computers and graphics calculators make it easier to draw several graphs at the same time. The use of technology or a sketch of the graphs can help you explore the relationship between a change in the function rule and a change in its graph.*

MAXIMUM
MINIMUM
Connection

In Chapter 2 you learned how to draw the graphs of several functions. One of those is the absolute value function, $f(x) = |x|$ or $y = |x|$. The *vertex* of this graph is where the V comes to a point. This vertex has its lowest, or minimum, point at the origin $(0, 0)$.

This graph is also *symmetric* with respect to the y-axis. The axis that passes vertically through the vertex of this function is the *axis of symmetry*. If you fold the graph along this axis, the left and right halves of the graph will match exactly.

x	|x|
−3	3
−2	2
−1	1
0	0
1	1
2	2
3	3

$y = |x|$

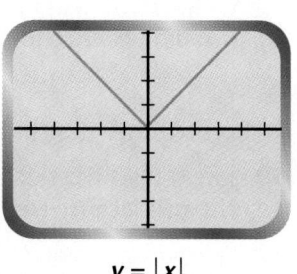

$y = |x|$

The original function $y = |x|$ is an example of a **parent function**. A parent function is the most basic of a family of functions. A variation such as a stretch, reflection, or shift is a **transformation** of the parent function.

There are various ways to stretch, reflect, and shift the graph of $y = |x|$.
Examine the graphs in a-f. The function $y = |x|$ has been

a. *stretched vertically by 2.*   **b.** *shifted vertically by* -2.   **c.** *reflected through the x-axis.*

d. *stretched vertically by 2 and shifted vertically by* -4.   **e.** *reflected through the x-axis and shifted vertically by 4.*   **f.** *stretched vertically by 3, reflected through the x-axis, and shifted horizontally by* -2.

•Exploration 1 *Transformations of Absolute Value Functions*

Graphics Calculator

1 Draw the graph of the parent function, $y = |x|$. If you have a graphics calculator, enter [ABS] X or ABS(X) to select the function.

2 Draw the graph of each of the following functions. Describe how the graph has been transformed from the parent function.

   a. $y = -|x|$ **b.** $y = 2|x|$ **c.** $y = |x| - 2$

   d. $y = -|x| + 4$ **e.** $y = 2|x| - 4$ **f.** $y = -3|x + 2|$

3 Compare your graphs a-f from Step 2 with the graphs a-f above. Match the functions and the graphs.

4 What must you do to the function rule for $y = |x|$ to make the new function rule represent a stretch? a reflection? a shift?

5 How can you tell if the graph of the absolute value has been stretched, reflected, or shifted?

6 Explain how to use the parent function $y = |x|$ to graph

   a. $y = 3|x|$.   **b.** $y = -5|x|$.   **c.** $y = |x| + 2$.

7 Make a conjecture about the graph of $y = |x + 1|$ as a transformation of $y = |x|$.

8 Test your conjecture by graphing

   a. $y = |x + 3|$.   **b.** $y = |x - 3|$. ❖

Exploration 2 *Transformations of Quadratic Functions*

 Draw the graph of the parent function $y = x^2$.

 Draw the graph of each of the following functions. Describe how the graph of each function has been transformed from the parent function.

   a. $y = 2x^2$ **b.** $y = -x^2$ **c.** $y = x^2 - 2$

 Compare your graphs to those drawn in a-f in Exploration 1. How are they similar?

Ⓐ What must you do to the function rule for $y = x^2$ to make the new function rule result in a stretch? a reflection? a shift?

Ⓔ Explain how to use the parent function $y = x^2$ to graph $y = -2x^2 + 3$. ❖

APPLICATION

GEOMETRY
Connection

Square photos are placed on a piece of cardboard backing. Each photo has a special border that adds a total of 2 inches to each side of the photo. The area of a photo including its border is a function of the length of a side of the photo. How can you model this application with a function?

Method A   Make a table.

Measure the lengths of the sides of the photo as they increase by 1 unit. Then find the corresponding areas.

Length of photo	x	0	1	2	3	4	5	6	x
Area with border	A(x)	4	9	16	25	36	49	64	$(x + 2)^2$

Notice that 2 is added to x and then squared. The function rule is $(x + 2)^2$.

*Graphics
Calculator*

Method B   Make a graph.

Graph the ordered pairs in the table.

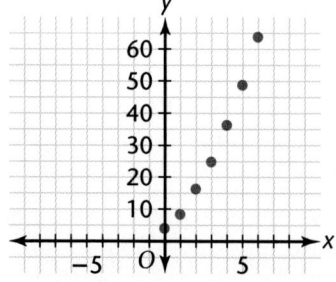

When a graphics calculator is used, the entire parabola is shown. Remember to match the graph with the original information in the problem.

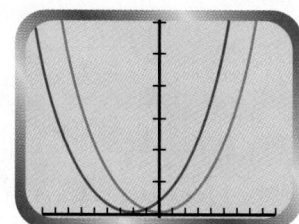

Compare the graph with the parent function $y = x^2$. Try various transformations until one fits. Notice that the x^2 function is shifted 2 units to the left. Thus, the function rule for the application is $(x + 2)^2$ and $A(x) = (x + 2)^2$ is the function that models the application data. ❖

The following is a summary of some parent functions from Chapter 2.

SUMMARY OF PARENT FUNCTIONS

FUNCTION	NAME	PAGE
$y = x$	Linear	60
$y = 2^x$ and $y = 10^x$	Exponential	60
$y = x^2$	Quadratic (parabola)	65
$y = \dfrac{1}{x}$	Reciprocal	70
$y = |x|$	Absolute value	79

In the following lessons you will examine some transformations of these parent functions. A guide to the parent functions and some of their transformations can be found on pages 704-707.

EXERCISES PROBLEMS

Communicate

1. Explain how to identify the vertex and axis of symmetry for the graph of the function $y = x^2$.

2. Describe how the graph changes when the graph of a parent function is stretched.

3. Describe how the graph changes when the graph of a parent function is reflected.

4. Describe how the graph changes when the graph of a parent function is shifted.

5. Explain how the vertex of $y = |x|$ and $y = x^2$ changes when the graph of the function is shifted.

6. Review the Summary of Parent Functions. How do you determine which of the functions have a vertex and which have symmetry through the y-axis?

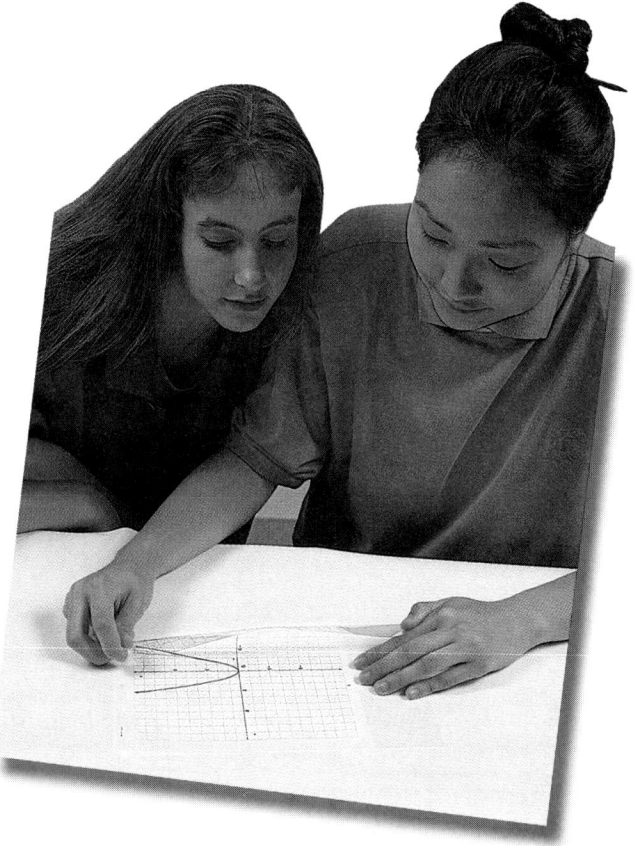

Practice & Apply

Give the parent function for each graph. Then describe the transformation using the terms stretch, reflect, and shift.

7.   **8.** **9.**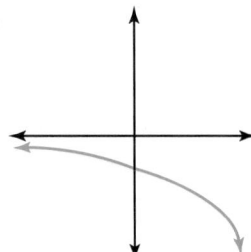

Graph each function. Then tell whether the graph of $y = |x|$ is stretched, reflected, or shifted when you graph the following functions.

10. $y = 4|x|$ **11.** $y = 4 + |x|$ **12.** $y = -|x|$

13. $y = 0.5|x|$ **14.** $y = |x - 5|$ **15.** $y = |x| - 3$

Graph each function. Then tell whether the graph of $y = \frac{1}{x}$ is stretched, reflected, or shifted, or give the combination of transformations when you graph the following functions.

16. $y = 5 + \frac{1}{x}$ **17.** $y = \frac{4}{x}$ **18.** $y = -\frac{1}{x}$

19. $y = \frac{1}{x} + \frac{1}{3}$ **20.** $y = \frac{1}{x + 2}$ **21.** $y = 3 - \frac{1}{x}$

Identify the parent function. Tell what transformation is applied to each, and then draw a carefully labeled graph.

22. $y = 5x^2$ **23.** $y = \sqrt{10}\, x$ **24.** $y = \text{INT}\, \frac{x}{2}$ **25.** $y = -|x - 2|$

A square picture is to be framed with a 2-inch mat around it. The area of the picture without the mat is $A = x^2$, where each side of the picture is x inches. The area of the picture including the mat is $A = (x + 4)^2$.

26. Make a table of values for $A = x^2$.

27. Make a table of values for $A = (x + 4)^2$.

28. Graph the two functions on the same axes.

29. Use transformations to tell how the graphs compare.

30. A group of students are weighed at the beginning of the year. What transformation of the data takes place if each student gains 3 pounds?

31. A teacher decides to double every student's score on a quiz because she used the wrong grading scale. What kind of transformation is this?

Geometry Tell what kind of transformation moves $\triangle ABC$ to $\triangle A'B'C'$, which is read A prime, B prime, C prime.

32.

33.

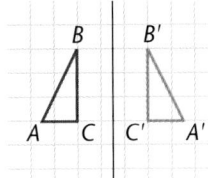

34. Find the equation of the axis of symmetry for this graph.

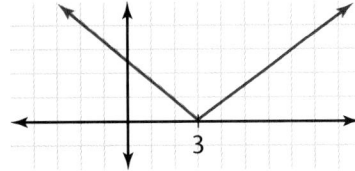

Look Back

35. What are the missing numbers in the sequence?   **[Lesson 1.2]**
3, 6, 11, 18, 27, _?_, _?_, 66, _?_

36. Calculate the value of $6 + 36 \div 3 - 1$.   **[Lesson 1.4]**

37. The cost of lunch was $6.09. The total amount for the food was $5.80. What percent of the total was the sales tax?   **[Lesson 4.5]**

Graph the solution of each inequality on a number line.
[Lesson 4.8]

38. $3x - 5 < 5x - 17$ **39.** $7x + 4 \geq -10$ and $2x + 3 < 13$

40. Find an equation of a line that contains the points $(3, -7)$ and $(-4, 3)$.
[Lessons 5.3, 5.4]

Solve the following systems of equations if possible. If the system has no solution, explain why.   **[Lessons 6.2, 6.3]**

41. $\begin{cases} 8x - 3y = -1 \\ 2x + y = 5 \end{cases}$ **42.** $\begin{cases} 4x - 3y = 6 \\ x = 12 \end{cases}$ **43.** $\begin{cases} 4x - 6y = 3 \\ 2x - 3y = 4 \end{cases}$

44. Given the following matrices, find the matrix products for both YA and XA.   **[Lesson 7.4]**

$$A = \begin{bmatrix} -6 & 3 \\ -8 & 7 \end{bmatrix} \qquad Y = \begin{bmatrix} -1 & 0 \\ 0 & 1 \end{bmatrix} \qquad X = \begin{bmatrix} 1 & 0 \\ 0 & -1 \end{bmatrix}$$

Look Beyond

Statistics Tell what happens to the mean of a set of data if each data point is

45. increased by 5.   **46.** doubled.   **47.** replaced by its opposite.

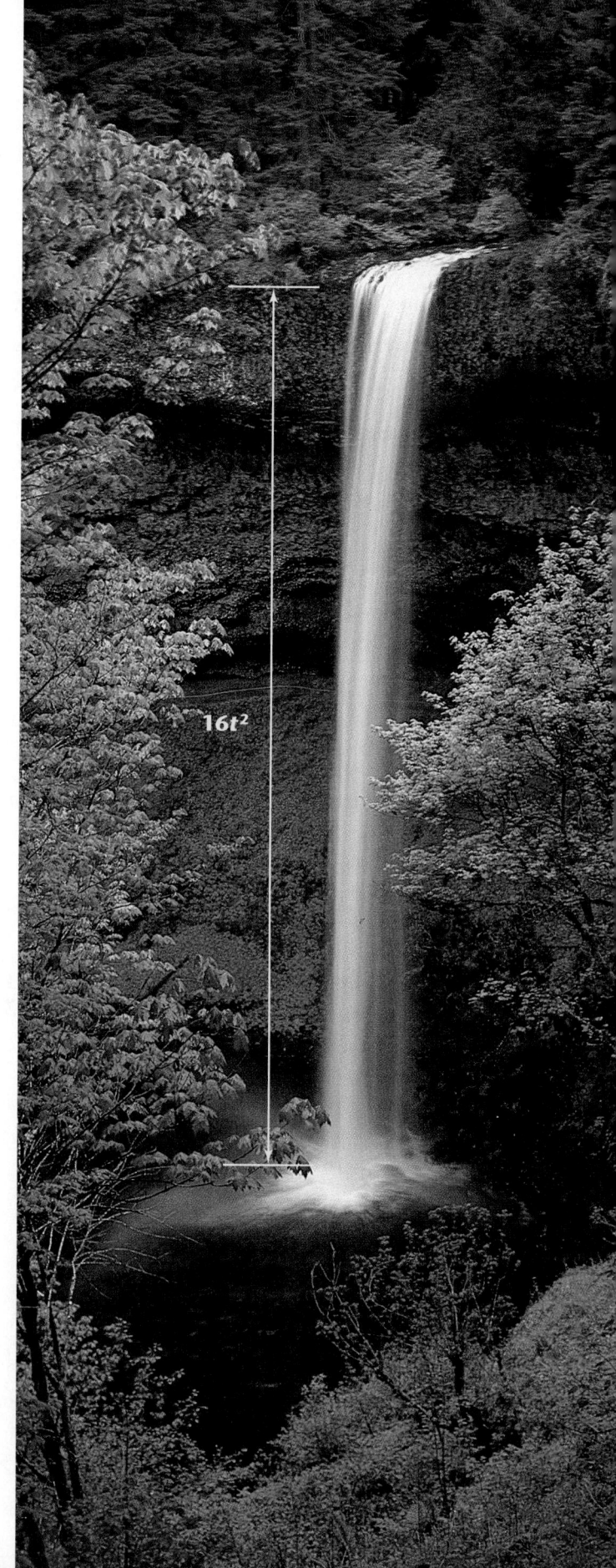

why *When you evaluate a function and plot the points, the function will have a distinctive shape. Changes in scale will change the shape of that graph. One such change is a vertical stretch. Once you know the effects of a change in scale on a parent function, it is easier to sketch or visualize the graph.*

$16t^2$

The height of a waterfall can be approximated if you know the time it takes the water to fall. This relationship can be modeled by the function $S = 16t^2$, where S is the distance in feet and t is the time in seconds. For example, if it takes the water 3 seconds to fall, the height is about $16(3)^2$ or 144 feet. What is the height if the time is 4 seconds?

The table shows the values of the function $S = 16t^2$ for times from 0 to 3 seconds.

How do these values compare to the values of the parent function $S = t^2$? That is, compare the functions $f(x) = x^2$ and $g(x) = 16x^2$.

$S = 16t^2$		
t	$16t^2$	S
0	$16(0)^2$	0
1	$16(1)^2$	16
2	$16(2)^2$	64
3	$16(3)^2$	144

What is the value of the parent function $f(x) = x^2$ when x is 5?

If $x = 5$, then $f(5) = 5^2 = 25$.

What is the value of the transformed function $g(x) = 16x^2$ when x is 5?

If $x = 5$, then $g(5) = 16(5)^2 = 400$.

The values of the function $y = 16x^2$ are 16 times greater than the corresponding values of the parent function $y = x^2$. The coefficient, 16, is called the **scale factor** of the function.

What does $g(5)$ represent in the waterfall problem?

Can you visualize the graph of $y = 16x^2$? Can you sketch it quickly without plotting the points? To do this you must understand how the scale factor affects the parent function.

Compare the graphs.

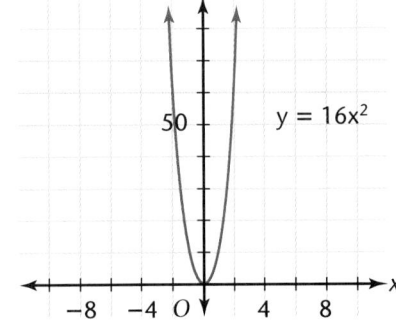

The graph of $y = x^2$ is *stretched* vertically to become the graph $y = 16x^2$.

When you apply the scale factor of 16, the point (2, 4) on the parent graph becomes the point (2, 64) on the transformed graph.

$$16(4) = 64$$

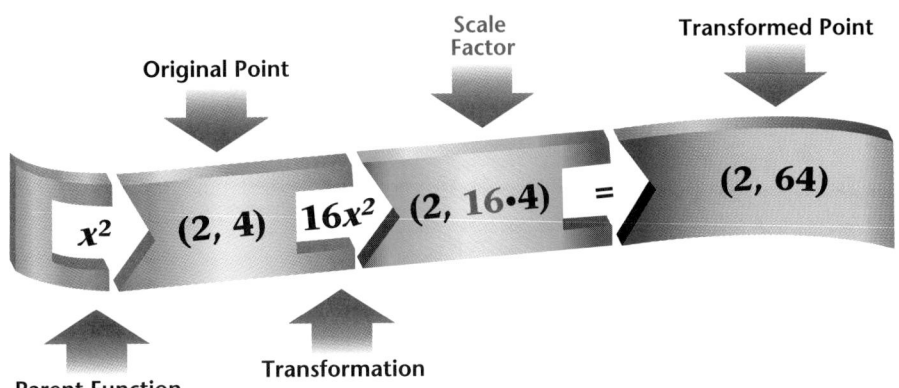

Original Point

Scale Factor

Transformed Point

x^2 (2, 4) $16x^2$ (2, 16·4) = (2, 64)

Parent Function

Transformation

EXAMPLE 1

Use the parent function to sketch the graph of $y = \frac{1}{2}|x|$.

Solution ➤

The parent function is $y = |x|$. The transformation has a scale factor of $\frac{1}{2}$.

Each of the y-values is $\frac{1}{2}$ the corresponding value of the parent function.

The graph will be stretched vertically to only $\frac{1}{2}$ the height of $y = |x|$, so the

graph of $y = \frac{1}{2}|x|$ is between the parent function and the x-axis.

x	$\frac{1}{2}|x|$	y
-2	$\frac{1}{2}|-2|$	1
-1	$\frac{1}{2}|-1|$	$\frac{1}{2}$
0	$\frac{1}{2}|0|$	0
1	$\frac{1}{2}|1|$	$\frac{1}{2}$
2	$\frac{1}{2}|2|$	1

Try This   Sketch the functions $f(x) = 4x^2$ and $g(x) = \frac{|x|}{4}$.

The graph of $y = 16x^2$ and $y = \frac{1}{2}|x|$ have been stretched from the graph of
their parent functions. When $a > 0$, $y = a \cdot f(x)$ is stretched vertically from
the graph of its parent function, $y = f(x)$, by the scale factor a.

EXAMPLE 2

Sketch the graph of $y = \frac{6}{x}$. How does it compare to its parent function?

Solution ➤

The function can be written $y = 6\left(\frac{1}{x}\right)$. Its parent function is the reciprocal

function, $y = \frac{1}{x}$. Each point of the parent graph is stretched vertically by a

scale factor of 6.

x	$6\left(\frac{1}{x}\right)$	y
-3	$6\left(\frac{1}{-3}\right)$	-2
-2	$6\left(\frac{1}{-2}\right)$	-3
1	$6\left(\frac{1}{1}\right)$	6
2	$6\left(\frac{1}{2}\right)$	3
3	$6\left(\frac{1}{3}\right)$	2

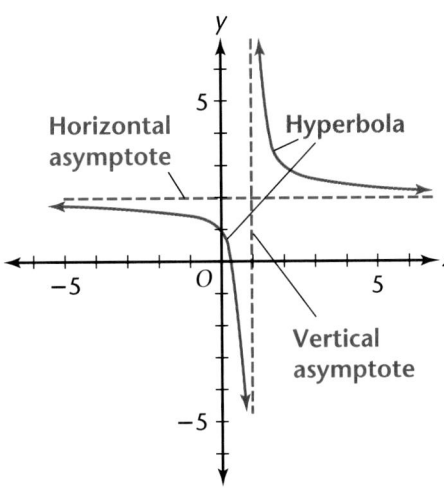

Horizontal asymptote

Hyperbola

Vertical asymptote

The graph of the reciprocal function forms a curve called a **hyperbola**. Notice that the graph of $y = \frac{1}{x}$ gets closer and closer to the x- and y-axes, but will never touch them. The lines that a hyperbola will approach, but not touch, are called **asymptotes**. The intersection of the asymptotes provides a reference point when transforming reciprocal functions. At what point do the asymptotes intersect in the graph at the left?

CRITICAL
Thinking

How does the change in the graph of $y = ax^2$ differ from the change in the graph of $y = a\left(\frac{1}{x}\right)$ as the scale factor a increases when $a > 0$?

EXAMPLE 3

Which functions result in stretches of the parent graph?

A $y = \frac{x^2}{3}$ **B** $y = x^2 + \frac{3}{4}$ **C** $y = 5|x|$

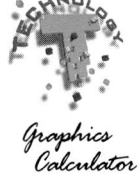
*Graphics
Calculator*

Solution ➤

A Since x^2 is multiplied by $\frac{1}{3}$, this is a vertical stretch by the scale factor $\frac{1}{3}$. The y-values will be only $\frac{1}{3}$ the y-values of the parent function.

B Since $\frac{3}{4}$ is *added* to x^2, this is *not* a stretch. The graph of the function has just moved up.

C Since $|x|$ is multiplied by the scale factor, 5, each value of the parent function stretches vertically by 5. ❖

SUMMARY

To graph a vertical stretch, you should

1. identify the parent function and
2. determine the scale factor.

The graph of the parent function will be stretched vertically by the amount of the scale factor.

Exercises & Problems

Communicate

1. Explain the pattern of changes to the graph of $y = ax^2$ for different values of a from 1 to 4.

2. Explain the pattern of changes to the graph of $y = ax^2$ for different values of a from 1 to 0.

3. Describe what is meant by vertical stretch.

4. Explain the pattern of changes to the graph of $y = b\left(\dfrac{1}{x}\right)$ for different values of b from 1 to 4.

5. Explain the pattern of changes to the graph of $y = b\left(\dfrac{1}{x}\right)$ for different values of b, such as $b = \dfrac{1}{2}, \dfrac{1}{3},$ and $\dfrac{1}{4}$.

6. If E is kinetic energy, m is mass, and v is velocity, explain the effect that $\dfrac{m}{2}$ has on the graph of $E = v^2$.

The formula for kinetic energy is $E = \dfrac{m}{2} v^2$.

Practice & Apply

Given the function $y = 5x^2$,

7. identify the parent function.

8. sketch the graphs of the function and the parent function on the same axes.

9. How does the graph of $y = 5x^2$ compare to the graph of the parent function?

Graph each of the following.

10. $y = 2\,|x|$ 11. $y = 0.5\,x^2$ 12. $y = 10\,\text{INT}(x)$

True or false?

13. The y-values of $y = 5\,|x|$ are five times as large as the y-values of $y = |x|$.

14. The x-values of $y = 5\,|x|$ can be $\dfrac{1}{5}$ as large as the x-values of $y = |x|$ in order to get the same y-values.

Given the function $y = \dfrac{1}{2x}$,

15. identify the parent function.

16. sketch the graphs of the function and the parent function on the same axes.

17. tell how the graph of the parent function was changed by the 2.

Which functions result in stretches of the parent function?

18. $y = 3x^2$ **19.** $y = |x| + 1$ **20.** $y = \dfrac{x^2}{5}$ **21.** $y = \dfrac{5}{x}$

Write an equation for each function.

22.

23.

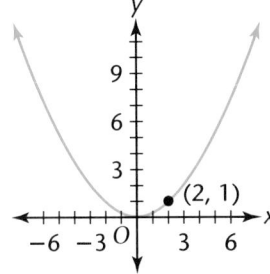

24. Art Pictures can be enlarged by using a grid. Cut a cartoon or other figure from a newspaper or magazine. Draw a grid on your picture. Make a larger grid off to the side and use it to make an enlargement of the picture.

25. The workers in an electronics plant got an "across the board" raise of 5%. Find the scale factor, the amount their salaries were "stretched."
HINT: Be careful. It isn't 0.05.

For each function, identify the parent function, draw the graph, and tell how the graph of the parent function is transformed by the 3.

26. $y = \text{ABS}(3x)$ **27.** $y = 2^{3x}$ **28.** $y = \left(\dfrac{x}{3}\right)^2$

29. 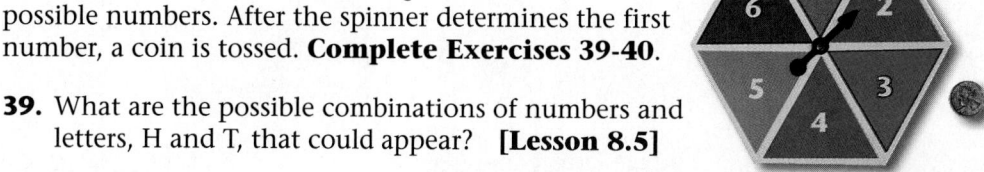 **Statistics** Draw a bar graph for this frequency distribution of Ms. Thompson's test.

Score	50	55	60	65	70	75	80	85	90	95	100
Frequency	1	0	2	5	10	6	2	0	0	0	0

30. Because the students didn't have enough time and the scores were low, Ms. Thompson decided to add 20 points to every student's score. Draw the new bar graph, and explain how it compares with the first graph.

31. Ms. Thompson also considered multiplying every student's original score by $\frac{5}{4}$. Draw the new bar graph, and explain how it compares with the graph you drew in Exercise 30.

Look Back

32. Which number indicates the strongest correlation? **[Lesson 1.6]**

   a. 0.57   **b.** 0.86   **c.** −0.01   **d.** −0.91

33. The cost for 2 assistants both working 12 hours of part-time work was $129.60. One was paid $0.30 more per hour because of experience. What is the hourly pay for each of the workers? **[Lesson 4.6]**

34. Solve the equation $5x - (7x + 4) = 3(x + 5) + 4(5 - 2x)$. **[Lesson 4.6]**

Solve the systems of linear equations. **[Lessons 6.2, 6.3]**

35. $\begin{cases} 3x + 7y = -6 \\ x - 2y = 11 \end{cases}$ **36.** $\begin{cases} 5y - 3x = -31 \\ 4y = 16 \end{cases}$ **37.** $\begin{cases} 1.25x + 2y = 5 \\ 3.75x + 6y = 15 \end{cases}$

38. The menu of a restaurant has 4 different burgers, 6 different soft drinks, and 2 extras, French fries or onion rings. How many different combinations of lunch can you select if each lunch consists of a burger, drink, and 1 extra? **[Lesson 8.5]**

Probability A spinner on a hexagonal base has 1-6 as possible numbers. After the spinner determines the first number, a coin is tossed. **Complete Exercises 39-40.**

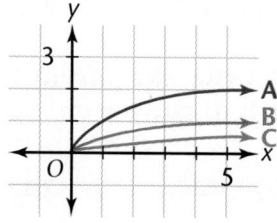

39. What are the possible combinations of numbers and letters, H and T, that could appear? **[Lesson 8.5]**

40. What is the probability of selecting any one of the pairs? **[Lesson 8.6]**

41. Two sets of numbers form the ordered pairs (3, 7), (5, −1), (1, 0), (2, 5), (16, −6), (0, 15), (1, −1), and (−2, 5). Is this relation a function? Why? **[Lesson 9.1]**

Look Beyond

Match each equation with its graph.

42. $y = \sqrt{x}$ **43.** $y = \sqrt{\dfrac{x}{4}}$ **44.** $y = \dfrac{\sqrt{x}}{4}$

LESSON 9.4 Reflections

Why *A mirror reflects a figure as its image on the other side. The graph of a function can be reflected through an axis or a given line. This reflection can be represented by a function rule.*

Exploration A Reflection

1 Use the values -2, -1, 0, 1, and 2 for x in this exploration. Make a table of values, and graph the parent function $f(x) = x^2$.

2 Reflect f through the x-axis. That is, create a mirror image of f as if the mirror were placed on the x-axis. Let g represent the function for this reflected image.

3 Make a table of values for the reflected function g. Use the same first coordinates or x-values that you used to graph f.

4 Compare the tables for the two graphs. The point $(2, 4)$ is on the parent graph. What is the y-value on the reflected graph when x is 2? If $(-3, 9)$ is a point on the parent graph, what is the corresponding y-value on the reflected graph?

5 Evaluate $g(x)$ for the given values of x.

  a. $g(3) = \underline{?}$ **b.** $g(0) = \underline{?}$ **c.** $g(-3) = \underline{?}$ **d.** $g(a) = \underline{?}$

6 Follow the same procedure for the following functions.

  a. $f(x) = |x|$

  b. $f(x) = 2^x$

  c. $f(x) = -2x^2$

  Examine the graphs, and describe in general the relationship between the graph of the function and the graph of its reflection.

7 Suppose the graph of $y = f(x)$ is reflected through the x-axis. What is the function for the reflected graph? ❖

CRITICAL
Thinking

Compare the graphs of $y = x$ and $y = -x$. Describe the transformation. How does the graph compare with $y = |x|$?

EXAMPLE 1

Describe each graph.   **A** $y = -|x|$ **B** $y = -3x^2$

Solution ➤

Graphics Calculator

A The parent function is $y = |x|$. The y-value of every point on the graph of the parent function changes to its opposite. Thus, the graph of $y = -|x|$ is a reflection of the graph of $y = |x|$ through the x-axis.

Notice that the vertex of the graph of $y = -|x|$ is now a maximum.

B The parent function is $y = x^2$. First, every point on the graph of the parent function is stretched vertically by a scale factor of 3. Then the negative sign in $y = -3x^2$ indicates a reflection of the graph $y = 3x^2$ through the x-axis.

The parabola for $y = -3x^2$ opens downward, so the vertex of the parabola is the maximum point on the graph. ❖

A **vertical reflection** through the x-axis results when each point (a, b) of the original graph is replaced by the point $(a, -b)$. Thus, the function $y = f(x)$ is transformed to $y = -f(x)$.

Let $f(x) = x + 2$. If a mirror image of f is created as if the mirror were placed on the y-axis, g would represent the function for the reflected image.

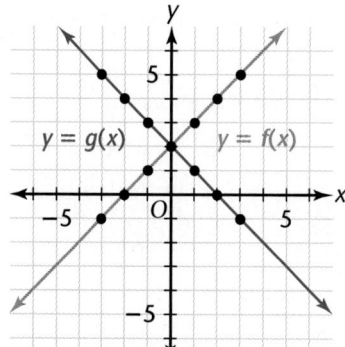

x	y = f(x)
3	5
2	4
1	3
0	2
−1	1
−2	0
−3	−1

x	y = g(x)
−3	5
−2	4
−1	3
0	2
1	1
2	0
3	−1

$(3, 5)$ is replaced by $(-3, 5)$.

A **horizontal reflection** through the y-axis results when each point (a, b) of the original graph is replaced by the point $(-a, b)$. Thus, the function $y = f(x)$ is transformed to $y = f(-x)$.

EXAMPLE 2

Identify which graphs are vertical reflections of parent graphs.

Ⓐ $y = -2^x$ **Ⓑ** $y = x^2 - 2$ **Ⓒ** $y = -\frac{1}{x}$

Solution ➤

Ⓐ The values of -2^x are opposite those for 2^x. This is a reflection through the x-axis of the parent graph $y = 2^x$.

Ⓑ This is not a reflection because -2 is *added* to x^2.

Ⓒ The y-values of $y = -\frac{1}{x}$ are the opposite of the y-values of the function $y = \frac{1}{x}$. This is a reflection through the x-axis of $y = \frac{1}{x}$. ❖

Remember, a reflection occurs when the values of the function are replaced by their opposites. For this reason, look for reflections when negative signs affect the function values. Plot a few points to check.

SUMMARY

To graph a vertical reflection

1. identify the parent function and
2. replace the rule with its opposite.

The graph of the parent function will be reflected through the x-axis.

EXERCISES & PROBLEMS

Communicate

1. Explain the effect of the negative sign in the formula $y = -ax^2$.

2. If $h(x) = a\frac{1}{x}$ is reflected through the x-axis, how would you change the original function to express the reflection?

3. Describe how to graph the reflection of the function $f(x) = 2x$ through the x-axis.

4. How does a reflection of $g(x)$ through the x-axis change the formula of the function?

5. What is the effect of reflecting the function through the y-axis when the function is symmetric about the y-axis? Give an example.

6. Identify the axis through which the graph $y = -(x)^2$ and the graph $y = (-x)^2$ are reflected. What effect does the location of the negative sign have on the reflection of the parent function?

7. Why are left and right reversed in a mirror image?

Practice & Apply

8. A vertical reflection is through which axis?

9. A horizontal reflection is a reflection through which axis?

10. If you see the reflection of a mountain in a lake, is this a vertical reflection or a horizontal reflection?

11. If the right half of a fir tree is the mirror image of the left half, is this a vertical reflection or a horizontal reflection?

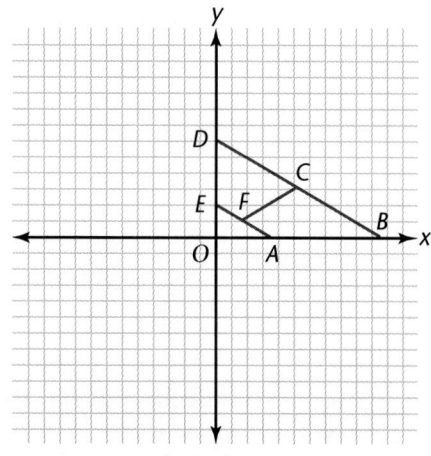

Graph each of the following.

12. $y = -x^2$ **13.** $y = -\text{INT}(x)$ **14.** $y = -2^x$

15. Compare the graphs of $y = -\dfrac{1}{x}$ and $y = \dfrac{1}{-x}$. Explain why they are the same or different.

16. What happens to the graph of $y = |x|$ when you reflect it through the x-axis? through the y-axis? Explain.

17. Graph $f(x) = 2x + 1$.   **18.** For $f(x) = 2x + 1$, find $f(3)$.

In each of the following, find $f(3)$, and use this information to tell which kind of reflection (horizontal or vertical) is applied to $y = 2x + 1$.

19. $y = -(2x + 1)$.   **20.** $y = 2(-x) + 1$

21. Show the result when a parabola is reflected vertically, then horizontally.

22. How can you describe the two reflections in Exercises 19 and 20 as a single transformation?

23. Draw a sketch that reflects $(2, 5)$ into $(5, 2)$, $(10, 3)$ into $(3, 10)$, etc. What happens to (a, b)?

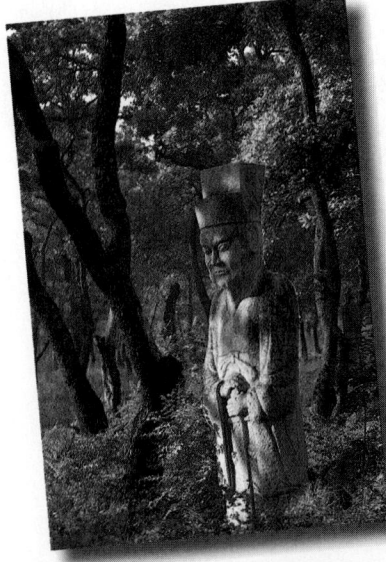

Stone figure in Kong Woods, family cemetery of Confucius, a famous Chinese philosopher

24. Cultural Connection: Asia

On a piece of graph paper, construct the line segments shown in bold. Point E is $(0, 2)$, and D is $(0, 6)$. Segment EA is twice the length of segment OE, and DB is twice OD. The segment FC connects the midpoints of the segments EA and DB. Reflect these lines through both the x- and y-axes into all 4 quadrants to complete the pattern. This basic pattern appears in a carving found in a tomb of the Han dynasty in China (206 B.C.E. to 222 C.E.). When the segments are reflected through both axes the shape is said to have *bilateral symmetry*.

25. Explain how the symmetry of a parabola can help you graph $y = x^2$ using fewer points.

26. Physics If the speed of sound through air is approximately 330 meters per second, how far away did the lightning strike? **[Lesson 2.3]**

27. Geometry The length of a rectangular garden needs to be 3 times the width. If the perimeter of the rectangular garden is 72 meters, what is the area? **[Lesson 4.6]**

Solve for the unknown. **[Lesson 4.6]**

28. $6(3 - 2r) = -(5r + 3)$ **29.** $\frac{x}{2} + 5 = \frac{3x - 2}{4}$ **30.** $3x = -9(x - 4)$

31. Graph the solution set for $x \geq 3$ or $x < -1$ on a number line. **[Lesson 4.7]**

32. **Statistics** Quality control measures 5 random samples of glass rods in a 15-minute production run. What is the average length of the glass rods tested in this run based on the following sample measurements? **[Lesson 8.3]**

Sample number	1	2	3	4	5
Measurement in cm	23.5	23.9	23.6	24.0	23.7

33. A number cube and a number tetrahedron (4 faces) are used for a probability experiment. How many possible outcomes are there when both are rolled once? **[Lesson 8.6]**

Look Beyond

Match each of these functions with the appropriate graphs.

34. $y = \sqrt{x}$ **35.** $y = \sqrt{-x}$ **36.** $y = -\sqrt{x}$ **37.** $y = -\sqrt{-x}$

A.

B.

C.

D.

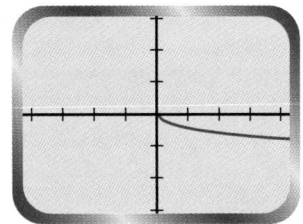

38. Graph $y = 5 - x^2$.

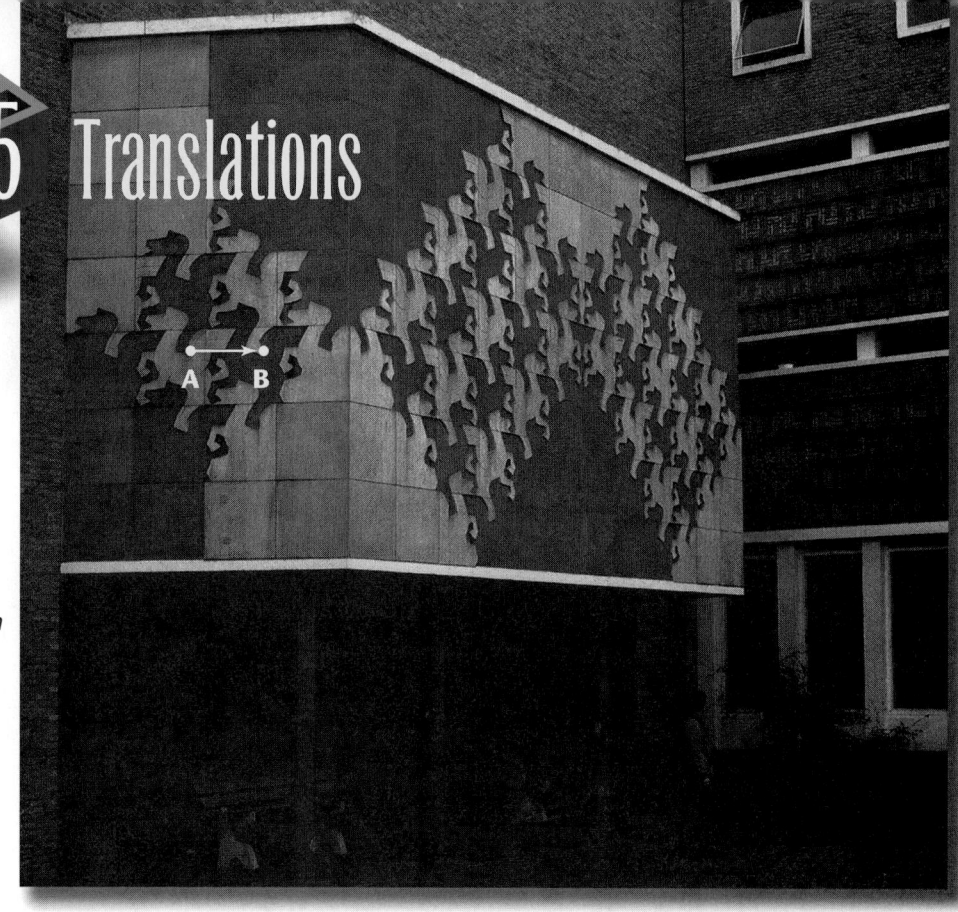

In 1960 Escher designed this tiled facade for the hall of a school in The Hague. It was executed in concrete in two colors.

WHY *Sometimes a graph looks like a parent function that has been shifted vertically or horizontally. When this occurs, there is a simple relationship between the rules for each function.*

A transformation that shifts the graph of a function horizontally or vertically is called a **translation**. In the Escher facade point A is translated to point B by shifting point A to the right.

EXAMPLE 1

Compare the graph of $E(x) = |x - 10|$ with the graph of its parent function.

Solution ➤

To graph the function $E(x) = |x - 10|$, start with a table of values. Include the number 10 in the table, as well as numbers greater than and less than 10.

Plot the points. Notice that the function is a variation of the absolute value function $y = |x|$.

Connect the points to make the familiar V-shaped graph. The graph of the function, $E(x) = |x - 10|$, is the graph of its parent function shifted 10 units to the right. ❖

x	0	2	5	8	10	13	15
E	10	8	5	2	0	3	5

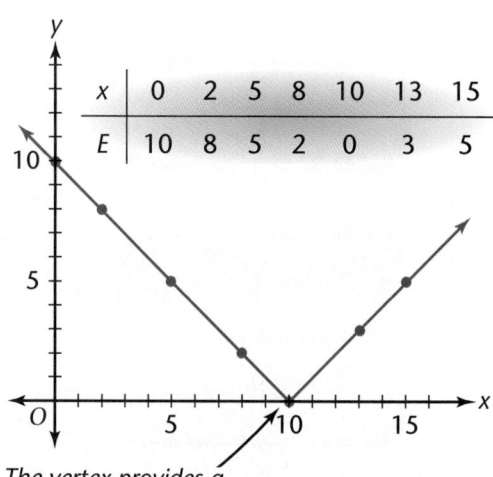

The vertex provides a convenient reference point.

Vertical Translation

A transformation that shifts the graph of a function up (positive direction) or down (negative direction) is a **vertical translation**.

EXAMPLE 2

Compare the graph $y = x^2 - 3$ with the graph of the parent function $y = x^2$. In what direction does the translation shift the vertex? How far?

Solution ➤

Evaluate $h(x) = x^2 - 3$ for several values of x, and plot the graph with $h(x)$ on the y-axis.

$h(x) = x^2 - 3$		
x	$(x)^2 - 3$	$h(x)$
-2	$(-2)^2 - 3$	1
-1	$(-1)^2 - 3$	-2
0	$(0)^2 - 3$	-3
1	$(1)^2 - 3$	-2
2	$(2)^2 - 3$	1

 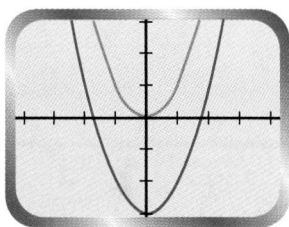

The vertex is vertically shifted downward 3 units. ❖

Try This   Sketch the graph of the function $y = |x| + 4$.

VERTICAL TRANSLATION

The graph of $y = f(x) + k$ is translated vertically by k units from the graph of $y = f(x)$.

The following translations show how the formula applies to basic parent functions. The graph of a function shifts vertically when you add a positive or negative constant to the function.

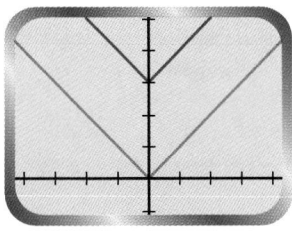

$y = |x| + 3$, $(k = 3)$

The graph is shifted up 3 units from the graph of $y = |x|$.

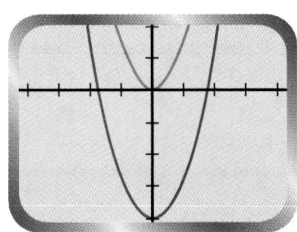

$y = x^2 - 4$, $(k = -4)$

The graph is shifted down 4 units from the graph of $y = x^2$.

$y = \dfrac{1}{x} + 5$, $(k = 5)$

The graph is shifted up 5 units from the graph of $y = \dfrac{1}{x}$.

Horizontal Translation

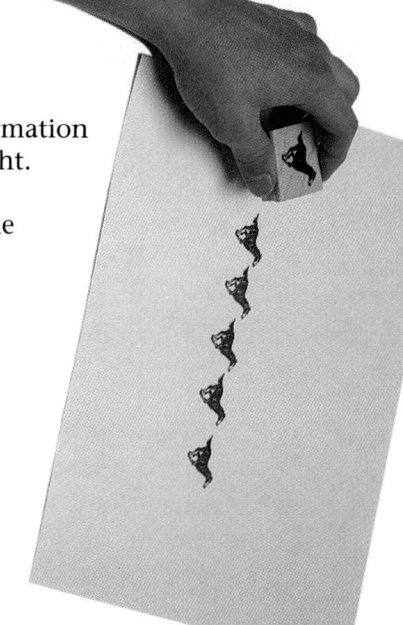

Horizontal translation is the transformation that shifts the function left or right. Remember, the positive direction is to the right and the negative direction is to the left.

EXAMPLE 3

Compare the graph of $y = (x - 3)^2$ with the graph of its parent function. Which direction does the translation shift the vertex and by how much?

Solution ➤

Graph the parent function $y = x^2$. The transformed function, $y = (x - 3)^2$, can be interpreted, "Subtract 3 from x, then square the quantity." The table shows some of the values of this function. Sketch this function and the parent function on the same axes.

$y = (x - 3)^2$		
x	$(x - 3)^2$	y
-2	$(-2 - 3)^2$	25
-1	$(-1 - 3)^2$	16
0	$(0 - 3)^2$	9
1	$(1 - 3)^2$	4
2	$(2 - 3)^2$	1
3	$(3 - 3)^2$	0
4	$(4 - 3)^2$	1

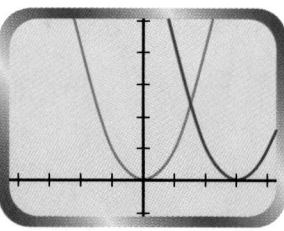

Notice that a translation does not change the shape of the parabola. If you subtract 3 from x *before* you square the quantity, you shift the graph to the right 3 units. The vertex is horizontally shifted to the right 3 units.

The x-values in the new function must be 3 units greater than the x-values for the parent function to get the same y-value. ❖

Try This   Sketch the graph of $y = |x - 6|$.

> ### HORIZONTAL TRANSLATION
> The graph of $y = f(x - \boldsymbol{h})$ is translated horizontally by h units from the graph of $y = f(x)$.

The following examples show how the graphs and the formulas represent the horizontal translation of various parent functions.

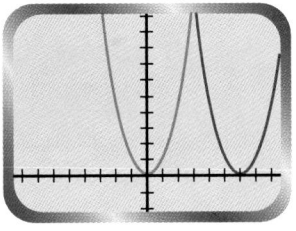

$y = (x - 6)^2$, $(h = 6)$

The graph shifts right 6 units from the graph of $y = x^2$.

$y = |x + 3|$, $(h = -3)$

The graph shifts left 3 units from the graph of $y = |x|$.

$y = \dfrac{1}{x + 2}$, $(h = -2)$

The graph shifts left 2 units from the graph of $y = \dfrac{1}{x}$.

EXAMPLE 4

Which of the functions represents translations of the parent graph?

Ⓐ $y = x^2 + 3$ **Ⓑ** $y = \dfrac{3}{4} + 5^x$ **Ⓒ** $y = \dfrac{|x|}{9}$ **Ⓓ** $y = \dfrac{1}{x - 10}$

Solution ➤

Ⓐ Since 3 is added to x^2, the parent function, $y = x^2$, is translated upward by 3 units.

Ⓑ Since $\dfrac{3}{4}$ is added to 5^x, the parent function is translated upward by $\dfrac{3}{4}$ of a unit.

Ⓒ Since $|x|$ is divided by 9, this is a stretch. It is not a translation.

Ⓓ Since 10 is subtracted from x in $\dfrac{1}{x - 10}$, this is a horizontal translation of $y = \dfrac{1}{x}$ to the right by 10 units. ❖

EXAMPLE 5

STATISTICS
Connection

Three salespeople have salaries of $21,000, $24,000, and $30,000. What happens to the average salary if each gets a $3000 raise?

Solution ➤

The average salary before the raise is $\dfrac{21,000 + 24,000 + 30,000}{3} = 25,000$.

If each gets a $3000 raise, the average will be $\dfrac{24,000 + 27,000 + 33,000}{3} = 28,000$.

If each salary increases by $3000, the average will increase by $3000. In general, if each salary is translated by a number, k, the average is translated by the same number, k. ❖

For example, consider the function $y = (x - 5)^2$. The parent function is $y = x^2$. When x is 5, $(x - 5)^2$ is 0. Thus, the vertex $(0, 0)$ of $y = x^2$ has been shifted to the right 5 units. What happens to the vertex of $y = x^2$ when $y = x^2 - 5$?

Cultural Connection: Africa The designs of African art for pottery and fabric show the use of translation and reflection for decoration. Such examples show that geometric transformations have existed for centuries in this as well as in many other cultures. Algebra has created a way to represent these geometric features by using symbols and expressions.

Contrast the two earthenware vessels. The one in the background dates from the early Twentieth Century, the other from 1200 C.E.

EXERCISES & PROBLEMS

Communicate

1. What is meant by a *translation* in mathematics?

2. A transformation of the parent quadratic function is $f(x) = (x - 3)^2 + 5$. What value tells you the amount and direction of the *horizontal* translation?

3. A transformation of the parent absolute value function is $g(x) = |x - 3| + 5$. What value tells you the amount and direction of the *vertical* translation?

4. Explain why the graph of $f(x) = (x + h)^2$ is translated horizontally when a constant, h, is added to x before the quantity is squared.

5. Explain why the graph of $f(x) = x^2 + k$ is translated vertically when a constant, k, is added after the variable is squared.

6. What is the relationship of the numbers 4 and 3 in the formula, and what is the location of the vertex of the graph for $y = (x - 4)^2 + 3$?

Practice & Apply

Identify the parent function. Tell what type of transformation is applied to each, then draw a carefully labeled graph. HINT: **Remember to plot a few points to help see what happens to the parent function.**

7. $y = -x^2$ **8.** $y = (-x)^2$ **9.** $y = x^2 + 3$ **10.** $y = (x + 3)^2$

11. $y = \text{INT}\left(\frac{x}{2}\right)$ **12.** $y = \text{INT}(-x)$ **13.** $y = \text{ABS}(x) - 1$ **14.** $y = \text{ABS}(x - 1)$

Consider the parent function $f(x) = x^2$. Note that $f(3) = 9$, so the point $(3, 9)$ is on the graph. In each of the following, use the given fact to tell which way the parent function is translated.

15. $f(x) = x^2 + 5$ contains the point $(3, 14)$.   **16.** $f(x) = x^2 - 5$ contains the point $(3, 4)$.

17. $f(x) = (x + 5)^2$ contains the point $(-2, 9)$.   **18.** $f(x) = (x - 5)^2$ contains the point $(8, 9)$.

The point $(5, 8)$ is on the graph of $f(x)$. Tell what happens to $(5, 8)$ when each of the following transformations are applied to the function.

19. vertical translation by 6   **20.** vertical translation by -2

21. vertical translation by -10 **22.** vertical stretch by 3

23. horizontal translation by 3   **24.** horizontal translation by -1

25. horizontal translation by -12 **26.** vertical stretch by 10

Tell whether the given graph is stretched or translated from the graph of the parent function.

27. $y = \frac{2}{3}|x|$ **28.** $y = \frac{2}{3} + |x|$

29. What happens to the average salary of the workers in Example 5 if all the salaries were increased by 10%?

30. Suppose one ball is thrown from level ground in a parabolic path as shown. A second ball is thrown in an identical path, but from a point on top of a cliff to a point on another cliff. What transformation relates the higher path to the lower one?

31. Simplify the expression $-3(8 - 4x) - 5(x + 2)$.   **[Lesson 4.2]**

32. 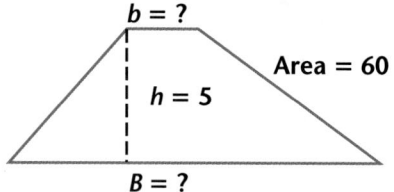 **Geometry** The area of a trapezoid can be determined by the formula $A = \frac{h(B + b)}{2}$. The area, A, of a given trapezoid is 60. The altitude, h, is 5, and the larger base, B, is twice the smaller base, b. What is the length of each base? **[Lesson 4.6]**

33. Solve for x. $7x - (24 + 3x) = 0$ **[Lesson 4.6]**

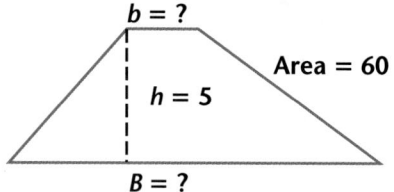 **Statistics** Investigate differences in reaction time. Place a ruler with the zero point between a person's thumb and index finger. Release the ruler. The person should catch the ruler as quickly as possible. Determine the point at which the ruler was caught.   **[Lesson 8.3]**

34. Find the average distance along the ruler where most people catch it.

35. Compare the average catch distance along the ruler for groups of males and females. Which group showed the shortest average catch distance? What does this say about the reaction time of each group?

36. Sketch the graph of $y = -\frac{10}{x}$.   **[Lesson 9.3]**

37. Determine a formula for the function $f(x)$ that will model the following data. Give the steps you used for finding the formula. **[Lesson 9.3]**

There should be a 4-centimeter distance between thumb and index finger with the ruler centered between them.

x	0	1	2	3	4	5	6	7
f(x)	0	3	12	27	48	75	108	147

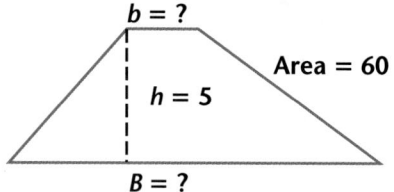 **Geometry** Notice the triangle on the coordinate graph with vertices at points $A(2, 1)$, $B(4, 4)$, and $C(7, 2)$.   **[Lesson 9.4]**

38. Reflect the triangle through the x-axis. What are the coordinates for the vertices of the transformed triangle?

39. Reflect the original triangle through the y-axis, and give the new coordinates.

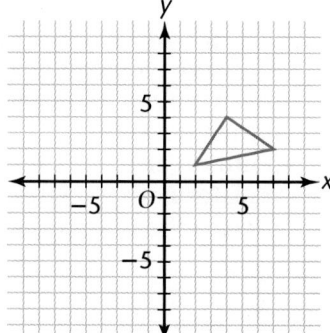

Look Beyond

Graph the parent function and the following transformations. How is each changed from the basic cubic function $y = x^3$?

40. $y = -x^3$ **41.** $y = (-x)^3$

LESSON 9.6 Combining Transformations

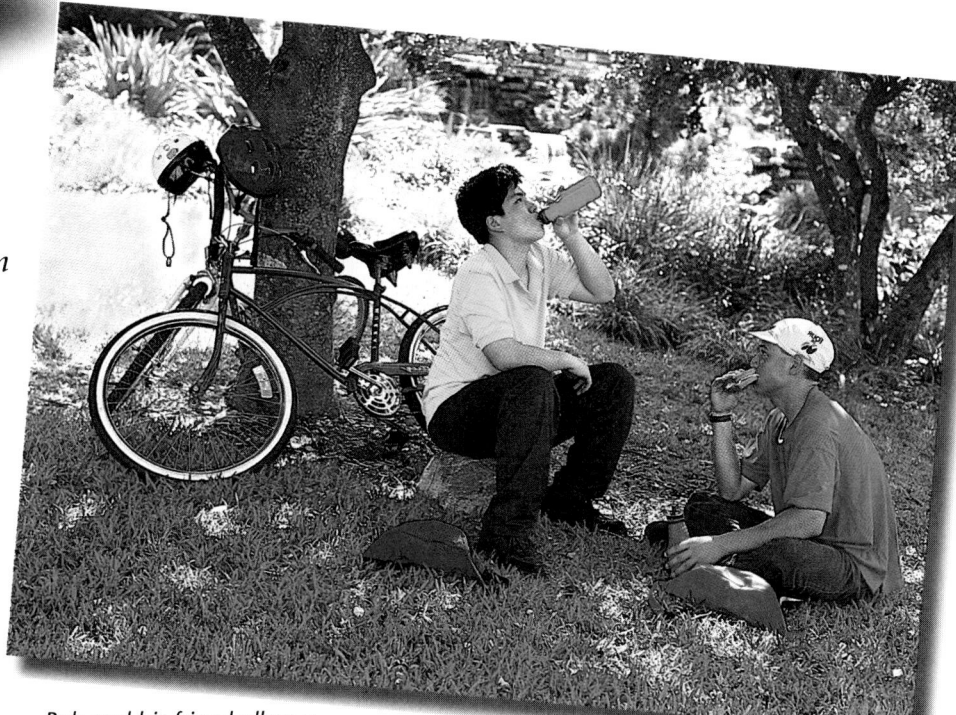

Once a function is recognized as related to a known parent function, the graph can be sketched by identifying and performing several transformations.

Bob and his friend allow a total of 2 hours for rest breaks.

Travel Bob and his friends are planning a 100-mile bike ride. How fast must they ride to complete the whole trip in 6 hours? 7 hours? *h* hours?

To finish the trip in 6 hours, the boys expect to ride 4 hours with 2 hours rest ($6 - 2 = 4$). That means they would have to ride 100 miles in 4 hours.

$$\text{rate} = \frac{100 \text{ miles}}{4 \text{ hours}} = \frac{25 \text{ miles}}{1 \text{ hour}}, \text{ or } 25 \text{ mph}$$

To finish the trip in 7 hours, they would need to ride for 5 hours and rest for 2 hours. They would have to ride 100 miles in 5 hours.

$$\text{rate} = \frac{100 \text{ miles}}{5 \text{ hours}} = \frac{20 \text{ miles}}{1 \text{ hour}}, \text{ or } 20 \text{ mph}$$

To finish the trip in *h* hours, they would need to ride $h - 2$ hours. They would have to ride 100 miles in $h - 2$ hours.

$$\text{rate} = \frac{100 \text{ miles}}{h - 2 \text{ hours}}, \text{ or } \frac{100}{h - 2} \text{ mph}$$

If *R* is the rate, and *h* is the total number of hours for the trip, the rate the boys would have to travel is expressed as

$$R = \frac{100}{h - 2}.$$

To sketch the graph of a transformed function, begin with the parent function, then perform the individual transformations one at a time.

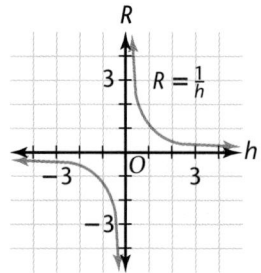

Sketch the graph of $R = \dfrac{100}{h-2}$.

First determine the parent function. Since h is in the denominator, you should first consider the parent reciprocal function $R = \dfrac{1}{h}$. Start with $R = \dfrac{1}{h}$ and find the transformations needed to end with $R = \dfrac{100}{h-2}$.

Next determine whether the function has been stretched or reflected. A graph has been stretched when its parent function is multiplied by a nonzero constant. If the constant is negative, the graph is also reflected.

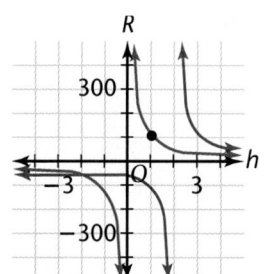

Examine $R = \dfrac{100}{h}$. Since $R = \dfrac{100}{h}$ can be written as $100\left(\dfrac{1}{h}\right)$, the parent function is stretched vertically by 100 and there is no reflection. Notice on the blue and red graphs that the y-axis is labeled in hundreds. Although the graphs of $R = \dfrac{1}{h}$ and $R = \dfrac{100}{h}$ look alike, each y-value in the graph of $R = \dfrac{100}{h}$ is 100 times the amount of the y-value in $R = \dfrac{100}{h}$.

Then determine whether the parent function has been translated. Remember that the graph is translated when you add or subtract a constant. Since 2 is subtracted from x in the function $y = \dfrac{100}{h-2}$, the x coordinate needs to be 2 units greater to get the same value as $y = \dfrac{100}{h}$. This indicates a horizontal translation of 2 to the right.

To summarize, the parent graph, $R = \dfrac{1}{h}$ is transformed by a vertical stretch of 100 followed by a translation of 2 to the right. The point $\left(3, \dfrac{1}{3}\right)$ on the original graph, is transformed to the point $\left(5, \dfrac{100}{3}\right)$.

From the function, determine how fast they must travel to finish in 8 hours.

Suppose the function $y = x^2$ represents the area of a square with side x. Interpret what the function $y = 3x^2 + 5$ would indicate about the change in the area of the original square.

EXAMPLE 1

Sketch the graph of $y = 2|x + 4|$.

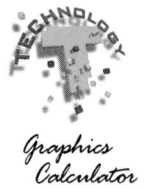

Graphics Calculator

Solution ➤

1. The parent graph is $y = |x|$.

2. Since the scale factor is 2, the y-values will be twice as great as the parent function. This is a vertical stretch by 2.

3. Because 4 is added to the variable, the x-values in the transformed function need to be 4 units less to produce the same y-values that the stretched parent function produces. The result is a translation to the left by 4.

The parent graph $y = |x|$ is stretched vertically by 2 and translated to the left by 4.

The vertex is at -4 on the x-axis. Since $y = 2|0| = 0$, $(-4, 0)$ is the new position of the vertex. ❖

The graphics calculator can plot the parent function and transformation quickly.

Try This Check several points for $y = 2|x + 4|$ to see how steep the V of the graph is. Is the graph symmetric? If so, what is the axis of symmetry?

Reflection   **Horizontal translation**

Summary
For many functions, the graph can be sketched directly from the information that appears in the function rule.

$$y = -a \cdot f(x - h) + k$$

Vertical stretch   **Vertical translation**

EXAMPLE 2

Use the information from the Summary to sketch the graph of the function $y = -(x + 3)^2 - 2$.

Solution ➤

The parent function is $y = x^2$, so the graph is a parabola. Since the sign is negative, the parabola is reflected through the x-axis. There is no constant following the negative sign. Thus, there is no vertical stretch. The value of h must be -3, since $x - (-3) = x + 3$. This indicates a horizontal translation to the left 3 spaces. Finally, $k = -2$, so the parabola is translated vertically downward 2 spaces. ❖

EXERCISES & PROBLEMS

Communicate

1. In the 100-mile bike ride problem, explain how to identify which number in the function represents the translation.

2. In the 100-mile bike ride problem, explain how to identify which number in the function represents a stretch of the graph.

3. Discuss the characteristics of various parent functions that will help you to identify them from their transformed functions.

4. Describe how to write the function for a translation of 4 units to the left if the parent function is the reciprocal function.

5. Given the function $y = -a(x - h)^2 + k$, identify and explain the effect of each transformation from the letters and symbols in the formula.

6. Explain the steps you would perform to sketch the graph of the function $y = |x + 7| - 3$.

Practice & Apply

Sketch the graph of each function.

7. $j(x) = 2(x + 5)^2$

8. $m(x) = 2x^2 + 5$

9. $t(x) = |x + 3| - 4$

10. $v(x) = 10|x - 4|$

11. $f(x) = \dfrac{3}{x - 2}$

12. $g(x) = \dfrac{3}{x} - 2$

13. $h(x) = 0.5(2^x) - 1$

14. $z(x) = \dfrac{3}{4} + \dfrac{2}{x}$

15. $p(x) = \dfrac{3}{4} + \dfrac{2}{4 + x}$

Tell what happens to the point (4, 10) on $y = \dfrac{5}{2}x$ when each of the following pairs of transformations are applied in the order given. HINT: If you are not sure, draw a sketch.

16. a vertical stretch by 2, followed by a vertical translation by 3

17. a vertical reflection, followed by a vertical stretch by 2

18. a vertical reflection, followed by a vertical translation by -3

19. a vertical stretch by -3, followed by a horizontal translation by 5

20. a vertical reflection, then a horizontal translation by -5, followed by a vertical stretch by $\dfrac{1}{2}$

21. Simplify the formula $y = 3(x^2 + 5)$ so that it can be sketched from the clues in the function rule.

22. In the figure at the right, an irregular triangle is shown in the first quadrant. Its reflection and translation are also shown. If the coordinates of A are $(5, 3)$, find the coordinates of A' and A''. Do the other vertices change using the same rule?

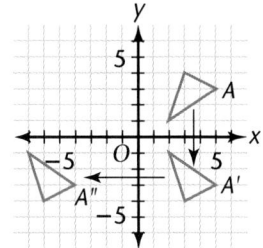

23. Use the information in the table to determine the transformation of a parent function that fits the data.

x	−3	−2	−1	0	1	2	3	4	5	6	7
y	21	16	11	6	1	−4	−9	−14	−19	−24	−29

24. **Portfolio Activity** Complete the problem in the portfolio activity on page 409.

Look Back

25. Sketch a scatter plot that could have a correlation of −0.95.
[Lesson 1.6]

26. **Packaging** A roll of tape contains 32 meters of tape. Each box requires 75 centimeters of tape to seal for delivery. If 35 centimeters of tape must be left for sealing other items, write and solve an equaton that shows how many boxes can be sealed with the available tape.
[Lesson 4.6]

27. Graph the function $4x - 5y = 3$. Describe the procedure for graphing this equation.   **[Lesson 5.4]**

28. If the graph of a line contains the points $A(-3, 7)$ and $B(2, -9)$, what is the equation for the line perpendicular to the given line at point A?
[Lesson 5.6]

Solve the system of equations. $\begin{cases} 2x - 3y = 8 \\ 6x + 5y = -4 \end{cases}$

29. by graphing   **[Lesson 6.1]**   **30.** by elimination   **[Lesson 6.3]**
31. by matrices   **[Lesson 7.5]**

32. **Probability** What is the theoretical probability of throwing 7 on a number cube?   **[Lesson 8.6]**

Look Beyond

33. The rocket function given in Chapter 2 can be written as $y = -16(x - 7)^2 + 784$. Graph the function.

34. Describe each transformation of the parent function in Exercise 33.

Plot the following graphs.

35. $y = 3 + \sqrt{x + 4}$ **36.** $y = -7 + 10\sqrt{x}$

Pick a Number

Try this with a classmate. Ask your classmate to

> *pick a number from 1 through 9, triple the number and add 6, then divide the result by 3.*

Ask for the number. If you mentally subtract 2, you can tell your classmate the original number.

Try other numbers. What do you notice?

Algebra can be used to see why this happens. Let x be the number chosen.

	Arithmetic	Algebra
Pick a number	5	x
Triple the number and add 6	$15 + 6 = 21$	$3x + 6$
Divide by 3	$21 \div 3 = 7$	$x + 2$

The algebra shows you how to go directly to the result. This is an example of finding a function of a function. In terms of a function machine, the output from the first machine is used as the input for the second machine.

When you find a function of a function, you are forming a **composite function**. If f is one function and g is another function, write $g(f(x))$ to indicate the value of the composite function for any x value.

If $f(x) = 3x + 6$ and $g(x) = \frac{x}{3}$, then

$$f(g(x)) = \frac{f(x)}{3} = \frac{3x + 6}{3} = x + 2.$$

The composition of two functions produces a new function with a new graph.

Composition of functions is not a commutative operation. $g(f(x)) = x + 2$, but

$$f(g(x)) = 3(g(x)) + 6 = 3\left(\frac{x}{3}\right) + 6 = x + 6.$$

Activity 1

Make up a two step computation game like that in the opener of the project. Show how the domain and range are related through the steps. Then show the simplified composition of the functions that lets you perform the calculation easily. Try the game on your friends.

Activity 2

Use what you have learned about composing functions to graph the composition of two functions. Begin by selecting a parent function from those you studied in Chapter 2. Replace the x in the function rule with a function, and graph the composition.

a. Identify the domains and ranges of the first function and the composite function.

b. Make a table of the input values, the intermediate values for input in the second step, and the final composite values.

c. Use a graphics calculator or draw the graph of the new composite function, and comment on the graph.

Chapter 9 Review

Vocabulary

Key Skills & Exercises

Lesson 9.1

➤ **Key Skills**

Test a graph to identify a function using the vertical line test.

Any vertical line will intersect the graph at the right no more then once, so the graph is a function.

Evaluate functions.

To evaluate the function $f(x) = x^2 + 1$ for $f(-3)$, substitute -3 for x, and simplify.

$$f(x) = x^2 + 1$$
$$f(-3) = (-3)^2 + 1$$
$$f(-3) = 10$$

➤ **Exercises**

Which of the following are functions?

1.   2.   3.

Evaluate the following functions when $x = 4$.

4. $f(x) = x - 8$ **5.** $g(x) = |x + 2|$ **6.** $h(x) = 3^x$

Lesson 9.2

➤ **Key Skills**

Identify which transformation, stretch, reflect, or shift, was applied to a parent function.

The parent function for the graph at the right is $y = x^2$. It has been reflected through the x-axis and shifted vertically.

Give the parent function for each graph. Then tell what kind of transformation was applied.

7.   8.   9.

Lesson 9.3

➤ *Key Skills*

Describe the effect of a stretch on the graph of a function.

The parent function of $y = 3|x|$ is $y = |x|$. Each of the y-values is 3 times the corresponding value of the parent function.

x	$3|x|$	y
-2	$3|-2|$	6
-1	$3|-1|$	3
0	$3|0|$	0
1	$3|1|$	3
2	$3|2|$	6

➤ *Exercises*

Graph each of the following.

10. $y = 4|x|$ **11.** $y = 3x^2$ **12.** $y = \dfrac{1}{4x}$ **13.** $y = 2\,\text{INT}(x)$

Lesson 9.4

➤ *Key Skills*

Describe the effect of a reflection on the graph of the parent function.

The parent function of $y = -|x|$ is $y = |x|$. The values of $-|x|$ are opposite of those for $|x|$. This is a reflection through the x-axis of the parent graph, $y = |x|$.

➤ *Exercises*

Sketch a graph of each of the following.

14. $y = -x$ **15.** $y = -\dfrac{1}{x}$ **16.** $y = -3^x$ **17.** $y = -3x^2$

Lesson 9.5

➤ *Key Skills*

Describe the effect of a translation on the graph of a function.

The parent function of $y = |x| + 2$ is $y = |x|$. The graph of $y = |x| + 2$ is shifted up by 2 from the graph of $y = |x|$. The graph of $y = |x + 1|$ is shifted to the left by 1 from the graph of $y = |x|$.

➤ *Exercises*

Identify the parent function. Tell what type of transformation is applied to each, and then draw a carefully labeled graph.

18. $y = x^2 + 1$ **19.** $y = (x + 1)^2$ **20.** $y = |x| - 4$ **21.** $y = |x - 4|$

Lesson 9.6

➤ *Key Skills*

Describe the effect of a combination of transformations on the graph of a function.

The parent function of $y = -2|x + 2|$ is $y = |x|$. Because of the factor -2, the y-values will be twice as great as the parent function and reflected through the x-axis. This is a reflection and a vertical stretch by 2. Because 2 is added to the variable, x-values are less than the same y-values for the parent function. This results in a horizontal translation to the left by 2.

➤ *Exercises*

Graph each function.

22. $f(x) = 3x^2 + 1$ **23.** $g(x) = |x - 6| + 2$ **24.** $h(x) = -\dfrac{2}{x} + 1$

Applications

25. Are meters a function of centimeters? (If you know the number of meters, do you necessarily know the number of centimeters?)

26. Are centimeters a function of meters? (If you know the number of centimeters, do you necessarily know the number of meters?)

27. **Temperature** Kelvins are often used to measure the temperature of substances in science. To change degrees Celsius to kelvins, add 273.16 to degrees Celsius. What transformation relates degrees Celsius to kelvins?

28. **Business** Tomas decides to have a sale at his art store. He decides to reduce the price of everything in the store by 20%. Find the scale factor that shows the amount by which the cost of the items were "stretched."

29. **Geometry** Belinda reflected the line segment with endpoints (2, 4) and (6, 6) through the x-axis and then shifted it to the left 5 units. What were the endpoints of the new line segment?

Chapter 9 Assessment

1. When is a relation a function?

2. Explain how to use the vertical line test to identify the graph of a function.

Evaluate the following functions when $x = -3$.

3. $f(x) = \text{INT}(x)$ **4.** $g(x) = -x^2$ **5.** $h(x) = 2|x|$

Graph each function. Then tell whether the graph of $y = |x|$ is stretched, reflected, of shifted.

6. $y = 3|x|$ **7.** $y = |x| - 5$ **8.** $y = -|x|$ **9.** $y = |x + 4|$

Given the function $y = \dfrac{2}{x}$,

10. identify the parent function.

11. sketch the graphs of the function and the parent function on the same axes.

12. tell how the graph of the parent function was changed by the 2.

Tell what happens to the point (3, 2) when each of the following is applied.

13. vertical translation by 5   **14.** horizontal translation by -3

15. vertical translation by -6 **16.** vertical stretch by 4

17. What happens to the average hourly wage of the workers at a restaurant if all hourly wages are increased by 5%?

18. Explain how you know when a parabola opens downward.

Graph each function.

19. $f(x) = (x - 3)^2$ **20.** $h(x) = 5|x + 3|$ **21.** $t(x) = \dfrac{-2}{x + 2}$

22. Business A local department store is having a sale. They marked down their clearance items by 25%, and for one day only they are offering an additional 30% off each clearance item. Miki decided to leave the original prices on the price tag and make a sign that shows 55% off. Is Miki's sign correct? Why or why not?

23. Explain how to use a graphics calculator to approximate the minimum point of a quadratic function like $y = 2(x + 4)^2 - 1$.

Explain how to use a graphics calculator to solve each of the following for x.

24. $\dfrac{4}{x} = 3 + x$ **25.** $|1 - 3x| = |x| + 5$

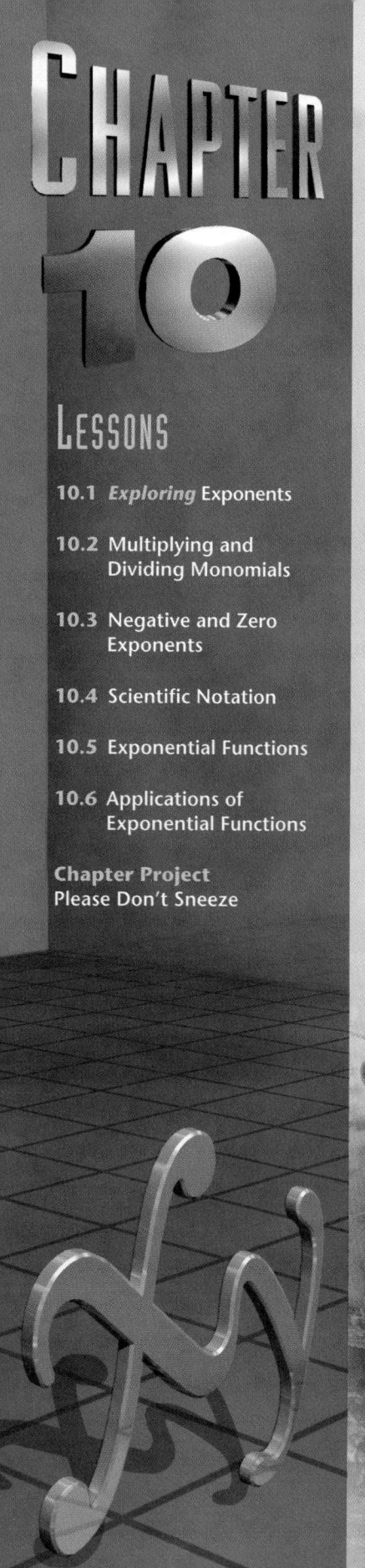

CHAPTER 10

Exponents

Exponential functions model a variety of important real-world activities. The growth of living organisms, population growth and decline, radioactive dating of fossils, and the spread of certain diseases are all examples of exponential growth or decay. You have already investigated some examples briefly in Chapter 2. This Chapter will give you the opportunity to explore them in greater depth.

One of the areas in which exponential functions play an important role is finance. The portfolio activity gives you an example of interest on savings and loans that typically behaves exponentially.

Exponential
Growth Function

$y = 2^x$

Exponential
Decay Function

$y = 2^{-x}$

PORTFOLIO ACTIVITY

A credit card company charges 1.5% on the unpaid balance and requires a minimum payment of $10 or 3% of the unpaid balance, whichever is greater. Suppose you borrow $1000 and pay it back making the minimum payment each month. How long will it take to pay it back? What is the total amount that you will pay back?

LESSON 10.1

Exploring Exponents

why *Numbers with many digits can be expressed in a simpler way by using exponents.*

By using an exponent, 2•2•2•2 is represented as 2^4.

$$2 \cdot 2 \cdot 2 \cdot 2 = 2^4 = 16$$

Base — Exponent

The fourth power of 2 is 16.

The expression 2^{1000} represents 2 to the thousandth power and is the product when a thousand 2s are multiplied.

Some powers have special names. The first power of 2 is 2^1. It is read *two to the first*. The second power of 2 is 2^2. It is read *two squared*. The third power of 2 is 2^3. It is read *two cubed*. The nth power of 2 is 2^n. It is read *two to the nth*.

PRINTOUT OF 2^1000

10715086071862673209484250490600018105
10715086071862673209484250490600018105
61404811705533607443750388303510511249
361224931983788815695858127594672917553
14682518714528569231404359845775746985
74813934567774824230985421074605062371
14187795418215304647498358194126739876
75591655439460770629145711964776865421
67660429837652624386837205668069376

EXPONENTS AND POWERS

If x is any number and a is an integer greater than 1, then

$$x^a = \underbrace{x \cdot x \cdot x \cdot \ \cdots \ \cdot x.}_{a \text{ factors}}$$

When $a = 1$, $x^a = x^1 = x$.

Our number system uses 10 as its base. A number written in the form **2315** is in *customary notation*. It can be expressed in *expanded notation* as

$$2 \cdot 1000 + 3 \cdot 100 + 1 \cdot 10 + 5$$

or in *exponential notation* as

$$2 \cdot 10^3 + 3 \cdot 10^2 + 1 \cdot 10^1 + 5.$$

How would you write 15,208 in exponential notation?

Exploration 1 *Multiplying Powers*

1 Examine the steps in the example.

Product Form → $10^3 \cdot 10^1$ → $1000 \cdot 10$ → $10{,}000$ → 10^4 → Simplified Form

Complete the steps as shown in Step 1.

a. $10^3 \cdot 10^2 = \quad 1000 \cdot 100 \quad = \quad \underline{?} \quad = \underline{?}$

b. $10^3 \cdot 10^3 = \quad \underline{(?)}\,\underline{(?)} \quad = \quad \underline{?} \quad = \underline{?}$

2 Complete the table for $10^4 \cdot 10^2$. Write the missing product as a power of ten. This product is now in simplest form.

Decimal form	$10{,}000 \cdot 100 = 1{,}000{,}000$
Exponential form	$10^4 \cdot 10^2 = \underline{?}$

3 Make a similar table for each of the following.

    a. $10^3 \cdot 10^6$ **b.** $10^1 \cdot 10^3$ **c.** $2^2 \cdot 2^4$ **d.** $3^2 \cdot 3^2$

4 How would you simplify $a^m \cdot a^n$ using exponents? Make a table to check your guess.

5 For any positive number a and any positive integers m and n, what is the simplified form for $a^m \cdot a^n$? ❖

The statement in Step 5 suggests that multiplying powers *with the same base* involves the addition of exponents. Guess what happens to the exponents when you divide powers.

•Exploration 2 Dividing Powers

1 Use the table to write the quotient for $\dfrac{10^6}{10^2}$ as a power of 10.

FORM	NUMERATOR	DENOMINATOR	QUOTIENT
Decimal	1,000,000	100	10,000
Exponential	10^6	10^2	?

2 Make a similar table for each of the following.

    a. $\dfrac{10^5}{10^2}$ **b.** $\dfrac{2^6}{2^2}$ **c.** $\dfrac{3^4}{3^3}$

3 Guess the simplified form in exponent notation for $\dfrac{10^5}{10^1}$. Make a table to check your guess.

4 For any positive number a and any positive integers m and n with $m > n$, what is the simplified form for $\dfrac{a^m}{a^n}$?

5 Explain how you would simplify $\dfrac{2^5}{2^3}$.

6 Can you use this procedure to simplify $\dfrac{5^2}{3^2}$? Make a conjecture to identify the restrictions on combining the exponents when dividing powers. ❖

CRITICAL
Thinking
    Explain why $(ab)^n = a^n b^n$ and $\left(\dfrac{a}{b}\right)^n = \dfrac{a^n}{b^n}$ when n is any positive integer.

You can multiply and divide powers. What is the result when you find a power of a power?

•Exploration 3 The Power of a Power

1 Write both 10^5 and 10^6 in customary notation.

2 Since $(10^3)^2$ means $10^3 \cdot 10^3$, change 10^3 to customary notation and square the number.

3 Change the resulting number back from customary notation to exponential notation. Does $(10^3)^2$ equal 10^5 or 10^6?

4 What was done with the exponents? Does $(5^2)^4$ equal 5^6 or 5^8?

5 For any positive number a and any positive integers m and n, what is the equivalent expression for $(a^m)^n$? ❖

Suppose a certain bacteria colony doubles every hour. If a colony contains 1000 bacteria at noon, the properties of exponents can be used to find how many bacteria the culture contains at 3 P.M. and 2 hours later.

At 3 P.M. there will be $1000 \cdot 2^3$ or 8000 bacteria. Two hours later the bacteria will double two more times. There will be $(1000 \cdot 2^3) \cdot 2^2 = 1000 \cdot 2^{3+2} = 1000 \cdot 2^5$ or 32,000 bacteria in the culture. ❖

Once you are familiar with the properties of exponents, you can simplify many exponential expressions.

a. $3^2 \cdot 3^3 = 3^5$ **b.** $5^4 \div 5^1 = 5^3$ **c.** $(2^2)^3 \div (2^3)^2 = 2^6 \div 2^6$
$9 \cdot 27 = 243$ $625 \div 5 = 125$ $64 \div 64 = 1$

EXTENSION

Calculator

Most computations with exponents can be performed easily on a calculator. Check the calculator you use to find how it performs these computations.

Most calculators will use the y^x or ⌃ key to indicate an exponent. Parentheses are used for more complicated numerators and denominators.

For example, to simplify $\dfrac{10^3 + 10^4}{10^2}$, the numerator contains addition of the powers, so it should be included in parentheses. On many calculators, you can use

[( ] 10 [y^x] 3 [+] 10 [y^x] 4 [ )] [÷] 10 [y^x] 2 [=] or

[( ] 10 [⌃] 3 [+] 10 [⌃] 4 [ )] [÷] 10 [⌃] 2 [ENTER] .

Notice the difference in each computation.

a. $\dfrac{10^3 + 10^4}{10^2} = \dfrac{10^3(1 + 10)}{10^2} = 10^1(11) = 110$

b. $\dfrac{10^3 \times 10^4}{10^2} = \dfrac{10^7}{10^2} = 10^5 = 100,000$ ❖

CRITICAL Thinking

Compare the way the exponents are treated when the powers are added with the way they are treated when they are multiplied. Make a conjecture about how to treat exponents when powers are added.

EXERCISES & PROBLEMS

Communicate

1. Explain the relationship between the exponent in a power of ten and the number of zeros in a number written in customary notation. Use 10^{12} as an example.

2. How can you tell how many times the base is used as a factor in 3^6?

3. When you multiply powers, what operation is applied to the exponents of the common base? Explain why.

4. When you divide powers, what operation is applied to the exponents of the common base? Explain why.

5. When you raise a power to a power, what operation is applied to the exponents of the common base? Explain why.

6. Explain how you can determine, without computing anything, which value is greatest: 5^{17}, 8^{16}, or 8^{17}.

Killer Whale

Weddell Seal

Explain why the expressions in each of the following equations are equivalent.

7. $a^r \cdot a^s = a^{r+s}$ 8. $\dfrac{a^p}{a^q} = a^{p-q}$ 9. $(a^s)^t = a^{st}$

Practice & Apply

Make a table like the one in Exploration 1 for each product.

10. $10^6 \cdot 10^4$ 11. $10^3 \cdot 10^5$ 12. $2^3 \cdot 2^4$ 13. $3^4 \cdot 3^2$

Make a table like the one in Exploration 2 for each quotient.

14. $\dfrac{10^3}{10^2}$ 15. $\dfrac{10^6}{10^3}$ 16. $\dfrac{2^6}{2^3}$ 17. $\dfrac{3^5}{3^4}$

Change each of the following to customary notation.

18. 2^5 19. 2^2

Squid

20. Use the evaluations of 2^5 and 2^2 to find $2^5 \cdot 2^2$.

21. Express the product of $2^5 \cdot 2^2$ as a single power of 2.

22. Use the evaluations of 2^5 and 2^2 to find $\dfrac{2^5}{2^2}$.

23. Express the quotient $\dfrac{2^5}{2^2}$ as a single power of 2.

24. Use the evaluation of 2^5 to find $(2^5)^2$.

25. Express the result in Exercise 24 as a single power of 2.

Evaluate each of the following using the rules for exponents when they apply.

26. $(6^2)^3$ 27. $\dfrac{4^3}{4^2}$ 28. $\dfrac{10^2 \cdot 10^5}{10^3}$ 29. $\dfrac{10^2 + 10^5}{10^3}$

Evaluate each of the following.

30. $(x^2)^3$ **31.** $\dfrac{c^4}{c^1}$ **32.** $\dfrac{r^3(r^4)}{r^2}$ **33.** $\dfrac{2^3 + 2^4}{2^2}$

Some of the following can be expressed as a single power of 10; others cannot. If you can, write the given expression as a power of ten. Then evaluate each expression.

34. $10^4 \cdot 10^2$ **35.** $10^4 + 10^2$ **36.** $10^4 \div 10^2$

37. $10^4 - 10^2$ **38.** $(10^2)^3$ **39.** 10^{2+3}

Technology For exponents, 2^3 is entered as 2 ⌃ 3 on some calculators and as 2 y^x 3 on others. Evaluate 2^3. Familiarize yourself with the way the calculator you are using handles exponents. Then use the calculator to evaluate the following.

40. 2^{20} **41.** 3^{10} **42.** 5^{10} **43.** 0.5^4

The table shows how to read certain large numbers.

Customary	Exponential	Word
1,000,000	10^6	million
1,000,000,000	10^9	billion
1,000,000,000,000	10^{12}	trillion

For example, the number 2,450,000,000 is read "two billion, four hundred fifty million," and 3×10^6 is read "three million." Write the following in words.

44. 7,000,000,000,000   **45.** 2,030,400,000,000   **46.** 5×10^9 **47.** 3×10^{12}

48. A newspaper article refers to a debt of 4.2 trillion dollars. Write this in customary notation.

Algae

Yellowtail Fish

Krill (zooplankton)

Biology The food chain explains why even small organisms affect much larger ones. It works in the following way.

- A killer whale eats about 10 pounds of seal per day.
- Each pound of seal requires about 10 pounds of fish or squid per day.
- Each pound of fish or squid requires about 10 pounds of zooplankton per day.
- Each pound of zooplankton requires about 10 pounds of algae per day.

Express each number of pounds per day as a power of 10.

49. seals needed to sustain a killer whale   **50.** fish or squid needed to sustain a killer whale

51. zooplankton needed to sustain a killer whale   **52.** algae needed to sustain a killer whale

53. **Statistics** Make a bar graph from a frequency distribution of the digits of 2^{1000} given at the beginning of this lesson. Count how many 0s, 1s, 2s, . . . , 9s there are. Would you say the distribution is about the same for each digit, or does it favor some particular digits?

54. Write 2^{10} in customary notation.

55. Technology A calculator shows the value 20^{10}. Compare this answer to the answer in Exercise 54. What do you think this notation means? Write out the answer in customary notation.

20^10

1.024E13

Look Back

Solve for x.   [Lesson 4.8]

56. $|x - 3| \leq 7$ **57.** $|2x + 4| > 8$

58. **Coordinate Geometry** A line passes through the points $P(14, -3)$ and $Q(-5, 27)$. Write an equation for the line through point P perpendicular to line PQ.   **[Lessons 5.4, 5.6]**

59. Find the inverse of $A = \begin{bmatrix} 3 & -1 \\ -2 & 1 \end{bmatrix}$. What is the product of $A \cdot A^{-1}$? **[Lesson 7.4]**

60. **Probability** A study of more than 50,000 nurses was conducted for over eight years. Those whose diets were rich in vitamin A were 40% less likely to have cataracts that impaired their vision than those whose diets were low in the vitamin. Is this an example of experimental probability or theoretical probability? Explain.   **[Lesson 8.1]**

61. How many different ways can you order an ice cream cone with a topping?   **[Lesson 8.5]**

62. **Geometry** The area in square centimeters of a square piece of cardboard is determined from the length of its sides. Write the function for the area if 3 centimeters is added to each side of the original square and that area is doubled.   **[Lesson 9.6]**

63. Describe the graph of the function in Exercise 62 without drawing it. Use your understanding of transformations.   **[Lesson 9.6]**

Super Scooper
flavors

1. vanilla	7. chocolate chip
2. chocolate	8. cookies & cream
3. strawberry	9. pineapple
4. banana	10. lime sherbet
5. rocky road	11. orange sherbet
6. pralines & cream	12. raspberry sherbet

toppings	cones
• sprinkles • shaved chocolate	plain
• malt • pecans • peanuts	sugar
• gummy bits • marshmallows	waffle

Look Beyond

64. Look at the pattern in the table for Exploration 1. Based on the table, 10^n can be written as a 1 followed by how many zeros?

65. What does Exercise 64 suggest for the value of 10^0?

Use a calculator to evaluate each of the following.

66. $4^{0.5}$ **67.** $9^{0.5}$ **68.** $81^{0.5}$ **69.** $100^{0.5}$

70. What do you think $b^{0.5}$, for any positive number b, means?

Computation with powers of variables can be simplified by the same properties you used for numerical powers.

In the previous lesson you discovered a property that lets you simplify the product of two powers with the same base. Just keep the base and add the exponents. For example,

$$x^3 \cdot x^5 = x^{3+5} = x^8.$$

PRODUCT-OF-POWERS PROPERTY

If x is any number and a and b are any positive integers, then

$$x^a \cdot x^b = x^{a+b}.$$

The Product-of-Powers Property can be used to find the product of more complex expressions such as $5a^2b$ and $-2ab^3$. Expressions like $5a^2b$ and $-2ab^3$ are *monomials*.

A **monomial** is an algebraic expression that is either a constant, a variable, or a product of a constant and one or more variables. A monomial can also contain powers of variables with positive integer exponents.

$$(5a^2b^2)(-2ab^3) = -10(a^2 \cdot a^1)(b^2 \cdot b^3) \quad \text{Multiply the coefficients.}$$

Then group powers of like bases.

$$= -10a^3b^5 \qquad \text{Product-of-Powers Property}$$

EXAMPLE 1

Find the products.

A $(6st)(-2s^2t)$

B $(-4a^2b)(-ac^2)(3b^2c^2)$

Solution ➤

Multiply the coefficients and group powers of the same base, then use the Product-of-Powers Property.

A $(6st)(-2s^2t) = -12s^3t^2$

B $(-4a^2b)(-ac^2)(3b^2c^2) = 12a^3b^3c^4$ ❖

CRITICAL Thinking

Explain how to use the Product-of-Powers Property to simplify each of the following expressions.

a. $x^{3c} \cdot x^{4c}$ **b.** $(x + c)^d \cdot (x + c)^{2d}$

The Product-of-Powers Property can be used to simplify $(a^2)^3$.

$$(a^2)^3 = a^2 \cdot a^2 \cdot a^2 = a^{2+2+2} = a^6$$

Notice that $(a^2)^3 = a^{2 \cdot 3}$. To raise a power to a power, multiply the exponents.

POWER-OF-A-POWER PROPERTY
If x is any number and a and b are any positive integers, then
$$(x^a)^b = x^{ab}.$$

EXAMPLE 2

Simplify.

A $(3^2)^4$

B $(p^2)^5$

Solution ➤

Use the Power-of-a-Power Property to multiply the exponents.

A $(3^2)^4 = 3^{2 \cdot 4} = 3^8$

B $(p^2)^5 = p^{2 \cdot 5} = p^{10}$ ❖

Sometimes an exponential expression has a monomial for a base.

EXAMPLE 3

Simplify $(xy^2)^3$.

Solution ➤

The exponent, 3, outside the parentheses indicates that the monomial xy^2 is used as a factor 3 times. Simplify by regrouping and multiplying.

$$(xy^2)^3 = (xy^2)(xy^2)(xy^2)$$
$$= (x \cdot x \cdot x)(y^2 \cdot y^2 \cdot y^2)$$
$$= x^3y^6 \quad \text{❖}$$

POWER-OF-A-PRODUCT PROPERTY
If x and y are any numbers and n is a positive integer, then

$$(xy)^n = x^n y^n.$$

If a is the numerical coefficient, then $(axy)^n = a^n x^n y^n$. For example,

$$(3xy)^2 = (3xy)(3xy) = (3^2)(x^2)(y^2) = 9x^2y^2.$$

Remember to apply the exponent outside of the parentheses to *each* factor of the monomial inside the parentheses. This includes the coefficient, 3. Thus, $(3xy)^2 = 9x^2y^2$.

EXAMPLE 4

Simplify each expression.

A $(3x^2y^3)^3$ **B** $(-t)^5$ **C** $(-t)^6$ **D** $-t^4$ **E** $(-5x)^3$

Solution ➤

A $(3x^2y^3)^3 = (3)^3(x^2)^3(y^3)^3 = (3^3)(x^{2 \cdot 3})(y^{3 \cdot 3}) = 27x^6y^9$

B $(-t)^5 = (-1 \cdot t)^5 = (-1)^5 \cdot t^5 = -1 \cdot t^5 = -t^5$

C $(-t)^6 = (-1 \cdot t)^6 = (-1)^6 \cdot t^6 = 1 \cdot t^6 = t^6$

D $-t^4$ is in simplest form. $(-1) \cdot t^4 = -t^4$
The exponent applies only to t and not to the -1.

E $(-5x)^3 = (-5 \cdot x)^3 = (-5)^3 \cdot x^3 = -125x^3$ ❖

CRITICAL Thinking

Explain whether $(-x)^m$ is positive or negative when x is positive and m is even. Do the same when m is odd. When are the exponents of a power added, and when are they multiplied?

Earlier you found that if you simplify quotients like $\frac{10^6}{10^2}$, the quotient can be written without a denominator. A similar property holds for variables and monomials.

QUOTIENT-OF-POWERS PROPERTY

If x is any number except 0, and a and b are any positive integers with $a > b$, then

$$\frac{x^a}{x^b} = x^{a-b}.$$

EXAMPLE 5

Simplify. **A** $\dfrac{-4x^2y^5}{2xy^3}$ **B** $\dfrac{c^4b}{c^2a}$

Solution ➤

Simplify any numerical coefficients. Then subtract the exponent in the denominator from the exponent in the numerator with the same base.

A $\dfrac{-4x^2y^5}{2xy^3} = \left(\dfrac{-4}{2}\right)(x^{2-1})(y^{5-3}) = -2xy^2$

B $\dfrac{c^4b}{c^2a} = (c^{4-2})\left(\dfrac{b}{a}\right) = \dfrac{c^2b}{a}$ ❖

Try This Simplify the expressions.

a. $\dfrac{6ab^2}{-2b}$ **b.** $\dfrac{(25^2)(-81t^3)}{45^3t}$

Exercises & Problems

Communicate

1. Explain why $3x^2$ is a monomial but $3x^{-2}$ is not.

2. Compare the Product-of-Powers Property with the Quotient-of-Powers Property.

3. Explain why the exponent of each factor of a monomial is multiplied by the outside exponent to calculate the power of a power.

4. Explain why $(x^a)^b = (x^b)^a$.

5. Explain why $y^2 \cdot y^4 \neq y^8$.

Practice & Apply

Evaluate each of the following.

6. $(3x)^2$ **7.** $\left(\dfrac{a}{b}\right)^3$ **8.** $\left(\dfrac{10x}{y^3}\right)^2$

Find each product.

9. $(8r^2)(4r^3)$ **10.** $(70x^4)(7x^3)$ **11.** $(-2a^5)(4a^2)$

12. $(-p^2)(10p^3)$ **13.** $(-2a^2)(-5a^4)$ **14.** $(-x^2y^5)(-x^3y^2)$

15. $(48a^4b^2)(-0.2ac^5)$ **16.** $(4.3d^2n^{10}k)(0.1n^2k^3)$

Find each quotient.

17. $\dfrac{8r^3}{4r^2}$ **18.** $\dfrac{70x^4}{7x^3}$ **19.** $\dfrac{-2a^5}{4a^2}$ **20.** $\dfrac{-p^4}{10p^3}$ **21.** $\dfrac{-2a^6b^7}{-5a^4b}$

22. $\dfrac{-x^4y^5}{-x^3y^2}$ **23.** $\dfrac{48a^4b^5}{-1.2ac^3}$ **24.** $\dfrac{0.8r^6}{0.004r^3}$ **25.** $\dfrac{4u^2v^{10}w^4}{0.1w^2v^3}$

Simplify each of the following.

26. $(2x^4)^3$ **27.** $(3b^2)^5$ **28.** $(-2r^3)^2$ **29.** $(-10m^4)^3$

30. $(5j^2k^3)^4$ **31.** $(3xy^4z^5)^2$ **32.** $2(3a^2)^3$ **33.** $10(-5b^5)^2$

34. $(8n^2p)^3$ **35.** $(7j^2)^3$ **36.** $(ab^5)(a^3)^2$ **37.** $(v^4w^3)^2(v^3)^4$

38. Evaluate the two monomials $2A^3$ and $(2A)^3$ for $A = 10$.

39. Evaluate each of the following. $(-1)^1,\ (-1)^2,\ (-1)^3,\ (-1)^4,\ (-1)^5$

40. Use the pattern in Exercise 39 to find the value $(-1)^{100}$. Explain what you notice about $(-1)^n$ for various values of n.

41. **Geometry** Find the area of a square if each side is twice as long as the side of the square shown.

42. **Geometry** Find the volume of a cube if each edge is double the edge of the cube shown.

Technology Calculators can be used to evaluate expressions involving monomials. Store 10 for A in memory and 2 for B in memory on a graphics calculator. Use a calculator to evaluate each of the following.

43. A^2 **44.** B^3 **45.** $A^2 + B^3$ **46.** A^2B^3 **47.** $\dfrac{A^2}{B^3}$ **48.** $\dfrac{1}{A^2 + B^3}$

Evaluate the following for $A = 9.8$ and $B = 2.1$.

49. A^2 **50.** B^3 **51.** $A^2 + B^3$ **52.** A^2B^3 **53.** $\dfrac{A^2}{B^3}$ **54.** $\dfrac{1}{A^2 + B^3}$

55. Does $(10^2)^3$ equal 10^8 or 100^3? Explain your reasoning.

Look Back

Solve the equations for n. [Lesson 4.6]

56. $(4n - 20) - n = n + 12$

57. $9(n + k) = 5n + 17k + 12$

58. Travel Distance, d, varies directly as time, t. If a car travels 195 miles in 3 hours, determine the constant, k, and write the equation for the direct variation. What is the distance the car will travel in 4 hours? **[Lesson 5.3]**

Solve the following systems. [Lessons 6.2, 6.3]

59. $\begin{cases} 5x - 3y = 32 \\ 2x = 12.8 \end{cases}$ **60.** $\begin{cases} 7x - 3y = 2 \\ 2x - y = -5 \end{cases}$

When traveling in foreign countries, the distances are usually given in kilometers.

Statistics

Use the data to find

61. the median.

62. the mean.

63. the mode. **[Lesson 8.3]**

64. A math class has 28 students. Of those students, 16 are girls. What is the probability that a random selection of one student will be a boy? **[Lesson 8.6]**

Look Beyond

Work all four exercises to discover a property of sums of squares.

65. Show that $1^2 + 8^2 = 4^2 + 7^2$. Show that $14^2 + 87^2 = 41^2 + 78^2$. How were the numbers 14, 87, 41, and 78 chosen?

66. Does $17^2 + 84^2 = 71^2 + 48^2$? How were the numbers 17, 84, 71, and 48 chosen?

67. Show that $0^2 + 5^2 = 3^2 + 4^2$. Find new numbers x and y so that $(03)^2 + 54^2 = x^2 + y^2$.

68. Find new numbers x and y so that $(04)^2 + 53^2 = x^2 + y^2$. Show that $4^2 + 6^2 = 3^2 + 7^2$ is *false*. Does $43^2 + 67^2 = 34^2 + 76^2$?

468 CHAPTER 10

why You have used positive exponents to represent numbers between 0 and 1. Negative exponents can also be used to represent the same numbers.

In chemistry, the approximate mass of a neutron or proton of an atom is 1.660×10^{-24} grams.

The expression 10^1 is defined to be 10. How should the expression 10^{-1} be defined?

Calculator

When the Quotient-of-Powers Property was defined, the exponent in the numerator was greater than the exponent in the denominator. For example, $\frac{10^4}{10^3} = 10^{4-3} = 10^1$. If the Quotient-of-Powers Property is applied when the exponent in the numerator is less than the exponent in the denominator, the result is $\frac{10^3}{10^4} = 10^{3-4} = \mathbf{10^{-1}}$.

If you write the powers as factors the result is

$$\frac{10^3}{10^4} = \frac{10 \cdot 10 \cdot 10}{10 \cdot 10 \cdot 10 \cdot 10} = \mathbf{\frac{1}{10}}.$$

The results above indicate that 10^{-1} should be defined as $\frac{1}{10}$. How can you find 10^{-1} on your calculator?

> **NEGATIVE EXPONENT**
>
> If *x* is any number except zero and *n* is any integer, then
>
> $$x^{-n} = \frac{1}{x^n}.$$

EXAMPLE 1

Simplify each expression.

A $2^{-3} \cdot 2^2$ **B** $\frac{10^3}{10^{-1}}$ **C** $10^2 \cdot 3^2$

Solution ➤

A Use the Product-of-Powers Property.

$$2^{-3} \cdot 2^2 = 2^{-3+2} = 2^{-1} = \frac{1}{2^1} = \frac{1}{2}$$

B Use the Quotient-of-Powers Property. Subtract the exponents, and simplify the expression.

$$\frac{10^3}{10^{-1}} = 10^{3-(-1)} = 10^4 = 10,000$$

C Since 10^2 and 3^2 do not have the same base, $10^2 \cdot 3^2$ cannot be simplified using the properties of exponents.

$$10^2 \cdot 3^2 = 100 \cdot 9 = 900 \; ❖$$

 CRITICAL *Thinking*

Compare the expressions $(-2)^3$ and $(-2)^{-3}$. Simplify each expression. How are the results related?

You have simplified expressions containing positive and negative exponents. What is the significance of an expression that contains a zero exponent?

•Exploration° *Defining x^0*

 1 Simplify each expression.

a. $\frac{10^5}{10^5}$ **b.** $\frac{10^3}{10^3}$ **c.** $\frac{10^1}{10^1}$

 2 Now use the Quotient-of-Powers Property to simplify each expression in Step 1.

 3 What is the simplified numerical value of 10^0?

 4 How would you define x^0 for any positive number *x*? ❖

Expressions containing variables with negative and zero exponents can be simplified if you use the properties of powers.

EXAMPLE 2

Simplify the following expressions.

(A) $c^{-4} \cdot c^4$ **(B)** $-3y^{-2}$ **(C)** $\dfrac{m^2}{n^{-3}}$

Solution ➤

(A) $c^{-4} \cdot c^4 = c^{-4+4} = c^0 = 1$
Does this agree with the definition you wrote in Step 4 of the Exploration?

(B) $-3y^{-2} = (-3)(y^{-2}) = \dfrac{-3}{y^2}$

(C) $\dfrac{m^2}{n^{-3}} = m^2 \cdot n^{-(-3)} = m^2 n^3$

Notice that $\dfrac{m^2}{n^{-3}} = \dfrac{m^2}{\dfrac{1}{n^3}} = m^2 n^3$. ❖

SUMMARY OF POWER PROPERTIES
Let a and b be any numbers with integer exponents m and n.

Product of Powers	Quotient of Powers $(b \neq 0)$	Power of a Power	Power of a Product
$b^m \cdot b^n = b^{m+n}$	$\dfrac{b^m}{b^n} = b^{m-n}$	$(b^m)^n = b^{mn}$	$(ab)^m = a^m b^m$

CRITICAL Thinking

If a and b are non-zero numbers and n is any integer, explain why the following is true.

$$\left(\frac{a}{b}\right)^{-n} = \left(\frac{b}{a}\right)^{n}$$

EXERCISES & PROBLEMS

Communicate

1. Explain why $5^{-2} = \dfrac{1}{5^2}$.

2. Use a pattern to show that $2^0 = 1$. For what other bases does this apply?

3. Can $5^2 \cdot 4^{-3}$ be simplified using the properties of exponents? Why or why not?

4. Explain why $(3a)^{-2}$ does not equal $\dfrac{3}{a^2}$.

5. What is the meaning of a negative exponent?

6. What is the meaning of a zero exponent?

Practice & Apply

Evaluate each expression.

7. -3^{-2} **8.** $(-3)^2$ **9.** 3^{-2} **10.** -3^2

11. Copy the table and fill in the remaining entries by continuing the pattern.

Decimal form	10,000	1000	100	10	?	?	?
Exponential form	10^4	10^3	10^2	10^1	?	?	?

12. The corresponding numbers 10,000 and 10^4 are equal. Are the corresponding numbers you wrote in the table equal?

Write each of the following without negative or zero exponents.

13. 2^{-3} **14.** 10^{-5} **15.** $a^3 b^{-2}$ **16.** $c^{-4} d^3$

17. $v^0 w^2 y^{-1}$ **18.** $(a^2 b^{-7})^0$ **19.** $r^6 r^{-2}$ **20.** $-t^{-1} t^{-2}$

21. $\dfrac{m^2}{m^{-3}}$ **22.** $\dfrac{2a^{-5}}{a^{-6}}$ **23.** $\dfrac{(2a^3)(10a^5)}{4a^{-1}}$ **24.** $\dfrac{b^{-2} b^4}{b^{-3} b^4}$

Probability The probability p of getting a multiple-choice question with four choices for each item correct by guessing is $\frac{1}{4}$, and the probability q of getting it wrong is $\frac{3}{4}$. On a test with 5 items, a formula for finding the probability of getting a certain number of items correct is shown in the chart.

25. Find the remaining values in the chart to the nearest thousandth.

No. Correct	Probability	Value
0	$p^0 q^5$	$\left(\frac{1}{4}\right)^0 \left(\frac{3}{4}\right)^5 \approx 0.237$
1	$5p^1 q^4$	?
2	$10p^2 q^3$	?
3	$10p^3 q^2$	?
4	$5p^4 q^1$	?
5	$p^5 q^0$	?

26. Which number of correct items is most likely?

27. Which number of correct items is least likely?

28. Find the sum of all six probabilities in Exercise 25.

Technology Tell whether each of the following is easier to do on a calculator or easier to do mentally. Explain why you think so. Then simplify.

29. $\dfrac{2.56^7}{2.56^6}$ **30.** $\dfrac{2.56^6}{2.56^7}$ **31.** 0^7

32. 7^0 **33.** $(2.992 \times 9.554)^0$ **34.** $(19.43 \times 0)^{18}$

35. Technology Try to compute 2^{236} on a calculator. What happens?

36. Try to compute 0^0 on a calculator. What happens?

37. Draw a graph of $Y_1 = 0^x$ for $x > 0$. What does this suggest the value of 0^0 should be?

38. Draw a graph of $Y_2 = x^0$ for $x > 0$. What does this suggest the value of 0^0 should be?

Look Back

39. What are the next three terms of the sequence 8, 11, 16, 23, 32 . . . ? **[Lesson 1.2]**

Solve each equation for y. How are the equations related? [Lesson 4.6]

40. $5(3y) = 3y - 24$ **41.** $aby = by - c$

42. Draw the number line graph for the values of x that solve the inequality $3x - 6 > -4x + 12$. Describe the solution set of this inequality in words. **[Lesson 4.8]**

43. How do the slopes of vertical and horizontal lines differ? **[Lesson 5.5]**

44. Transformations The graph of the quadratic parent function is stretched vertically by a scale factor of 2. It is then translated -3 units horizontally and 4 units vertically. Finally, it is reflected through the x-axis. Draw a graph and write a formula for this transformation. **[Lesson 9.6]**

45. Let $A = 3 \cdot 3 \cdot 3 \cdot 3 \cdot 3 \cdot x \cdot x \cdot x \cdot x \cdot x \cdot x \cdot y \cdot y \cdot y \cdot y$ and $B = 2 \cdot 2 \cdot 2 \cdot 2 \cdot 3 \cdot 3 \cdot 3 \cdot x \cdot x \cdot x \cdot x \cdot y$. Simplify each expression. **[Lesson 10.2]**

46. Write the product $A \cdot B$ from Exercise 45 using powers. **[Lesson 10.2]**

Look Beyond

Cultural Connection: Europe Pierre de Fermat (1601–1665) was a French mathematician who helped develop early number theory and probability. He studied numbers of the form $2^{2^n} + 1$, now called Fermat numbers.

47. Copy the table and complete the Fermat numbers.

n	Fermat number
0	$2^{2^0} + 1 = 2^1 + 1 = 3$
1	$2^{2^1} + 1 = 2^2 + 1 = 5$
2	$2^{2^2} + 1 = 2^4 + 1 = \underline{\ ?\ }$
3	$2^{2^3} + 1 = 2^8 + 1 = \underline{\ ?\ }$
4	$2^{2^4} + 1 = \underline{\ ?\ } = \underline{\ ?\ }$
5	$2^{2^5} + 1 = \underline{\ ?\ } = \underline{\ ?\ }$

48. Number Theory Fermat observed that each of the numbers was prime. He made the conjecture that all such numbers were prime. Though an unusually gifted mathematician, he was wrong in thinking the sixth number, $2^{2^5} + 1$, was prime, since it has a factor of 641. Use a calculator to find the other factor. Later research suggested that *no* more Fermat numbers are prime, so even great mathematicians can be wrong.

EYEWITNESS MATH

All Mixed Up?

You are playing a card game and it's your turn to deal. How many times do you shuffle the deck? Is that enough to be sure the cards are thoroughly mixed? Is there such a thing as shuffling too much? Read the article below to learn how two mathematicians, one of them a magician as well, found some answers that surprised shufflers around the world.

In Shuffling Cards, 7 Is Winning Number

By Gina Kolata

It takes just seven ordinary, imperfect shuffles to mix a deck of cards thoroughly, researchers have found. Fewer are not enough and more do not significantly improve the mixing.

The mathematical proof, discovered after studies of results from elaborate computer calculations and careful observation of card games, confirms the intuition of many players that most shuffling is inadequate.

The finding has implications for everyone who plays cards and everyone who has a stake in knowing whether a shuffle is random. . . .

No one expected that the shuffling problem would have a simple answer, said Dr. Dave Bayer, a mathematician and computer scientist at Columbia who is coauthor of the recent discovery. Other problems in statistics, like analyzing speech patterns to identify speakers, might be amenable to similar approaches, he said. . . .

Dr. Persi Diaconis, a mathematician and statistician at Harvard University who is another author of the discovery, said the methods used are already helping mathematicians analyze problems in abstract mathematics that have nothing to do with shuffling or with any known real-world phenomena.

Dr. Diaconis has been carefully watching card players shuffle for the past 20 years. The researchers studied the dovetail or riffle shuffle, in which the deck of cards is cut in half and the two halves are riffled together. They said this was the most commonly used method of shuffling cards. But, Dr. Diaconis said it produces a card order that is "far from random." . . .

Dr. Diaconis began working with Dr. Jim Reeds at Bell Laboratories and showed that a deck is perfectly mixed if it is shuffled between 5 and 20 times.

Next, Dr. Diaconis worked with Dr. Aldous and showed that it takes 5 to 12 shuffles to perfectly mix a deck.

In the meantime, he also worked on "perfect shuffles," those that exactly interlace the cards. He derived a mathematical proof showing that if a deck is perfectly shuffled eight times, the cards will be in the same order as they were before the shuffling.

To find out how many ordinary shuffles were necessary to mix a deck, Dr. Diaconis and Dr. Bayer watched players shuffle. He also watched Las Vegas dealers to see how perfectly they would interlace the cards they shuffled . . .

The researchers did extensive simulations of shuffling on a computer. To get the proof, the researchers looked at a lot of shuffles, guessed that the answer was seven, and finally proved it by finding an abstract way to describe what happens when cards are shuffled.

"When you take an honest description of something realistic and try to write it out in mathematics, usually it's a mess," Dr. Diaconis said. "We were lucky that the formula fit the real problem. That is just miraculous, somehow."

Getting Lost in the Shuffle

Even after a deck of cards is cut and shuffled, fragments of the original arrangement remain. In this example, the hearts are cut into two sequences: ace through six, and seven through king. Then the sequences are shuffled.

The cards are no longer ordered, but the sequence ace through six remains, with cards from the other fragment interspersed, also in sequence. The larger the number of cards in play, the more shuffles are required to mix them.

Curved lines show the odds that the first card in a deck will occupy any other position in the deck after one to seven shuffles. After one shuffle, for example, the first card is very likely to be one of the first few cards in the deck and very unlikely to be even five or six cards back. After four shuffles, it is still far more likely to be at the beginning of the deck than at the end. Only after seven shuffles does the card have about the same odds of being in any given position.

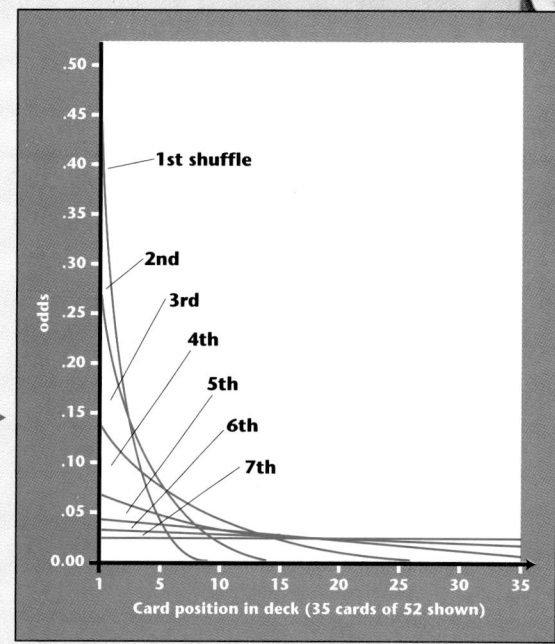

Cooperative Learning

1. Based on the graph in the article, after 4 shuffles, how many times more likely is the first card to be in the first position than in the 35th position? Explain.

2. **a.** What is the difference between a perfect shuffle and an ordinary shuffle?

   b. Is the shuffle in the diagram above perfect? Explain.

3. **a.** Use a diagram or model to find out how many perfect shuffles it takes for a deck to return to its original order if the deck consists of
   a. 2^2 cards.   **b.** 2^3 cards.   **c.** 2^4 cards.

4. **a.** How many perfect shuffles do you think it would take for a deck of 32 cards to return to its original order? Why?

   b. Use a diagram or model to check your prediction. What did you find?

5. Would a machine that always made perfect shuffles be useful when you play a card game? Why or why not?

LESSON 10.4 Scientific Notation

Orion

Betelgeuse

why

The study of the universe requires measurements of very large and very small quantities. The distance from the Earth to one of the stars, for instance, is about 6,000,000,000,000,000 miles. The wavelength of an X-ray is approximately 0.0000000001 meters.

These numbers with all their zeros are hard to read and to use for calculations. They do not even fit on a calculator display. A special way to express these large and small quantities in a compact number form is to use *scientific notation*.

Astronomy Betelgeuse is one of the stars in the constellation Orion, named for the great hunter of Greek and Roman mythology. The approximate distance in miles from Earth to Betelgeuse is written as a 6 followed by 15 zeros. In scientific notation this distance is written 6×10^{15}.

A number written in **scientific notation** is written with two factors, a number from 1 to 10, but not including 10, and a power of ten.

$$6 \times 10^{15}$$

First factor
Number from 1 to 10

Second factor
Power of 10

EXAMPLE 1

The distance from earth to Proxima Centauri, the nearest star other than the sun, is about 24,000,000,000,000 miles. It is read as 24 trillion miles. Write this number using scientific notation.

Solution ➤

To express the first factor, place the decimal point after the 2. This moves the decimal point 13 places to the left. To compensate for moving the decimal point, multiply the expression 2.4 by 10^{13}. This moves the decimal point 13 places to the right of 2. The two representations are now equivalent.

Customary Notation	**Scientific Notation**
24,000,000,000,000	2.4×10^{13} ❖

Try This   Write 875,000 in scientific notation.

EXAMPLE 2

Space Travel   If you travel at a speed of 1000 miles per hour, how long will it take to get to the star Betelgeuse?

Solution A ➤

Find $6 \times 10^{15} \div 1000$. If you use long division,

$$1000 \overline{)6,000,000,000,000,000} = 6 \times 10^{12}.$$

where the quotient is $6,000,000,000,000$.

The flight will take about 6 trillion hours!

Solution B ➤

You can also use the properties of exponents.

$$6 \times 10^{15} \div 1000 = \frac{6 \times 10^{15}}{10^3} = 6 \times 10^{15-3} = 6 \times 10^{12}$$ ❖

Try This   How many years are there in 6 trillion hours?

Numbers for small quantities can also be written in scientific notation. Recall that a decimal such as 0.01 can be written with a negative exponent, 10^{-2}.

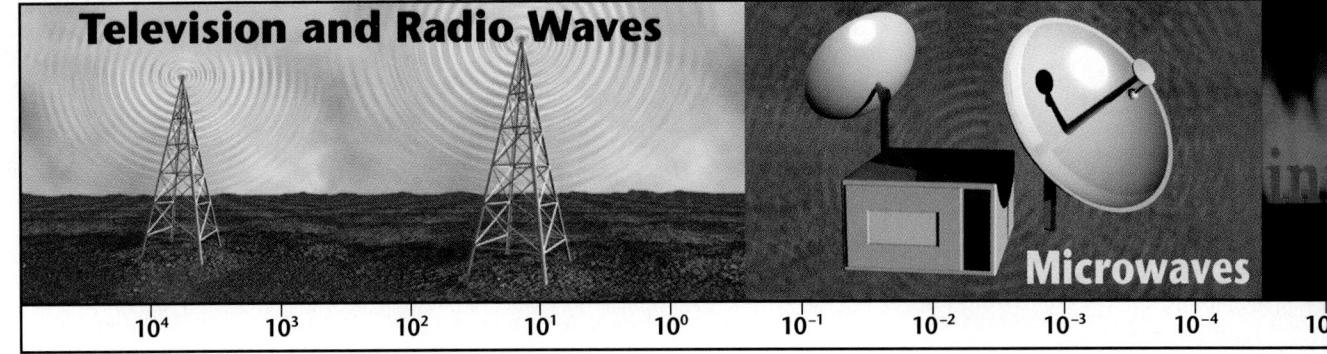

Television and Radio Waves

Microwaves

| 10^4 | 10^3 | 10^2 | 10^1 | 10^0 | 10^{-1} | 10^{-2} | 10^{-3} | 10^{-4} | 10 |

EXAMPLE 3

A typical X-ray has a wavelength of about 0.0000000001 meter. Write this number in scientific notation.

Solution ➤

To express the first factor, place the decimal after the 1. This, in effect, moves the decimal point 10 places to the right.

To compensate for moving the decimal point, multiply 1 by 10^{-10}. This is equivalent to dividing by 10^{10}, which moves the decimal point 10 places to the left, where it belongs.

Customary Notation	**Scientific Notation**
Thus, 0.0000000001 =	1×10^{-10} ❖

Try This Write 0.0000000402 in scientific notation.

Cultural Connection: Europe About 300 years ago, the English mathematician, Sir Isaac Newton, passed a ray of sunlight through a prism. Newton observed the colors of the rainbow. Today we know that visible light is only part of a larger set of waves that makes up the electromagnetic spectrum.

EXAMPLE 4

Physics How many times longer is a microwave of 10^{-2} meters than an X-ray of 10^{-10} meters?

Solution ➤

Divide the length of the microwave by the length of the X-ray.

$$\frac{10^{-2}}{10^{-10}} = 10^{-2-(-10)} = 10^{-2+10} = 10^8$$

The microwave is 10^8, or 100 million, times longer than the X-ray. ❖

CRITICAL Thinking

How can you change 12×10^{-2} and 0.12×10^{-2} to scientific notation using mental arithmetic?

When you multiply or divide two numbers written in scientific notation, you are working with pairs of numbers that are each represented by two factors. Rearrange the factors to simplify the expression.

EXAMPLE 5

Express the product and quotient in scientific notation.

A $(3 \times 10^3)(4 \times 10^{-5})$ **B** $\dfrac{2 \times 10^6}{5 \times 10^2}$

Solution ➤

A
$$(3 \times 10^3)(4 \times 10^{-5}) = (3 \cdot 4)(10^3 \times 10^{-5})$$
$$= 12 \times 10^{-2}$$
$$= 1.2 \times 10^{-1}$$

B
$$\frac{2 \times 10^6}{5 \times 10^2} = \left(\frac{2}{5}\right)(10^6 \times 10^{-2})$$
$$= 0.4 \times 10^4$$
$$= 4 \times 10^3 \; ❖$$

Most computations with scientific notation can be done on a calculator. Check to see if the calculator you use has an [EE] key or [EXP] key.

EXAMPLE 6

Calculator

The speed of light is about 2.98×10^5 kilometers per second. If the sun is about 1.49×10^8 kilometers from the Earth, how long does it take sunlight to reach Earth?

Solution ➤

Depending on your calculator use the [EE] or [EXP] key.

1.49 [EE] 8 [÷] 2.98 [EE] 5 [ENTER]

1.49 [EXP] 8 [÷] 2.98 [EXP] 5 [=]   [500]

It takes about 500 seconds for sunlight to reach Earth. ❖

> 1.49E8/2.98E5
>   500

EXERCISES & PROBLEMS

Communicate

1. Why is it important to be able to write a number with many zeros in scientific notation?

2. Explain what moving the decimal point 8 places to the left does to the value of a number. By what power of 10 must you multiply to compensate for that decimal point move?

3. Explain how to write the distance in miles from the Earth to the sun using scientific notation.

4. Explain what moving the decimal point 12 places to the right does to the value of a number. By what power of 10 must you multiply to compensate for that decimal point move?

5. Explain how to use the properties of exponents to multiply two numbers written in scientific notation. Use 240,000 and 0.006 in your discussion.

Practice & Apply

Write each number in scientific notation.

6. 2,000,000   **7.** 8,000,000,000   **8.** 340,000   **9.** 58,000

10. 0.00008   **11.** 0.0000005   **12.** 0.000234   **13.** 0.000000082

Write each number in customary notation.

14. 3×10^4 **15.** 4×10^8 **16.** 6.7×10^{10} **17.** 9.01×10^5

18. 4×10^{-7} **19.** 5×10^{-9} **20.** 8.8×10^{-12} **21.** 7.2×10^{-10}

Social Studies The following information shows how large numbers appear when studying different aspects of American culture. Write each in scientific notation.

22. 125,000: the number of passengers bumped on United States airlines each year due to overbooking

23. 5 trillion: the total value of outstanding stock in the stock market

24. 5.4 million: the number of American businesses owned by women

25. 3.7 million: the number of square feet in the Pentagon, the world's largest office building

Perform the following computations. Answer using scientific notation.

26. $(2 \times 10^4)(3 \times 10^5)$

27. $(6 \times 10^8)(1 \times 10^6)$

28. $(8 \times 10^6)(2 \times 10^{10})$

29. $(9 \times 10^6)(7 \times 10^6)$

30. $(9 \times 10^6) + (7 \times 10^6)$

31. $(9 \times 10^6) - (7 \times 10^6)$

32. $(8.2 \times 10^6)(3.1 \times 10^6)$

33. $(1.9 \times 10^8)(2 \times 10^{10})$

34. $\dfrac{8 \times 10^6}{2 \times 10^2}$

35. $\dfrac{9 \times 10^8}{2 \times 10^4}$

36. $\dfrac{3 \times 10^{10}}{6 \times 10^4}$

37. $\dfrac{2 \times 10^6}{5 \times 10^5}$

38. $\dfrac{(2 \times 10^6)(9 \times 10^4)}{6 \times 10^4}$

39. $(2 \times 10^{10})(6 \times 10^7)(4 \times 10^{-8})$

Economics The national debt for the United States in 1992 was listed as 4,064.6 billion dollars.

40. Write 4,064.6 billion in scientific notation.

41. This can be expressed as about 4 _?_ dollars.

42. Assuming the population of the United States was about 250 million in 1992, how much debt was this per person?

43. Write your age at your last birthday in *seconds*. Use scientific notation.

44. **Astronomy** Write the distance in miles from Earth to the moon in scientific notation.

The distance from Earth to the moon is about 248,000 miles.

Astronomy The distance light travels in one year, 5.87×10^{12} miles, is called a light-year.

45. Write this number in customary notation.

46. If the distance from Betelgeuse to Earth is 6×10^{15} miles, how long does it take light to travel to Earth from that star?

47. If the distance from Proxima Centauri to Earth is 2.4×10^{13} miles, how long does it take light to travel to Earth from Proxima Centauri?

48. **Science** According to Nobel Prize-winning physicist Leon Lederman, the universe is 10^{18} seconds old. How many years is this? Write your answer in words.

Sometimes scientific notation is used to indicate a level of accuracy. For example, an attendance figure for a rock concert, given as 1.2×10^4, (or 12,000), would be assumed to be given to the nearest *thousand*, while a figure of 1.20×10^4 (also 12,000) would be assumed to be given to the nearest hundred. This is because the appearance of the 0 in 1.20 would indicate that it is significant. Similarly, 1.2000×10^4 would indicate that all the zeros are significant, so it is accurate to the nearest *one*.

Write the following in scientific notation to indicate which digits are significant.

49. 76,000 to the nearest thousand

50. 76,000 to the nearest hundred

51. 8,000,000 to the nearest million

52. 8,000,000 to the nearest thousand

53. 8,000,000 to the nearest hundred

54. 8,000,000 to the nearest ten

55. 0.002 to the nearest thousandth

56. 0.0020 to the nearest ten-thousandth

57. A **googol** is the number formed by writing 1 followed by 100 zeros. Write this number in scientific notation.

Technology Write the following numbers in customary notation.

58. 2.3 E 04   **59.** 5.6 E 03   **60.** 7.22 E −03

61. 1.01 E −04   **62.** −2.8 E 02   **63.** −9.303 E −04

64. Technology What is the largest number that can be shown on your calculator without using scientific notation?

65. What is the largest number that can be shown on your calculator using scientific notation?

66. What is the smallest number that can be shown on your calculator without using scientific notation?

67. What is the smallest number that can be shown on your calculator using scientific notation?

Look Back

68. Statistics The scores on a test and a measure of student behavior have a correlation coefficient of $r = -0.90$. If there are 10 subjects, draw the scatter plot as you think it might look.   **[Lesson 1.6]**

69. Find the value of the expression. $3 + (-4) - (-9) - [-8 + 2(3 - 5)]$
[Lessons 3.2, 4.1]

70. Find the product. $(5)(-3)[-2(4)](-12)$ **[Lesson 4.1]**

71. Simplify. $(6x + 4y - 8) - 5(-7x + y)$ **[Lesson 4.2]**

72. Coordinate Geometry Given points on the coordinate graph $A(3, 7)$ and $B(-3, 12)$, find the slope of the line joining the points.   **[Lesson 5.1]**

73. Graph the following inequalities, and describe the region that solves both inequalities.   **[Lesson 6.5]**

$$\begin{cases} x - y > 5 \\ -3x + y \le 9 \end{cases}$$

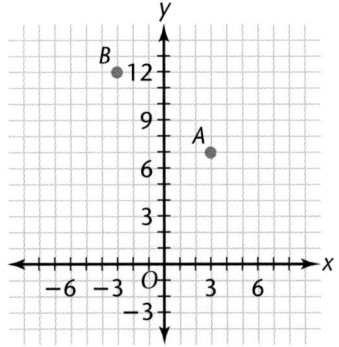

Look Beyond

Technology Use the constant feature of your calculator to write the following sequences in customary notation.

74. $0.9^1, 0.9^2, 0.9^3, 0.9^4, 0.9^5, \ldots$

75. $1.1^1, 1.1^2, 1.1^3, 1.1^4, 1.1^5, \ldots$

76. Compare what happens when these numbers are raised to higher and higher powers.

An exponential function models problems of growth and decline in population, finance, and science.

Social Studies   To better understand the effects of population growth it is helpful to understand the underlying mathematics. This involves a look back to the exponential function introduced in the second chapter.

According to the headline above, how many people are added to the population in one year? How many are added in one minute?

$$1.7\% \times 5.5 \text{ billion} = 0.017 \times 5,500,000,000$$
$$= 93,500,000 \text{ (in one year)}$$
$$\approx 256,000 \text{ (in one day)}$$
$$\approx 10,700 \text{ (in one hour)}$$
$$\approx 178 \text{ (in one minute)}$$

In the time it took you to read this far, the world population probably increased by several dozen people.

CRITICAL Thinking   Explain how the increase for one day was obtained from the increase for one year.

EXAMPLE 1

Ⓐ Estimate the world's population in the year 2000.

Ⓑ If population growth continues at the same rate, how many people will be added to the population each minute in the year 2000?

Solution ➤

B3	=B2*1.017	
	A	**B**
1	Year	Population
2	1990	5.50
3	1991	**5.59**
4	1992	5.69
5	1993	5.79
6	1994	5.88
7	1995	5.98
8	1996	6.09
9	1997	6.19
10	1998	6.29
11	1999	6.40
12	2000	6.50

Ⓐ Start with the 1990 population of 5.5 billion. Since the population increases by 1.7% each year, the new population will be 101.7%, or 1.017, times the old population. Use 1.017 as a multiplier to get the population for 1991, 1992, 1993, and so on.

If you round to the nearest 0.1 billion, the population in the year 2000 will be about 6.5 billion.

Ⓑ If this new population of 6.5 billion increases by 1.7%, the number of people added is calculated as follows.

$$
\begin{aligned}
1.7\% \times 6.5 \text{ billion} &= 0.017 \times 6{,}500{,}000{,}000 \\
&= 110{,}500{,}000 \text{ (in one year)} \\
&\approx 303{,}000 \text{ (in one day)} \\
&\approx 12{,}600 \text{ (in one hour)} \\
&\approx 210 \text{ (in one minute)}
\end{aligned}
$$

This means that while the population was increasing by 178 people per minute in 1990, it will be increasing by 210 people per minute in 2000. ❖

What is the formula for finding the population if you know its rate of growth? Examine this population pattern.

Years	Population (billions)	Exponential notation
0	5.5	$5.5(1.017)^0$
1	5.5(1.017)	$5.5(1.017)^1$
2	5.5(1.017)(1.017)	$5.5(1.017)^2$
3	5.5(1.017)(1.017)(1.017)	$5.5(1.017)^3$

750 CE   1000 CE   1250 CE

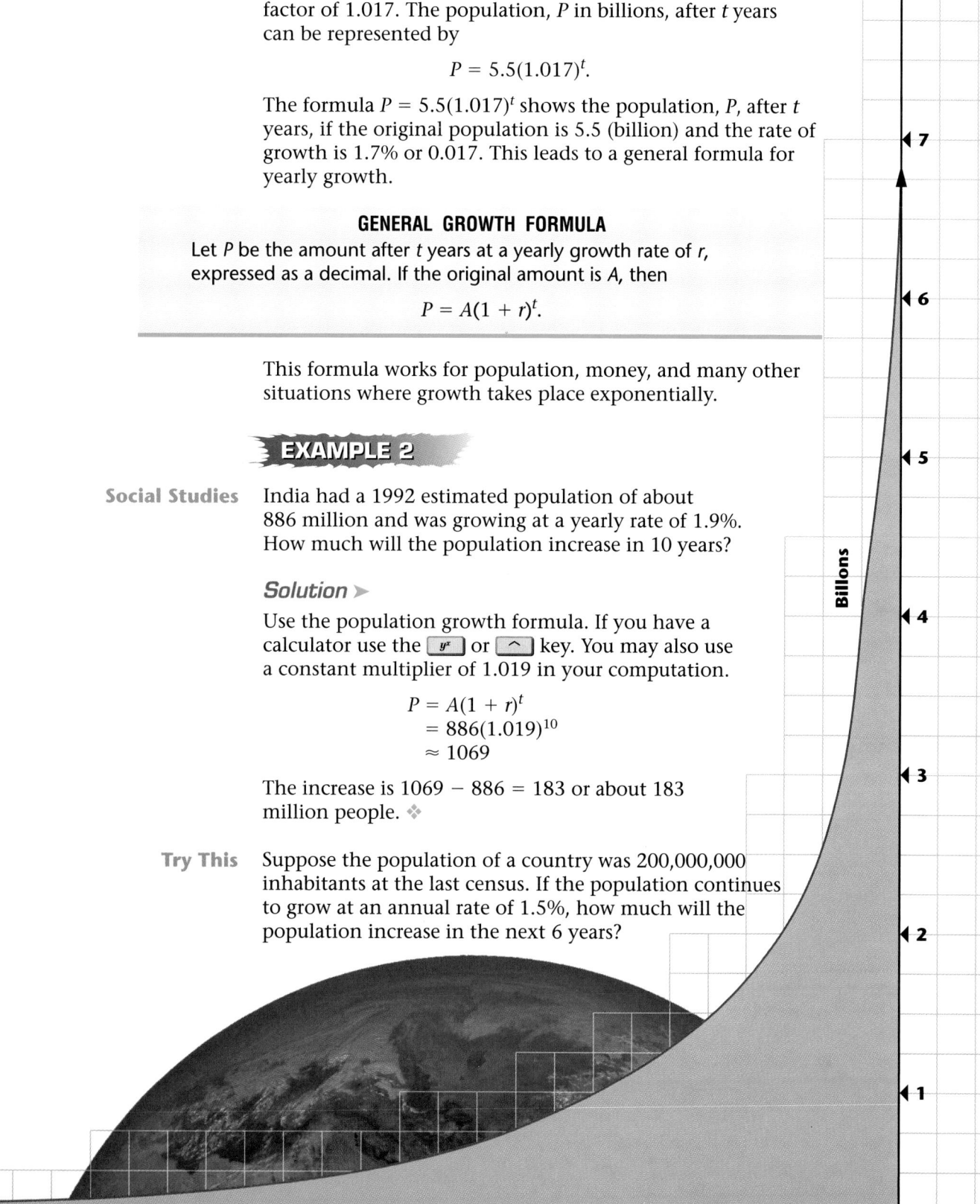

Each year the world population increases by another factor of 1.017. The population, P in billions, after t years can be represented by

$$P = 5.5(1.017)^t.$$

The formula $P = 5.5(1.017)^t$ shows the population, P, after t years, if the original population is 5.5 (billion) and the rate of growth is 1.7% or 0.017. This leads to a general formula for yearly growth.

GENERAL GROWTH FORMULA

Let P be the amount after t years at a yearly growth rate of r, expressed as a decimal. If the original amount is A, then

$$P = A(1 + r)^t.$$

This formula works for population, money, and many other situations where growth takes place exponentially.

EXAMPLE 2

Social Studies India had a 1992 estimated population of about 886 million and was growing at a yearly rate of 1.9%. How much will the population increase in 10 years?

Solution ➤

Use the population growth formula. If you have a calculator use the [y^x] or [$\wedge$] key. You may also use a constant multiplier of 1.019 in your computation.

$$\begin{aligned} P &= A(1 + r)^t \\ &= 886(1.019)^{10} \\ &\approx 1069 \end{aligned}$$

The increase is $1069 - 886 = 183$ or about 183 million people. ❖

Try This Suppose the population of a country was 200,000,000 inhabitants at the last census. If the population continues to grow at an annual rate of 1.5%, how much will the population increase in the next 6 years?

Billons

8

7

6

5

4

3

2

1

1500 CE   1750 CE   2000 CE

485

EXAMPLE 3

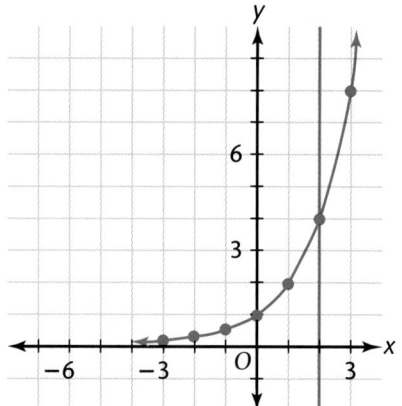

Make a table of values. Graph and describe the function $f(x) = 2^x$. Show why it is a function.

Solution ➤

x	-3	-2	-1	0	1	2	3
2^x	0.125	0.25	0.5	1	2	4	8

The function $f(x) = 2^x$ shows the characteristics of an exponential function. The graph remains above the x-axis for all values of x. When x is 0 the function equals 1. After crossing the y-axis the function increases rapidly.

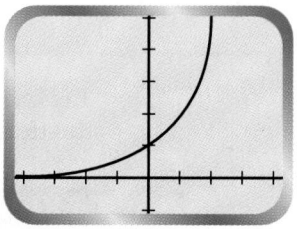

The vertical line test shows that $f(x) = 2^x$ is a function.

A calculator produces the same distinctive curve. ❖

EXAMPLE 4

Graph $y = 3^x$ and $y = 4^x$ on the same axes. Explain how they are similar and how are they different.

Graphics Calculator

Solution ➤

You can graph each function using a graphics calculator, or you can make a table of values and graph them by hand. Include positive and negative values of x.

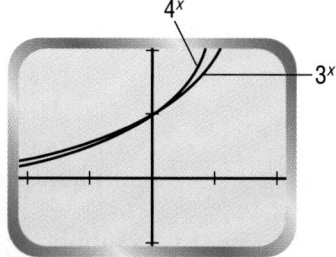

x	-3	-2	-1	0	1	2	3
3^x	0.04	0.11	0.33	1	3	9	27
4^x	0.0156	0.0625	0.25	1	4	16	64

The graphs of the functions are similar in that each of them rises from left to right. Each crosses the y-axis at $y = 1$. Each gets closer and closer to the *x-axis* as x decreases.

They are different in that the graph of $y = 4^x$ rises faster than the graph of $y = 3^x$ for values of x greater than 0. ❖

Continue graphing powers on the same graph using other bases, such as 5^x, 6^x, and so on. What do you notice about the pattern?

EXERCISES & PROBLEMS

Communicate

1. The population of Zimbabwe in 1995 was about 11 million and was growing at a rate of about 1.9% per year. Explain how to find the constant multiplier.

2. Explain how mathematics can give you a clearer understanding of the effects of population growth.

3. Identify and explain each of the variables in the growth formula, $P = A(1 + r)^t$.

4. Describe how the exponential function changes as the base increases.

5. Describe the relationship between the constant multiplier on a calculator and the exponent in mathematics.

6. Explain why population growth and invested money can be represented by the same formula.

Practice & Apply

Graph each of the following.

7. $y = 2^x$ **8.** $y = 4^x$ **9.** $y = 5^x$ **10.** $y = 10^x$ **11.** $y = 0.5^x$ **12.** $y = 0.1^x$

Generalize the results in Exercises 7–12.

13. For what values of b does the graph of $y = b^x$ rise from left to right?

14. For what values of b does the graph of $y = b^x$ fall from left to right?

Graph each of the following. In each case describe the effect of the 3 as a transformation of the parent function $y = 2^x$.

15. $y = 2^x + 3$ **16.** $y = 2^{x+3}$ **17.** $y = 3 \cdot 2^x$ **18.** $y = 2^{3x}$

19. Is the graph of $y = x^4$ an exponential function? Explain.

20. Is the graph of $y = 4^x$ an exponential function? Explain.

If $f(x) = 2^x$, find the following values for the function.

21. $f(3)$ **22.** $f(0)$ **23.** $f(-1)$ **24.** $f(-4)$

The population of Spain was about 39 million in 1992 and was growing at a rate of about 0.3% per year.

25. What multiplier is used to find the new population each year?

26. Use this information to estimate the population for 1993.

27. Estimate the population for the year 2000.

Demographics The population of Italy was about 58 million in 1995 and was growing at a rate of about 0.1% per year. The population of France was about 57.8 million in 1995 and was growing at a rate of about 0.4% per year. Use this information to estimate the population of

28. Italy in 1996.

29. Italy in 1998.

30. Italy in 2000.

31. France in 1996.

32. France in 1998.

33. France in 2000.

34. In what year does the population of France surpass that of Italy?

35. **Portfolio Activity** Complete the problem in the portfolio activity on page 455.

~~~ *Look Back*

Refer to matrix A. $A = \begin{bmatrix} 1 & -3 \\ 5 & 2 \end{bmatrix}$ **[Lesson 7.3]**

36. What is A^2?   **37.** What is A^3?

38. Explain the characteristic that a relation must have to be a function. **[Lesson 9.1]**

39. What must happen to the parent function for it to be reflected through the x-axis? **[Lesson 9.4]**

Evaluate each of the following. [Lesson 10.3]

40. 10^{-2} **41.** -10^2 **42.** $(-10)^2$ **43.** $(-10)^{-2}$

Look Beyond

Cultural Connection: Europe Maria Gaetana Agnesi was born in Italy in 1718. She is considered to be the first woman in modern times to achieve a reputation for mathematical works. She wrote her best known work in 1748 in which she first discussed a curve that is now known as the Agnesi curve.

This curve is represented by the function $f(x) = \dfrac{a^3}{x^2 + a^2}$. Let the domain of f be $\{-3, -2, -1, 0, 1, 2, 3\}$.

44. Graph the equation for $a = 1$.

45. Graph the equation for $a = 5$.

46. How do the graphs compare?

LESSON 10.6 Applications of Exponential Functions

Why *How old is a dinosaur bone? How much money should be invested to reach a savings target? These are a few of the many questions that can be answered with the use of exponential functions.*

Exponential functions can be used to determine the age of fossils and archeological finds. This method is based on the fact that the amount of radioactivity in an object decreases, or decays, over time.

Science Carbon-14 dating is a reliable method of determining the age of objects up to 40,000 years old. Carbon-14 is a radioactive form of carbon that has a half-life of 5700 years. This means that half of the substance is converted to nonradioactive carbon every 5700 years. For example, it can be concluded that an object with half as much carbon-14 as its living counterpart died 5700 years ago. Then, each 5700 years, half of the remaining amount of carbon-14 decays.

Years	Fraction of Carbon-14 Remaining
0	$1 = \left(\frac{1}{2}\right)^0$
5,700	$\frac{1}{2} = \left(\frac{1}{2}\right)^1$
11,400	$\frac{1}{4} = \left(\frac{1}{2}\right)^2$
17,100	$\frac{1}{8} = \left(\frac{1}{2}\right)^3$
22,800	$\frac{1}{16} = \left(\frac{1}{2}\right)^4$
28,500	$\frac{1}{32} = \left(\frac{1}{2}\right)^5$
34,200	$\frac{1}{64} = \left(\frac{1}{2}\right)^6$

	flute	gourd canteen	agave knife	shell ornament	charred wood	bone	teeth
Fraction of Carbon-14 Remains	1	$\frac{1}{2}$	$\frac{1}{4}$	$\frac{1}{8}$	$\frac{1}{16}$	$\frac{1}{32}$	$\frac{1}{64}$
Years ago	0	5,700	11,400	17,100	22,800	28,500	34,200

EXAMPLE 1

Use the table to estimate the age of an object which has 10% of its original carbon-14.

Solution ➤

Since 10% or $\frac{1}{10}$ is between $\frac{1}{8}$ and $\frac{1}{16}$, the age is between 17,100 and 22,800 years ago. A reasonable estimate is about 20,000 years. ❖

EXAMPLE 2

Finance   Suppose that you wish to save $1000 for a down payment on a car and that you want to buy the car when you graduate in 3 years. The current interest rate for savings is 5% compounded annually. How much will you need to invest in a savings account now to buy the car in 3 years?

Solution ➤

If A represents the original amount, then $P = A(1 + r)^t$ is the formula for compound interest. Let P represent the amount of money after t years. Let r represent the yearly rate of growth (expressed as a decimal) compounded once each year. Substitute the values in the formula, and solve for A.

$$P = A(1 + r)^t$$
$$1000 = A(1.05)^3$$
$$1000 = A(1.157625)$$
$$\frac{1000}{1.157625} = A$$

Since $863.84 \approx A$, you will need to invest $863.84. ❖

The example can also be solved by working backward. Think of the $1000 as being invested 3 years ago, and use -3 as the number of years. Substitute the values into the formula, then solve for A.

$$P = A(1 + r)^t$$
$$P = 1000(1.05)^{-3}$$
$$P \approx 863.84$$

EXAMPLE 3

Investment   If the sculpture is now worth $14,586.08, what was the value of the investment 4 years ago?

The value of a sculpture has been growing at a rate of 5% per year for 4 years.

Solution ➤

Use the formula. Since you are trying to find the value 4 years ago, substitute −4 for *t*.

$$P = A(1 + r)^t$$
$$= 14,586.08(1.05)^{-4}$$
$$\approx 14,586.08(0.8227)$$
$$\approx 12,000$$

The value was $12,000. ❖

Try This   An art museum recently sold a painting for $4,500,000. The directors claimed that its value had been growing at a rate of 7% for the past 50 years. What did the museum pay for the painting 50 years ago?

EXAMPLE 4

PROBABILITY
Connection

Ginny has a 90% probability of getting a given step correct in an algebra problem. She does 100 problems each with two steps.

A In how many problems can you expect her to get both steps correct?

B Calculate the probability if the problems have three steps.

C Calculate the probability if the problems have ten steps.

Solution ➤

A In 100 problems you can expect her to get the first step correct 90 times. Of these 90 times, she would get the second step correct 90% of the time. Since 90% of 90 is 81, she would get both steps correct 81 times out of 100. Notice that the answer is $100(0.90)^2$.

B From the solution for A, you know that in 100 problems she would have the first two steps correct 81 times. She would also get the third step correct 90% of these times or about 73 times. Notice that the answer is $100(0.90)^3$.

C By the same reasoning, the answer is $100(0.90)^{10}$. If you use the power key on a calculator, you get approximately 35. Thus, she gets only about 35% correct when doing 10-step problems, even though she has a 90% chance of getting any particular step correct. ❖

CRITICAL *Thinking*

How many steps would a problem have to have for Ginny to have less than a 20% probability of getting the problem correct?

In this lesson you have seen several examples of exponential growth and decay. The general formula for exponential growth gives the amount, P, when A is the initial amount, r is the percent of increase expressed in decimal form, and t is the time expressed in years, or

$$P = A(1 + r)^t.$$

For exponential growth, r is positive, but when decay or decrease takes place, the value of r is negative.

EXAMPLE 5

Demographics   The population of a city in the United States was 152,494 in 1990 and was *decreasing* at a rate of about 1% per year. When will the city's population fall below 140,000? Solve the problem

A graphically.   **B** using a table.

Solution ➤

Each year the population will be 99% of what it was the previous year, so the multiplier is $1.00 - 0.01$ or 0.99. The population x years after 1990 is given by $Y = 152,494\,(0.99)^x$.

Graphics Calculator

A Enter the formula $Y = 152,494\,(0.99)^x$ into a graphics calculator and graph it. Use X-values between 0 and 10 and Y-values between 130,000, and 160,000. Trace the function to find when it falls below 140,000.

X=8.5106383 .Y=139992.73.

It falls below 140,000 in about $8\frac{1}{2}$ years.

B The population decreases by a factor of 0.99 each year. Use the constant feature on a calculator to generate a table of values for the population. Continue the table until you reach the year that the value passes below 140,000.

x	$Y = 152{,}494(0.99)^x$ Population	x	$Y = 152{,}494(0.99)^x$ Population
0	152,494	5	145,020
1	150,969	6	143,570
2	149,459	7	142,134
3	147,965	8	140,713
4	146,485	9	139,306

According to the table, the population falls below 140,000 between 8 and 9 years. ❖

In Chapter 9 you learned how to use transformations to translate the graph of a parent function. Graphs that represent the exponential function $y = a^x$ can be translated in the same way.

EXAMPLE 6

Compare the graphs of the transformations with the graph of the parent function $y = 2^x$.

A $y = 2^x - 3$ **B** $y = 2^{x-3}$

Solution ➤

Graph each function.

Parent function
$y = 2^x$ **A** $y = 2^x - 3$ **B** $y = 2^{x-3}$

Note that all of the graphs have the same shape.

A The graph of $y = 2^x - 3$ is shifted *down* 3 units from the parent graph $y = 2^x$. Each y-value will be 3 units less than the corresponding point in the parent graph.

B The graph of $y = 2^{x-3}$ is shifted 3 units *to the right* of the parent graph $y = 2^x$. Subtracting 3 before you raise 2 to the power means that to get the same y-value as in the parent graph, you have to substitute an x-value that is 3 units greater. ❖

EXERCISES & PROBLEMS

Communicate

1. List several examples of situations that are described by exponential functions.

2. Describe the formula for exponential decay.

3. Explain why r is negative in the formula for exponential decay.

4. Tell how a horizontal translation is shown in the function rule for exponential functions.

5. Tell how a vertical translation is shown in the rule for exponential functions.

6. Choose a base and form an exponential sequence of 5 or 6 terms. Describe what you find when you examine the differences.

Practice & Apply

Carbon Dating Use the method of carbon-14 dating to estimate the age of an object which has

7. 25% of its original carbon-14 remaining.

8. 1% of its original carbon-14 remaining.

9. Technology The population of the metropolitan area of Corpus Christi, Texas, was 361,280 in 1993 and was *decreasing* at a rate of about 0.1% per year. Draw a graph of the population after *x* years using a "friendly" window.

Demography Trace the graph to estimate the population of Corpus Christi, Texas,

10. in the year 1995.   **11.** in the year 1998.   **12.** in the year 2000.

Investment An investment is growing at a rate of 8% per year and now has a value of $8200. Find the value of the investment

13. in 5 years.   **14.** in 10 years.   **15.** 5 years ago.   **16.** 10 years ago.

Investment An investment is losing money at a rate of 2% per year and now has a value of $94,000. Find the value of the investment

17. in 5 years.   **18.** in 10 years.   **19.** 5 years ago.   **20.** 10 years ago.

21. **Probability** If Ginny (Example 4) does 100 four-step problems, in how many problems will she get *all* of the steps correct?

22. **Probability** If *x* is chosen randomly from the list 1, 2, 3, 4, 5, find the probability that 2^x is greater than 10.

23. Repeat Exercise 22 if *x* is chosen from the list 1, 2, 3, . . . , 100.

The digits of a repeating decimal like $\frac{1}{3} = 0.33333$ . . . can be described in terms of an exponential function. Evaluate each of the following.

24. $3\left(\frac{1}{10}\right)^1$ **25.** $3\left(\frac{1}{10}\right)^2$ **26.** $3\left(\frac{1}{10}\right)^3$ **27.** $3\left(\frac{1}{10}\right)^4$

Write the digit in the sixth decimal place of

28. $\frac{1}{3}$.

29. $\frac{2}{3}$.

30. $\frac{4}{9}$.

Technology Graph each of the following, and describe the effect of the 5 as a transformation of the parent function $y = 10^x$.

31. $y = 10^x + 5$ **32.** $y = 10^{x+5}$ **33.** $y = 5 \cdot 10^x$ **34.** $y = 10^{5x}$

Look Back

35. Calculate. $5(3 - 6) \div 3 \times 2$ **[Lesson 1.4]**

36. Jamie bought tapes and cassettes and paid the cashier $42.75. If the cost of the items before tax was $39.33, what was the percent of the tax? **[Lesson 4.5]**

37. **Technology** What formula would you use in a calculator or spreadsheet to produce random 0s and 1s? **[Lesson 8.1]**

38. What formula would produce random numbers from 0 to 9 in a calculator or spreadsheet? **[Lesson 8.1]**

39. There are 44 students that signed up for band or art. Of those students, 21 chose band, and 29 chose art. How many students are taking both band and art? **[Lesson 8.4]**

40. Tamitra flips a coin, records the result, and then flips it again. She wants to know the probability of heads on both tosses. Are the events independent? How do you know? **[Lesson 8.7]**

41. Describe the differences between the graph of $y = 2^x$ and the graph of $y = 2^{-x}$. **[Lesson 10.3]**

Look Beyond

42. **Technology** A number that is often used to describe growth of natural phenomena and money is e, which is about 2.718. Find e^x on your calculator. Write the first 8 digits of the decimal expansion. Use the $\boxed{e^x}$ key and 1 for x. Then draw the graph of $y = e^x$.

43. Compare.

     a. The value of $1000 invested at 8% interest compounded *once a year* is given by $Y = 1000(1.08)^x$, where x is the number of years. Find the value after 10 years.

     b. The value of $1000 invested at 8% interest compounded *continuously* is given by $y = 1000e^{0.08x}$. Find the value after 10 years.

Please Don't Sneeze

Suppose that one person in your class gets the flu. While contagious, this person makes contact with two other people so that two more people are exposed to the virus. If any have not yet had the flu, assume that each of them will now get it and make contact with two more people. If these people already have had contact, they are immune and will not pass it on. Use a simulation to find out how many people in the class will get the flu before the epidemic is over.

To do this, assign each person in your class a number. Then use random numbers from a calculator or computer spreadsheet or use numbers drawn from a hat to see who is exposed to the flu virus according to the conditions given above.

The formula INT(RAND*30)+1 will generate a random number from 1 to 30, where RAND is a random number from 0 to 1 including 0, but not 1. (If your class has 28 students, replace the 30 in the formula by 28, etc.) Summarize the results in a tree diagram.

In the example, the following thirteen random numbers were initially generated. 10, 23, 12, 25, 29, 25, 4, 5, 29, 23, 9, 3, 5 . . .

The first random number was 10, so student 10 was the first to get the flu.

The next two random numbers were 23 and 12. This means that while contagious, student 10 had contact with students 23 and 12 who also got the flu.

The next two random numbers were 25 and 29, indicating that student 23 passed it on to students 25 and 29.

The next two random numbers were 25 and 4, indicating that student 12 had contact with 25 and 4. Since student 25 already had the virus, this branch ends.

The next random numbers generated were 5, 29, 23, 9, 3, and 5, so the tree diagram so far looks like this. Remember that a circle around a number indicates that the branch ends.

When the above simulation was continued, all students except 1, 16, 17, 18, 22, 27, 28, and 30 got the virus before all the branches ended. Thus 22 of the 30 students got the flu.

Conduct a simulation of an epidemic. Then pool your data for the class, and summarize the results. What is the average number of people who are infected before the simulated epidemic comes to an end?

Chapter 10 Review

Vocabulary

Key Skills & Exercises

Lesson 10.1

> **Key Skills**

Use the definition of exponent to simplify expressions.

$$3^5 = 3 \cdot 3 \cdot 3 \cdot 3 \cdot 3 = 243 \qquad 4^3 = 4 \cdot 4 \cdot 4 = 64$$

The exponent in a power of 10 tells you the number of zeros to use when writing the number in decimal notation.

$$10^4 = 10{,}000$$

> **Exercises**

Simplify each of the following.

1. 6^3 **2.** 5^5 **3.** 10^8 **4.** 10^6

Lesson 10.2

> **Key Skills**

Use Product-of-Powers Property or Quotient-of-Powers Property to simplify expressions.

Product of Powers
$$(6x^2y^3)(-3x^4y) = (6 \cdot -3)(x^{2+4})(y^{3+1})$$
$$= -18x^6y^4$$

Quotient of Powers
$$\frac{15x^4y^3z}{3xy^2} = (5)(x^{4-1})(y^{3-2})(z) = 5x^3yz$$

Use Power-of-a-Power Property or Power-of-a-Product Property to simplify expressions.

Power of a Power
$$(x^3)^5 = x^{3 \cdot 5} = x^{15}$$

Power of a Product
$$(4x^4y^3)^2 = (4^2)(x^{4 \cdot 2})(y^{3 \cdot 2}) = 16x^8y^6$$

> **Exercises**

Simplify each of the following.

5. $(2t^4)(3t^2)$ **6.** $(-a^2b^3)(4ab^2)$ **7.** $(-0.1w^2x^2y)(-0.5w^5)$

8. $\dfrac{10w^5}{2w^2}$ **9.** $\dfrac{-21wx^2y^4}{-3xy^3}$ **10.** $(r^3)^4$

11. $(-3c^2)^3$ **12.** $(2p^2q^3)^2$ **13.** $(m^4n^2)^2(m^3)^4$

Lesson 10.3

➤ *Key Skills*

Simplify expressions containing negative exponents.

$$5x^{-3}y^2 = \frac{5y^2}{x^3}$$

$$\frac{m^{-3}p^2}{n^{-2}p^2} = \frac{n^2}{m^3}$$

Simplify expressions containing zero exponents.

$$(3x)^0 = 1$$

$$7a^0b^2 = 7b^2$$

➤ *Exercises*

Write each of the following without negative or zero exponents.

14. 3^{-2}

15. a^2b^{-3}

16. $(-3)^0$

17. $a^0b^{-1}c^2$

18. $(2c^2d^{-1})^0$

19. $-q^{-3}q^2$

20. $\frac{b^{-2}}{b^3}$

21. $\frac{t^{-2}u}{t^{-4}u^2}$

Lesson 10.4

➤ *Key Skills*

Write numbers in scientific notation.

Customary Notation	Scientific Notation
32,000,000,000	3.2×10^{10}
0.000000784	7.84×10^{-7}

Perform computations with numbers written in scientific notation.

$$(3 \times 10^{-3})(4 \times 10^8) = (3 \cdot 4)(10^{-3} \times 10^8)$$
$$= 12 \times 10^5$$
$$= 1.2 \times 10^6$$

$$\frac{4 \times 10^6}{2 \times 10^3} = \left(\frac{4}{2}\right)(10^{6-3}) = 2 \times 10^3$$

➤ *Exercises*

Write each number in scientific notation.

22. 5,900,000

23. 368,000,000,000

24. 0.0000075

Write each number in decimal notation.

25. 2×10^3

26. 7.9×10^{-5}

27. 8.34×10^9

Perform the following computations. Answer using scientific notation.

28. $(3 \times 10^2)(5 \times 10^5)$

29. $(2.1 \times 10^5)(4 \times 10^{-3})$

30. $(8 \times 10^2) + (2 \times 10^2)$

31. $(9 \times 10^5) - (3 \times 10^5)$

32. $\frac{9 \times 10^7}{3 \times 10^2}$

33. $\frac{8 \times 10^4}{2 \times 10^{-2}}$

Lesson 10.5

➤ Key Skills

Use exponential functions to estimate population growth.

The population of a country was estimated to be about 750 million in 1990 and was growing at a yearly rate of about 1.7% At this rate of growth, the population of this country in 10 years will be $P = 750(1 + 0.017)^{10} \approx 888$, or about 888 million people.

➤ Exercises

The population of New Zealand was about 3.3 million in 1991 and was growing at a rate of about 0.8% per year. Estimate the population

34. in the year 1998.   **35.** in the year 2000.

Lesson 10.6

➤ Key Skills

Use exponential functions to estimate growth or decay.

An investment is losing money at a rate of 3% per year and now has a value of $6500. The value of the investment in 5 years will be $y = 6500(1 - 0.03)^5 \approx \$5581.77$. The value of the investment 5 years ago was $y = 6500(1 - 0.03)^{-5} \approx \$7569.28$.

➤ Exercises

An investment is growing at a rate of 5.5% per year and now has a value of $6300. Find the value of the investment

36. in 5 years.   **37.** in 10 years.

38. 5 years ago.   **39.** 10 years ago.

Applications

40. Geometry Suppose the radius of a circle has length r. Then the area of the circle is πr^2. Find the area of a circle if the radius is 5 times as long as the radius of the first circle.

41. Astronomy The sun has a diameter of about 864,000 miles. Write this number in scientific notation.

42. The interior temperature of the sun is about 3.5×10^7 °F. Write this number in decimal notation.

43. Time The calendar year is increasing at a rate of about 1.7×10^{-6} minutes each year. How many seconds is this? Write your answer in scientific notation.

Probability Suppose x is chosen randomly from the list $-1, -2, -3, -4, -5$. Find the probability that

44. 2^x is less than 0.5.   **45.** 3^x is greater than $\frac{1}{27}$.

46. 10^x is greater than 0.001.   **47.** $\frac{1}{2^x}$ is less than 16.

Chapter 10 Assessment

1. What is a monomial?
2. The product of two powers can be found by adding the exponents. What must be true about these powers?

Simplify each of the following.

3. $(2y)^3$
4. $(2f^3)(3f^4)$
5. $(-x^3y)(5x^2y^2z)$
6. $\dfrac{36t^5}{6t^2}$
7. $\dfrac{-2w^3z^4}{-3wz^5}$
8. $\dfrac{ab^4}{-2b^3}$
9. $(g^3)^4$
10. $(4c^2)^5$
11. $6(2cd^3)^2$

12. What is the difference between an exponent of 0 and an exponent of 1?

Write each of the following without negative or zero exponents.

13. 2^{-4}
14. $a^{-3}b^2$
15. $m^{-4}n^0p^4$
16. $-2t^{-2}$
17. y^5y^{-3}
18. $\dfrac{3h^{-2}}{h^{-1}}$

19. What is the meaning of a negative exponent in a number written in scientific notation?

Taxes In 1970, the IRS collected about 196 million dollars in taxes. In 1992, about 1.1 billion dollars was collected.

20. Write 196 million in scientific notation.
21. Write 1.1 billion in scientific notation.
22. How much more did the IRS collect in 1992 than in 1970? Answer using scientific notation and decimal notation.

Perform the following computations. Answer using scientific notation.

23. $(3 \times 10^{-2})(5 \times 10^5)$
24. $\dfrac{3 \times 10^6}{5 \times 10^4}$

Demographics The population of Peru in 1992 was about 23 million and growing at a rate of 2% per year.

25. What is the multiplier to get the new population each year?
26. Estimate the population for the year 1997.
27. Estimate the population for the year 2000.

Investment An investment is losing money at a rate of 0.5% per year and had a value of $8500 five years ago. Find the value of the investment

28. 10 years ago.
29. now.
30. in 5 years.
31. in 10 years.

Chapters 1 – 10
Cumulative Assessment

College Entrance Exam Practice

Quantitative Comparison   For Questions 1–4, write

A if the quantity in Column A is greater than the quantity in Column B;
B if the quantity in Column B is greater than the quantity in Column A;
C if the two quantities are equal; or
D if the relationship cannot be determined from the information given.

	Column A	Column B	Answers
1.	INT(4.5)	ABS(-4.5)	(A) (B) (C) (D) **[Lesson 2.4]**
2.	5×10^{-3}	0.005	(A) (B) (C) (D) **[Lesson 10.4]**
3.	slope of $y = -2$	slope of $y = x - 2$	(A) (B) (C) (D) **[Lesson 5.1]**
4.	7^0	$56[9 + (-9)]$	(A) (B) (C) (D) **[Lesson 10.3]**

5. What is the sum of $-14 + (-6)$?   **[Lesson 3.1]**
   a. -20 **b.** -8 **c.** 20   **d.** 8

6. What is 65% of 20?   **[Lesson 4.5]**
   a. 1300   **b.** 0.0325   **c.** 13   **d.** 3.25

7. Which command generates a random number from the list 3, 4, 5?
   [Lesson 8.2]
   a. INT(RAND() * 5) + 5   **b.** INT(RAND() * 3) + 5
   c. INT(RAND() * 5) + 3   **d.** INT(RAND() * 3) + 3

8. There are 3 blue chips, 7 red chips, and 2 yellow chips in a
   bag. If you select a chip, replace it, and then select another
   chip, what is the probability of selecting a red chip and a
   yellow chip?   **[Lesson 8.6]**
   a. $\frac{3}{4}$ **b.** $\frac{7}{72}$ **c.** $\frac{7}{66}$ **d.** $\frac{1}{14}$

9. What is the parent function for the graph?   **[Lesson 2.5]**
   a. $y = x^2$ **b.** $y = x$
   c. $y = |x|$ **d.** $y = $ INT(x)

10. Cathi has $17.40 in dimes and quarters. If she has a total of 81 coins, how many of each type of coin does she have?   **[Lesson 6.3]**

11. Solve $\dfrac{x}{-10} = \dfrac{-3}{5}$.   **[Lesson 4.3]**

12. The first three terms of a sequence are 2, 4, and 9. The second differences are a constant 3. What are the next three terms?   **[Lesson 1.2]**

13. In 1992, a total of 10,924,000 individual tax returns were filed electronically. Write this number in scientific notation.   **[Lesson 10.4]**

14. Find the equation of the line passing through the origin and the point $(3, -4)$.   **[Lesson 5.3]**

15. Find the inverse for $\begin{bmatrix} -2 & 3 \\ 4 & -1 \end{bmatrix}$ if it exists.   **[Lesson 7.4]**

16. Identify the parent function for $y = \dfrac{-2}{x}$. Tell what type of transformation is applied and then draw a carefully labeled graph.   **[Lesson 9.3]**

17. There are 48 students in the choir and 50 students in the orchestra. If 15 students are in both the choir and the orchestra, what is the total number of students in the choir and the orchestra?   **[Lesson 8.4]**

Free-Response Grid   The following questions may be answered using a free-response grid commonly used by standardized test services.

18. Simplify $3(4 + 5) \cdot (4 - 3)2 \div 5 \cdot 3$.   **[Lesson 1.4]**

19. The perimeter of a rectangle is 30 centimeters and its length is twice its width. Find the length of this rectangle.   **[Lesson 4.6]**

20. Suppose you roll a number cube. What is the probability that you will roll a number greater than 4?   **[Lesson 8.6]**

21. What is the slope of a line parallel to the line passing through $(-7, -5)$ and $(1, -3)$?   **[Lesson 5.6]**

CHAPTER 11

Polynomials and Factoring

$$(a+b)^3$$

$$a^3 + 3a^2b + 3ab^2 + b^3$$

Punnett Squares are used in genetics to determine the possible characteristic of offspring. Assume that each parent has both the recessive and the dominant form of the gene for height. Show how to determine the probability that the offspring from a cross between these two parents will be tall if tall is dominant and short is recessive. How is this related to polynomials?

505

LESSON 11.1 Adding and Subtracting Polynomials

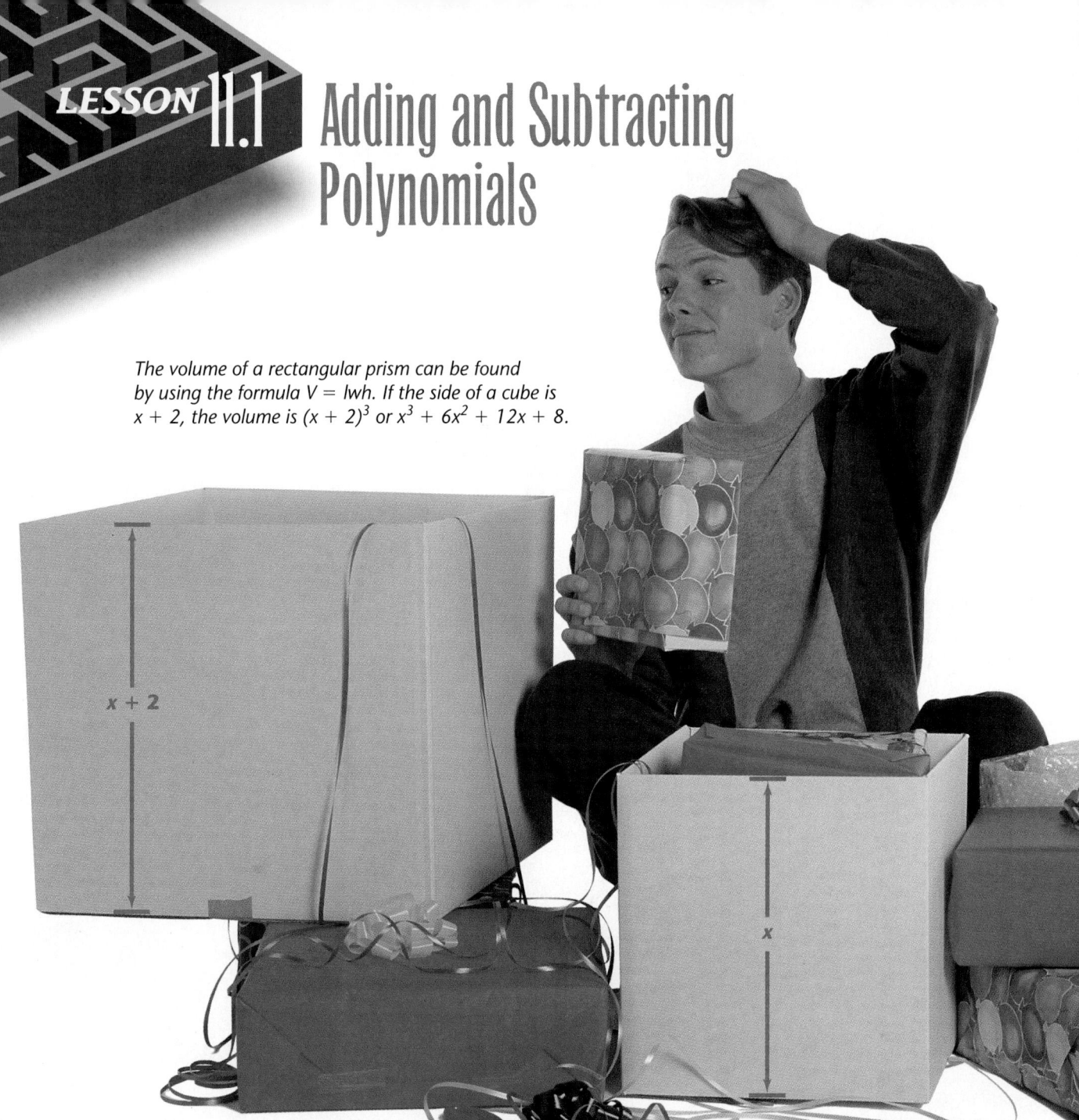

The volume of a rectangular prism can be found by using the formula $V = lwh$. If the side of a cube is $x + 2$, the volume is $(x + 2)^3$ or $x^3 + 6x^2 + 12x + 8$.

$x + 2$

x

Why The arc of a football, the distance travelled by an accelerating car, and a formula for volume are examples that can be modeled by a polynomial.

A **polynomial** is a monomial or the sum or difference of two or more monomials. An example of a polynomial in one variable is $x^3 + 6x^2 + 12x + 8$.

The **degree** of a polynomial in **one variable** is the *exponent with the greatest value of any of the polynomial's terms.* The degree of $9 - 4x^2$ is 2. What is the degree of $2a^2 + 5a^4 - 3a^3 + 1$?

The terms of a polynomial may appear in any order. However, in **standard form**, the terms of a polynomial are ordered from left to right, from the greatest to the least degree of the variable. This is called **descending order**.

The polynomial $3x^2 - 4x + 6$ is in standard form. The degrees of the x-terms are in order from greatest to least. The polynomial $9 - 4x^2$ written in standard form is $-4x^2 + 9$. Sometimes it is convenient to write $-4x^2 + 9$ as $(-1)(4x^2 - 9)$.

Write $5x^2 + x^3$ in standard form.

Some polynomial expressions have special names that are dependent on either their *degree* or their *number of terms*, as illustrated in the table.

Polynomial	# of terms	Name by # of terms	Degree	Name by degree
12	1	monomial	0	constant
$8x$	1	monomial	1	linear
$9 - 4x^2$	2	binomial	2	quadratic
$5x^2 + x^3$	2	binomial	3	cubic
$3x^2 - 4x + 6$	3	trinomial	2	quadratic
$3x^4 - 4x^3 + 6x^2 - 7$	4	polynomial	4	quartic

Second degree polynomials can be modeled by algebra tiles.

1. represents $3x^2 - 4x + 6$.

2. represents $5x^2 + x - 3$.

3. represents $9 - 4x^2$.

Adding Polynomials

Algebra tiles can be used to model polynomial addition. To add $x^2 + x + 1$ and $2x^2 + 3x + 2$, model each polynomial and group like tiles.

Polynomials can also be added in vertical or horizontal form. In vertical form align the like terms and add.

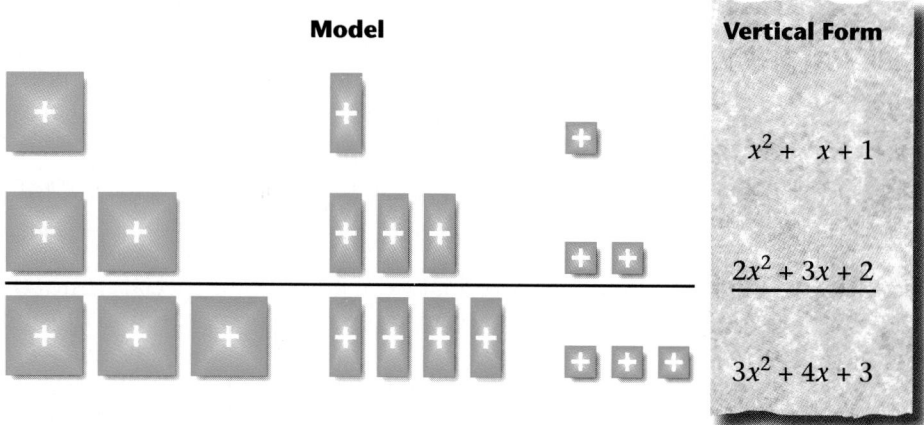

Model

Vertical Form

$x^2 + x + 1$

$2x^2 + 3x + 2$

$3x^2 + 4x + 3$

In horizontal form, write $(x^2 + x + 1) + (2x^2 + 3x + 2)$ and add like terms.

EXAMPLE 1

Use algebra tiles to model $(2x^2 - 3x + 5) + (4x^2 + 7x - 2)$. Find the sum.

Solution A ➤

Model each polynomial, and group like tiles. Notice that when an equal number of negative and positive tiles of the same type are put together, the combination represents zero.

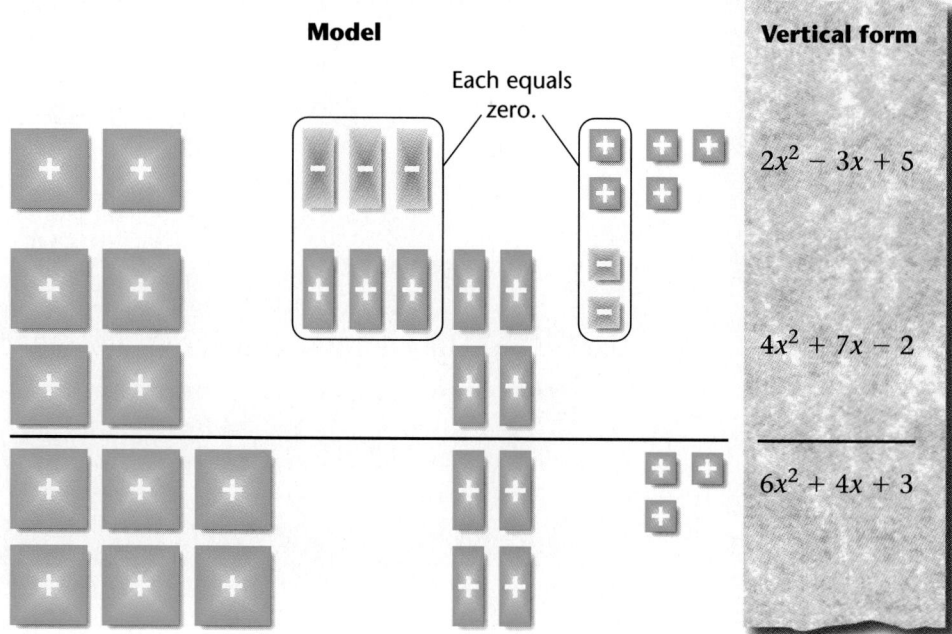

Model

Each equals zero.

Vertical form

$2x^2 - 3x + 5$

$4x^2 + 7x - 2$

$6x^2 + 4x + 3$

Solution B ➤

In horizontal form, add like terms. $(2x^2 - 3x + 5) + (4x^2 + 7x - 2)$
$= (2x^2 + 4x^2) + (-3x + 7x) + (5 - 2)$
$= 6x^2 + 4x + 3$ ❖

Try This   Use algebra tiles to model $(3x^2 + 5x) + (4 - 6x - 2x^2)$. Find the sum using the horizontal form.

EXAMPLE 2

Find the perimeter of each polygon. Check the addition by substituting a value for each variable.

GEOMETRY
Connection

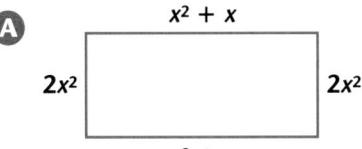

Ⓐ $x^2 + x$

$2x^2$ $2x^2$

$x^2 + x$

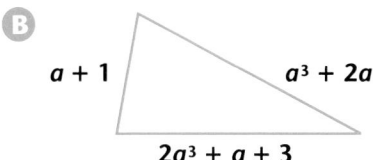

Ⓑ $a + 1$ $a^3 + 2a$

$2a^3 + a + 3$

Solution ➤

Use vertical form to add like terms. Substitute 10 for each variable to check.

Ⓐ **Vertical form**
$x^2 + x$
$2x^2$
$2x^2$
$x^2 + x$
_____
$6x^2 + 2x$

Check: $x = 10$
$10^2 + \quad\;\; 10 = 110$
$2 \cdot 10^2 \qquad\quad = 200$
$2 \cdot 10^2 \qquad\quad = 200$
$10^2 + \quad\;\; 10 = 110$
_____
$6 \cdot 10^2 + 2 \cdot 10 = 620$

Ⓑ **Vertical form**
$a^3 + \;\; 2a$
$\qquad\quad a + 1$
$2a^3 + \quad a + 3$
_____
$3a^3 + \;\; 4a + 4$

Check: $x = 10$
$10^3 + 2 \cdot 10 \qquad = 1020$
$\qquad\qquad 10 + 1 = \quad 11$
$2 \cdot 10^3 + \qquad 10 + 3 = 2013$
_____
$3 \cdot 10^3 + 4 \cdot 10 + 4 = 3044$ ❖

The perimeter can also be found using horizontal form.

Ⓐ $(x^2 + x) + 2x^2 + 2x^2 + (x^2 + x) = 6x^2 + 2x$

Ⓑ $(a^3 + 2a) + (a + 1) + (2a^3 + a + 3) = 3a^3 + 4a + 4$ ❖

CRITICAL
Thinking

Explain how using the horizontal form allows you to add polynomials mentally.

Subtracting Polynomials

Recall that the definition of subtraction is to add the opposite.

EXAMPLE 3

Subtract $x^2 - 4$ from $3x^2 - 2x + 8$.

Solution ➤

The binomial $x^2 - 4$ can be written as $x^2 + 0x - 4$, where $0x$ is a "placeholder" for the missing x term. Remember to change signs when subtracting.

Vertical form

$$3x^2 - 2x + 8$$
$$- \ (x^2 + 0x - 4)$$
$$\overline{2x^2 - 2x + 12}$$

Horizontal form

$$3x^2 - 2x + 8 - (x^2 - 4) = 3x^2 - 2x + 8 - x^2 + 4$$
$$= 2x^2 - 2x + 12$$

Study the two examples. How is the addition of whole numbers related to the addition of polynomials?

1. $485 = 4 \cdot 10^2 + 8 \cdot 10 + 5$
$\underline{113} = \underline{1 \cdot 10^2 + 1 \cdot 10 + 3}$
$598 = 5 \cdot 10^2 + 9 \cdot 10 + 8$

2. $4x^2 + 8x + 5$
$\underline{x^2 + x + 3}$
$5x^2 + 9x + 8$

EXERCISES & PROBLEMS

Communicate

1. Explain how to use algebra tiles to represent $5x^2 - 2x + 3$.

2. How do you determine the degree of a polynomial?

3. Give two different names for the polynomial $3x^3 + 4x^2 - 7$.

4. Draw an algebra-tile model of $(2x^2 + 3) + (x^2 - 1)$. Find the sum.

5. Explain how to find $(3b^3 - 2b + 1) - (b^3 + b - 3)$.

6. Explain how to write a polynomial in standard form. Use $5x^2 - 2x + 3x^4 - 6$ as a model.

Practice & Apply

7. Use algebra tiles to represent $x^2 + 4x + 4$.

8. Draw an algebra-tile model of $(3y^2 + 6) + (2y^2 - 1)$.

Rewrite each polynomial in standard form.

9. $6 + c + c^3$

10. $5x^3 - 1 + 5x^4 + 5x^2$

11. $10 + p^7$

Write the degree of each polynomial.

12. $4r + 1$

13. $x^3 + x^4 + x - 1$

14. $y + y^3$

Identify each polynomial by name from the number of terms and degree.

15. $3x + 1$

16. $8x^2 - 1$

17. $8x^2 - 2x + 3$

Give an example of a polynomial that is a

18. quadratic trinomial.

19. linear binomial.

20. cubic monomial.

Use vertical form to add.

21. $3x^2 + 4x^4 - x + 1$ and $3x^4 + x^2 - 6$

22. $2y^3 + y^2 + 1$ and $3y^3 - y^2 + 2$

23. $4r^4 + r^3 - 6$ and $r^3 + r^2$

24. $2c - 3$ and $c^2 + c + 4$

Use horizontal form to add.

25. $y^3 - 4$ and $y^2 - 2$

26. $x^3 + 2x - 1$ and $3x^2 + 4$

27. $3s^2 + 7s - 6$ and $s^3 + s^2 - s - 1$

28. $w^3 + w - 2$ and $4w^3 - 7w + 2$

Use vertical form to subtract.

29. $x^2 + x$ from $x^3 + x^2 + 7$

30. $3y^2 - 4$ from $4y^2 - y + 6$

31. $4c^3 - c^2 - 1$ from $5c^3 + 10c + 5$

32. $8x^3$ from $x^3 - x + 4$

Use horizontal form to subtract.

33. $y^2 + 3y + 2 - (3y - 2)$

34. $3x^2 - 2x + 10 - (2x^2 + 4x - 6)$

35. $3x^2 - 5x + 3 - (2x^2 - x - 4)$

36. $2x^2 + 5x - (x^2 - 3)$

Simplify. Express all answers in standard form.

37. $(1 - 4x - x^4) + (x - 3x^2 + 9)$

38. $5 - 3x - 1.4x^2 - (13.7x - 62 + 5.6x^2)$

39. Subtract $x^5 - 2x^2 + 3x + 5$ from $4x^5 - 3x^3 - 3x - 5$.

$$34{,}276 = 3(10^4) + 4(10^3) + 2(10^2) + 7(10^1) + 6$$

40. 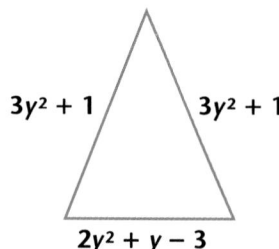 **Geometry** Find the perimeter of the triangle.

$3y^2 + 1$ $3y^2 + 1$

$2y^2 + y - 3$

 Geometry Exercises 41–43 refer to the dimensions of the figure at the right.

$5x + 4$

$3x$

$2x + 5$

$3x + 2$

41. Find the polynomial expression that represents the perimeter of the large rectangle.

42. Find the polynomial expression that represents the perimeter of the smaller rectangle.

43. Find the difference between the perimeters of the two rectangles.

Look Back

44. State the pattern, and find the next three terms of the sequence.
1, 3, 9, 27, 81, ___, ___, ___ **[Lesson 1.1]**

45. Divide $\dfrac{6x^2 - 12x + 18}{3}$. **[Lesson 4.2]**

46. Solve $-2(a + 3) = 5 - 6(2a - 7)$. **[Lesson 4.6]**

47. Solve. $\begin{cases} 6x = 4 - 2y \\ 12x - 4y = 16 \end{cases}$ **[Lesson 6.3]**

48. Without repeating any digits, how many different 3-digit codes can be made with the digits 1–9? **[Lesson 8.5]**

Multiply.

49. $6x(x^2 - 4x)$ **50.** $8m^3(m^2 + 6m)$ **[Lesson 10.2]**

51. Simplify $(3x^5)(-2x^3)(-x)$.

Write each number in scientific form.

52. 7,100,000   **53.** 8,900,000,000 **[Lesson 10.4]**

Look Beyond

54. Use algebra tiles to model a square whose area can be represented as $x^2 + 6x + 9$. What is the length of a side of the square?

Exploring Products and Factors

Factor
tiles

$x \cdot x = x^2$

Product
tile

*Multiplication of polynomials can be
represented by a geometric model.*

WHY Algebra tiles can be used to model polynomials,
their sums, and differences. These tiles can also be used to
explore products and factors.

To model $x \cdot x$, the factor tiles are placed
above and next to the crossed lines. A
product tile is placed within the crossed
lines. The rules for the signs of the factor
tiles are the same as those developed for
integers. If the signs of the factor tiles are
the same, the product tile is positive. If the
signs are opposite, the product tile is
negative.

$(1)(1) = (1)$ $(-1)(-1) = (1)$ $(1)(-1) = (-1)$ $(-1)(1) = (-1)$

513

•Exploration 1 Modeling With Tiles

1 What product is shown in the diagram that models $(x)(-x)$?

2 Use tiles to model $(x)(x)$, $(-x)(x)$, and $(-x)(-x)$. What is the product for each?

3 The diagram models $(-x)(1)$ and $(1)(-x)$. What product is shown for each?

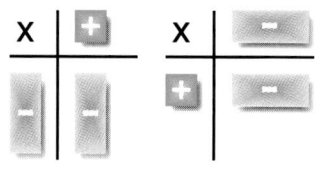

4 Use tiles to model $(x)(-1)$, $(x)(1)$, $(-x)(-1)$, $(1)(x)$, $(1)(-x)$, and $(-1)(-x)$. What is the product for each? ❖

Either tiles or the Distributive Property can be used to find the product of 3 and $2x + 1$.

Algebra-Tile Model	Distributive Property

$2x + 1$

X

$3\{$

Count the tiles for the product.
$6x + 3$

$$3(2x + 1) = 3(2x) + 3(1)$$
$$= 6x + 3$$

The model shows that the product of 3 and $2x + 1$ is $6x + 3$. The tiles modeling 3 and $2x + 1$ represent two **factors** of $6x + 3$. Notice that applying the Distributive Property results in the same algebraic product.

APPLICATION

Suppose each of three students has 2 boxes of pencils and 1 loose pencil. If b represents the number of pencils in a box, $2b + 1$ represents the number of pencils each student has. Since there are 3 students, $3(2b + 1)$ represents the total number of pencils. This product can be rewritten $6b + 3$, that is, $3(2b + 1) = 6b + 3$. ❖

•Exploration 2 Finding Products

$x + 1$

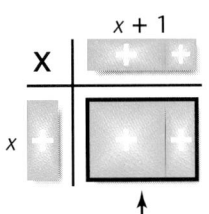

↑
Product rectangle

Examine the product rectangle.

1 What are the two factors?

2 What is the product?

3 Use the Distributive Property to find the product $x(x + 1) = \underline{?}$.

4 Use tiles to model each multiplication problem.
   a. $2x(x + 3)$ **b.** $3(3x + 2)$

5 Use the Distributive Property to find each product.
   a. $2x(x + 3)$ **b.** $3(3x + 2)$

6 Explain how to use the Distributive Property to find the product $2x(x + 1)$. ❖

Rectangles that model products can be built using positive and negative tiles. When you build *product rectangles* be careful to obey the laws of signs for multiplying integers.

•Exploration 3 Using Negative Tiles

$x - 1$

$2x$

Examine the product rectangle.

1 What are the two factors?

2 What is the product?

3 Use the Distributive Property to find the product $2x(x - 1) = \underline{?}$.

4 Use tiles to model each multiplication.
   a. $4(-x + 2)$ **b.** $3(2x - 1)$

5 Use the Distributive Property to find each product.
   a. $4(-x + 2)$ **b.** $3(2x - 1)$

6 Explain how to use the Distributive Property to find $-3(-x + 4)$. ❖

Once a product rectangle is formed, you can tell what two factors have been multiplied to form the product.

 Exploration 4 *From Product Rectangles to Factors*

Begin with the tiles for $4x^2$ and $6x$.

1️⃣ Form a product rectangle with $2x$ as one factor.

2️⃣ What is the second factor?

3️⃣ Start with the tiles for $3x^2$ and $6x$. Form a product rectangle with $3x$ as one factor.

4️⃣ What is the second factor?

5️⃣ Explain how to use the Distributive Property to complete $3x^2 + 6x = 3x(\underline{\ ?\ } + \underline{\ ?\ })$.

6️⃣ Explain how to use a product rectangle and the Distributive Property to complete $5x^2 + 15x = 5x(\underline{\ ?\ } + \underline{\ ?\ })$. ❖

The **greatest common factor**, or **GCF**, of $4x^2$ and $6x$ is $2x$. For the polynomial $4x^2 + 6x$, the factored form is $2x(2x + 3)$. Here are some examples.

Polynomial	GCF	Factored Form
$5x^2 + 10x$	$5x$	$5x(x + 2)$
$3x^2 - 9x$	$3x$	$3x(x - 3)$
$8x^3 - 4x$	$4x$	$4x(2x^2 - 1)$
$-2x - 6x^2$	$-2x$	$-2x(1 + 3x)$
$6x^4 - 9x^2$	$3x^2$	$3x^2(2x^2 - 3)$

 CRITICAL Thinking

Given the tiles that model a second-degree polynomial, can you always form a rectangle from the tiles? Explain.

EXERCISES PROBLEMS

Communicate

1. Explain how to use algebra tiles to show the area of a rectangle with sides $2x$ and $x + 5$.

2. Explain how to use the Distributive Property to find the area of a rectangle with sides $2x$ and $x + 5$.

3. Draw a tile model to find the sides of a rectangle that has a product of $4x^2 + 2x$.

4. Tell what it means to write the polynomial $x^3 + x$ in factored form.

Practice & Apply

5. Draw a tile model to show the area of a rectangle that has factors of $x + 7$ and x.

6. Draw a tile model to find the sides of a rectangle that has a product of $2x^2 + 2x$.

Geometry Find the factors for the rectangle modeled with positive tiles and write the product shown by the tiles.

7.   **8.**   **9.**

10.   **11.**   **12.**

Use the Distributive Property to find each product.

13. $4(x + 2)$ **14.** $6(2x + 7)$ **15.** $5(y + 10)$ **16.** $3(m + 8)$

17. $x(x + 2)$ **18.** $2y(y - 4)$ **19.** $3r(r^2 - 3)$ **20.** $p^2(p + 7)$

21. $8(3x - 4)$ **22.** $4y(y - 4)$ **23.** $5x(4x + 9)$ **24.** $3w(w^2 - w)$

25. $2y^2(y^2 + y)$ **26.** $2t^2(2t^2 + 8t)$ **27.** $3y^2(3y - 6)$ **28.** $z(11z^2 + 22z)$

Write each polynomial in factored form.

29. $3x^2 + 6$ **30.** $5x^2 - 20$ **31.** $y^2 - y^3$ **32.** $4p + 12p^3$

33. $7y^4 + 49y$ **34.** $m^4 - 6m^2$ **35.** $-4r^2 - 6r$ **36.** $9n^2 - 27n^4$

Geometry Mary wants a rectangular flower bed and lawn in her backyard.

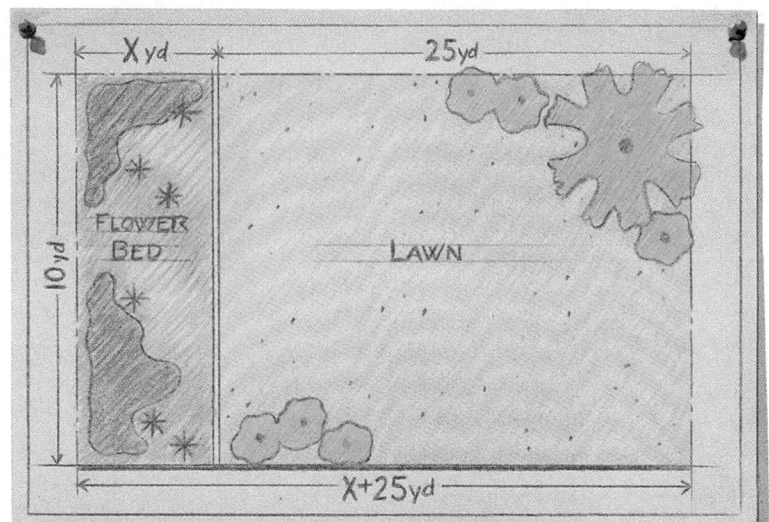

A landscape artist is designing Mary's backyard.

37. Write an expression for the area of the flower bed.

38. Write an expression for the area of the lawn.

39. Write an expression for the area of her backyard.

40. What conjecture can you make about the areas of the flower bed and the lawn compared with the area of her backyard?

41. Discount Sales Wholesale Grocery has pieces of dinnerware for sale in the houseware department. Glasses are $1.50, plates are $2.50, cups are $1.25, and saucers are $1.00. A 25% discount is given for orders over $10.00. If a customer buys x glasses and y plates and receives a discount, write an expression for how much discount the customer receives.

42. **Geometry** Find the perimeter of a rectangle with a width of $2x + 3$ and a length of $3x$.

Look Back

43. **Coordinate Geometry** Find the equation of the line between $A(2, -7)$ and $B(2, 11)$. **[Lesson 5.2]**

44. Advertising In how many different orders can four different models of bicycles be displayed in a store's window? **[Lesson 8.5]**

Let $f(x) = 2^x$.   **45.** Find $f(7)$.   **46.** Find $f(4)$. **[Lesson 9.1]**

Use the Distributive Property to find each polynomial. **[Lesson 10.1]**

47. $2a(a - 6b)$ **48.** $x(z - x^2)$ **49.** $9y^2(y - 3)$

Write each number in scientific notation. **[Lesson 10.4]**

50. 1,900,000   **51.** 0.0000000001

Look Beyond

52. Explain how factoring $x^2 - x - 12$ will help simplify $(x^2 - x - 12) \div (x + 3)$.

LESSON 11.3 Multiplying Binomials

$(x+4)(x+5)=$

why

The Distributive Property can be used to multiply a monomial and a binomial. It can also be used to multiply two binomials.

•Exploration 1 *Multiplying With Tiles*

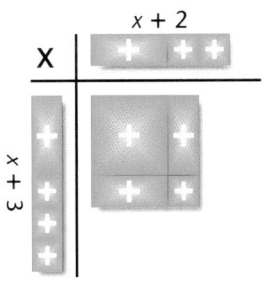

1. Complete the algebra-tile model.

2. What two factors are multiplied?

3. What is the product represented by the completed model?

4. If a rectangle has sides of $x + 3$ and $x + 2$, what is the area?

5. Explain how to use an algebra-tile model to show that $(x + 4)(x + 5) = x^2 + 9x + 20$. ❖

The Distributive Property can be used to show that $(x + 3)(x + 2) = x^2 + 5x + 6$.

$(x + 3)(x + 2)$	Given
$= x(x + 3) + 2(x + 3)$	Distribute the first binomial to *each term of the second binomial*.
$= x^2 + 3x + 2x + 6$	Distribute *each term* of the second binomial *to each term* of the first binomial.
$= x^2 + \quad 5x \quad + 6$	Add like terms.

Use the Distributive Property to show that $(x + 4)(x + 5) = x^2 + 9x + 20$.

Exploration 2 Multiplying Binomials

$$(x + 3)(x + 5) = x^2 + 8x + 15$$

1 What is the relationship of the first term of each binomial factor to the first term of the trinomial?

$$(x + 3)(x + 5) = x^2 + 5x + 3x + 15$$

2 What is the relationship of the outside terms and the inside terms of the binomial factors to the middle term of the trinomial?

$$(x + 3)(x + 5) = x^2 + \quad 8x \quad + 15$$

$$(x + 3)(x + 5) = x^2 + 8x + 15$$

3 What is the relationship of the last term of each binomial factor to the last term of the trinomial?

4 Explain how to find $(x + 3)(x + 5)$ using what you learned in the first 3 steps.

5 Use this shortcut to write each product. Check the result by using a tile model or the Distributive Property.

    a. $(x + 3)(x - 5)$ **b.** $(x - 3)(x + 5)$ **c.** $(x - 3)(x - 5)$

6 Show how you would use this shortcut to complete $(x + a)(x + b) = \underline{\ ?\ }$ ❖

FOIL Method

In Exploration 2 you used a method for multiplying two binomials that is usually referred to as **FOIL**.

- Multiply **F**irst terms.

- Multiply **O**utside terms. Multiply **I**nside terms. Add outside and inside products.

- Multiply **L**ast terms.

1. First terms product **3. Last terms product**

2. Add Outside terms product to Inside terms product

EXAMPLE 1

Use the FOIL method to find each product.

A $(3x + 1)(3x - 2)$ **B** $(2a - 1)(5a + 3)$ **C** $(5m - 2)(5m + 2)$

Solution ➤

Multiply to find each term, then simplify.

A $(3x + 1)(3x - 2) = 9x^2 - 6x + 3x - 2 = 9x^2 - 3x - 2$

B $(2a - 1)(5a + 3) = 10a^2 + 6a - 5a - 3 = 10a^2 + a - 3$

C $(5m - 2)(5m + 2) = 25m^2 + 10m - 10m - 4 = 25m^2 - 4$ ❖

CRITICAL
Thinking

Draw an algebra-tile diagram to show that $(2x + 3)(x + 1) = 2x^2 + 5x + 3$. Explain how you can check your diagram by the Distributive Property and the FOIL method.

EXAMPLE 2

Small Business Cruz's Frame Shop makes a mat by cutting out the inside of a rectangular mat board.

Find the length and width of the original mat if the area of the mat board is 148 square inches.

$2x - 1$

$2x - 5$

$x + 2$

$x + 6$

Solution ➤

Find the area of each rectangle.

Area of the outside rectangle	**Area of the inside rectangle**
$(2x - 1)(x + 6)$	$(x + 2)(2x - 5)$
$2x^2 + 12x - x - 6$	$2x^2 - 5x + 4x - 10$
$2x^2 + 11x - 6$	$2x^2 - x - 10$

Substitute each area into the formula, and then solve for x.

Area outside rectangle − Area inside rectangle = Area of the mat

$2x^2 + 11x - 6$	$-$	$(2x^2 - x - 10)$	$= 148$
$2x^2 + 11x - 6$	$-$	$2x^2 + x + 10$	$= 148$
		$12x + 4$	$= 148$
		$12x$	$= 144$
		x	$= 12$

The length of the original rectangle is $2x - 1$. Substitute.

$$2x - 1 = 2(\mathbf{12}) - 1 = 23$$

The width is $x + 6$. Substitute. $x + 6 = \mathbf{12} + 6 = 18$

The original mat was 23 inches by 18 inches. ❖

PROBABILITY
Connection

The Punnett Square shows the possible results of crossing two flowers, each containing a purple gene and a white gene. Since the purple gene dominates, only flowers with a ww genetic makeup are white.

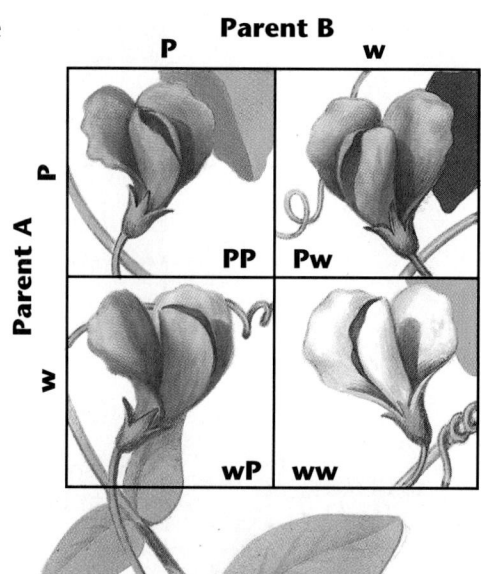

Parent B

P   w

Parent A

P

PP   Pw

w

wP   ww

EXAMPLE 3

Show how squaring a binomial models the probability that crossing two hybrid purple (Pw) flowers results in a flower that has a 25% probability of being white.

Solution ➤

Square the binomial $P + w$.

$(P + w)^2 = P^2 + 2Pw + w^2 = PP + Pw + Pw + ww$

Since there are four possibilities and one is w^2 or ww, the probability that the offspring will be white is $\frac{1}{4}$ or 25%. ❖

EXERCISES & PROBLEMS

Communicate

1. Show how to model each product.
   a. $(x + 1)(x + 2)$
   b. $(x - 1)(x - 2)$

2. Explain how to use the Distributive Property to find each product.
   a. $(x + 1)(x + 2)$
   b. $(x - 1)(x - 2)$

3. Describe how to find $(2x + 3)(x - 4)$ using the FOIL method.

4. Use algebra tiles to model the areas of $(x + 2)^2$ and $x^2 + 2^2$. How are the two models different?

Practice & Apply

Use the Distributive Property to find each product.

5. $(x + 2)(x + 5)$
6. $(a - 3)(a + 4)$
7. $(b - 1)(b - 3)$
8. $(y + 3)(y - 2)$
9. $(c + 5)(c + 5)$
10. $(d + 5)(d - 5)$

Use the FOIL method to find each product.

11. $(y + 5)(y + 3)$ **12.** $(w + 9)(w + 1)$ **13.** $(b - 7)(b + 3)$

14. $(3y - 2)(y - 1)$ **15.** $(5p + 3)(p + 1)$ **16.** $(2q - 1)(2q + 1)$

17. $(2x + 5)(2x - 3)$ **18.** $(4m + 1)(5m - 3)$ **19.** $(2w - 9)(3w - 8)$

20. $(3x - 5)(3x - 5)$ **21.** $(7s + 2)(2s - 3)$ **22.** $\left(y - \frac{1}{2}\right)\left(y + \frac{1}{2}\right)$

23. $\left(y - \frac{1}{3}\right)\left(y + \frac{5}{9}\right)$ **24.** $(c^2 + 1)(c^2 + 2)$ **25.** $(2a^2 + 3)(2a^2 + 3)$

26. $(a + c)(a + 2c)$ **27.** $(p + q)(p + q)$ **28.** $(a^2 + b)(a^2 - b)$

29. $(x^2 + y)(x + y)$ **30.** $(c + d)(2c + d)$ **31.** $(1.2m + 5)(0.8m - 4)$

 Geometry A rectangular garden has a length of $x + 8$ units and a width of $x - 4$ units.

32. Draw a diagram, and label the dimensions.

33. Find the area.

34. **Geometry** Find the area of a square rug that is $y + 6$ units on a side.

35. **Geometry** Which has the greater area, a square with sides $(x + 1)$ units long or a rectangle with length $(x + 2)$ units and width x units? How much greater is it?

36. **Geometry** Which has a greater area, a square with sides $(x + 1)$ units long or a rectangle with length x units and width $(x - 2)$ units? How much greater is it?

Look Back

Evaluate.   [Lesson 2.4]

37. INT(6.5) **38.** INT$\left(2\frac{3}{4}\right)$ **39.** INT(7.91)

Find the slope of the following lines.   [Lesson 5.3]

40. $4x + 3y = 12$ **41.** $y = 4x$

42. Write the equation of the line through $(-2, -1)$ and parallel to $y = -2x - 4$.   **[Lesson 5.6]**

Graph each system of equations. Check by the substitution method.   [Lesson 6.2]

43. $\begin{cases} 2x + y = 9 \\ x - 2y = 2 \end{cases}$ **44.** $\begin{cases} x + y = 4 \\ x + y = 10 \end{cases}$

Look Beyond

Where does the graph of each function cross the x-axis?

45. $f(x) = (x + 3)(x - 2)$ **46.** $g(x) = (x + 2)^2$ **47.** $h(x) = x^2 - 4$

Common Factors

why

The product of two factors is found by multiplication. Sometimes the product and one of the two factors are known and it is necessary to find the second factor.

Two unit squares can be used to form a rectangle of height 1 and length 2. A rectangle of height 2 and length 1 can also be formed.

However, the two rectangles will be considered the same, because the second rectangle is a rotation of the first.

Four unit squares form two different rectangles, a 2-by-2 and a 1-by-4. A 4-by-1 rectangle is the same as a 1-by-4 rectangle.

Exploration · Prime Numbers

1 Use unit squares to form rectangles. Record the number of different rectangles you can make.

	Number of Unit Squares	Number of Rectangles		Number of Unit Squares	Number of Rectangles
a.	2	1	g.	8	?
b.	3	1	h.	9	?
c.	4	2	i.	10	?
d.	5	?	j.	11	?
e.	6	?	k.	12	?
f.	7	?	l.	13	?

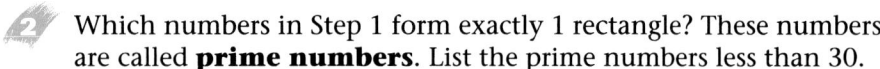 Which numbers in Step 1 form exactly 1 rectangle? These numbers are called **prime numbers**. List the prime numbers less than 30.

 The numbers in the list that are not prime are called **composite numbers**. List the composite numbers less than 30.

4 The number 1 is not considered to be prime or composite. Why do you think this is so?

5 Give a definition of a prime number using the word factor. ❖

When two numbers are multiplied they form a product. Each number is called a **factor** of that product. Thus, 3 and 8 are factors of 24 because $3 \cdot 8 = 24$, and one way to express 24 in factored form is $3 \cdot 8$. Sometimes individual factors can be factored further. The process ends when all the factors are prime. Name the prime factors of 24.

When you are asked to factor $3x^2 + 12x$, the first step is to examine the terms for a common monomial factor. Since $3x$ is common to both terms, write

$$3x^2 + 12x = 3x(x + 4).$$
$$\uparrow$$
$$\textbf{common monomial factor}$$

The polynomial $3x^2 + 12x$ has been factored over the integers because each term in $3x(x + 4)$ has an integer coefficient. When a polynomial has no polynomial factors with integral coefficients except itself and 1, it is a **prime polynomial** with respect to the integers. In this text, to factor means to factor over the integers.

EXAMPLE 1

Factor each polynomial.

A $5am - 5an$ B $5x^3 - 3y^2$ C $2c^4 - 4c^3 + 6c^2$

Solution ➤

Factor out the greatest common factor, or GCF, from each term.

A The GCF is $5a$. $5am - 5an = 5a(m - n)$

B The GCF is 1. $5x^3 - 3y^2$ is prime.

C The GCF is $2c^2$. $2c^4 - 4c^3 + 6c^2 = 2c^2(c^2 - 2c + 3)$

To check each factorization, multiply the two factors to find the original expression. ❖

Substitute numbers for the variables in the previous example and simplify. Compare the results. What do you find?

CRITICAL
Thinking

Even though the expression $2x + 3y$ can be written as the product of a monomial and a binomial, $6\left(\frac{1}{3}x + \frac{1}{2}y\right)$, the binomial $2x + 3y$ is considered prime. Explain why.

Sometimes a polynomial has a common factor that contains two terms. This expression is called a **common binomial factor**.

EXAMPLE 2

Factor $r(t + 1) + s(t + 1)$.

Solution ➤

Notice that $r(t + 1) + s(t + 1)$ contains the common binomial factor $t + 1$. Think of $t + 1$ as c.

$$r(t + 1) + s(t + 1) = rc + sc$$
$$= (r + s)c$$
$$= (r + s)(t + 1)$$

The expression can also be factored mentally if you apply the Distributive Property.

$$r(t + 1) + s(t + 1) = (r + s)(t + 1) \;❖$$

GEOMETRY
Connection

You can visualize this factoring procedure with an area model. It shows the same total area in different arrangements.

$$r(t + 1) \quad + \quad s(t + 1) \quad = \quad (r + s)(t + 1)$$

Try This Factor $a(x - 3) + 6(x - 3)$.

When a polynomial has terms with more than one common factor, the expression can sometimes be factored by grouping terms.

EXAMPLE 3

Factor $x^2 + x + 2x + 2$ by grouping.

Solution ➤

There are several ways to factor this expression by grouping. One way is to group the terms that have a common coefficient or variable. Treat $x^2 + x$ as one expression, and treat $2x + 2$ as another expression.

$x^2 + x + 2x + 2 = (x^2 + x) + (2x + 2)$	Group terms.
$= x(x + 1) + 2(x + 1)$	Factor the GCF from each group.
$= (x + 2)(x + 1)$	Write the expression as the product of two binomials.

To check, multiply the factors. ❖

Try This Factor $ax + bx + ay + by$ by grouping.

EXERCISES & PROBLEMS

Communicate

1. Define a *prime* polynomial.

2. What property of mathematics is being used when you factor by removing a GCF?

3. Once you have found the GCF, how do you find the remaining factor of the polynomial?

Each of the given pairs contains a GCF. Find the GCF, and explain how you find it.

4. 60 and 150

5. x^3y^5 and x^5y^2

6. $25(x + y)$ and $39(x + y)$

7. Explain how to group the terms to factor $y^2 + 2y + 3y + 6$.

Practice & Apply

Identify each polynomial as prime or composite.

8. $4x^2 - 16$

9. $r^2 + 10$

10. $n^2 + 4$

Factor each polynomial by removing the GCF.

11. $2x^2 - 4$

12. $5n^2 - 10$

13. $3x^2 + 6x$

14. $x^9 - x^2$

15. $k^5 + k^2$

16. $4a^8 - 20a^6 + 8a^4$

17. $4x^2 + 2x - 6$

18. $7x^2 - 28x - 14$

19. $27y^3 + 18y^2 - 81y$

20. $3m^3 - 9m^2 + 3m$

21. $90 + 15a^5 - 45a$

22. $2x^3y - 18x^2y^2 + 17xy^3$

Write each as the product of two binomials.

23. $x(x + 1) + 2(x + 1)$

24. $5(y + 3) - x(y + 3)$

25. $a(x + y) + b(x + y)$

26. $(4 + p)3q - 4(4 + p)$

27. $x(x - 1) + 2(x - 1)$

28. $r(x - 4) + t(x - 4)$

29. $5a(a - 3) + 4(a - 3)$

30. $2w(w + 4) - 3(w + 4)$

31. $2(x - 2) + x(2 - x)$

32. $8(y - 1) - x(y - 1)$

33. $2r(r - s)^2 - 3(r - s)^2$

34. $ax(u - v)^n + bz(u - v)^n$

Factor.

35. $2x + 2y + ax + ay$

36. $nu + nv + 3u + 3v$

37. $12ab - 15a - 8b + 10$

38. $ax + ay + 12x + 12y$

39. $3(x + y) + 12(x + y)$

40. $x^2 + 3x + 4x + 12$

41. $2n^2 - 6n + 14n - 42$

42. $6pq + 12p^2 - 8qp + 2p^2$

43. $5x(2d + 3)^3 - 10(2d + 3)^3$

An annulus is an object shaped like a washer.

 Geometry The shaded area between the two concentric circles is called an **annulus**. The formula for the area of the *annulus* is $\pi R^2 - \pi r^2$ where the radius of the larger circle is R and the radius of the smaller is r.

44. Write the formula for the area of an annulus in factored form.

45. Use the factored form to find the area of an annulus formed by concentric circles of radii 8 and 5.

46. **Portfolio Activity**
Complete the problem in the portfolio activity on page 505.

Look Back

47. Is the given sequence 3, 6, 12, 24, 48 representative of a linear relationship? How do you know?
[Lesson 2.1]

48. Evaluate $8 - 5(2 - 6) - 3^2$. **[Lesson 3.1]**

49. Solve the formula $A = \frac{1}{2} h(b_1 + b_2)$ for h. **[Lesson 4.6]**

50. Solve $3(x - 5) - 2(x + 8) = 7x + 3$. **[Lesson 4.6]**

51. Graph $x - 4y = 8$. **[Lesson 5.2]**

52. Solve. $\begin{cases} 4x = 11 + 15y \\ 6x + 5y = 0 \end{cases}$ **[Lesson 6.3]**

53. **Probability** Find the probability of drawing a heart or a queen from a deck of cards. **[Lesson 8.5]**

Given $y = 3x - 5$ as the original function, write the equation that would represent each of the following transformations.

54. A reflection across the x-axis **[Lesson 9.4]**

55. A translation of 4 units to the right and 2 units down **[Lesson 9.5]**

56. Evaluate $\dfrac{2^2 \cdot 2^4}{2^5} - 2^0$. **[Lesson 10.3]**

Look Beyond

57. **Physics** When a projectile is fired vertically into the air, its motion can be modeled by the equation $h = -16t^2 + 320t$, where h is the height in feet at time t in seconds. From the graph of the equation, find the time when $h = 1200$.

Factoring Special Polynomials

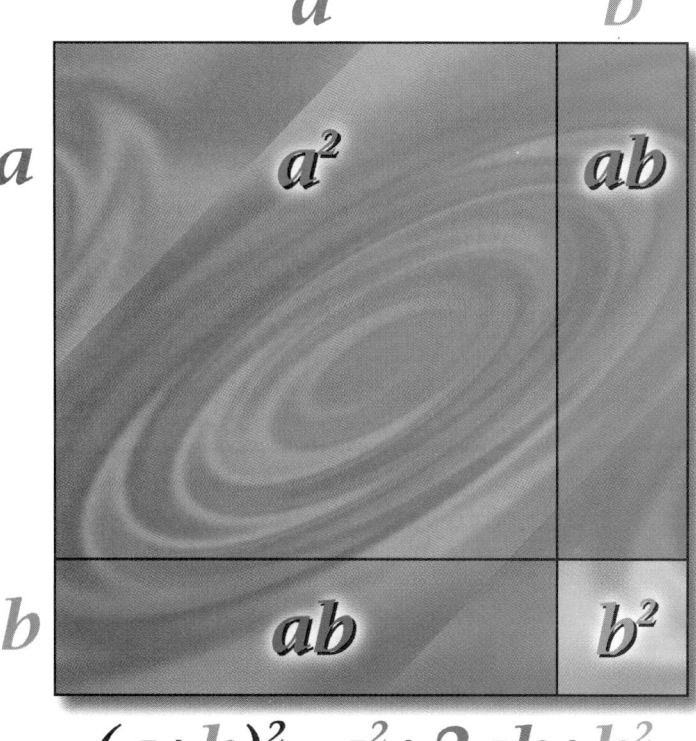

a

b

a a^2 ab

b ab b^2

why *If the length of one side of a square is a, the expression for its area is a^2. If the length of the side is increased, then the new expression for the area exhibits a simple pattern.*

$$(a+b)^2 = a^2 + 2ab + b^2$$

Perfect-Square Trinomial

GEOMETRY *Connection*

You can see from the geometry of the figure that the area of the large square is $(a + b)^2$, which is $a^2 + 2ab + b^2$. This shows why the product $a^2 + 2ab + b^2$ is called a **perfect-square trinomial**.

The same pattern is used to square the expression $a - b$. Square the first term, subtract twice the product of the binomial terms, and add the square of the last term.

$$(a - b)^2 = a^2 - 2ab + b^2$$

Explain how to find the perfect-square trinomials for $(2x + 6)^2$ and $(2x - 6)^2$. How do their products compare?

To see if a trinomial such as $4x^2 + 12x + 9$ is a perfect square, use this pattern.

- The first term $4x^2$ is a perfect square $(2x)^2$.
- The last term 9 is a perfect square $(3)^2$.
- The middle term is $12x$ or $2(2x)(3)$.

Thus, $4x^2 + 12x + 9$ is a perfect square trinomial.

EXAMPLE 1

Examine each expression. Is the expression a perfect-square trinomial? If not, why not?

Ⓐ $x^2 + 8x + 16$ **Ⓑ** $3x^2 + 16x + 16$ **Ⓒ** $4x^2 - 3xy - y^2$

Ⓓ $4a^2 + 12ab + 9b^2$ **Ⓔ** $9y^2 + 6xy + x^2$ **Ⓕ** $16x^2 + 12xy + y^2$

Solution ➤

Ⓐ Yes.   **Ⓑ** No, $3x^2$ is not a perfect square.

Ⓒ No, $4x^2$ and y^2 are perfect squares, but the sign in front of y^2 is negative, not positive.

Ⓓ Yes.   **Ⓔ** Yes.

Ⓕ No, $16x^2$ and y^2 are perfect squares, but the middle term is not $2(4x)(y)$. ❖

The perfect-square trinomial pattern can be used to factor expressions in the form of $a^2 + 2ab + b^2$ or $a^2 - 2ab + b^2$.

For example, $9m^2 + 12mn + 4n^2$ fits the pattern.
The first term is a perfect square, **$(3m)^2$**.
The last term is a perfect square, **$(2n)^2$**.
The middle term is twice the product of ($3m$ and $2n$), **$(12mn)$**.
Thus, **$9m^2 + 12mn + 4n^2 = (3m + 2n)^2$**.

FACTORING A PERFECT-SQUARE TRINOMIAL
For all numbers a and b,

1. $a^2 + 2ab + b^2 = (a + b)(a + b) = (a + b)^2$.
2. $a^2 - 2ab + b^2 = (a - b)(a - b) = (a - b)^2$.

EXAMPLE 2

Factor each expression.

Ⓐ $x^2 - 10x + 25$ **Ⓑ** $9s^2 + 24s + 16$

Ⓒ $64a^2 - 16ab + b^2$ **Ⓓ** $49y^4 + 14y^2 + 1$

Solution ➤

Ⓐ $x^2 - 10x + 25 = (x - 5)^2$

Ⓑ $9s^2 + 24s + 16 = (3s + 4)^2$

Ⓒ $64a^2 - 16ab + b^2 = (8a - b)^2$

Ⓓ $49y^4 + 14y^2 + 1 = (7y^2 + 1)^2$

To check, multiply the factors. ❖

CRITICAL
Thinking

The area of a square is $x^2 + 6x + 9$ square units. If the length of one side is 5 units, find the value of x by factoring.

Difference of Two Squares

When you multiply $(a + b)(a - b)$, the product is $a^2 - b^2$. This product is called the **difference of two squares**.

$$(\boldsymbol{a} + \boldsymbol{b})(\boldsymbol{a} - \boldsymbol{b}) = a(a - b) + b(a - b)$$
$$= a^2 - ab + ab - b^2$$
$$= \boldsymbol{a}^2 \qquad - \qquad \boldsymbol{b}^2$$

EXAMPLE 3

Examine each expression. Is the expression a difference of two squares? If not, why not?

A $4a^2 - 25$ **B** $9x^2 - 15$ **C** $b^2 + 49$

D $c^2 - 4d^2$ **E** $a^3 - 9$

Solution ➤

A Yes.

B No, 15 is not a perfect square.

C No, $b^2 + 49$ is not a difference; it is a sum.

D Yes.

E No, a^3 is not a perfect square. ❖

The difference-of-two-squares pattern can be used to factor expressions in the form $a^2 - b^2$.

For example, $4c^2 - 81d^2$ fits the pattern.
The first term is a perfect square, $(\boldsymbol{2c})^2$.
The second term is a perfect square, $(\boldsymbol{9d})^2$.
The terms are subtracted.

Thus, $\boldsymbol{4c^2 - 81d^2} = (2c + 9d)(2c - 9d)$.

FACTORING A DIFFERENCE OF TWO SQUARES
For all numbers a and b, $a^2 - b^2 = (a + b)(a - b)$.

EXAMPLE 4

Factor each expression.

A $x^2 - 4$ **B** $36a^2 - 49b^2$ **C** $16x^2 - 25$ **D** $m^4 - n^4$

Solution ➤

A $x^2 - 4 = (x + 2)(x - 2)$ **B** $36a^2 - 49b^2 = (6a + 7b)(6a - 7b)$

C $16x^2 - 25 = (4x + 5)(4x - 5)$ **D** $m^4 - n^4 = (m^2 + n^2)(m^2 - n^2)$
$$= (m^2 + n^2)(m + n)(m - n)$$ ❖

Try This Factor $25w^2 - 81$.

CRITICAL Thinking Explain why $a^8 - b^8$ has more than two factors. What are they?

EXAMPLE 5

Find each product by using the difference of two squares.

A $31 \cdot 29$　　**B** $17 \cdot 13$　　**C** $34 \cdot 26$

Solution ➤

A Think of $31 \cdot 29$ as $(30 + 1)(30 - 1)$.
The product is $30^2 - 1^2 = 900 - 1 = 899$.

B Think of $17 \cdot 13$ as $(15 + 2)(15 - 2)$.
The product is $15^2 - 2^2 = 225 - 4 = 221$.

C Think of $34 \cdot 26$ as $(30 + 4)(30 - 4)$.
The product is $30^2 - 4^2 = 900 - 16 = 884$. ❖

Graphics Calculator

Graph $y = x^2 - 10x + 25$. Explain how the graph of a perfect-square trinomial can give you the factors of the expression. Use your graphics calculator to factor $x^2 - 16x + 64$.

EXERCISES & PROBLEMS

Communicate

1. Describe the process for determining the factors for the perfect-square trinomial $x^2 + 20x + 100$.

2. Explain how to factor $4x^2 - 12x + 9$.

3. Explain how to factor the difference of two squares, $p^2 - 121$.

Practice & Apply

Use the generalization of a perfect-square trinomial or the difference of two squares to find each product.

4. $(p + 3)^2$ 　　　　**5.** $(2x - 1)^2$ 　　　　**6.** $(a - 4)(a + 4)$

7. $(7y - 9)(7y + 9)$ 　　**8.** $(8x - 3y)(8x + 3y)$ 　　**9.** $(5z - 12)^2$

Find the missing terms in each perfect-square trinomial.

10. $x^2 - 14x + \underline{\ ?\ }$ 　　**11.** $16y^2 + \underline{\ ?\ } + 9$ 　　**12.** $25a^2 + 60a + \underline{\ ?\ }$

13. $9x^2 + \underline{\ ?\ } + 25$ 　　**14.** $x^2 - 12x + \underline{\ ?\ }$ 　　**15.** $\underline{\ ?\ } - 36y + 81$

Factor each polynomial completely.

16. $x^2 - 4$ **17.** $x^2 + 4x + 4$ **18.** $y^2 - 100$

19. $y^2 + 8y + 16$ **20.** $16c^2 - 25$ **21.** $4t^2 - 1$

22. $81 - 4m^2$ **23.** $25x^2 - 9$ **24.** $r^2 - 18r + 81$

25. $4x^2 - 20x + 25$ **26.** $100 - 36q^2$ **27.** $36d^2 + 12d + 1$

28. $p^2 - q^2$ **29.** $9c^2 - 4d^2$ **30.** $16x^2 + 72xy + 81y^2$

31. $9a^2 - 12a + 4$ **32.** $49x^2 - 42xy + 9y^2$ **33.** $a^2x^2 + 2axb + b^2$

34. $4m^2 + 4mn + n^2$ **35.** $81a^4 - 9b^2$ **36.** $x^4 - y^4$

37. $x^2(25 - x^2) - 4(25 - x^2)$ **38.** $(x - 1)x^2 - 2x(x - 1) + x - 1$

39. $(3x + 5)(x^2 - 3) - (3x + 5)$ **40.** $(x^2 - y^2)(x^2 + 2xy) + (x^2 - y^2)(y^2)$

41. Cultural Connection: Africa Abu Kamil, known as "The Egyptian Calculator," used geometric models in 900 C.E. to solve problems. You can make his models with algebra tiles or paper rectangles that you can cut.

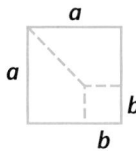

Use a sheet of paper to create a model of $a^2 - b^2$ as shown. Remove the b^2 from the corner. Next cut along the dotted line, and move the pieces to form a rectangle. Explain how this shows the factorization of $a^2 - b^2$.

Geometry Exercises 42–47 refer to the square figure shown.

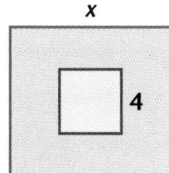

42. What is the area of the large square?

43. What is the area of the smaller square?

44. If the smaller region is removed, what is the area of the blue region?

45. Factor the polynomial that represents the area of the blue region.

46. Draw a rectangle whose dimensions are the factors you just found.

47. Show that the area of this rectangle equals the original blue area.

Geometry The area of a square is represented by $n^2 - 12n + 36$.

48. Find the length of each side.

49. Find the perimeter of the square.

50. **Geometry** Within a large square whose side is a units is a smaller square whose side is 3 units. What is the area of the large square? What is the area of the smaller square? What is the area of the yellow surface between the two squares? Factor the expression for the yellow area.

51. **Geometry** A smaller circle whose radius is *r* units is drawn within a large circle whose radius is *R* units. What is the area of the large circle? What is the area of the shaded part? Factor the expression for the shaded area. HINT: First, factor out the common monomial factor, π. Then factor the binomial.

Look Back

52. There is exactly one line that connects any two points. Use the table and the number pattern to find how many lines can be drawn that connect two points at a time for 10 points. **[Lesson 1.1]**

points	2	3	4	5	6	. . .	10
lines	1	3	6	?	?	. . .	?

53. **Statistics** Mark has scores of 78, 83, and 92 on his first 3 tests. What must his score be on the next test to have an average of 87? **[Lesson 4.1]**

54. Graph the solution of $|x| < 4$ on a number line. **[Lesson 4.8]**

A line is represented by the equation $3y = 5 - 4x$.

55. Find the slope of a line parallel to the given line. **[Lesson 5.6]**

56. Find the slope of a line perpendicular to the given line. **[Lesson 5.6]**

57. Graph the solution common to the following system. Label the solution clearly. $y < x - 5$ and $y \geq 3$ **[Lesson 6.5]**

58. Find the inverse matrix of $\begin{bmatrix} 1 & 2 \\ -2 & 1 \end{bmatrix}$. **[Lesson 7.4]**

59. **Probability** Two coins are tossed. Find the probability of getting at least one head. **[Lesson 8.1]**

Determine whether the ordered pairs represent a function. [Lesson 9.1]

60. {(1, 2), (2, 2), (3, 2), (4, 2)}   **61.** {(2, 7), (3, 5), (−2, 4), (3, −2)}

62. What is the graph of the equation $y = (x - 5)^2$ called? **[Lesson 9.5]**

63. What is the name for the point where the graph of a quadratic function changes direction? **[Lesson 9.4]**

Look Beyond

64. Given the trinomial $x^2 + x - 42$, what are possible factor pairs for 42? Which factors result in the sum of 1? What does the sign before the last term determine?

LESSON 11.6 Factoring Trinomials

The area of a rectangular garden can be found by using the formula A = lw.

Why

Factoring can be thought of as undoing multiplication. If you examine certain products carefully, you will discover patterns that enable you to recognize the factors.

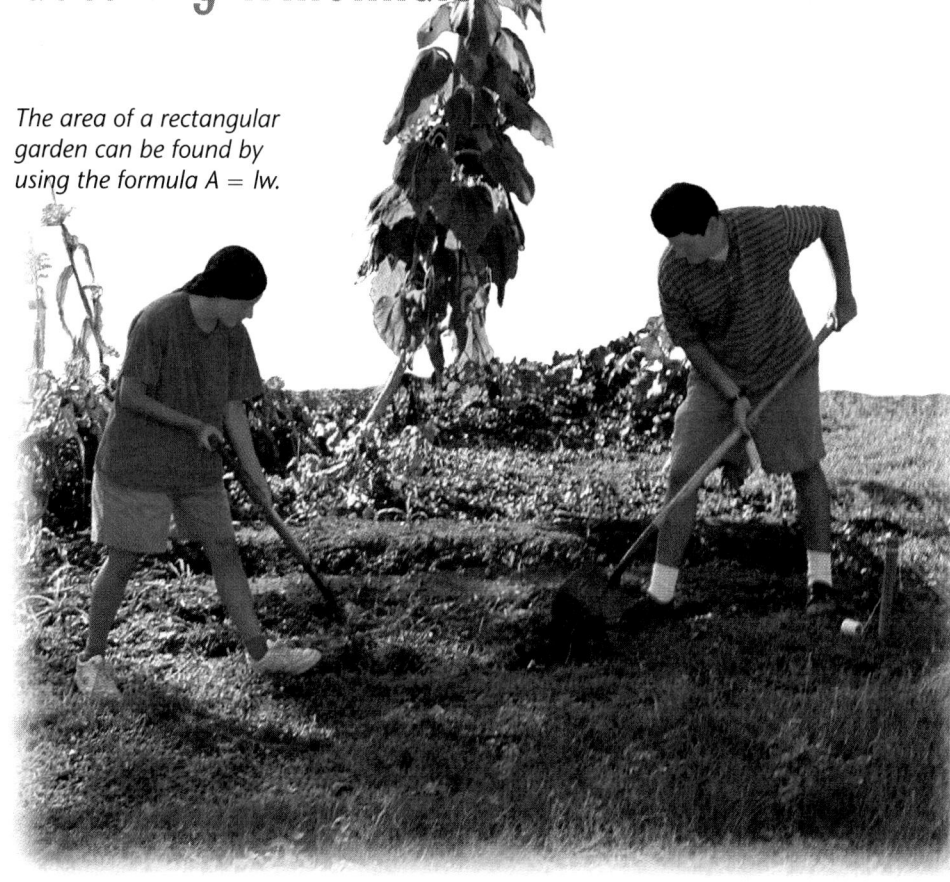

Factoring With Tiles

GEOMETRY
Connection

Suppose you know that the area of a rectangle is represented by $x^2 + 6x + 8$. How can you find the representation of its length and width? One way to find the length and width is to use algebra tiles. If the tiles can be arranged to form a rectangle, the length and width can be determined.

Start with the tiles that model $x^2 + 6x + 8$.

$x^2 + 6x + 8$

Arrange the tiles in a rectangle.

The length is represented by $x + 4$, and the width is represented by $x + 2$. To check, multiply the factors.

$$(x + 2)(x + 4) = x^2 + 2x + 4x + 8$$
$$= x^2 + 6x + 8$$

x $x + 4$

$x + 2$

Lesson 11.6   Factoring Trinomials   **535**

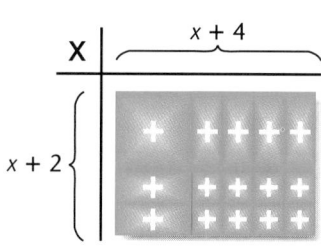

Sometimes using tiles may take more than one try. Examine the tiles for $x^2 + 2x - 8$.

Place the x^2-tile. Try a 1-by-8 arrangement of the negative 1-tiles. It is impossible to complete a rectangle.

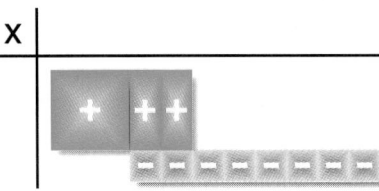

Try a 2-by-4 arrangement of the negative 1-tiles.

Complete a rectangle by adding 2 positive and 2 negative x-tiles.

Finally, decide on the factor tiles.

Thus, $x^2 + 2x - 8 = (x - 2)(x + 4)$. To check, multiply the factors.

$$(x - 2)(x + 4) = x^2 - 2x + 4x - 8$$
$$= x^2 + 2x - 8$$

Try This **a.** Explain how to use tiles to factor $x^2 - 2x - 8$. Check by multiplication.

b. Explain how to use tiles to factor $x^2 - 6x + 8$. Check by multiplication.

Factoring by Guess-and-Check

Examine the following trinomial patterns and their factors.

a. $x^2 + 6x + 8 = (x + 2)(x + 4)$ **c.** $x^2 + 2x - 8 = (x - 2)(x + 4)$

b. $x^2 - 6x + 8 = (x - 2)(x - 4)$ **d.** $x^2 - 2x - 8 = (x + 2)(x - 4)$

In each trinomial, the coefficient of x^2 is 1, and the last term is either 8 or -8. Since factoring is related to multiplication, a trinomial can be factored by working backward using the FOIL method. You can use guess-and-check to write the correct factors and signs in the FOIL model.

Follow the steps to factor $x^2 + x - 12$.

1. Examine the last term of the trinomial. Since there is a negative sign before the 12 and the coefficient of x is positive, $x^2 + x - 12$ is an example like trinomial **c** above. Write the appropriate signs in the FOIL model.

2. To find the values in ■ and ◆, remember that their product must be 12. Possible factor pairs of 12 are $1 \cdot 12$, $2 \cdot 6$, or $3 \cdot 4$. Choose the factors 3 and 4, since the sign for the middle term of the trinomial is positive and the coefficient is 1. Let 3 replace ■ and 4 replace ◆, so that $-3x + 4x$ will equal x.

Thus, the result of factoring $x^2 + x - 12$ is $(x - 3)(x + 4)$.

3. Explain why the other factor pairs do not work.

4. Check the factors. $(x - 3)(x + 4) = x^2 + 4x - 3x - 12$
$$= x^2 + x - 12$$

EXAMPLE

Factor each trinomial.

Ⓐ $x^2 - x - 20$ Ⓑ $x^2 - 10x + 16$

Ⓒ $x^2 + 4x - 21$ Ⓓ $x^2 + 9x + 18$

Solution ➤

Ⓐ Use pattern **d**.
$x^2 - x - 20 = (x + 4)(x - 5)$

Ⓑ Use pattern **b**.
$x^2 - 10x + 16 = (x - 2)(x - 8)$

Ⓒ Use pattern **c**.
$x^2 + 4x - 21 = (x - 3)(x + 7)$

Ⓓ Use pattern **a**.
$x^2 + 9x + 18 = (x + 6)(x + 3)$ ❖

CRITICAL
Thinking

Sometimes a trinomial has a common monomial factor. Explain why $2x^3 + 16x^2 + 24x$ written in factored form is $2x(x + 2)(x + 6)$.

There are some trinomials such as $x^2 - 6x - 8$ that cannot be factored by any method. For the middle term to equal $-6x$, the factors for the constant term, 8, would have to be -2 and -4. The constant term would then equal $+8$. There are no signs for the factor pair that fit the pattern of signs for the trinomial. The polynomial $x^2 - 6x - 8$ is prime or *irreducible* over the integers.

EXERCISES & PROBLEMS

Communicate

Explain how to use algebra tiles to factor these polynomials.

1. $x^2 - 5x + 4$ **2.** $x^2 - 4x - 12$ **3.** $x^2 + 6x + 9$

Write each trinomial in factored form. Tell why the signs of the factors are the same or opposite.

4. $x^2 + x - 6$ **5.** $x^2 - 7x + 10$ **6.** $x^2 + 2x - 15$

7. Explain how to use the guess-and-check method to factor $x^2 - 5x - 24$.

8. If the third term of a trinomial is 36, write the possible factor pairs.

Practice & Apply

9. Use algebra tiles to factor $x^2 + 5x + 6$.

Write each trinomial in factored form. Are the signs of the factors the same or opposite?

10. $x^2 - 13x + 36$ **11.** $x^2 + 11x + 24$
12. $x^2 + 10x - 24$ **13.** $x^2 - 35x - 36$

For each polynomial, write all the factor pairs of the third term, then circle the pair that would successfully be used to factor the polynomial.

Example: $x^2 - 3x - 4$ has factor pairs of $\boxed{1 \cdot 4}$ and $2 \cdot 2$.

14. $y^2 - 9y - 36$ **15.** $x^2 - 21x + 54$
16. $x^2 + 19x + 48$ **17.** $z^2 + 10z - 144$

538   CHAPTER 11

Write each trinomial as a product of its factors. Use factoring patterns or algebra tiles to assist you in your work.

18. $a^2 - 2a - 35$ **19.** $p^2 + 4p - 12$ **20.** $y^2 - 5y + 6$ **21.** $b^2 - 5b - 24$

22. $n^2 - 11n + 18$ **23.** $z^2 + z - 20$ **24.** $x^2 - 3x - 28$ **25.** $s^2 - 24s + 63$

This set of exercises includes all factoring patterns used throughout the chapter. Write each polynomial as a product of its factors.

26. $4x^3y - 20x^2y + 16xy$ **27.** $6y^3 - 18y^2 + 12y$ **28.** $x^2 - 18x + 81$

29. $(a + 3)(a^2 + 5a) - 6(a + 3)$ **30.** $5x^3 - 50x^2 + 45x$ **31.** $x^3 + 2x^2 - 36x - 72$

32. $-x^4 + 2x^2 + 8$ **33.** $64p^4 - 16$ **34.** $z^2 - 5z - 36$

35. $x^2 - 2x + 1$ **36.** $125x^2y - 5x^4y$ **37.** $2ax + ay + 2bx + by$

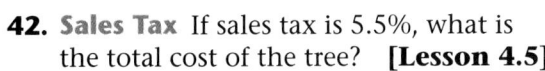

Look Back

38. Write the next three terms in the sequence 3, 9, 19, 33, . . . **[Lesson 1.3]**

Solve. [Lesson 4.4]

39. $\frac{3}{4}a = 363$ **40.** $\frac{z}{-8} = \frac{9}{12}$ **41.** $w - \frac{7}{9} = 93$

42. Sales Tax If sales tax is 5.5%, what is the total cost of the tree? **[Lesson 4.5]**

43. A line with slope -2 passes through point $A(4, -1)$. Write the equation of the line that passes through the point and is perpendicular to the original line. **[Lesson 5.6]**

44. Represent the inequality $-4 < x < 5$ on the number line. **[Lesson 6.5]**

45. Write the system of equations as a matrix, and solve using the matrix method. **[Lesson 7.5]** $\begin{cases} 3x + y = 7 \\ 2x - y = 3 \end{cases}$

A freshman class donates a tree to the childrens' hospital.

46. List the integers from 1 to 25, inclusive, that are multiples of 2. **[Lesson 8.4]**

Look Beyond

Cultural Connection: Americas De Padilla, a Guatemalan mathematician, wrote this problem 250 years ago.

47. Find two numbers, given that the second is triple the first. If you multiply the first number by the second number and the second number by 4, the sum of the products is 420.

48. Graph $y = |x|$ and $y = x^2$. If these graphs are "folded" together along a vertical line, what is the equation of the line that divides them both equally?

Powers, Pascal, & Probability

cultural connection: ASIA If a great mathematical discovery were made today, the discoverer and his work would be on the evening news. This has not been the case throughout history. Pascal's Triangle, named after the French mathematician Blaise Pascal, appeared in work published in Europe in 1665. However, historical writings show that mathematicians from the ancient Islamic civilization were already using the remarkable properties of Pascal's Triangle in the tenth century.

ACTIVITY 1

Look at the different powers of the binomial $(a + b)$, and compare the expanded form to the Pascal Triangle.

POWER	EXPANDED FORM	PASCAL'S TRIANGLE
$(a+b)^0 =$	1	1
$(a+b)^1 =$	$1a + 1b$	$1 \quad 1$
$(a+b)^2 =$	$1a^2 + 2ab + 1b^2$	$1 \quad 2 \quad 1$
$(a+b)^3 =$	$1a^3 + 3a^2b + 3ab^2 + 1b^3$	$1 \quad 3 \quad 3 \quad 1$
$(a+b)^4 =$	$1a^4 + 4a^3b + 6a^2b^2 + 4ab^3 + 1b^4$	$1 \quad 4 \quad 6 \quad 4 \quad 1$
$(a+b)^5 =$	$1a^5 + 5a^4b + 10a^3b^2 + 10a^2b^3 + 5ab^4 + 1b^5$	$1 \quad 5 \quad 10 \quad 10 \quad 5 \quad 1$

1. Start with the expanded form of $(a + b)^2$, and multiply it by $a + b$. What do you get?
2. Do the same with the expanded forms of $(a + b)^3$ and $(a + b)^4$. What do you get?
3. How does the number of terms in each row of the expanded form compare with the exponent of the binomial?
4. How many terms would you expect in the expansion of $(a + b)^7$?
5. Recalling that the triangle's elements are created by adding pairs of elements from a previous row, extend the rows of the triangle through row 10.
6. Write the coefficients for each term in the expansion of $(a + b)^7$.
7. Look at the algebraic terms of the expanded form. Describe the pattern in the exponents as you read them from left to right. What happens to the exponents of a? What happens to the exponents of b?
8. Write the complete algebraic expansion of $(a + b)^7$ without multiplying.

A Chinese version of the triangle appeared in 1303 C.E.

ACTIVITY 2

1. Copy the table.
2. Flip three coins a total of 32 times. Record the results in the table.
3. Divide each total by 4, and round the answer to the nearest whole number. What are the results for each column?

ACTIVITY 3

1. Look at the tree diagram for the coin-tossing activity.
2. How many ways are there to toss 3 heads?
3. How many ways are there to toss 3 tails?
4. How many ways are there to toss 2 heads and 1 tail?
5. How many ways are there to toss 1 head and 2 tails?
6. What is the probability of each outcome?

ACTIVITY 4

1. Write the expanded form of $(h + t)^3$.
2. The probability of getting a head when tossing a coin is $\frac{1}{2}$. The probability of getting a tail when tossing a coin is $\frac{1}{2}$. Let h and t each equal $\frac{1}{2}$, and find the value of *each term* of the expanded form of $(h + t)^3$.

3. How are the values for each term related to
   a. the probabilities you calculated in Activity 3?
   b. the results you obtained in Activity 2?
4. If you tossed 5 coins at a time, what would be the probability of getting 5 heads? 4 heads and 1 tail?

Chapter 11 Review

Vocabulary

Key Skills & Exercises

Lesson 11.1

➤ **Key Skills**

Use vertical form to add polynomials.

Add $4x^2 + 6x - 2$ and $x^3 - 2x^2 + 7$.

$$
\begin{array}{r}
4x^2 + 6x - 2 \\
+ (x^3 - 2x^2 \quad + 7) \\
\hline
x^3 + 2x^2 + 6x + 5
\end{array}
$$

Use vertical form to subtract polynomials.

Subtract $5m^2 - 4$ from $m^2 - 5m - 10$.

$$
\begin{array}{r}
m^2 - 5m - 10 \\
-(5m^2 \qquad - 4) \\
\hline
-4m^2 - 5m - 6
\end{array}
$$

Use horizontal form to add polynomials.

Add $5y^2 - 3y + 8$ and $y^2 + y - 9$.

$(5y^2 - 3y + 8) + (y^2 + y - 9) = 6y^2 - 2y - 1$

Use horizontal form to subtract polynomials.

Subtract $6b^2 + 4b - 8$ from $10b^2 - 13$.

$(10b^2 - 13) - (6b^2 + 4b - 8)$
$= 10b^2 - 13 - 6b^2 - 4b + 8$
$= 4b^2 - 4b - 5$

➤ **Exercises**

Simplify. Express all answers in standard form.

1. $(3x^2 - 4x + 2) + (2x^2 + 3x - 2)$

2. $(c^3 + 4c^2 + 6) + (c^2 + 3c - 5)$

3. $(8d^2 - d) - (2d^2 + 4d - 5)$

4. $(w^3 - 3w + 9) - (8w^3)$

5. $(10m^2 - m + 4) - (2m^2 + m)$

6. $(7c + 3) + (3c^2 - 7c - 2)$

Lesson 11.2

➤ **Key Skills**

Use the Distributive Property to find products.

$3x(2x - 3) = 3x(2x) - 3x(3) = 6x^2 - 9x$

Write a polynomial in factored form.

$-15z^2 - 20z^3 = -5z^2(3 + 4z)$

> **Exercises**

Use the Distributive Property to find each product.

7. $5(x - 5)$ **8.** $y(y + 4)$ **9.** $4t(t^2 + 7)$ **10.** $2r^2(r^2 - 3r)$

11. $b(12b^2 + 11b)$ **12.** $4y(y + 5)$ **13.** $5x^2(2x^2 - x)$ **14.** $6d^2(d^2 - 1)$

Write each polynomial in factored form.

15. $6x^2 + 8$ **16.** $5c^3 - 25c$ **17.** $n^4 + 2n^3$ **18.** $-9w^2 - 21w^4$

19. $8y^2 - 3y^2$ **20.** $-8p - 14p^2$ **21.** $z^4 + 5z^2$ **22.** $16y^5 - 4y^3$

Lesson 11.3

> **Key Skills**

Use the Distributive Property to find a product of two binomials.

$$(x + 6)(x + 2)$$
$$x(x + 6) + 2(x + 6)$$
$$x^2 + 6x + 2x + 12$$
$$x^2 + 8x + 12$$

Use the FOIL method to multiply binomials.

$$(2x + 3)(x - 1)$$

$$(2x + 3)\ (x - 1) = 2x^2 - 2x + 3x - 3$$
$$2x^2 + x - 3$$

> **Exercises**

Use the Distributive Property to find a product of two binomials.

23. $(x + 6)(x - 1)$ **24.** $(y + 9)(y - 2)$ **25.** $(z - 3)(z - 6)$

26. $(3m + 5)(m + 5)$ **27.** $(2p - 9)(p + 5)$ **28.** $(2d + 7)(2d - 6)$

Use the FOIL method to find each product.

29. $(x + 3)(x - 4)$ **30.** $(5d - 8)(d - 1)$ **31.** $(4w + 3z)(w + z)$

32. $(y + 4)(y + 5)$ **33.** $(x + 2)(x - 4)$ **34.** $(3z + 1)(4z - 1)$

Lesson 11.4

> **Key Skills**

Factor a polynomial by removing the GCF.

$$8m^6 + 4m^4 - 2m^2$$
$$= 2m^2(4m^4 + 2m^2 - 1)$$

Find a binomial as the greatest common factor.

$$2x^2 + 2 + x^3 + x$$
$$= (2x^2 + 2) + (x^3 + x)$$
$$= 2(x^2 + 1) + x(x^2 + 1)$$
$$= (2 + x)(x^2 + 1)$$

> **Exercises**

Factor each polynomial by removing the GCF.

35. $16x^3 + 8x^2$ **36.** $9y^7 + 6y^3 + 3y$ **37.** $b^6 + 15b^3 - 30b^2$

38. $24m^9 - 16m^4 + 8m^3$ **39.** $60a^4 + 20a^3 + 10a^2$ **40.** $100p^8 - 50p^6 - 25p$

Factor the common binomial factor from each polynomial.

41. $d(f + 1) + h(f + 1)$ **42.** $3y(y - 3) - 4(y - 3)$ **43.** $5x - 5y + x^2 - xy$

44. $x(z - 4) + y(z - 4)$ **45.** $10 - 5t - 2t + t^2$ **46.** $6c^2 + 6 + c^3 + c$

Lesson 11.5

➤ Key Skills

Use a generalization of special polynomials to find each product.

$(x + 7)^2 = (x + 7)(x + 7) = x^2 + 14x + 49$

$(p + 8)(p - 8) = p^2 - 64$

Factor perfect-square trinomials.

$x^2 - 18x + 81 = (x - 9)^2$

Factor the difference of two squares.

$4x^2 - 9 = (2x + 3)(2x - 3)$

➤ Exercises

Use a generalization of special polynomials to find each product.

47. $(c - 9)^2$ **48.** $(b - 10)(b + 10)$ **49.** $(5a - 2)(5a + 2)$

Factor each polynomial.

50. $a^2 + 6a + 9$ **51.** $w^2 - 16w + 64$ **52.** $36p^2 + 12p + 1$

53. $y^2 - 81$ **54.** $16r^2 - 25$ **55.** $9z^2 - 1$

Lesson 11.6

➤ Key Skills

Use sign and number patterns to determine the factors of a trinomial.

$x^2 - x - 12 = (x + \underline{?})(x - \underline{?})$ Factor pairs: $1 \cdot 12, 2 \cdot 6, 3 \cdot 4$

The factors are $(x + 3)(x - 4)$.

➤ Exercises

Write each polynomial as a product of its factors.

56. $n^2 - 2n - 24$ **57.** $h^2 + 14h + 40$ **58.** $y^2 + 3y - 18$

59. $a^2 - 7a + 12$ **60.** $x^2 - 5x + 4$ **61.** $2a^2 + 8a + 6$

Applications

Geometry Given the rectangle with sides as shown, solve the following.

62. What is the area of the large rectangle?

63. What is the area of the small rectangle?

64. If the smaller region is removed, what is the area of the shaded region?

65. Factor the polynomial that represents the area of the shaded region.

Crafts Marcus works at a local craft store mounting pictures and posters on mats for framing. He is mounting a picture on a mat that has an area of 99 square inches.

66. The length of the mat is 2 inches greater than the width of the mat. The equation representing the area of the mat is $x(x + 2) = 99$, or $x^2 + 2x - 99 = 0$. What are the dimensions of the mat?

67. An equal amount of mat will be showing on each side of a picture. Let x represent the amount of mat showing on each side. What are the dimensions of the picture? What is the area of the picture?

Chapter 11 Assessment

Simplify. Express all answers in standard form.

1. $(2 + 4x + 2x^2) + (4x^2 - 6)$

2. $9 - 5v^2 + 5v^4 - (10v^2 + 7v - 11)$

Use the Distributive Property to find each product.

3. $8(x - 7)$

4. $r(r - 6)$

5. $3a^2(a - 1)$

6. $q(2q^2 + q)$

Write each polynomial in factored form.

7. $6x^6 + 3x^4$

8. $25a^8 - 15a^4$

9. $8x^5 - 4x^4 + x^3$

10. $25y^9 + 15y^7 - 5y^2$

11. Describe how to find $(x + 4)(x - 8)$ using the FOIL method.

12. Write the degree of the polynomial $6y^2 + 3y^3 - 1$.

Factor the common binomial factor from each polynomial.

13. $5(r + 1) - t(r + 1)$

14. $p^2 + 2p + 3p + 6$

Use the Distributive Property or the FOIL method to find each product.

15. $(y + 3)(y + 2)$

16. $(c + 3)(c - 4)$

17. $\left(\dfrac{1}{2}a + 1\right)(a - 1)$

18. $(4x - y)(3x + 2y)$

19. $(w - 9)(w - 6)$

20. $(2n + 1)(n + 1)$

Use a generalization of special polynomials to find each product.

21. $(x + 11)^2$

22. $(p + 12)(p - 12)$

23. $(m - 9)^2$

24. Find the area of the floor of a square garage that is $t + 5$ units on a side.

25. Can $7c^2 + 45c - 28$ be factored? Explain why or why not.

Use any method to factor each polynomial completely.

26. $4r + 4s + rt + st$

27. $14x^3 + 7x - 8x^2 - 4$

28. $4v^2 - 144$

29. $a^4 - 81$

30. $n^2 - 12n + 36$

31. $15z^2 - z - 2$

32. $w^4 - 7w^3 - 18w^2$

33. $d^4 + 5d^2 + 6$

Find the greatest common factor.

34. $y^3 - y^2 + 4y$

35. $6w^3 + 3w^2 + 3$

36. $7m^3 - m^2$

Within a large rectangle there is a smaller rectangle.

37. What is the area in square units of the large rectangle?

38. What is the area in square units of the smaller rectangle?

39. What is the area in square units of the shaded area between the two figures?

CHAPTER 12

Quadratic Functions

You have seen that the height of a projectile can be modeled by a quadratic function. Often the trajectory of a projectile will resemble a parabola, the graph of a quadratic function. The cost of an automobile can be modeled using a quadratic function. In this chapter you will learn different methods for analyzing quadratic functions and quadratic equations.

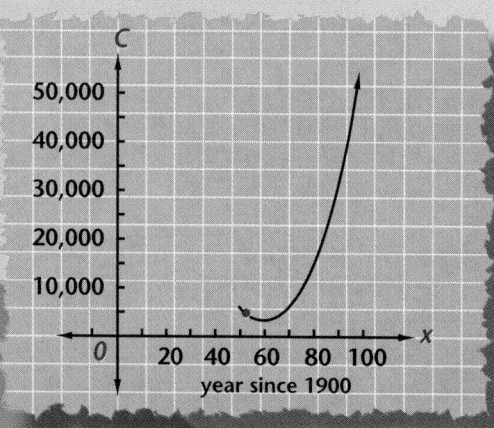

year since 1900

Sometimes the cost of a product can be approximated by a quadratic function.

1953   1962

PORTFOLIO ACTIVITY

This quadratic function approximates the annual cost of a popular car from 1953-1995.

$$C = 34x^2 - 4206x + 132{,}539$$

The variable C represents the cost. The variable x represents the number of years since 1900. For example, x is 80 for the year 1980.

To find the cost of the car for 1980, substitute 80 for x and compute the value of C.

$C = 34x^2 - 4206x + 132{,}539$
$C = 34(80)^2 - 4206(80) + 132{,}539$
$ = 13{,}659$

The cost of this sports car for 1980 was approximately $13,659.

Use the formula to project the cost of this sports car for the year 2000. In what year did the car cost $25,000?

1989

1990

1995

LESSON 12.1

Exploring Parabolas

why *You have already seen that a graph that represents a quadratic function is a parabola. When you understand the geometric properties of the parabola, you gain additional mathematical tools for modeling and solving problems related to quadratic functions.*

•Exploration 1 *Parabolas and Constant Differences*

The table shows some points on the graph of the function $y = f(x)$.

x	0	1	2	3	4	5	6
$f(x)$	10	0	−6	−8	−6	?	?

1 Calculate the first and second differences for $f(x)$.

2 What do you notice about the second differences?

3 Work backward using the pattern, and predict $f(5)$ and $f(6)$.

4 Plot $y = f(x)$. Connect the points with a smooth curve.

5 What kind of curve did you draw? Explain the relationship between second differences and the curve that you drew. ❖

In chapter 9, you learned that the function $g(x) = a(x − h)^2 + k$ transforms the parent quadratic function $f(x) = x^2$ by stretching the parent function by a factor of a and moving its vertex from $(0, 0)$ to (h, k). The axis of symmetry is $x = h$. If $a > 0$ the minimum value of g is k.

548 CHAPTER 12

Exploration 2 Parabolas and Transformations

1 Consider the graph of the function $g(x) = 2(x - 3)^2 - 8$. What is its vertex?

2 Find 4 more points on the graph.

3 Plot the points and sketch the graph.

4 What kind of curve did you draw? How does your graph compare with the graph you drew in Exploration 1? ❖

The graph of a quadratic function is a parabola. If the parabola crosses the *x*-axis, its *x*-value at the intersection is called a **zero of the function**. How many zeros can a quadratic function possibly have? Explain your answer.

Exploration 3 Parabolas and Polynomials

**MAXIMUM
MINIMUM**
Connection

1 Graph the function $t(x) = 2x^2 - 12x + 10$.

2 Find the zeros of the function.

3 Find the average, *h*, of the two zeros.

4 Find $t(h)$.

5 What is the vertex of the graph?

6 What is the axis of symmetry of the graph?

7 What is the minimum value of the function?

8 Compare the graphs for Exploration 2 and 3.

9 Simplify $2(x - 3)^2 - 8$. Compare $g(x)$ with $t(x)$. ❖

APPLICATION

In Lesson 1.2, you used a table and differences to find the maximum height of the flight of a small rocket. The flight can be modeled with the function $f(x) = -16(x - 7)^2 + 784$. In this form, the vertex, the axis of symmetry and the maximum value can be read directly from the equation. The zeros can be read from the graph.

function	$f(x) = -16(x - 7)^2 + 784$
vertex	(7, 784)
axis of symmetry	$x = 7$
maximum value	784
zeros	0, 14

The zeros of *f* are 0 and 14. Check by substitution. ❖

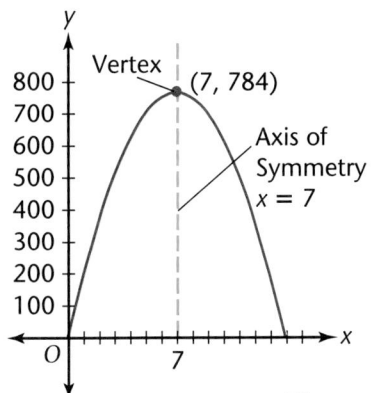

**CRITICAL
Thinking**

How can the zeros of $f(x) = -16(x - 7)^2 + 784$ be interpreted in terms of the flight of the rocket?

EXERCISES & PROBLEMS

Communicate

1. Explain how you can tell from second differences that the values of a function represent a quadratic.

x	-4	-3	-2	-1	0
$f(x)$	0	-1	0	3	8

2. Discuss how the graph of the quadratic function $g(x) = 2(x - 3)^2 - 8$ differs from the graph of the parent function $y = x^2$.

Refer to the function $g(x) = a(x - h)^2 + k$ to complete Exercises 3-4.

3. Explain how to determine the vertex of $g(x) = (x - 3)^2 - 4$.

4. Explain how to determine the axis of symmetry.

5. Discuss how to determine the zeros of $g(x) = x^2 - 8x + 16$.

6. Describe how to determine the axis of symmetry for $g(x) = x^2 - 8x + 16$ from the zeros of the polynomial.

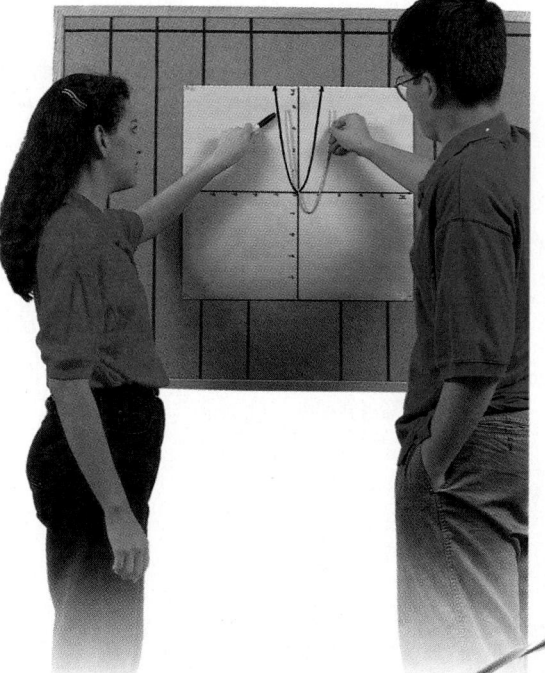

Practice & Apply

7. Do the values represent a quadratic function? Graph to check.

x	0	1	2	3	4	5	6
$f(x)$	-5	0	3	4	3	-5	0

Compare each graph with the graph of $y = x^2$. Write the change in units and direction that takes place to the left, right, up, or down.

8. $y = (x - 2)^2 + 3$

9. $y = 3(x - 5)^2 - 2$

10. $y = -(x - 2)^2 + 1$

11. $y = (x + 1)^2$

12. $y = -3(x - 2) + 1$

13. $y = \frac{1}{2}(x - 2)^2 + 3$

Determine the vertex and axis of symmetry for each function, and then sketch a graph from the information.

14. $y = -2(x + 4)^2 - 3$

15. $y = \frac{1}{2}(x - 2)^2 + 3$

16. $y = (x - 3)^2 - 7$

17. $y = -3(x - 5)^2 + 2$

18. $y = -(x + 3)^2 - 2$

19. $y = 2(x + 5)^2 + 7$

Find the zeros of each function by graphing.

20. $y = x^2 + 8x - 9$

21. $y = x^2 - 20x + 100$

22. $y = x^2 - x - 72$

23. $y = x^2 + 6x - 7$

24. $y = x^2 + 4x - 5$

25. $y = x^2 + 2x - 24$

26. $y = 2x^2 - 2x - 144$

27. $y = 3x^2 + 9x - 12$

28. $y = 5x^2 + 15x - 20$

First graph the function. Then find the zeros, the axis of symmetry, and the minimum value of the functions.

29. $f(x) = x^2 + 18x + 81$

30. $f(x) = x^2 + 6x - 7$

31. $f(x) = x^2 - 5x + 6$

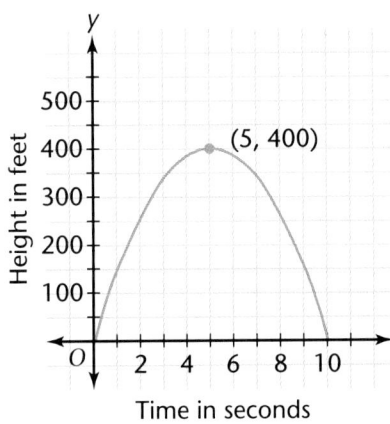

Time in seconds

Physics The graph represents a relationship between the time when a projectile is propelled vertically into the air and the height that it reaches.

32. What is the maximum height reached by the projectile?

33. How long does it take for the projectile to reach its maximum height?

34. How long does it take for the projectile to return to Earth?

35. What is the equation of the axis of symmetry of the graph?

36. Which equation was used to graph the parabola?

    a. $y = -16(x + 5)^2 + 400$ **b.** $y = -16(x + 5)^2 - 400$

    c. $y = -16(x - 5)^2 + 400$ **d.** $y = -16(x - 5)^2 - 400$

37. Physics Use the formula in the caption to find the height in meters after 5 seconds.

The formula $h = 40t - 5t^2$ can be used to find the height in meters of the arrow after t seconds.

Look Back

The expression $2(x - 3)^2 + 1$ involves multiplication, an operation inside parentheses, an exponent, and addition. **[Lesson 1.4]**

38. Which of these operations should be done first? second? third?

Perform each matrix operation. [Lesson 7.2]

39. $\begin{bmatrix} 6 & -2 \\ 8 & 3 \end{bmatrix} + \begin{bmatrix} -2 & 4 \\ 3 & 6 \end{bmatrix}$

40. $\begin{bmatrix} -4 & 5 \\ 6 & 2 \end{bmatrix} - \begin{bmatrix} 3 & -1 \\ -5 & 6 \end{bmatrix}$

41. $\begin{bmatrix} -3 & 7 \\ 4 & -5 \end{bmatrix} + \begin{bmatrix} 6 & 8 \\ -1 & -3 \end{bmatrix}$

Factor. [Lesson 11.6]

42. $x^2 + 10x + 25$

43. $x^2 + 14x + 49$

44. $x^2 - 18x + 81$

Look Beyond

45. Several students are able to collect 100,000 cans to recycle. What is the minimum space the cans will take? Assume that a can has a circumference of 21 centimeters and a height of 12.5 centimeters.

h = 185 ft

1 Sec.
h = 169 ft

2 Secs.
h = 129 ft

ITALY

Pisa

Adriatic Sea

Rome

Tyrrhenian
Sea

Mediterranean
Sea

3 Secs.
h = 41 ft

Ground Level 0

Galileo Galilei, born in Pisa in 1564, proved theoretically that objects fall according to the law of uniformly accelerated motion.

why *You already know how to find approximate solutions to a quadratic equation by using technology or graphing. Algebraic techniques can be used to find exact solutions.*

An object falls from the top of the Leaning Tower of Pisa, 185 feet above ground level. Its height, h, after t seconds is given by the function $h = -16t^2 + 185$. How long will it take the object to reach the ground?

When the object reaches the ground, the height is 0 feet. Thus, we substitute 0 for h in the given quadratic function to obtain the quadratic equation $0 = -16t^2 + 185$.

Tabular Method

Begin by making a table of values for the function $h = -16t^2 + 185$. Substitute integer values of t into $h = -16t^2 + 185$ to find h. According to the table, h will be 0 somewhere between $t = 3$ and $t = 4$ seconds. A new table can be constructed where t changes by tenths.

time (t)	height (h)	
0	185	
1	169	
2	121	
3	41	
4	−71	← Stop here.

time (t)	height (h)	
3.0	41.00	
3.1	31.24	
3.2	21.16	
3.3	10.76	
3.4	0.04	
3.5	−11.00	← Stop here.

Closer investigation using the table shows that 3.4 is an approximate solution to the nearest tenth for the equation $0 = -16t^2 + 185$.

How can you get a solution to the nearest hundredth for the equation $0 = -16t^2 + 185$?

Graphing Method

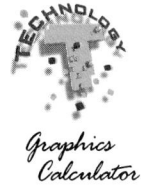

Graphics Calculator

Graph the function $h = -16t^2 + 185$ with t on the x-axis and h on the y-axis. Use the trace function to find a point where $h \approx 0$. According to the graph, $t \approx 3.4$.

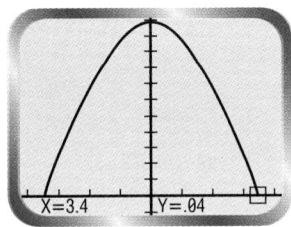

X=3.4   Y=.04

In fact, the quadratic equation $0 = -16t^2 + 185$ has 2 solutions. A more complete picture of the graph of $0 = -16t^2 + 185$ suggests the solutions are approximately 3.4 or -3.4. Since the falling-object problem is about height and time, only the positive time values make sense. In this case, ignore the negative time values.

Finding Square Roots

Finding the exact solutions for $0 = -16t^2 + 185$ involves finding square roots. Every positive number has two square roots. For example,

The square root of 9 is 3. $\sqrt{9} = 3$.
The negative square root of 9 is -3. $-\sqrt{9} = -3$.

If $x^2 = 9$, $x = 3$ or $x = -3$. The two solutions can be combined using the symbol $\pm$, read plus or minus.

$$\text{If } x^2 = 9, x = \pm\sqrt{9} = \pm 3.$$

EXAMPLE 1

Show that $\sqrt{\dfrac{4}{9}} = \dfrac{\sqrt{4}}{\sqrt{9}}$.

Solution ➤

Simplify the square root under the radical signs.

$\sqrt{\dfrac{4}{9}} = \dfrac{2}{3}$ because $\dfrac{2}{3} \cdot \dfrac{2}{3} = \dfrac{4}{9}$.

$\dfrac{\sqrt{4}}{\sqrt{9}} = \dfrac{2}{3}$ because $\sqrt{4} = 2$ and $\sqrt{9} = 3$.

Since $\sqrt{\dfrac{4}{9}} = \dfrac{2}{3}$ and $\dfrac{\sqrt{4}}{\sqrt{9}} = \dfrac{2}{3}$, $\sqrt{\dfrac{4}{9}} = \dfrac{\sqrt{4}}{\sqrt{9}}$ by substitution. ❖

Example 1 shows two ways to find the square root of $\dfrac{4}{9}$. But, it does not show how you find the solution or solutions to $x^2 = \dfrac{4}{9}$.

EXAMPLE 2

Solve the following equations.

A $x^2 = \frac{4}{9}$ **B** $x^2 = 1.44$ **C** $x^2 = 10$

Solution >

A There are two solutions: $\sqrt{\frac{4}{9}} = \frac{2}{3}$ and $-\sqrt{\frac{4}{9}} = -\frac{2}{3}$.

B There are two solutions: $\sqrt{1.44} = 1.2$ and $-\sqrt{1.44} = -1.2$.

C The solutions are $\sqrt{10}$ and $-\sqrt{10}$. There is no rational number answer. An approximation can be found by using [√] on your calculator.

$$\boxed{\sqrt{\;}}\; 10 \;\boxed{=}\; 3.16227766$$

The approximate solutions are 3.16 and −3.16.

The results of Example 2 lead to a generalization for solving a quadratic equation of the form $x^2 = k$. ❖

SOLVING $x^2 = k$ WHEN $k \geq 0$

If $x^2 = k$, $k \geq 0$, then

1. $x = \pm \sqrt{k}$ and

2. the solutions are $\sqrt{k}$ and $-\sqrt{k}$.

For example, if $x^2 = 16$, Then $x = \pm\sqrt{16} = \pm 4$. The solutions are 4 and −4.

Algebraic Method

Now the time can be found for the object falling from the Tower of Pisa.

$$-16t^2 + 185 = 0$$
$$-16t^2 = -185$$
$$t^2 \approx 11.56$$
$$t \approx \pm\sqrt{11.56}$$

The time is approximately $\sqrt{11.56}$ seconds.

Why is $-\sqrt{11.56}$ not considered as a solution?

A calculator gives the solution 3.4, which agrees with the tabular and graphing methods.

 CRITICAL *Thinking*   Explain why the algebraic method for solving the quadratic equation $-16t^2 + 185 = 0$ gives another way to find the zeros for the quadratic function $h = -16t^2 + 185$.

EXAMPLE 3

Solve the following equations.

A $(a - 2)^2 - 9 = 0$ **B** $(x - 2)^2 = 11$

Solution ➤

A In this equation, the expression $a - 2$ plays the role of x in the statement $x^2 = k$.

$$(a - 2)^2 - 9 = 0$$
$$(a - 2)^2 = 9$$
$$a - 2 = \pm 3$$
$$a = 2 \pm 3$$

The solutions are 5 and -1.

Check each solution in the original equation.

B $(x - 2)^2 = 11$
$$x - 2 = \pm \sqrt{11}$$

The approximate solutions are 5.32 and -1.32.

Check each solution in the original equation. ❖

The solution to Example 3a will help you sketch the graph of the function $f(x) = (x - 2)^2 - 9$.

COORDINATE GEOMETRY *Connection*

In the form $y = (x - 2)^2 - 9$, the vertex is $(2, -9)$ and the axis of symmetry is $x = 2$. Since the coefficient of the quadratic term is positive, the parabola opens upward and has a minimum value. The values where $y = 0$ are the zeros of the function. When $(x - 2)^2 - 9 = 0$, x is 5 or x is -1. Thus, the graph is a parabola that crosses the x-axis at 5 and -1.

EXERCISES & PROBLEMS

Communicate

1. Explain how to make a table of values for the function $y = -12x^2 + 300$ to determine where the function crosses the x-axis. Why would it be important to know when the values of the function are no longer positive?

2. Discuss how to find the square roots of 100. What is meant by the sign $\pm$?

Explain how to solve the following equations.

3. $x^2 = 64$ **4.** $x^2 = 8$ **5.** $x^2 = \dfrac{16}{100}$

Discuss how to solve each equation using square roots.

6. $(x + 3)^2 - 25 = 0$ **7.** $(x - 8)^2 = 2$

8. Explain how to sketch a graph from the vertex, axis of symmetry, and zeros for $f(p) = (p + 4)^2 - 5$.

Practice & Apply

9. Make a table of values for the function $f(t) = -14t^2 + 300$. Where does the function cross the x-axis?

Find each square root. Round answers to the nearest hundredth when necessary.

10. $\sqrt{121}$ **11.** $\sqrt{144}$ **12.** $\sqrt{625}$ **13.** $\sqrt{36}$

14. $\sqrt{44}$ **15.** $\sqrt{90}$ **16.** $\sqrt{88}$ **17.** $\sqrt{19}$

Solve each equation. Round answers to the nearest hundredth when necessary.

18. $x^2 = 25$ **19.** $x^2 = 169$ **20.** $x^2 = 81$ **21.** $x^2 = 625$

22. $x^2 = 12$ **23.** $x^2 = 24$ **24.** $x^2 = 18$ **25.** $x^2 = 54$

26. $x^2 = \dfrac{25}{81}$ **27.** $x^2 = \dfrac{49}{121}$ **28.** $x^2 = \dfrac{36}{49}$ **29.** $x^2 = \dfrac{4}{100}$

30. $x^2 = 32$ **31.** $x^2 = 45$ **32.** $x^2 = 28$ **33.** $x^2 = 63$

34. $(x + 4)^2 - 25 = 0$ **35.** $(x - 5)^2 - 9 = 0$ **36.** $(x + 1)^2 - 1 = 0$ **37.** $(x - 2)^2 - 6 = 0$

38. $(x + 7)^2 - 5 = 0$ **39.** $(x - 3)^2 - 2 = 0$ **40.** $(x + 3)^2 = 36$ **41.** $(x - 2)^2 = 144$

42. $(x - 8)^2 = 81$ **43.** $(x - 1)^2 = 11$ **44.** $(x + 5)^2 = 10$ **45.** $(x + 6)^2 = 15$

Find the vertex, axis of symmetry, and zeros of each function. Sketch a graph.

46. $f(x) = (x - 4)^2 - 9$ **47.** $g(x) = (x + 2)^2 - 1$ **48.** $h(x) = (x - 4)^2 - 3$

Refer to the function $f(x) = (x + 4)^2 - 4$.

49. Find the vertex.

50. Find the axis of symmetry.

51. Find the zeros.

52. Sketch a graph.

53. Technology Check by using a graphics calculator.

54. **Geometry** The area of a square garden is 169 square feet. Find the length of each side. ($A = s^2$)

55. 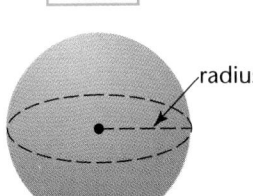 **Geometry** Use the formula $S = 4\pi r^2$ to find the radius of a sphere with a surface area of 90 square meters. (Use $\pi = 3.14$.)

The height of the Met Life Building in New York City is 700 feet.

Physics If an object falls from the Met Life Building, and its height, h, after t seconds is given by the function $h = -16t^2 + 700$, find

56. how long it will take for the object to reach the ground.

57. how long it will take for the object to reach a height of 100 feet.

Look Back

Find the products. [Lesson 7.3]

58. $\begin{bmatrix} 4 & -1 \\ 6 & 0 \end{bmatrix}\begin{bmatrix} 4 & -3 \\ -7 & 2 \end{bmatrix}$

59. $\begin{bmatrix} 9 & -4 \\ 2 & -4 \end{bmatrix}\begin{bmatrix} 6 & 6 \\ -2 & -4 \end{bmatrix}$

Simplify each expression. [Lesson 10.2]

60. $\dfrac{-6x^2y^2}{2x}$

61. $\dfrac{b^3c^4}{bc}$

62. $(p^2)^4(2a^2b^3)^2$

Add or subtract each polynomial. [Lesson 11.1]

63. $(3x^2 - 2x + 1) + (2x^2 + 4x + 6)$

64. $x^2 + 2x - 1 - (2x^2 - 5x + 7)$

65. Find the factors of $a^2 - b^2$, $a^4 - b^4$, and $a^8 - b^8$. How many factors does $a^{64} - b^{64}$ have? Explain.

Look Beyond

66. Find $\sqrt{5^2 - 4 \cdot 1 \cdot 2}$.

67. Find $\sqrt{6^2 - 4(1)(9)}$.

If you rewrite the quadratic function $y = x^2 + bx + c$ in the form $y = (x - h)^2 + k$, you can find the vertex and axis of symmetry of the parabola. A technique called completing the square will enable you to do this.

In the previous chapter you used algebra tiles to factor perfect-square trinomials such as $x^2 + 6x + 9$. How can a square be constructed when you begin with the following tiles?

To make a square, start with the x^2-tile and arrange the six x-tiles into 2 groups of 3.

How many 1-tiles will it take to fill in the corner?

$x^2 + 6x$

$x^2 + 6x + 9$

factor
(x + 3)

factor
(x + 3)

$x^2 + 6x + 9 = (x + 3)^2$

completed square $x^2 + 6x + 9$

Why do you think the process of adding the nine 1-tiles is called completing the square?

·Exploration How Many Ones?

Use algebra tiles to complete the square for $x^2 + 10x$.

1 Into how many groups should the x-tiles be divided?

2 How many 1-tiles must you add?

3 Write the algebraic expression for the area of the square.

4 Fill in the blanks to complete the square for $x^2 + 10x$.

  a. Divide the x-tiles into 2 sets of __?__ x-tiles.

  b. Add __?__ 1-tiles.

  c. Write $x^2 + 10x + $ __?__ $= (x + 5)^2$.

5 Model the expression $x^2 + 16x$ with algebra tiles. What is the length of the side of your new square? How does the constant in the expression compare with the number 16?

  a. Write the area of your new square in the form $x^2 + bx + c$.

  b. Write the area of your new square in the form $(x + $ __?__ $)^2$.

6 Suppose you have 1 x-square tile and 74 x-tiles. How many 1-tiles will you need to complete the square? Explain how you get your answer. ❖

In the Exploration, the coefficient of x is an even number. Thus, it is easy to divide the x-tiles into 2 sets. What happens if the coefficient of x is odd? What happens if the coefficient of x is negative?

EXAMPLE 1

Complete the square for $x^2 + 5x$.

Solution ➤

It is not possible to use actual tiles to complete the square. However, a picture will serve as a good model in this case.

Split the 5 x-tiles into 2 groups, each with $2\frac{1}{2}$ x-tiles. Then add 1-tiles to complete the square.

You need $\left(\frac{5}{2}\right)^2$, or $\frac{25}{4}$, 1-tiles.

By adding $6\frac{1}{4}$ 1-tiles, the result is a square whose area is $x^2 + 5x + 6\frac{1}{4}$, or $\left(x + \frac{5}{2}\right)^2$. ❖

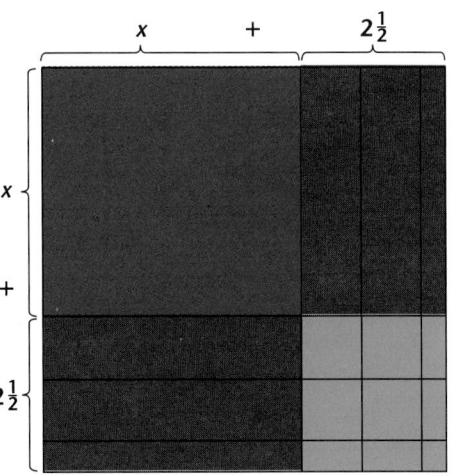

$x^2 + 5x + 6\frac{1}{4} = (x + 2\frac{1}{2})^2$

EXAMPLE 2

Complete the square for $x^2 - 6x$.

Solution ➤

One-half the coefficient of x is -3.
Add $(-3)^2$ or 9 to form a perfect square.
The perfect square is $x^2 - 6x + 9 = x^2 - 6x + (-3)^2 = (x - 3)^2$. ❖

How do you decide what to add to $x^2 + 6x$ and $x^2 - 6x$
to form a perfect square?

Completing the square gives a technique for rewriting the quadratic
$y = x^2 + 6x$ in the form $y = (x - h)^2 + k$.

Start with the quadratic.	$y = x^2 + 6x$
Find what you need to add to complete the square.	$\frac{1}{2} \cdot 6 = 3; \quad 3^2 = 9$
Add and subtract the square from the quadratic.	$y = x^2 + 6x + 9 - 9$
Group the terms.	$y = (x^2 + 6x + 9) - 9$
Write in the form $y = (x - h)^2 + k$.	$y = (x + 3)^2 - 9$

Why can you add $9 - 9$ to $x^2 + 6x$ without changing the value of the
expression?

MAXIMUM
MINIMUM
Connection

*The **minimum** value
for a quadratic function
$y = (x - h)^2 + k$
is the y-value of the
vertex.*

Minimum

The vertex of the graph of
$y = (x + 3)^2 - 9$ is $(-3, -9)$.
Thus, the minimum value
of $y = x^2 + 6x$ is -9.

EXAMPLE 3

Find the minimum value for the function $f(x) = x^2 - 3x$.

Solution ➤

The vertex of the parabola can be found by completing the square. The
coefficient of x is -3. Add and subtract the square of $-\frac{3}{2}$ to the equation.

$$f(x) = x^2 - 3x + \left(-\frac{3}{2}\right)^2 - \left(-\frac{3}{2}\right)^2$$

$$= \left(x - \frac{3}{2}\right)^2 - \frac{9}{4}$$

The graph of $f(x) = \left(x - \frac{3}{2}\right)^2 - \frac{9}{4}$ is a parabola with vertex $\left(\frac{3}{2}, -\frac{9}{4}\right)$ and axis

of symmetry $x = \frac{3}{2}$. Thus, the minimum value is $-\frac{9}{4}$. ❖

You can now complete the square for a quadratic of the form $y = x^2 + bx$. The next step is to complete the square for a quadratic of the form $y = x^2 + bx + c$.

EXAMPLE 4

A Rewrite $y = x^2 - 8x + 7$ in the form $y = (x - h)^2 + k$.

B Find the vertex of the parabola.

Solution ➤

A The new feature in this problem is the constant 7.

| Original function | $y = x^2 - 8x + 7$ |
| Group the x^2 and x terms. | $y = (x^2 - 8x) + 7$ |

First complete the square using the coefficient of the x-term, -8.

$$\frac{1}{2}(-8) = -4 \quad \text{and} \quad (-4)^2 = 16$$

| Now add and subtract **16**. | $y = (x^2 - 8x + 16) + 7 - 16$ |
| Write in the form $y = (x - h)^2 + k$. | $y = (x - 4)^2 - 9$ |

B The vertex is $(4, -9)$. Check by graphing. ❖

Try This   Find the minimum value for $y = x^2 - 6x + 11$.

EXERCISES & PROBLEMS

Communicate

1. Describe how to use algebra tiles to complete the square for $x^2 + 4x$.
2. Discuss how to complete the square algebraically for $x^2 - 7x$.
3. How do you rewrite $y = x^2 + 10x + 25 - 25$ in the form $y = (x - h)^2 + k$?

Describe how to find the minimum value for each quadratic function.

4. $y = x^2 + 2$ **5.** $y = x^2 - x$

6. Explain how to rewrite $y = x^2 - 10x + 11$ in the form $y = (x - h)^2 + k$. How do you find the vertex from this form of the equation?

Practice & Apply

Use algebra tiles to complete the square for each of the following. Write the area of each new square in the form $x^2 + bx + c$.

7. $x^2 + 14x$ **8.** $x^2 - 14x$ **9.** $x^2 + 8x$ **10.** $x^2 - 8x$

Complete the square.

11. $x^2 + 6x$ **12.** $x^2 - 2x$ **13.** $x^2 + 12x$ **14.** $x^2 - 12x$

15. $x^2 + 7x$ **16.** $x^2 - 10x$ **17.** $x^2 + 15x$ **18.** $x^2 - 5x$

19. $x^2 + 16x$ **20.** $x^2 + 20x$ **21.** $x^2 - 9x$ **22.** $x^2 + 40x$

Rewrite each function in the form $y = (x - h)^2 + k$.

23. $y = x^2 + 8x + 16 - 16$ **24.** $y = x^2 - 4x + 4 - 4$

25. $y = x^2 - 10x + 25 - 25$ **26.** $y = x^2 + 14x + 49 - 49$

27. $y = x^2 - 16x + 64 - 64$ **28.** $y = x^2 + 20x + 100 - 100$

Refer to the equation $y = x^2 + 10x$ to complete Exercises 29-31.

29. Complete the square and rewrite it in the form $y = (x - h)^2 + k$.

30. Find the vertex.

31. Find the maximum or minimum value.

Find the minimum value for each quadratic function.

32. $f(x) = x^2 + 3$ **33.** $f(x) = x^2 - 4$ **34.** $f(x) = x^2 + 2$

35. $f(x) = x^2 - 1$ **36.** $f(x) = x^2 + 6$ **37.** $f(x) = x^2 - 5$

Rewrite each function in the form $y = (x - h)^2 + k$. Find each vertex.

38. $y = x^2 + 4x - 1$ **39.** $y = x^2 - 2x - 3$ **40.** $y = x^2 + 10x - 12$

41. $y = x^2 - x + 2$ **42.** $y = x^2 - 6x - 5$ **43.** $y = x^2 + \frac{1}{3}x - 3$

A water rocket is shot vertically into the air with an initial velocity of 192 feet per second.

Physics   The relationship between time, t, and height, h, is given by the formula $h = -16t^2 + 192t$.

44. Graph the function. (If you use graphing technology you will need to use x in place of t and y in place of h.)

45. Complete the square. Find the vertex of the parabola.

46. What is the maximum height reached by the projectile?

47. How long does it take for the projectile to reach its maximum height?

A tennis player throws a ball vertically into the air and watches as it begins to descend. He extends his racket to make contact with the ball.

Physics The motion of the ball can be modeled by the function $h = -(t - 1)^2 + 9$, where h is the height of the ball in feet after t seconds have elapsed.

48. What is the maximum height reached by the tennis ball?

49. How long did it take for the tennis ball to reach its maximum height?

50. The player can make contact with the ball at a height of 9 feet. How much time will elapse before he hits the tennis ball?

51. If the tennis ball is allowed to fall to the ground, how much time will elapse?

 Look Back

Solve each system by graphing. [Lesson 6.1]

52. $\begin{cases} x - 3y = 3 \\ 2x - y = -4 \end{cases}$ **53.** $\begin{cases} x - 2y = 0 \\ x + y = 3 \end{cases}$

54. Can you find the inverse of $\begin{bmatrix} 4 & 6 \\ 2 & 3 \end{bmatrix}$? Explain. **[Lesson 7.4]**

55. **Probability** Find the probability that an even number will be drawn from a bag containing the whole numbers from 1 to 20 inclusive. **[Lesson 8.6]**

Write each of the following without negative exponents. [Lesson 10.3]

56. $a^2 b^{-3}$ **57.** $\dfrac{m^5}{m^{-2}}$ **58.** $\dfrac{2n^{-6}}{n^{-4}}$

59. **Geometry** Find the area of the square rug. **[Lesson 11.3]**

$\longleftarrow x + 4 \longrightarrow$

Look Beyond

The equation for the height in feet of a model rocket after t seconds is given by $h(t) = -16t^2 + 200t$. Find t when the rocket has a height of 400 feet

60. using the graphing method.

61. using the algebraic method.

Rescue at 2000 Feet

A Miraculous Sky Rescue

The jump began as a routine skydiving exercise, part of a convention of 420 parachutists sponsored by Skydive, Arizona, but it quickly turned into a test of nerve, instinct and courage...

Moments after he went out the open hatch of a four-engine DC-4 airplane near Coolidge, Ariz., Sky Diver Gregory Robertson, 35, could see that Debbie Williams, 31, a fellow parachutist with a modest 50 jumps to her credit, was in big trouble. Instead of "floating" in the proper stretched-out position parallel to the earth, Williams

was tumbling like a rag doll. In attempting to join three other divers in a hand-holding ring formation, she had slammed into the backpack of another chutist, and was knocked unconscious.

From his instructor's position above the other divers, Robertson reacted with instincts that had been honed by 1,700 jumps during time away from his job as an AT&T engineer in Phoenix. He straightened into a vertical dart, arms pinned to his body, ankles crossed, head aimed at the ground in what chutists call a "no-lift" dive, and plummeted toward Williams... <u>At 3,500 ft., about ten seconds before impact, Robertson caught up with Williams,</u> almost hitting her but slowing his own descent by assum-

ing the open-body frog-like position. He angled the unconscious sky diver so her chute could open readily and <u>at 2,000 ft., with some six seconds left, yanked the ripcord</u> on her emergency chute, then pulled his own ripcord. The two sky divers floated to the ground. Williams, a fifth-grade teacher from Post, Texas, landed on her back, suffering a skull fracture and a perforated kidney—but alive. In the history of recreational skydiving, there has never been such a daring rescue in anyone's recollection.

Robertson explains how he saved a life at 2,000 ft.

Does the article give you a complete picture of the rescue? How can you tell if the numbers in the article are accurate?

You can use what you know about distance, time, and speed to check the facts in the article and to get a fuller sense of what took place during this amazing feat.

First, study these diagrams to get an idea of some of the forces that affect sky divers in free fall.

When a parachutist jumps out of a plane, the force of gravity causes the diver to accelerate downward.

As the sky diver's speed increases, the force of air resistance (or drag) plays a larger and larger role.

At some point, the two forces balance and the sky diver no longer accelerates, but falls at a steady rate, called the terminal velocity.

Cooperative Learning

1. Use the fact that 88 feet per second equals
    60 miles per hour to convert. Answer Part 1 to
    see if the underlined part of the article makes sense.

    Based on the data in the article, how long did it take Robertson to pull
    Williams's ripcord after he caught up to her? During that time, how far did
    the two divers fall according to the article? Given that the maximum speed
    of a person in free fall is about 190–200 miles per hour, is the data in the
    article accurate? Explain.

2. After Robertson caught up with Williams, the two divers were probably
    falling at a speed between 125 and 150 miles per hour. Suppose they
    fell at 135 miles per hour. You can use that value to find more plausible
    time estimates for the rescue.

    About how many seconds from impact would the two sky divers be at 3500
    feet? at 2000 feet? Suppose Robertson did catch up to Williams at 3500 feet and
    did pull her ripcord at 2000 feet. About how much time would have elapsed
    between those two events if the divers were falling at 135 miles per hour?

3. While Robertson was catching up to Williams, he changed his body position
    to speed up and then slow down. Such motion is difficult to calculate exactly,
    but you can make estimates for this part of the rescue by using a single
    average velocity for Robertson. Assume that Robertson was at 8500 feet and
    Williams was at 8300 feet when Robertson started trying to catch up.

    Write an equation for t in terms of v_r and v_w, where

    v_r is Robertson's average velocity in feet per second while catching
    up to Williams.

    v_w is Williams' velocity in feet per second.

    t is the time in seconds from when Robertson starts to go after
    Williams to when he reaches her.

4. Use your equation to find t if v_r = 206 (about 140 miles per hour) and
    v_w = 198 (about 135 miles per hour). Based on your results, at what altitude
    would Robertson catch up to Williams?

5. Imagine you are a journalist writing about the rescue. Use your data from
    Parts 2, 3, and 4 to create a paragraph that will give your readers a reasonably
    accurate sense of the event.

Solving Equations of the Form $x^2 + bx + c = 0$

You have already solved equations of the form $x^2 = k$. Equations of the form $x^2 + bx + c = 0$ can be solved by completing the square.

In Lesson 12.3 you found the vertex of $y = x^2 - 8x + 7$ by completing the square. When the function is rewritten in the form $y = (x - 4)^2 - 9$, the vertex $(4, -9)$ can be noted by inspection.

$$(x - 4)^2 - 9 = 0$$

$$(x - 4)^2 = 9$$

$$x - 4 = \pm 3$$

$$x - 4 = 3 \quad \text{or} \quad x - 4 = -3$$

The solutions are 7 and 1. These two solutions are the x-intercepts of $y = x^2 - 8x + 7$ and the zeros of $f(x) = x^2 - 8x + 7$.

$$f(7) = 7^2 - 8(7) + 7 = 49 - 56 + 7 = 0$$

$$f(1) = 1^2 - 8(1) + 7 = 1 - 8 + 7 = 0$$

Thus, the parabola intersects the x-axis at $(7, 0)$ and $(1, 0)$.

CRITICAL Thinking

Why are the zeros of the function $f(x) = x^2 - 8x + 6$ and the solutions to the equation $x^2 - 8x + 6 = 0$ the same?

EXAMPLE 1

A Solve $x^2 - 6x + 8 = 0$.

B Find the zeros of $h(x) = x^2 - 6x + 7$.

Solution ➤

For each problem, write the expression in the form $(x - h)^2 + k$ by completing the square. Then use the $x^2 = k$ form to find the solution. Check by substitution.

A $x^2 - 6x + 8 = 0$
$x^2 - 6x = -8$
$x^2 - 6x + 9 = -8 + 9$
$(x - 3)^2 = 1$
$x - 3 = \sqrt{1}$ or $x - 3 = -\sqrt{1}$
$x = 4$ or $x = 2$

The solutions are 4 and 2.

B $x^2 - 6x + 7 = 0$
$x^2 - 6x = -7$
$x^2 - 6x + 9 = -7 + 9$
$(x - 3)^2 = 2$
$x - 3 = \sqrt{2}$ or $x - 3 = -\sqrt{2}$
$x = 3 + \sqrt{2}$ or $x = 3 - \sqrt{2}$

The zeros are $3 + \sqrt{2}$ and $3 - \sqrt{2}$. ❖

Try This Solve $x^2 - 7x + 10 = 0$.

When a quadratic can be factored mentally you can quickly solve the quadratic equation. For example, factor $x^2 - 6x + 8$ into $(x - 4)(x - 2)$.

Thus, $x^2 - 6x + 8 = 0$ is the same as $(x - 4)(x - 2) = 0$.

If the product of two factors is equal to zero, then at least one of the factors must be zero. This generalization is called the **Zero Product Property.**

ZERO PRODUCT PROPERTY
If a and b are real numbers such that $ab = 0$, then $a = 0$ or $b = 0$.

The Zero Product Property can be used to solve $(x - 4)(x - 2) = 0$.

$x - 4 = 0$ or $x - 2 = 0$
$x = 4$ or $x = 2$

The solutions are 4 and 2.

Check

$$x^2 - 6x + 8 = (4)^2 - 6(4) + 8 = 16 - 24 + 8 = 0 \quad \text{True}$$

or

$$x^2 - 6x + 8 = (2)^2 - 6(2) + 8 = 4 - 12 + 8 = 0 \quad \text{True}$$

EXAMPLE 2

Solve $x^2 + x = 2$ by the indicated method.

A Completing the square   **B** Factoring

Solution ▸

A
$$x^2 + x = 2$$
$$x^2 + x + \left(\frac{1}{2}\right)^2 = 2 + \frac{1}{4}$$
$$\left(x + \frac{1}{2}\right)^2 = \frac{9}{4}$$

$$x + \frac{1}{2} = \frac{3}{2} \quad \text{or} \quad x + \frac{1}{2} = -\frac{3}{2}$$
$$x = \frac{3}{2} - \frac{1}{2} \quad \text{or} \quad x = -\frac{3}{2} - \frac{1}{2}$$
$$x = 1 \quad \text{or} \quad x = -2$$

B
$$x^2 + x = 2$$
$$x^2 + x - 2 = 0$$
$$(x + 2)(x - 1) = 0$$

$$x + 2 = 0 \quad \text{or} \quad x - 1 = 0$$
$$x = -2 \quad \text{or} \quad x = 1$$

The solutions are 1 and -2. ❖

How would you decide which method to use?

EXAMPLE 3

Let $f(x) = x^2 + 3x - 15$. Find the value of x when $f(x)$ is -5.

Graphics Calculator

Solution A ▸

Graph $y = x^2 + 3x - 15$ and $y = -5$ on the same set of axes. Use a graphics calculator if you have one. Find the point or points of intersection. You see that when $x = 2$ and $x = -5$, then $f(x) = -5$.

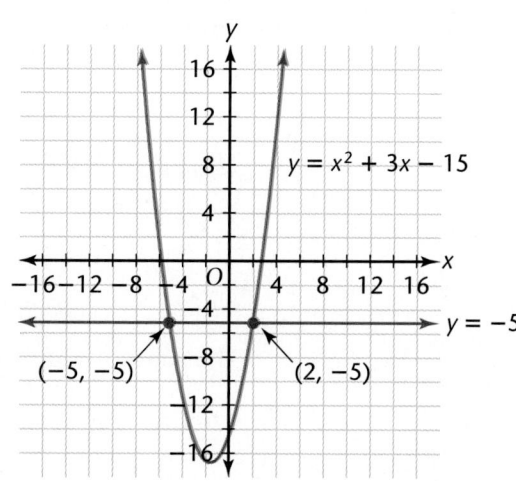

Solution B ➤

Since we want $x^2 + 3x - 15$ to equal -5, write $x^2 + 3x - 15 = -5$. Add 5 to both sides of the equation. Now, $x^2 + 3x - 10 = 0$. Factor the quadratic, and use the Zero Product Property to find the possible values of x.

$$(x - 2)(x + 5) = 0$$
$$x - 2 = 0 \quad \text{or} \quad x + 5 = 0$$
$$x = 2 \quad \text{or} \quad x = -5 \qquad \text{Check by substitution. } ❖$$

Try This   Solve $x^2 - 8x + 22 = 10$.

EXAMPLE 4

GEOMETRY
Connection

A box with a square base and no top is to be made by cutting out 2-inch squares from each corner of a square piece of cardboard and folding up the sides. The volume of the box is to be 8 cubic inches. What size should the piece of cardboard be?

Solution ➤

Let x be the length (and width) of the piece of cardboard.

$V = 8$ cubic inches

The volume of a rectangular box is
$V = \text{length} \cdot \text{width} \cdot \text{height}$.

$8 = (x - 4)(x - 4)(2)$	$V = lwh$
$4 = (x - 4)(x - 4)$	Divide both sides by 2.
$4 = x^2 - 8x + 16$	Simplify.
$x^2 - 8x + 12 = 0$	Subtract 4 from both sides.
$(x - 6)(x - 2) = 0$	Factor.
$x - 6 = 0 \quad \text{or} \quad x - 2 = 0$	Zero Product Property
$x = 6 \quad \text{or} \quad x = 2$	Addition Property of Equality

Can you cut 2-inch squares from each corner of a 2-inch square?

The value of x cannot equal 2. The piece of cardboard should be 6 inches by 6 inches. ❖

EXAMPLE 5

Find the points where the graph of $y = x - 3$ intersects the graph of $y = x^2 - 10x + 21$.

Solution A ➤

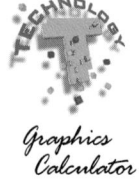

*Graphics
Calculator*

Graph the equations, and find the points of intersection.

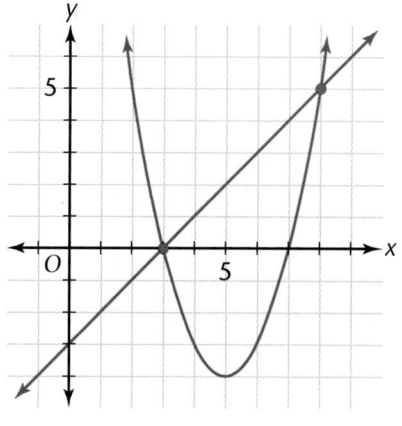

The graphs intersect at (8, 5) and (3, 0).

Solution B ➤

Solve the system. $\begin{cases} x^2 - 10x + 21 = y \\ y = x - 3 \end{cases}$

First, substitute $x - 3$ (from the second equation) for y (in the first equation) and simplify.

$x^2 - 10x + 21 = x - 3$
$x^2 - 11x + 24 = 0$

Then solve the new quadratic by factoring.

$(x - 8)(x - 3) = 0$
$x - 8 = 0$ or $x - 3 = 0$
$\quad x = 8$ or $\quad x = 3$

Substitute each value for x into the equation $y = x - 3$.

$y = x - 3$	$y = x - 3$
$y = 8 - 3$	$y = 3 - 3$
$y = 5$	$y = 0$

The graphs intersect at (8, 5) and (3, 0). ❖

Try This Find the points where the graph of $y = -2x + 4$ intersects the graph of $y = x^2 - 4x + 1$.

EXERCISES & PROBLEMS

Communicate

1. How are x-intercepts related to the zeros of a function?

2. Explain how to find the zeros of $f(x) = x^2 - 6x + 8$ by completing the square.

3. Describe how to solve $x^2 - 2x = 15$ by completing the square.

4. Discuss how to solve $x^2 + 10x = 24$ by factoring.

Which method would you choose to solve the following problems? Explain why.

5. $x^2 + 12x + 36 = 0$

6. $x^2 - 8x - 9 = 0$

Graph $f(x) = x^2 - 7x + 12$.

7. Explain how to find the value of x when $f(x)$ is 2.

8. Describe how to find where the graph of $y = x - 1$ intersects the graph of $y = x^2 - 3x + 2$.

Practice & Apply

Find the x-intercepts or zeros of each function. Graph to check.

9. $y = x^2 - 2x - 8$

10. $y = x^2 + 6x + 5$

11. $y = x^2 - 4x + 4$

Find the solution of each equation by factoring.

12. $x^2 - 2x - 3 = 0$

13. $x^2 + 4x - 5 = 0$

14. $x^2 + 7x + 12 = 0$

15. $x^2 - 10x + 24 = 0$

16. $x^2 - 3x - 10 = 0$

17. $x^2 - 8x + 15 = 0$

18. $x^2 - 6x + 9 = 0$

19. $x^2 + 10x + 25 = 0$

20. $x^2 - 2x + 1 = 0$

Find the solution of each equation by completing the square.

21. $x^2 - 2x - 15 = 0$

22. $x^2 + 4x - 5 = 0$

23. $x^2 - x - 20 = 0$

24. $x^2 + 2x - 8 = 0$

25. $x^2 - 4x - 12 = 0$

26. $x^2 + x - 6 = 0$

27. $x^2 + 2x - 5 = 0$

28. $x^2 + 4x - 1 = 0$

29. $x^2 + 8x + 13 = 0$

Lesson 12.4   Solving Equations of the Form $x^2 + bx + c = 0$ **571**

Solve each equation by factoring or completing the square.

30. $b^2 + 10b = 0$ **31.** $r^2 - 10r + 24 = 0$ **32.** $x^2 - x - 3 = 0$ **33.** $x^2 + 4x - 12 = 0$

34. $x^2 + 3x - 5 = 0$ **35.** $x^2 - 3x - 1 = 0$ **36.** $x^2 - 5x - 3 = 0$ **37.** $x^2 - x = 0$

**Refer to the equation $f(x) = x^2 - 4x + 3$ to complete Exercises 38-41.
Graph the equation to find a value for x when $f(x)$ is the
following value. Check by substituting the given value of $f(x)$.**

38. $f(x) = 3$ **39.** $f(x) = 8$ **40.** $f(x) = -1$ **41.** $f(x) = 15$

**Find the point or points where the graphs intersect. Graph to
check.**

42. $\begin{cases} y = 9 \\ y = x^2 \end{cases}$ **43.** $\begin{cases} y = 4 \\ y = x^2 - 2x + 1 \end{cases}$ **44.** $\begin{cases} y = x - 1 \\ y = x^2 - 3x + 3 \end{cases}$ **45.** $\begin{cases} y = x + 3 \\ y = x^2 - 4x + 3 \end{cases}$

46. Find two consecutive even integers whose product is 224.

47. **Geometry** The length of a rectangle is 4 yards longer than the width. Find the length and the width.

$x + 4$

x

$A = 60$ square yards

48. Photography A perimeter of a photograph is 80 centimeters. Find the dimensions of the photo for a frame if the area is 396 square centimeters.

A marching band uses a rectangular formation with 8 columns and 10 rows.

49. An equal number of columns and rows were added to the formation when 40 new members joined at the start of a new school year. How many rows and columns were added?

50. The difference of two numbers is 3. What are the numbers if the sum of their squares is 117?

51. Physics A boat fires an emergency flare that travels upward at an initial velocity of 25 meters per second. To find the height in meters, h, of the flare at t seconds, use the formula $h = 25t - 5t^2$. After how many seconds will the flare be at a height of 10 meters? Round answers to the nearest tenth.

Look Back

Write each system in its matrix equation form. [Lesson 7.5]

52. $\begin{cases} x - 2y = 1 \\ 4x + 2y = -1 \end{cases}$

53. $\begin{cases} 5x - 2y = 11 \\ 3x + 5y = 19 \end{cases}$

Statistics Find the mean and the mode for the following sets of numbers. [Lesson 8.3]

54. 10, 15, 20, 36, 10, 25

55. 8, 10.5, 30, 15, 12, 10.5

Write each number in decimal form. [Lesson 10.4]

56. $4 \cdot 10^4$

57. $6.5 \cdot 10^7$

58. $9.6 \cdot 10^{-5}$

Write each polynomial in factored form. [Lesson 11.2]

59. $4b^2 + 6b - 36$

60. $6w(w^2 - w)$

61. $8p^4 - 16p$

Look Beyond

62. Factor $2x^2 + x - 6$.

63. Earlier you learned a pattern that simplifies the process of determining the powers of binomials. Can you find a pattern to determine the powers of a trinomial, $(a + b + c)^n$? What are the first 3 powers?

64. Imagine that someone invented a number for $\sqrt{-1}$ and called it i. Consider i^0 to be 1. If $i = \sqrt{-1}$, then what is i^2 or i^3? What is i^6? Can you find a pattern that will let you predict the value of i^{92}?

LESSON 12.5 The Quadratic Formula

In the Shot-put event in track and field, a shot is tossed into the air from a position about 6 feet above the ground.

Why An object thrown in the air can be modeled by a quadratic function. The zeros of the quadratic function tell when the object is at ground level. When you derive the quadratic formula by completing the square, the formula will provide a way to find those zeros.

If the shot is tossed with an initial horizontal and vertical velocity of 31 feet per second, the time that the shot travels can be found by solving the quadratic equation $-16t^2 + 31t + 6.375 = 0$. You can then use t to solve for the horizontal distance, d, in the formula $d = 31t$.

Cultural Connection: Asia Since early Babylonian times scholars have known how to solve the standard quadratic equation $ax^2 + bx + c = 0$ by using a formula.

To see how this quadratic formula can be derived, follow the steps as the equations $2x^2 + 5x + 1 = 0$ and $ax^2 + bx + c = 0$ are solved.

Divide both sides of each equation by the coefficient of x^2.

$$2x^2 + 5x + 1 = 0 \qquad\qquad ax^2 + bx + c = 0$$

$$x^2 + \frac{5}{2}x + \frac{1}{2} = 0 \qquad\qquad x^2 + \frac{b}{a}x + \frac{c}{a} = 0$$

Subtract the constant term from both sides of each equation.

$$x^2 + \frac{5}{2}x = -\frac{1}{2} \qquad\qquad x^2 + \frac{b}{a}x = -\frac{c}{a}$$

Complete the square.

$$x^2 + \frac{5}{2}x + \left(\frac{5}{4}\right)^2 = -\frac{1}{2} + \left(\frac{5}{4}\right)^2 \qquad x^2 + \frac{b}{a}x + \left(\frac{b}{2a}\right)^2 = -\frac{c}{a} + \left(\frac{b}{2a}\right)^2$$

$$\left(x + \frac{5}{4}\right)^2 = -\frac{1}{2} + \frac{25}{16} \qquad\qquad \left(x + \frac{b}{2a}\right)^2 = -\frac{c}{a} + \frac{b^2}{4a^2}$$

Work with the right side of each equation to simplify.

$$\left(x + \frac{5}{4}\right)^2 = -\frac{8 \cdot 1}{8 \cdot 2} + \frac{25}{16} \qquad\qquad \left(x + \frac{b}{2a}\right)^2 = -\frac{4a \cdot c}{4a \cdot a} + \frac{b^2}{4a^2}$$

$$\left(x + \frac{5}{4}\right)^2 = \frac{25 - 8}{16} \text{ or } \frac{17}{16} \qquad\qquad \left(x + \frac{b}{2a}\right)^2 = \frac{b^2 - 4ac}{4a^2}$$

Use $x^2 = k$ to solve each equation.

$$x + \frac{5}{4} = \pm\sqrt{\frac{17}{16}} \qquad\qquad x + \frac{b}{2a} = \pm\sqrt{\frac{b^2 - 4ac}{4a^2}}$$

$$x + \frac{5}{4} = \pm\frac{\sqrt{17}}{\sqrt{16}} \qquad\qquad x + \frac{b}{2a} = \pm\frac{\sqrt{b^2 - 4ac}}{\sqrt{4a^2}}$$

$$x = -\frac{5}{4} \pm \frac{\sqrt{17}}{4} \qquad\qquad x = -\frac{b}{2a} \pm \frac{\sqrt{b^2 - 4ac}}{2a}$$

$$x = \frac{-5 \pm \sqrt{17}}{4} \qquad\qquad x = \frac{-b \pm \sqrt{b^2 - 4ac}}{2a}$$

You have derived the quadratic formula.

THE QUADRATIC FORMULA

The solutions of the quadratic equation $ax^2 + bx + c = 0$ are

$$\frac{-b \pm \sqrt{b^2 - 4ac}}{2a}.$$

Notice that if you substitute the values $a = 2$, $b = 5$, and $c = 1$ from $2x^2 + 5x + 1 = 0$ into the formula $x = \dfrac{-b \pm \sqrt{b^2 - 4ac}}{2a}$, you get the following result.

$$x = \frac{-5 \pm \sqrt{25 - (4 \cdot 2 \cdot 1)}}{2 \cdot 2} = \frac{-5 \pm \sqrt{17}}{4}$$

EXAMPLE 1

Use the quadratic formula to solve $x^2 - 10 + 3x = 0$.

Solution ➤

Rewrite $x^2 - 10 + 3x = 0$ in the form $x^2 + 3x - 10 = 0$. Notice that $a = 1$, $b = 3$, and $c = -10$. Substitute and simplify.

$$x = \frac{-3 \pm \sqrt{(3)^2 - 4(1)(-10)}}{2(1)}$$

$$x = \frac{-3 \pm \sqrt{9 + 40}}{2}$$

$$x = \frac{-3 + 7}{2} \quad \text{or} \quad x = \frac{-3 - 7}{2}$$

$$x = \frac{4}{2} = 2 \quad \text{or} \quad x = \frac{-10}{2} = -5$$

The solutions are 2 and -5. ❖

Try This   Use the quadratic formula to solve $x^2 + x - 12 = 0$.

You now have three different methods for solving a quadratic equation.

1. Factoring

2. Completing the Square

3. The Quadratic Formula

CRITICAL
Thinking

Use all three methods to solve $2x^2 - 7x - 4 = 0$. Which method do you think is best for solving real applications? When is factoring the best way to solve a quadratic equation?

EXAMPLE 2

Find the zeros of $h(x) = 3x^2 + 2x - 4$.

Solution ➤

Use the quadratic formula, where $a = 3$, $b = 2$, and $c = -4$.

$$x = \frac{-2 \pm \sqrt{4 - (-48)}}{6}$$

$$x = \frac{-2 \pm \sqrt{52}}{6}$$

A calculator shows that $x \approx 0.87$ or $x \approx -1.54$. Check by substitution. ❖

How would you use the quadratic formula to solve $3n^2 - 2n = 2$?

You can now solve the quadratic equation
$-16t^2 + 31t + 6.375 = 0$ that models the
shot-put problem.

Let $a = -16$, $b = 31$ and $c = 6.375$.

$$t = \frac{-31 \pm \sqrt{(31)^2 - (4)(-16)(6.375)}}{2(-16)} = \frac{-31 \pm \sqrt{961 + 408}}{-32}$$

$t = -0.1875$ or $t = 2.125$ and $d = 31t$, so
$d = 31(2.125) = 65.875$

The shot travels a horizontal distance of
about 66 feet.

•Exploration • *The Discriminant*

Examine each graph.

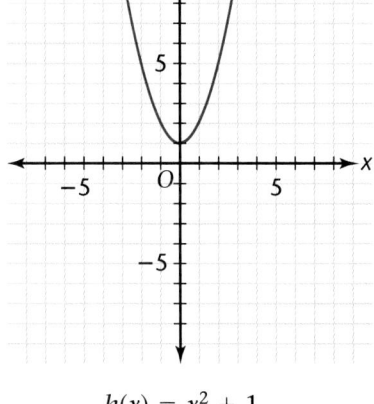

$f(x) = x^2 + 6x + 5$ $\qquad$ $g(x) = x^2 - 10x + 25$ $\qquad$ $h(x) = x^2 + 1$

Each expression is written in the form $ax^2 + bx + c$.

 Find the values of a, b, and c, and complete the chart. The expression $b^2 - 4ac$ is called the **discriminant** of the quadratic formula.

Function	a	b	c	b² − 4ac	Number of x-intercepts
f	?	?	?	?	?
g	?	?	?	?	?
h	?	?	?	?	?

 Explain the role of the discriminant in determining the number of solutions to a quadratic equation. ❖

EXAMPLE 3

What does the discriminant tell you about the solution to each of the following equations?

A $3x^2 - 2x + 1 = 0$ **B** $4x = 4x^2 + 1$ **C** $2x^3 + 3x + 1 = 0$

Solution ➤

A Find $b^2 - 4ac$ where $a = 3$, $b = -2$, and $c = 1$. The discriminant is $(-2)^2 - 4 \cdot 3 \cdot 1 = 4 - 12 = -8$. If the discriminant is negative, there are no x-intercepts, and $3x^2 - 2x + 1 = 0$ has no real solutions.

B Rewrite the equation in standard form to find a, b, and c. If $4x^2 - 4x + 1 = 0$, then $a = 4$, $b = -4$, and $c = 1$. The discriminant is $4^2 - 4 \cdot 4 \cdot 1 = 16 - 16 = 0$. If the discriminant equals 0, there is exactly one real solution, and $4x = 4x^2 + 1$ has one real solution.

C The equation is not a quadratic. There is no discriminant. ❖

When the discriminant is a perfect square, the quadratic equation can be factored over the integers. You can use the quadratic formula to find the factors.

EXAMPLE 4

Factor $6x^2 + 23x + 20$.

Solution ➤

Evaluate the discriminate.

$$b^2 - 4ac = (23)^2 - 4(6)(20) = 49 \text{ which is a perfect square.}$$

Find the solution to $6x^2 + 23x + 20 = 0$.

$$x = \frac{-b \pm \sqrt{b^2 - 4ac}}{2a} = \frac{-23 \pm \sqrt{49}}{12} = \frac{-23 \pm 7}{12}$$

Simplify.

$$x = -\frac{30}{12} = -\frac{5}{2} \text{ or } x = -\frac{16}{12} = -\frac{4}{3}$$

Work backward to find the factors.

$x = -\dfrac{5}{2}$ or	$x = -\dfrac{4}{3}$	Given
$2x = -5$ or	$3x = -4$	Multiplication Property of Equality
$2x + 5 = 0$ or	$3x + 4 = 0$	Addition Property of Equality

The factors of $6x^2 + 23x + 20$ are $2x + 5$ and $3x + 4$. Check by multiplying the factors. ❖

EXERCISES & PROBLEMS

Communicate

1. Discuss how to identify the coefficients a, b, and c in the equation $2y^2 + 3y = 7$ in order to use the quadratic formula. Why is the standard form of the equation important?

2. How do you find the discriminant for $2y^2 + 3y - 7 = 0$?

Discuss how to use the discriminant to determine the number of solutions to each equation.

3. $2x^2 - x - 2 = 0$

4. $x^2 - 2x = -7$

5. Describe how to use the quadratic formula to find the solution to $4y^2 + 12y = 7$.

6. If the solutions to $4x^2 + 12x + 5 = 0$ are $-\frac{1}{2}$ and $-\frac{5}{2}$, discuss how to work backward to find the factors of $4x^2 + 12x + 5$.

Practice & Apply

Identify a, b, and c for each quadratic equation.

7. $x^2 - 4x - 5 = 0$

8. $r^2 - 5r - 4 = 0$

9. $5n^2 - 2n - 1 = 0$

10. $-p^2 - 3p + 7 = 0$

11. $3t^2 - 45 = 0$

12. $-3m + 7 - m^2 = 0$

Find the value of the discriminant for each quadratic equation.

13. $x^2 - x - 3 = 0$

14. $x^2 + 2x - 8 = 0$

15. $x^2 + 8x + 13 = 0$

16. $z^2 + 4z - 21 = 0$

17. $2y^2 - 8y - 8 = 0$

18. $8y^2 - 2 = 0$

19. $4k^2 - 3k = 5$

20. $3x^2 - x + 4 = 0$

21. $4x^2 - 12x + 9 = 0$

Use the quadratic formula to solve each equation. Give answers to the nearest hundredth when necessary. Check by substitution.

22. $a^2 - 4a - 21 = 0$

23. $t^2 + 6t - 16 = 0$

24. $m^2 + 4m - 5 = 0$

25. $w^2 - 4w = 0$

26. $x^2 + 9x = 0$

27. $x^2 - 9 = 0$

28. $x^2 + 2x = 0$

29. $3m^2 = 2m + 1$

30. $-x^2 + 6x - 9 = 0$

31. $2r^2 - r - 3 = 0$

32. $2w^2 - 5w + 3 = 0$

33. $x^2 + 6x + 3 = 0$

34. $10y^2 + 7y = 12$

35. $3x^2 - 2x - 7 = 0$

36. $8y^2 + 7y - 2 = 0$

Choose any method to solve the following quadratic equations. Report answers to the nearest hundredth when necessary.

37. $x^2 + 4x - 12 = 0$ **38.** $x^2 - 13 = 0$ **39.** $2x^2 + 5x - 3 = 0$

40. $x^2 + 7x + 10 = 0$ **41.** $2a^2 - 5a - 12 = 0$ **42.** $3x^2 - 2x - 2 = 0$

43. $3x^2 - 7x - 3 = 0$ **44.** $8x^2 + 10x + 3 = 0$ **45.** $3x^2 + 10x - 5 = 0$

46. Accounting To approximate profits per day for her business, Mrs. Haag uses the following formula.

$$p = -x^2 + 50x - 350$$

The profit, p, is derived from selling x cases of decorator napkins. How many cases of napkins must she sell to make a maximum profit? Find the maximum profit.

47. The number of seats in each row of a theater is 16 fewer than the number of rows. How many seats are in each row of a 1161-seat theater?

48. **Portfolio Activity** Complete the problem in the portfolio activity on page 547.

Look Back

49. Probability Find the probability of guessing at least 60% of the correct answers on a 5-question multiple-choice quiz in which each item has 4 possible responses. **[Lesson 8.2]**

Write each of the following numbers in scientific notation. [Lesson 10.4]

50. 3,000,000   **51.** 67,000   **52.** 0.006   **53.** 0.0000009

Graph. [Lesson 10.5]

54. $y = 2^x + 4$ **55.** $y = 2 \cdot 3^x$

56. Geometry Write the expression for the perimeter of the rectangle. **[Lesson 11.2]**

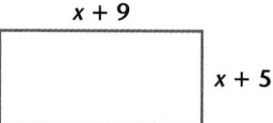

Multiply. [Lesson 11.3]

57. $(x - 5)(x + 3)$ **58.** $(2x + 4)(2x - 2)$

Look Beyond

59. From what you know about graphing linear inequalities, make a conjecture about how you might graph $x^2 + 3x + 2 < 0$.

LESSON 12.6 Graphing Quadratic Inequalities

Why *Not all problems involving quadratics use equations. Sometimes you will need to solve an inequality to get an answer. Solving a quadratic inequality includes techniques you already know with one more important step.*

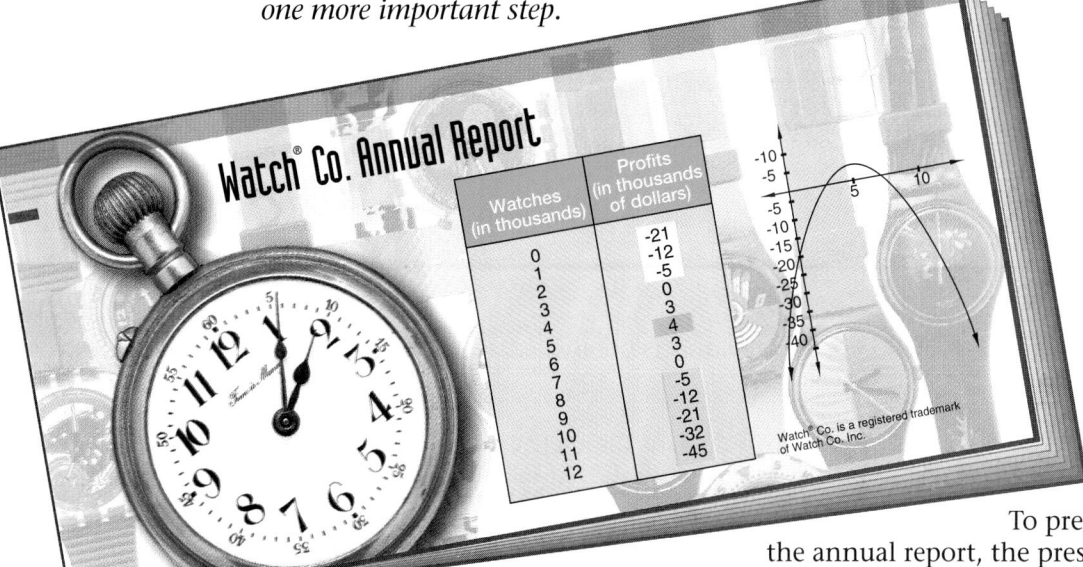

Watch® Co. Annual Report

Watches (in thousands)	Profits (in thousands of dollars)
0	-21
1	-12
2	-5
3	0
4	3
5	4
6	3
7	0
8	-5
9	-12
10	-21
11	-32
12	-45

Watch® Co. is a registered trademark of Watch Co. Inc.

To prepare the annual report, the president of a watch company gathers data about the number of watches produced per week and the corresponding profit. He uses the model $p = -x^2 + 10x - 21$ in which p represents profit and x represents the number of watches produced per week.

Exploration *Production and Profit*

MAXIMUM MINIMUM *Connection*

Use the graph of the profit function and the table of values to answer each question.

1. At what levels of production is the profit 0?

2. At what levels of production is the profit greater than 0?

3. At what levels of production is the profit less than 0?

4. How many watches should be produced to maximize profit?

5. According to the consultant's model, what is the maximum profit that the company can make? ❖

In the Exploration, you solved a quadratic equation and two quadratic inequalities.

$-x^2 + 10x - 21 = 0$	$-x^2 + 10x - 21 > 0$	$-x^2 + 10x - 21 < 0$
quadratic equation	**quadratic inequality**	**quadratic inequality**

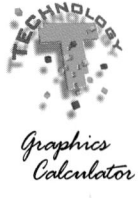

EXAMPLE 1

Solve $x^2 - 2x - 15 > 0$.

Solution A ➤

A graphics calculator can be used to graph
$Y = x^2 - 2x - 15$. Trace the graph to see where
the values are greater than 0.

Graphics Calculator

Solution B ➤

Use the x-axis as a number line.

1. Solve the related quadratic equation using the Zero Product Property.

$$x^2 - 2x - 15 = 0$$
$$(x - 5)(x + 3) = 0$$
$$x - 5 = 0 \quad \text{or} \quad x + 3 = 0$$
$$x = 5 \quad \text{or} \quad x = -3$$

2. Plot the solutions on an x-axis. When would you darken the points at -3 and 5?

3. The two solutions divide the x-axis into three intervals, to the left, between, and to the right of the two points. Test each interval.

$$\begin{array}{ccccccc} & & & & \circ & & & & & \circ & \\ \hline -6 & -4 & -2 & 0 & 2 & 4 & 6 \end{array}$$

Choose a number from each interval. Substitute those numbers for x in the inequality.

$x^2 - 2x - 15 > 0$	$x^2 - 2x - 15 > 0$	$x^2 - 2x - 15 > 0$
$(-5)^2 - 2(-5) - 15 \, ? \, 0$	$(0)^2 - 2(0) - 15 \, ? \, 0$	$(6)^2 - 2(6) - 15 \, ? \, 0$
$25 + 10 - 15 \, ? \, 0$	$0 - 0 - 15 \, ? \, 0$	$36 - 12 - 15 \, ? \, 0$
$20 > 0$	$-15 > 0$	$9 > 0$
Solution interval	**No**	**Solution interval**

4. Graph the solutions.

$$\begin{array}{ccccccc} & & \circ & & & & & \circ & \\ \hline & & -3 & & 0 & & & 5 \end{array}$$

$$x < -3 \quad \text{or} \quad x > 5 \quad \diamond$$

Try This Solve $x^2 + x - 12 \geq 0$.

CRITICAL *Thinking*

Explain how you could tell which intervals should be shaded without testing numbers from each interval.

EXAMPLE 2

Graph $y > -x^2 + 1$.

Solution ➤

1. Graph $y = -x^2 + 1$. Draw the parabola with a dashed line to show the solutions of $y = x^2 + 1$ are *not* on the graph.

2. To graph $y > -x^2 + 1$, shade the points above the parabola.

3. As a check, test points inside and outside the parabola.

Test $(0, 0)$ inside.	Test $(0, 3)$ outside.
$y > -x^2 + 1$	$y > -x^2 + 1$
$0 > 0 + 1$?	$3 > 0 + 1$?
$0 > 1$? No	$3 > 1$? Yes

4. Shade the region outside the parabola.

The graph of $y > -x^2 + 1$ is the **set of points outside the parabola**. ❖

You can test points inside and outside the graph of $y = -x^2 + 1$ to show that the shaded region on each graph below represents the given set of points.

$y < -x^2 + 1$

$y \geq -x^2 + 1$

$y \leq -x^2 + 1$

EXAMPLE 3

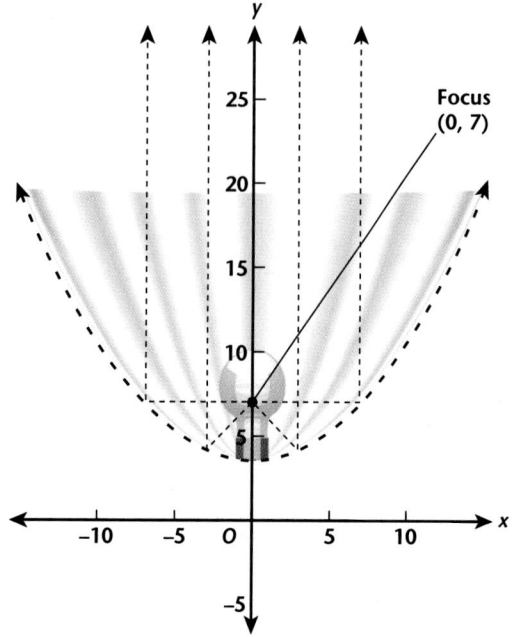

A cross section of the parabolic light reflector at the right is described by the equation $y = \frac{1}{14}x^2 + \frac{7}{2}$. The bulb inside the reflector is located at the point (0, 7). Light bounces off the reflector in parallel rays. This quality allows flashlights to direct light in a narrow beam. Determine which of the inequalities indicate the region in which the light bulb is located.

A $y > \frac{1}{14}x^2 + \frac{7}{2}$ **B** $y < \frac{1}{14}x^2 + \frac{7}{2}$

Solution ➤

A $y > \frac{1}{14}x^2 + \frac{7}{2}$

Test (0, 7).

$7 > \left(\frac{1}{14}\right)(0) + \frac{7}{2}$

$7 > \frac{7}{2}$ Yes

B $y < \frac{1}{14}x^2 + \frac{7}{2}$

Test (0, 7).

$7 < \left(\frac{1}{14}\right)(0) + \frac{7}{2}$

$7 < \frac{7}{2}$ No

The inequality $y > \frac{1}{14}x^2 + \frac{7}{2}$ contains the point of focus inside the parabola. ❖

EXERCISES & PROBLEMS

Communicate

1. Explain how to solve $x^2 - 9x + 18 > 0$ using the Zero Product Property.

2. Describe how to graph and check the solution intervals to Exercise 1 on a number line.

Explain how to determine if the boundary line is solid or dashed for the graph of each inequality.

3. $y < x^2 - 3$ **4.** $y \leq x^2 - 3$ **5.** $y \geq x^2 - 3$

Discuss how to graph each inequality. Explain how to decide whether to shade inside or outside the parabola for each solution.

6. $y < x^2 - 3$ **7.** $y \leq x^2 - 3$ **8.** $y \geq x^2 - 3$

Practice & Apply

Solve each quadratic inequality using the Zero Product Property. Graph the solution on a number line.

9. $x^2 - 6x + 8 > 0$
10. $x^2 + 3x + 2 < 0$
11. $x^2 - 9x + 14 \leq 0$

12. $x^2 + 2x - 15 > 0$
13. $x^2 - 7x + 6 < 0$
14. $x^2 - 8x - 20 < 0$

15. $x^2 - 8x + 15 \leq 0$
16. $x^2 - 3x - 18 \leq 0$
17. $x^2 - 11x + 30 \geq 0$

18. $x^2 - 4x - 12 \geq 0$
19. $x^2 + 5x - 14 > 0$
20. $x^2 + 10x + 24 \geq 0$

Graph each quadratic inequality. Shade the solution region.

21. $y \geq x^2$
22. $y \geq x^2 + 4$
23. $y < -x^2$

24. $y > x^2$
25. $y > x^2 - 2x$
26. $y \geq 1 - x^2$

27. $y > -2x^2$
28. $y \leq x^2 - 3x$
29. $y \leq x^2 + 2x - 8$

30. $y \geq x^2 - 2x + 1$
31. $y \leq x^2 - 4x + 4$
32. $y \leq x^2 - 5x - 6$

33. $y > x^2 + 5x + 6$
34. $y < x^2 - 3x$
35. $y < x^2 - x - 6$

A projectile is fired vertically into the air. Its motion is described by $h = -16t^2 + 320t$, where h is its height (in feet) after t seconds.

Physics Use a quadratic inequality to answer each of the following questions.

36. During what time intervals will the height of the projectile be below 1024 feet?

37. During what time intervals will the height of the projectile be above 1024 feet?

38. Technology Use graphing technology to graph $h = -16t^2 + 320t$. Trace the curve to check your answers to Exercises 36 and 37.

Look Back

Find the products. [Lesson 10.2]

39. $(-ab^2c)(a^2bc^3)$
40. $(3p^2q^3r^4)(-2pqr)$

Write each number in scientific notation. [Lesson 10.4]

41. 0.8
42. 0.000001
43. 0.0000074

Factor each polynomial by first removing a greatest common factor. [Lesson 11.4]

44. $12y^2 - 2y$
45. $2ax + 6x + ab + 3b$
46. $4x^2 - 24x + 32$

Look Beyond

47. Three identical rectangles with sides of positive integer length, each have a perimeter of 24 inches. Find the area of the square formed when the rectangles are joined.

WHAT'S THE DIFFERENCE?

If n represents the dots on a side, the sequence shows the total number of dots in a square pattern.

A function whose domain is consecutive positive integers can produce a sequence. Many sequences have a function rule that you can discover.

A function that will produce this sequence is $A(n) = n^2$. The variable n is usually used when working with sequences, and it represents the number of the term in the sequence.

How do you determine the function rule? Sometimes the pattern in the sequence is clear enough to guess and check to find the rule. When the sequence is more complicated, other methods are needed. If a sequence eventually produces a constant difference, one way to find the general rule for that sequence is to use **finite differences**.

Examine the sequence 0, 5, 12, 21, 32, 45, 60. You want to find a general rule that will generate any term of this sequence. Begin by finding differences until they are constant.

$$P(n) = an^2 + bn + c = a + b + c$$
$$P(1) = a(1)^2 + b(1) + c = a + b + c$$
$$P(2) = a(2)^2 + b(2) + c = 4a + 2b + c$$
$$P(3) = a(3)^2 + b(3) + c = 9a + 3b + c$$
$$P(4) = a(4)^2 + b(4) + c = 16a + 4b + c$$

Table 1

n	1	2	3	4
P(n)	0	5	12	21

1st Term
1st Difference   5   7   9
2nd Difference   2   2

If the sequence produces constant second differences, you expect a quadratic equation. Write the quadratic in the form of a polynomial function, $P(n) = an^2 + bn + c$. Evaluate the function for n from 1 to at least 3. Next, make a table of the terms and differences using the expressions you get from evaluating the general quadratic polynomial.

The sequence in Table 1 is equal to the corresponding general expressions in Table 2. Match the first expression in each row of Table 2 with the corresponding expression in Table 1.

Table 2		Table 1
$a + b + c$	$=$	0
$3a + b$	$=$	5
$2a$	$=$	2

Use *substitution* to find the values of a, b, and c.

Use $2a = 2$ to solve for a.
$$a = 1$$

Use $3a + b = 5$ and the value for a to solve for b.
$$3 + b = 5, \text{ so } b = 2$$

Use $a + b + c = 0$ and the values for a and b to solve for c.
$$1 + 2 + c = 0, \text{ so } c = -3$$

Table 2

Finally, replace the letters in the quadratic polynomial, $P(n) = an^2 + bn + c$ with the values $a = 1$, $b = 2$, and $c = -3$.

$$P(n) = n^2 + 2n - 3$$

Check your rule by substitution to see if it generates the appropriate sequence. 0, 5, 12, 21, 32, 45, 60, . . .

Thus, $P(n) = n^2 + 2n - 3$ is a rule that generates the sequence.

Activity

Find a Rule

In groups of 2 to 5 people, the leader begins by choosing a quadratic equation with integer coefficients and generating a sequence of 5 numbers. The leader then shows only the sequence to the other people in the group. The group then tries to find the function rule that generated the sequence. Finally, the group uses the rule they found to generate any 2 terms between 6 and 20. The leader confirms the results by showing the original function rule and sequence. Let the group members take turns being the leader.

Chapter 12 Review

Vocabulary

Key Skills & Exercises

Lesson 12.1

➤ **Key Skills**

Examine quadratic functions to find the vertex, axis of symmetry, and zeros of the quadratic function to sketch a graph.

$$g(x) = (x - 1)^2 - 4$$

The vertex is $(1, -4)$ and the axis of symmetry is $x = 1$.

$$g(x) = (x - 1)^2 - 4$$
$$= x^2 - 2x - 3$$
$$= (x - 3)(x + 1)$$

The zeros of $g(x)$ are 3 and -1.

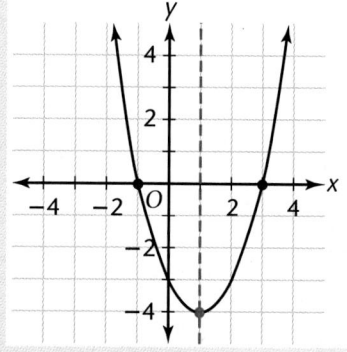

➤ **Exercises**

Determine the vertex, axis of symmetry, and zeros for each function. Then sketch a graph from the information.

1. $y = (x + 1)^2 - 4$ **2.** $y = (x - 5)^2 - 1$ **3.** $y = 2(x - 3)^2 - 2$

Lesson 12.2

➤ **Key Skills**

Use square roots to solve quadratic equations.

$x^2 = 36 \rightarrow x = \sqrt{36}$ or $-\sqrt{36} \rightarrow x = 6$ or -6

$(x - 3)^2 = 15 \rightarrow x - 3 = \pm\sqrt{15} \rightarrow x = 3 + \sqrt{15} \approx 6.87$ or $3 - \sqrt{15} \approx -0.87$

➤ **Exercises**

Solve each equation. Give answers to the nearest hundredth when necessary.

4. $x^2 = 225$ **5.** $x^2 = \dfrac{9}{144}$ **6.** $x^2 = 50$

7. $(x + 1)^2 - 4 = 0$ **8.** $(x + 7)^2 - 81 = 0$ **9.** $(x - 3)^2 - 3 = 0$

Lesson 12.3

➤ Key Skills

Complete the square.

To complete the square, find one half of the coefficient of $-6x$, which is -3. Square -3, and add it to $x^2 - 6x$.

$$x^2 - 6x + (-3)^2 = x^2 - 6x + 9$$

Write quadratic equations in the form $y = (x - h)^2 + k$. Find the vertex.

To find the vertex of $y = x^2 - 6x + 2$, complete the square. Add and subtract 9.

$$y = (x^2 - 6x + 9) + 2 - 9 = (x - 3)^2 - 7$$

The vertex is $(3, -7)$.

➤ Exercises

Complete the square.

10. $x^2 + 4x$ **11.** $x^2 - 16x$ **12.** $x^2 - x$ **13.** $x^2 + 3x$

Rewrite each function in the form $y = (x - h)^2 + k$. Find each vertex.

14. $y = x^2 + 4x + 3$ **15.** $y = x^2 - 8x + 18$ **16.** $y = x^2 - 12x + 32$

Lesson 12.4

➤ Key Skills

Solve quadratic equations by completing the square.

$$x^2 + 2x = 24$$
$$x^2 + 2x + 1 = 24 + 1$$
$$(x + 1)^2 = 25$$
$$x + 1 = 5 \quad \text{or} \quad x + 1 = -5$$
$$x = 4 \quad \text{or} \quad x = -6$$

Solve quadratic equations by factoring.

$$x^2 + 2x = 24$$
$$x^2 + 2x - 24 = 0$$
$$(x + 6)(x - 4) = 0$$
$$x + 6 = 0 \quad \text{or} \quad x - 4 = 0$$
$$x = -6 \quad \text{or} \quad x = 4$$

➤ Exercises

Solve each equation by factoring or completing the square.

17. $x^2 + 3x - 10 = 0$ **18.** $x^2 + 5x + 6 = 0$ **19.** $x^2 - 8x + 16 = 0$

Lesson 12.5

➤ Key Skills

Evaluate the discriminant. Determine the number of solutions.

$$3x^2 + 4x + 7 = 0$$

Substitute $a = 3$, $b = 4$, and $c = 7$ in $b^2 - 4ac$. Simplify. If the discriminant is negative, the equation has no real solutions. If the discriminant is 0, there is exactly one real solution. If the discriminant is positive, the equation has 2 real solutions.

$$b^2 - 4ac = (4)^2 - (4 \cdot 3 \cdot 7) = -68$$

The equation has no real solutions.

Use the quadratic formula to find solutions to quadratic equations.

For the equation $x^2 + 3x - 14 = 0$, $a = 1$, $b = 3$, and $c = -14$. Substitute these values in the quadratic formula and simplify.

$$x = \frac{-3 \pm \sqrt{9 - (4 \cdot 1 \cdot -14)}}{2 \cdot 1}$$

$$x = \frac{-3 \pm \sqrt{65}}{2}$$

$$x \approx 2.53 \text{ or } x \approx -5.53$$

Evaluate the discriminant to determine the number of solutions. Solve. Give answers to the nearest hundredth.

20. $x^2 - 2x + 9 = 0$ **21.** $4x^2 - 5x - 4 = 0$ **22.** $4x^2 - 4x + 1 = 0$

23. $x^2 - 9x + 18 = 0$ **24.** $6x^2 - 7x - 3 = 0$ **25.** $8x^2 + 2x - 1 = 0$

Lesson 12.6

➤ **Key Skills**

Solve and graph quadratic inequalities.

Use factoring and the Zero Product Property to solve $x^2 + 3x - 4 > 0$.

$$x^2 + 3x - 4 = 0$$
$$(x + 4)(x - 1) = 0$$
$$x = -4 \text{ or } x = 1$$

Test numbers in the inequality to determine the solutions. Since numbers less than -4 or numbers greater than 1 produce a true inequality, the solutions are $x < -4$ or $x > 1$.

To graph $y \le x^2 + 2$, graph $y = x^2 + 2$. Draw a solid parabola to show that the solutions are on the graph of $y = x^2 + 2$. Since $y \le x^2 + 2$, shade the points below the parabola.

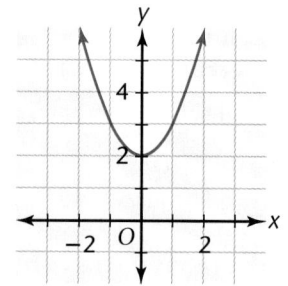

➤ **Exercises**

Solve each quadratic inequality using the Zero Product Property. Graph the solution on a number line.

26. $x^2 + 6x + 8 \ge 0$ **27.** $x^2 - 3x - 10 < 0$ **28.** $x^2 + 9x + 18 > 0$

Graph each inequality. Shade the solution region.

29. $y \le x^2$ **30.** $y > x^2 - 4x$ **31.** $y \ge x^2 - x - 10$

Applications

Sports The graph represents a relationship between the number of yards a football is thrown and its height above the ground.

Distance in yards

32. What is the maximum height reached by the football?

33. How far does the football travel before it reaches its maximum height?

34. If a receiver catches the ball 30 yards from where the ball was thrown, what was the height of the ball?

35. Which equation was used to graph the parabola?

  a. $y = -\dfrac{1}{30}(x - 15)^2 + 12$ **b.** $y = -\dfrac{1}{30}(x + 15)^2 + 12$

  c. $y = -\dfrac{1}{30}(x - 15)^2 - 12$ **d.** $y = -\dfrac{1}{30}(x + 15)^2 - 12$

36. Find two numbers whose sum is 55 and whose product is 600.

37. A piece of wrapping paper has an area of 480 square inches. If Orin cuts out a square piece from the paper and the new piece of paper has an area of 455 square inches, what is the length of each side of the square?

Chapter 12 Assessment

1. Compare the graph of $y = (x + 2)^2 - 4$ to the graph of $y = x^2$. Write the change in units and direction that takes place to the left, right, up, or down.

Determine the vertex, axis of symmetry, and zeros for each function. Then sketch a graph from the information.

2. $y = (x - 6)^2 - 4$ 3. $y = (x + 3)^2 - 1$

4. Why does an equation of the form $x^2 = k$ have two solutions?

Solve each equation. Give answers to the nearest hundredth when necessary.

5. $x^2 = 900$ 6. $x^2 = 12$ 7. $(x + 5)^2 - 16 = 0$

8. Use the formula $A = \pi r^2$ to find the radius of a circle with an area of 100 square feet. (Use $\pi = 3.14$.)

9. An object is dropped and its height, h, after t seconds is given by the function $h = -16t^2 + 550$. About how long will it take for the object to reach the ground?

Complete the square.

10. $x^2 + 8x$ 11. $x^2 - 30x$ 12. $x^2 + 11x$

Rewrite each function in the form $y = (x - h)^2 + k$. Find each vertex.

13. $y = x^2 + 4x + 1$ 14. $y = x^2 - 14x + 50$

Solve each equation by factoring or completing the square.

15. $x^2 - 35 - 2x = 0$ 16. $x^2 + 6x - 12 = 0$

17. Why does a quadratic equation need to be written in standard form in order to use the quadratic formula to solve it?

Find the value of the discriminant for each quadratic equation. Describe the number of solutions.

18. $x^2 + 10x - 13 = 0$ 19. $5x^2 - 4x + 2 = 0$

Use the quadratic formula to solve each equation. Give answers to the nearest hundredth when necessary.

20. $x^2 - 4x - 60 = 0$ 21. $2x^2 + 3x - 4 = 0$

Solve each quadratic inequality using the Zero Product Property. Graph the solution on a number line.

22. $x^2 + 4x - 5 > 0$ 23. $x^2 + 10x + 21 \leq 0$

Graph each quadratic inequality. Shade the solution region.

24. $y \leq -x^2 + 2$ 25. $y > x^2 + 5x + 6$

Chapters 1 – 12
Cumulative Assessment

College Entrance Exam Practice

Quantitative Comparison   For Questions 1–4, write
A if the quantity in Column A is greater than the quantity in Column B;
B if the quantity in Column B is greater than the quantity in Column A;
C if the two quantities are equal; or
D if the relationship cannot be determined from the information given.

	Column A	Column B	Answers
1.	$-6 \cdot 9$	$6 \cdot -9$	Ⓐ Ⓑ Ⓒ Ⓓ **[Lesson 4.1]**
2.	correlation of a line of best fit that rises	correlation of a line of best fit that falls	Ⓐ Ⓑ Ⓒ Ⓓ **[Lesson 1.6]**
3.	70% of a number	90% of a number	Ⓐ Ⓑ Ⓒ Ⓓ **[Lesson 4.5]**
4.	$\dfrac{5 \times 10^{-2}}{7 \times 10^{-5}}$	$\dfrac{5 \times 10^{2}}{7 \times 10^{5}}$	Ⓐ Ⓑ Ⓒ Ⓓ **[Lesson 10.3]**

5. What is the reciprocal of 4?   **[Lesson 2.3]**
   a. -4 **b.** $\frac{1}{4}$ **c.** 4   **d.** $-\frac{1}{4}$

6. What is the simplest form of the expression $5r - (3r + 2)$?   **[Lesson 4.2]**
   a. $8r + 2$ **b.** $2r + 2$ **c.** $2r - 2$ **d.** $8r - 2$

7. What is the solution to $6 - 4t > 18$?   **[Lesson 4.7]**
   a. $t < -3$ **b.** $t > -3$ **c.** $t > -6$ **d.** $t < -6$

8. What is the slope of a line passing through $(-1, 2)$ and $(4, -3)$?   **[Lesson 5.3]**
   a. -2 **b.** 2   **c.** 1   **d.** -1

9. Which equation is graphed at the right?
   a. $y = 2x^2 - 1$
   b. $y = -2x^2 - 1$
   c. $y = \frac{1}{2}x^2 - 1$
   d. $y = -\frac{1}{2}x^2 - 1$

   [Lesson 12.1]

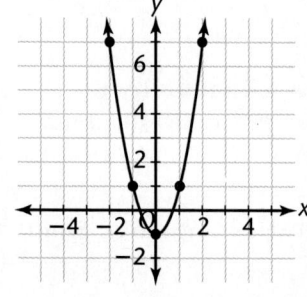

10. What is the factored form of $4x^2 + 25$?   **[Lesson 11.4]**
   a. $(2x - 5)^2$ **b.** $(2x + 5)(2x - 5)$
   c. $(2x + 5)^2$ **d.** Cannot be factored.

11. If tickets to a concert cost $24 per person, plus a $7 handling charge per order, regardless of how many tickets are ordered, how many tickets could you order for $199?   **[Lesson 4.2]**

12. Write an inequality that describes the points graphed on the number line.   **[Lesson 3.6]**

13. Simplify $\dfrac{-100a^8b^{10}}{-20a^5b^9}$.   **[Lesson 10.2]**

14. Vernon scored 86 on his first two tests, 90 on his third test, and 78 on his fourth test. Find the median test score.   **[Lesson 8.3]**

15. Write the equation with slope -2 and passing through $(-1, 0)$.   **[Lesson 5.3]**

16. Write an equation for a line that is perpendicular to the line in Item 15.   **[Lesson 5.6]**

17. Factor $3a(2c + d)^2 - 5(2c + d)^2$.   **[Lesson 11.4]**

18. Solve the system. $\begin{cases} 3x + y = -2 \\ 2x + 3y = 8 \end{cases}$ **[Lesson 6.2]**

19. Does the matrix $\begin{bmatrix} -2 & 2 \\ 1 & -1 \end{bmatrix}$ have an inverse?   **[Lesson 7.4]**

20. Simplify $(3x^2y^3)^0$.   **[Lesson 10.3]**

21. Suppose you choose a card from an ordinary deck of 52 cards. What is the probability that you will choose an ace or an eight?   **[Lesson 8.7]**

Free-Response Grid   The following questions may be answered using a free-response grid commonly used by standardized test services.

22. How many ways are there to arrange 6 different letters?   **[Lesson 8.5]**

23. Solve $\dfrac{2x}{3} = 12$.   **[Lesson 4.6]**

24. Evaluate $h(x) = 2x^2 + 7x - 20$ for $x = 5$.   **[Lesson 12.4]**

25. Find the slope of the line for the equation $y = 2x + 5$.   **[Lesson 5.1]**

CHAPTER 13

Radicals and Coordinate Geometry

In this chapter you can examine square roots and their relationship to the process of squaring. Square roots are used whenever the Quadratic formula or the distance formula are used. For this reason, they play a central and useful role in both algebra and geometry.

Square roots appear in many applications. For example,

• You know the length of a pendulum and want to know how long it will take for one complete swing.

• You know the braking distance of a car and want to know the speed of the car.

$$t = 2\pi\sqrt{\frac{l}{g}}$$

$$d = \sqrt{(x_2-x_1)^2+(y_2-y_1)^2}$$

$A = 25\pi$

$r = \sqrt{\dfrac{A}{\pi}}$

$s = \sqrt{9}$

$a = \sqrt{c^2 - b^2}$

$c = \sqrt{a^2 + b^2}$

$b = \sqrt{c^2 - a^2}$

PORTFOLIO ACTIVITY

The data in the table below comes from a Driver's Manual.

Braking Distance on Dry Concrete in feet	Speed in mph
22	20
50	30
88	40
138	50
198	60
270	70

a. Use the data from the table to create a mathematical model that describes speed as a function of braking distance.

b. Use your model to answer the following questions:

   1. If the braking distance is 100 feet, what is the speed?

   2. If the speed is 55 miles per hour, what is the braking distance?

c. A skid mark at the scene of an accident which occurred on dry concrete is 36 feet long. Assume that this number represents the braking distance. Note that 36 is midway between 22 and 50 feet, the braking distances for 20 and 30 miles per hour, respectively.

$s = a\sqrt{d}$

The speed limit at the location of the accident is 25 miles per hour.

The driver says, "I was going below the speed limit."
A passenger says, " We were going at exactly the speed limit."
A policeman says, "You were going above the speed limit."

You be the judge. Decide who is correct and present a mathematical argument to support your conclusion.

Square Root Functions

the length of a square is the same as finding the positive square root of a number.

An artist is working on a square table-top mosaic with an area covered by 36 square tiles. What is the length of one side of that table?

Since the mosaic is square, count the tiles on any side to find the length. The length of each side is 6 tiles.

$$\text{Area} = 36 \text{ square tiles}$$
$$\text{Side} = 6 \text{ tile lengths}$$

You can check your answer by multiplying the length of the side by itself.

$$6 \cdot 6 = 6^2, \text{ or } 36$$

Since the 36 tiles can be arranged in a square with the measure of each side equal to the same integer, the number 36 is called a **perfect square**.

List the first ten perfect-square numbers.

Estimating Square Roots

How can you determine the length of the side of the blue square if its area is 12 square centimeters?

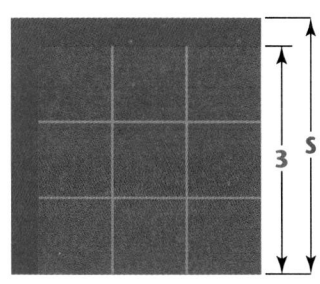

Side s is between 3 and 4.

 1 What is the area of the large green square?

 2 What is the area of the large red square?

 3 Estimate the length of a side of the blue square.

 4 Multiply the length you find in Step 3 by itself.

 5 Compare this product with the actual area, 12 square centimeters. If the product is greater than the area of the blue square, select a new number that is slightly less for the length. If the product is less than the area, try a new number that is slightly greater. Again test by multiplying.

 6 Continue to guess and check until you have an estimate with 2 decimal places. You have estimated $\sqrt{12}$.

 7 Use this procedure to estimate the decimal value for the length of a side of a square with area 20 square centimeters. Describe how to find $\sqrt{20}$ to the nearest hundredth. ❖

Irrational Numbers

Calculator

Every *rational number* can be represented by a terminating or repeating decimal. A repeating decimal is written with a bar over the numbers that repeat. On the other hand, when you calculate $\sqrt{12}$, the decimal part will never terminate or repeat. Numbers such as $\sqrt{12}$ or π are called *irrational numbers*.

Rational Numbers	Irrational Numbers
$\frac{5}{8} = 0.\mathbf{625}$	$\sqrt{2} = 1.\mathbf{4142}\ldots \quad \pi = 3.\mathbf{1415926}\ldots$
$\frac{2}{3} = 0.\overline{6}$ (6 repeats forever.)	$\sqrt{12} = 3.\mathbf{4641}\ldots \quad a = 0.121221222\ldots$

To save time and effort, many people use a calculator to find the square root. The square root key on most calculators is identified by the **radical sign**, √ . When the [√] is pressed, the square root is given as an exact number or as an approximation. On some calculators, √2 is represented by 1.414213562. This, however, is only an approximation. When a calculator displays an irrational number, it cannot keep producing digits indefinitely, so it rounds the value.

The *real numbers* include all the rational and irrational numbers. The real numbers form a **dense set**. That is, between any two real numbers there is another real number. Explain why the rationals are dense but the integers are not.

The Principal Square Root Function

In Lesson 12.3 you found the positive and negative square roots of a positive number. For example, 2 and -2 are square roots of 4 because $(-2)^2 = 4$ and $2^2 = 4$. The positive square root, sometimes called the **principal square root**, is indicated by the radical sign √ .

Thus, $\sqrt{4} = 2$.

The function $f(x) = \sqrt{x}$ represents the parent function of the principal square root.

What are the domain and range of $f(x) = \sqrt{x}$?

The squaring function and the principal square root function are related. Plot $f(x) = \sqrt{x}$ and $g(x) = x^2$ on the same axes. Compare the two graphs. What is the relationship? Give as many similarities and differences for the two functions as you can.

•Exploration 2 *Transforming the Square Root Function*

1 Graph each function on the same axes.
   a. $g(x) = \sqrt{x} + 2$ **b.** $h(x) = \sqrt{x} + 5$ **c.** $p(x) = \sqrt{x} - 3$

2 Compare each function with $f(x) = \sqrt{x}$. What transformation results when $f(x) = \sqrt{x}$ is changed to $q(x) = \sqrt{x} + a$?

3 Graph each function on the same axes.
   a. $g(x) = \sqrt{x + 2}$ **b.** $h(x) = \sqrt{x + 5}$ **c.** $p(x) = \sqrt{x - 3}$

4 Compare each function with $f(x) = \sqrt{x}$. What transformation results when $f(x) = \sqrt{x}$ is changed to $q(x) = \sqrt{x - a}$?

5 Graph each function on the same axes.
   a. $g(x) = 2\sqrt{x}$ **b.** $h(x) = 3\sqrt{x}$ **c.** $p(x) = 5\sqrt{x}$

6 Compare each function with $f(x) = \sqrt{x}$. What transformation results when $f(x) = \sqrt{x}$ is changed to $g(x) = a\sqrt{x}$, when $a > 0$?

7 Graph each function on the same axes.
   a. $g(x) = -2\sqrt{x}$ **b.** $h(x) = -3\sqrt{x}$ **c.** $p(x) = -5\sqrt{x}$

8 Compare each function to $f(x) = \sqrt{x}$. What transformation results when $f(x) = \sqrt{x}$ is changed to $g(x) = a\sqrt{x}$, when $a < 1$?

9 Summarize your findings by explaining the results of statements 2, 4, 6, and 8. ❖

Compare the graph of $y = \sqrt{x}$ with the graph of $y = -\sqrt{x}$. The graph of $y = -\sqrt{x}$ shows that it is also a function. However, the relation $y = \pm\sqrt{x}$ is not a function.

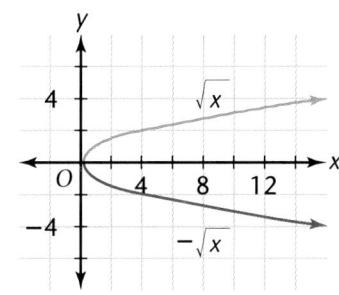

Explain the transformations that were applied to $f(x) = \sqrt{x}$ to produce g and h graphed below.

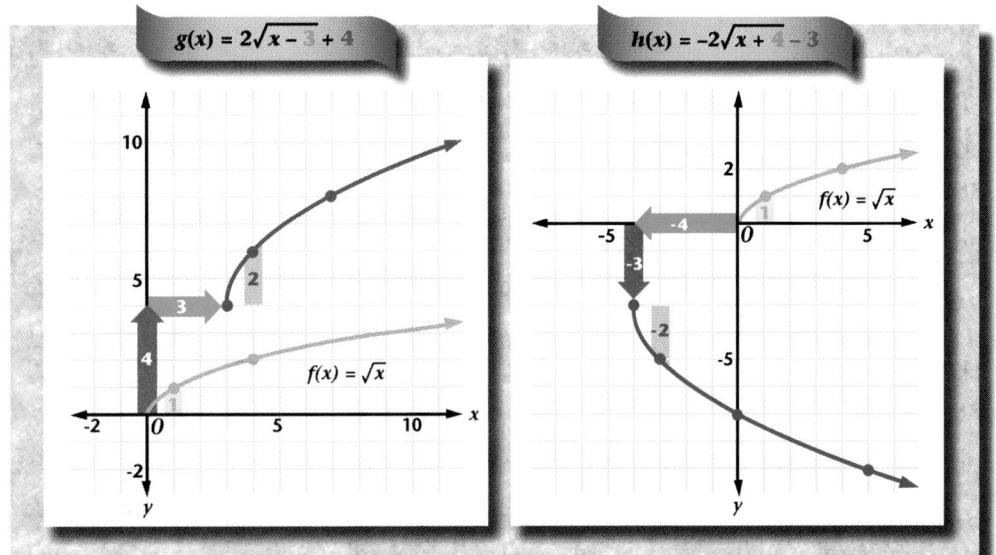

The transformations in Exploration 2 can be combined to graph functions such as $f(x) = 3 - \sqrt{x + 2}$. First, write the function in the form $f(x) = -\sqrt{x + 2} + 3$.

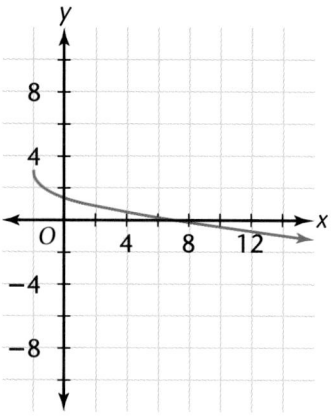

The transformation $\sqrt{x + 2}$ shifts the graph of $\sqrt{x}$ left 2 units. Then $-\sqrt{x + 2}$ reflects the graph through the x-axis. Finally, $-\sqrt{x + 2} + 3$ shifts the graph up 3 units.

Thus, the transformation $f(x) = -\sqrt{x + 2} + 3$ shifts the graph to the left 2 units, reflects it through the x-axis, then shifts it up 3 units. ❖

Janice Brown made the first solar-powered flight in the Solar Challenger on December 3, 1980.

APPLICATION

The approximate distance to the horizon (in kilometers) is given by the formula $D = 3.56\sqrt{x}$. D represents the distance to the horizon in kilometers and x is the altitude in meters. Find the approximate distance in kilometers to the horizon. Use the formula $3.56\sqrt{x}$ with $x = 20$.

$D = 3.56\sqrt{20}$

$D \approx 3.56 \cdot 4.472$

$D \approx 15.92$

The distance to the horizon is approximately 15.92 kilometers. ❖

20 meters

Horizon

CRITICAL Thinking

Explain why the square root of a negative number does not make sense for real numbers.

EXERCISES & PROBLEMS

Communicate

1. Describe how you can determine the square root of a perfect square with graph paper.

2. How can you estimate the square root of a number that is not a perfect square with graph paper?

3. What are the characteristics of an irrational number? Give three examples.

4. Explain how to estimate $\sqrt{7}$.

5. Explain how to write the transformation that moves the graph of the square root function up 3 and right 2.

6. Explain why every positive number has two square roots.

Practice & Apply

Estimate each square root. If the square root is irrational, find the value to the nearest hundredth.

7. $\sqrt{225}$ **8.** $-\sqrt{169}$ **9.** $\sqrt{11}$ **10.** $\sqrt{\dfrac{4}{9}}$ **11.** $-\sqrt{40}$

12. $-\sqrt{27}$ **13.** $\sqrt{1000}$ **14.** $\sqrt{10,000}$ **15.** $-\sqrt{0.04}$ **16.** $\sqrt{0.059}$

Geometry Find the length of the side of each square that has the given area.

17. 250 square meters   **18.** 144 square centimeters   **19.** 28 square miles

20. The Smiths have a yard in the shape of a square. If the area of the yard is 676 square feet, what is the length of each side?

21. Use the formula, $D = 1.22\sqrt{A}$ to find the distance in miles to the horizon if an airplane is flying at an altitude of 30,000 feet.

22. **Technology** Use a calculator for the following activities. Choose five positive numbers greater than 50. Find the square of each positive number. Then find the square root of the result. Record your results in a table with the following headings. One example is started for you.

Positive Number	(Positive Number)^2	$\sqrt{\text{(Positive Number)}^2}$
52	2704	?

Formulate a conjecture based on the data in your table.

23. Technology Choose five negative numbers. Find the square of each negative number, and then find the square root of the result. Record your results in a table with the following headings.

Negative Number	(Negative Number)^2	$\sqrt{\text{(Negative Number)}^2}$
?	?	?

Formulate a conjecture based on the data in your table.

Visualize the transformations of the principal square root function. Use a calculator or graph paper to graph the following.

24. $y = \sqrt{x} + 6$ **25.** $y = 4\sqrt{x}$ **26.** $y = \sqrt{x} - 2$

27. $y = -\sqrt{x} + 1$ **28.** $y = 2\sqrt{x + 1} - 3$ **29.** $y = -2\sqrt{x - 1} + 6$

30. Write equations for two radical functions whose graphs create a parabola with the vertex at (0, 3).

31. Find the vertex and three additional ordered pairs for each pair of functions. Plot the ordered pairs on the same coordinate axes. State the domain and range.

  a. $y = \sqrt{4 - x}$ and $y = -\sqrt{4 - x}$ **b.** $y = \sqrt{-x} + 3$ and $y = -\sqrt{-x} + 3$

32. Test each statement with numbers. Then determine which statements are always true, and explain your reasoning. Give a counterexample if the statement is false.

  a. $\sqrt{k^2} = k$, for any k.   **b.** $\sqrt{k^2} = |k|$, for any k.

  c. $\sqrt{k^2}$ must be written two different ways, depending on the value of k.

33. For what values of x is $\sqrt{x}$ defined?

34. For what values of x is $\sqrt{x + 5}$ defined?

35. What is the domain of the function $y = \sqrt{x + 5}$?

Identify the transformation(s), and describe the effect on the parent function.

36. $y = -\sqrt{x} - 4$ **37.** $y = 5\sqrt{x + 3} - 4$ **38.** $y = 2.00 - 3.50\sqrt{x + 3.75}$

Write the equation of the function graphed.

39.

40.

41.

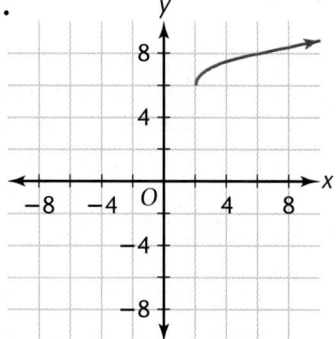

42. Technology Copy and complete the table below. Use a calculator to obtain decimal approximations for the square roots to the nearest tenth.

x	0	1	2	3	4	5	6	7	8	9	10
$\sqrt{x}$											

43. Plot the data in your table on graph paper with x on the horizontal axis and $\sqrt{x}$ on the vertical axis.

Is it possible to have a point on the graph in Exercise 43 in the

44. second quadrant?   **45.** third quadrant?   **46.** fourth quadrant?

47. Technology Use graphing technology to graph $y = \sqrt{x}$. Trace the function. Do the values in your table agree with trace values on the calculator?

Look Back

Without using graphing technology, sketch the graph from what you know about transformations.   [Lesson 9.2]

48. $y = 3x + 2$ **49.** $y = x^2 - 4$ **50.** $y = 3 - 2^x$ **51.** $y = -x^2 + 6$

Find the vertex of each absolute value function. Use graphing technology to check your answer.   [Lesson 9.6]

52. $y = |x - 2|$ **53.** $y = |x| + 5$

54. $y = 2|x + 3| - 4$ **55.** $y = -|x - 1| + 3$

56. Solve the equation $x^2 - 5x - 7 = 0$.   **[Lesson 12.3]**

Look Beyond

Decide whether the following statements are true or false. If you believe that a statement is false, give a counterexample.

57. If $a^2 = b^2$, then $a = b$.

58. If $a = b$, then $a^2 = b^2$.

59. If $a^2 = b^2$ and both a and b have the same sign, then $a = b$.

60. In a long-distance race of 4 runners, Caitlin finished ahead of Andrew. David finished with a time 0.1 second more than Barbara, who finished 1 second after the winner. David was faster than only 1 other runner. Who won the race, and what was the order of finish?

LESSON 13.2 Operations With Radicals

why *Radicals often appear when computing in statistics or evaluating formulas in physics. Radicals are sometimes used with numbers, but they are also used in algebraic expressions. Several methods are used to simplify radicals.*

The area model is restricted to positive values. What happens when $A = x^2$ and x is not restricted? Recall that 9 has two square roots, 3 and -3.

DEFINITION OF SQUARE ROOT

If a is a number greater than or equal to zero, $\sqrt{a}$ and $-\sqrt{a}$ represent the square roots of a. Each square root of a has the following property.

$$\sqrt{a} \cdot \sqrt{a} = a \qquad (-\sqrt{a})(-\sqrt{a}) = a$$

When you work with radicals, you should know the name for each part.

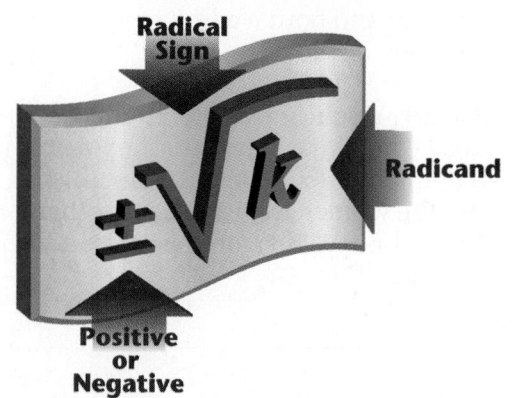

Radical Sign

Radicand

Positive or Negative

EXAMPLE 1

Find each square root.   **A** $\sqrt{25}$ **B** $-\sqrt{144}$ **C** $\sqrt{7}$

Solution ➤

A $\sqrt{25} = 5$ because $5 \cdot 5 = 25$.

B $-\sqrt{144} = -12$ because $-(12 \cdot 12) = -144$.

C $\sqrt{7}$ is not a perfect square. It is approximately 2.646. ❖

Try This Find each square root.   **a.** $\sqrt{196}$ **b.** $-\sqrt{225}$

Simplifying Radicals and Radical Expressions

Radical expressions that are simplified allow you to manipulate algebraic expressions more easily. In some cases the calculations can be done without technology. A radical expression is in **simplest radical form** if the expression within the radical sign

1. contains no perfect squares greater than 1,

2. contains no fractions, and

3. is not in the denominator of a fraction.

EXAMPLE 2

Simplify.

A $\sqrt{20}$ **B** $\dfrac{\sqrt{12}}{4}$ **C** $\dfrac{\sqrt{18}}{\sqrt{2}}$

Solution ➤

A $\sqrt{20}$ is not a perfect square, but it can be factored so the radicand contains a perfect square. Simplify further.
$$\sqrt{20} = \sqrt{2^2 \cdot 5} = \sqrt{2^2}\sqrt{5} = 2\sqrt{5}$$

B $\dfrac{\sqrt{12}}{4} \cdot \dfrac{\sqrt{2^2 \cdot 3}}{4} = \dfrac{2\sqrt{3}}{4} = \dfrac{\sqrt{3}}{2}$

C $\dfrac{\sqrt{18}}{\sqrt{2}} = \sqrt{\dfrac{18}{2}} = \sqrt{9} = 3$ ❖

When the radicand contains variables such as a^4 or b^5, rewrite the expression as a product of even powers. Then simplify.
$$\sqrt{a^4 \cdot b^5} = \sqrt{a^2 \cdot a^2 \cdot b^2 \cdot b^2 \cdot b}$$
$$= \sqrt{(a^2)^2 \cdot (b^2)^2 \cdot b}$$
$$= a^2 b^2 \sqrt{b}$$

The procedures used in Example 2 use two important properties of square roots.

MULTIPLICATION PROPERTY OF SQUARE ROOTS

For all numbers a and b, such that a and $b \geq 0$,

$$\sqrt{ab} = \sqrt{a}\sqrt{b}.$$

EXAMPLE 3

Simplify.

  Ⓐ $\sqrt{12}$ **Ⓑ** $\sqrt{400}$ **Ⓒ** $\sqrt{a^5 b^{12}}$

Solution ➤

Look for perfect square factors, and apply the Multiplication Property of Square Roots. Then find the square roots of the perfect squares. Leave the factor that is not a perfect square in radical form.

  Ⓐ $\sqrt{12} = \sqrt{4 \cdot 3} = \sqrt{4} \cdot \sqrt{3} = 2\sqrt{3}$

  Ⓑ $\sqrt{400} = \sqrt{4 \cdot 100} = \sqrt{4} \cdot \sqrt{100} = 2 \cdot 10 = 20$

  Ⓒ $\sqrt{a^5 b^{12}} = \sqrt{(a^2)^2 \cdot a \cdot (b^6)^2} = a^2 b^6 \sqrt{a}$ ❖

Try This   Simplify $\sqrt{72m^2 n^5}$.

DIVISION PROPERTY OF SQUARE ROOTS

For all numbers $a \geq 0$ *and* $b > 0$,

$$\sqrt{\frac{a}{b}} = \frac{\sqrt{a}}{\sqrt{b}}.$$

EXAMPLE 4

Simplify.

  Ⓐ $\sqrt{\dfrac{16}{25}}$ **Ⓑ** $\sqrt{\dfrac{7}{16}}$ **Ⓒ** $\sqrt{\dfrac{9}{5}}$ **Ⓓ** $\sqrt{\dfrac{a^2 b^3}{c^2}}$

Solution ➤

Rewrite the square root using the Division Property of Square Roots. Then simplify the numerator and denominator separately.

  Ⓐ $\sqrt{\dfrac{16}{25}} = \dfrac{\sqrt{16}}{\sqrt{25}} = \dfrac{4}{5}$

  Ⓑ $\sqrt{\dfrac{7}{16}} = \dfrac{\sqrt{7}}{\sqrt{16}} = \dfrac{\sqrt{7}}{4}$

C $\sqrt{\dfrac{9}{5}} = \dfrac{\sqrt{9}}{\sqrt{5}} = \dfrac{3}{\sqrt{5}}$ You can can simplify this expression by multiplying

the fraction by $\dfrac{\sqrt{5}}{\sqrt{5}}$ or 1. $\quad \dfrac{3}{\sqrt{5}} \cdot \dfrac{\sqrt{5}}{\sqrt{5}} = \dfrac{3\sqrt{5}}{\sqrt{5}\sqrt{5}} = \dfrac{3\sqrt{5}}{\sqrt{25}} = \dfrac{3\sqrt{5}}{5}$

This is known as **rationalizing the denominator**.

D For $a, b \geq 0$, and $c > 0$, $\sqrt{\dfrac{a^2 b^3}{c^2}} = \dfrac{\sqrt{a^2 b^2 b}}{\sqrt{c^2}} = \dfrac{ab\sqrt{b}}{c}$. ❖

Operations and Radical Expressions

You know that $\sqrt{16 \cdot 9} = \sqrt{16} \cdot \sqrt{9}$.
What is the relationship between
$\sqrt{16 + 9}$ and $\sqrt{16} + \sqrt{9}$? Complete
the computations to find the answer.

$$\sqrt{16 + 9} = \sqrt{25} \text{ or } 5$$
$$\sqrt{16} + \sqrt{9} = 4 + 3 \text{ or } 7$$

Since 5 does not equal 7, $\sqrt{16 + 9}$
does not equal $\sqrt{16} + \sqrt{9}$. In general,
$\sqrt{a + b} \neq \sqrt{a} + \sqrt{b}$.

Radicands must be the same to add
expressions that contain radicals.
Like terms that contain $\sqrt{5}$, for
example, can be added by the
Distributive Property.

$$2\sqrt{5} + 4\sqrt{5} = (2 + 4)\sqrt{5} = 6\sqrt{5}$$

EXAMPLE 5

Simplify.

A $5\sqrt{6} - 2\sqrt{6}$ **B** $5 + 6\sqrt{7} - 2\sqrt{7} - 3$

C $8\sqrt{3} + 6\sqrt{2} - \sqrt{3} + 2\sqrt{2}$ **D** $a\sqrt{x} + b\sqrt{x}$

Solution ➤

A Use the Distributive Property to combine like terms. In this case, $\sqrt{6}$ is the common factor. Then simplify.

$$5\sqrt{6} - 2\sqrt{6} = (5 - 2)\sqrt{6} = 3\sqrt{6}$$

B Rearrange and combine like terms.

$$5 + 6\sqrt{7} - 2\sqrt{7} - 3 = (5 - 3) + (6 - 2)\sqrt{7} = 2 + 4\sqrt{7}$$

C Rearrange and combine like terms.

$$8\sqrt{3} + 6\sqrt{2} - \sqrt{3} + 2\sqrt{2} = (8 - 1)\sqrt{3} + (6 + 2)\sqrt{2} = 7\sqrt{3} + 8\sqrt{2}$$

D Literal expressions are treated in a similar way.

$$a\sqrt{x} + b\sqrt{x} = (a + b)\sqrt{x} ❖$$

Does $\sqrt{a} - \sqrt{b}$ equal $\sqrt{a-b}$ for any *a* and *b*? Substitute values to test the idea. Explain the results of the test. Are there any values of *a* and *b* that allow the statement to be true?

There are several ways that radicals may appear in multiplication.

EXAMPLE 6

Simplify.

A $(5\sqrt{3})^2$

B $\sqrt{3}\sqrt{6}$

C $\sqrt{2}(6 + \sqrt{12})$

D $(3 - \sqrt{2})(4 + \sqrt{2})$

Solution ➤

A Recall from the properties of exponents that the second power means the product of two identical factors. Rearrange the factors and multiply.

$$(5\sqrt{3})^2 = (5\sqrt{3})(5\sqrt{3}) = (5 \cdot 5)(\sqrt{3}\sqrt{3}) = 25 \cdot 3 = 75$$

B The Multiplication Property of Square Roots allows you to multiply separate radicals. You can then factor the new product in a different way and simplify.

$$\sqrt{3}\sqrt{6} = \sqrt{3 \cdot 6} = \sqrt{18} = \sqrt{9 \cdot 2} = \sqrt{9}\sqrt{2} = 3\sqrt{2}$$

C Use the Distributive Property to multiply, then factor again and simplify the result. Notice that the radical is usually written last in each term.

$$\sqrt{2}(6 + \sqrt{12}) = \sqrt{2} \cdot 6 + \sqrt{2} \cdot \sqrt{12}$$
$$= 6\sqrt{2} + \sqrt{2 \cdot 12}$$
$$= 6\sqrt{2} + \sqrt{24}$$
$$= 6\sqrt{2} + \sqrt{4 \cdot 6}$$
$$= 6\sqrt{2} + 2\sqrt{6}$$

D To multiply differences and sums, multiply the two binomials.
$$(3 - \sqrt{2})(4 + \sqrt{2}) = 12 + 3\sqrt{2} - 4\sqrt{2} - 2 = 10 - \sqrt{2} \diamondsuit$$

Calculator

When you enter more complicated expressions into a calculator, you may need to use parentheses. To find the square root of $3^2 + 4^2$, enter $\sqrt{(3^2 + 4^2)}$ into the calculator.

| $\sqrt{}$ | (| 3 | x^2 | + | 4 | x^2 |) | = | | 5 |

What do you think you will get on a calculator if the parentheses are not used?

EXERCISES & PROBLEMS

Communicate

1. Explain when $\sqrt{a^2} = a$.

2. How is factoring used to simplify expressions that contain radicals? What is simplest radical form?

3. For what values of a can you find $\sqrt{-a+3}$?

4. Explain in your own words what the Multiplication and Division Properties of Square Roots allow you to do.

5. When is it possible to add two radical expressions?

6. Describe a procedure for multiplying $(\sqrt{3} + 2)(\sqrt{2} + 3)$.

Practice & Apply

Simplify the radicals by factoring.

7. $\sqrt{49}$　　　　8. $\sqrt{196}$　　　　9. $\sqrt{576}$　　　　10. $\sqrt{3600}$

11. $\sqrt{75}$　　　　12. $\sqrt{98}$　　　　13. $\sqrt{1620}$　　　　14. $\sqrt{264}$

Decide whether the given statement is true or false. Assume that $a, b \geq 0$.

15. $\sqrt{a+b} = \sqrt{a} + \sqrt{b}$　　　　16. $\sqrt{a-b} = \sqrt{a} - \sqrt{b}$

17. $\sqrt{ab} = \sqrt{a}\sqrt{b}$　　　　18. $\sqrt{\dfrac{a}{b}} = \dfrac{\sqrt{a}}{\sqrt{b}}$, $b \neq 0$

Express in simplest radical form.

19. $\sqrt{3}\sqrt{12}$　　　　20. $\sqrt{8}\sqrt{18}$　　　　21. $\sqrt{48}\sqrt{3}$　　　　22. $\sqrt{54}\sqrt{6}$

23. $\sqrt{\dfrac{64}{16}}$　　　　24. $\sqrt{\dfrac{96}{2}}$　　　　25. $\dfrac{\sqrt{50}}{\sqrt{8}}$　　　　26. $\dfrac{\sqrt{150}}{\sqrt{6}}$

27. $\sqrt{5}\sqrt{15}$　　　　28. $\sqrt{98}\sqrt{14}$　　　　29. $\sqrt{\dfrac{56}{8}}$　　　　30. $\dfrac{\sqrt{96}}{\sqrt{8}}$

Simplify each of the following. Assume that all variables are non-negative and that all denominators are non-zero.

31. $\sqrt{a^4 b^6}$　　　　32. $\sqrt{x^8 y^9}$　　　　33. $\sqrt{\dfrac{p^9}{q^{10}}}$　　　　34. $\sqrt{\dfrac{x^3}{y^6}}$

If possible, perform the indicated operation, and simplify your answer.

35. $3\sqrt{5} + 4\sqrt{5}$

36. $\sqrt{10} + \sqrt{15}$

37. $\sqrt{7} + \sqrt{29}$

38. $4\sqrt{5} + 2\sqrt{5} - 5\sqrt{5}$

39. $\sqrt{6} + 2\sqrt{3} - \sqrt{6}$

40. $(4 + \sqrt{3}) + (1 - \sqrt{2})$

41. $\dfrac{6 + \sqrt{18}}{3}$

42. $\dfrac{\sqrt{15} + \sqrt{10}}{\sqrt{5}}$

Simplify.

43. $(3\sqrt{5})^2$

44. $(4\sqrt{25})^2$

45. $\sqrt{12}\sqrt{6}$

46. $\sqrt{72}\sqrt{32}$

47 $3(\sqrt{5} + 9)$

48. $\sqrt{5}(6 - \sqrt{15})$

49. $\sqrt{6}(6 + \sqrt{18})$

50. $(\sqrt{5} - 2)(\sqrt{5} + 2)$

51. $(\sqrt{3} - 4)(\sqrt{3} + 2)$

52. $(\sqrt{5} + 7)(\sqrt{2} - 8)$

53. $(\sqrt{6} + 2)^2$

54. $\sqrt{3}(\sqrt{3} + 2)^2$

55. **Geometry** Copy and complete the table. Each entry should be simplified.

56. Add three more columns to your table according to the number pattern established in the Area of Square column.

Area of Square	6	12	24	48	96
Length of side	?	?	?	?	?

Look Back

57. Find the solution to the system of equations. [Lesson 6.2]
$$\begin{cases} y = -x + 20 \\ y = 2x - 1 \end{cases}$$

Simplify. [Lesson 10.2]

58. $(-a^2b^2)^3(a^4b)^2$

59. $\dfrac{x^5y^7}{x^2y^3}$

60. $\left(\dfrac{20x^3}{-4x^2}\right)^3$

Find each product. [Lesson 11.3]

61. $(2x - 4)(2x - 4)$

62. $(3a + 5)(2a - 6)$

63. $(6b + 1)(3b - 1)$

Solve each equation by completing the square. [Lesson 12.3]

64. $y^2 - 8y + 12 = 0$

65. $x^2 + 14x - 15 = 0$

66. $r^2 - 24r + 63 = 0$

Look Beyond

67. Notice that $2 \cdot 2 \cdot 2 = 2^3 = 8$. Then $\sqrt[3]{8} = 2$ (read the cube root of 8 equals 2). Copy and complete the following table.

x	1	2	3	4	5	6	7	8	9	10
x^3	1	8	27	?	?	?	?	?	?	?

Based on the data in the table, answer Exercises 68–71.

68. $\sqrt[3]{125} = $?

69. $\sqrt[3]{1000} = $?

70. $\sqrt[3]{100}$ is between what two whole numbers?

71. **Technology** Use a calculator to approximate $\sqrt[3]{100}$ to the nearest tenth.

LESSON 13.3 Solving Radical Equations

Why *Equations that contain radicals are often used in science to model natural phenomena. Scientists also use radicals to solve equations that contain squares of numbers and variables. The mathematics of radicals provides a necessary tool to solve many problems that occur in science and other areas.*

The relationship of the length of a pendulum to the number of swings per minute is one example of an equation that contains a radical.

Physics In a physical science class, six groups are given six strings of different sizes, each tied to a washer. They are asked to time one complete swing of each pendulum. Then they are asked to put all their data together, make a table, and draw a graph of the data. The point (0, 0) represents the time for a string of zero length.

Examine the data in the table and on the graph.

Let l represent the length of the string in centimeters and t represent the time of one swing in seconds.

ℓ	t
0	0
6	0.5
25	1.0
58	1.5
100	2.0
155	2.5

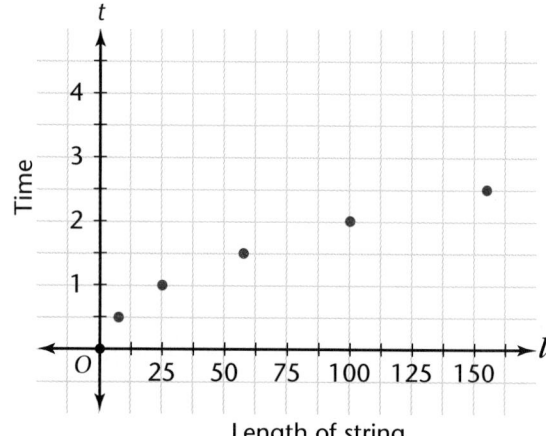

What conclusions can be drawn from the information? How can you use the graph to predict the time that it will take for a string of any length to make a complete swing?

First, notice that the curve resembles a transformation of the parent function $y = \sqrt{x}$. In this case, the function is written $t = \sqrt{l}$. The transformation for a stretch of the function is $t = a\sqrt{l}$. The physical science class found that $a = \frac{1}{5}$ worked for their data. Check to see if the class was right.   Use the function to predict the time it will take for a string 225 centimeters long to make a complete swing.

The equation that models the motion of a pendulum is $t = 2\pi \sqrt{\dfrac{l}{g}}$. The variable l is the length in centimeters, t is the time in seconds and g is acceleration due to gravity (980 centimeters per second per second). This formula can be used to see how close the physical science class's data is to the theoretical value determined by the formula.

EXAMPLE 1

A Find the time it takes a 75-centimeter pendulum to make a complete swing.

B Find the length of a pendulum that takes 3 seconds to make a complete swing.

Solution ➤

Scientific Calculator

A Substitute into the pendulum formula.

$$t = 2\pi \sqrt{\frac{75}{980}}$$

Use a calculator to help in the calculation.

$$t \approx 1.74$$

One swing takes about 1.7 seconds.

B Substitute the known information into the formula, and solve for l.

$3 = 2\pi \sqrt{\dfrac{l}{980}}$	Given
$3\sqrt{980} = 2\pi\sqrt{l}$	Multiply both sides by $\sqrt{980}$.
$\dfrac{3\sqrt{980}}{2\pi} = \sqrt{l}$	Divide both sides by 2π.
$\left(\dfrac{3\sqrt{980}}{2\pi}\right)^2 = l$	Use the definition of square root.
$223.4 \approx l$	Use a calculator.

The length of the pendulum is about 223.4 centimeters. ❖

Solving Equations Containing Radicals

EXAMPLE 2

Solve $\sqrt{x + 2} = 3$.　　**A** graphically　　**B** algebraically

Solution ➤

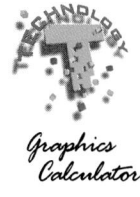

Graphics Calculator

A You can solve the equation on a graphics calculator. Set a *friendly window*. Enter the left side of the equation in the graphics calculator as $Y_1 = \sqrt{x + 2}$. Then enter the right side of the equation, $Y_2 = 3$.

The calculator will graph both equations on the same set of axes. You can then use to find the point(s) of intersection. Since the point of intersection is $(7, 3)$, $x = 7$.

B The square root and the squaring operations are inverses and can be used to solve equations.

$$\sqrt{x + 2} = 3 \qquad \text{Given}$$
$$(\sqrt{x + 2})^2 = 3^2 \qquad \text{Square both sides.}$$
$$x + 2 = 9 \qquad \text{Simplify.}$$
$$x = 7 \qquad \text{Subtraction Property of Equality}$$

Check by substituting. Since $\sqrt{7 + 2} = \sqrt{9} = 3$, the solution is 7. ❖

Try This Solve $\sqrt{2x - 3} = 4$.

What is the domain of the function $y = \sqrt{x + 2}$? Explain.

EXAMPLE 3

Solve $\sqrt{x + 6} = x$.

Solution ➤

Notice that the variable appears on both sides of the equation.

$$\sqrt{x + 6} = x \qquad \text{Given}$$
$$(\sqrt{x + 6})^2 = x^2 \qquad \text{Square both sides of the equation.}$$
$$x + 6 = x^2 \qquad \text{Simplify.}$$
$$0 = x^2 - x - 6 \qquad \text{Form a quadratic equal to 0.}$$
$$0 = (x - 3)(x + 2) \qquad \text{Factor.}$$
$$x - 3 = 0 \quad \text{or} \quad x + 2 = 0 \qquad \text{Set each factor equal to zero.}$$

The possible solutions are 3 and -2. You **must** check each possible solution by substituting it into the original equation to determine whether it is, in fact, a solution.

Check

For $x = 3$, $\sqrt{3 + 6} = \sqrt{9} = 3$, so 3 is a solution.

For $x = -2$, $\sqrt{-2 + 6} = \sqrt{4} \neq -2$, so -2 is *not* a solution. ❖

The graphs of $y = \sqrt{x + 6}$ and $y = x$ show that the equation $\sqrt{x + 6} = x$ has only one solution. How do you know this?

Solving Equations by Using Radicals

You can solve the equation $x^2 = 225$ by using the generalization on page 554. Recall that the square root may have two answers, since the values of x can be positive or negative.

$$x^2 = 225$$
$$x = \pm\sqrt{225}$$
$$x = \pm 15$$

EXAMPLE 4

Solve the following equations for x.

A $x^2 = 150$ **B** $x^2 = 4^2 + 3^2$ **C** $x^2 = y^2 + z^2$

Solution ➤

For each problem, use the generalization on page 554.

A $x^2 = 150$

$x = \pm\sqrt{150} = \pm\sqrt{25 \cdot 6} = \pm 5\sqrt{6}$

Thus, $x = 5\sqrt{6}$ or $x = -5\sqrt{6}$.

Check
$$5\sqrt{6} \cdot 5\sqrt{6} = 25 \cdot 6 = 150$$
$$(-5\sqrt{6}) \cdot (-5\sqrt{6}) = 25 \cdot 6 = 150$$

B $x^2 = 4^2 + 3^2$

$x = \pm\sqrt{4^2 + 3^2} = \pm\sqrt{25} = \pm 5$

C For a literal equation, follow the same procedure that you would with numbers. Remember to include everything on the right side of the equation under one radical. Account for positive and negative values as possible solutions.

$$x^2 = y^2 + z^2$$
$$x = \pm\sqrt{y^2 + z^2}$$
$$x = \sqrt{y^2 + z^2} \text{ or } -\sqrt{y^2 + z^2} \;❖$$

Problems that have two possible answers must also be considered in context. For example, if a geometric problem relates to distance, the answer must be positive. In this case, the negative value should be disregarded.

EXAMPLE 5

$A = \pi r^2$

The area of the circular flower garden is 23 square yards. Find the radius.

Solution ➤

Substitute 23 for A, and solve for r.
Since $\pi r^2 = 23$ and $r^2 = \dfrac{23}{\pi}$,

$$r = \pm\sqrt{\frac{23}{\pi}} \approx \pm\sqrt{\frac{23}{3.14}} \approx \pm\sqrt{7.325} \approx \pm 2.706.$$

Only the positive square root can be used. The length of the radius is approximately 2.7 yards. ❖

EXAMPLE 6

Solve the equation $x^2 - 8x + 16 = 36$ using radicals.

Solution ➤

Recall the methods from the previous two chapters. The trinomial $x^2 - 8x + 16$ is a perfect square. When it is factored it becomes $(x - 4)^2$. Write the equation in this form, and solve.

$$(x - 4)^2 = 36$$
$$x - 4 = \pm\sqrt{36}$$
$$x - 4 = \pm 6$$
$$x = 10 \text{ and } x = -2$$

Since x can be any number, you must account for both positive and negative values. ❖

Check by substitution to see if the values actually fit the original equation.

Solve the equation $x^2 - 8x + 16 = 36$ by graphing it as a transformation. How does the graph justify 2 solutions for x?

EXERCISES & PROBLEMS

Communicate ⌇⌇⌇⌇⌇

1. Explain why you must consider both positive and negative values if you take the square root of a number.

2. Describe a step-by-step method for finding the solution to an equation of the form $\sqrt{x + 7} = 114$.

3. Does $\sqrt{-x} = -3$ have a solution? Explain.

4. Explain the difference between solving $a^2 = b^2 + c^2$ and solving $a^2 = (b + c)^2$, for a.

5. Give a step-by-step procedure for finding the solution to the equation $x^2 + 2bx + b^2 = c^2$ for x. Then test the procedure by substituting numbers for b and c.

6. Find the explanation of the quadratic formula in Lesson 12.5, and identify how radicals are used to develop the formula.

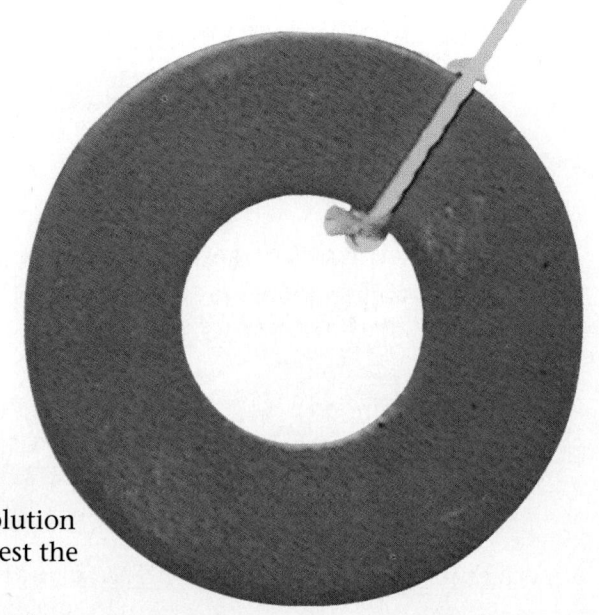

Practice & Apply

Solve each equation algebraically. Be sure to check your solution.

7. $\sqrt{x-5} = 2$ **8.** $\sqrt{x+7} = 5$ **9.** $\sqrt{2x} = 6$

10. $\sqrt{10-x} = 3$ **11.** $\sqrt{2x+9} = 7$ **12.** $\sqrt{2x-1} = 4$

Solve each equation algebraically. Be sure to check your solution.

13. $\sqrt{x+2} = x$ **14.** $\sqrt{6-x} = x$ **15.** $\sqrt{5x-6} = x$

16. $\sqrt{x-1} = x - 7$ **17.** $\sqrt{x+3} = x + 1$ **18.** $\sqrt{2x+6} = x - 1$

19. $\sqrt{x^2 + 3x - 6} = x$ **20.** $\sqrt{x-1} = x - 1$ **21.** $\sqrt{x^2 + 5x + 11} = x + 3$

Make separate graphs from the left and right sides of the equation. Use graphing technology or sketch a graph. Solve the equations that have solutions by finding the intercepts of the graphs.

22. $\sqrt{x+12} = x$ **23.** $\sqrt{x-2} = x$ **24.** $\sqrt{x} = \frac{1}{3}x + \frac{2}{3}$

Physics   The motion of a pendulum can be modeled by $t = 2\pi\sqrt{\dfrac{l}{32}}$, where l is the length of the pendulum in feet and t is the number of seconds required for one complete swing.

25. A grandfather clock is based on the motion of a pendulum. About how long should a pendulum be so that the time required for 1 complete swing is 1 second?

26. If the time required for 1 complete swing is doubled, by what is the length multiplied?

27. If the time is multiplied by 3, by what is the length multiplied?

28. The time required for 1 complete swing of a pendulum is multiplied by c. By what will the corresponding length be multiplied? Generalize.

Sports   The formula for kinetic energy is $E = \frac{1}{2}mv^2$. E is the kinetic energy in joules, m is the mass of the object in kilograms, and v is the velocity a moving object in meters per second.

29. Solve the formula for v.

30. The temperature is 70°F. If a baseball is thrown with 50 joules of energy and has a mass of 0.14 kilograms, what is its velocity?

Solve each equation and simplify the solution.

31. $x^2 = 90$ **32.** $4x^2 = 7$ **33.** $3x^2 - 27 = 0$

34. $2x^2 = 48$ **35.** $x^2 - 8x + 16 = 0$ **36.** $x^2 - 12x + 36 = 0$

Solve each equation if possible. If not possible, explain why.

37. $x^2 = 9$ **38.** $\sqrt{x} = 9$ **39.** $|x| = 9$

40. $x^2 = -9$ **41.** $\sqrt{x} = -9$ **42.** $|x| = -9$

Physics   A beam is balanced when the product of the weight and the distance from the fulcrum on one side is equal to the product of the weight and the distance from the fulcrum on the right side.

$W_l d_l = W_r d_r$ or $\dfrac{W_l}{W_r} = \dfrac{d_r}{d_l}$

To find the weight of the metal cube, W, first place W on the right, and balance it with weight W_l. Then switch sides, and balance W with weight W_r.

43. The ratios $\dfrac{W_l}{W} = \dfrac{W}{W_r}$ are equal. So $W_l W_r = W^2$. Express the value of W in terms of the other weights.

44. A *geometric mean* is the square root of the product of two factors. If 25 and 49 are two factors, what is the geometric mean?

45. If W_l is 9 pounds, and W_r is 16 pounds, what is the true weight of W?

Look Back

46. Solve $-2x + 3 < 11$.   **[Lesson 4.7]**

47. Find the inverse of the matrix $A = \begin{bmatrix} 6 & -1 \\ 2 & -7 \end{bmatrix}$.   **[Lesson 7.4]**

Graph each pair of functions on the same set of axes. Describe the relationship between the two graphs.   [Lesson 9.6]

48. $y = x + 3$
     $y = (x - 2) + 3$

49. $y = x^2 + 3$
     $y = (x - 2)^2 + 3$

50. $y = |x| + 3$
     $y = |x - 2| + 3$

51. Factor $4x^4 - 16y^4$ completely.   **[Lesson 11.5]**

52. What conditions for the discriminate indicates that a quadratic has no real solution? Give an example.   **[Lesson 12.5]**

Solve and graph each inequality.   [Lesson 12.6]

53. $x^2 - x - 12 > 0$ **54.** $y^2 + 5y + 6 < 0$ **55.** $a^2 - 6a - 16 < 0$

Look Beyond

56. What is the center of the circle $x^2 + y^2 = 4$?

LESSON 13.4 The "Pythagorean" Right-Triangle Theorem

Right Triangle ABC

hypotenuse *c*

B

leg *a*

right angle *C*

leg *b*

A

why The "Pythagorean" Right-Triangle theorem is one of the most familiar and significant theorems in mathematics. It makes extensive use of squares and square roots. You can use this theorem to find the length of one side of a right triangle if you know the lengths of the other two.

Copy and complete the table for the given sides of a right triangle.

Side a	Side b	Side c	a^2	b^2	c^2	$a^2 + b^2$
3	4	5				
5	12	13				
16	30	34				
13	84	85				

Compare the last two columns of the table. What do you notice about the relationship between c^2 and $a^2 + b^2$?

Cultural Connection: Asia The relationships of the sides of right triangles have been studied in many parts of the world throughout history. The ancient Chinese, Babylonians, and Hindus had all discovered the right-triangle relationship. Throughout history, many people have discovered clever proofs of the relationship.

European culture provided the name we currently associate with this right-triangle relationship. It is usually referred to as the "Pythagorean" Right-Triangle Theorem, although Babylonians knew of the relationship over 1000 years earlier.

"PYTHAGOREAN" RIGHT-TRIANGLE THEOREM

Given a right triangle with legs of length a and b and hypotenuse of length c, then $c^2 = a^2 + b^2$ or $a^2 + b^2 = c^2$.

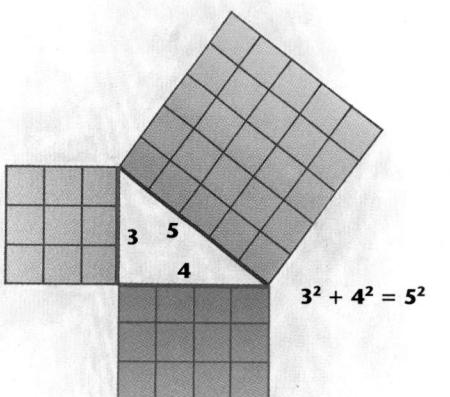

Over the years this illustration has been used to show that the sum of the squares of the legs of a right triangle is equal to the square of the hypotenuse.

EXAMPLE 1

Find the length of x, y, and z.

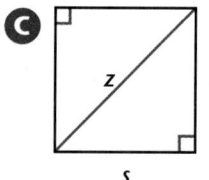

Solution ➤

A Since x is opposite the right angle, x represents the length of the hypotenuse.

$$x^2 = 7^2 + 10^2$$
$$x^2 = 149$$
$$x = \pm\sqrt{149} \approx \pm 12.21$$

Since x represents the length of a segment, it must be positive. Therefore, $x = \sqrt{149} \approx 12.21$.

B The length of the hypotenuse is 6.

$$y^2 + 4^2 = 6^2$$
$$y^2 + 16 = 36$$
$$y^2 = 20$$
$$y = \pm\sqrt{20} \text{ or } \pm 2\sqrt{5}$$
$$y \approx 4.47 \qquad \text{The negative root is not a solution.}$$

C The diagonal, z, of a square is the hypotenuse. Each leg is s.

$$z^2 = s^2 + s^2$$
$$z^2 = 2s^2$$
$$z = \sqrt{2s^2}$$
$$z = s\sqrt{2} \qquad \text{Only positive numbers represent a length.}$$

Thus, the formula for the diagonal of a square with sides s is $s\sqrt{2}$. ❖

EXAMPLE 2

Boating Navigation

Your boat is traveling due north at 20 miles per hour. Your friend's boat left at the same time from the same location headed due west at 15 miles per hour. After an hour you get a call from your friend that he has engine trouble. How far must you travel to reach your friend?

Solution ➤

The legs will be 20 miles and 15 miles. The distance you must travel is the length of the hypotenuse. Use the "Pythagorean" Right-Triangle Theorem to find the distance. Add and use the definition of square root to solve.

$$c^2 = a^2 + b^2$$
$$c^2 = 20^2 + 15^2$$
$$c^2 = 625$$

Since distance is involved, only positive values are used.

$$c = \sqrt{625} \rightarrow c = 25 \text{ miles} \; ❖$$

Try This Find the hypotenuse of a right triangle if the legs are 36 and 48.

What do you notice about multiples of the sides of a 3-by-4-by-5 right triangle? If you do not know, make a table of multiples of the legs and see if you can guess the hypotenuse. Then check your results.

Use algebra to solve the equation $a^2 + b^2 = c^2$ for a or b. Then you can use the new equation to determine the unknown side of the right triangle. All you need to know is the hypotenuse and the other side.

EXAMPLE 3

GEOMETRY
Connection

The height of a pyramid with a square base can be determined from measurements of the base and the slant height. The measurement of the base is 40 meters, and the slant height is 52 meters. How high is the pyramid?

Solution ➤

Use half the base of the pyramid, 20 meters, as one side of the triangle. Use the slant height, 52 meters, as the hypotenuse.

If you solve for side a, the formula from the "Pythagorean" Right-Triangle Theorem will be $a = \sqrt{c^2 - b^2}$. Let the height, a, represent the leg whose length we do not know. Since distance is positive, you can disregard the negative value. Substitute the values for b and c to find the height of the pyramid.

$$a^2 = 52^2 - 20^2$$
$$a = \sqrt{52^2 - 20^2}$$
$$a = \sqrt{2304} = 48$$

The height of the pyramid is 48 meters. ❖

Height (a)

Slant height

Half the base

CRITICAL *Thinking*

An important theorem in geometry states that for all non-zero numbers a and b, $\sqrt{a^2 + b^2} < a + b$. Use this theorem to explain why the path along the hypotenuse is always shorter than the path along the two legs.

EXERCISES & PROBLEMS

Communicate

1. Name the sides of a right triangle.
2. State the "Pythagorean" Right-Triangle Theorem in your own words.
3. Explain how to find the hypotenuse of a right triangle if you know the lengths of the legs.
4. Describe how to change the basic equation of the Pythagorean theorem to let you find the length of one of the legs.
5. How would you show someone that the "Pythagorean" Right-Triangle Theorem is true?
6. Explain how to use the "Pythagorean" Right-Triangle Theorem to find the length of a diagonal of a rectangle.

622 CHAPTER 13

Practice & Apply

Technology Copy and complete the table. Use a calculator and round the answer to the nearest tenth.

	Leg	Leg	Hypotenuse			Leg	Leg	Hypotenuse
7.	24	45	?		**8.**	10	?	26
9.	?	8	17		**10.**	5	9	?
11.	12	?	25		**12.**	?	6	25
13.	30	40	?		**14.**	0.75	?	1.25

Solve for *x*.

15.

16.

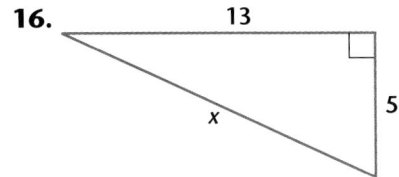

17.

x

2.5 3.5

Landscaping A garden in the shape of a right triangle is shown in the figure.

18. How many feet of fencing must be bought to enclose the garden? Assume that fencing is sold by the foot.

14 feet   10 feet

19. If the cost of fencing is $4.98 per foot, how much will it cost to enclose the garden?

A baseball diamond is a square with sides of 90 feet.

20. **Sports** To the nearest foot, how long is a throw from third base to first base?

21. Find the diagonal of a square that is 5 inches on each side.

22. A hiker leaves camp and walks 5 miles east. He then walks 10 miles south. How far from camp is the hiker?

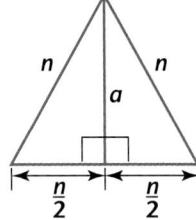

Geometry An equilateral triangle has an axis of symmetry from each vertex that is perpendicular to the opposite side at its midpoint. The axis of symmetry divides the equilateral triangle into two right triangles that have the same size and shape. A representative equilateral triangle is shown where *a* is the altitude of the triangle.

23. Fill in the table. Simplify each answer.

Equilateral Triangles

Side	4	6	8	10	12	20	*n*	?
Half of a Side	2	?	?	?	?	?	?	?
Altitude	?	?	?	?	?	?	?	$17\sqrt{3}$

Lesson 13.4   The "Pythagorean" Right-Triangle Theorem   **623**

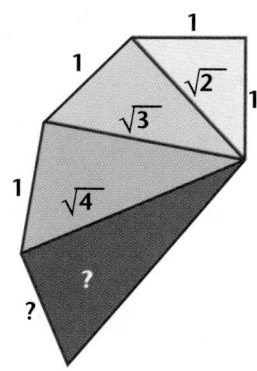

Cultural Connection: Europe The given figure is known as the wheel of Theodorus, a Greek mathematician who lived in the mid-fifth century B.C.

24. Continue the spiral by extending the length of the hypotenuse of each new right triangle. Check the numbers with the "Pythagorean"Right-Triangle Theorem. Make a table and extend the number to 9 right triangles.

Right Triangle Number	1	2	3	4	5	6	7	8	9
Hypotenuse	$\sqrt{2}$	?	?	?	?	?	?	?	?

25. Predict the length of the hypotenuse for the tenth right triangle.

26. Describe the length of the hypotenuse for the nth right triangle in terms of n.

27. The hypotenuse of a right triangle is 8 centimeters. Find the lengths of the two legs if one leg is 1 centimeter longer than the other. Give both an exact and an approximate answer rounded to the nearest hundredth.

Look Back

Darnell is paid a weekly fee of $20 plus $0.50 for every T-shirt that he sells. Copy and complete the table. [Lesson 4.6]

T-Shirts Sold	0	1	2	3	4
Total	28. ___	29. ___	30. ___	31. ___	32. ___

33. Write an equation that describes w, weekly salary, in terms of t, the number of T-shirts sold.

34. If Darnell sells 29 T-shirts in a given week, what will his salary be?

35. If Darnell needs to earn $49 in a given week in order to buy a particular sweater, how many T-shirts must he sell?

Decide whether the lines are parallel, perpendicular, or neither. [Lesson 5.6]

36. $y = \frac{1}{2}x + 3$

$y = 4x + 3$

37. $y = 3x - 4$

$y = -\frac{1}{3}x + 2$

38. $-2x + y = 8$

$-6x + 3y = 15$

Look Beyond

Plot each pair of points on graph paper, and find the distance between them.

39. $(3, 2), (10, 2)$ **40.** $(-7, -1), (5, -1)$ **41.** $(4, 1), (4, 8)$ **42.** $(-6, -4), (-6, -1)$

LESSON 13.5 The Distance Formula

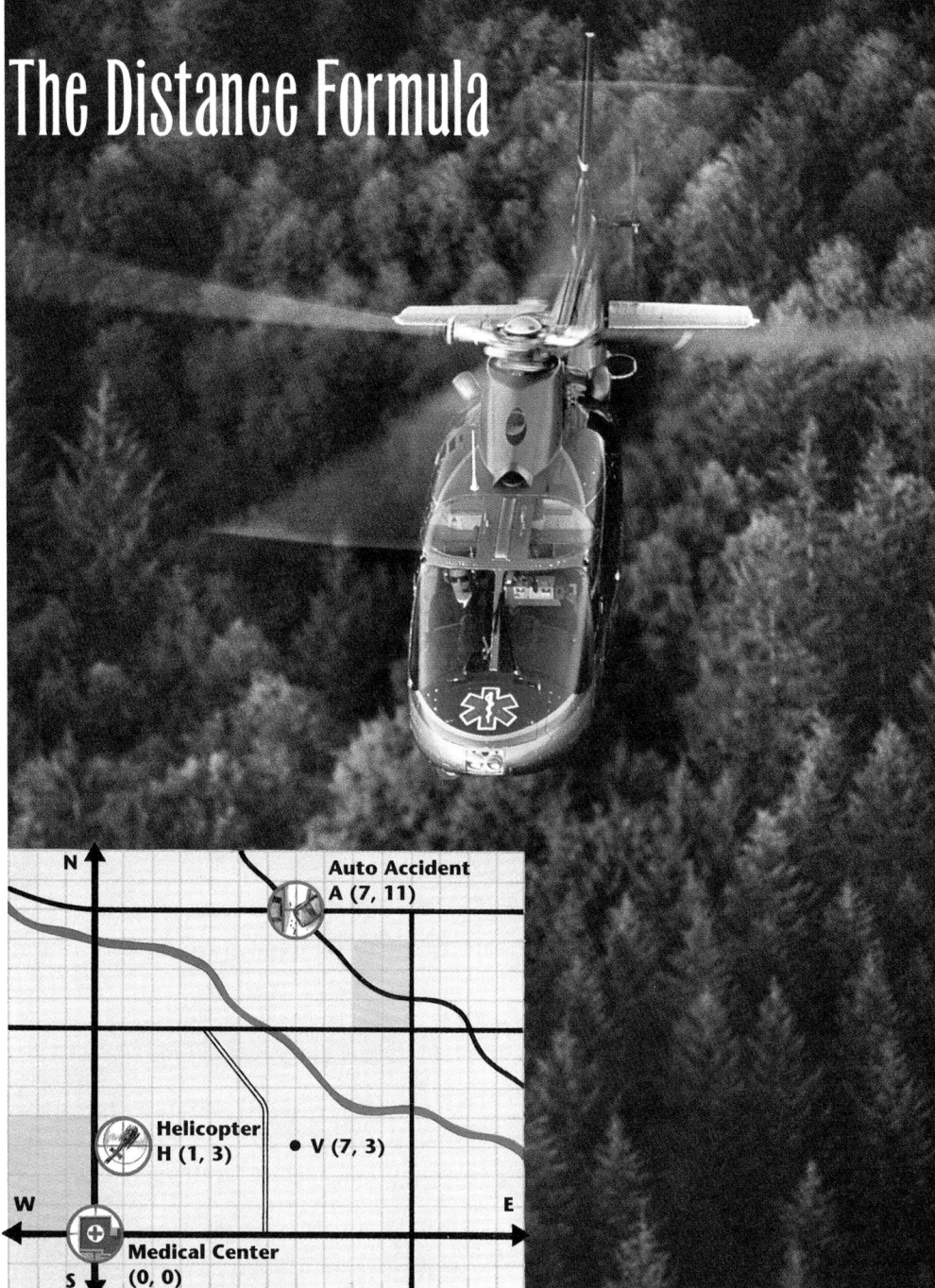

Auto Accident
A (7, 11)

Helicopter
H (1, 3)

• V (7, 3)

N

W

E

S

Medical Center
(0, 0)

The ability to find the distance between two points is of critical importance to sailors, surveyors, mathematicians, architects, astronomers, and draftsmen. Countless others, taxi drivers, carpenters, and emergency medical technicians, must possess the ability to measure indirect distance.

Emergency Service A medical center learns of an auto accident located 7 miles east and 11 miles north of its location. If the medical center's helicopter is now located 1 mile east and 3 miles north of the medical center, what distance must the helicopter travel to get to the scene of the accident?

Visualize a segment drawn from H to A. This segment represents the distance the helicopter must travel. Now visualize a line through A perpendicular to the west-east axis and a line through H, perpendicular to the north-south axis. The intersection of these two lines meet at point V (7, 3). The 3 lines form the right triangle AVH with the right angle at V. How can you find the length of HA?

**COORDINATE
GEOMETRY**
Connection

1. You can find the length of $\overline{HV}$ by subtracting the first coordinates of V and H.

length of $\overline{HV} = 7 - 1 = 6$

2. You can find the length of $\overline{AV}$ by subtracting the second coordinates of A and V.

length of $\overline{AV} = 11 - 3 = 8$

3. Now use the "Pythagorean" Right-Triangle Theorem.

$$c^2 = a^2 + b^2$$
$$(\text{length of } \overline{AH})^2 = 6^2 + 8^2 = 100$$
$$\text{length of } \overline{AH} = \sqrt{100} = 10$$

The medical center's helicopter must travel 10 miles.

•Exploration 1 *Finding the Distance Formula*

**COORDINATE
GEOMETRY**
Connection

1 Use graph paper and the method illustrated in the medical center problem to find the distance between each pair of given points.

a. $K(11, 2)$, $L(14, 6)$ **b.** $M(2, 3)$, $N(7, 15)$

c. $R(-4, 3)$, $S(5, 8)$ **d.** $T(-2, 3)$, $U(-6, -2)$

The **distance** between two points A and B is the **length** of $\overline{AB}$. This is the segment with A and B as endpoints.

2 Follow the same procedure as before to find the distance from A to B, but use the letters for the coordinates of the points $A(x_1, y_1)$ and $B(x_2, y_2)$ instead of numbers. Compare your results with the distance formula.

DISTANCE FORMULA

The distance, d, between two points $A(x_1, y_1)$ and $B(x_2, y_2)$ is given by

$$d = \sqrt{(x_2 - x_1)^2 + (y_2 - y_1)^2}.$$

The Converse of the Theorem

When the *if* and *then* portions of a theorem are interchanged, the new statement is called a **converse**.

> **CONVERSE OF THE "PYTHAGOREAN" RIGHT-TRIANGLE THEOREM**
>
> If a, b, and c are the lengths of the sides of a triangle with $a^2 + b^2 = c^2$, $c > a$ and $c > b$, then the triangle is a right triangle.

CRITICAL Thinking Is the *converse* of an if-then statement always true? If not, give an example.

EXAMPLE 1

COORDINATE GEOMETRY *Connection*

Given points $P(1, 2)$, $Q(3, -1)$, and $R(-5, -2)$, determine whether triangle PQR is a right triangle.

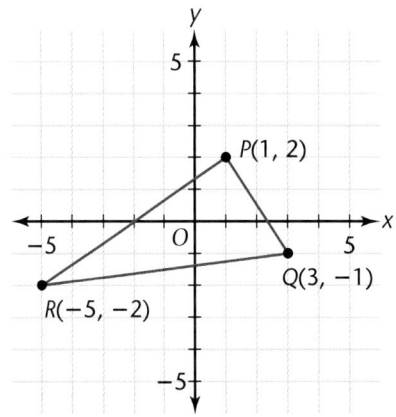

Solution ➤

Draw triangle PQR on graph paper.

Use the Distance Formula to find the lengths of each side.

$$\text{Length of } \overline{PR} = \sqrt{(-5-1)^2 + (-2-2)^2} = \sqrt{(-6)^2 + (-4)^2} = \sqrt{52}$$
$$\text{Length of } \overline{PQ} = \sqrt{(3-1)^2 + (-1-2)^2} = \sqrt{2^2 + (-3)^2} = \sqrt{13}$$
$$\text{Length of } \overline{RQ} = \sqrt{[3-(-5)]^2 + [-1-(-2)]^2} = \sqrt{8^2 + 1^2} = \sqrt{65}$$

Now use the converse of the "Pythagorean" Right-Triangle Theorem to test whether the three lengths can be the sides of a right triangle.

Does $(\sqrt{52})^2 + (\sqrt{13})^2 = (\sqrt{65})^2$?

Since $52 + 13 = 65$, the triangle is a right triangle. ❖

Try This The sides of a triangle are 60, 90, and 109 centimeters. Is it a right triangle? Is a triangle with sides 54, 71, 90 a right triangle?

CRITICAL Thinking How can you use the idea of slope to show that the triangle PQR in Example 1 is a right triangle?

The Midpoint Formula

Exploration 2 Finding the Midpoint Formula

COORDINATE GEOMETRY *Connection*

1 Locate the points on graph paper. Guess the midpoint of $\overline{PQ}$.

a. $P(1, 6), Q(7, 10)$ **b.** $P(-2, 3), Q(6, 7)$ **c.** $P(-4, -1), Q(8, 7)$

2 Use the distance formula to verify that your proposed midpoint is the midpoint of the given segment.

3 What rule can be applied to the *x*-coordinates of the endpoints of a segment to obtain the *x*-coordinate of the midpoint of the segment?

4 What rule can be applied to the *y*-coordinates of the endpoints of a segment to obtain the *y*-coordinate of the midpoint of the segment?

5 Apply your rule in Steps 3 and 4 to find the coordinates of the midpoint of $\overline{AB}$, with $A(12, 63)$ and $B(43, 20)$.

6 Use the distance formula to verify your answer. ❖

MIDPOINT FORMULA

Given: $\overline{PQ}$ with $P(x_1, y_1)$ and $Q(y_2, y_2)$.

The coordinates of the midpoint, M, of $\overline{PQ}$, are $\left(\dfrac{x_1 + x_2}{2}, \dfrac{y_1 + y_2}{2}\right)$.

EXAMPLE 2

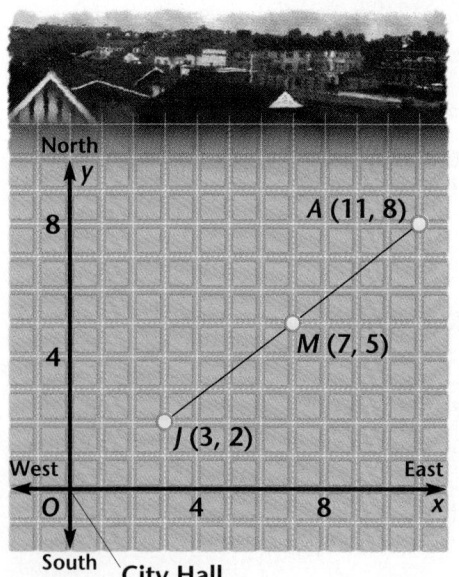

North

South   City Hall

The streets of a city are laid out like a coordinate plane with the origin at City Hall. Jacques lives 3 blocks east and 2 blocks north of City Hall. His friend Alise lives 11 blocks east and 8 blocks north of City Hall. They want to meet at the point midway between their two locations. At what point should they meet?

Solution ➤

Plot $J(3, 2)$ and $A(11, 8)$ on a coordinate plane as shown in the figure. To find the point midway between J and A, use the midpoint formula, which gives the "average" coordinate, $(\bar{x}, \bar{y})$.

$$\bar{x} = \frac{x_1 + x_2}{2} = \frac{3 + 11}{2} = \frac{14}{2} = 7$$

$$\bar{y} = \frac{y_1 + y_2}{2} = \frac{2 + 8}{2} = \frac{10}{2} = 5$$

The midpoint $M = (7, 5)$ is where they should meet. ❖

EXAMPLE 3

GEOMETRY
Connection

The midpoint of a diameter of a circle is $M(3, 4)$. If one endpoint of a diameter is $A(-3, 6)$, what are the coordinates of the other endpoint, $B(x_2, y_2)$?

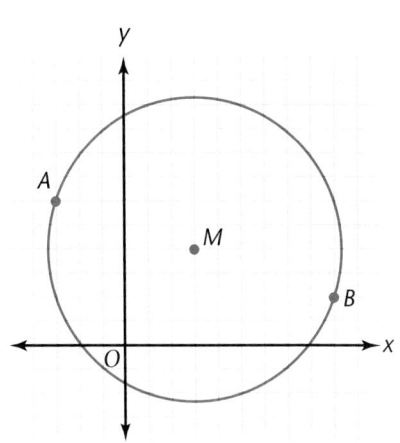

Solution ➤

Substitute the information into the midpoint formula, and solve.

$$\frac{x_1 + x_2}{2} = \bar{x} \qquad \frac{y_1 + y_2}{2} = \bar{y}$$

$$\frac{-3 + x_2}{2} = 3 \qquad \frac{6 + y_2}{2} = 4$$

$$2\left(\frac{-3 + x_2}{2}\right) = 2(3) \qquad 2\left(\frac{6 + y_2}{2}\right) = 2(4)$$

$$-3 + x_2 = 6 \qquad 6 + y_2 = 8$$

$$x_2 = 9 \qquad y_2 = 2$$

The other endpoint is $B(9, 2)$. ❖

EXERCISES & PROBLEMS

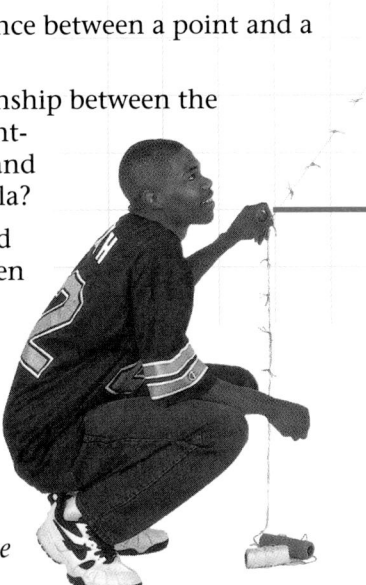

Communicate

1. What is the difference between a point and a coordinate?

2. What is the relationship between the "Pythagorean" Right-Triangle Theorem and the distance formula?

3. Explain how to find the distance between points on a coordinate plane.

4. Describe a way to find the midpoint of a segment on a coordinate plane.

5. Explain the *converse* of an if-then statement.

Practice & Apply

For each figure, find the coordinates of Q and the lengths of the three sides of the triangle.

6.

7.

8.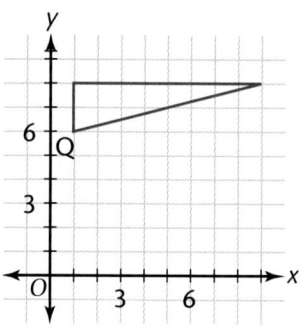

Find the distance between the two given points.

9. $A(4, 7)$, $B(1, 3)$

10. $P(5, 6)$, $Q(17, 11)$

11. $R(-5, -2)$, $S(-9, 3)$

Plot $\triangle PQR$ on graph paper. Decide which description(s) apply to $\triangle PQR$: scalene (no sides equal), isosceles (2 sides equal), equilateral (3 sides equal), or right triangle. Justify your answer.

12. $P(-1, 3)$, $Q(4, 6)$, $R(4, 0)$

13. $P(-2, 2)$, $Q(2, 5)$, $R(8, -3)$

14. Rescue Services A Coast Guard communications center is located at $(0, 0)$. A rescue helicopter located 3 miles west and 2 miles north of the center must respond to a ship located 5 miles east and 7 miles south of the center. What distance must the helicopter travel to reach the ship?

Plot each segment in a coordinate plane, and find the coordinates of the midpoint of the segment.

15. $\overline{AB}$, with $A(2, 5)$ and $B(9, 5)$

16. $\overline{EF}$, with $E(-4, -1)$ and $F(-2, -7)$

17. $\overline{GH}$, with $G(-5, 6)$ and $H(-1, -2)$

18. $\overline{IJ}$, with $I(4, 3)$ and $J(9, -4)$

The locations of two points are given relative to the origin, $(0, 0)$. Find the coordinates of the midpoint of the segment connecting the two points.

19. A is 2 units east, 3 units north of 0.
B is 4 units west, 5 units north of 0.

20. C is 2 units west, 5 units south of 0.
D is 6 units west, 1 unit south of 0.

21. Use the method in Example 2 to find the coordinates of the midpoint of $\overline{PQ}$ with $P(10, 7)$ and $Q(2, 3)$.

22. If the coordinates of P are (a, b) and the coordinates of Q are (c, d), what are the coordinates of the midpoint of $\overline{PQ}$?

The midpoint of $\overline{PQ}$ is M. Find the missing coordinates.

23. $P(2, -5)$, $Q(6, 7)$, $M(\underline{?}, \underline{?})$

24. $P(4, 8)$, $Q(\underline{?}, \underline{?})$, $M(10, 7)$

25. $P(\underline{?}, \underline{?})$, $Q(6, -2)$, $M(9, 4)$

26. $P(3, \underline{?})$, $Q(\underline{?}, 5)$, $M(2, 8)$

Complete Exercises 27–32.

27. Plot $\overline{RS}$ with $R(2, 1)$ and $S(6, 3)$.

28. Find the coordinates M, the midpoint of $\overline{RS}$, and plot M.

29. Find the slope of $\overline{RS}$.

30. What is the slope of any line perpendicular to $\overline{RS}$?

31. Find an equation of k, the line that is perpendicular to $\overline{RS}$ and contains M. Draw k on your graph. The line k is called the *perpendicular bisector* of $\overline{RS}$ because it is perpendicular to $\overline{RS}$ and contains the midpoint of $\overline{RS}$.

32. Choose any two points on k other than M. Name them A and B. Compute the distance from A to R and the distance from A to S. Compute the distance from B to R and the distance from B to S.

33. **Portfolio Activity** Complete the problem in the portfolio activity on page 595.

34. Use right triangles and the "Pythagorean" Right-Triangle Theorem to show that $(4, 7)$ is the midpoint between $(1, 3)$ and $(7, 11)$.

Look Back

Factor each trinomial.   [Lesson 11.5]

35. $y^2 + 35y + 300$ **36.** $x^2 + 30x + 216$

Technology Find the vertex of each parabola. Use graphing technology to assist you.   **[Lesson 12.1]**

37. $y = (x - 2)^2 + 3$ **38.** $y = (x + 3)^2 - 1$

39. $y = -2(x - 1)^2 - 4$ **40.** $y = -(x + 4)^2 + 1$

41. Use an algebraic technique to find the x-intercept(s) of the graph of the function $y = x^2 - 6x - 10$. Give an exact solution(s).   **[Lesson 12.4]**

Use properties of radicals to simplify each expression. Use a calculator to check your answers.   [Lesson 13.2]

42. $\sqrt{3}(\sqrt{12} - \sqrt{75})$ **43.** $\dfrac{\sqrt{75}}{\sqrt{3}}$ **44.** $\dfrac{\sqrt{36} + \sqrt{81}}{\sqrt{9}}$

Look Beyond

Technology Use a scientific calculator to evaluate each expression.
NOTE: To evaluate 9^(1/2), use the key strokes of 9 $\boxed{y^x}$ $\boxed{(}$ 1 $\boxed{\div}$ 2 $\boxed{)}$ or 9 $\boxed{y^x}$ .5. If your calculator has an exponent key, $\boxed{\wedge}$, you may use 9 $\boxed{\wedge}$ $\boxed{(}$ 1 $\boxed{\div}$ 2 $\boxed{)}$ or 9 $\boxed{\wedge}$ .5.

45. 9^(1/2)   **46.** 16^(1/2)   **47.** 39^(1/2)   **48.** 2^(1/2)

49. Make a conjecture about the meaning of the exponent $\dfrac{1}{2}$.

Exploring Geometric Properties

WHY *In coordinate geometry, algebra is used to find relationships between the line segments that form geometric figures. The distance formula and the midpoint formula are useful in connecting geometry with algebra. This connection can then help you solve problems.*

$r = 5$

$(0, 0)$

COORDINATE GEOMETRY *Connection*

A *circle* is the set of all points in a plane that are the same distance from a given point called the *center*. You can use the distance formula to find the equation of the circle with *center* $(0, 0)$ and *radius* 5.

If (x, y) is on the circle, then its distance from $(0, 0)$ must be 5.

$$d = \sqrt{(x - 0)^2 + (y - 0)^2} = 5$$
$$\sqrt{x^2 + y^2} = 5$$

Square both sides and simplify.

$$x^2 + y^2 = 5^2 \text{ or}$$
$$x^2 + y^2 = 25$$

The equation $x^2 + y^2 = 25$ represents all possible points that form the circle.

Exploration 1 The Equation of a Circle

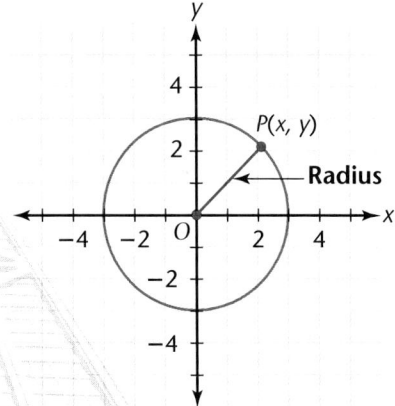

1 Construct a circle with center (0, 0) and radius 3. Then repeat with a radius of 1.

2 Find the equation of each circle.

3 What is the equation of a circle with center (0, 0) and radius *r*?

4 What is the equation of a circle with center (1, 1) and radius 1? ❖

Exploration 2 The Diagonals of a Rectangle

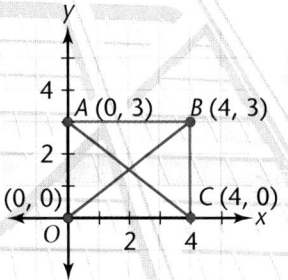

1 Use the distance formula to find the length of each diagonal of rectangle *ABCO*.

2 Construct two other rectangles and find the lengths of the diagonals for each rectangle.

3 Make a conjecture about the diagonals of a rectangle. ❖

When you change from numbers to letters, the conjecture becomes more general. It can now fit whatever numbers are specified for the letters.

Exploration 3 Finding the Formula

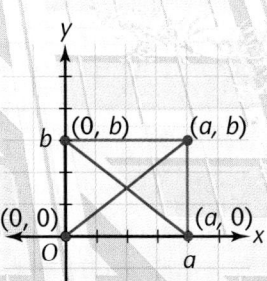

1 Use the 0, *a*, and *b* labels in the diagram to write a formula for finding the length of a diagonal of a rectangle.

2 Explain the general case from your discoveries and your conjecture. ❖

The midpoint formula is also helpful for showing geometric relationships algebraically.

Exploration 4 *Segments and Midpoints*

1 Construct three triangles on a coordinate plane like the one shown. For each, find the midpoints of two sides, and draw a segment to connect those midpoints.

What do you notice about the length of the segment *AB* and the length of the side parallel to that segment?

2 Make a conjecture that describes this relationship.

3 Test the other triangles to see if your conjecture is true for them.

4 Use coordinate geometry with letters to represent the nonzero coordinates to show that your conjecture is true. ❖

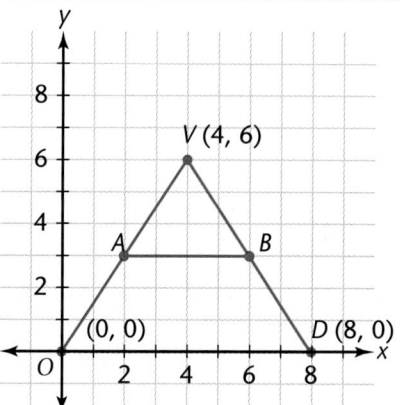

One of the interesting properties of a parabola is that it represents the set of all points the same distance from a given line and a given point. This is illustrated in the following application.

APPLICATION

Surveying   Engineers plan to locate a maintenance facility at a point that is the same distance from a TV tower and a road. The TV tower is located 2 miles from the road. Point *P* is the same distance from the TV tower, *T*, and the road, *X*. Is there an equation that describes the possible placement of the maintenance facility?

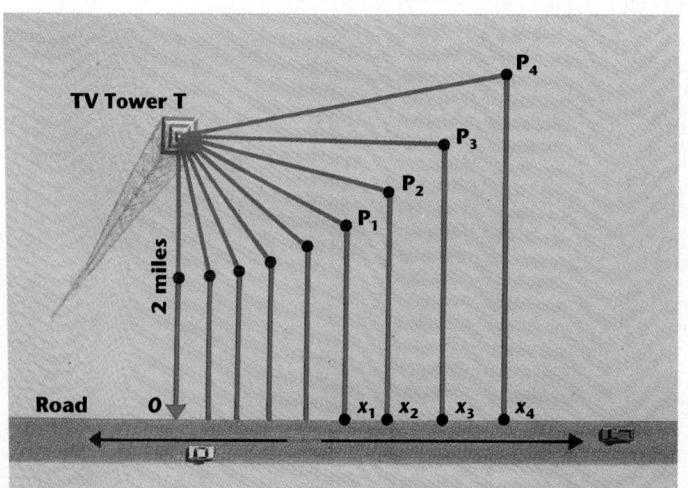

A coordinate grid can be used to model the problem. Let $T(0, 2)$ be the TV tower and $X(x, 0)$ be a point on the road x units away from the origin. Let any point $P(x, y)$ be any point equidistant from T and X. You can use coordinate geometry to find the equation of a graph that is equidistant from the x-axis $(y = 0)$ and the point $(0, 2)$.

Let $P(x, y)$ be any point that is equidistant from $(x, 0)$ and $(0, 2)$.

$$d_1 = \sqrt{(x - x)^2 + (y - 0)^2} = \sqrt{y^2} = y$$
$$d_2 = \sqrt{(x - 0)^2 + (y - 2)^2} = \sqrt{x^2 + (y - 2)^2}$$

Since d_1 and d_2 are equal distances, the two expressions are equal. Substitute the values, square both sides of the equation, and simplify. This is the function that will produce the graph.

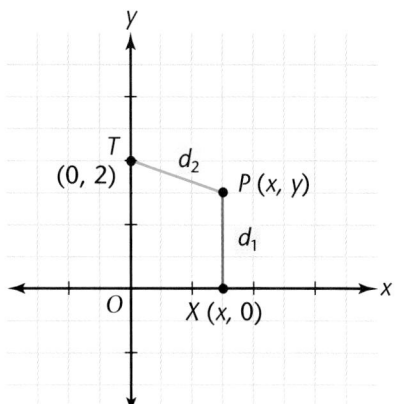

$d_1 = d_2$	Given
$y = \sqrt{x^2 + (y - 2)^2}$	Substitute the values.
$y^2 = x^2 + (y - 2)^2$	Square both sides.
$y^2 = x^2 + y^2 - 4y + 4$	Simplify.
$4y = x^2 + 4$	Addition Property of Equality
$y = \dfrac{1}{4} x^2 + 1$	Division Property of Equality

The equation that describes the possible placement of the maintenance facility is $y = \dfrac{1}{4} x^2 + 1$. ❖

CRITICAL Thinking What is the name for the graph of $y = \dfrac{1}{4} x^2 + 1$? What is the vertex? What is the line of symmetry?

Cultural Connection: Africa In ancient Egypt, African surveyors used geometry to solve algebraic problems. They noticed that by using the diagonal of a 1-cubit square as the side of a larger square, the area of the new square would be double the original square. Recall that $(\sqrt{2})^2 = 2$. They made the diagonal of a 1-cubit square part of their system of measurement. It was called a *double-remen* and measured $\sqrt{2}$ cubits.

1. Square B is 1 square unit in area. How can you determine that square C has an area of 2 square units?

2. Since the area of square C is 2, how would you explain why the side of square C has length $\sqrt{2}$?

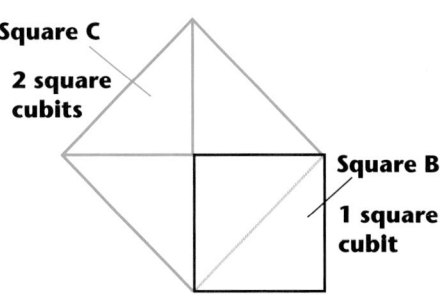

Square C

2 square cubits

Square B

1 square cubit

Exercises & Problems

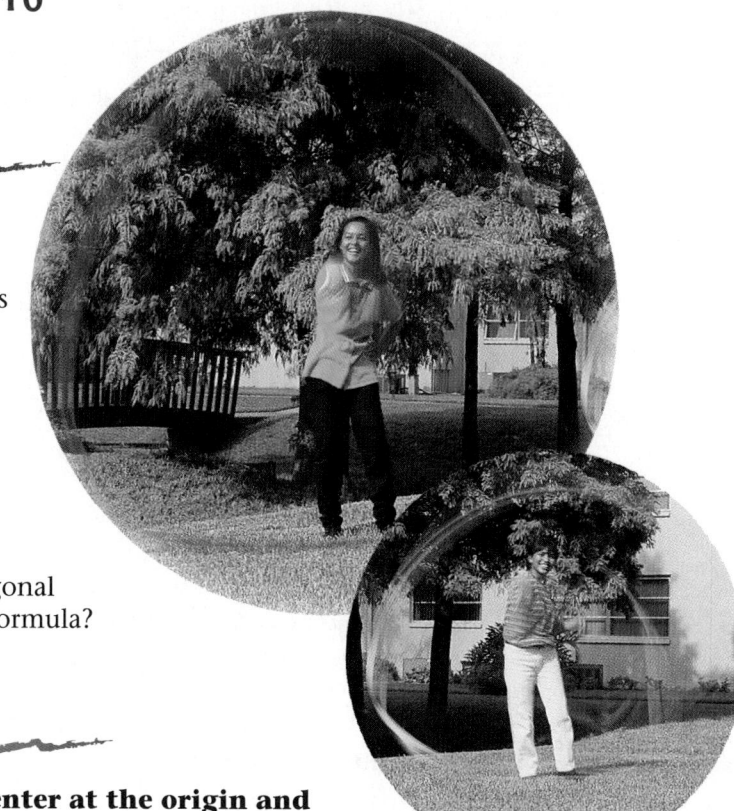

Communicate

1. Discuss the ways that geometry can help solve problems in algebra.

2. What is meant by the set of all points equidistant from a given point?

3. Describe how to find the distance between two points using the distance formula.

4. Explain how the distance formula is used to express the set of all points equidistant from a given point.

5. How do you find the length of a diagonal of a rectangle by using the distance formula?

Practice & Apply

Find an equation of a circle with center at the origin and having the given radius.

6. radius = 3   7. radius = 7   8. radius = 15   9. radius = r

10. Copy and complete the table of ordered pairs that satisfy the equation $x^2 + y^2 = 100$.

x	0	6	−6	8	−8	10	−10	?	?	?	?	?	?	?
y	?	?	?	?	?	?	?	0	6	−6	8	−8	10	−10

11. Plot the ordered pairs on graph paper.

12. What are the x-intercepts?

13. What are the y-intercepts?

14. What shape is the graph?

Use graph paper to sketch each circle below. Find an equation of a circle with

15. center (2, 3) and radius = 5.

16. center (− 2, 3) and radius = 5.

17. center at (2, − 3) and radius = 5.

18. center (− 2, − 3) and radius = 5.

19. Rewrite the equations for Exercises 15 – 18 with a radius of 7.

20. Generalize. Find an equation of a circle with center (h, k) and radius, r.

Use your generalization to identify the center and radius for each circle.

21. $(x − 1)^2 + (y − 5)^2 = 9$

22. $(x + 3)^2 + (y − 4)^2 = 16$

23. $(x − 2)^2 + (y + 4)^2 = 50$

24. $(x + 5)^2 + (y + 10)^2 = 10$

Geometry **Plot triangle *PQR* on graph paper, with *P*(2, 1), *Q*(4, 7), and *R*(12, 3).**

25. Find the coordinates of *M*, the midpoint of $\overline{PQ}$, and plot *M*.

26. Find the coordinates of *N*, the midpoint of $\overline{PR}$, and plot *N*.

27. Draw $\overline{MN}$ and find the length of $\overline{MN}$ (simplify your answer).

28. Find the length of $\overline{QR}$.

29. Compare the lengths of $\overline{MN}$ and $\overline{QR}$.

30. Find the slopes of $\overline{MN}$ and $\overline{QR}$.

31. What does the slope information tell you about $\overline{MN}$ and $\overline{QR}$?

32. Use the distance formula to find an equation for all points *P*(*x*, *y*) such that the sum of the distances from *P* to *A*(− 8, 0) and from *P* to *B*(8, 0) is 20.

33. Find an equation for all points *P*(*x*, *y*) such that the sum of the distances from *P* to *A*(0, − 4) and from *P* to *B*(0, 4) is 10.

 Look Back

34. Identify which of the following correctly represents the distance between 2 and 5 on a number line.   **[Lesson 4.8]**

    a. $2 + 5$ **b.** $2 - 5$ **c.** $5 + 2$ **d.** $5 - 2$

    e. $|2 + 5|$ **f.** $|2 - 5|$ **g.** $|5 + 2|$ **h.** $|5 - 2|$

35. Evaluate $\dfrac{3^0 \cdot 3^2}{3^2} - 3^2$.   **[Lesson 10.3]**

Write each number in scientific form.   [Lesson 10.4]

36. 2,300,000   **37.** 0.00000125

Add or subtract each polynomial.   [Lesson 11.1]

38. $(x^2 + 3x + 5) + (7x^2 - 5x - 10)$ **39.** $(8b^2 - 15) - (2b^2 + b + 1)$

Find the value of the discriminate for each equation.
[Lesson 12.5]

40. $3x^2 - 6x + 3 = 0$ **41.** $2x^2 + 3x - 2 = 0$

Graph each quadratic inequality.   [Lesson 12.6]

42. $y < 2x^2 - 3x + 4$ **43.** $y > -4x^2 + 6x - 5$

Look Beyond

 Geometry In geometry, you can prove several theorems by using coordinates. Prove each of the following.

44. The diagonals of a rectangle bisect each other.

45. The diagonals of an isosceles trapezoid have the same length.

Trigonometric Functions

why *The tangent function is used for solving problems by indirect measure. If you know the measurement of an acute angle and one leg of a right triangle, you can determine the length of the other side without measuring it. In reverse, you have a way to determine the acute angle of the triangle if you know the lengths of the two legs.*

Les works on a milling machine in a machine shop. A blueprint calls for a part with the dimensions as shown. At what angle should Les make the cut from the rectangular metal piece?

Consider the triangle that is to be removed. Examine the following ratio with respect to angle B.

$$\frac{\text{length of the leg opposite the angle}}{\text{length of the leg adjacent to the angle}} = \frac{AC}{BC} = \frac{2}{3}$$

In a right triangle, this ratio is called the **tangent** of angle B or tan B. The goal is to find an angle whose tangent is $\frac{2}{3}$. "The angle whose tangent is $\frac{2}{3}$" is written $\tan^{-1}\left(\frac{2}{3}\right)$. $\tan^{-1}x$ is also read "the inverse tangent of x."

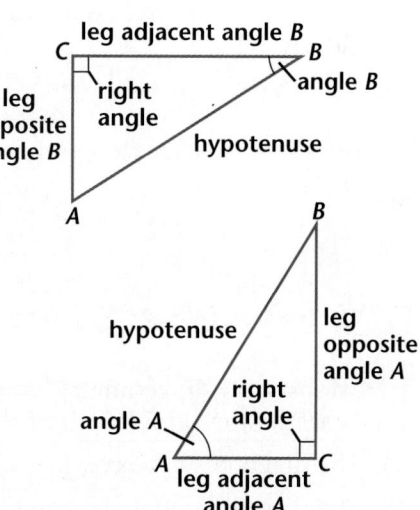

•Exploration• Angles and Tangents

$\overline{CA}$ is the base of all 7 triangles. For example, one triangle is CAB_1. As angle A increases in increments of $10°$, the length of the side opposite angle A for each triangle also increases.

 Find the tangent ratio for each given angle measure. Round the ratio to the nearest 10th.

a. $10°$ (Angle CAB_1)

b. $20°$ (Angle CAB_2)

c. $30°$ (Angle CAB_3)

d. $40°$ (Angle CAB_4)

e. $50°$ (Angle CAB_5)

f. $60°$ (Angle CAB_6)

g. $70°$ (Angle CAB_7)

Refer to your answers for **a–g** to answer the following questions.

 How does the value of the tangent ratio increase as the measure of the angle doubles? (Compare, for example, tan $10°$ and tan $20°$, and tan $30°$ and tan $60°$.)

 For which of the angles in Step 1 is the tangent ratio less than 1?

 The measure of angle CAB_S is $45°$. What can you say about the length of the side opposite the angle and the length of the side adjacent to the angle?

 How can you find the length of the side opposite angle A if you know that the measure of angle A is $45°$ and you know the length of the side adjacent to A? Explain. ❖

The function $f(x) = \tan x$ is an example of a trigonometric function. The domain of f is a set of angle measures. Thus $f(45°) = \tan 45° = 1$.

Calculator

You can use a calculator to find the tangent. Since you are using degrees to measure the angles, be sure the calculator mode is set for degrees. Enter the angle and use the TAN key. For example, for $f(x) = $ TAN x, press $\boxed{\text{TAN}}$ 45 $\boxed{\text{ENTER}}$ or $\boxed{=}$, and the calculator will return the value 1.

You can also find an acute angle if you know the tangent value. To find the angle x, where x represents the degree measure of the acute angle, use the TAN⁻¹ key. For example, if $\tan x = 1$, press TAN⁻¹ 1 ENTER and the calculator will show 45.

4 units

X

3 units

7 units

Calculator

Now you can solve Les's problem. That is, find the angle whose tangent is $\frac{2}{3}$ or approximately 0.67.

$$\boxed{\text{TAN}^{-1}} \left(\frac{2}{3}\right) \approx \boxed{\text{TAN}^{-1}} \, 0.67 \approx 34$$

Les should cut the metal at a 34° angle.

EXAMPLE 1

Find the tangent of each acute angle of each triangle.

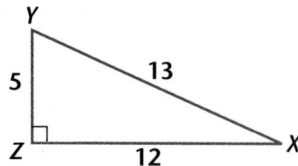

Solution ➤

Ⓐ $\tan A = \frac{4}{3}$, $\tan B = \frac{3}{4}$ Ⓑ $\tan Y = \frac{12}{5}$, $\tan X = \frac{5}{12}$ ❖

EXAMPLE 2

An airplane climbs to 30,000 feet at a steady rate. If the altitude is reached after the airplane has covered a distance of 35 miles on the ground, what is its angle of elevation?

Solution ➤

Draw a picture to model the problem. Change 35 miles to feet by multiplying by 5280.

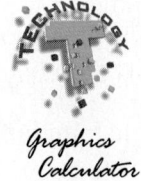

Graphics Calculator

$$\tan A = \frac{30,000}{184,800}$$

In degree mode, use TAN⁻¹ to find the angle whose tangent is $\frac{30,000}{184,800}$.

$$A = \tan^{-1} \frac{30,000}{184,800} \approx 9°$$

The airplane is climbing at an angle of 9°. The angle of elevation is 9°. ❖

C

30,000 Feet

A

184,800 Feet

B

Angle of Elevation

A group of skateboarders wants to build a ramp with an angle of 12°. How much should the ramp rise over each 10 meters of run to achieve a 12° angle?

Solution ➤

To find the rise, find the tangent of 12°.

$$\tan 12° = \frac{x}{10}$$

$$0.2126 = \frac{x}{10}$$

$$(10)(0.2126) = x$$

$$2.126 = x$$

The ramp should rise about 2.1 meters for each 10 meters of run. ❖

EXERCISES & PROBLEMS

Communicate

1. Explain how the tangent function is related to the slope of a line.

2. How is the length of the opposite leg of a right triangle determined from the acute angle A and the length of the other leg?

3. How is the angle determined from the ratio of the opposite over the adjacent legs of a right triangle?

4. Why is finding the tangent considered indirect measure?

Practice & Apply

Technology Use a calculator to find the tangents of the following angles to the nearest tenth.

5. 45°   6. 30°   7. 60°   8. 0°

9. 38°   10. 57°   11. 89°   12. 89.9°

What are the corresponding angles for the following tangent approximations?

13. 1.732   **14.** 1.192   **15.** 0.268   **16.** 3.732

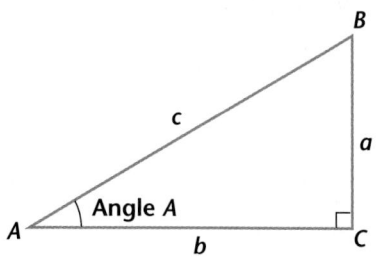

The small letters represent the sides, and the capital letters represent the angles of a general right triangle. Use the given information to find the indicated side to the nearest hundredth.

17. If angle A is 30° and side b is 12 meters, find a.

18. If angle B is 60° and side a is 10 feet, find b.

19. If side a is 3 centimeters and tan A is 0.75, what is the length of side b?

20. If side b is 23 feet long and tan A is 1.0, what is the length of side a?

21. If side a is 6 inches and angle A is approximately 31°, how long is side b?

22. If angle A is approximately 35° and side b is 20 millimeters, what is the length of side a?

23. House Construction Mr. Fernandez wants the roof of his house to have an angle of 20°. The slope of the roof is the same on both sides and has the same length on either side of the center. How high does the roof rise if the whole length of the roof is 14 meters long?

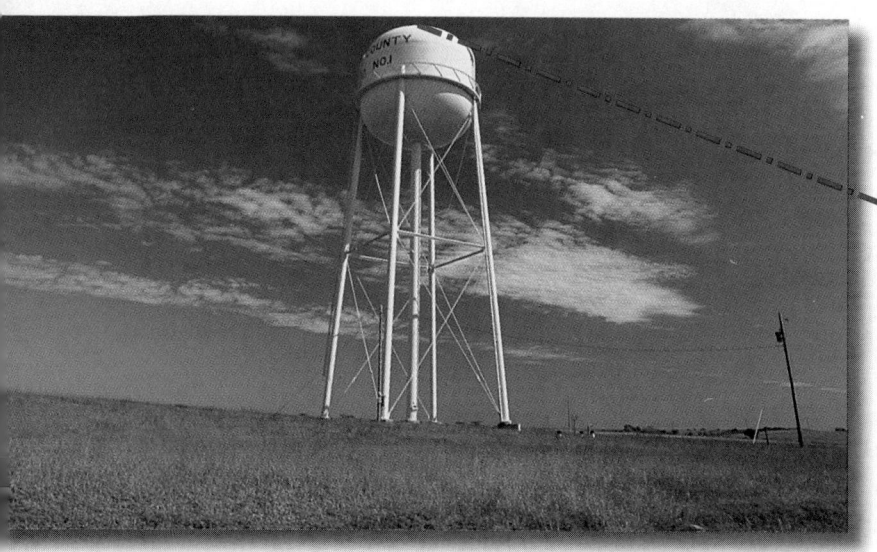

24. A water tower is 600 feet from the observer. Sighting the top of the water reservoir forms an angle of 28°. A sighting at the bottom of the reservoir has an angle of 26°. How tall is the reservoir part of the water tower?

25. Highways If a road has a 7% grade, its slope is $\frac{7}{100}$. What is the measure of the angle that the road inclines?

26. Model Airplanes A model airplane and a car begin at the same place. After 15 seconds the car has driven $\frac{1}{4}$ mile, and the elevation angle from the starting point to the plane is 30°. What is the altitude of the plane?

Look Back

27. Solve the equation $4x - 5 = 7(x - 3) + 2$, for x. **[Lesson 4.6]**

28. What is the probability that after 1 roll of 2 number cubes the sum of the dots equals 7? **[Lesson 8.6]**

29. Determine whether the quadratic equation $2x^2 + x - 15 = 0$ has a real solution. If it does, use the quadratic formula to determine the solution. **[Lesson 12.5]**

30. Solve $x^2 - 5x + 6 \geq 0$. **[Lesson 12.6]**   **31.** Graph $x^2 - 16 < 0$. **[Lesson 12.6]**

32. 🖳 **Geometry** The 45-45-90 degree triangle and the 30-60-90 degree triangle have sides with special relationships.
[Lesson 13.4]

Use the "Pythagorean" Right-Triangle Theorem to show that the sides actually have the relationship shown.

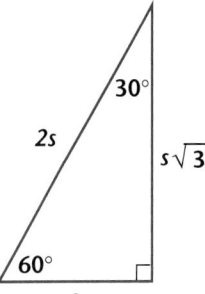

33. Points $P(4, 9)$ and $Q(10, 5)$ are given on a coordinate plane. Find the coordinates of M, the midpoint of segment PQ.   **[Lesson 13.5]**

Find the length of each segment.   **[Lesson 13.5]**

34. PQ.   **35.** PM.   **36.** MQ.

Look Beyond

Besides the tangent (tan), the sine (sin) and cosine (cos) are two other basic trigonometric functions. They provide additional tools for solving problems by indirect measure.

$f(x) = \tan (x)$ is $\dfrac{b}{a}$,

$g(x) = \sin (x)$ is $\dfrac{b}{c}$,

$h(x) = \cos (x)$ is $\dfrac{a}{c}$

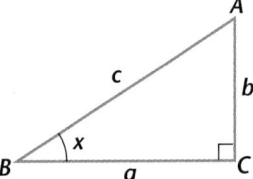

37. If the tangent of angle x is $\dfrac{3}{4}$, what is the sine and the cosine of x?

HINT: Use the "Pythagorean" Right-Triangle Theorem.

38. Explain why $(\sin x)^2 + (\cos x)^2 = 1$ is true for any right triangle.

RADICAL
RECTANGLES

a

b

D ID YOU KNOW THAT SOME RECTANGLES ARE considered more beautiful than others? Greek architects of the fifth century B.C.E. believed that the golden rectangle was the most beautiful rectangle possible. Great artists all over the world have used this rectangle in their creations.

What is a golden rectangle? It is a rectangle whose length and width form a special ratio. If *a* is the longer side of the rectangle and *b* is the shorter side, then the golden ratio, *m*, is the proportion

$$\frac{b}{a} = \frac{a}{a = b}.$$

This proportion can be rewritten as $a^2 = b(a + b)$ and can be solved by the quadratic formula. The ratio is

$$m = \frac{\sqrt{5} - 1}{2} \approx 0.618.$$

The Italian renaissance artist, Masaccio's fresco, *The Tribute Money*, (detail) compositionally exploits properties of The "Golden Rectangle".

Photo: Scala/ Art Resource, NY

A golden rectangle will produce a similar rectangle when a square is cut from the end of the original rectangle. A similar rectangle is one that has an identical shape, but has a different size.

ACTIVITY 1

Construct an approximate golden rectangle with a 3-by-5 inch note card. Since the golden mean has the ratio of b to a or 0.61803, this is approximately $\frac{6}{10}$ or $\frac{3}{5}$.

You can see that the approximation is close. If you fold one side and draw a vertical line to form the square portion, the remaining rectangle will have a ratio of $\frac{2}{3} = 0.667$ which is close to 0.61803.

ACTIVITY 2

There is another way you can create a golden rectangle that is more accurate. Begin with a square. Find the midpoint of the base. Use a compass to measure the distance from, the midpoint, M, to the vertex, V, of the opposite side. Extend the base of the square, and draw an arc to the point of the intersection with the base, B. Use this to complete the rectangle.

ACTIVITY 3

Construct a large golden rectangle. Fold a corner of the rectangle and draw the line to create a smaller rectangle. Repeat the process with each of the resulting rectangles. Continue this process several times. You can produce a spiral by connecting the corresponding vertex of each square.

Examine the decreasing rectangles. Explore and describe the mathematical relationships that result. For example, consider the length of the side of the smallest square as 1. Use the "Pythagorean" Right-Triangle Theorem to determine the lengths of the diagonals.

The spiral pattern appears frequently in nature. It is shown here in the flower.

Chapter 13 Review

Vocabulary

Key Skills & Exercises

Lesson 13.1

➤ **Key Skills**

Estimate square roots. If the square root is irrational, find the value to the nearest hundredth.

a. $\sqrt{16} = 4$ **b.** $\sqrt{\frac{9}{25}} = \frac{3}{5}$ **c.** $\sqrt{18} = 4.24$ **d.** $-\sqrt{0.012} = -0.11$

➤ **Exercises**

Estimate each square root. If the square root is irrational, find the value to the nearest hundredth.

1. $\sqrt{20}$ **2.** $\sqrt{115}$ **3.** $\sqrt{134}$ **4.** $-\sqrt{67}$ **5.** $\sqrt{\frac{9}{16}}$ **6.** $-\sqrt{0.09}$

Lesson 13.2

➤ **Key Skills**

Simplify radical expressions.

a. $\sqrt{32} = \sqrt{16 \cdot 2} = \sqrt{16} \cdot \sqrt{2} = 4\sqrt{2}$ **b.** $\sqrt{\frac{4}{25}} = \frac{\sqrt{4}}{\sqrt{25}} = \frac{2}{5}$

c. $6\sqrt{3} + 5\sqrt{2} - 3\sqrt{3} + \sqrt{2} = (6 - 3)\sqrt{3} + (5 + 1)\sqrt{2} = 3\sqrt{3} + 6\sqrt{2}$

d. $(4\sqrt{5})^2 = (4\sqrt{5})(4\sqrt{5}) = (4 \cdot 4)(\sqrt{5} \cdot \sqrt{5}) = 16 \cdot 5 = 80$

➤ **Exercises**

Simplify.

7. $\sqrt{150}$ **8.** $\sqrt{\frac{48}{27}}$ **9.** $\sqrt{a^2 b^7}$ **10.** $\sqrt{2} + 3\sqrt{7} - 3\sqrt{2}$

11. $(2\sqrt{3})^2$ **12.** $\sqrt{3}(2 - \sqrt{12})$ **13.** $(\sqrt{2} + 2)^2$ **14.** $(\sqrt{3} - 2)(\sqrt{5} + 6)$

Lesson 13.3

➤ **Key Skills**

Solve radical equations. Use radicals to solve equations.

$\sqrt{x + 2} = 5 \longrightarrow (\sqrt{x + 2})^2 = 5^2 \longrightarrow x + 2 = 25 \longrightarrow x = 23$

$2x^2 = 36 \longrightarrow x^2 = 18 \longrightarrow x = \pm\sqrt{18} \longrightarrow x = \pm\sqrt{9 \cdot 2} = \pm 3\sqrt{2}$

➤ **Exercises**

Solve each equation. Be sure to check your solution.

15. $\sqrt{x - 7} = 2$　　　　**16.** $\sqrt{3x + 4} = 1$　　　　**17.** $\sqrt{x + 6} = x$

18. $\sqrt{x^2 - 2x + 1} = x - 5$　　**19.** $x^2 = 40$　　　　**20.** $2x^2 - 32 = 0$

Lesson 13.4

➤ *Key Skills*

Use the "Pythagorean" Right-Triangle Theorem to find the length of a missing side of a right triangle.

Find the length of the hypotenuse of the triangle.

$$a^2 + b^2 = c^2$$
$$8^2 + 11^2 = c^2$$
$$64 + 121 = c^2$$
$$185 = c^2$$
$$c = \pm\sqrt{185} \approx \pm 13.6$$

Since c represents the length of a segment, it must be positive. Therefore, $c \approx 13.6$ meters.

➤ *Exercises*

Use the "Pythagorean" Right-Triangle Theorem to find the length of the missing side of each right triangle.

21.　　　　　　　　**22.**　　　　　　　　**23.**

Lesson 13.5

➤ *Key Skills*

Use the Distance formula to find the distance between two points.

Use the Distance formula to find the distance between $(2, 3)$ and $(-4, -1)$.

$$d = \sqrt{(x_2 - x_1)^2 + (y_2 - y_1)^2}$$
$$d = \sqrt{(-4 - 2)^2 + (-1 - 3)^2}$$
$$d = \sqrt{36 + 16}$$
$$d = \sqrt{52} = 2\sqrt{13} \approx 7.2$$

Use the Midpoint formula to find the midpoint of a segment.

Use the Midpoint formula to find the midpoint of the segment whose endpoints are $(-2, 6)$ and $(3, 4)$.

$$\left(\frac{x_1 + x_2}{2}, \frac{y_1 + y_2}{2}\right) = \left(\frac{-2 + 3}{2}, \frac{6 + 4}{2}\right) = \left(\frac{1}{2}, 5\right)$$

➤ **Exercises**

Find the distance between the two given points.

24. $A(0, 3)$, $B(2, 8)$　　　　**25.** $X(-1, 5)$, $Y(3, -8)$　　　　**26.** $G(7, 2)$, $F(-6, -1)$

Find the midpoint of each segment whose endpoints are given.

27. $M(5, 2)$, $N(3, 6)$　　　　**28.** $P(-3, 4)$, $Q(-7, -1)$　　　　**29.** $R(8, -2)$, $S(-1, 4)$

Lesson 13.6

➤ Key Skills

Write the equation of a circle given the center and the radius.

The equation of a circle with center at the origin and radius 4 is $x^2 + y^2 = 16$.

The equation of a circle with center $(2, -3)$ and radius 5 is $(x - 2)^2 + (y + 3)^2 = 25$.

➤ Exercises

Write the equation of each circle with the given center and radius.

30. center $(0, 0)$, radius 2   **31.** center $(0, 2)$, radius 6   **32.** center $(-2, 5)$, radius 3

  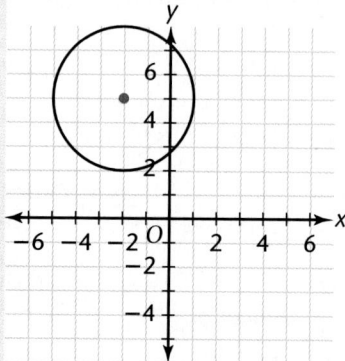

Lesson 13.7

➤ Key Skills

Find the tangent of an acute angle of a right triangle.

$$\tan M = \frac{15}{8} \qquad \tan N = \frac{8}{15}$$

➤ Exercises

Find the tangent ratio for the given angle of the right triangle.

33. $\tan R$ **34.** $\tan S$

Applications

35. The bottom of a circular swimming pool has an area of 80 square feet. Find the radius to the nearest tenth. Use 3.14 for π.

36. Physics The number of seconds it takes an object to fall d feet is given by the formula $s = \sqrt{\dfrac{d}{16}}$. About how long does it take an object dropped from the top of a 50-foot building to hit the ground? Give your answer to the nearest tenth.

Sports At the end of a ski lift, riders are 50 feet above the ground. The ski lift travels a horizontal distance of 120 feet.

37. How long is the cable?

38. What angle does the cable make with the ground?

Chapter 13 Assessment

Simplify.

1. $\sqrt{50}$ 2. $\sqrt{\dfrac{72}{2}}$ 3. $\sqrt{\dfrac{m^8}{n^{10}}}$

4. $6\sqrt{6} + 5\sqrt{6} - \sqrt{3}$ 5. $(\sqrt{10} - 3)^2$

Solve each equation. Be sure to check your solution.

6. $\sqrt{4x + 2} = 8$ 7. $\sqrt{x + 2} = x + 2$ 8. $4x^2 = 48$

Find the length of the missing side of each right triangle to the nearest hundredth.

9.

21 meters

7 meters

x

10.

10 inches

8 inches

x

11. One boat travels 9 miles south. Another boat travels 6 miles west and meets the first boat. How far apart were the two boats when they started?

Find the indicated tangent ratio for the right triangles.

12. tan M
13. tan K
14. tan A
15. tan B
16. tan R
17. tan T

M

$15\sqrt{2}$ 15

K 15 L

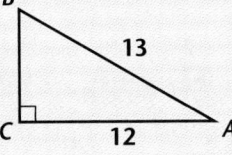

B

13

C 12 A

T

40

S 50 R

18. A ladder is leaning against a house. It forms an angle of 55° with the house. If the ladder touches the house at a point 12 feet high, how far is the bottom of the ladder from the bottom of the house?

Write the equation of each circle with the given center and radius.

19.

20.

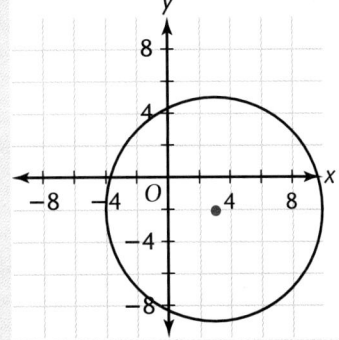

Rational Functions

Have you ever noticed how it takes you longer to row upstream than downstream? Have you ever used a lever to lift a heavy object? Do you know what a gear ratio is?

The rate of the current affects your time spent rowing. The distance from the balance point determines the amount of lift for a lever. The gear ratio affects how fast you ride a bicycle.

Rational expressions can be used to model these relationships. Look carefully at the words *rational* and *irrational*. At the heart of each is the word *ratio*. A **ratio** compares two quantities, and a **rational** expression is a fraction that compares two algebraic expressions.

PORTFOLIO ACTIVITY

A bicycle controls the speed and effort of pedaling by the relationship of the teeth on the pedal gears to the teeth on the wheel gears. For example, if the pedal gear has 54 teeth and the wheel gear has 16, the gear ratio is $\frac{16}{54} = 0.296$. The reciprocal is $\frac{54}{16} = 3.375$ or about 3.4. This represents the number of turns of the wheel for every full turn of the pedal. Assume the circumference of a bicycle tire is approximately 2.19 meters, how far will the rider travel during one full turn of the pedal?

a. The wheel gears on a bicycle might have 13, 14, 15, 16, 18, 21, and 24 teeth and the pedal gears, 49 or 54 teeth. Write a function for each pedal gear that will calculate the gear ratios. Then find the reciprocals.

b. Assume a bicycle rider turns the pedal 60 full turns per minute. Choose 2 gear combinations and calculate the miles per hour for each. Note: 1 Kilometer per hour is approximately 0.62 miles per hour.

Rational Expressions

The art institute charges a $55 membership fee and $4.50 for each pottery lesson.

why People are constantly comparing data. Baseball players compare their batting averages. Business owners compare profit to total income. Students figure their average expenses. Special expressions are needed to model these comparisons.

When Saul's *total* expenses for institute membership and pottery lessons are considered, the following formula represents the average cost per lesson.

$$\text{Average cost} = \frac{\text{total expense}}{\text{number of lessons}}$$

If Saul takes only 1 lesson, his cost will be $55 + $4.50 or $59.50. If Saul takes 2 lessons and then stops, his average cost per lesson will be $\dfrac{\$55 + \$4.50(2)}{2}$ or $32.00.

If x is the number of lessons taken, then $55 + 4.50x$ is the total expense for all lessons. So,

$$\text{Average cost} = \frac{55 + 4.50x}{x}.$$

Spreadsheet

How can Saul compare the average cost per lesson for taking 2 lessons with the average cost for taking 3 lessons? Remember, he has 10 weeks to take lessons.

To compare the average cost per lesson, evaluate the expression

$$\frac{55 + 4.50x}{x}$$

for 20 lessons and 30 lessons. A spreadsheet or graphics calculator can quickly evaluate the expression for any value of *x*.

	A	B	C
1	x	55 + 4.50x	(55 + 4.50x)/x
2	0	55.00	error
3	1	59.50	59.50
4	2	64.00	32.00
5	3	68.50	22.83
6	4	73.00	18.25
7	5	77.50	15.50
8	6	82.00	13.67
9	20	145.00	7.25
10	30	190.00	6.33

Saul's average cost per lesson for 20 lessons is $7.25. If Saul takes 30 lessons, his average cost is $6.33 per lesson. Why does the entry in row 2, column C, read "error"?

The expression $\frac{55 + 4.50x}{x}$ is an example of a *rational expression*.

RATIONAL EXPRESSION

If *P* and *Q* are polynomials and $Q \neq 0$, then an expression in the form $\frac{P}{Q}$ is a rational expression.

Since $55 + 4.50x$ and *x* are polynomials and $x \neq 0$, $\frac{55 + 4.50x}{x}$ is a rational expression.

A function of the form $y = \frac{P}{Q}$, or $f(x) = \frac{P(x)}{Q(x)}$, where $\frac{P}{Q}$ is a rational expression is called a **rational function**. Thus, $g(x) = \frac{55 + 4.50x}{x}$ is a rational function. Since you cannot divide by zero, *g* is *undefined* when $x = 0$. A rational function is undefined or restricted when its denominator is equal to zero.

EXAMPLE 1

For what value or values is each rational function undefined?

A $f(x) = \dfrac{1}{x}$ **B** $y = \dfrac{a^2 - b^2}{a - b}$ **C** $t = \dfrac{m + n}{m^2 - 4m + 3}$

Solution ➤

A rational function is undefined when its denominator is equal to zero.

A $\dfrac{1}{x}$ is undefined when $x = 0$.

B $\dfrac{a^2 - b^2}{a - b}$ is undefined when $a - b = 0$. That is, when $a = b$.

C $\dfrac{m + n}{m^2 - 4m + 3}$ is undefined when $m^2 - 4m + 3 = 0$.

Since $m^2 - 4m + 3 = (m - 1)(m - 3)$, $\dfrac{m + n}{m^2 - 4m + 3}$ is undefined when $m = 1$ or $m = 3$. ❖

Try This For what value or values is each rational function undefined?

a. $\dfrac{y - 2}{y^2 - 5y + 6}$ **b.** $\dfrac{3x + 6x}{x - y}$

Recall that the parent function for a variety of rational functions is $f(x) = \dfrac{1}{x}$. In Chapter 9 you found its graph to be a hyperbola with vertical asymptote $x = 0$.

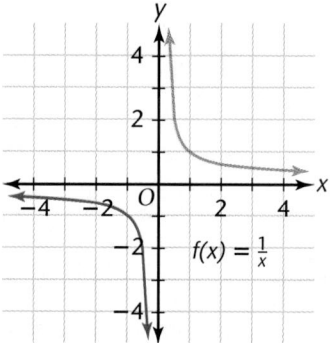

EXAMPLE 2

A Describe the transformations applied to the parent function $f(x) = \dfrac{1}{x}$ to produce the graph of $g(x) = \dfrac{55 + 4.50x}{x}$.

B Use the graph to find the number of pottery lessons that can be taken for an average cost of $5.60 per lesson.

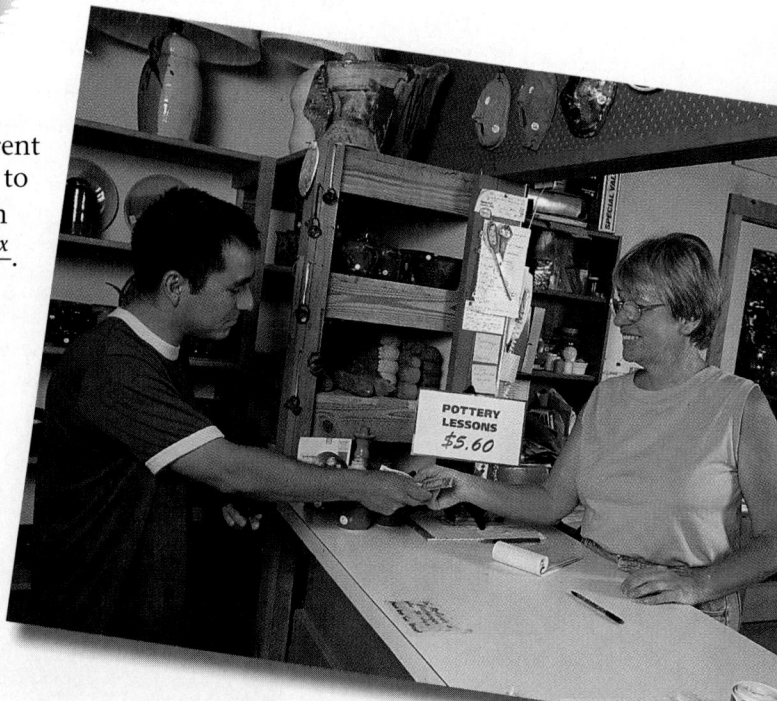

Solution ➤

A The parent function is $f(x) = \frac{1}{x}$. To compare g with f, rewrite the expression using the Distributive Property and simplify.

$$\frac{55 + 4.50x}{x} = \frac{55}{x} + \frac{4.50x}{x}$$

$$= \frac{55}{x} + 4.5$$

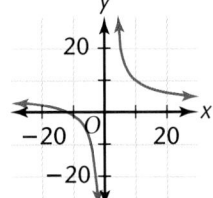

The graph of the function $g(x) = \frac{55}{x} + 4.5$ shows that the parent function has been transformed by a vertical stretch of 55 and a translation up by 4.5.

B Examine the graph, or use the trace function to find the value of x so that $f(x) = 5.60$.

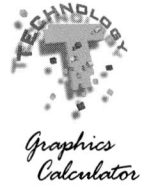

Graphics Calculator

The x value represents 50 lessons.

Therefore, in a 10-week period, the average cost for 50 lessons is $5.60. ❖

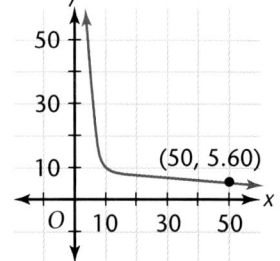

(50, 5.60)

EXAMPLE 3

Describe the transformation applied to the parent function $f(x) = \frac{1}{x}$ to produce the graph of $g(x) = \frac{5}{x-3} + 2$.

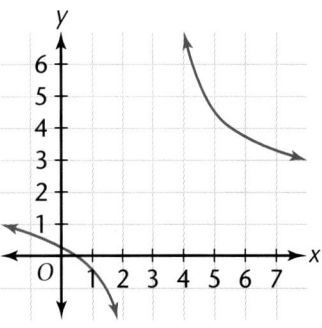

Solution ➤

The information from the function $g(x) = \frac{5}{x-3} + 2$ shows that the parent function has been transformed by a vertical stretch of 5, a shift to the right by 3, and a shift up by 2. ❖

Try This Describe the transformations applied to the parent function $f(x) = \frac{1}{x}$ to produce the graph of $g(x) = \frac{3}{2x+3} - 5$.

CRITICAL Thinking

Begin with the function $f(x) = a\left(\frac{1}{x-h}\right) + k$. First vary the values of a, then h, then k, keeping the other 2 values constant, and graph the results. Describe the effects on the graph of varying a, h, and k.

EXERCISES & PROBLEMS

Communicate

1. Explain how a rational expression is defined. Give two examples of a rational expression.

2. What condition causes a rational expression to be undefined?

Graph the function $h(x) = \dfrac{12x - 7}{x}$.

3. Discuss how to express $h(x)$ in a form that shows that it is a transformation of the parent function $g(x) = \dfrac{1}{x}$.

4. Describe the behavior of $h(x)$ when x is 0.

5. Describe all the transformations to $g(x) = \dfrac{1}{x}$ that produce the graph of $h(x)$.

Practice & Apply

For what values are these rational expressions undefined?

6. $\dfrac{3x + 1}{x}$

7. $\dfrac{6}{y - 2}$

8. $\dfrac{2m - 5}{6m^2 - 3m}$

9. $\dfrac{x^2 + 2x - 3}{x^2 + 4x - 5}$

Evaluate each rational expression for $x = 1$ and $x = -2$. Write "undefined" if appropriate.

10. $\dfrac{5x - 1}{x}$

11. $\dfrac{4x}{x - 1}$

12. $\dfrac{x^2 - 1}{x^2 - 4}$

13. $\dfrac{x^2 + 2x}{x^2 + x + 2}$

Describe all the transformations to $f(x) = \dfrac{1}{x}$ needed to graph each of the following rational functions.

14. $g(x) = \dfrac{1}{x - 5}$

15. $g(x) = \dfrac{2}{x}$

16. $g(x) = \dfrac{-1}{x} + 3$

17. $g(x) = \dfrac{1}{2x + 3}$

18. $g(x) = \dfrac{4}{x + 2} - 5$

19. $g(x) = \dfrac{-2}{x - 1} + 4$

Graph each rational function. Label the vertical asymptotes. List the value for which the function is undefined.

20. $h(x) = \dfrac{1}{x} + 1$

21. $h(x) = \dfrac{1}{x + 2}$

22. $h(x) = \dfrac{-1}{x} + 3$

23. $h(x) = \dfrac{3}{x - 1}$

24. $h(x) = \dfrac{2}{x - 3} + 4$

25. $h(x) = \dfrac{-2}{x - 4} + 3$

Transportation   The air pressure, P, in pounds per square inch, in the tires of a car can be modeled by the rational equation $P = \dfrac{(1.1)10^5}{V}$, where V is volume in cubic inches.

26. Graph the rational function represented by P.

27. The owner's manual calls for a pressure of 32 pounds per square inch. What volume of air does a tire hold at this pressure?

28. Your class orders 500 mugs with all the names of the class members. In the first week, 356 mugs were sold. What percent of the mugs sold the first week? **[Lesson 4.5]**

29. Graph a line through the points (2, 1) and (−3, 5). **[Lesson 5.3]**

30. Find the inverse matrix to $\begin{bmatrix} 0.3 & 0.4 \\ 0.5 & 0.6 \end{bmatrix}$. **[Lesson 7.4]**

31. How many different 4-digit numbers are possible if the digits 1-9 can be used with repetition? **[Lesson 8.5]**

32. What transformation is applied to $f(x) = x^2$ to obtain the graph of $y = (x - 5)^2$? **[Lesson 9.5]**

33. Solve $x^2 - 20 = -x$. **[Lesson 12.2]**

34. Cultural Connection: Americas Benjamin Banneker, the first African American astronomer, is most famous for the almanacs he created from 1791 to 1797. Banneker also enjoyed solving mathematical problems, such as this one from his journal. **[Lesson 13.4]**

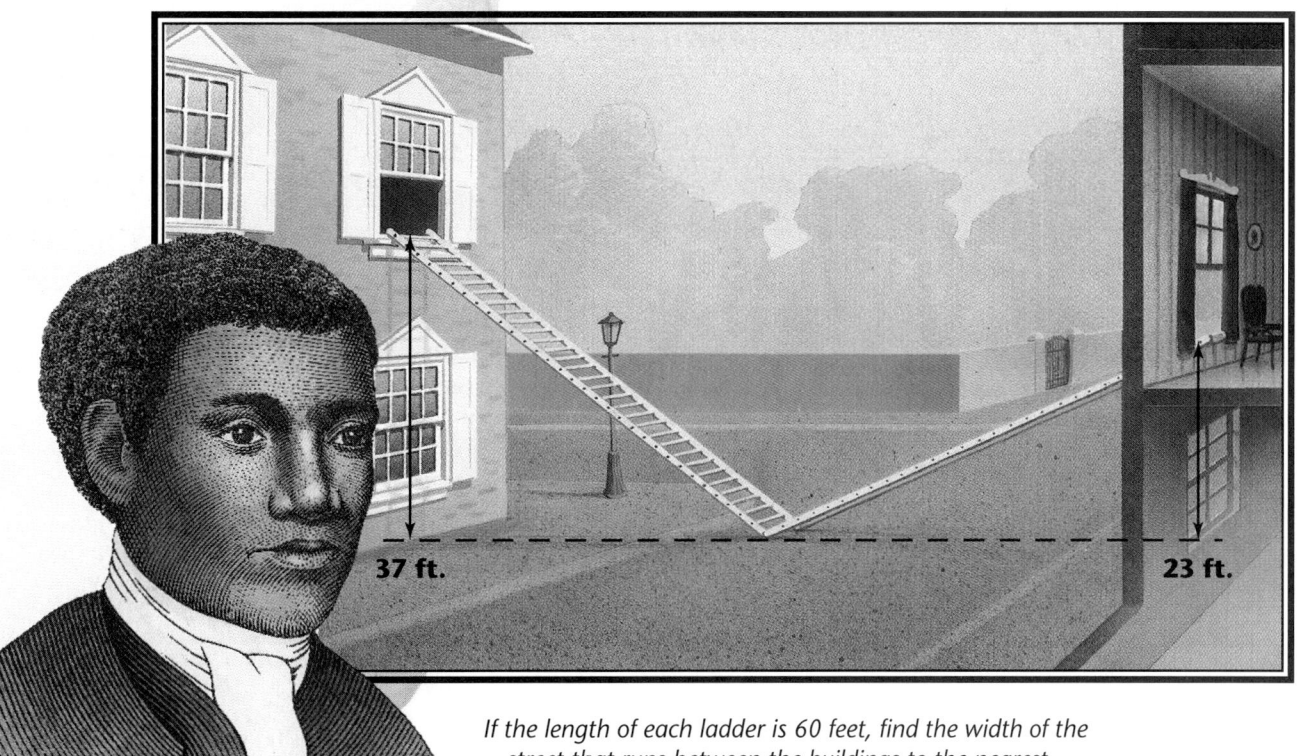

37 ft. 23 ft.

If the length of each ladder is 60 feet, find the width of the street that runs between the buildings to the nearest hundredth.

35. Describe the transformations that produce $y = \dfrac{5}{x - 1} + 2$ from $y = \dfrac{1}{x}$.

36. Sketch the graph.

LESSON 14.2 Inverse Variation

It takes a person 6 minutes to plant each tree for this project.

WHY

Variation describes how a change in one quantity changes another quantity. You have seen how quantities vary directly. They can also vary inversely.

In Chapter 2 you solved a fund-raising problem about students raising $1000 to help a needy family. You found that the number of contributors needed decreased as the amount of money each person gave increased. This relationship is an example of inverse variation.

> **INVERSE VARIATION**
>
> If $x \cdot y = k$, or $y = \dfrac{k}{x}$, when $x \neq 0$ and k is a constant, then y varies inversely as x. The constant k is called the constant of variation.

EXAMPLE 1

Ecology A volunteer organization has 5000 trees to plant to beautify an area. If 5 people need 100 hours to plant all the trees, how many hours are needed when 50 people help?

Solution ➤

Let *n* represent the number of people and *t* represent the number of hours to plant all the trees. You can determine the constant by the formula for inverse variation. Since $n \cdot t = k$, $k = 5 \cdot 100$ or $k = 500$. After you determine the constant of variation, you can use it and the remaining information in the problem to find the number of hours, *t*.

$$nt = 500$$
$$50t = 500$$
$$t = 10$$

Examine the table for various numbers of volunteers.

Persons *n*	1	2	4	5	40	50	125	250	500
Hours *t*	500	250	125	100	12.5	10	4	2	1
nt = *k*	500	500	500	500	500	500	500	500	500

To plant 500 trees, 50 volunteers must work 10 hours planting one tree every 6 minutes. ❖

Try This   If *y* is 4 when *x* is 12, write an equation that shows how *y* is related to *x*. Assume that *y* varies inversely as *x*.

CRITICAL *Thinking*

In Example 1, *k* is the constant of variation. What physical quantity does *k* represent?

Cultural Connection: Africa One of the earliest references to variation comes from an ancient Egyptian papyrus that describes the commodity exchange system used in some Egyptian cities. Bread was used as a form of currency, and its value was based on the amount of wheat that was used to make each loaf. A rating system was created to value the bread.

Loaves	1000	1000	1000	1000	1000
*Rating	10	20	25	40	50
**Value	100	50	40	25	20

* Rating is the number of loaves from 1 hekat.
** Value is the total number of hekats used.

The table shows the value of 1000 loaves of bread at various ratings.

Notice that as the rating for each 1000 loaves *increases* the *value* for each 1000 loaves *decreases*. This is an example of **inverse variation**.

Algebraically, the variables of rating and value are related by the following equation.

$$\text{Rating} \cdot \text{Value} = \text{Loaves, as a constant, } k$$
$$20 \, \frac{loaves}{hekat} \cdot 50 \, hekats = 1000 \text{ loaves}$$

The bread's rating is said to vary inversely as its value. ❖

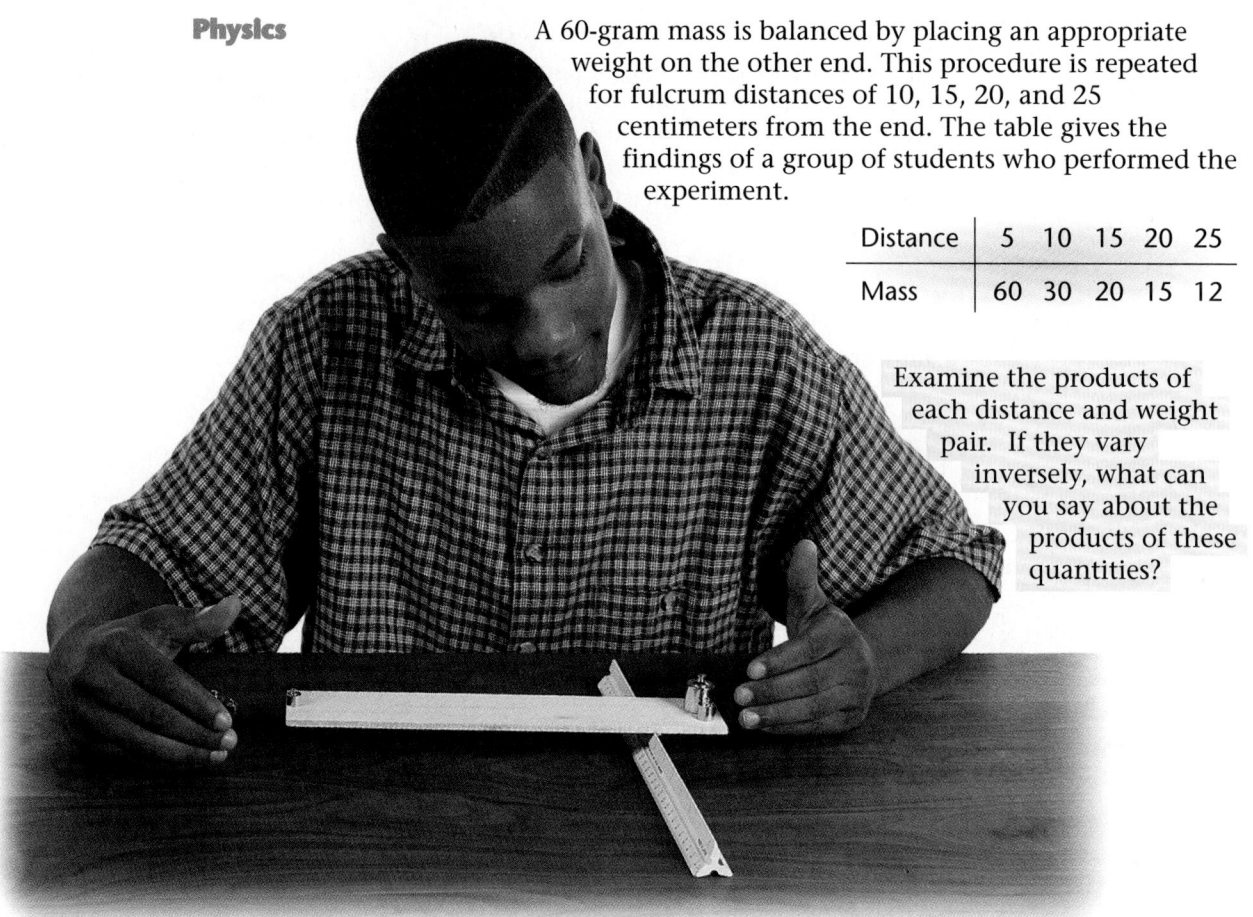

Physics

A 60-gram mass is balanced by placing an appropriate weight on the other end. This procedure is repeated for fulcrum distances of 10, 15, 20, and 25 centimeters from the end. The table gives the findings of a group of students who performed the experiment.

Distance	5	10	15	20	25
Mass	60	30	20	15	12

Examine the products of each distance and weight pair. If they vary inversely, what can you say about the products of these quantities?

EXAMPLE 2

Analyze the experiment just described.

A Write a sentence that specifies the variation between the variables.

B Write an equation that relates the variables.

C Construct a graph of the equation, and analyze the type of function that the graph represents.

Solution ➤

This experiment illustrates inverse variation because the product of the distance from the balance point and the balance weight is constant.

A The balance weight varies inversely as the distance to the balance point.

B The inverse variation relationship means that

$$\text{weight} \cdot \text{fulcrum distance} = k.$$

For the values of 15 grams and 20 centimeters,

$$15 \cdot 20 = 300 = k.$$

The constant of variation is 300.

C The equation for this inverse variation is $m \cdot d = 300$.

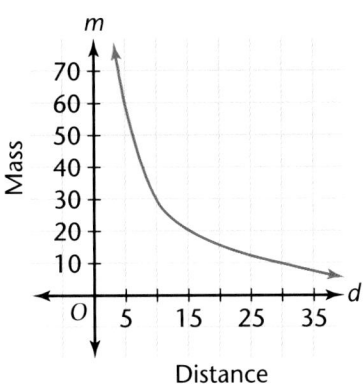

Distance

If $md = 300$, then $m = \dfrac{300}{d}$. The equation represents a rational function that is a transformation of the parent function $y = \dfrac{1}{x}$ by a vertical stretch of 300. ❖

Investment The amount of time it takes to double your money at *compound* interest can be approximated by the function $t = \dfrac{72}{r}$, where r is the interest rate written as a whole number. In business this function is usually referred to as the Rule of 72.

EXAMPLE 3

A How long does it take to double your money at 6% compound interest?

B Graph the Rule of 72 function.

Solution ➤

A To use the Rule of 72, substitute 6 for r so $t = \dfrac{72}{6} = 12$. It takes about 12 years to double your money at 6%.

B Choose interest rates for r. Make a table and sketch a graph from the points.

$r\%$	t
4	18
6	12
8	9
12	6

❖

Describe the graph of the Rule of 72 as a transformation of its parent graph. What are the domain and range of the Rule of 72 function? What effect does the constant of variation have on its parent function?

EXERCISES & PROBLEMS

Communicate

1. Give two equivalent equations expressing inverse variation if y varies inversely as x.
2. Explain what inverse variation between rate and time means. Give a real-life example of inverse variation.
3. If y is 3 when x is 8, and y varies inversely as x, explain how to find y when x is 2.
4. Explain how to use the Rule of 72 to find how long it would take to double your money at 4% interest compounded yearly.
5. Describe the parent function for inverse variation functions.

Practice & Apply

Determine if each of the following is an example of inverse variation.

6. $rt = 400$
7. $y = \dfrac{-28}{x}$
8. $x - y = 10$
9. $\dfrac{n}{5} = \dfrac{3}{m}$
10. $\dfrac{x}{y} = \dfrac{1}{2}$
11. $x = 10y$
12. $a = \dfrac{42}{b}$
13. $r = t$

For Exercise 14-19, y varies inversely as x. If y is

14. 8 when x is 6, find x when y is 12.
15. 9 when x is 12, find y when x is 36.
16. 3 when x is 32, find x when y is 4.
17. 3 when x is -8, find x when y is -4.
18. $\dfrac{3}{5}$ when x is -60, find y when x is 2.
19. $\dfrac{3}{4}$ when x is 12, find y when x is 27.

20. **Geometry** If the area is constant, the base of a triangle varies inversely as the height. When the base is 22 centimeters, the height is 36 centimeters. Find the length of the base when the height is 24 centimeters.

The speeds are inversely proportional to the number of teeth.

21. **Physics** If a gear with 12 teeth revolves at a speed of 500 revolutions per minute, at what speed should a gear with 16 teeth revolve?

22. Music A harp string vibrates to produce sound. The number of these vibrations varies inversely to its length. If a string 28 centimeters long vibrates 510 times per second, how long is a string that vibrates 340 times per second?

23. Time varies inversely as the average rate of speed over a given distance. When Mike travels 6 hours, his average rate is 80 kilometers per hour. Find Mike's time when his rate is 90 kilometers per hour.

24. The frequency of a radio wave varies inversely as its wavelength. If a 200-meter wave has a frequency of 3000 kilocycles, what is the wavelength of a wave that has a frequency of 2000 kilocycles?

25. Transportation Time varies inversely as the average speed over a given distance. A jet takes about 2.7 hours to fly from Boston to Paris averaging 2200 kilometers per hour. How long would it take a plane traveling at an average speed of 1760 kilometers per hour to make the same trip?

26. Investment If Harold invests $500 of his 4-H prize money at 5% interest compounded yearly, how long will it take him to double his money?

27. 🖳 **Geometry** The area of a rectangle is 36 square centimeters. What is the length of a rectangle of the same area that has twice the original width? Explain why the formula for the area of a rectangle is a model for inverse variation.

Look Back

28. The first three terms of a sequence are 1, 5, and 12. The second difference is a constant 3. Find the next three terms of the sequence. **[Lesson 1.1]**

29. The 5-by-7 picture of your award winning cat, Kat, is surrounded by a frame of width $3x$. Find the dimensions of the frame and picture unit. **[Lesson 3.3]**

30. The total cost for your order of computer discs, including the $1.95 handling charge, is $57.60. The discs are sold in boxes of 10 and were on sale for $7.95 a box. How many discs were in the order? **[Lesson 4.6]**

31. What are the coordinates of the vertex of $y = -x^2 - 4x$? **[Lesson 12.1]**

Look Beyond

32. Subtract. $\dfrac{1}{x-3} - \dfrac{x}{x^2-9}$

33. The intensity, I, of light from a light source varies inversely as the square of the distance, d^2, from the light source. If I is 30 units when d is 6 meters, find I when d is 3 meters.

Simplifying Rational Expressions

In arithmetic, a fraction is simplified when it is expressed in lowest terms. In algebra, rational expressions are simplified in much the same way. Algebraic operations on rational expressions follow the same mathematical steps that you use to simplify fractions.

$$\frac{3(x^2 + 4x + 4)}{(4x + 8)(x + 2)} \qquad \frac{3}{4}$$

Recall the steps to reduce $\frac{18}{24}$ to lowest terms.

$\frac{18}{24} = \frac{2 \cdot 3 \cdot 3}{2 \cdot 2 \cdot 2 \cdot 3}$ Write both the numerator and denominator as a product of prime factors.

$= \frac{3}{2 \cdot 2} \cdot \frac{2}{2} \cdot \frac{3}{3}$ Rewrite to show fractions equal to one, $\frac{2}{2} = 1$ and $\frac{3}{3} = 1$.

$= \frac{3}{4}$ Simplify.

The same procedure is used to simplify or reduce a rational expression to lowest terms. A rational expression is in simplest form when the numerator and denominator have no common factors other than 1 or −1.

EXAMPLE 1

Simplify $\dfrac{5a + 10}{5a}$.

Solution ➤

$$\dfrac{5a + 10}{5a} = \dfrac{5(a + 2)}{5a}, \; a \neq 0$$ Factor the numerator and denominator.

$$= \dfrac{5}{5} \cdot \dfrac{a + 2}{a}, \; a \neq 0$$ Rewrite to show fractions equal to 1, $\dfrac{5}{5} = 1$.

$$= \dfrac{a + 2}{a}, \; a \neq 0$$ Simplify.

One way to check your work is to replace the variable by a convenient value. For example, let $a = 10$.

Let $a = 10$.

Original form $\dfrac{5a + 10}{5a} = \dfrac{50 + 10}{50} = \dfrac{60}{50} = \dfrac{6}{5}$

Reduced form $\dfrac{a + 2}{a} = \dfrac{12}{10} = \dfrac{6}{5}$

The check shows that the value of the original form is the same as that of the reduced form. ❖

Why is it important that the check for the original fraction and the check for the reduced fraction be the same?

STEPS TO REDUCE RATIONAL EXPRESSIONS

1. Factor the numerator and denominator.
2. Express, or think of, each common factor pair as 1.
3. Simplify.

Be sure that only common factors are eliminated when you simplify. Notice the common factors in the examples below.

Expression	Common factor	Reduced form
$\dfrac{x(a + 3)}{2x}$	x	$\dfrac{a + 3}{2}, \; x \neq 0$
$\dfrac{x + 4}{x}$	none	Cannot be simplified further.
$\dfrac{2x(x + 1)}{4x^2}$	$2x$	$\dfrac{x + 1}{2x}, \; x \neq 0$
$\dfrac{2x^2 + 3}{x}$	none	Cannot be simplified further.

CRITICAL
Thinking

Explain why $\dfrac{x}{x^2 + x} = \dfrac{1}{x + 1}$, $x \neq 0, -1$, is a true statement.

Sometimes a rational expression can be simplified by first factoring the numerator and denominator. Remember to check the restrictions on the denominator.

EXAMPLE 2

Simplify $\dfrac{3x - 6}{x^2 + x - 6}$.

Solution ➤

$$\dfrac{3x - 6}{x^2 + x - 6} = \dfrac{3(x - 2)}{(x + 3)(x - 2)}$$ Factor numerator and denominator.

$$= \dfrac{3}{x + 3} \cdot \dfrac{x - 2}{x - 2}; \; x \neq 2, -3$$ Rewrite to show fractions equal to 1.

$$= \dfrac{3}{x + 3}; \; x \neq 2, -3$$ Simplify. Write restrictions from original expression. ❖

Try This   Simplify $\dfrac{x^2 + 3x - 4}{x - 1}$.

Simplifying rational expressions can be tricky. Many steps may be needed. Examine each expression carefully to see if it can be factored.

EXAMPLE 3

Simplify $\dfrac{x^2 + x - 20}{16 - x^2}$.

Solution ➤

Rewrite the denominator with the variable first. In this case, factor -1 from the denominator.

$$\dfrac{x^2 + x - 20}{16 - x^2} = \dfrac{x^2 + x - 20}{(-1)(x^2 - 16)}$$ Factor -1 from the denominator.

$$= \dfrac{(x + 5)(x - 4)}{(-1)(x + 4)(x - 4)}, \; x \neq 4, -4$$ Factor the numerator and denominator.

$$= \dfrac{x + 5}{(-1)(x + 4)}, \; x \neq 4, -4$$ The common factor is $x - 4$. Simplify.

$$= \dfrac{x + 5}{-x - 4}, \; x \neq 4, -4$$ Multiply $x + 4$ in the denominator by -1. ❖

Explain why the opposite of $(x - 4)$ can be written $(4 - x)$.

Try This   Simplify $\dfrac{5x^3 - 20x}{12 - 3x^2}$.

CRITICAL
Thinking

Explain why you cannot write $\dfrac{x - 1}{1 - x} = -1$ without adding restrictions. What restrictions must be added?

EXAMPLE 4

Find the ratio of the volume of a right circular cylinder with radius r and height h to its surface area.

GEOMETRY
Connection

Solution ➤

$$\frac{\text{Volume}}{\text{Surface area}} = \frac{\pi r^2 h}{2\pi r^2 + 2\pi rh} \qquad \text{Given}$$

$$= \frac{\pi \mathbf{r} rh}{2\pi \mathbf{r}(r+h)} \qquad \text{Factor.}$$

$$= \frac{rh}{2(r+h)} \qquad \text{The common factors are } \pi \text{ and } r. \text{ Simplify.}$$

Therefore, the ratio $\dfrac{\text{Volume}}{\text{Surface area}}$ is $\dfrac{rh}{2(r+h)}$. ❖

Try This Find the ratio of the surface area of a cube to its volume.

EXERCISES & PROBLEMS

Communicate

1. Explain what makes a rational expression undefined.

2. Discuss how to find the restrictions of the denominator of a rational expression.

3. Explain what is meant by a common factor.

4. Describe the process that reduces $\dfrac{x+1}{x^2+2x+1}$ to lowest terms.

5. Explain how the expression $x - 7$ is related to $7 - x$.

Practice & Apply

For what values of the variable is each rational expression undefined?

6. $\dfrac{10x}{x-3}$

7. $\dfrac{10}{5-y}$

8. $\dfrac{r-6}{r}$

9. $\dfrac{7p}{p-4}$

10. $\dfrac{k-3}{3-k}$

11. $\dfrac{(a+3)(a+4)}{(a-3)(a+4)}$

12. $\dfrac{c-4}{2c-10}$

13. $\dfrac{3}{y(y^2-5y+6)}$

Write the common factors.

14. $\dfrac{9}{12}$

15. $\dfrac{3(x+4)}{6x}$

16. $\dfrac{x-y}{(x+y)(x-y)}$

17. $\dfrac{r+3}{r^2+5r+6}$

Simplify.

18. $\dfrac{16(x+1)}{30(x+2)}$

19. $\dfrac{3(a+b)}{6(a-b)}$

20. $\dfrac{4(c+2)}{10(2+c)}$

21. $\dfrac{3(x+y)(x-y)}{6(x+y)}$

22. $\dfrac{6m+9}{6}$

23. $\dfrac{7t+21}{t+3}$

24. $\dfrac{12+8x}{4x}$

25. $\dfrac{3d^2+2d}{3d+1}$

26. $\dfrac{b+2}{b^2-4}$

27. $\dfrac{x-2}{x^2+2x-8}$

28. $\dfrac{-(a+1)}{a^2+8a+7}$

29. $\dfrac{4-k}{k^2-k-12}$

30. $\dfrac{c^2-9}{3c+9}$

31. $\dfrac{3n-12}{n^2-7n+12}$

32. $\dfrac{y^2+2y-3}{y^2+7y+12}$

33. $\dfrac{a^2-b^2}{(a+b)^2}$

34. **Geometry** The area of a certain rectangle is represented by x^2+6x+9 and its length by $x+3$. Find the width.

35. Biology Two equal populations of animals, one a predator, the other a prey were released into the wild. Five years later a survey showed the prey population had multiplied 3 times, while the predators had multiplied by 8 times the original population. Then 32 predators and 12 prey were released into the area. What is the numerical ratio of prey to predator after the release?

Geometry Exercises 36-37 refer to the cube below.

36. To find the volume of a cube, multiply the length by the width by the height. Write an expression to represent the volume of this cube.

37. To find the total surface area of a cube, multiply the area of one face by 6. Write an expression to represent the total surface area of this cube.

x

x

x

38. Savings How much does Heather have to deposit in the bank at a 4.5% interest rate, compounded yearly, to have $1500 for the stereo on her sixteenth birthday? **[Lesson 2.1]**

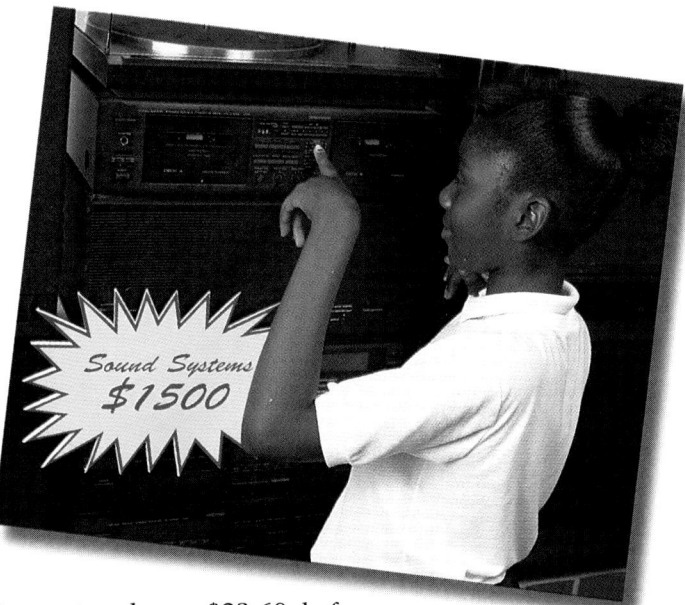

In 2 years, Heather wants to buy a stereo for her sixteenth birthday.

39. Chin-Lyn buys a new sweater at a 20% discount and pays $23.60, before tax. What is the sweater's original price? **[Lesson 4.5]**

40. Find the equation of the line segment connecting the points $\left(\frac{2}{3}, -2\right)$ and $\left(-\frac{5}{6}, 3\frac{1}{3}\right)$. **[Lesson 5.4]**

41. Multiply $\begin{bmatrix} -3 & 2 \\ -1 & 4 \end{bmatrix} \cdot \begin{bmatrix} 5 & -3 \\ 3 & -6 \end{bmatrix}$. **[Lesson 7.3]**

42. **Probability** How many different 3-letter arrangements can be formed from the letters in the word PHONE? **[Lesson 8.5]**

43. If $f(x) = 3^x$, then explain how $g(x) = 3^{x+2} - 5$ compares with $f(x)$. **[Lesson 10.6]**

44. Multiply $(x + 2)(x - 2)$. **[Lesson 11.3]**

45. Write the equation of the parabola whose vertex is (2, 3) and passes through the point (0, −1). **[Lesson 12.1]**

46. Which of the parabolas open downward? **[Lesson 12.2]**
   a. $y = 3x^2 - 8$ **b.** $y = 9 - x^2$ **c.** $y = -x^2 + 4$ **d.** $y = \frac{3}{4}x^2 - 6$

47. Simplify $\sqrt{\frac{2}{3}} - \sqrt{96}$. **[Lesson 13.2]**

Look Beyond

48. Eight squares are the same size, but each is a different color. The squares are stacked on top of each other as shown. Identify the order of the squares numerically from top to bottom. The top square is number 1.

LESSON 14.4 Operations With Rational Expressions

Operations with rational expressions are like operations with ordinary fractions. Your arithmetic skills will be needed to add, subtract, multiply, and divide rational expressions.

Anna's rowing rate is 4 times the rate of the river's current.

Each year a local club sponsors River Challenge Day at the riverfront to feature safety and recreational activities. One of the most popular activities is the Current Challenge. The challenge requires a participant to paddle a kayak up and down a 1-mile portion of the river. Is there a simple rational expression that models Anna's round-trip time?

Write a formula for the total time it takes to paddle upstream and downstream.

$$\underbrace{\text{Time upstream}}_{\text{against current}} + \underbrace{\text{Time downstream}}_{\text{with current}} = \text{Total time}$$

Write the formula for time in terms of distance and rate.

If distance = rate · time, then time = $\dfrac{\text{distance}}{\text{rate}}$.

Thus, $\dfrac{\text{distance}}{\text{rate against current}} + \dfrac{\text{distance}}{\text{rate with current}} = \text{total time}$.

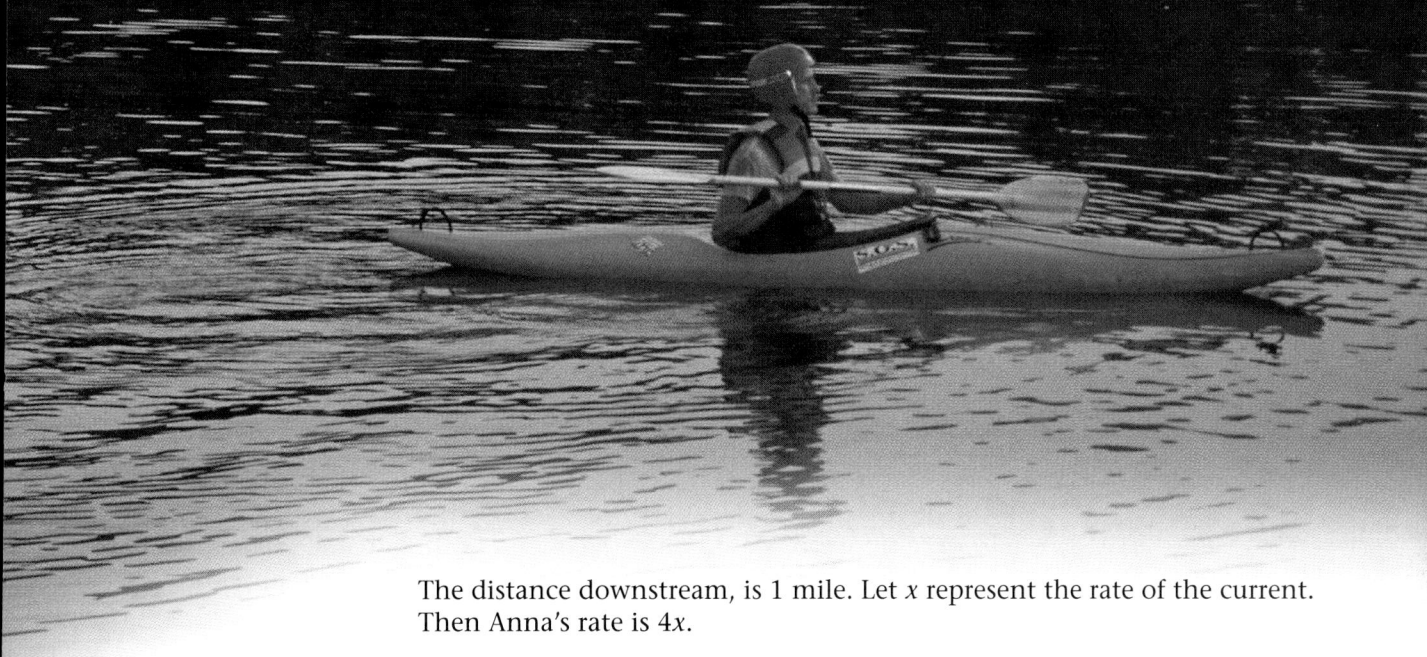

The distance downstream, is 1 mile. Let x represent the rate of the current. Then Anna's rate is $4x$.

So, $\dfrac{1 \text{ mile}}{4x \text{ mph} - x \text{ mph}} + \dfrac{1 \text{ mile}}{4x \text{ mph} + x \text{ mph}} = \text{total time.}$

The equation that models the relationship is

$$\frac{1}{3x} + \frac{1}{5x} = \text{total time.}$$

To find a simple rational form for $\dfrac{1}{3x} + \dfrac{1}{5x}$, add the rational expressions in the same way you add rational numbers with unlike denominators. Compare the methods for adding $\dfrac{1}{30} + \dfrac{1}{50}$ and $\dfrac{1}{3x} + \dfrac{1}{5x}$.

Rational Numbers	Rational Expressions	
$\dfrac{1}{30} + \dfrac{1}{50}$	$\dfrac{1}{3x} + \dfrac{1}{5x}$	Given
$\dfrac{5}{5} \cdot \dfrac{1}{30} + \dfrac{3}{3} \cdot \dfrac{1}{50}$	$\dfrac{5}{5} \cdot \dfrac{1}{3x} + \dfrac{3}{3} \cdot \dfrac{1}{5x}$	Find a common denominator.
$\dfrac{5}{150} + \dfrac{3}{150}$	$\dfrac{5}{15x} + \dfrac{3}{15x}$	Change to common denominators.
$\dfrac{8}{150}$	$\dfrac{8}{15x}$	Simplify.

Anna's time in terms of the rate of the current is $\dfrac{8}{15x}$ where $x \neq 0$.

In working with rational expressions, why is it important to remember to check for the restrictions on the denominator in the original expression?

In adding the rational expressions on page 671, it is necessary to find a common denominator by multiplying $\frac{5}{5} \cdot \frac{1}{3x} = \frac{5}{15x}$. This illustrates the method used for multiplying any two rational expressions.

EXAMPLE 1

Multiply $\frac{x - 2}{x + 3} \cdot \frac{x + 3}{x - 5}$.

Solution ➤

$$\frac{x - 2}{x + 3} \cdot \frac{x + 3}{x - 5} = \frac{(x - 2)(x + 3)}{(x + 3)(x - 5)}$$ Multiply numerators and denominators.

$$= \frac{x - 2}{x - 5} \cdot \frac{x + 3}{x + 3}$$ Find fractions that equal 1.

$$= \frac{x - 2}{x - 5}, \, x \neq 5, \, -3$$ Simplify and write restrictions to original problem. ❖

Try This Multiply $\frac{7}{x} \cdot \frac{x^2}{14} \cdot \frac{5}{x}$.

One way to add or subtract two rational expressions such as $\frac{1}{x + 2}$ and $\frac{1}{x - 1}$ is to rewrite the expressions to have a common denominator. This can be done by multiplying each expression by the appropriate name for 1.

$$\frac{x}{x + 2} + \frac{x}{x - 1} = \frac{x - 1}{x - 1} \cdot \frac{x}{x + 2} + \frac{x + 2}{x + 2} \cdot \frac{x}{x - 1}$$

Since the expressions now have common denominators, the numerators can be added.

$$\frac{(x - 1) x}{(x - 1)(x + 2)} + \frac{(x + 2) x}{(x + 2)(x - 1)} = \frac{(x^2 - x) + (x^2 + 2x)}{(x - 1)(x + 2)}$$

$$= \frac{2x^2 + x}{(x - 1)(x + 2)} \quad \text{or} \quad \frac{2x^2 + x}{x^2 + x - 2}, \, x \neq 1, \, -2$$

How do you choose the appropriate names for 1?

EXAMPLE 2

Simplify $\frac{a}{a^2 - 4} + \frac{2}{a + 2}$.

Solution ➤

Since $a^2 - 4 = (a + 2)(a - 2)$, multiply the second expression by $\frac{a - 2}{a - 2}$ and add.

$$\frac{a}{a^2 - 4} + \frac{2}{a + 2} = \frac{a}{(a + 2)(a - 2)} + \frac{2(a - 2)}{(a + 2)(a - 2)}$$

$$= \frac{a + 2a - 4}{(a + 2)(a - 2)}$$

$$= \frac{3a - 4}{a^2 - 4}, \, a \neq 2, \, -2 \, ❖$$

EXAMPLE 3

Simplify. $\dfrac{x-4}{3x-3} - \dfrac{x-3}{2x-2}$

Solution ➤

To find the common denominator, factor the denominators.

$$\frac{x-4}{3(x-1)} - \frac{x-3}{2(x-1)}$$

A common denominator is $3 \cdot 2(x-1) = 6(x-1)$, $x \neq 1$.

$\dfrac{x-4}{3(x-1)} - \dfrac{x-3}{2(x-1)} = \dfrac{2(x-4)}{6(x-1)} - \dfrac{3(x-3)}{6(x-1)}$, $x \neq 1$.   Change to common denominators.

$= \dfrac{2x-8-3x+9}{6(x-1)}$, $x \neq 1$ Subtract the numerators.

$= \dfrac{-x+1}{6(x-1)}$, $x \neq 1$ Simplify.

$= \dfrac{-1(x-1)}{6(x-1)}$, $x \neq 1$ Factor -1 from the numerator.

$= \dfrac{-1}{6}$, $x \neq 1$ Simplify. ❖

CRITICAL Thinking

Explain when to factor -1 from an expression. How does factoring -1 simplify a rational expression?

EXERCISES & PROBLEMS

Communicate

Refer to the expressions $\dfrac{x}{x+1}$ and $\dfrac{3}{x}$ for Exercises 1–5.

1. Explain how a common denominator may be found for the two rational expressions.

2. Describe how to find the sum of these rational expressions.

3. Describe how to find the difference of these rational expressions.

4. Describe how to find the product of these rational expressions.

5. How do you think you can find the quotient of the first expression divided by the second expression?

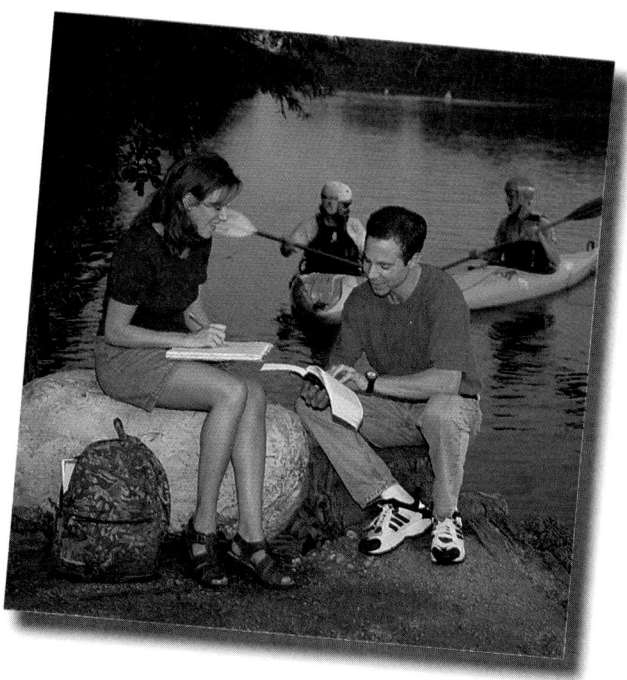

Practice & Apply

Perform the indicated operations. Simplify and state restrictions for variables.

6. $\dfrac{5}{3x} + \dfrac{2}{3x}$

7. $\dfrac{8}{x+1} - \dfrac{5}{x+1}$

8. $\dfrac{2x}{y+4} - \dfrac{5x}{y+4}$

9. $\dfrac{5x+4}{a-c} - \dfrac{7+3x}{a-c}$

10. $\dfrac{2}{a} + \dfrac{3}{b}$

11. $\dfrac{7}{3t} - \dfrac{8}{2t}$

12. $\dfrac{t}{2rs} + \dfrac{3s}{rt}$

13. $\dfrac{5}{pq} - \dfrac{m}{p^2q}$

14. $\dfrac{-2}{x+1} + \dfrac{3}{2(x+1)}$

15. $x - \dfrac{x-4}{x+4}$

16. $\dfrac{-3-d}{d-1} + 2$

17. $\dfrac{a}{b} + \dfrac{c}{d}$

18. $\dfrac{5+m}{3+n} \cdot \dfrac{3+n}{a+b}$

19. $\dfrac{x+2}{x(x+1)} \cdot \dfrac{x^2}{(x+2)(x+3)}$

20. $\dfrac{q^2-1}{q^2} \cdot \dfrac{q}{q+1}$

21. $\dfrac{1}{x+1} \cdot \dfrac{2}{x}$

22. $\dfrac{3}{x-1} \cdot \dfrac{5}{x}$

23. $\dfrac{y}{y-4} - \dfrac{1}{y}$

24. $\dfrac{-2}{x+1} + \dfrac{3}{x}$

25. $\dfrac{a-2}{a+1} + \dfrac{5}{a+3}$

26. $\dfrac{m+5}{m+2} \cdot \dfrac{m-3}{m-1}$

27. $\dfrac{2r}{(r+5)(r+1)} - \dfrac{r}{r+5}$

28. $\dfrac{3}{b^2+b-6} - \dfrac{5}{b-2}$

29. $\dfrac{4}{y-2} + \dfrac{5y}{y^2-4y+4}$

30. $\dfrac{2}{y-5} - \dfrac{5y-3}{y^2+y-30}$

31. $\dfrac{x}{1-x} + \dfrac{1}{x-1}$

32. $\dfrac{2}{x^2-9} - \dfrac{1}{2x+6}$

Ecology Students at an aquarium are creating an exhibit of a bay ecosystem. They determine that the water in the bay contains 4.0% salt-by-mass concentration. The water in the aquarium tanks for the sea exhibits is 6.0% salt-by-mass concentration. The students plan to dilute 750 kilograms of the water in the aquarium by adding enough pure water to get the 4.0% salt-by-mass concentration. The concentration can be determined by

$$\text{Concentration} = \frac{\text{sea-water salt concentration} \cdot \text{mass of sea water}}{\text{mass of sea water} + \text{mass of pure water}}.$$

33. Write a rational function for this problem to represent the resulting concentration when a mass of pure water, *w*, is added.

34. Graph the function.

35. How much pure water needs to be added to reach a concentration of 4.0% salt-by-mass?

36. State the pattern and find the next three terms of the following sequence. $-1, 6, 15, 26, \_\_\_, \_\_\_, \_\_\_$ **[Lesson 1.1]**

37. Solve $0.3(x - 90) = 0.7(2x - 70)$. **[Lesson 4.6]**

38. Graph the solution of $y < -2x + 6$ or $y \le 3x - 4$. **[Lesson 6.5]**

39. Add $\begin{bmatrix} -3 & 7 \\ 5 & 11 \end{bmatrix} + \begin{bmatrix} 7 & -13 \\ 22 & 2 \end{bmatrix}$. **[Lesson 7.2]**

Cultural Connection: Americas The Blackfoot people of Montana played a game of chance in which they threw 4 decorated bones made from buffalo ribs. Three were blank on one side, but the fourth, called the chief, was engraved on both sides like a coin with heads and tails. The highest score was 6 points, for a throw of 3 blanks and the chief landing heads up. For 3 blanks and the chief reversed, the score was 3 points.

40. What is the probability of throwing 3 blanks and the chief landing heads up? **[Lesson 8.1]**

41. What is the probability of throwing 3 blanks and the chief reversed? **[Lesson 8.1]**

42. **Probability** If a bag contains 4 red marbles, 3 black marbles, and 1 white marble, what is the probability that you will select a white marble? **[Lesson 8.1]**

43. Express 25,000,000,000 in scientific notation. **[Lesson 10.5]**

44. Graph $y = (x - 1)^2 - 4$. **[Lesson 12.1]**

Look Beyond

A proportion is the equality of two ratios such as $\frac{5}{10} = \frac{2}{4}$, since both ratios equal $\frac{1}{2}$. This reduced fraction is called the constant of the proportion. Find the missing term in each proportion and state the constant.

45. $\frac{26}{39} = \frac{x}{9}$ **46.** $\frac{14}{x} = \frac{49}{28}$ **47.** $\frac{15}{40} = \frac{27}{x}$

How Worried Should You Be?

Now, the testing question

Concerns about iron's role in heart disease, if ultimately confirmed, surely will touch off a debate over iron tests. Who should be tested? and should testing be a matter of public-health policy? . . .

The virtue of such broad screening would be to help ferret out many of the estimated 32 million Americans who carry a faulty gene that prompts their body to harbor too much iron. There is no test for the gene itself, but anyone with a high blood level of iron is a suspect. Siblings and chil-

dren would be candidates for testing as well, since the condition is inherited . . .

Two widely used blood tests for iron generally cost from $25 to $75 each. The first measures ferritin, a key iron-storing protein . . .

The results should always be discussed with a doctor, since some physicians take recent studies seriously enough to worry about even moderately elevated ferritin levels. Moreover, the test might bear repeating. Erroneously high readings are relatively common.

CENTRAL HIGH HEALTH FAIR

Although people submit to medical tests because they want to find out if something is wrong, the results may cause them to worry. Read the following excerpt about an imaginary case.

Assume that there is a test for cancer which is 98 percent accurate; i.e., if someone has cancer, the test will be positive 98 percent of the time, and if one doesn't have it, the test will be negative 98 percent of the time. Assume further that 0.5 percent—one out of two hundred people—actually have cancer. Now imagine that you've taken the test and that you've tested positive. The question is: How concerned should you be?

To find out whether your reaction would be justified, you need to look at the numbers more closely.

Cooperative Learning

Activity 1. Use the data in the excerpt to complete a–g. Explain each answer.

  a. Of the 10,000 people tested, how many would you expect to have cancer?

  b. How many of those people that have cancer would you expect to test positive? (We'll call such cases *true positives*.)

  c. Of the 10,000 people tested, how many would you expect to not have cancer?

  d. How many of those people that do not have cancer would you expect to test positive? (We'll call such cases *false positives*.)

  e. Add your results in b and d to find the total number of people expected to test positive.

  f. What is the probability (**P**) that if you tested positive, you actually have cancer?

  g. How does your answer in **f** compare with your response to the question, "How concerned should you be?"

> **Remember:**
>
> $$\text{Probability} = \frac{\text{considered outcome}}{\text{total number of outcomes}}$$
>
> $$P = \frac{\text{number of true positives}}{\text{total number expected to test positive}}$$

Activity 2. The data you used in Activity 1 are made up. To explore what happens with different sets of data, you can find a general formula for **P**, the probability that if you tested positive, you actually have cancer.

  Let **N** = the number of people tested
  a = the accuracy of the test
      (if test is 98% accurate, then **a** = 0.98)
  r = the portion of people tested that actually have what's being tested for (in the example), **r** = 0.005)

  a. Write a formula for the number of true positives.
  HINT: Look back at a and b of Activity 1.

  b. Write a formula for the number of false positives.
  HINT: Look back at c and d of Activity 1.

  c. Write a formula for **P**. HINT: Look back at e and f of Activity 1.

3. Look at your formula from 2c.

  a. Find **P** when **r** = 1. Explain why that result makes sense.

  b. In the imaginary case, suppose **r** = 0.5 instead of 0.005. Would the results in f of Activity 1 have been as surprising? Explain.

Why

Many relationships such as time, work, and average costs are modeled using rational expressions in equations. Graphing and algebraic methods can be used to solve rational equations.

Sam rode his bike 6 miles before his chain broke, and he had to walk home.

Sports During the walk home, Sam estimated that he bikes 5 times as fast as he walks. If the entire trip took him 2 hours and 24 minutes, find his biking rate.

The time for each part of the trip can be represented using the formula $\frac{\text{Distance}}{\text{Rate}} = \text{Time}$. The total time can be represented by the formula

$$\frac{\text{Distance biked}}{\text{Rate}} + \frac{\text{Distance walked}}{\text{Rate}} = \text{Total time}.$$

If x is Sam's walking rate, then $5x$ is his biking rate.

$$\frac{6 \text{ miles}}{5x \text{ mph}} + \frac{6 \text{ miles}}{x \text{ mph}} = 2\frac{2}{5} \text{ hours}$$

The equation that models the problem is $\frac{6}{5x} + \frac{6}{x} = \frac{12}{5}$. It can be solved by either of two methods.

Common Denominator Method Multiply each of the terms of both sides of the original equation by a common denominator. This will clear the equation of fractions. Using the lowest common denominator saves extra steps in the solution.

$\frac{6}{5x} + \frac{6}{x} = \frac{12}{5}$	Given
$5x\left(\frac{6}{5x} + \frac{6}{x}\right) = 5x\left(\frac{12}{5}\right)$	Multiply both sides by $5x$, the lowest common denominator.
$6 + 30 = 12x$	Simplify.
$12x = 36$	Simplify.
$x = 3$	Division Property of Equality

Graphing Method Graph
$Y_1 = \frac{6}{5x} + \frac{6}{x}$ and $Y_2 = \frac{12}{5}$. You find that
their intersection is at $x = 3$.

Sam walks at 3 miles per hour. His biking
rate is 5 times that, or 15 miles per hour.

CRITICAL
Thinking

What is the "real-life" domain for the
function in the previous example? What
are the undefined value(s) for the
rational expressions in the equations?

EXAMPLE 1

Solve $\frac{6}{x-1} + 2 = \frac{12}{x^2 - 1}$ using the common denominator method and the
graphing method.

Solution A ➤

Common Denominator Method
Factor. Notice that this equation is undefined for $x = 1$ and $x = -1$.

$$\frac{6}{x-1} + 2 = \frac{12}{(x+1)(x-1)}$$

Multiply both sides by the lowest common denominator $(x + 1)(x - 1)$.

$$(x+1)(x-1)\left[\frac{6}{x-1} + 2\right] = (x+1)(x-1)\left[\frac{12}{(x+1)(x-1)}\right], \; x \neq 1, -1$$

$$6(x+1) + 2(x+1)(x-1) = 12 \qquad \text{Divide out common factors.}$$
$$6x + 6 + 2x^2 - 2 = 12 \qquad \text{Simplify.}$$
$$2x^2 + 6x - 8 = 0 \qquad \text{Standard quadratic form}$$
$$2(x+4)(x-1) = 0 \qquad \text{Factor.}$$
$$x + 4 = 0 \text{ or } x - 1 = 0 \qquad \text{Zero Product Property}$$
$$x = -4 \text{ or } x = 1 \qquad \text{Solve.}$$

Reject 1 as a solution since it makes the *original* equation undefined.
Thus, -4 is the only solution. It is very important to check all possible
solutions to be sure that the original equation is defined for them.

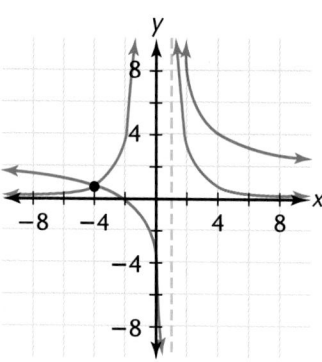

Solution B ➤

Graphing Method

Graph $Y_1 = \dfrac{6}{x-1} + 2$ and $Y_2 = \dfrac{12}{x^2-1}$. The two functions have a common asymptote at $x = 1$ and a clear point of intersection at $x = -4$.

The solution is $x = -4$. ❖

Explain the effect of an undefined value on the graph of a rational expression in simplest form.

EXAMPLE 2

Juanita is starting a lawn-mowing business in her neighborhood. The startup cost for the business is $400.

There are expenses of $6 per lawn that include a helper and the cost of gasoline. Juanita has the $6 expense each time she mows a lawn. How many lawns must she mow before the average cost per lawn is $20?

Solution ➤

Let x = the number of lawns mowed. Substitute the known values.

$$\begin{array}{c}\text{average} \\ \text{cost per} \\ \text{lawn}\end{array} = \dfrac{\begin{array}{c}\text{Start-up} \\ \text{costs}\end{array} + \left(\begin{array}{c}\text{Costs per} \\ \text{lawn}\end{array}\right) \times \left(\begin{array}{c}\text{Number} \\ \text{of lawns}\end{array}\right)}{\text{Number of lawns}}$$

$$20 = \dfrac{400 + 6x}{x}$$

Common Denominator Method

$$\dfrac{20}{1} = \dfrac{400 + 6x}{x}$$
$$20x = 400 + 6x$$
$$14x = 400$$
$$x = 28\tfrac{4}{7}$$

Graphing Method

$$y = \dfrac{400 + 6x}{x}$$
$$y = 20$$

Juanita would have to mow 29 lawns before the average cost per lawn is $20. ❖

Solve the rational equation by finding the lowest common denominator. Verify the solution by graphing. $\dfrac{20}{x^2 - 4} + 6 = \dfrac{10}{x - 2}$

EXAMPLE 3

Max and his younger brother Carl are raising extra money by delivering papers. If the boys work together, how long will it take them to finish this job?

Solution ➤

Let t represent the number of minutes the boys work together. The completed job is equal to the sum of all the fractional parts and is represented by 1 complete job.

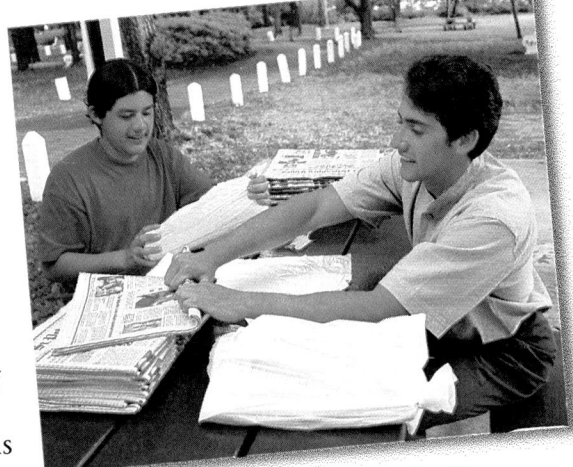

Max can deliver papers in 80 minutes, but it takes Carl 2 hours to deliver the same number of papers.

Fractional part of the job done by Max	+	Fractional part of the job done by Carl	=	The complete job, or 1 whole unit
$\left(\begin{array}{c}\text{Rate} \\ \text{of Max's} \\ \text{work}\end{array}\right) \cdot \left(\begin{array}{c}\text{Time} \\ \text{worked}\end{array}\right)$	+	$\left(\begin{array}{c}\text{Rate} \\ \text{of Carl's} \\ \text{work}\end{array}\right) \cdot \left(\begin{array}{c}\text{Time} \\ \text{worked}\end{array}\right)$	=	1
$\dfrac{1}{80} \cdot t$	+	$\dfrac{1}{120} \cdot t$	=	1
$\dfrac{t}{80}$	+	$\dfrac{t}{120}$	=	1

Continue to solve $\dfrac{t}{80} + \dfrac{t}{120} = 1$. Multiply each term by the lowest common denominator of 80 and 120.

$$240 \cdot \dfrac{t}{80} + 240 \cdot \dfrac{t}{120} = 240 \cdot 1$$
$$3t + 2t = 240$$
$$5t = 240$$
$$t = 48$$

Working together it would take Carl and Max 48 minutes to deliver all the papers.

In 48 minutes, Carl does $\dfrac{48}{80}$ of the job and Max does $\dfrac{48}{120}$ of the job.

$$\dfrac{48}{80} + \dfrac{48}{120} = \dfrac{6}{10} + \dfrac{4}{10} = \dfrac{10}{10} = 1 \; ❖$$

EXERCISES & PROBLEMS

Communicate

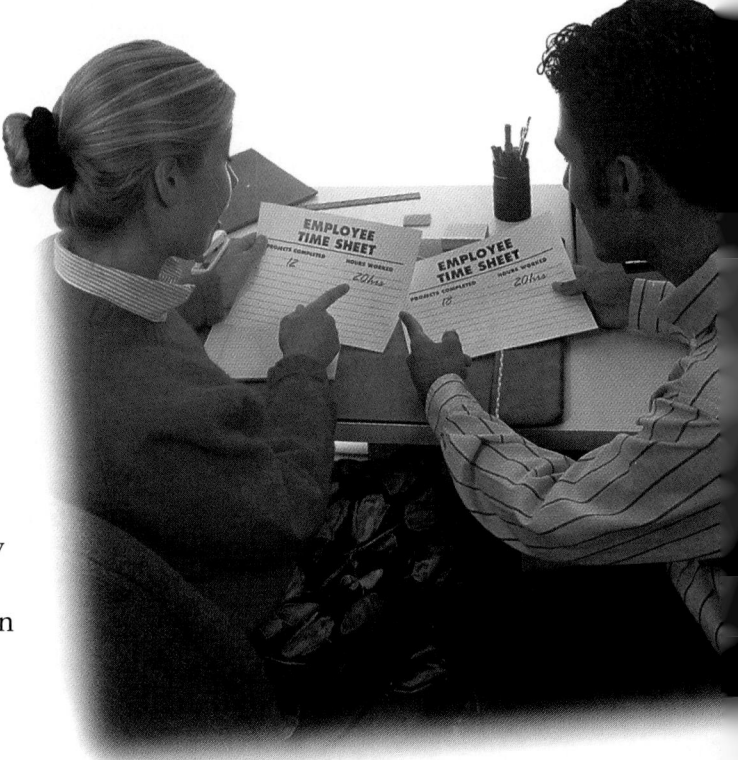

1. Discuss what "undefined variable values of rational equations" means.

2. How is the "realistic domain" of a rational equation determined in applications?

Refer to the equation $\dfrac{3}{c} - \dfrac{2}{c-1} = \dfrac{6}{c}$ **for Exercises 3–5.**

3. Describe the steps needed to find the lowest common denominator.

4. Explain how to solve the rational equation by graphing.

5. How are the undefined values for the equation found? What are they?

6. Explain how to write the rate of work on a particular job as a fraction.

Practice & Apply

Solve the following rational equations by finding the lowest common denominator.

7. $\dfrac{x-1}{x} + \dfrac{7}{3x} = \dfrac{9}{4x}$

8. $\dfrac{10}{2x} + \dfrac{4}{x-5} = 4$

9. $\dfrac{x-3}{x-4} = \dfrac{x-5}{4+x}$

10. $\dfrac{x+3}{x} + \dfrac{6}{5x} = \dfrac{7}{2x}$

11. $\dfrac{x}{x-3} = 5 + \dfrac{x}{x-3}$

12. $\dfrac{x+3}{x^2-9} - \dfrac{6}{x-3} = 5$

13. $\dfrac{4}{y-2} + \dfrac{5}{y+1} = \dfrac{1}{y+1}$

14. $\dfrac{10}{x+3} - \dfrac{3}{5} = \dfrac{10x+1}{3x+9}$

15. $\dfrac{3}{x-2} - \dfrac{6}{x^2-2x} = 1$

Solve the following rational equations by the graphing method.

16. $\dfrac{12x-7}{x} = \dfrac{17}{x}$

17. $\dfrac{3}{7x+x^2} = \dfrac{9}{x^2+7x}$

18. $\dfrac{20-x}{x} = x$

19. $\dfrac{2}{x} + \dfrac{1}{3} = \dfrac{4}{x}$

20. $\dfrac{1}{2} + \dfrac{1}{x} = \dfrac{1}{2x}$

21. $\dfrac{x-3}{x-4} = \dfrac{x-5}{x+4}$

22. $\dfrac{3}{x-2} - \dfrac{6}{x^2-2x} = 1$

23. $\dfrac{1}{x-2} + \dfrac{16}{x^2+x-6} = -3$

24. $\dfrac{1}{x-2} + 3 = \dfrac{-16}{x^2+x-6}$

25. The sum of two numbers is 56. If the larger number is divided by the smaller, the quotient is 1 with a remainder of 16. Find the two numbers.

Marisa pedals downstream at a still-water rate of 20 meters per minute for 500 meters before turning around to pedal upstream to where she started.

Recreation If Marisa maintains the same still-water rate, she will only pedal 300 meters upstream in the same time it takes her to pedal 500 meters downstream.

26. Write rational expressions that represent the time it takes Marisa to pedal downstream and upstream in the current.

27. What is the rate of the current?

28. **Politics** Jeff, Phil, and Allison decide to support their favorite candidate by stuffing envelopes for a candidate's political campaign. They find that they all work at different rates. Phil takes 2 hours longer than Jeff to stuff 1000 envelopes, and Allison takes 3 hours less than Phil to do the same. One day they work together for 4 hours. Write rational expressions that represent the rate at which each works.

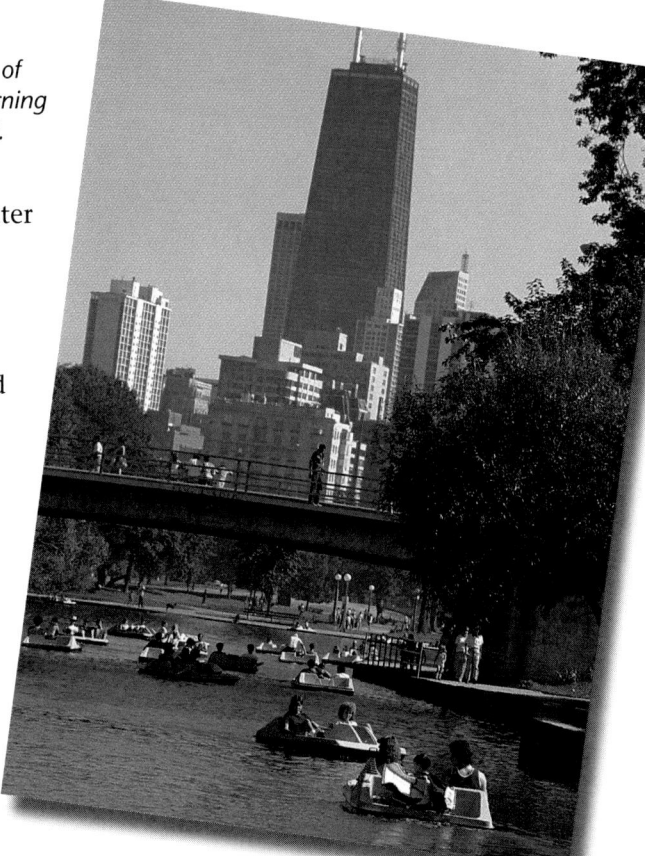

Auto Painting In their shop, it takes Tom about 5 hours to paint a car working alone. It takes Shelley about 7 hours to paint a car.

29. Write rational expressions that represent the fractional parts that Tom and Shelley complete when painting a car together.

30. Write and solve a rational equation to find the time it would take to paint a car if they work together.

Look Back

31. Solve $x - 4 = \frac{-3}{4}(x + 2)$. **[Lesson 4.7]**

32. Graph $|2x - 7| \le 7$. **[Lesson 4.8]**

33. Solve graphically $x + 5y = 9$ and $3x - 2y = 10$. **[Lesson 6.1]**

34. A club has 20 members. In how many ways can 3 officers be selected if all members are eligible for these positions? **[Lesson 8.1]**

35. Multiply $(3x - 8)(5x + 2)$. **[Lesson 11.3]**

36. If $f(x) = \frac{x^2 - 3x - 10}{(x - 4)^2}$, find $f(-2)$. **[Lesson 14.1]**

Look Beyond

37. As a game, Antonio climbs up 6 steps in 3 seconds and hops back 4 steps in 2 seconds. How long will it take him to reach step 100?

Exploring Proportions

why *You have used ratios and proportions to solve percent problems. Since a ratio containing a variable is a rational expression, proportions can be solved using the methods in the previous lesson.*

In the first round of a free-throw contest Al made 28 out of 40 shots. Alicia made 21 out of 30. Who wins the first round?

There are several methods to determine the winner. One way is to use fractions.

Al	Alicia
$\dfrac{28}{40} = \dfrac{7}{10}$	$\dfrac{21}{30} = \dfrac{7}{10}$

Since both fractions reduce to $\dfrac{7}{10}$, $\dfrac{28}{40}$ is equal to $\dfrac{21}{30}$, and

the contest ends in a tie. The equation $\dfrac{28}{40} = \dfrac{21}{30}$ is called a proportion.

A **proportion** is an equation that states that 2 ratios are equal.

A proportion $\dfrac{a}{b} = \dfrac{c}{d}$ is sometimes written $a : b = c : d$.

MEANS
EXTREMES

Exploration 1 — Solving Proportions

1. For each statement, identify the extremes and the means.

a. $\dfrac{1}{2} = \dfrac{5}{10}$ **b.** $\dfrac{3}{4} = \dfrac{9}{12}$ **c.** $\dfrac{4}{6} = \dfrac{6}{9}$ **d.** $\dfrac{6}{2\sqrt{2}} = \dfrac{3\sqrt{2}}{2}$

2. For each statement find the product of the extremes and the product of the means.

a. $\dfrac{1}{2} = \dfrac{5}{10}$ **b.** $\dfrac{3}{4} = \dfrac{9}{12}$ **c.** $\dfrac{4}{6} = \dfrac{6}{9}$ **d.** $\dfrac{6}{2\sqrt{2}} = \dfrac{3\sqrt{2}}{2}$

3. Make a conjecture about the relationship between the product of the extremes and the product of the means in a proportion.

4. In $\dfrac{a}{b} = \dfrac{c}{d}$, ad and bc are often called the *cross products*. Restate your conjecture from Step 3 using cross products. Use your conjecture to tell which of these equations are proportions.

a. $\dfrac{6}{10} = \dfrac{20}{25}$ **b.** $\dfrac{2 + 4\sqrt{3}}{-4} = \dfrac{11}{2 - 4\sqrt{3}}$ **c.** $\dfrac{24}{15} = \dfrac{160}{100}$

5. Explain how to use your conjecture to solve the proportion $\dfrac{x}{4} = \dfrac{42}{24}$.

6. Use your conjecture to solve each of the following proportions.

a. $\dfrac{4}{b} = \dfrac{6}{b + 3}$ **b.** $\dfrac{c - 6}{7} = \dfrac{1}{c}$

7. How are the proportions in Step 6 the same? different? ❖

Exploration 2 *True or False?*

1 Suppose $\frac{a}{b} = \frac{c}{d}$. Try various values for a, b, c, and d. Find the cross products in each proportion.

a. $\frac{b}{a} = \frac{d}{c}$ **b.** $\frac{a+b}{c} = \frac{a-b}{c}$ **c.** $\frac{a+b}{b} = \frac{c+d}{d}$

d. $\frac{a}{c} = \frac{b}{d}$ **e.** $\frac{a-b}{b} = \frac{c-d}{d}$ **f.** $\frac{a}{b} = \frac{a+c}{b+d}$

2 Which of the proportions a-f seem to be true for any replacements you make?

3 State a rule that gives the relationship between the original proportion and each true proportion. ❖

APPLICATION

Sports In competing for the best free-throw percentage, Al made 28 out of 40 free throws for 70%. To win the contest, Al needs to shoot 76%. If Al makes the rest of his free throws, how many free throws must he make?

Let x represent how many more successful free throws Al must make. Write a proportion using Al's new percentage.

$$70\% = \frac{28}{40} = \frac{\text{successes}}{\text{number attempted}} \qquad 76\% = \frac{\text{successes} + x}{\text{number made} + x}$$

$\dfrac{76}{100} = \dfrac{28+x}{40+x}$	Given, $76\% = \dfrac{76}{100}$
$76(40+x) = 100(28+x)$	Cross products are equal.
$3040 + 76x = 2800 + 100x$	Distributive Property
$240 = 24x$	Simplify.
$10 = x$	Division Property of Equality

Al must make 10 more consecutive free throws to shoot 76%. ❖

APPLICATION

GEOMETRY *Connection*

The measures of the three angles of a triangle are in the ratio $1:2:3$. Find the measure of each angle.

Draw a picture. Let the angle measures be x, $2x$, and $3x$. Recall that the sum of the degree measures of the angles of a triangle is $180°$.

$$x + 2x + 3x = 180$$
$$6x = 180$$
$$x = 30 \qquad 2x = 60 \qquad 3x = 90$$

The angle measures are $30°$, $60°$, and $90°$. ❖

A theorem in geometry states that if a line is parallel to one side of a triangle and intersects the other two sides, then it divides the two sides proportionately. Illustrate the theorem with a proportion.

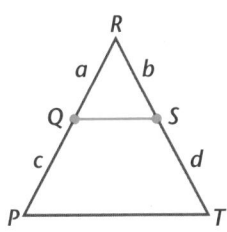

EXERCISES & PROBLEMS

Communicate

1. What is a proportion? Give an example.
2. Explain how to find the means and the extremes of a proportion.
3. Make a statement relating the cross products of the means and the extremes of a proportion.
4. Discuss how to solve for x in the proportion $\frac{1}{2} = \frac{2x + 1}{4}$.
5. Explain how to write a proportion to solve the following. A car travels 320 kilometers on 40 liters of gas. How far can the car travel on 75 liters?

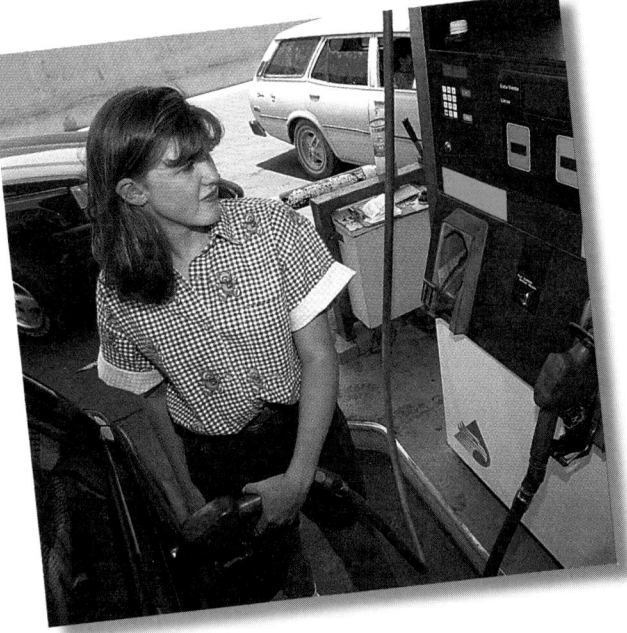

Practice & Apply

Identify the means and extremes of each proportion.

6. $\frac{x}{3} = \frac{7}{4}$ **7.** $\frac{x}{10} = \frac{2}{5}$ **8.** $\frac{3}{4} = \frac{y}{5}$ **9.** $\frac{17}{2} = \frac{8.5}{x}$

10. $\frac{10}{m - 2} = \frac{m}{7}$ **11.** $\frac{2w - 3}{5} = \frac{w}{3}$ **12.** $\frac{3}{n + 6} = \frac{2}{n}$ **13.** $\frac{c + 3}{5} = \frac{c - 2}{4}$

Solve for the variable indicated.

14. $\frac{8}{14} = \frac{4}{x}$ **15.** $\frac{33}{n} = \frac{3}{1}$ **16.** $\frac{p}{4} = \frac{4}{16}$ **17.** $\frac{65}{13} = \frac{x}{5}$

18. $\frac{2}{m - 3} = \frac{5}{m}$ **19.** $\frac{w}{3} = \frac{w + 4}{7}$ **20.** $\frac{x - 1}{3} = \frac{x + 1}{5}$ **21.** $\frac{1}{y - 3} = \frac{3}{y - 5}$

22. $\frac{p}{4} = \frac{10}{p - 3}$ **23.** $\frac{12}{c} = \frac{c + 4}{1}$ **24.** $\frac{x}{2} = \frac{30}{x - 4}$ **25.** $\frac{r}{3} = \frac{56}{r + 2}$

26. $\frac{2}{x} = \frac{x}{9}$ **27.** $\frac{m - 5}{4} = \frac{m + 3}{3}$ **28.** $\frac{x}{3} = \frac{6}{2x}$ **29.** $\frac{2n}{3} = \frac{16}{n + 2}$

For Exercises 30–32, use a variable to represent the unknown. Write a proportion to solve each problem.

30. Twenty-five feet of copper wire weighs 1 pound. How much does 325 feet of copper wire weigh?

31. The distance between two cities is 750 kilometers. Find how far apart the cities are on a map with a scale that reads "1 centimeter = 250 kilometers."

32. In a mixture of concrete, the ratio of sand to cement is 1 : 4. How many bags of cement are needed for 100 bags of sand?

Solve each problem.

33. **Baking** A recipe for making blueberry muffins requires 1600 grams of flour to make 40 muffins. How many muffins can be made using 1000 grams of flour?

34. **Geometry** Find the measures of two supplementary angles if their measures are in a ratio of 3 : 6. The sum of the measures of supplementary angles is 180°.

35. **Transportation** Mrs. Sanchez used 15 gallons of gasoline to drive on a trip of 450 miles. How far can she drive on a full tank of 20 gallons?

It costs $8 to cover a 25-square foot area with wallpaper.

36. **Decorating** How much will it cost to cover a wall with an area of 165 square feet?

37. A recent poll found that 4 out of 6 people nationwide use Klean toothpaste. How many people can be expected to use this toothpaste in a city of 30,000?

38. **Travel** A plane has a cruising speed of 650 miles per hour. At this rate, how long does it take the plane to travel 3250 miles on a cross-country flight?

39. **Portfolio Activity** Complete the problem in the portfolio activity on page 651.

Look Back

40. Solve. $2(x - 3) = 6x$ **[Lesson 4.6]**

41. Find the slope of the line parallel to $6x - 3y = 24$. **[Lesson 5.6]**

42. Simplify $\begin{bmatrix} 2 & -1 \\ 3 & 0 \end{bmatrix}\begin{bmatrix} 1 & 4 \\ -3 & 2 \end{bmatrix}$. **[Lesson 7.3]**

43. If x is chosen at random from {3, 4, 5, 7, 9}, what is the probability that the area of this rectangle will exceed 40 square units? **[Lesson 8.1]**

Complete the square.

44. $x^2 + 20x$ **45.** $y^2 - 10y$ **46.** $a^2 - 14a$ **47.** $c^2 + c$ **[Lesson 12.3]**

LESSON 14.7 Proof in Algebra

why *Throughout this book statements have been made that are assumed to be true. Many of these statements can be proven by using a logical argument.*

IF $a + 6 < 9$ Statement p / Hypothesis

THEN $a < 3$ Statement q / Conclusion

A statement that has been proven is called a **theorem**. In the *if p-then q* form of a theorem, the **hypothesis**, p, is the *if* part, and is assumed to be true. A **proof** provides a logical justification that shows that the **conclusion**, q, the *then* part, is also true.

Each step in a proof must be based on established definitions, on axioms or postulates (statements assumed to be true), or on previously proven theorems. Study the proof of Theorem A.

Theorem A   If $a + 6 < 9$, then $a < 3$.

$a + 6 < 9$	Hypothesis or Given
$a + 6 - 6 < 9 - 6$	Subtraction Property of Inequality
$a < 3$	Simplify.

The conclusion follows logically from the hypothesis, and the theorem is proven.

If you interchange the hypothesis and the conclusion, the new statement is called the **converse** of the original statement. "If q, then p" is the converse of "if p, then q." What is the converse of Theorem A? Is it true?

In Lesson 14.6 on proportions, one conjecture made is

$$\text{if } \frac{a}{b} = \frac{c}{d}, \text{ then } \frac{a+b}{b} = \frac{c+d}{d}.$$

Lesson 14.7   Proof in Algebra   **689**

Hypothesis: $\dfrac{a}{b} = \dfrac{c}{d}$ **Conclusion:** $\dfrac{a+b}{b} = \dfrac{c+d}{d}$

If $\dfrac{a}{b} = \dfrac{c}{d}$, then $\dfrac{a+b}{b} = \dfrac{c+d}{d}$.

The conjecture can be proven by adding 1 to each side of the original proportion.

$\dfrac{a}{b} = \dfrac{c}{d}$	Hypothesis
$\dfrac{a}{b} + 1 = \dfrac{c}{d} + 1$	Addition Property of Equality
$\dfrac{a}{b} + \dfrac{b}{b} = \dfrac{c}{d} + \dfrac{d}{d}$	$\dfrac{b}{b} = 1$ and $\dfrac{d}{d} = 1$, Substitution
$\dfrac{a+b}{b} = \dfrac{c+d}{d}$	Definition of Addition for Rational Numbers

EXAMPLE

Assume that you know all of the rules for operating with integers except the rule for multiplying two negatives. Prove the following statement.

$$\text{If } -3(2) = -6, \text{ then } -3(-2) = 6.$$

Solution ➤

Use the Definition of Opposites. Recall that if the sum of two numbers is zero, the numbers are opposites.

$-3(0) = 0$	Zero Property
$-3(-2 + 2) = 0$	$-2 + 2 = 0$, Substitution, Definition of Opposites
$-3(-2) + (-3)(2) = 0$	Distributive Property
$-3(-2) + (-6) = 0$	Hypothesis $-3(2) = -6$, Substitution
$-3(-2) + (-6) + 6 = 0 + 6$	Simplify. Addition Property of Equality
$-3(-2) + 0 = 0 + 6$	$-6 + 6 = 0$, Substitution, Definition of Opposites
$-3(-2) = 6$	Simplify. Addition Property of Zero ❖

DEFINITION OF EVEN NUMBERS
An even number is of the form $2n$, where n is an integer. For example, 16 is even because $16 = 2(8)$.

DEFINITION OF ODD NUMBERS
An odd number is of the form $2n + 1$ where n is an integer. For example, 17 is odd because $17 = 2(8) + 1$.

From the definition of even numbers, you can show that $2k + 2$ is even if k is an integer.

By the Distributive Property, $2k + 2 = 2(k + 1)$. Since k is an integer, $k + 1$ is also an integer. If $n = k + 1$, then $2k + 2$ can be written in the form $2n$, which is an even integer.

From the definition of odd numbers, you can show that $2k + 3$ is odd if k is an integer.

Notice that $2k + 3 = (2k + 2) + 1$. We just showed that $2k + 2$ is even if k is an integer. So, $(2k + 2) + 1$ can be written as $2n + 1$ and is an odd integer. Thus, $2k + 3$ is odd.

Now you can prove a theorem about odd numbers.

THEOREM B
If two numbers are odd, then their sum is even.

Hypothesis: **a and b are odd.** **Conclusion:** **a + b is even.**

If 2 numbers are odd, **then** their sum is even.

Proof:

a and b are two odd numbers.	Hypothesis or Given
Let $a = 2k + 1$ and $b = 2p + 1$.	Definition of Odd Numbers (k and p are integers.)
$a + b = (2k + 1) + (2p + 1)$	Addition of Real Numbers
$\quad = 2k + 2p + 2$	Simplify.
$\quad = 2(k + p + 1)$	Distributive Property

$2(k + p + 1)$ is in the form $2n$, where n is equal to $k + p + 1$ and $k + p + 1$ is an integer. Thus, $2(k + p + 1)$ is an even number and Theorem B is proved.

Try This Prove: If $a + c = b + c$, then $a = b$. Use the Addition Property of Equality and the Definition of Opposites, $-c + c = 0$.

CRITICAL Thinking Suppose you are given the following statement: If I live in Texas, then I live in the United States. Is the converse true? Make up your own *if, then*-statement. Is the converse true? Explain your answer.

Cultural Connection: Africa The use of logic and proof was first formalized in the study of geometry. Sometime around 300 B.C.E., Euclid, a professor of mathematics at the University of Alexandria, wrote *The Elements*. This text used axioms and postulates to prove theorems. His method of proof incorporated what we know today as the rules of formal logic.

Exercises & Problems

Communicate

1. What does a proof in algebra provide?
2. Name 3 types of reasons that can be used in proving theorems.
3. Identify the part of an *if, then*-statement that is assumed to be true. What part has to be proven?
4. Explain what is meant by the converse of a statement.
5. Explain how to prove that if $7x + 9 = -5$, then $x = -2$. Give a reason for each step.

Practice & Apply

Give a reason for each step.

Proof	Hypothesis or Given
$3m - 4 = -19$	
6. $3m - 4 + 4 = -19 + 4$	_____
7. $3m + 0 = -15$	_____
8. $3m = -15$	_____
9. $\dfrac{3m}{3} = \dfrac{-15}{3}$	_____
10. $1 \cdot m = -5$	_____
11. $m = -5$	_____

Give a reason for each step in the proof that $(a + b) + (-a) = b$.

Proof	
12. $(a + b) + (-a)$	_____
13. $= (b + a) + (-a)$	_____
14. $= b + [a + (-a)]$	_____
15. $= b + 0$	_____
16. $= b$	_____

Prove each of the following. Give a reason for each step.

17. For all y, $2y + 3y = 5y$.   **18.** For all x, $5x - 3x = 2x$.   **19.** For all a, $(-3a)(-2a) = 6a^2$.

20. If $7n - 3n = 32$, then $n = 8$.   **21.** If $3(x + 2) = -15$, then $x = -7$.

22. If $\dfrac{2}{3}x + 5 = -9$, then $x = -21$.   **23.** If $\dfrac{3}{4}m + 8 = -1$, then $m = -12$.

Write a proof for each statement. Let all variables represent real numbers. Complete Exercises 24–32.

24. If $a + (b + c)$, then $(a + b) + c$.

25. If $a = b$, then $a + c = b + c$.

26. $a(b - c) = ab - ac$

27. $(ax + b) + ay = a(x + y) + b$

28. The square of an even number is even.

29. The square of an odd number is odd.

30. For all real numbers a and b, $a - b$ and $b - a$ are opposites. HINT: Prove that $(a - b) + (b - a) = 0$ is true. Then refer to the Definition of Opposites.

31. If $x \neq 0$, then $(xy)\frac{1}{x} = y$.

32. Let a, b, c, and d be integers with $b, d \neq 0$. Prove that the arithmetic sum $\frac{a}{b} + \frac{c}{d} = \frac{da + bc}{bd}$.

Look Back

33. **Statistics** What is the average length in centimeters of the hand span from the tip of the right thumb to the tip of the little finger from this sample? 16.2, 18.5, 23.0, 21.2, 22.9, 21.1, 20.6, 19.2 **[Lesson 4.1]**

21 cm span

34. Solve $x = 16x + 45$. **[Lesson 4.6]**

35. Write the equation of the line perpendicular to $2x - 3y = 9$ and passing through the point $(-6, 12)$. **[Lesson 5.6]**

36. Solve for x and y. $\begin{bmatrix} 1 & 3 \\ 2 & 1 \end{bmatrix} \cdot \begin{bmatrix} x \\ y \end{bmatrix} = \begin{bmatrix} 5 \\ 0 \end{bmatrix}$ **[Lesson 7.5]**

37. Describe the transformations to $f(x) = x^2$ for $g(x) = -2(x - 1)^2 + 3$. **[Lesson 9.2]**

38. **Geometry** Find the area of a triangle whose dimensions are represented as base with length $2(x - 4)$ and height $x - 6$. **[Lesson 11.3]**

39. Solve $x^2 - 8 = 2x$. **[Lesson 12.5]**

40. Solve $\sqrt{x - 5} = -3$. **[Lesson 13.3]**

41. **Geometry** Find the midpoint of the segment joining the points $(-2.5, 4)$ and $(2.5, 6)$. **[Lesson 13.5]**

42. Simplify $\frac{x^2 - 9}{x + 3}$. **[Lesson 14.3]**

A Different Dimension

DIMENSION

When you apply mathematics to solve problems, the numbers are only one part of the problem to consider. Basic dimensions such as mass, length, force, time, and temperature contribute essential information about the problem. Some dimensions are derived from powers or combinations of dimensions. The measurement of these dimensions uses various units such as feet or centimeters to identify relative size.

Part of dimensional analysis is knowing how to convert measurements from one unit to another.

1 mile	= 5280 feet
1 hour	= 60 minutes
1 minute	= 60 seconds

A test car travels on the track at 200 *miles per hour*. What is the equivalent speed in feet per second? Express the original speed in terms of its dimensions.

$$\frac{200 \text{ miles}}{1 \text{ hour}}$$ The *per* indicates that hour is in the denominator.

Express the equivalent dimensions as a fraction, and multiply.

$$\frac{1 \text{ hour}}{60 \text{ minutes}} \cdot \frac{200 \text{ miles}}{1 \text{ hour}}$$

Since units can be treated as factors, the expression can be simplified by dividing identical expressions.

$$\frac{1 \text{ hour}}{60 \text{ minutes}} \cdot \frac{200 \text{ miles}}{1 \text{ hour}}$$

Now simplify, and multiply by the next equivalent expression.

$$\frac{200 \text{ miles}}{60 \text{ minutes}} \cdot \frac{1 \text{ minute}}{60 \text{ seconds}} = \frac{200 \text{ miles}}{3600 \text{ seconds}} = \frac{1 \text{ mile}}{18 \text{ seconds}}$$

Finally, multiply the result by the equivalent fraction to change from miles to feet.

$$\frac{1 \text{ mile}}{18 \text{ seconds}} \cdot \frac{5280 \text{ feet}}{1 \text{ mile}} = \frac{5280 \text{ feet}}{18 \text{ seconds}} = \frac{293.333 \text{ feet}}{1 \text{ second}}$$

Thus, 200 miles per hour is approximately equivalent to 293.333 feet per second.

Dimensional Analysis

Some derived dimensions are powers of a basic dimension. The centimeter is a unit for measuring length. Length squared is area. The units are square centimeters. Length cubed is volume. Its units are cubic centimeters.

You can use dimensional analysis to find the number of square feet in 16 square yards. Begin with the basic conversion 3 feet = 1 yard. This can be written $\frac{3 \text{ feet}}{1 \text{ yard}} = 1$.

$$\frac{16 \text{ yards}^2}{1} \cdot \left(\frac{3 \text{ feet}}{1 \text{ yard}}\right)^2 = \frac{16 \text{ yd}^2}{1} \cdot \frac{9 \text{ ft}^2}{1 \text{ yd}^2} = 144 \text{ square feet}$$

Since 1 inch is approximately 2.54 centimeters, you can use dimensional analysis to relate cubic centimeters and liters to cubic inches.

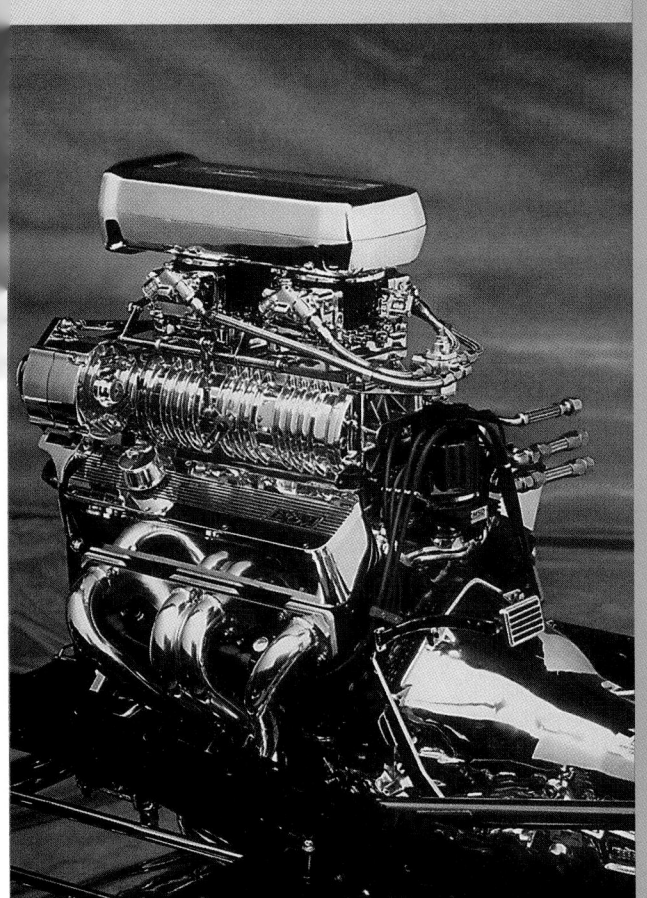

ACTIVITY 1

Use dimensional analysis to convert 220 kilometers per hour into meters per second. Find at least 3 other ways to express 220 kilometers using different units. Show the steps for converting to those units.

ACTIVITY 2

A racing automobile has a 7.1 liter engine. Use dimensional analysis to determine the cubic inch displacement of the engine. Consider a liter as 1000 cubic centimeters. Find the engine size of 3 different kinds of car and compare the displacement of these engines with the size of a racing engine.

ACTIVITY 3

Find two change-of-dimension problems that you might encounter in your activities, and show how to change the dimensions.

Chapter 14 Review

Vocabulary

Key Skills & Exercises

Lesson 14.1

➤ **Key Skills**

Identify transformations to $f(x) = \dfrac{1}{x}$ needed to graph a rational function. List the value(s) for which the function is undefined.

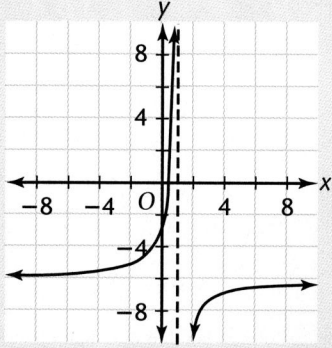

The transformation of $g(x) = \dfrac{-3}{x-1} - 6$ is a reflection over the x-axis, a vertical stretch by 3, a shift to the right by 1, and a shift down by 6. The vertical asymptote is $x = 1$. The rational expression is undefined when x is 1.

➤ **Exercises**

Describe the transformations to $f(x) = \dfrac{1}{x}$ needed to graph each of the following rational functions. List any values for which the function is undefined.

1. $g(x) = \dfrac{-2}{x} + 4$ **2.** $h(x) = \dfrac{1}{x-3} + 2$ **3.** $g(x) = \dfrac{-3}{x+4} - 1$

Lesson 14.2

➤ **Key Skills**

Identify and use inverse variation.

If y varies inversely as x, and y is 10 when x is 3, solve the equation $10 = \dfrac{k}{3}$ to find the constant of variation, k.

$$10 = \frac{k}{3} \qquad 30 = k$$

Use the constant of variation, 30, to find x when $y = 5$.

$$5 = \frac{30}{x} \qquad 5x = 30 \qquad x = 6$$

4. If x varies inversely as y, and $x = 75$ when $y = 20$, find y when $x = 25$.

5. If p varies inversely as q, and $q = 36$ when $p = 15$, find p when $q = 24$.

6. If m varies inversely as n, and $m = 0.5$ when $n = 5$, find m when $n = 50$.

Lesson 14.3

➤ **Key Skills**

Simplify rational expressions.

$$\frac{-x^2 + x + 2}{x^2 + 3x + 2} = \frac{(-1)(x + 1)(x - 2)}{(x + 1)(x + 2)} = \frac{(-1)(x - 2)}{x + 2} = \frac{-x + 2}{x + 2}, x \neq -1, -2$$

➤ **Exercises**

7. $\dfrac{8(x + 1)}{10(x + 1)^2}$
 8. $\dfrac{2x^2 + 6x}{6x - 30}$
 9. $\dfrac{2x^2 - x - 1}{1 - x^2}$

Lesson 14.4

➤ **Key Skills**

Add, subtract, and multiply rational expressions.

a. $\dfrac{5}{x - 3} + \dfrac{2}{x - 5} = \dfrac{5(x - 5)}{(x - 3)(x - 5)} + \dfrac{2(x - 3)}{(x - 5)(x - 3)}$

$$= \frac{5x - 25 + 2x - 6}{x^2 - 8x + 15} = \frac{7x - 31}{x^2 - 8x + 15}$$

b. $\dfrac{3x + 6}{x + 5} \cdot \dfrac{x^2 - 25}{x + 2} = \dfrac{3(x + 2) \cdot (x + 5)(x - 5)}{(x + 5)(x + 2)}$

$$= 3 \cdot \frac{x + 2}{x + 2} \cdot \frac{x + 5}{x + 5} \cdot (x - 5) = 3(x - 5)$$

➤ **Exercises**

Perform the indicated operations, and simplify.

10. $\dfrac{2}{5x} + \dfrac{5}{10y}$
 11. $\dfrac{16}{y^2 - 16} - \dfrac{2}{y + 4}$
 12. $\dfrac{t^2 - 9}{6} \cdot \dfrac{9}{3 - t}$

Lesson 14.5

➤ **Key Skills**

Solve rational equations. $\dfrac{1}{x + 3} + \dfrac{3}{4x} = \dfrac{2}{x}$

$$4x(x + 3)\left(\frac{1}{x + 3} + \frac{3}{4x}\right) = 2 \cdot 4\left(\frac{x}{x}\right)(x + 3)$$

$$\frac{4x(x + 3)}{x + 3} + \frac{3(4x)(x + 3)}{4x} = \frac{x}{x} \cdot 4(x + 3)(2)$$

$$4x + 3(x + 3) = 8(x + 3)$$
$$7x + 9 = 8x + 24$$
$$x = -15$$

➤ **Exercises**

Solve.

13. $\dfrac{z + 2}{2z} + \dfrac{z + 3}{z} = 5$
 14. $\dfrac{6 - x}{6x} = \dfrac{1}{x + 1}$
 15. $\dfrac{5}{w + 6} - \dfrac{2}{w} = \dfrac{9w + 6}{w^2 + 6w}$

Lesson 14.6

➤ Key Skills

Solve proportions.

$$\frac{x}{8} = \frac{-2}{x+10}$$
$$x(x+10) = 8(-2)$$
$$x^2 + 10x = -16$$
$$x^2 + 10x + 16 = 0$$
$$(x+8)(x+2) = 0$$
$$x+8 = 0 \quad \text{or} \quad x+2 = 0$$
$$x = -8 \qquad\qquad x = -2$$

➤ Exercises

Solve.

16. $\dfrac{x}{3} = \dfrac{3}{9}$ **17.** $\dfrac{x+1}{2} = \dfrac{x-1}{3}$ **18.** $\dfrac{7}{x+3} = \dfrac{x}{4}$

Lesson 14.7

➤ Key Skills

Prove algebraic statements.

Prove that $a + 2(b + a) = 3a + 2b$.

    Proof

1. $a + 2(b + a)$ Hypothesis
2. $a + 2b + 2a$ Distributive Property
3. $3a + 2b$ Simplify.

➤ Exercises

Prove each of the following. Give a reason for each step.

19. If $2a + 3 = 5$, then $a = 1$.   **20.** For all a and b, $(a - b) - (b - a) = 2a - 2b$.

Applications

21. Fitness Millie joins a fitness club. There is a $300 fee to join and then a monthly charge of $20. Find a rational expression to represent Millie's average cost per month for belonging to the fitness club.

22. Geometry The area of a certain triangle is represented by $x^2 - 2x - 15$ and its height is $x - 5$. Find the expression for the length of the base.

23. It takes Andrea 3 days to hoe her garden. Helen can hoe the same garden in 4 days. Find the time it would take them to hoe the garden working together.

24. Travel A trip that is 189 miles long can take one hour less if Gerald increases the average speed of his car by 12 miles per hour. Find the original average speed of his car.

25. A tree that is 75 feet tall casts a shadow that is 40 feet long. Find the height of a tree that casts a shadow 15 feet long at the same time of day.

Chapter 14 Assessment

1. For what value is $f(x) = \dfrac{1}{x - 2}$ undefined?

2. Describe all the transformations to $f(x) = \dfrac{1}{x}$ needed to graph $g(x) = \dfrac{-4}{x - 5} + 1$. Write the restriction for the variable.

If x varies inversely as y, and $x = 0.2$ when $y = 7$, find

3. y when $x = 4$.   4. x when $y = 10$.

5. If 6 feet of wire weigh 0.7 kilograms, how much does 50 feet of wire weigh?

6. Write the common denominator for $\dfrac{3}{x - 2}$ and $\dfrac{2}{x + 4}$.

Simplify.

7. $\dfrac{4x^2(x - 1)}{6x}$

8. $\dfrac{x - y}{x^2 - y^2}$

9. $\dfrac{5}{6y} + \dfrac{7}{4y}$

10. $\dfrac{6}{x - 8} - \dfrac{7}{8 - x}$

11. $\dfrac{4 - x^2}{6x} \cdot \dfrac{3x^2 - x}{5x + 10}$

12. $\dfrac{a}{a - 3} - \dfrac{2a}{a + 4}$

13. The area of a certain rectangle is represented by $6x^2 + x - 35$ and its length is $2x + 5$. Find the width.

Solve.

14. $\dfrac{5}{2y} - \dfrac{3}{10} = \dfrac{1}{y}$

15. $\dfrac{x^2 + 5}{3} - \dfrac{6x}{3} = 0$

16. $\dfrac{2(y - 1)}{y + 2} + \dfrac{1}{y + 2} = 1$

17. $\dfrac{4}{7} = \dfrac{x}{49}$

18. $\dfrac{x + 4}{2} = \dfrac{7}{4}$

19. $\dfrac{x}{3} = \dfrac{4}{x - 1}$

20. A large truck has a load capacity of 3 tons more than a smaller truck. If the ratio of their load capacities is 5 to 2, find the capacity of both trucks.

21. The ratio of a number to 5 more than the same number is 2 to 5. Find the number.

22. A blueprint for a house shows a room as a square that has sides of 3 centimeters. The scale indicates that the actual length of a side of the square room is 12 meters. The actual length of another square room is 16 meters. What is the length of this room in centimeters on the blueprint?

23. Sean and Sam are bricklayers. Sean can build a wall with bricks in 5 hours, and Sam can build a wall of the same size in 4 hours. How long will it take them to build the wall if they work together?

24. A boat can travel 90 kilometers downstream in the same time that it can travel 60 kilometers upstream. If the boat travels 15 kilometers per hour in still water, find the rate of the current.

25. Prove that $3b - (b - a) = 2b + a$. Give a reason for each step.

Chapters 1–14
Cumulative Assessment

College Entrance Exam Practice

Quantitative Comparison   For Questions 1–4 write

A if the quantity in Column A is greater than the quantity in Column B;

B if the quantity in Column B is greater than the quantity in Column A;

C if the two quantities are equal; or

D if the relationship cannot be determined from the information given.

	Column A	Column B	Answers
1.	the number of $15 tickets you can buy for $90	the number of $12 tickets you can buy for $84	(A) (B) (C) (D) **[Lesson 4.3]**
2.	$-18 \cdot -4$	$18 \cdot -2$	(A) (B) (C) (D) **[Lesson 4.1]**
3.	$\dfrac{2\sqrt{2}}{\sqrt{5}}$	$\dfrac{2\sqrt{10}}{5}$	(A) (B) (C) (D) **[Lesson 13.2]**
4.	$\cos Y$	$\sin Y$	(A) (B) (C) (D) **[Lesson 13.7]**

(For Question 4: right triangle with Y at top, X at bottom-left right angle, Z at bottom-right; side $YX = 12$, hypotenuse $YZ = 20$, side $XZ = 16$.)

5. Find $|-3.6|$.   **[Lesson 2.4]**
   a. 3.6   **b.** -3.6 **c.** 4   **d.** -4

6. Simplify $3\sqrt{2} - \sqrt{5} + 6\sqrt{2} + 9\sqrt{5}$.   **[Lesson 13.2]**
   a. $9\sqrt{2} + 9$ **b.** $-3\sqrt{2} + 8\sqrt{5}$ **c.** $-3\sqrt{2} + 9$ **d.** $9\sqrt{2} + 8\sqrt{5}$

7. What is the solution to $x^2 = 144$?   **[Lesson 13.3]**
   a. 12   **b.** 72   **c.** ± 12 **d.** ± 72

8. What is the slope of a line parallel to $2x + 3y = -5$?   **[Lesson 5.6]**
   a. 2   **b.** $-\dfrac{2}{3}$ **c.** $\dfrac{3}{2}$ **d.** 3

9. What is the probability of choosing a blue chip from a bag containing 4 blue chips and 3 red chips?   **[Lesson 8.5]**

   a. $\dfrac{3}{4}$ **b.** $\dfrac{4}{3}$ **c.** $\dfrac{4}{7}$ **d.** $\dfrac{3}{7}$

10. What is the simplified form of $(-2m^2n^4)^2$? **[Lesson 10.2]**
   a. $-4m^4n^6$ **b.** $4m^4n^6$ **c.** $-4m^4n^8$ **d.** $4m^4n^8$

11. Name the type of function whose graph is shown at the right.
   [Lesson 2.4]

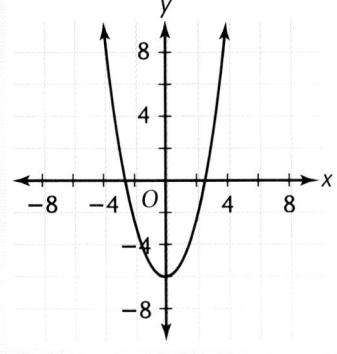

12. Explain how to use the FOIL method to multiply $(x + 4)(2x - 5)$.
   Find the product.   **[Lesson 11.3]**

13. Factor $4x^2 - 16$.   **[Lesson 11.2]**

14. If $A = \begin{bmatrix} 4 & -2 \\ 5 & -3 \end{bmatrix}$ and $B = \begin{bmatrix} -2 \\ 6 \end{bmatrix}$, find AB.   **[Lesson 7.3]**

15. Simplify $3w(4w + 5) - (w + 10)$.   **[Lesson 4.2]**

16. Solve $(x + 1)^2 - 4 = 0$.   **[Lesson 12.3]**

17. The first three terms of a sequence are 4, 7, and 12. The second
   differences are a constant 2. What are the next three terms?
   [Lesson 1.2]

18. Find two numbers whose sum is 32 and whose product is 255.
   [Lesson 6.6]

19. A right triangle has legs with lengths 16 inches and 18 inches. Find
   the length of the hypotenuse to the nearest tenth of an inch.
   [Lesson 13.4]

20. Find the distance between $(-2, 3)$ and $(4, 7)$. Give your answer as a
   radical in simplest form.   **[Lesson 13.5]**

Free-Response Grid   The following questions may be answered using a
free-response grid commonly used by standardized test services.

21. Solve $\frac{5}{2} = \frac{2x}{5}$.   **[Lesson 14.6]**

22. Evaluate $g(x) = x^3 - 4x + 7$ for $x = 3$.   **[Lesson 12.4]**

23. 70% of 45 is _?_.   **[Lesson 4.5]**

24. Find the slope of the line through $(-1, -2)$ and $(3, 4)$.
   [Lesson 5.3]

25. Find INT(5.6).   **[Lesson 2.4]**

INFO BANK

Functions and Their Graphs

Throughout the text a variety of functions were studied—linear, exponential, absolute value, quadratic, and reciprocal. The simplest form of any function is called the parent function. Each parent function has a distinctive graph. Changes made to the parent function will alter the graph but retain the distinctive features of the parent function. Some changes produce a horizontal or vertical shift, others produce a stretch, and still others a reflection. Pages 704–707 summarize some parent functions and the changes that transform them.

Linear Function

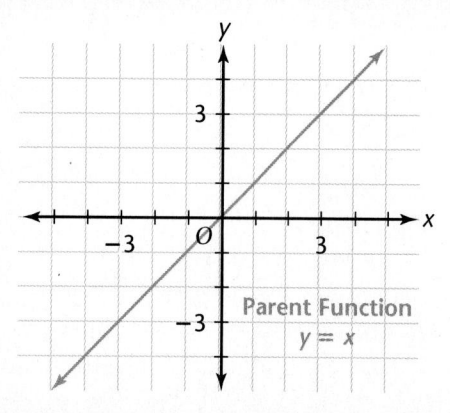

Parent Function
$y = x$

Stretch, Reflection
Vertical Shift
$y = -2x + 3$

Reflection
$y = -x$

Vertical Shift
$y = x - 3$

Stretch
$y = 2x$

Exponential Function

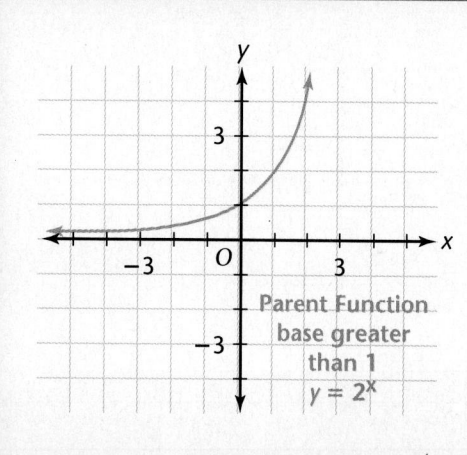

Parent Function
base greater
than 1
$y = 2^x$

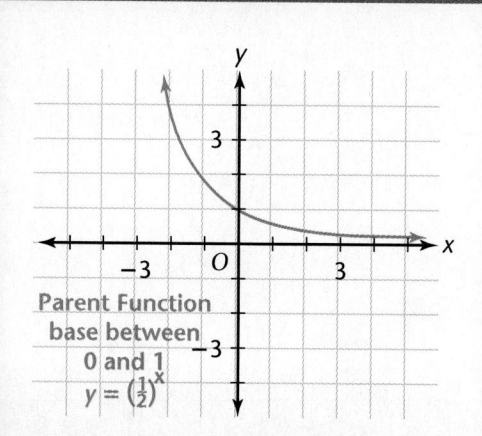

Parent Function
base between
0 and 1
$y = \left(\frac{1}{2}\right)^x$

Absolute Value Function

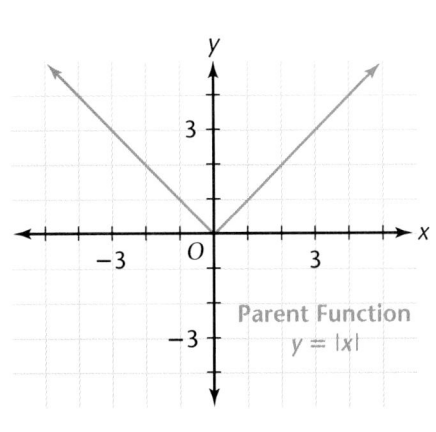

Parent Function
$y = |x|$

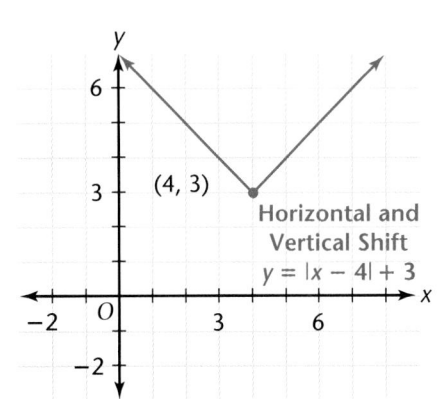

$(4, 3)$

Horizontal and
Vertical Shift
$y = |x - 4| + 3$

Stretch
$y = 2|x|$

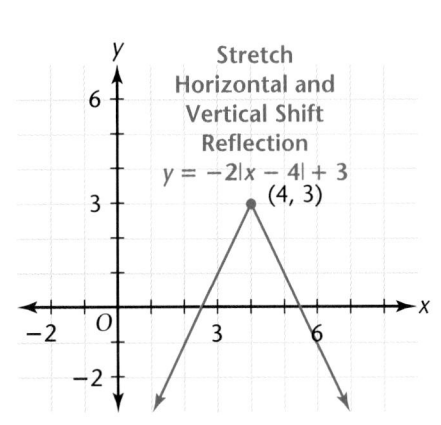

Stretch
Horizontal and
Vertical Shift
Reflection
$y = -2|x - 4| + 3$
$(4, 3)$

Quadratic Function

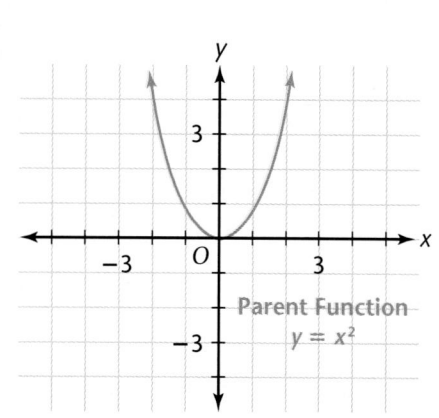

Parent Function
$y = x^2$

(4, 3)

Horizontal and
Vertical Shift
$y = (x - 4)^2 + 3$

Stretch
$y = 2x^2$

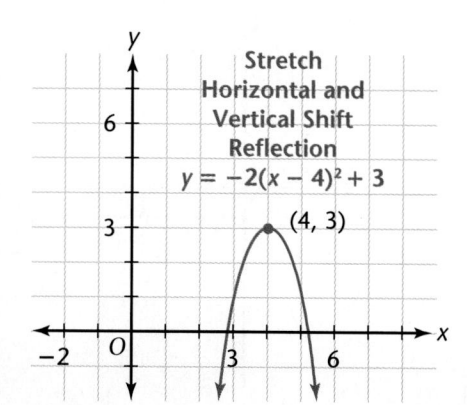

Stretch
Horizontal and
Vertical Shift
Reflection
$y = -2(x - 4)^2 + 3$

(4, 3)

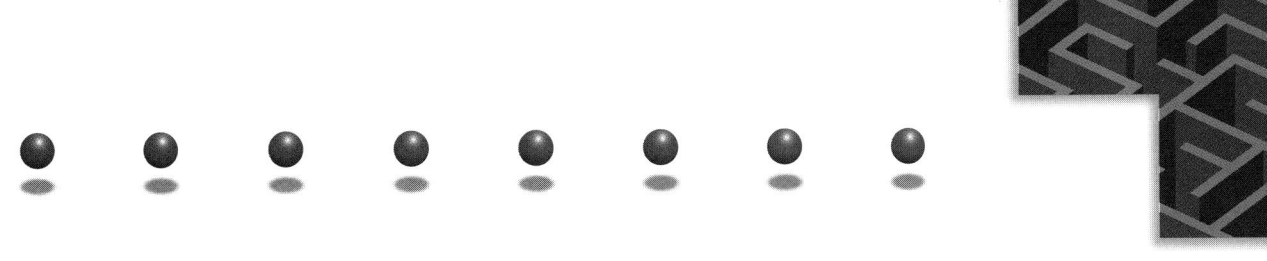

Reciprocal Function

y

3

-3 O 3 x

-3

Parent Function
$y = \frac{1}{x}$

y

6

3 $(4, 3)$

Horizontal and Vertical Shift
$y = \frac{1}{x-4} + 3$

O 3 6 x

y

3 **Reflection**
$y = -\frac{1}{x}$

-3 O 3 x

-3

y

Horizontal and Vertical Shift Reflection
$y = -\frac{1}{x-4} + 3$
$(4, 3)$

6

3

-2 O 3 6 x

-2

For additional transformations of the reciprocal function, see pages 654–655.

Table of Squares, Cubes, and Roots

No.	Squares	Cubes	Square Roots	Cube Roots	No.	Squares	Cubes	Square Roots	Cube Roots
1	1	1	1.000	1.000	51	2,601	132,651	7.141	3.708
2	4	8	1.414	1.260	52	2,704	140,608	7.211	3.733
3	9	27	1.732	1.442	53	2,809	148,877	7.280	3.756
4	16	64	2.000	1.587	54	2,916	157,464	7.348	3.780
5	25	125	2.236	1.710	55	3,025	166,375	7.416	3.803
6	36	216	2.449	1.817	56	3,136	175,616	7.483	3.826
7	49	343	2.646	1.913	57	3,249	185,193	7.550	3.849
8	64	512	2.828	2.000	58	3,364	195,112	7.616	3.871
9	81	729	3.000	2.080	59	3,481	205,379	7.681	3.893
10	100	1,000	3.162	2.154	60	3,600	216,000	7.746	3.915
11	121	1,331	3.317	2.224	61	3,721	226,981	7.810	3.936
12	144	1,728	3.464	2.289	62	3,844	238,328	7.874	3.958
13	169	2,197	3.606	2.351	63	3,969	250,047	7.937	3.979
14	196	2,744	3.742	2.410	64	4,096	262,144	8.000	4.000
15	225	3,375	3.873	2.466	65	4,225	274,625	8.062	4.021
16	256	4,096	4.000	2.520	66	4,356	287,496	8.124	4.041
17	289	4,913	4.123	2.571	67	4,489	300,763	8.185	4.062
18	324	5,832	4.243	2.621	68	4,624	314,432	8.246	4.082
19	361	6,859	4.359	2.668	69	4,761	328,509	8.307	4.102
20	400	8,000	4.472	2.714	70	4,900	343,000	8.367	4.121
21	441	9,261	4.583	2.759	71	5,041	357,911	8.426	4.141
22	484	10,648	4.690	2.802	72	5,184	373,248	8.485	4.160
23	529	12,167	4.796	2.844	73	5,329	389,017	8.544	4.179
24	576	13,824	4.899	2.884	74	5,476	405,224	8.602	4.198
25	625	15,625	5.000	2.924	75	5,625	421,875	8.660	4.217
26	676	17,576	5.099	2.962	76	5,776	438,976	8.718	4.236
27	729	19,683	5.196	3.000	77	5,929	456,533	8.775	4.254
28	784	21,952	5.292	3.037	78	6,084	474,552	8.832	4.273
29	841	24,389	5.385	3.072	79	6,241	493,039	8.888	4.291
30	900	27,000	5.477	3.107	80	6,400	512,000	8.944	4.309
31	961	29,791	5.568	3.141	81	6,561	531,441	9.000	4.327
32	1,024	32,768	5.657	3.175	82	6,724	551,368	9.055	4.344
33	1,089	35,937	5.745	3.208	83	6,889	571,787	9.110	4.362
34	1,156	39,304	5.831	3.240	84	7,056	592,704	9.165	4.380
35	1,225	42,875	5.916	3.271	85	7,225	614,125	9.220	4.397
36	1,296	46,656	6.000	3.302	86	7,396	636,056	9.274	4.414
37	1,369	50,653	6.083	3.332	87	7,569	658,503	9.327	4.431
38	1,444	54,872	6.164	3.362	88	7,744	681,472	9.381	4.448
39	1,521	59,319	6.245	3.391	89	7,921	704,969	9.434	4.465
40	1,600	64,000	6.325	3.420	90	8,100	729,000	9.487	4.481
41	1,681	68,921	6.403	3.448	91	8,281	753,571	9.539	4.498
42	1,764	74,088	6.481	3.476	92	8,464	778,688	9.592	4.514
43	1,849	79,507	6.557	3.503	93	8,649	804,357	9.644	4.531
44	1,936	85,184	6.633	3.530	94	8,836	830,584	9.695	4.547
45	2,025	91,125	6.708	3.557	95	9,025	857,375	9.747	4.563
46	2,116	97,336	6.782	3.583	96	9,216	884,736	9.798	4.579
47	2,209	103,823	6.856	3.609	97	9,409	912,673	9.849	4.595
48	2.304	110,592	6.928	3.634	98	9,604	941,192	9.899	4.610
49	2,401	117,649	7.000	3.659	99	9,801	970,299	9.950	4.626
50	2,500	125,000	7.071	3.684	100	10,000	1,000,000	10.000	4.642

Table of Random Digits

Line\Col	(1)	(2)	(3)	(4)	(5)	(6)	(7)	(8)	(9)	(10)	(11)	(12)	(13)	(14)
1	10480	15011	01536	02011	81647	91646	69179	14194	62590	36207	20969	99570	91291	90700
2	22368	46573	25595	85393	30995	89198	27982	53402	93965	34095	52666	19174	39615	99505
3	24130	48360	22527	97265	76393	64809	15179	24830	49340	32081	30680	19655	63348	58629
4	42167	93093	06243	61680	07856	16376	39440	53537	71341	57004	00849	74917	97758	16379
5	31570	39975	81837	16656	06121	91782	60468	81305	49684	60672	14110	06927	01263	54613
6	77921	06907	11008	42751	27756	53498	18602	70659	90655	15053	21916	81825	44394	42880
7	99562	72905	56420	69994	98872	31016	71194	18738	44013	48840	63213	21069	10634	12952
8	96301	91977	05463	07972	18876	20922	94595	56869	69014	60045	18425	84903	42508	32307
9	89579	14342	63661	10281	17453	18103	57740	84378	25331	12566	58678	44947	05585	56941
10	85475	36857	53342	53988	53060	59533	38867	62300	08158	17983	16439	11458	18593	64952
11	28918	69578	88231	33276	70997	79936	56865	05859	90106	31595	01547	85590	91610	78188
12	63553	40961	48235	03427	49626	69445	18663	72695	52180	20847	12234	90511	33703	90322
13	09429	93969	52636	92737	88974	33488	36320	17617	30015	08272	84115	27156	30613	74952
14	10365	61129	87529	85689	48237	52267	67689	93394	01511	26358	85104	20285	29975	89868
15	07119	97336	71048	08178	77233	13916	47564	81056	97735	85977	29372	74461	28551	90707
16	51085	12765	51821	51259	77452	16308	60756	92144	49442	53900	70960	63990	75601	40719
17	02368	21382	52404	60268	89368	19885	55322	44819	01188	65225	64835	44919	05944	55157
18	01011	54092	33362	94904	31273	04146	18594	29852	71585	85030	51132	01915	92747	64951
19	52162	53916	46369	58586	23216	14513	83149	98736	23495	64350	94738	17752	35156	35749
20	07056	97628	33787	09998	42698	06691	76988	13602	51851	46104	88916	19509	25625	58104
21	48663	91245	85828	14346	09172	30168	90229	04734	59193	22178	30421	61666	99904	32812
22	54164	58492	22421	74103	47070	25306	76468	26384	58151	06646	21524	15227	96909	44592
23	32639	32363	05597	24200	13363	38005	94342	28728	35806	06912	17012	64161	18296	22851
24	29334	27001	87637	87308	58731	00256	45834	15398	46557	41135	10367	07684	36188	18510
25	02488	33062	28834	07351	19731	92420	60952	61280	50001	67658	32586	86679	50720	94953
26	81525	72295	04839	96423	24878	82651	66566	14778	76797	14780	13300	87074	79666	95725
27	29676	20591	68086	26432	46901	20849	89768	81536	86645	12659	92259	57102	80428	25280
28	00742	57392	39064	66432	84673	40027	32832	61362	98947	96067	64760	64584	96096	98253
29	05366	04213	25669	26422	44407	44048	37937	63904	45766	66134	75470	66520	34693	90449
30	91921	26418	64117	94305	26766	25940	39972	22209	71500	64568	91402	42416	07844	69618
31	00582	04711	87917	77341	42206	35126	74087	99547	81817	42607	43808	76655	62028	76630
32	00725	69884	62797	56170	86324	88072	76222	36086	84637	93161	76038	65855	77919	88006
33	69011	65795	95876	55293	18988	27354	26575	08625	40801	59920	29841	80150	12777	48501
34	25976	57948	29888	88604	67917	48708	18912	82271	65424	69774	33611	54262	85963	03547
35	09763	83473	73577	12908	30883	18317	28290	35797	05998	41688	34952	37888	38917	88050
36	91567	42595	27958	30134	04024	86385	29880	99730	55536	84855	29080	09250	79656	73211
37	17955	56349	90999	49127	20044	59931	06115	20542	18059	02008	73708	83517	36103	42791
38	46503	18584	18845	49618	02304	51038	20655	58727	28168	15475	56942	53389	20562	87338
39	92157	89634	94824	78171	84610	82834	09922	25417	44137	48413	25555	21246	35509	20468
40	14577	62765	35605	81263	39667	47358	56873	56307	61607	49518	89656	20103	77490	18062
41	98427	07523	33362	64270	01638	92477	66969	98420	04880	45585	46565	04102	46880	45709
42	34914	63976	88720	82765	34476	17032	87589	40836	32427	70002	70663	88863	77775	69348
43	70060	28277	39475	46473	23219	53416	94970	25832	69975	94884	19661	72828	00102	66794
44	53976	54914	06990	67245	68350	82948	11398	42878	80287	88267	47363	46634	06541	97809
45	76072	29515	40980	07391	58745	25774	22987	80059	39911	96189	41151	14222	60697	59583
46	90725	52210	83974	29992	65831	38857	50490	83765	55657	14361	31720	57375	56228	41546
47	64364	67412	33339	31926	14883	24413	59744	92351	97473	89286	35931	04110	23726	51900
48	08962	00358	31662	25388	61642	34072	81249	35648	56891	69352	48373	45578	78547	81788
49	95012	68379	93526	70765	10592	04542	76463	54328	02349	17247	28865	14777	62730	92277
50	15664	10493	20492	38391	91132	21999	59516	81652	27195	48223	46751	22923	32261	85653

GLOSSARY

absolute error The absolute value of the difference between the actual measure, x, and the specified measure. (192)

absolute value For any number x, if x is greater than or equal to 0, $|x| = x$, and if x is less than 0, $|x| = -x$. (101)

Addition Principle of Counting Suppose there are m ways to make a first choice, n ways to make a second choice, and t ways that have been counted twice. Then there are $m + n - t$ ways to make the first choice OR the second choice. (378)

Addition Property of Equality If equal amounts are added to the expressions on both sides of an equation, the expressions remain equal. (126)

Addition Property of Inequality If equal amounts are added to the expressions on both sides of an inequality, the resulting inequality is still true. (132)

Addition Property of Zero For any number a, $a + 0 = 0 + a$. (107)

additive identity The number that adds to a second number to give a sum that equals the second number. Zero is the additive identity for addition. (107)

additive inverses Two numbers, such as a and $-a$, whose sum is 0. (102)

address Each entry in a matrix can be located by its matrix address. (309)

approximate solution A reasonable estimate for a point of intersection for a system of equations. (264)

Associative Property of Addition For all numbers a, b, and c, $(a + b) + c = a + (b + c)$. (104)

asymptotes The lines that a hyperbola will approach, but not touch. (427)

axis of symmetry The line along which a figure can be folded so that the two halves of the figure match exactly. (418)

base In an expression of the form x^a, x is the base. (456)

coefficient The number multiplied by a variable. (112)

common binomial factor A binomial factor that is common to all terms of a polynomial. (526)

common monomial factor A monomial factor that is common to all terms of a polynomial. (525)

Commutative Property for Addition For any numbers a and b, $a + b = b + a$. (114)

composite number A whole number greater than 1 that is not a prime number. (525)

compound interest Money paid or earned on a given amount of money. (60)

conjecture A statement about observations that is believed to be true. (10)

consistent system A system of equations that has one or more solutions. (279)

constant A term in an algebraic expression that represents a fixed amount. (112) In the constant function $y = b$, b is the constant. (241)

constant of variation In direct variation of the form $\frac{y}{x} = k$ or $y = kx$, if y varies directly as x, k is called the constant of variation. (228)

correlation coefficient The measure of how closely a set of data point falls along a line. (47)

degree of a polynomial The degree of a polynomial in one variable is the exponent with the greatest value of any of the polynomial's terms. (507)

dependent events Two events are dependent if the occurrence of the first event does affect the probability of the second event occurring. (396)

dependent system A system of equations that has an infinite number of solutions. The graph of each equation is the same line with the same y-intercepts. A dependent system is called a consistent system. (280)

descending order The order of the terms of a polynomial when they are ordered from left to right, from the greatest to the least degree of the variable. (507)

difference of two squares A polynomial of the form $a^2 - b^2$ that can be written as the product of two factors, $a^2 - b^2 = (a - b)(a + b)$. (531)

direct variation If y varies directly as x, then $\frac{y}{x} = k$ or $y = kx$. The k is called the constant of variation. (228)

discriminant The expression $b^2 - 4ac$ in the quadratic formula. (577)

Distance formula The distance, d, between two points $A(x_1, y_1)$ and $B(x_2, y_2)$ is given by $d = \sqrt{(x_2 - x_1)^2 + (y_2 - y_1)^2}$. (626)

Distributive Property For all numbers a, b, and c, $a(b + c) = ab + ac$ and $(b + c)a = ba + ca$. (113)

Distributive Property Over Subtraction For all numbers a, b, and c, $a(b - c) = ab - ac$, and $(b - c)a = ba - ca$. (153)

Division Property of Equality See Multiplication and Division Property of Equality.

Division Property of Inequality See Multiplication and Division Property of Inequality.

Division Property of Square Roots For all numbers $a \geq 0$ and $b > 0$, $\sqrt{\frac{a}{b}} = \frac{\sqrt{a}}{\sqrt{b}}$. (606)

domain The set of first coordinates, or x-values, in an ordered pair of a function. (413)

elimination method A method used to solve a system of equations in which one variable is eliminated by adding or subtracting opposites. (272)

equally-likely outcomes Outcomes of an experiment that have the same probability of happening. (389)

equation Two equivalent expressions separated by an equal sign. (23)

experimental probability Let t be the number of trials in an experiment. Let f be the number of times an event occurs. The experimental probability, P, of the event is given by $P = \frac{f}{t}$. (354)

exponent The number that tells how many times a number is used as a factor. In an expression of the form x^a, a is the exponent. (457)

exponential decay A situation in which a number is repeatedly multiplied by a number between 0 and 1. (60)

exponential growth A situation in which a number is repeatedly multiplied by a number greater than 1. (60)

expression In algebra, variables combined with numbers and operations. (23)

extremes In the proportion $\frac{a}{b} = \frac{c}{d}$, a and d are the extremes. (684)

factor Numbers or variables that are multiplied. (112) Numbers or polynomials that are multiplied to form a product. (525)

FOIL method A method for multiplying two binomials.

1. Multiply **F**irst terms.

2. Multiply **O**utside terms. Multiply **I**nside terms. Add products from 2 and 3.

3. Multiply **L**ast terms. (520)

formula An equation that describes a numerical relationship. (127)

function A set of ordered pairs for which no two pairs have the same first coordinate. (411)

greatest common factor (GCF) The greatest factor that is common to all terms of a polynomial. (516)

horizontal translation The graph of $y = f(x - h)$ is translated horizontally by h units from the graph of $y = f(x)$. (438)

hyperbola The graph of the reciprocal function. (427)

hypotenuse The side opposite the right angle in a right triangle. (619)

identity matrix for addition A matrix that is added to a second matrix to give a sum that equals the second matrix. The identity matrix for addition is the zero matrix. (317)

identity matrix for multiplication A matrix that is multiplied by a second matrix to give a product that equals the second matrix. (327)

inconsistent system A system of equations that has no solution. The graphs of the equations are parallel lines with the same slope but different y-intercepts. (279)

independent events Two events are independent if the occurrence of the first event does not affect the probability of the second event occurring. (396)

independent system A system of equations that has one solution. The graphs of the equations are lines that intersect at one point and have different slopes. Independent systems are called consistent systems. (280)

inequality A statement containing one of the signs $<$, $>$, $\leq$, $\geq$, or $\neq$. (132)

integer function A function written in the form $\text{INT}(x)$. The integer function rounds the number x down to the nearest integer. (80)

integers Whole numbers and their opposites. (100)

inverse variation If $x \cdot y = k$, or $y = \frac{k}{x}$ and k is a constant, then y varies inversely as x. The constant k is called the constant of variation. (658)

irrational numbers Numbers whose decimal part never terminates or repeats. Irrational numbers cannot be represented by the ratio of two integers. (597)

like terms Terms that contain the same form of a variable. (112)

line of best fit Represents an approximation of the data on a scatter plot. (46)

linear equation An equation whose graph is a straight line. (37)

linear relationship A relationship in which a number grows by a fixed amount. (60)

literal equation An equation that contains a number of different letters. Many formulas are examples of literal equations. (127)

matrix Data arranged in a table of rows and columns and enclosed by brackets [ ]. The plural of matrix is matrices. (309)

matrix equality Two matrices are equal when their dimensions are the same and their corresponding entries are equal. (311)

matrix equation An equation for a system of linear equations of the form $AX = B$ where A is the coefficient matrix, X is the variable matrix, and B is the constant matrix. (338)

maximum value The y-value of the vertex of a parabola that opens down. (560)

mean A measure of central tendency in which all pieces of data are added and the sum is divided by the number of pieces of data. (368)

means In the proportion $\frac{a}{b} = \frac{c}{d}$, b and c are the means. (684)

median The middle number in a set of data arranged in order. If there are two middle numbers, the median is the mean of the two numbers. (369)

Midpoint Formula The midpoint, M, of segment PQ where $P(x_1, y_1)$ and $Q(x_2, y_2)$ is $M\left(\frac{x_1 + x_2}{2}, \frac{y_1 + y_2}{2}\right)$. (628)

minimum value The y-value of the vertex of a parabola that opens upward. (560)

mode The piece of data that occurs most often. There can be more than one mode for a set of data. (369)

monomial An algebraic expression that is either a constant, a variable, or a product of a constant and one or more variables. (464)

Multiplication Principle of Counting If there are m ways to make a first choice and n ways to make a second choice, then there are $m \cdot n$ ways to make a first choice AND a second choice. (383)

Multiplication and Division Property of Equality If both sides of an equation are multiplied or divided by equal numbers, the results are equal. (161–162)

Multiplication and Division Property of Inequality If both sides of an inequality are multiplied or divided by the same positive number, the resulting inequality has the same solution. If both sides of an inequality are multiplied or divided by the same negative number and the inequality sign is reversed, the resulting inequality has the same solution. (188)

Multiplication Property of Square Roots For all numbers a and b, such that a and $b \geq 0$, $\sqrt{ab} = \sqrt{a}\sqrt{b}$. (606)

multiplicative inverse The reciprocal of a number. (169)

multiplicative inverse matrix The matrix A has an inverse matrix, A^{-1}, if $AA^{-1} = I = A^{-1}A$, where I is the identity matrix. (331)

Multiplicative Property of −1 For all numbers a, $-1(a) = -a$. (153)

negative exponent If x is any number except zero and n is a positive integer, then $x^{-n} = \frac{1}{x^n}$. (470)

opposites Two numbers that are on opposite sides of zero and the same distance from zero on a number line. (100)

order of operations The set of rules for computation.

1. Perform all operations enclosed in symbols of inclusion (parentheses, brackets, braces, and bars) from innermost outward.
2. Perform all operations with exponents.
3. Perform all multiplications and divisions in order from left to right.
4. Perform all additions and subtractions in order from left to right. (29)

ordered pair The address of a point in the rectangular coordinate system, indicated by two numbers in parentheses, (x, y). (35)

origin The point of intersection of the x- and y-axes in the rectangular coordinate system. (35)

parabola The graph of a quadratic function. (66)

parallel lines If two different lines have the same slope, the lines are parallel. If two different lines are parallel, they have the same slope. Two parallel, vertical lines have undefined slope. (247)

parent function The most basic of a family of functions. (418)

percent A ratio that compares a number with 100. (173)

perfect square A number whose square root is a positive integer. (596)

perfect-square trinomial A trinomial of the form $a^2 + 2ab + b^2$ or $a^2 - 2ab + b^2$. The factored form of $a^2 + 2ab + b^2$ is $(a + b)(a + b) = (a + b)^2$, and of $a^2 - 2ab + b^2$ is $(a - b)(a - b) = (a - b)^2$. (529)

perpendicular lines If the slopes of two lines are m and $-\frac{1}{m}$, the lines are perpendicular. If the slope of a line is m, then the slope of a line perpendicular to it is $-\frac{1}{m}$. (250)

point-slope form The form $y - y_1 = m(x - x_1)$ is the point-slope form for the equation of a line. The coordinates x_1 and y_1 are taken from a given point (x_1, y_1), and the slope is m. (235)

polynomial The sum or difference of two or more monomials. (506)

Power-of-a-Power Property If x is any number and a and b are any positive integers, then $(x^a)^b = x^{ab}$. (464)

Power-of-a-Product Property If x and y are any numbers and n is a positive integer, then $(xy)^n = x^n y^n$. (465)

prime number A whole number greater than 1 whose only factors are itself and 1. (525)

prime polynomial A polynomial that has no polynomial factors with integral coefficients except itself and 1. (525)

principal square root The positive square root of a number that is indicated by the radical sign. (598)

product matrix The result of multiplying two matrices. (325)

Product-of-Powers Property If x is any number and a and b are any positive integers, then $x^a \cdot x^b = x^{a+b}$. (463)

Properties of Zero Let a represent any number.

1. The product of any number and zero is zero. $a \cdot 0$ and $0 \cdot a = 0$
2. Zero divided by any nonzero number is zero. $\frac{0}{a} = 0$, $a \neq 0$
3. A number divided by zero is undefined. That is, never divide by zero. (148)

Property of Opposites For any number a, $-a$ is its opposite, and $a + (-a) = 0$. (102)

proportion An equation that states that two ratios are equal. (684)

"Pythagorean" Right–Triangle Theorem Given a right triangle with legs of length a and b and hypotenuse of length c, $c^2 = a^2 + b^2$ or $a^2 + b^2 = c^2$. (620)

quadrant One of the four regions in a coordinate plane. A horizontal and vertical number line divide a coordinate plane into 4 quadrants. (35)

quadratic formula The solutions of the quadratic equation $ax^2 + bx + c = 0$ are $\frac{-b \pm \sqrt{b^2 - 4ac}}{2a}$. (575)

quadratic inequality An inequality that contains one or more quadratic expressions. (581)

Quotient-of-Powers Property If x is any number except 0, and a and b are any positive integers with $a > b$, then $\frac{x^a}{x^b} = x^{a-b}$. (466)

radical expression An expression that contains a square root. (605)

radical sign The sign $\sqrt{}$ used to denote the square root. (598)

radicand The number under a radical sign. (604)

range The difference between the highest and lowest numbers in a set of data. (369) The set of second coordinates, or y-values, in an ordered pair of a function. (413)

rate of change The amount of increase or decrease of a function. (221)

ratio The comparison of two quantities. (650)

rational expression If P and Q are polynomials and $Q \neq 0$, then an expression in the form $\frac{P}{Q}$ is a rational expression. (653)

rational function A function of the form $y = \frac{P}{Q}$, or $f(x) = \frac{P(x)}{Q(x)}$. (653)

rational numbers A number that can be expressed as the ratio of two integers with 0 excluded from the denominator. (168)

real numbers The combination of all rational and irrational numbers. (598)

Reciprocal Property For any nonzero number r, there is a number $\frac{1}{r}$ such that $r \cdot \frac{1}{r} = 1$. (168)

rectangular coordinate system A system in which a plane is divided into four regions by a horizontal and a vertical number line and the addresses, or coordinates, of points are given by ordered pairs. (35)

reflection A transformation that flips a figure over a given line. (431)

regression line The line of best fit for a set of data. (227)

relation Pairs elements from one set with elements of another set. (410)

rise The vertical change in a line. (208)

run The horizontal change in a line. (208)

scalar The number by which each entry in a matrix is multiplied. (324)

scale factor The number by which a parent function is multiplied to create a vertical stretch. (425)

scatter plot A display of data that has been organized into ordered pairs and graphed on the coordinate plane. (40)

scientific notation A number written in scientific notation is written with two factors, a number from 1 to 10, but not including 10, and a power of 10. (476)

simplest form of a radical expression A radical expression is in simplest form if the radicand contains no perfect square factors greater than 1, contains no fractions, and if there is no radical sign in the denominator of a fraction. (605)

simulation An experiment with mathematical characteristics that are similar to the actual event. (360)

slope Measures the steepness of a line by the formula, slope $= \dfrac{\text{rise}}{\text{run}}$. (209)

slope formula Given two points with coordinates (x_1, y_1) and (x_2, y_2), the slope is
$$m = \frac{\text{change in } y}{\text{change in } x} = \frac{\text{difference in } y}{\text{difference in } x} = \frac{y_2 - y_1}{x_2 - x_1}$$
(226)

slope-intercept form The slope-intercept formula or form for a line with slope m and y-intercept b is $y = mx + b$. (225)

square matrix A matrix with equal row and column dimensions. (310)

square root If a is greater than or equal to zero, $\sqrt{a}$ and $-\sqrt{a}$ represent the square root of a. Each square root of a has the following property. $\sqrt{a} \cdot \sqrt{a} = a$ $(-\sqrt{a}) \cdot (-\sqrt{a}) = a$

standard form An equation in the form $Ax + By = C$ is in standard form when A, B, and C are integers, A and B are not both zero, and A is not negative. (233)

standard form of a polynomial A polynomial is in standard form when the terms of the polynomial are ordered from left to right, from the greatest to the least degree of the variable. (507)

stretch A transformation in which the graph of a parent function is stretched vertically by the amount of the scale factor. (424)

substitution method A method used to solve a system of equations in which variables are replaced with known values or algebraic expressions. (267)

Subtraction Property of Equality If equal amounts are subtracted from the expressions on both sides of an equation, the expressions remain equal. (125)

Subtraction Property of Inequality If equal amounts are subtracted from the expressions on both sides of an inequality, the resulting inequality is still true. (132)

symmetric A figure is symmetric with respect to a line called the axis of symmetry if the figure can be folded along this axis and the left and right halves of the figure match exactly. (418)

system of equations Two or more equations in two or more variables. (261)

system of linear inequalities Two or more linear inequalities in two or more variables. The solution for a system of linear inequalities is the intersection of the solution sets for each inequality. (285)

tangent function In a right triangle, the ratio of the length of the leg opposite an acute angle of the triangle to the length of the leg adjacent to the acute angle. (638)

term Each number in a sequence. (10) The numbers, variables, or product or quotient of numbers and variables that are added or subtracted in an algebraic expression. (112)

theoretical probability Let *n* be the number of equally-likely outcomes in an event. Let *s* be the number of successful outcomes in the same event. Then, the theoretical probability that the event will occur is $P = \frac{s}{n}$. (389)

transformation A variation such as a stretch, reflection, or shift of a parent function. (418)

translation A transformation that shifts the graph of a function horizontally or vertically. (436)

tree diagram A diagram that is used to find all the possible choices in a situation in which one choice AND another choice need to be made. (382)

trial The number of times an experiment is conducted. (353)

variable A letter or other symbol that can be replaced by any number or other expression. (23)

variable matrix A matrix containing the variables of a system of equations. (338)

Venn diagram A diagram used to represent the relationships among different sets of data. (376)

vertex The point where a parabola or absolute value function changes direction. (66)

vertical line test A test used to determine whether the graph of a relation is a function. A relation is a function if any vertical line intersects the graph of the relation no more than once. (412)

vertical translation The graph of $y = f(x) + k$ is translated vertically by *k* units from the graph of $y = f(x)$. (437)

x-axis The horizontal number line in the rectangular coordinate system. (35)

x-coordinate A point on the *x*-axis in the rectangular coordinate system. (35)

y-axis The vertical number line in the rectangular coordinate system. (35)

y-coordinate A point on the *y*-axis in the rectangular coordinate system. (35)

y-intercept The point where a line crosses the *y*-axis. (224)

zero exponent If *x* is any number, $x^0 = 1$. (470)

zero matrix A matrix that is filled with zeros. The zero matrix is the identity matrix for matrix addition. (317)

Zero Product Property If *a* and *b* are real numbers such that $ab = 0$, then $a = 0$ or $b = 0$. (567)

INDEX

Definitions of boldface entries can be found in the glossary.

Ratio, 170, 173, 650
Rational equations
    common denominator method,
       678–679
    graphing method, 678–679
    solving, 678–681
Rational expression(s), 650,
    652–655
    operations with, 670–673
    simplifying, 664–667
Rational function, 653–655
    reducing, 665
    undefined, 653–654
Rational numbers, 168, 597
Rationalizing the denominator,
    607
Real numbers, 598
Reasoning (See *Critical Thinking*)
Reciprocal, 71, 168–170
Reciprocal function, 70–73, 87
Reciprocal Property, 168–70
**Rectangular coordinate
    system,** 35
Reflections, 431–433
    vertical, 432
Regression line, 227
Relation, 410–412
Right triangle, 619–620
    converse of the "Pythagorean"
       Right-Triangle Theorem,
       627
    hypotenuse, 619–620
    legs, 619–620
    "Pythagorean" Right-Triangle
       Theorem, 620–622
Rise, 208–212
Run, 208–212

Scalar, 324
Scalar multiplication of matrices,
    324
Scale factor, 425
Scatter plot, 40–43, 46–48
Scientific notation, 476–479
Sequences, 10
    diminishing, 24
    generated by a quadratic, 66–67,
       586–587

Sequences, *(cont.)*
    number, 10
    terms of, 10, 66–67
Sets
    disjoint, 376
    intersection of, 376
    union of, 376
**Simplest form of a radical
    expression,** 605
Simulation, 360–363
    basketball, 361
    designing a, 361
    weather, 362
Slope, 208–212
    geometric interpretation of, 209
    graphic interpretation of, 209
    negative, 209–210
Slope formula, 226
Slope-intercept form, 224–227,
    237
Spreadsheets, 12, 19, 78–79, 83,
    151
Square matrix, 310
Square root(s), 553, 596–600,
    604–608
    division property of, 606
    estimating, 597
    multiplication property of, 606
    negative, 604
    positive, 604
    principal, 598
Standard form, 233–235, 237
**Standard form of a
    polynomial,** 507
Statistics
    deceptive, 370
    descriptive, 368–371
Stretches, 424–427
    scale factor, 425
Substitution method, 267–269
Subtraction, 109
    equations, 126, 164
    of expressions, 118–121
    of integers, 108–109
    of matrices, 315–317
    of polynomials, 510
    of rational expressions, 672
**Subtraction Property of
    Equality,** 125
**Subtraction Property of
    Inequality,** 132
Successful event, 353

Symbols of inclusion, 29, 30
Symmetric, 418
Systems of equations
    consistent, 279–280
    dependent, 280
    elimination method, 272–274
    graphing, 260–264
    inconsistent, 279–280
    independent, 280
    matrix solution of, 338–339
    substitution method, 267–269
Systems of linear inequalities
    boundary line, 286
    graphs of, 285–288

Tangent, 638–641
Tangent function, 538
Technology, 13, 19, 20, 28, 63, 83,
       198, 232, 238, 244, 266, 314,
       334, 335, 339–340, 343, 358,
       359, 373, 393, 461, 462, 467,
       472, 473, 482, 494, 495, 602,
       603, 631, 641 (Also see
       *Calculators, Computers,
       Graphics Calculators,* and
       Spreadsheets.)
Term, 10, 112
Theodorus, 624
Theorem, 689
Theoretical probability,
    388–391
Transformations, 418–421
    combining, 443–445
    of absolute value functions,
       419
    of quadratic functions, 420
    of square root function,
       599–600
    reflections, 431–433
    stretches, 424–427
    translations, 436–440
Translations, 436–440
    horizontal, 438
    vertical, 437
Tree diagram, 382–384, 541
Trial, 353
Trigonometric functions, 638–641
    tangent, 638–641

CREDITS

PHOTOS

Abbreviations used: (t) top, (c) center, (b) bottom, (l) left, (r) right, (bckgd) background, (bdr) border.

FRONT COVER: (l), James W. Kay/Daniel Schaefer; (c) SuperStock; (r), Viesti Associates, Inc. **TABLE OF CONTENTS:** Page v(tr), Jerry Jacka; v(br), Sam Dudgeon/ HRW Photo; vi(tr), Michelle Bridwell/HRW Photo; vi(cl), Larry Stevens/Nawrocki Stock Photo, Inc.; vi(br), Sam Dudgeon/HRW Photo; vii(tr), Tony Stone Images; vii(cl), Ron Kimball; vii(br), Nawrocki Stock Photo, Inc.; viii(tr), Dennis Fagan/HRW Photo; viii(cl), John Langford/HRW Photo; viii(br) Sam Dudgeon/HRW Photo; ix(tr), Douglas Dawson Gallery/Chicago, Illinois; ix(cl), Scott Van Osdol/HRW Photo; ix(br), Andrew Leonard/The Stock Market; x(tr), Michelle Bridwell/HRW Photo; x(b), David Madison;xi(tr), David Phillips/HRW Photo; xi(br), Dennis Fagan/HRW Photo. **CHAPTER ONE:** Page 2(bl)(tr), Jerry Jacka; 2-3(bckgd), Steve Vidler/Nawrocki Stock Photo, Inc.; 3(tr)(br), Michelle Bridwell/Frontera Fotos; 3(c), Jim Newberry/HRW Photo; 4(tr), Scott Van Osdol/HRW Photo; 5(tl), Fred Griffen; 5(cr), The Stock Market; 6(tl), Tony Stone Images; 6(br), Scott Van Osdol/HRW Photo; 7(tr), Flip McCririck; 8(cl), Dennis Fagan/HRW Photo; 8(c), Art Matrix; 9(tl), (tr),(cr), Michelle Bridwell/Frontera Fotos; 12-13(br), Dennis Fagan/HRW Photo;13 (tr), Jonathan Daniel/Allsport; 14(tl), Bettmann Archive; 15(tr), Alex Bartel/FPG International; 16(tr), Warren Faidley/Weatherstock; 16(cl), Dennis Fagan/HRW Photo; 18(c), Sam Dudgeon/HRW Photo; 19(br), Jim Newberry/HRW Photo; 20-21(bckgd), NASA; 22(c), Sam Dudgeon/HRW Photo; 24(l), Advanced Satellite Productions; 25(l), Sam Dudgeon/HRW Photo; 26(t), Dennis Fagan/HRW Photo; 28(t),(bc), Michelle Bridwell/Frontera Fotos; 29(bc), 31(t), 33(t), Sam Dudgeon/HRW Photo; 34(t), Dennis Fagan/HRW Photo; 34(tr), Sam Dudgeon/HRW Photo/Printed with permission of State of Minnesota; 36(t), Sam Dudgeon/HRW Photo; 37(cl), Jay Thomas/International Stock; 38(t), Dennis Fagan/HRW Photo; 39(tr), Sam Dudgeon/HRW Photo; 39(bl), Kathleen Campbell/Allstock; 40(tr), 43(t), Sam Dudgeon/HRW Photo; 45(t), David Madison; 46(bl), (tr), (bl), Sam Dudgeon/HRW Photo; 48(b), Dennis Fagan/HRW Photo; 50-51(bckgd), The Quilt Complex Courtesy of the Espirit Quilt Collection. **CHAPTER TWO:** Page 57(r), Michelle Bridwell/ HRW Photo; 56-57(bckgd), Sam Dudgeon/HRW Photo; 58(tl), 59(bl), Dennis Fagan/HRW Photo; 60(br), Phil Schermeister/AllStock; 62(tr), Dennis Fagan/HRW Photo; 63(tr), Harold E. Edgerton/Palm Press; 64(tc),(bl),(bc),(b), Sam Dudgeon/HRW Photo; 65(b), Michelle Bridwell/HRW Photo; 67(tr), Dennis Fagan/HRW Photo; 68(tr), (br), Michelle Bridwell/HRW Photo; 69(cr), Laurence Parent; 70(c), Michelle Bridwell/HRW Photo; 72(bl), Sam Dudgeon/HRW Photo; 73(tr), Tony Freeman/PhotoEdit; 74(tr),(bl),75(cr), Sam Dudgeon/HRW Photo; 78(t), 83(br), Michelle Bridwell/HRW Photo; 84(t), Richard T. Bryant/The Stock Source/Atlanta; 85(tr), Michelle Bridwell/HRW Photo; 88(br), Robert E. Daemmrich/Tony Stone Images; 89(br), Babylonian Sillabary, 442 B.C., British Museum, London, England/Art Resource, NY; 90(tr), Index Stock Photography Stock; 90(cl),(bl), Sam Dudgeon/HRW Photo; 91(br), Nawrocki Stock Photo, Inc.; 91(tr), Sam Dudgeon/HRW Photo; 91(b), Index Stock Photography Stock. **CHAPTER THREE:** Page 98(bckgd), Martin Rogers/Tony Stone Images; 99(tc), Ray Massey/Tony Stone Images; 99(br), Barbara Adams/FPG International; 98(br), SuperStock; 99(bckgd), Paluan/Art Resource.; 100(r), NASA; 101(tc), Michelle Bridwell/Frontera Fotos; 101(cr), Sam Dudgeon/HRW Photo; 102(tr), Michelle Bridwell/HRW Photo; 104(b), Sam Dudgeon/HRW Photo; 105(tl), Michelle Bridwell/Frontera Fotos; 107(tr), Sam Dudgeon/HRW Photo; 108(bl), Michelle Bridwell/HRW Photo; 109(tr), Sam Dudgeon/HRW Photo; 110(tr), Michelle Bridwell/HRW Photo; 111(bckgd), Lobl-Schreyer/FPG International; 112(tl), Michelle Bridwell/HRW Photo; 112(t), 115-116(bckgd), Sam Dudgeon/HRW Photo; 118(t),121(b), Michelle Bridwell/HRW Photo; 123(bckgd), Trevor Wood/Tony Stone Images; 124(tr),125(tr), Michelle Bridwell/Frontera Fotos; 126(cl), 126(br), 128(r), Sam Dudgeon/HRW Photo; 129(bckgd), SuperStock; 131(c), 132(l),135(tr), Michelle Bridwell/Frontera Fotos; 136-137(bckgd),Tom McCarthy/PhotoEdit;138(bckgd), Matthew Neal McVay/Tony Stone Images; 139(tl), (cr), Bob Daemmrich; 139(cl), (br), Sam Dudgeon/HRW Photo. **CHAPTER FOUR:** Page 144(bc),145(cl,cr), SuperStock;144-145(bckgd), Scott Van Osdol/HRW Photo;146(r), Michelle Bridwell/HRW Photo; 148(cl), (cr), David Phillips/HRW Photo; 149(r), Sam Dudgeon/HRW Photo; 150(cl), Mike Valeri/FPG International; 151(tr), Michelle Bridwell/HRW Photo; 152(bl) 153(tr), Sam Dudgeon/HRW Photo; 155(br), Michelle Bridwell/HRW Photo; 156(bl),157(tl), Sam Dudgeon/HRW Photo; 157(br), David Phillips/HRW Photo 158(bc), (br), Sam Dudgeon/HRW Photo; 158-159(bckgd), Ron Sanford/Tony Stone Images; 160(tr), Michelle Bridwell/HRW Photo; 162-163, Stock Editions; 162(cl), Robert E. Daemmrich; 164(br), Dennis Fagan/HRW Photo; 165(cr), Sam Dudgeon/HRW Photo; 165(bckgd), John W. Warden/SuperStock; 166(bckgd), Philip Habib/Tony Stone Images; 167(t), 168(tr), 169(c), Sam Dudgeon/HRW Photo; 171(tr), Dennis Fagan/HRW Photo; 171(br), (br), 172(tl), Sam Dudgeon/HRW Photo; 172(cr), Larry Stevens/Nawrocki Stock Photo Inc.; 172(bl), Robert Wolf/Robert Wolf Photography; 173(tr), Colin Prior/Tony Stone Images; 174(br), Sam Dudgeon/HRW Photo; 175(cr), 177(tl), Michelle Bridwell/HRW Photo; 178(tl), Dennis Fagan/HRW Photo; 180(tr), Michelle Bridwell/HRW Photo; 181(tr), Dennis Fagan/HRW Photo; 183(bl), Michelle Bridwell/HRW Photo; 184(tr), Dennis Fagan/HRW Photo; 185(c), SuperStock; 186(c), John Terence Turner/FPG International; 189(tr), Maria Taglienti/The Image Bank; 190(bl), The VNR Concise Encyclopedia of Mathematics/Van Nostrand Reinhold Company; 190(br), Photri; 191(t), SuperStock; 190-191(bckgd), John Terence Turner/FPG International; 192(t), Mark Segal/Tony Stone Images; 194(t), Dennis Fagan/HRW Photo; 197(br), David Phillips/HRW Photo; 198(bckgd), Robert E. Daemmrich/Tony Stone Images; 199(cl), "Egyptian Equation Solving:/Image #27 Limestone/Dunasty XII. c.1900 B.C./Planet Art Classic Graphics, Ancient Egypt; 199(tr), Courtesy of The National Council of Teachers of Mathematics: The Rhind Mathematical Papyrus by Arnold Buffum Chace. **CHAPTER FIVE:** Page 206(c), Robert Harding Associates, London; 206(b), John Langford/HRW Photo; 207(bckgd) Hugh Sitton/Tony Stone Images; 208(tc), (inset) Michelle Bridwell/HRW Photo; 209(l), 212(cr), Sam Dudgeon/HRW Photo; 211(bckgd), E. Nagele/ FPG International; 212-213(bckgd), 213(cr), Birdseye, Ltd.; 213(br), Michelle Bridwell/HRW Photo; 214(tl), Tony Stone Images; 214(br), William R. Saliaz/Duomo; 215(t), Mark Reinstein/FPG International; 216,(bc)SuperStock: 216-217(bckgd), Lee Balteman/FPG International; 218(c), Joe Towers/George Hall/Check Six; 220(bl), 221(bckgd), Sam Dudgeon/HRW Photo; 222(tr), Dennis Fagan/HRW Photo; 222(bl), SuperStock; 224(r), Steven E. Sutton/Duomo; 226-227(bc), J. Zimmerman/FPG International; 227(tc), Richard Dole/Duomo; 228(bl), 229(cr), Michelle Bridwell/HRW Photo; 230(tc), James Newberry/HRW Photo; 231(br), Michelle Bridwell/HRW Photo; 232(tl), Sam Dudgeon/HRW Photo; 233(tc), Shooting Star International; 234(tl), Michelle Bridwell/Frontera Fotos; 236(cl), Sam Dudgeon/HRW Photo; 237(br), Dennis Fagan/HRW Photo; 238(cr), Michelle Bridwell/Frontera Fotos; 238-239(bckgd), Bob Daemmrich; 240(r)(l), David Madison; 241(bckgd), Chris Harvey/Tony Stone Images; 242-243(bckgd), Michael Hart/FPG International; 243(bl), Dennis Fagan/HRW Photo; 244(br), Sam Dudgeon/HRW Photo; 245(tr), 246(tr), SuperStock; 246(tc), Michael Keller/FPG International; 248(l), 248(r), Sam Dudgeon/HRW photo; 252(tr), The Bettmann Archive; 250-51(bckgd), Superstock; 252-253(bckgd), Carl Yarbrough; 253(tr), Chris Marona. **CHAPTER SIX:** Page 258(cr), John Banagan/Image Bank; 258(b),259(r), SuperStock; 258-259(bckgd), SuperStock; 260(t), David Phillips/HRW Photo; 261(c), Jeff Zaruba/The Stock Market; 262(br), Michelle Bridwell/HRW Photo; 264-265(bckgd), A.Upitis/The Image Bank; 267(tr), SuperStock; 268(br), David Phillips/HRW Photo; 270(tr), Dennis Fagan/HRW Photo; 270(bl), Robert Reiff/FPG International; 272(t), 276(cl), David Phillips/HRW photo; 276(br), Sam Dudgeon/HRW Photo; 276(bckgd), Michelle Bridwell/HRW Photo; 278(t), 279(br),Dennis Fagan/HRW Photo; 284(t),Superstock; 287(tl), (tr), Michelle Bridwell/HRW Photo; 289(tr), David Phillips/HRW Photo; 291(tr), Michelle Bridwell/HRW Photo; 292(tr), (cl), Ron Kimball; 292(bc), Dennis Fagan/HRW Photo; 293(bl), Sam Dudgeon/HRW Photo; 295(bl), Nawrocki Stock Photo, Inc.; 296(bc), Michelle Bridwell/HRW Photo; 297(t), Joe Towers/The Stock Market; 297(c), A. Edgeworth/The Stock Market; 297(bckgd), Luis Villota/The Stock Market; 298(l), Michelle Bridwell/HRW Photo; 298(tl), Sam Dudgeon/HRW Photo; 298-299(bckgd), Superstock. **CHAPTER SEVEN:** 306(bl), Winfield Parks/National Geographic; 306-307 (bckgd), Joseph Scherschel/National Geographic; Page 306(r)(tr), Erich Lessing/Art Resource; 307(c), Art Resource; 307(tl), Werner Forman Archive/Art Resource; 308(tc), Michelle Bridwell/HRW Photo; 309(tc), Sam Dudgeon/HRW Photo; 310(tc), Tom Van Sant/The Stock Market; 312(br), Sam Dudgeon/HRW Photo; 315(r), David Phillips/HRW Photo; 318(c), Sam Dudgeon/HRW Photo; 318(cr), Dennis Fagan/HRW Photo; 319(br), Michelle Bridwell/HRW Photo; 320(tr), Jonathan Daniel/AllSport USA; 321(tr), Michelle Bridwell/HRW Photo; 322(tr), Mike Anich/Adventure Photo; 322(bl), D. R. Fernandez & M. L. Peck/Adventure Photo; 323(bckgd), Sam Dudgeon/HRW Photo; 324(t), David Phillips/HRW Photo; 325(br), Michelle Bridwell/HRW Photo; 328-329(bckgd), Sam Dudgeon/HRW Photo; 334(br), 336(t), Dennis Fagan/HRW Photo; 337(t), Sam Dudgeon/HRW Photo; 340(tl), 341(c), David Phillips/HRW Photo; 343(c), Michelle Bridwell/HRW Photo; 344(t), UPI/Bettmann; 345(br), FPG International. **CHAPTER EIGHT:** Page 350-351(bckgd), Sam Dudgeon/HRW Photo; 350(c),351(tr),(c), John Langford/HRW Photo; 352(t), David Phillips/HRW Photo; 352(bc), Sam Dudgeon/HRW Photo; 353(tr), John Langford/HRW Photo; 354(tl), Michelle Bridwell/HRW Photo; 354(bl), Erich Lessing/Art Resource, NY; 357(br), Dennis Fagan/HRW Photo; 358(tr), (bl), (bc), Sam Dudgeon/HRW Photo; 360(cl), 362(l), 363(tr), 364(tr), David Phillips/HRW Photo; 366(c),Ron Chapple/FPG International; 366-367(bckgd), J. Zimmermann/FPG International; 368(t), R. Stewart/AllSport USA; 372(t), Warren Morgan/Westlight; 372(c), Dick Reed/The Stock Market; 372(b), SuperStock; 374(c), Bruce Coleman; 375(r), 379(tr), David Phillips/HRW Photo; 380(cr), Mel Digiacomo/The Image Bank; 381(b), Michelle Bridwell/HRW; 382(tl), David Phillips/HRW; 382(tc), Ken Lax/HRW Photo 382(br), Ken Karp/HRW Photo; 383(tl), (tc), Sam Dudgeon/HRW Photo; 383(tr), Coronado Rodney Jones/HRW Photo; 384(t), 385(br), Sam Dudgeon/HRW Photo; 386(br), Michelle Bridwell/HRW Photo; 387(tr), Sam Dudgeon/HRW Photo; 388(tc), Mary Kate Denny/Photo Edit; 389(tr),(br), 390(cl),(c), Sam Dudgeon/HRW Photo; 391(tl), Michelle Bridwell/HRW Photo; 391(br), David Phillips/HRW Photo; 392(tl), SuperStock; 392(cr), Photo Researchers, Inc.; 394(bl), David Phillips/HRW Photo; 396(tr), 398-399(bckgd), 400(c), Michelle Bridwell/HRW Photo; 401(c), Sam Dudgeon/HRW Photo. **CHAPTER NINE:** Page 408(b), 408-409(t), (l), Scott Van Osdol/HRW Photo; 410(t), Tibor Bognar/The Stock Market; 411(t), Jonathan A. Meyers/FPG International; 415(br), David Phillips/HRW Photo; 416(bckgd), Keith Gunnar/FPG International; 418(t), Michelle Bridwell/HRW Photo; 418(cl), Sam Dudgeon/HRW Photo; 421(br), David Phillips/HRW Photo; 424(r), Ron Thomas/FPG International; 428(tr), David Phillips/HRW Photo; 430(br),

Sam Dudgeon/HRW Photo; 431(t),(cl), 433(br), Scott Van Osdol/HRW Photo; 434(bl), Dennis Cox/FPG International; 435(tc), Sam Dudgeon/HRW Photo; 436(t), MC Escher Heirs/Cordon Art, Baarn, Holland; 435(tr), Tom Ives/The Stock Market; 438(t), Sam Dudgeon/HRW Photo; 440(inset), Janet Brooks, artist; 440(tl), Douglas Dawson Gallery/Chicago, Ill. c.1920, Earthenware Vessel, Bamileke Culture, Cameroon grasslands; 440(tl), Douglas Dawson Gallery/Chicago, Ill. c.1200 A.D., Earthenware slip decorated vessel, D'jenne Culture, Mali; 440(br), Michelle Bridwell/HRW Photo; 442(l), Sam Dudgeon/HRW Photo; 443(t), 444(b), 446(t), David Phillips/HRW Photo; 448(r), 449(t), Scott Van Osdol/HRW Photo. **CHAPTER TEN:** Page 454(c), Thomas Craig/FPG International; 455(tr), James King Holmes/SPL/Photo Researchers, Inc.; 454(r), Frithfoto/ Bruce Coleman, Inc; 454(b), Stan Osolinski/FPG International; 454-455 (bckgd), Lea Kuhn/FPG International; 455(b), Douglas Faulkner/ Photo Researchers, Inc; 455(r), Joe McDonald/Bruce Coleman, Inc; 456, Scott Van Osdol/HRW Photo; 459(tl), Visuals Unlimited; 459(cr), 468(tr), Sam Dudgeon/HRW Photo; 466(t), David Phillips/HRW Photo; 469 (bckgd), Dr. Mitsuo Ohtsuki/SPL/Photo Researchers, Inc.; 469(t), Telegraph Colour Library/FPG International; 472(br), David Phillips/HRW Photo; 473(br), The VNR Concise Encyclopedia of Mathematics/Van Nostrand Reinhold Company; 474(inset), Rick Friedman/New York Times; 474-475(bckgd), Sam Dudgeon/HRW Photo; 475(inset), Scott Van Osdol/HRW Photo; 475(c), Robert Friedman/New York Times; 476(l), Scott Van Osdol/HRW Photo; 476(bckgd), John Sanford/Science Photo Library/Photo Researchers; 478(t), Dr. E. R. Degginger; 480(t), Scott Van Osdol/HRW Photo; 481(c), NASA; 483(b), Jack Zehrt/FPG International; 484(b), Telegraph Colour Library/FPG International; 485(bc), NRSC LTD/Science Photo Library/Photo Researchers; 487(t), Michelle Bridwell/HRW Photo; 488(tl), Alese & Mort Pechter/The Stock Market; 488(cl), Ulf Sjostedt/FPG International; 488(br), New York Public Library; 489(t), Ken Reid/FPG International; 489(cl), James King Holmes/Science Photo Library/Photo Researchers, Inc.; 491(t), Matt Bradley/Bruce Coleman, Inc.; 493(b), Michelle Bridwell/HRW Photo; 494-495(bc), Richard Stockton; 496-497(bckgd), Andrew Leonard/The Stock Market; 497(c), Michelle Bridwell/HRW Photo. **CHAPTER ELEVEN:** Page 506(bl), 510(b), Scott Van Osdol/HRW Photo; 513(t), Dennis Fagan/HRW Photo; 514-515(b), 517(tr), Sam Dudgeon/HRW Photo; 519(t), Scott Van Osdol/HRW Photo; 521(cr), 524(tc), Michelle Bridwell/HRW Photo; 527(tr), Scott Van Osdol/HRW Photo; 528(bckgd), Sam Dudgeon/HRW Photo; 532(bc), Dennis Fagan/HRW Photo; 533(t), David Phillips/HRW Photo; 535(t), Michelle Bridwell/HRW Photo; 534(bckgd), John Langford/HRW Photo; 536(b), Michelle Bridwell/HRW Photo; 539(cr), Robert Wolff/ Robert Wolff Photography; 541(bl), Michelle Bridwell/HRW Photo; 541(bckgd) John Langford/HRW Photo. **CHAPTER TWELVE:** Page 546(bckgd), Ron Kimball; 546(c), Michael Brohm/Nawrocki; W.S. Nawrocki/ Nawrocki; 548(t), Bob Burch/Bruce Coleman, Inc. 550(t), David Phillips/HRW Photo; 551(br),

Sam Dudgeon/HRW Photo; 552(cl), Shinichi Kanno/FPG International; 554(inset), Superstock; 555(tr), David Phillips/HRW Photo; 557(cl), Carlos V. Causo/Bruce Coleman; 558(tl), 558(tr), (bl), (br), Sam Dudgeon/HRW Photo; 562(bl), Michelle Bridwell/HRW Photo; 563(tr), David Phillips/HRW Photo; 564(c), Nancy Engebretson/Phoenix Gazette/Used with permission. Permission does not imply endorsement.; 565(c), Superstock; 566(br), Michelle Bridwell/HRW Photo; 568(bl), Sam Dudgeon/HRW Photo; 569(cr), David Phillips/HRW Photo; 571(tr), 572(b), Michelle Bridwell/HRW Photo; 573(tc), Robert Frerch/Tony Stone Images; 573(tr), Peter Brandt; 574(l), (r), (c), David Madison; 576(t), David Madison; 577(tl), David Madison; 579(tr), Michelle Bridwell/HRW Photo; 581(cl), FPG International; 585(r), Dave Gleiter/FPG International; 586(inset), SuperStock. **CHAPTER THIRTEEN:** Page 594(tc), W. Warren/Weslight; 595(c), Superstock; 596(t), Michelle Bridwell/HRW Photo; 600(cl), Rhonda Bishop/Contact Press/Woodfin Camp & Associates; 603(br),609(t), Michelle Bridwell/HRW Photo; 604(t), 607(tr), 611(br), Dennis Fagan/HRW Photo; 612(tl), Comstock; 612(br), John Langford/HRW Photo; 615(bl), SuperStock; 616(b), 617(br), 619(t), Comstock; 619(bl), Dave Bartruff/Nawrocki Stock Photo, Inc.; 619(bckgd), Sam Dudgeon/HRW Photo; 621(t), Tony Stone Images; 623(cr), SuperStock; 624(b), David Phillips/HRW Photo; 625(t), Matthew McVay/Tony Stone Images; 626(tc), Nawrocki Stock Photo, Inc.; 627(tr), Michael J. Howell/SuperStock; 629(br), Dennis Fagan/HRW Photo; 630(c), Index Stock Photography, Inc.; 632(t), Comstock; 633(bckgd), Bill Ross/Tony Stone Images; 634(t), Dallas & John Heaton/Westlight; 635(bl), Alan Bolesta/Index Stock Photography, Inc.; 636(t), David Phillips/HRW Photo; 636-37(bckgd), Michael Hans/Agence Vandystadt/Allsport; 638(t), Michelle Bridwell/HRW Photo; 641(t), SuperStock; 641(br), Michelle Bridwell/HRW Photo; 642(c),Superstock; 643(bl), Dennis Fagan/HRW Photo; 644(c), Scala/Art Resource, NY; 645(tl), Paul Chelsey/Tony Stone Images; 645(cr), Dennis Fagan/HRW Photo; 645(br), Ron Rovtar/Photonica. **CHAPTER FOURTEEN:** Page 651(t), David Philips/HRW Photo; 652(t), 653(b), 654(b), 656(bckgd), Michelle Bridwell/HRW Photo; 658(t), David Phillips/HRW Photo; 660(t), Dennis Fagan/HRW Photo; 662(br), 663(tr), SuperStock; 664(t), Dennis Fagan/HRW Photo; 667(tl), Index Stock Photography Stock; 667(br), Dennis Fagan/HRW Photo; 668(bl), Robert A. Lubeck/Animals Animals; 668-669(bckgd), Nicholas DeVore/Tony Stone; 669(tr), David Phillips/HRW Photo; 670-671(t), 673(br), 674(b), Michelle Bridwell/Frontera Fotos; 675(cr), Museum of the Rockies/Montana State Univ.; 676(bl), David Phillips/HRW Photo; 677(bckgd), SuperStock; 678(t), Dennis Fagan/HRW Photo; 679(b), Sam Dudgeon/HRW Photo; 680(c), 681(t), 682(tr), David Phillips/HRW Photo; 683(tr), Robert Frerck/Tony Stone; 684-685(t), David Phillips/HRW Photo; 687(tr), Michelle Bridwell/HRW Photo; 688(l), Dennis Fagan/HRW Photo; 692(tr), 693(t), Sam Dudgeon/HRW Photo; 694(tc), Hot Rod Magazine; 695(bl), Ron Kimball.

ILLUSTRATIONS

Abbreviations used: (t) top, (c) center, (b) bottom, (r) right, (l) left.

BB&K Design pages 50, 90,138, 199, 252, 298, 344, 400, 448, 496, 540, 586, 644, 694

Bindon, John pages 158 , 159

Boston Graphics pages 12, 21, 27, 30, 42 (r)(l) 44,46, 58, 72, 84, 86, 105, 111, 125, 129, 173, 185, 191, 218, 225, 236, 239, 260 (t)(b), 266, 272, 290, 312, 313 (t)(b), 330, 342

Fisher, David pages 199 (t)(b), 377, 388, 411, 441, 461, 468, 469, 476, 483, 484, 490, 518, 522, 525, 552, 600, 620, 622, 625, 634, 638, 657,

Holcomb, Rhonda pages 129, 224, 234, 235, 238, 376, 425, 445, 457, 512, 516, 520, 599, 599, 604, 669, 684, 689

Kumar, Nishi pages 8, 146, 149, 176, 308, 327, 329, 331, 336, 342, 355, 369, 394, 394, 422, 434, 439, 483, 505, 529, 546, 547, 563, 598, 668

Longacre, Jimmy pages 14, 657

Morrow, Mike pages 10, 18, 22, 23, 25, 27, 30, 33, 34, 43, 47(t)(b), 64, 70, 79, 81, 82, 89, 103, 106, 110, 116, 117, 118, 123, 124, 147, 160, 179, 186, 192, 224, 227, 228, 245, 246, 265, 266, 267, 269, 271, 275, 277, 278, 281(t)(c)(b), 283, 283, 294, 295, 314, 319, 337, 338, 368, 370(t)(b), 371(t)(b), 373(t)(b), 392, 398, 410, 413, 414, 429, 448, 456, 462, 463, 478, 520, 584, 607 , 628, 661

Ray, Julie page 181

Rivers, Joan page 210

Wong, Ophelia pages 76 , 216, 322, 366 , 474, 564 , 676

PERMISSIONS

Selected Answers

Practice & Apply and Look Back

Lesson 1.1

7. 55 **9.** 10 **11.** 28 **13.** 1275; Form a rectangle 50 by 51 and divide by 2. **15.** 52, 61, 70; The difference is 9. **17.** 64, 128, 256; Each term is twice the one before. **19.** 729, 2187, 6561; Each term is three times the one before. **21.** $\frac{1}{2}, \frac{1}{4}, \frac{1}{8}$; Each term is half of the one before. **23.** 45 **25.** 49 **27.** 5050 **29.** 400 **31.** 25 **33.** 210 **35.** $100 = 45 + 55$ **37.** 10,000 **39.** 3025 **41.** $1.80 **43.** $77 **45.** 529 square centimeters **47.** 78 **49.** $1.63

Lesson 1.2

7. First differences: 7, 9, 11, 13; Second differences: 2, 2, 2 **9.** 113, 129 **11.** 65, 81 **13.** 29, 26 **15.** 141, 207 **17.** Second differences constant; 2 **19.** 4 **21.** 20, 30, 42 **23.** 2, 9, 19, 32, 48 **25.** 48 seconds **27.** $D5 = C6 - C5$ **29.** Yes **33.** $1080 **35.** 13 **37.** 171

Lesson 1.3

7. $\frac{1}{49}$ **9.** $0.00 **11.** $2.00 **13.** Enter 6 and repeatedly add 2. **15.** Enter 100 and repeatedly subtract 10.

17.

x	4x
1	4
2	8
3	12
4	16
5	20

19.

s	7s + 4
1	11
2	18
3	25
4	32
5	39

21. She divided both sides of the equation by 2. **23.** 11 **25.** $30a = 99$; a is the number of apples. **27.** 6 **29.** $6.40 **31.** 3 **33.** 142 milliliters **35.** Sample answers: $4 = 1 + 3$, $9 = 6 + 3$, $16 = 10 + 6$, $25 = 10 + 15$, $36 = 20 + 16$ **37.** $1 + 2 + 3 + 4 + 5 +$

$6 = 21 = \frac{6 \cdot 7}{2}$; $1 + 2 + 3 + 4 + 5 +$ $6 + 7 = 28 = \frac{7 \cdot 8}{2}$; The sum of the first n numbers should be $\frac{n(n + 1)}{2}$.

Lesson 1.4

7. $(28 - 2) \cdot 0 = 0$ **9.** 1742 **11.** 1742 **13.** 7473 **15.** 1.088 **17.** 19.565 **19.** 19 **21.** 423 **square feet 23.** $610.53 **25.** 2 **27.** 6 **29.** 4 **31.** 16 **33.** 18 **35.** 2100.3636 **37.** 55 **39.** constant second difference; 46, 61 **41.** $l = w = 4$

Lesson 1.5

7. Yes **9.** No **11.** Yes **13.** $(0, 7)$ **15.** $(5, 4)$ **17–18.**

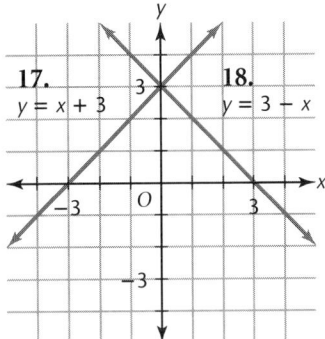

17. $y = x + 3$
18. $y = 3 - x$

19–22.

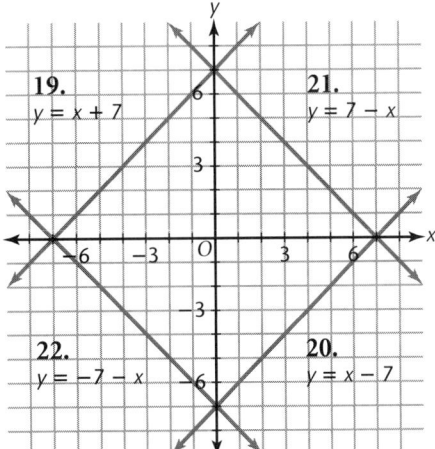

19. $y = x + 7$
21. $y = 7 - x$
22. $y = -7 - x$
20. $y = x - 7$

23.

x	y
1	4
2	5
3	6
4	7
5	8

25.

x	y
1	2
2	4
3	6
4	8
5	10

27. $d = 3h$

29.

31. $19

33. Yes

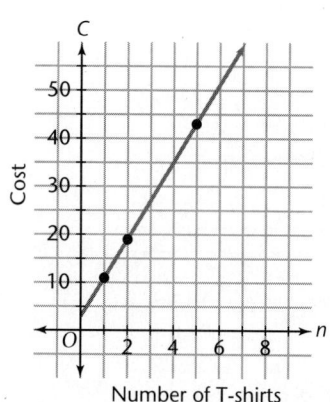

35. 31, 46, 64   **37.** 13, 21; Differences do not help. Each of the differences repeats the same sequence.   **39.** $x = 5\frac{1}{7}$ exactly   **41.** 4   **43.** 4

Lesson 1.6

9. strong positive
11. little or no

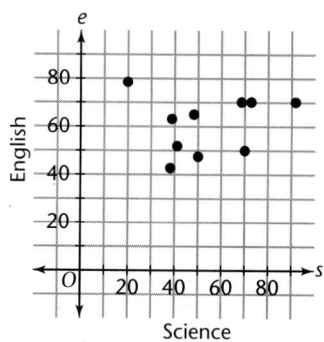

13. Yes   **15.** strong negative   **17.** strong negative   **19.** graph b   **21.** By stretching the scale, the steepness of the line increases.
23. 12   **25.** 13   **27.** 15

29–32.

Lesson 1.7

7. positive   **9.** rises, not a good fit   **11.** rises; good fit   **13.** 1   **15.** Line 1   **17.** $x = 1.5$
19. C; 0.9

21.

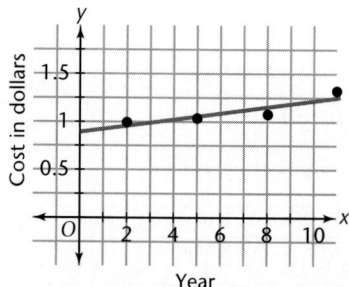

Regression line: $y = 0.4x + 0.89$ (1980 is 0 on the x-axis.)

23. 37, 50 **27.** 64 square centimeters

Chapter

Lesson 2.1

9. 5, 10, 20, 40, 80 **11.** 1100, 1300, 1500
13. 15, 10, 5 **15.** 11, 13 **17.** $y = 3, 9, 27, 81,$
243 **19.** 15 years **21.** $2\frac{1}{2}$ feet **23.** $\frac{5}{8}$ feet
25. linear **27.** A constant amount is added
to or subtracted from each successive term.
29. 256 **31.** Add 4 to each successive term. 23,
27 **33.** Subtract 3 from each successive term.
18, 15 **35.** 24, 18

Lesson 2.2

11. b **13.** a **15.** b **17.** The area becomes 100
times as large. **19.** 15-inch: 176.63 square
inches; 10-inch: 78.5 square inches **21.** c **23.** a
25. twice as large **27.** upward **29.** 64 feet

31.

33. 2673 **35.** $\frac{64}{80}$ **37.** 6 **39.** 84

41.

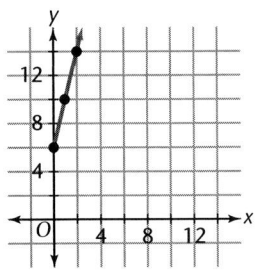

Lesson 2.3

9. $\frac{1}{6}, \frac{1}{5}, \frac{1}{3}, 1, 3, 5, 6$

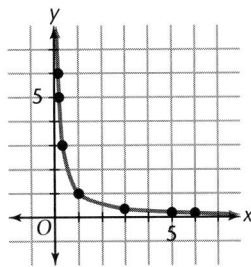

11. $\frac{1}{7}$ **13.** $\frac{1}{25}$ **15.** 10 **17.** Division by zero
is not possible. **19.** 600 vps **21.** 200 vps
23. 1320 vps

25.

27.

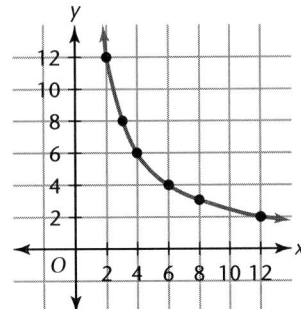

29. $5 **31.** $0.10 **33.** 15 cars **35.** 40 miles
per hour **37.** 60 miles per hour **39.** 240 miles
per hour **41.** 5 hours **43.** $-8, -7, -8, -6;$
No **45.** 3

47–49.

51.

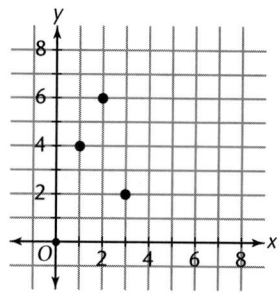

Lesson 2.4

11. under **13.** 8 **15.** 33 **17.** 7.11 **19.** 3.2
21. 5 **23.** 5.5 **25.** 8 **27.** 7 **29.** 8 **31.** −8
33. 6 **35.** 5 **37.** −1, 0, 11, −4, −2, 10, −2,
2, 8, 14, 10, 13, 5 **41.** 22, 26, 30 **43.** 5

44–47.

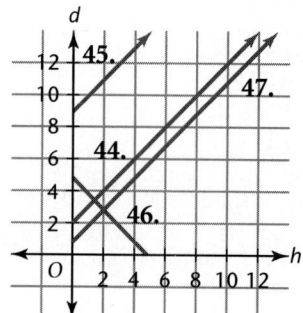

Lesson 2.5

11. a **13.** b **15.** f **17.** neither **19.** quadratic
21. quadratic

23.

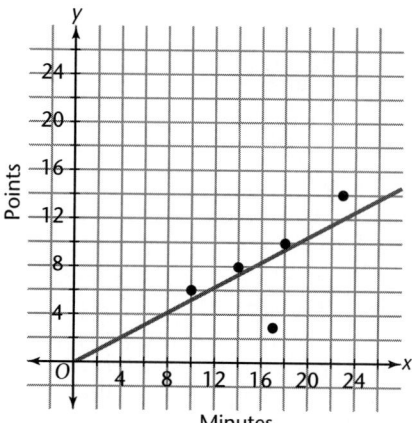

There is a positive correlation between time
played and points scored.

25.

27.

29. 18 years **31.** 8 years **33.** reciprocal
35. 19 **37.** 8 **39.** end of year 1: $1200.00
41. end of year 3: $1728.00 **43.** end of year 5:
$2488.32; 4 years to double the amount.

Chapter 3

Lesson 3.1

7. -17 **9.** 0 **11.** -7 **13.** -4 **15.** 5 **17.** 4
19. 22 **21.** -34 **23.** 58 **25.** -2 **27.** 6-yard
gain **29.** \$33 **31.** 4 **33.** 4 **35.** 4 **37.** $28, 25$
39. $15, 21$ **41.** 1 **43.** 24

44–47.

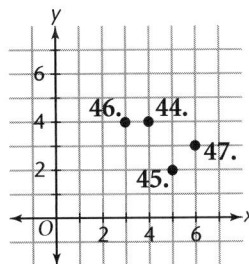

Lesson 3.2

7. 64 **9.** 11 **11.** 66 **13.** 1145 **15.** -63
17. 82 **19.** 8 **21.** 18 **23.** -8 **25.** -6
27. 5 units **29.** 24 units **31.** \$108
33. $-10°F$ **35.** $30°F$ **37.** quadratic function
39. 18.5 **41.** quadratic function

Lesson 3.3

7. $8a - 8$ **9.** $5x + 14y$ **11.** $\frac{5}{6}x$
13. $9a + 10c$ **15.** $8c + 4$ (8 cartons and 4 loose
tiles) **17.** 26 inches **19.** 11 inches **21.** $x + 1$
in. **23.** $14c - 14$ (14 cases − 14 single cookies)
25. $m + 1, m + 2$ **27.** $0.30a + 0.25b + 0.20p$
29. $6a + 10b - 2c$ **31.** $xy + xz; xy + xz =$
$x(y + z)$ **33.** $6x - 6 \cdot 3$ **35.** $4(3 + 5)$
37. $(a + b)z$ or $z(a + b)$ **39.** quadratic function
41. $(4, 8)$ **43.** reciprocal function **45.** absolute value function **47.** -1 **49.** -15

Lesson 3.4

9. 13 **11.** 6 **13.** $-5a - 3b$ **15.** $-9c + 5d$
17. $7r - 4s$ **19.** $j + k$ **21.** $6y$ **23.** $8d - 1$
25. $11k + 9j$ **27.** $v + w$ **29.** $-5y + 17$
31. False, it should be $x - y + z$. **33.** True if
$x > 0$ and false if $x < 0$. **35.** approximately 75
million **37.** approximately 110 million

39. approximately 220 million **41.** True
43. \$735 **45.** \$918.75 **47.** \$185.00 **49.** 5
cases of juice and 3 extra cans of juice

51.

55. $16 \div (5 + 3) \div 2 = 1$ **57.** 1 **59.** the
shape of a V

Lesson 3.5

7. Subtraction Property of Equality, -18
9. Addition Property of Equality, -70
11. Addition Property of Equality, 167
13. Subtraction Property of Equality, 140
15. Subtraction Property of Equality, -2.2
17. Addition and Subtraction Properties of
Equality, $-\frac{3}{8}$ **19.** 10 **21.** -10 **23.** $a = b + c$
25. $a = b - c$ **27.** 2.6 **29.** 131 **31.** Let $y =$
the number of yards the second-string running
back ran. Then $89 + y = 94; y = 5$ **33.** Let $c =$
the combined yardage of both running backs.
Then $94 + 89 = c; c = 183$ **35.** Let $w =$
amount of water to be added in cups. Then
$w + \frac{7}{8} = 1\frac{1}{2}; w = \frac{5}{8}$ **37.** Let $m =$ the
odometer reading at the end of the trip in
miles. Then $m - 149 = 23{,}580; m = 23{,}729$
miles **39.** Let $L =$ the lowest score. Then
$28 = 47 - L; L = 19$ **41.** Let $a =$ the measure
of the third angle in degrees. Then $2(50) +$
$a = 180; a = 80$ **43.** Let $x =$ the measure of
angle SRT in degrees. Then $x + 18 = 70; x = 52$
45. Let $y =$ the measure of angle UVZ in
degrees. Then $28 + y = 90; y = 62$
47. solve$(A + B = C + D, D)$; **49.** Subtract 6 from
the first term to get the second. Add 12 to the
second term to get the third. Subtract 6 from
the third term to get the fourth, and so on;
24, 18, 30 **51.** 6 pieces of poster board
53. $2x - 3$ **55.** $-n + 4$ **57.** 6

Lesson 3.6

11. False **13.** True **15.** $x > -9$ **17.** $x < \frac{1}{4}$
19. $x > 0.56$ **23.** $x < -1$

$$\begin{array}{c}\leftarrow\!\!+\!\!\!+\!\!\!+\!\!\!+\!\!\!+\!\!\!+\!\!\!\circ\!\!\!+\!\!\!+\!\!\!+\!\!\!+\!\!\rightarrow\\ -7\,-6\,-5\,-4\,-3\,-2\,-1\ \ 0\ \ 1\ \ 2\end{array}$$

25. $-1 < x \le 3.5$ **29.** $T \ge 54$ **31.** $70 \le T +$
$5 \le 80$ or $65 \le T \le 75$ **33.** $P + 74 \le 450$ or
$0 \le P \le 376$ **35.** $41.8 \le M \le 42.8$ **37.** $78 \le B$
< 89 **39.** 5 rulers **41.** negative correlation
43. integer function **45.** 9 **47.** 86

Chapter 1

Lesson 4.1

7. -6 **9.** -2.2 **11.** 4 **13.** 22 **15.** 0.2
17. -168 **19.** 6210 **21.** 45 **23.** -17
25. -2 **27.** 24 **29.** -200 **31.** false
33. false **35.** \$80 **37.** \$50 **39.** 2 yards
41. 2.5 yards **43.** 101; yes **45.** -9
47. 7 **49.** 17 **51.** 6 **53.** $x \geq 3$

Lesson 4.2

9. 19.5 **11.** 6.5 or $\frac{13}{2}$ **13.** 18 **15.** 2.5 or $\frac{10}{4}$
or $\frac{5}{2}$ **17.** $-12x$ **19.** $-33x$ **21.** $-12x - 6$
23. $-4x$ **25.** $-21x + 9$ **27.** $19x$ **29.** $16x^2$
31. $-63x^2$ **33.** $-3x$ **35.** $13x + 6$
37. $2x - 18y$ **39.** $23x - 23y$ **41.** $1 + 2w$
43. $-5x + 5y - 36$ **45.** $1 - 3y$
47. $-32x + 18y$ **49.** \$21 **51.** $5.25h$
53. $11.5 + 2h$ **55.** 96 in.^{3} **57.** \$125
59. \$110 **61.** $70h + 40$ **63.** $x^2 - 2x + 5y + 6$
65. $x \geq -7$ **67.** 4 **69.** $24 - 6x$

Lesson 4.3

11. multiplication **13.** addition
15. multiplication **17.** subtraction
19. division **21.** multiplication
23. 0.875 or $\frac{7}{8}$ **25.** 1665 **27.** -56
29. -22 **31.** -3.75 or $-\frac{15}{4}$ **33.** 40
35. $2\frac{1}{5}$ **37.** -19 **39.** -12 **41.** 0 **43.** 1
45. 1405 **47.** Let t represent the number of
rolls of tape Max can buy. $1.15t = 6$; 5 **49.** Let
t represent the number of tape measures that
can be purchased. $4.95t = 19$; 3 **51.** Let b
represent the cost of 1 battery. $4b = 2.52$; \$0.63
53. 225 miles **55.** $rt = d$ **57.** $\frac{V}{lh} = w$
59. 90° **61.** 135° **63.** $n = \frac{2A}{m}$ **65.** 3, 6, 12,
24, 48 **67.** 3 **69.** $30 \cdot 7 - (4 \cdot 10 \div 2 \div 2) =$
200 **71.** 5 **73.** 9

Lesson 4.4

7. $-\frac{1}{2}$ **9.** $\frac{5}{21}$ **11.** 1 **13.** 42 **15.** 12
17. -4.8 **19.** 20 **21.** -120 **23.** $\frac{-6}{5}$ or
$-1\frac{1}{5}$ **25.** $\frac{12}{7}$ or $1\frac{5}{7}$ **27.** 15 **29.** -12

31. $\frac{15}{4}$ or $3\frac{3}{4}$ **33.** 45 **35.** Let s represent the
number of seconds it will take Mary to run the
75 yards. $\frac{s}{75} = \frac{9}{50}$; 13.5 seconds **37.** Let p
represent the number of pesos exchanged for
63 crowns. $\frac{7}{9} = \frac{p}{63}$; 49 pesos **39.** 16 **41.** 6
43. $7x + 5$ **45.** $7b + 1$ **47.** $y \leq 10.55$

Lesson 4.5

11. 0.55 **13.** 0.08 **15.** 0.005 **17.** 1.9%
19. 0.1%

21.

23.

25.

27. less **29.** more **31.** less **33.** 20 **35.** 40
37. 50 **39.** 28 **41.** 0.1% **43.** 800 **45.** 0.05%
47. 26

49.

Clinton	43.24%
Bush	37.73%
Perot	19.02%

51.   **1992 Presidential Popular Vote**

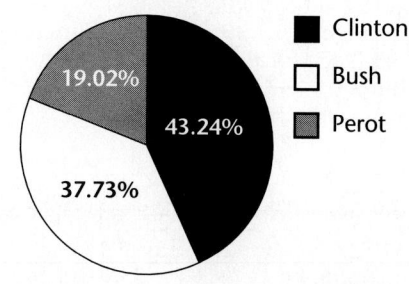

53. 28,476 **55.** about 16.7%

56–59.

61. $-6x$ **63.** $-2x - 5$

Lesson 4.6

7. -2 **9.** 9 **11.** -11 **13.** 24 **15.** 10
17. -300 **19.** $\frac{11}{12}$ **21.** $-1\frac{2}{5}$ **23.** 6 **25.** 12
27. $t = \frac{r}{3}$ **29.** $a = \frac{q}{m}$ **31.** $b = \frac{c - 2a}{2}$ **33.** Let
n represent the first number, then $n + (n + 1)+$
$(n + 2) = 48$; $n = 15$, $n + 1 = 16$, $n + 2 = 17$
35. $2(x) + 2(x + 60) = 180$; The width is 15
feet and the length is 75 feet. **37.** 3.1 hours or
$3\frac{1}{10}$ hours **39.** 1750 miles **41.** 36 square
inches **43.** $s - t$ **45.** 6 **47.** 23 **49.** -10.4

Lesson 4.7

9. $L > W$ **11.** $3.1 \leq V \leq 3.2$ **13.** $m > 0$
15. true **17.** false **19.** false **21.** true
23. $x \geq 3$ **25.** $G \leq 15$ **27.** $x < 7$ **29.** $x < 8$
31. $b > \frac{3}{5}$ **33.** $d < 2$ **35.** $x < 18$ **37.** $x \leq 1\frac{1}{2}$
39. $255 \leq C \leq 300$ **41.** $5 < 6$ **43.** $97 < 178$
45. $l \geq 25$ centimeters **47.** 1, 2, or 3 **49.** 103
51. -13 **53.** positive **55.** negative
57. $-8y + 4$

Lesson 4.8

7. 7, -5 **9.** 13, 3 **11.** $1\frac{1}{2}$, $-5\frac{1}{2}$ **13.** 4, -2
15. $x > 4$ or $x < -12$

17. $x \geq 7$ or $x \leq 3$

19. $-8 \leq x \leq 12$

21. $x < 2$ or $x > 6$

23. Let w represent Margo's weight. Then
$|w - 118| = 5$; 113 pounds and 123 pounds.

25.

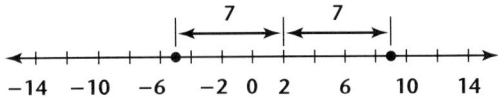

27. The distance between x and 3 is less than 4.
29. Any number between, but not including,
-1 and 7 **31.** The distance between x and 3 is
greater than 4. **33.** All numbers greater than
7 or less than -1. **35.** 53%; 47% **37.** yes;
Candidate A could receive a percentage of
the votes in the range $47\% \leq A < 50\%$.
39. $y = -x + 5$ is a line with a slope -1 and a
y-intercept, 5. $y = |-x + 5|$ is a V that opens
upward with a vertex at (5, 0). The portion of
$y = -x + 5$ above the x-axis is the left side of
the V. **41.** 7 **43.** 16 **45.** $63b + 18$
47. $x < 5$ **49.** $m < 12$

Lesson 5.1

7. positive **9.** neither **13.** $\frac{1}{7}$ **15.** 0 **17.** $-\frac{5}{2}$
19. $\frac{1}{4}$ **21.** 1 **23.** $\frac{1}{10}$ **25.** $-\frac{17}{59}$ **27.** 0
29. undefined **31.** 7.56 feet per mile **33.** The
slope of the line is $\frac{3}{1}$. **35.** 1.08 **37.** The
points fit the line very closely. The slope of the
line should be between 3.1 and 3.2. This ratio
of circumference to diameter is very near 3.14
or $\approx \pi$. **39.** Slope of the line of best fit is
approximately $\frac{4}{7} \approx 0.57$. There is approximately one representative per 570,000 people.
41. $\frac{d - b}{c - a}$ or $\frac{b - d}{a - c}$ **43.** little or no **45.** 7
47. -9 **49.** $12.25 **51.** 8

Lesson 5.2

9.

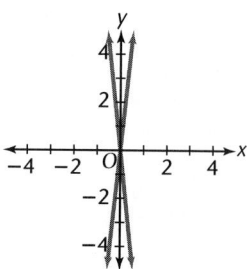

The lines both have y-intercept at $(0, 0)$. They have the same steepness but one has a positive slope and the other has a negative slope.

11.

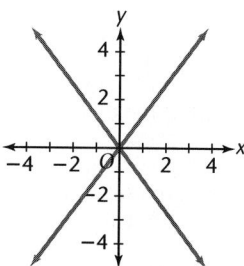

The lines both have y-intercept at $(0, 0)$. They have the same steepness but one has a positive slope and the other has a negative slope.

13.

The lines both have y-intercept at $(0, -7)$. They have the same steepness but one has a positive slope and the other has a negative slope.

15. $T = m + 30$

17. $y = \dfrac{8}{5}x$

19. $y = \dfrac{1}{2}x$

21.

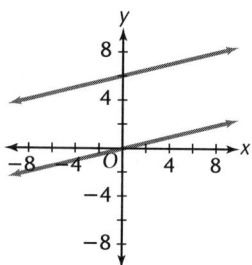

They have the same slope, $\dfrac{1}{4}$, and they have different y-intercepts , $(0, 0)$ and $(0, 6)$.

23. The line will have the same slope but different y-intercepts.

25. $l = 34 - 0.5d$; 16 days

27.

29.

31.

33.

35.

37.

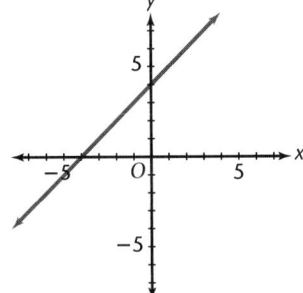

39. $E = 8n$; $264 **41.** $y = -\frac{3}{4}x + 6$
43. neither   **45.** positive

9. 4; When x is 0, the y-coordinate is where the line crosses the y-axis.   **11.** (0, 5)   **13.** (0, 7)
15. k **17.** $-\frac{6}{5}$ **19.** -1 **21.** $-\frac{4}{13}$ **23.** $-\frac{3}{2}$
25. $-\frac{4}{3}$ **27.** $y = -4x - 4$ **29.** $y = -5x + 7$
31. $y = -3x - 1$ **33.** $y = \frac{2}{3}x + 2$
35. $y = 3x + 3$ **37.** $y = \frac{9}{5}x - \frac{12}{5}$
39. $y = \frac{1}{2}x - \frac{7}{2}$ **41.** $y = 3x + 4$
43. $y = 6.50x - 43.80$

45.

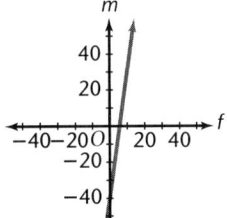

49. The population will reach 20,000 between 11 and 12 years.

51.

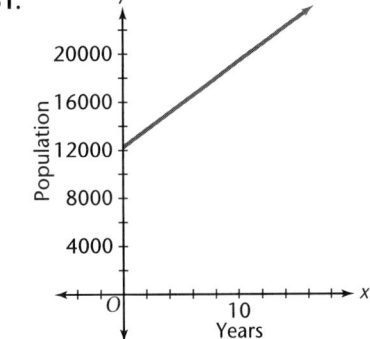

53. $x + 5y$ **55.** $x \geq 10$

7. $4x + 3y = 24$ **9.** $6x + 4y = -12$
11. $6x - 2y = 14$ **13.** $4x + 14y = -12$
15. (0, 10), (10, 0)   **17.** (0, 5), (4, 0)

19.

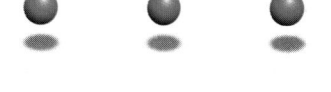

21. -3 **23.** x is the number of adult tickets, 122; 147 adult tickets, 294 student tickets
25. $y = -\dfrac{2}{7}x + 2$ **27.** $y = 2$ **29.** $x + y = 5$
31. slope-intercept form, $y = mx + b$
33. $y = 3x - 19$ **35.** $y = 3x - 6$
37. $y = 5x - 98$ **39.** $5(40)$ **41.** $3s + 5a$
43. 86 adult tickets sold **45.** October 1, 1908; 10-01-08 **47.** July 20, 1969; 690720 **49.** 6
51. 0 **53.** The data points are close to the line of best fit, sloping downward from left to right. The line of best fit has a negative slope.
55. 140 **57.** 216 **59.** $m = 4; y = 4x$
61. $m = \dfrac{7}{5}, y = \dfrac{7}{5}x$

Lesson 5.5

9. m is undefined; vertical **11.** $m = \dfrac{1}{5}$
13. Horizontal is related to the word horizon, the line where the sky meets the earth.
15. F **17.** A **19.** D

Slope-intercept Form	Standard Form
21. $y = 4$	$0x + 1y = 4$
23. $y = 4x$	$4x - 1y = 0$

25. horizontal **27.** horizontal **29.** Press *Graph, Draw,* and select *Vertical.* **31.** 4
33. more than 13 times **35.** season-pass option **37.** points **39.** 100 **41.** $8a + 9$
43. $-16 - 4p$ **45.** $1.09 **47.** $y = x + 2$

Lesson 5.6

9. The slope must be $-\dfrac{1}{5}$ to be perpendicular, but the y-intercept can vary. **11.** 0
13. undefined **15.** $5x - 2y = 25$
17. $3x - y = -13$ **19.** $y = 4$ **21.** 3 **23.** 2
25. 3 **27.** -1 **31.** $2(7 + 35) \div 7 - 10 = 2$
33. -3 **35.** $5x^2 + 7y$ **37.** $5x^2 + 10xy + 2x$

Lesson 6.1

Approximations may vary for 7–13.
7. $(34.3, -12.9)$ **9.** There is no solution since the lines are parallel. **11.** $(200, 100)$

13. $(2, 5)$ **15.** yes **17.** no **19.** $(4, 4)$
21. $\begin{cases} y = -120x + 3800 \\ y = 40x + 520 \end{cases}$

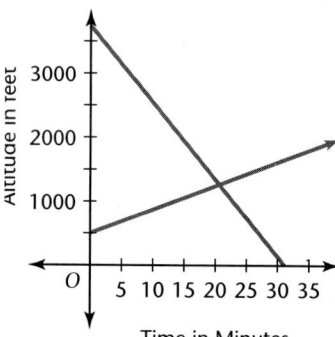

20.5 minutes

23. $y = -\dfrac{1}{3}x + 6$ **25.** $5 = 5$ True; $5 = 5$ True
27. The lines of best fit are $y = 301x + 2623.4$ and $y = -210x + 7667.4$ **35.** zero correlation
37. 3 **39.** $-\dfrac{3}{4}$ **41.** -3

Lesson 6.2

5. $(-2, 9)$ **7.** $(-14, 19)$ **9.** $(1, 3)$
11. $\left(\dfrac{1}{3}, \dfrac{1}{3}\right)$ **13.** $\left(\dfrac{1}{3}, \dfrac{1}{3}\right)$ **15.** $(-18, -28)$
17. $\left(\dfrac{57}{10}, \dfrac{78}{10}\right)$ **19.** $(4, 2)$ **21.** Let x represent the larger number and y represent the smaller number.
$\begin{cases} x + y = 27 \\ x = y + 3 \end{cases}$ $x = 15$ and $y = 12$

23. width is about 34.7 m; length is about 69.4 m **25.** $a + c = 210$ **27.** Graphing and substitution; Choose substitution because the variables in the second equation both have coefficients of 1. Graphing would involve very large numbers. **29.** 7 people; 53 coins
31. Tom; Sam; Rob; Mark; Joe **33.** $\dfrac{1}{5}$ **35.** 45
37. perpendicular **39.** perpendicular

Lesson 6.3

11. $(-4, 4)$ **13.** $\left(\dfrac{3}{4}, \dfrac{2}{3}\right)$ **15.** $\left(-\dfrac{49}{12}, -\dfrac{11}{2}\right)$
17. Price of 1 pizza is $6.99. Price of 1 bottle of soda is $0.79. **19.** There are an infinite number of solutions because both of the equations represent the same line. One example is $(1, -2)$. **21.** $(0, -1)$
23. $(-9, -2)$

25. Let p represent the overtime pay per hour in dollars and s represent the weekly salary in dollars. $\begin{cases} s + 10p = 240 \\ s + 7.5p = 213.7 \end{cases}$ Solve by elimination, $p = \$10.50$ and $s = \$135$. **27.** Let b represent the amount of money invested at the bank in dollars and s represent the amount of money invested in stocks in dollars.

$\begin{cases} b + s = 6000 \\ 0.05b + 0.09s = 380 \end{cases}$ Solve by elimination, $b = \$4000$ and $s = \$2000$. **29.** Let l represent the length of the rectangle in centimeters and w represent the width of the rectangle in centimeters. $\begin{cases} 2l + 2w = 24 \\ l = 3w \end{cases}$

Solve by substitution, $l = 9$ centimeters and $w = 3$ centimeters. **31.** $(-2, -4)$

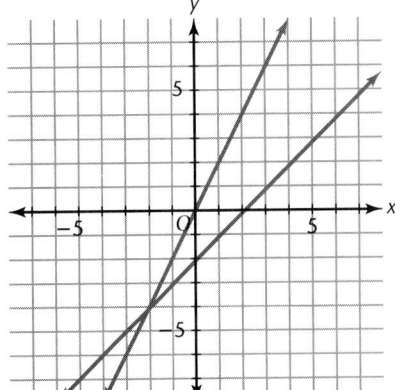

35. Add 3 to each term. 20, 23, 26 **37.** 31
39. c. exponential **41.** $\dfrac{7}{3}$ **43.** 2

Lesson 6.4

7. parallel lines, no solution **9.** $(-87, -66)$
11. $(-2, 5)$ **13.** $(2, 0)$ **15.** parallel lines, no solution **17.** independent **19.** dependent
21. independent **23.** inconsistent; The system is inconsistent because the slopes of both lines are equal and the y-intercepts are not equal. The lines are parallel. **29.** 13 **31.** One angle is 50 degrees and the other angle is 130 degrees.
33. Center City: $y = 500x + 40,070$, Bay City:
$y = 500x + 43,750$ **35.** $\begin{cases} -x + 4y = 7 \\ 2x + 7y = 16 \end{cases}$ $(1, 2)$
37. at the vertex **39.** 7 **41.** 10 **43.** -3

9.

11.

13.

15.

17. $2(4x) + 2(2y) \le 30$
  $2x + 2y \ge 12$

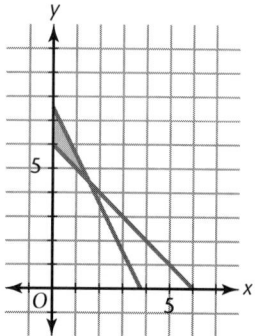

19. B **21.** A (3, 8); B (9, 5); C (8, 2); D (1, 2)

23.

Baskets	0	1	2	3	4	5	6	7	8
Free Throws	16	14	12	10	8	6	4	2	0

25. $l \ge w + 5$ **29.** 8; 8; 8 **31.** \$60
33. $8a + 2b^2 - b + z$ **35.** neither
37. perpendicular

Lesson 6.6

7. Creek's Current Rate, 1 mph; Ramona's Paddling Rate, 5 mph **9.** 26 **11.** May is 25 years old and David is 20 years old.
13. Sabrina is 12 years old, and Wymon is 32 years old. **15.** 5 **17.** 3 **19.** 16
21. $11x - 11y$ **23.** $-\dfrac{3}{4}$

Chapter

Lesson 7.1

5. 1; There was 1 mechanic absent on Thursday. **7.** mechanics

9.

	A	B	C	D
A	0	2	4	8
B	4	0	2	4
C	2	5	0	2
D	1	3	5	0

11. 4 rows by 3 columns
13.

	Atlantis	Columbia	Discovery	Total
1988	1	0	1	2
$L =$ 1989	2	1	2	5
1990	2	2	2	6
1991	3	1	3	7

1991
15. Chula Vista: 4,500; Abilene: 1,000
17. 1984 **19.** linear **21.** a. Property of Opposites **23.** e. Distributive Property
25. \$160 **27.** $x \le -4$ and $x \ge -4$

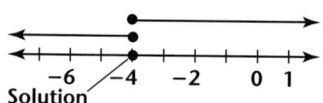

Solution

Lesson 7.2

9. $\begin{bmatrix} 10 & 6 \\ -5 & 17 \\ 13 & 5 \end{bmatrix}$ **11.** These 2 matrices cannot be subtracted because the first one has 2 rows and 3 columns, and the second one has 3 rows and 2 columns.

13. $\begin{bmatrix} 1.1 & -11.4 & 7.1 \\ -8 & -8.1 & 7.7 \\ -1.3 & 8.4 & 5.84 \end{bmatrix}$

15. $\begin{bmatrix} -4.3 & 1.1 & 7.8 \\ -4.7 & 16.7 & -1.8 \end{bmatrix}$

17. $\begin{bmatrix} 16 & 145 & 133 \\ -23 & -75 & -115 \\ -84 & 220 & 99 \end{bmatrix}$

19. 744 girls ; 756 boys **21.** 1822 students
23. Waters **25.** $-2; 6; 1; -15$ **27.** Yes, because all of the matrices have the same dimensions. The identity matrix would be $\begin{bmatrix} 0 & 0 \\ 0 & 0 \end{bmatrix}$.

29. $-A = \begin{bmatrix} -12 & -8 \\ -7 & 3 \end{bmatrix}$; $-B = \begin{bmatrix} 13 & 1 \\ 8 & 6 \end{bmatrix}$;

$-C = \begin{bmatrix} 1 & -7 \\ 1 & 9 \end{bmatrix}$

31. $(9, 9), (-9, 9)$
33. $4a + 2b$
35. $4(4x + 7)$
37. $10s; 6s^2$
39. $\left(2, -\dfrac{3}{2}\right)$

Lesson 7.3

7. Not Possible **9.** 4 by 1 **11.** Not Possible

13. $\begin{bmatrix} 15 & -5 & -20 \\ 10 & -10 & -30 \end{bmatrix}$ **15.** Not Possible

17. Not Possible **19.** Not Possible **21.** -24

23. $\begin{bmatrix} 32 & 48 & -24 \\ 4 & -4 & 2 \\ 12 & 14 & -7 \end{bmatrix}$ **25.** 3 by 2; 2 by 2;

2 by 1; AB, AC, and BC are possible. **27.** The same Associative Property of Multiplication that is true for numbers is also true for matrices; For example,

$\begin{bmatrix} 4 & 3 \\ 7 & 1 \end{bmatrix}\begin{bmatrix} 2 \\ 5 \end{bmatrix}$ = [8 6 2]. **29.** No; The Commutative Property of Multiplication does not apply to matrices; No **31.** [22];

$\begin{bmatrix} 3 & 6 & 9 \\ 2 & 4 & 6 \\ 5 & 10 & 15 \end{bmatrix}$ **33.** Sam $\begin{array}{c} \text{Cost} \\ \begin{bmatrix} 63.68 \\ 56.52 \end{bmatrix} \end{array}$

35. $\begin{array}{c} \text{Sales Tax} \\ \text{Cars} \begin{bmatrix} 8.25 \\ 7.50 \end{bmatrix} \\ \text{Trucks} \end{array}$

37.

	Cost	Sale Price
Freshmen	163.75	450.00
Sophomore	145.00	400.00
Junior	90.00	250.00
Senior	237.50	650.00

39. Answers may vary. Seniors have graduation activities and trips that require more funds to be raised. **41.** $12y^2 + 26y + 5$ **43.** -51

45. $\dfrac{2A}{b_1 + b_2}$

47.

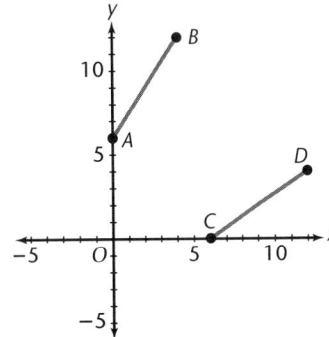

$y = -3x - 11$ **49.** $\dfrac{1}{2}$ mph

Lesson 7.4

7. $\begin{bmatrix} 3 & 7 \\ 2 & 5 \end{bmatrix}$ **9.** $\begin{bmatrix} -1.025 & 0.041 \\ -0.615 & 1.025 \end{bmatrix}$ or

$\begin{bmatrix} \frac{-125}{122} & \frac{5}{122} \\ \frac{-75}{122} & \frac{125}{122} \end{bmatrix}$ **11.** $\begin{bmatrix} 0 & 0.5 \\ -0.5 & -0.25 \end{bmatrix}$ **13.** Yes

15. No **17.** No **19.** None

21. $\begin{bmatrix} 0 & -0.2 & 0.2 \\ -0.2 & 0 & 0.2 \\ 1.2 & 1.2 & -1.4 \end{bmatrix}$ **23.** Answers

may vary. Yes, both of the products for $ad = bc$ for each matrix are equal. **25.** $-49 = -56$; This matrix has an inverse.

27.

The slopes are reciprocals (inverses) of each other. **29.** Answers will vary. For example,

$\begin{bmatrix} 4 & 2 \\ 6 & 7 \end{bmatrix} \cdot T = \begin{bmatrix} 2 & 4 \\ 7 & 6 \end{bmatrix}$.

The slope of each line segment is the inverse of the other. **31.** linear

33.

x	-3	-2	-1	0	1	2	3
$f(x)$	$-\frac{2}{5}$	$-\frac{1}{2}$	$-\frac{2}{3}$	-1	-2	undefined	2

Yes **35.** Addition Property of Equality; 78
37. Multipication Property of Equality; -60

39.

$C = \begin{array}{c} \\ \text{Job 1} \\ \text{Job 2} \end{array} \begin{array}{ccccc} \text{M} & \text{T} & \text{W} & \text{Th} & \text{F} \\ \begin{bmatrix} 8 & 4 & 4 & 8 & 6 \\ 4 & 12 & 8 & 10 & 5 \end{bmatrix} \end{array}$

Lesson 7.5

5. $\begin{bmatrix} -6 & 4 \\ 5 & -2 \end{bmatrix}\begin{bmatrix} x \\ y \end{bmatrix} = \begin{bmatrix} -12 \\ 10 \end{bmatrix}$

7. $\begin{bmatrix} 2 & 3 & 4 \\ 6 & 12 & 16 \\ 4 & 9 & 8 \end{bmatrix}\begin{bmatrix} x \\ y \\ z \end{bmatrix} = \begin{bmatrix} 8 \\ 31 \\ 17 \end{bmatrix}$ **9.** $(2, 0)$

11. c. $(-4, 12)$ **13.** 10 thousand **15.** 1982
17. $3.50, $1.75 **9.** Cashmere: $25.00;
Cotton: $15.00 ; Wool: $20.00 **21.** Kevin:
$5.25/hr; Doug: $4.75/hr **23.** $(-1, -4)$
25. 8 C-Group sites; 5 Pan-grave sites;
5 Transitional sites **27.** The first differences for
the y-values are 2, 4, 6, 8 and the second
differences are 2, 2, 2. Since the second
differences are constant, the function is
quadratic. **29.** $y = 5$ **31.** 8 ounces

Lesson 8.1

7. Toss 4 coins, recording the result for each
coin. One toss of all 4 coins represents 1 trial.
Repeat for 10 trials. Count the number of trials
where 4 heads or 4 tails show up. Divide this
number by 10. The quotient represents the
experimental probability. **9.** Yes, both Fred
and Ted will have the same result if all trials
were successful events or if all trials were
unsuccessful events. $\frac{15}{15} = \frac{16}{16}$ or $\frac{0}{15} = \frac{0}{16}$
11. $\frac{1}{4}$ **13.** $\frac{1}{10}$ **15.** $\frac{3}{5}$ Exercises 17–23
answers may vary. **17.** Toss a coin twice,
recording the result for each toss. Repeat for 10
trials. Divide the number of trials where 2 tails
show up by 10. **19.** Roll a number cube;
record the result. Repeat for 20 trials. Divide
the number of trials where 3 or 6 shows up by
20. **21.** Roll 2 number cubes; record the result.
Repeat for 40 trials. Divide the number of trials
where the same numbers show up on both
cubes by 40. **23.** Toss a coin 5 times, recording
the result after each toss. Repeat for 40 trials.
Divide the number of trials where heads and
tails alternate by 40. **25.** integers 0 and 1
27. integers 100 and 200 **29.** 10
31. INT(RAND*3) **33.** INT(RAND*6)+1
35. The correlation may be any value between
-1 and 1, but should be 0 if the numbers are
truly random.

37. 32 **39.** 33.725 **41.** $-\frac{1}{10}$ **43.** 57 hours

45. $\begin{bmatrix} 3 & 5 \\ 1 & -3 \end{bmatrix}\begin{bmatrix} x \\ y \end{bmatrix} = \begin{bmatrix} 2 \\ -4 \end{bmatrix}$; $\begin{bmatrix} -1 \\ 1 \end{bmatrix}$

Lesson 8.2

7. 13 **9.** 0; INT(RAND()*2) **11.** Let the
numbers 1 to 8 represent the number of
occurrences of rain, and the number 9 and 10
mean no rain. Generate 2 random numbers, let
the results represent the weather on day 1 and
day 2, respectively. Repeat for 10 trials. Divide
the number of trials where rain occurred on at
east one day by 10. The quotient is the
experimental probability. Exercises 13–15
answers may vary. **13.** Use a calculator to
generate a number from 1 to 7; Use
INT(RAND*7)+1. **15.** Use a calculator to
generate a number from 1 to 24; Use
INT(RAND*24)+1. **23.** the nth term is $4n$
25. $l = w + 74$

27.
```
←|—|ο|—|—|—|—|—|—|ο|—|—|→
  -10      0      10
```

29. $3x - 2y = 13$ **31.** yes

Lesson 8.3

9. 24 **11.** 14 **13.** 14.35 **15.** 2.9 **17.** Arrange in order, then find the mean or average
of the 2 middle values. **19.** 68 **21.** Impossible; one of the 2 remaining numbers must be
less than 18; 20 would not be the median.
23. Miami **27.** The averages dropped the first
year, then improved. The averages improved by
a mean of 0.006. **29.** The correlation, r, for
the National League is 0.52. The correlation for
the National League is weaker. **33.** 1.7 times
as large **35.** False **37.** 20.5% **39.** 1 more
41. 0.6 **43.** Houston-Juneau: 15°
  Houston-Duluth: 32°
  Juneau-Duluth: 17°
45. Let d represent the number of dimes.
$0.05(231) + 0.1d = 36.45$; $d = 249$ **47.** -1

Lesson 8.4

9. 2, 4, 6, 8, 10 **11.** 6 **13.** 4 **15.** 28 **17.** 19
19. 26 **21.** 5, 10, 15, 20 **23.** 15 **25.** 51
people **27.** row 29 **29.** 20 **31.** 361
33. 111 **35.** 97 **37.** 241 **39.** 198 **41.** 29
43. $h = \frac{3V}{\pi r^2}$ **45.** $x = \frac{3}{2}$

Lesson 8.5

7. 36 **9.** 5 **11.** 360 **13.** tar **15.** 120
17. 720 **19.** cares, races, acres, scare **21.** 6
23. 4 **25.** 50 **27.** The slopes are the same.

Lesson 8.6

7. $\dfrac{13}{25}$ **9.** $\dfrac{9}{25}$ **11.** $\dfrac{16}{25}$ **13.** $\dfrac{3}{13}$ **15.** $\dfrac{4}{13}$

17. $\dfrac{2}{9}$ **19.** experimental **21.** 4.86 or about 5

times **23.** $\dfrac{1}{9}$ **25.** $\dfrac{1}{4}$ **27.** As x approaches 0,
the graph rises and approaches the y-axis. As x
gets larger, the graph decreases and approaches
the x-axis. **29.** If $b = \dfrac{a}{0}$ and $a \neq 0$, then there
is no value of b such that $b \cdot 0 = a$.

31. $\dfrac{3.18}{53.00} = \dfrac{x}{100}$, $x = \dfrac{3.18}{53.00}(100)$, $x = 6$; The
sales tax is 6%. **33.** 32 **35.** 21

Lesson 8.7

11. $\dfrac{1}{3}$ **13.** $\dfrac{1}{12}$ **15.** $\dfrac{4}{25}$ **17.** $\dfrac{6}{25}$ **19.** The
outcome of the second toss is independent of
the outcome of the first toss. **21.** For two
heads to occur in a row, the first toss must be a
head. Therefore, the outcome of the first toss
affects the success of the event. **23.** $\dfrac{1}{6}$ **25.** $\dfrac{2}{3}$
27. $\dfrac{1}{6}$ **29.** $\dfrac{4}{9}$

31.

x	-2	-1	0	1	2
y	-2	-4	undefined	4	2

reciprocal function

35.

x	-2	-1	0	1	2
y	8	4	0	4	8

absolute value function
35. 20 **37.** 59 **39.** 15

Lesson 9.1

7. function; All of the x-coordinates are
unique. **9.** function; Any vertical line drawn
through the graph will only intersect it once.
11. D: 3, 4, 5; R: 4 **13.** D: $-2 \leq x \leq 2$;
R: $0 \leq y \leq 4$ **15.** 10 **17.** camping and hiking
19. None **21.** Grand Canyon **23.** Grand
Canyon, Mesa Verde, Rocky Mountain, Chaco
Culture, Arches, Bryce Canyon, Canyonlands,
Capitol Reef and Zion **25.** Grand Canyon
27. function **29.** not a function **31.** D: 0.1,
0.2, 0.3; R: 1, 2, 3 **33.** D: all numbers;
R: $-1 \leq y \leq 1$ **35.** 0 **37.** -30 **39.** 8 **41.** $\dfrac{1}{3}$
43. satisfies **45.** satisfies **47.** satisfies
49. yes; yes **51.** 2 **53.** $B = \dfrac{2A}{h} - b$

55. $5x - 3y = -15$ **57.** $\begin{bmatrix} 2 & 5 \\ 1 & 3 \end{bmatrix}$ **59.** $\dfrac{1}{3}$

Lesson 9.2

7. $y = x^2$; reflection through the x-axis
9. $y = 2^x$ or $y = 10^x$; reflection through the
x-axis
11.

13.

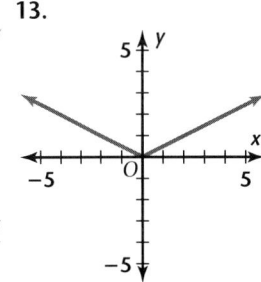

shifted vertically by 4   stretched vertically
by 0.5

15.

17.

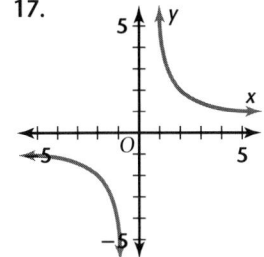

shifted vertically   stretched vertically
by -3 by 4

Selected Answers   **743**

19.

21.

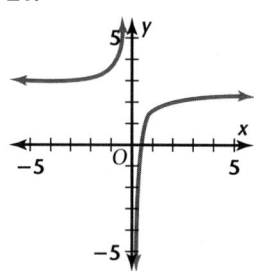

shifted vertically by $\frac{1}{3}$ reflected through the x-axis and shifted vertically by 3

23.

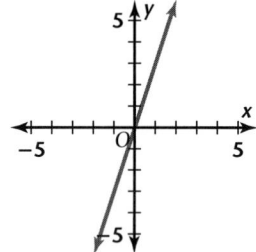

$y = x$; stretched vertically by $\sqrt{10}$
25. $y = |x|$; reflected through the x-axis and shifted horizontally by 2

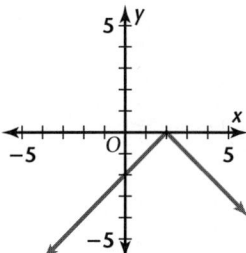

27.

x	$A = (x + 4)^2$
0	16
-1	9
-2	4
-3	1
-4	0
-5	1
-6	4

29. $A = (x + 4)^2$ is the graph of $A = x^2$ shifted 4 units to the left.   **31.** vertical stretch of 2   **33.** reflected through a vertical line   **35.** 38, 51, 83   **37.** 5%   **39.**

41. (1, 3)   **43.** no solution; inconsistent set of equations; the lines are parallel

Lesson 9.3

7. $y = x^2$ **9.** The graph of $y = 5x^2$ is a stretch of $y = x^2$, by a factor of 5.

11.

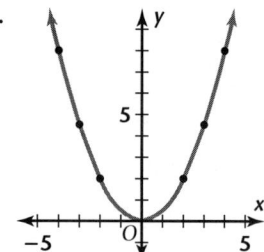

13. true   **15.** $y = \frac{1}{x}$ **17.** The graph of $y = \frac{1}{2x}$ comes closer to the x-axis than $y = \frac{1}{x}$ because its y-values are $\frac{1}{2}$ as large.   **19.** no   **21.** yes
23. $y = \frac{1}{4} x^2$ **25.** 1.05
27. a. $y = 2^x$ b.

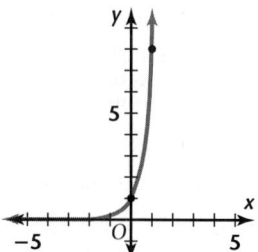

c. The graph of $y = 2^{3x}$ has been stretched by a factor of 3.

29.

31.

TEST SCORE DISTRIBUTION

Score

Each bar is shifted to the right. Although grades of 80 become 100 in both cases, grades less than 80 are not shifted as far to the right as when 20 points were added to each graph.
33. less experienced worker: $5.25; more experienced worker: $5.55 **35.** (5, −3)
37. infinitely many solutions: dependent system of equations **39.** 1H, 1T, 2H, 2T, 3H, 3T, 4H, 4T, 5H, 5T, 6H, 6T **41.** no; $x = 1$ is paired with two y-values, 0 and −1.

Lesson 9.4

9. y-axis **11.** horizontal

13.

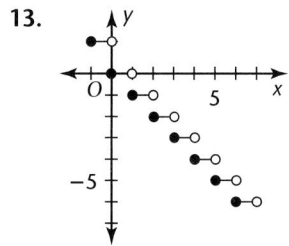

15. They are the same because $-\left(\dfrac{1}{x}\right)$ is the same as $\dfrac{1}{-x}$.

17.

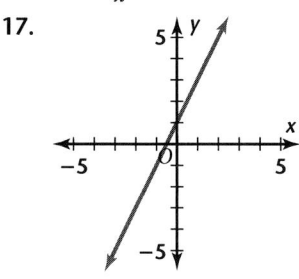

19. − 7; vertical **21.** $(x, y) \rightarrow (-x, -y)$ **23.** It becomes (b, a). **25.** Since the y-values are the same for a given x on either side of the y-axis, plot a y-value for a positive x and use that same y-value for negative x. **27.** 243 m^2 **29.** 22

31.

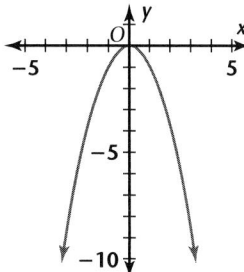

33. 24

Lesson 9.5

7. $y = x^2$; vertical reflection;

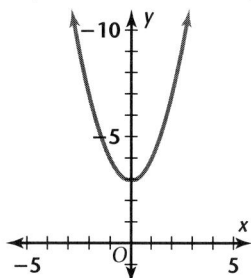

9. $y = x^2$; translation up 3 units;

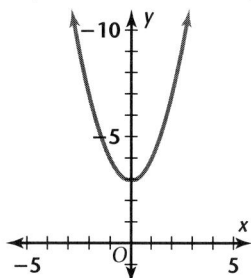

11. $y = \text{INT}(x)$; stretched vertically by $\dfrac{1}{2}$

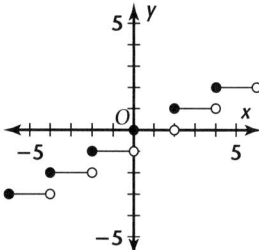

13. $y = \text{ABS}(x)$; translation down 1 unit;

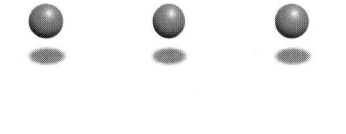

15. translated up 5 units **17.** translated left
5 units **19.** The *y*-value changes to 14 giving
(5, 14). **21.** The *y*-value changes to -2 giving
(5, -2). **23.** The *x*-value changes to 8 giving
(8, 8) **25.** The *x*-value changes to -7 giving
(-7, 8). **27.** stretched **29.** The average salary
will increase by 10% making it $27,500.
31. $7x - 34$ **33.** 6 **35.** Answers may vary.
37. $y = 3x^2$; Answers may vary. For example,
from examining the first 2 points, it seems that
y could be 3*x*. The third point, however, makes
this impossible. Divide both the third and
fourth *y*-values by 3 and it becomes apparent
that what remains is the *x*-value multiplied by
itself. Hence, *y* could be $3x^2$. Substituting this
for the fifth and following points confirms that
this is the correct answer.

39.

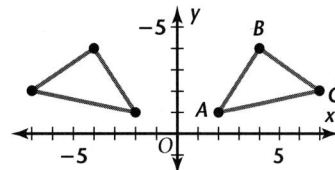

(-2, 1), (-4, 4), and (-7, 2)

Lesson 9.6

7.

9.

11.

13.

15.

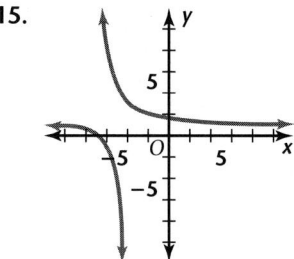

17. (4, -20) **19.** (9, -30). **21.** $y = 3x^2 + 15$
23. $y = -5x + 6$

25. Graphs may vary.

27.

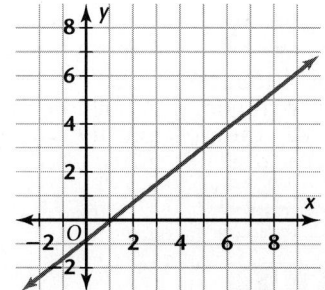

First write the equation in slope-intercept form
giving $y = \frac{4}{5}x - \frac{3}{5}$. Plot the *y*-intercept of
$\left(0, -\frac{3}{5}\right)$. Apply the slope of $\frac{4}{5}$ (up 4 units and
right 5 units) to this point to generate a second

point $\left(5, \dfrac{17}{5}\right)$ Draw a line through these two points.

29. $(1, -2)$ **31.** $(1, -2)$

Chapter

Lesson 10.1

11. Decimal form $1000 \cdot 100{,}000 = 100{,}000{,}000$; Exponential form $10^3 \cdot 10^5 = 10^8$
13. Decimal form $81 \cdot 9 = 729$; Exponential form $3^4 \cdot 3^2 = 3^6$

	Form	Num	Denom	Quot
15.	Decimal	1,000,000	1000	1000
	Exponential	10^6	10^3	10^3
17.	Decimal	243	81	3
	Exponential	3^5	3^4	3^1

19. 4 **21.** 2^7 **23.** 2^3 **25.** 2^{10} **27.** 4
29. 100.1 **31.** c^3 **33.** 6 **35.** 10,100
37. 9900 **39.** $10^5 = 100{,}000$ **41.** 59,049
43. 0.0625 **45.** two trillion, thirty billion, four hundred million **47.** three trillion **49.** 10^1 or 10 **51.** 10^3 **53.** The distribution is fairly flat.

55. The calculator shows 20^{10} as 1.024 E13. Since 20^{10} is the same as $10^{10} \times 2^{10}$, it is 10^{10} times larger than the previous answer, or 10,240,000,000,000. Thus, 1.024 E13 means 1.024×10^{13}. **57.** $x > 2$ or $x < -6$
59. $A^{-1} = \begin{bmatrix} 1 & 1 \\ 2 & 3 \end{bmatrix}$; $A \cdot A^{-1} = \begin{bmatrix} 1 & 0 \\ 0 & 1 \end{bmatrix}$ **61.** 252
63. $y = 2(x + 3)^2$ is a parabola shifted 3 units to the left and stretched vertically by a factor 2.

Lesson 10.2

7. $\dfrac{a^3}{b^3}$ **9.** $32r^5$ **11.** $-8a^7$ **13.** $10a^6$
15. $-9.6a^5b^2c^5$ **17.** $2r$ **19.** $\dfrac{-a^3}{2}$ **21.** $\dfrac{2a^2b^6}{5}$
23. $\dfrac{-40a^3b^5}{c^3}$ **25.** $40u^2v^7w^2$ **27.** $243b^{10}$
29. $-1000m^{12}$ **31.** $9x^2y^8z^{10}$ **33.** $250b^{10}$
35. $343j^6$ **37.** $v^{20}w^6$ **39.** $-1; 1; -1; 1; -1$
41. $4s^2$ **43.** 100 **45.** 108 **47.** 12.5
49. 96.04 **51.** 105.301 **53.** 10.37
55. $(10^2)^3 = 100^3$ or $(10^2)^3 = 10^{2\cdot3} = 10^6$, not 10^8 **57.** $2k + 3$ **59.** $(6.4, 0)$ **61.** 3 **63.** 3

Lesson 10.3

7. $-\dfrac{1}{9}$ **9.** $\dfrac{1}{9}$
11.

Decimal form	10,000	1000	100	10	1	0.1	0.01
Exponential form	10^4	10^3	10^2	10^1	10^0	10^{-1}	10^{-2}

13. $\dfrac{1}{2^3}$ **15.** $\dfrac{a^3}{b^2}$ **17.** $\dfrac{w^2}{y}$ **19.** r^4 **21.** m^5
23. $5a^9$ **25.** 0.396; 0.264; 0.088; 0.015; 0.001
27. 5 **29.** mentally: subtract exponents to obtain 2.56^1 or 2.56. **31.** mentally: any non-zero power of 0 is 0, so $0^7 = 0$. **33.** mentally: the 0 exponent makes the value of the product 1. **35.** On some calculators an error code appears because the value is too large. On other calculators the answer is given in scientific notation. **37.** A horizontal line, lying on the x-axis, beginning just to the right of the origin. It suggests that the value of 0^0 should be 0.

39. 43, 56, 71 **41.** $y = \dfrac{-c}{ab - b}$ If $a = 5$, $b = 3$, and $c = 24$, then $y = \dfrac{-c}{ab - b} = -2$, the solution for Exercise 40. **43.** vertical: undefined slope; horizontal: slope 0.   **45.** $A = 243x^6y^4$; $B = 432x^4y$

Lesson 10.4

7. 8×10^9 **9.** 5.8×10^4 **11.** 5×10^{-7}
13. 8.2×10^{-8} **15.** 400,000,000   **17.** 901,000
19. 0.000000005   **21.** 0.00000000072
23. 5×10^{12} **25.** 3.7×10^6 **27.** 6×10^{14}
29. 6.3×10^{13} **31.** 2×10^6 **33.** 3.8×10^{18}
35. 4.5×10^4 **37.** 4×10^0 **39.** 4.8×10^{10}
41. trillion   **43.** Answers may vary. 1 year = 3.1536×10^7 seconds   **45.** 5,870,000,000,000
miles   **47.** 4.088 years   **49.** 7.6×10^4
51. 8×10^6 **53.** 8.0000×10^6 **55.** 2×10^{-3}
57. 1×10^{100} **59.** 5600   **61.** 0.000101
63. -0.0009303 **64–67.** Answers may vary
depending on your calculator: check its
manual.   **69.** 20   **71.** $41x - y - 8$

73.

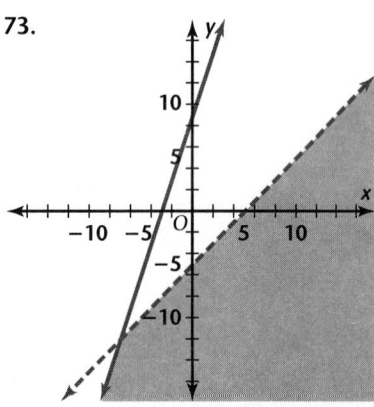

The solution is all the points below the line
$y = x - 5$ and to the right of $y = 3x + 9$.

Lesson 10.5

7.

9.

11.

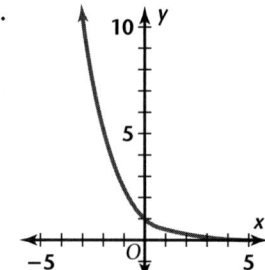

13. $b > 1$

15.

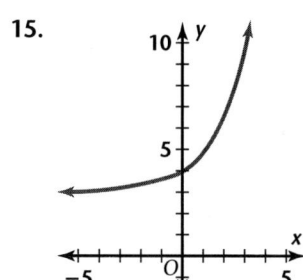

$y = 2^x + 3$ shifts the parent graph $y = 2^x$
vertically up 3 units.

17.

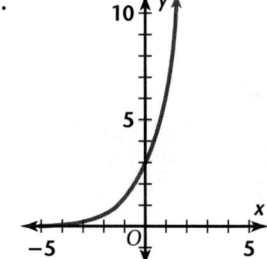

$y = 3 \cdot 2^x$ stretches the parent graph vertically
by a factor of 3.   **19.** no; variable is not in the
exponent   **21.** 8   **23.** $\frac{1}{2}$ **25.** 1.003   **27.** about
39.95 million   **29.** about 58.174 million
31. about 58.031 million   **33.** about 58.965
million   **37.** $\begin{bmatrix} -59 & 24 \\ -40 & -67 \end{bmatrix}$ **39.** $-f(x)$ is a
reflection of $f(x)$ through the x-axis.   **41.** -100
43. $\frac{1}{100}$

Lesson 10.6

7. about 11,400 years

9.

11. 359,477 **13.** \$12,048 **15.** \$5581
17. \$84,969 **19.** \$103,991 **21.** 66 **23.** $\frac{97}{100}$
25. 0.03 **27.** 0.0003 **28.** 3 **29.** 6

31.

shifts parent graph up 5 units

33.

stretched vertically by a factor of 5
35. -10 **37.** INT(RAND*2) or INT(RAND( )*2)
39. 6 students **41.** $y = 2^x$ increases from left to right; $y = 2^{-x}$ decreases from left to right.

Lesson 11.1

7.

9. $c^3 + c + 6$ **11.** $p^7 + 10$ **13.** 4 **15.** binomial; linear **17.** trinomial; quadratic
21. $7x^4 + 4x^2 - x - 5$ **23.** $4r^4 + 2r^3 + r^2 - 6$
25. $y^3 + y^2 - 6$ **27.** $s^3 + 4s^2 + 6s - 7$
29. $x^3 - x + 7$ **31.** $c^3 + c^2 + 10c + 6$

33. $y^2 + 4$ **35.** $x^2 - 4x + 7$ **37.** $-x^4 - 3x^2 - 3x + 10$ **39.** $3x^5 - 3x^3 + 2x^2 - 6x - 10$
41. $16x + 12$ **43.** $6x + 2$ **45.** $2x^2 - 4x + 6$
47. $(1, -1)$ **49.** $6x^3 - 24x^2$ **51.** $6x^9$
53. 8.9×10^9

Lesson 11.2

5. $x^2 + 7x$

7. $2x$ and $3x + 1$; $6x^2 + 2x$ **9.** $2x$ and $x + 5$; $2x^2 + 10x$ **11.** 2 and $3x + 1$; $6x + 2$
13. $4x + 8$ **15.** $5y + 50$ **17.** $x^2 + 2x$
19. $3r^3 - 9r$ **21.** $24x - 32$ **23.** $20x^2 + 45x$
25. $2y^4 + 2y^3$ **27.** $9y^3 - 18y^2$ **29.** $3(x^2 + 2)$
31. $y^2(1 - y)$ **33.** $7y(y^3 + 7)$ **35.** $-2r(2r + 3)$
37. $10x$ yd^2 **39.** $(10x + 250)$ yd^2
41. $0.25(1.5x + 2.5y)$ or $0.375x + 0.625y$
43. $x = 2$ **45.** 128 **47.** $2a^2 - 12ab$
49. $9y^3 - 27y^2$ **51.** 1×10^{-10}

Lesson 11.3

5. $x^2 + 7x + 10$ **7.** $b^2 - 4b + 3$ **9.** $c^2 + 10c + 25$ **11.** $y^2 + 8y + 15$ **13.** $b^2 - 4b - 21$
15. $5p^2 + 8p + 3$ **17.** $4x^2 + 4x - 15$
19. $6w^2 - 43w + 72$ **21.** $14s^2 - 17s - 6$
23. $y^2 + \frac{2}{9}y - \frac{5}{27}$ **25.** $4a^4 + 12a^2 + 9$
27. $p^2 + 2pq + q^2$ **29.** $x^3 + x^2y + xy + y^2$
31. $0.96m^2 - 0.8m - 20$ **33.** $x^2 + 4x - 32$
35. the square; 1 unit **37.** 6 **39.** 7 **41.** 4
43. $(4, 1)$;

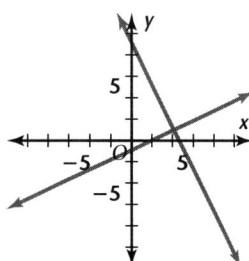

Lesson 11.4

9. prime **11.** $2(x^2 - 2)$ **13.** $3x(x + 2)$
15. $k^2(k^3 + 1)$ **17.** $2(2x^2 + x - 3)$
19. $9y(3y^2 + 2y - 9)$ **21.** $15(6 + a^5 - 3a)$
23. $(x + 2)(x + 1)$ **25.** $(a + b)(x + y)$
27. $(x + 2)(x - 1)$ **29.** $(5a + 4)(a - 3)$
31. $(x - 2)(2 - x)$ or $-(x - 2)^2$ **33.** $(2r - 3)$ $(r - s)^2$ **35.** $(2 + a)(x + y)$ **37.** $(3a - 2)$

Selected Answers **749**

$(4b - 5)$ **39.** $15(x + y)$ **41.** $2(n + 7)(n - 3)$
43. $5(x - 2)(2d + 3)^3$ **45.** 122.522 **47.** no;
The first difference is not constant. **49.** $\dfrac{2A}{b_1 + b_2}$
51. $x - 4y = 8$

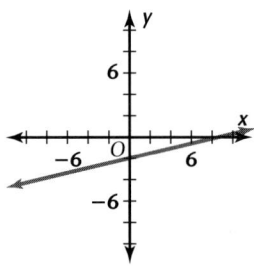

53. $\dfrac{4}{13}$ **55.** $y = 3x - 19$

Lesson 11.5

5. $4x^2 - 4x + 1$ **7.** $49y^2 - 81$ **9.** $25z^2 - 120z + 144$ **11.** $24y$ **13.** $30x$ **15.** $4y^2$
17. $(x + 2)^2$ **19.** $(y + 4)^2$ **21.** $(2t + 1)(2t - 1)$
23. $(5x + 3)(5x - 3)$ **25.** $(2x - 5)^2$
27. $(6d + 1)^2$ **29.** $(3c + 2d)(3c - 2d)$
31. $(3a - 2)^2$ **33.** $(ax + b)^2$
35. $9(3a^2 + b)(3a^2 - b)$ **37.** $(x + 2)(x - 2)$
$(5 + x)(5 - x)$ **39.** $(3x + 5)(x + 2)(x - 2)$
41. The rectangle formed has dimensions $a + b$
and $a - b$, therefore, comparing areas,
$a^2 - b^2 = (a + b)(a - b)$.

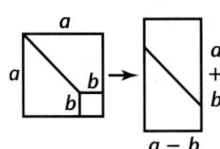

43. 16 **45.** $(x + 4)(x - 4)$ **47.** The area of
the rectangle is $(x + 4)(x - 4)$ or $x^2 - 16$ which
is the area of the original shaded area.
49. $4n - 24$ **51.** πR^2; $\pi R^2 - \pi r^2$; $\pi(R + r)$
$(R - r)$ **53.** 95 **55.** $-\dfrac{4}{3}$

57.

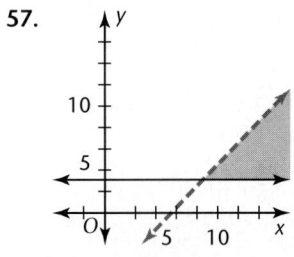

59. $\dfrac{3}{4}$ **61.** no **63.** vertex

Lesson 11.6

9. $(x + 2)(x + 3)$ **11.** $(x + 3)(x + 8)$; same
13. $(x + 1)(x - 36)$; opposite **15.** 1, 54; 2, 27;
3, 18; 6, 9; [circle 3, 18] **17.** 1, 144; 2, 72; 3,
48; 4, 36; 6, 24; 8, 18; 9, 16; 12, 12;
[circle 8, 18] **19.** $(p + 6)(p - 2)$ **21.** $(b - 8)$
$(b + 3)$ **23.** $(z - 4)(z + 5)$ **25.** $(s - 21)(s - 3)$
27. $6y(y - 1)(y - 2)$ **29.** $(a + 6)(a - 1)(a + 3)$
31. $(x + 6)(x - 6)(x + 2)$ **33.** $16(2p^2 + 1)$
$(2p^2 - 1)$ **35.** $(x - 1)^2$ 37. $(a + b)(2x + y)$
39. 484 **41.** $93\dfrac{7}{9}$ **43.** $y = \dfrac{1}{2}x - 3$

45. $\begin{bmatrix} 3 & 1 \\ 2 & -1 \end{bmatrix} \begin{bmatrix} x \\ y \end{bmatrix} = \begin{bmatrix} 7 \\ 3 \end{bmatrix}$; 2, 1

Lesson 12.1

7. yes; the second differences are constant + 2.
9. Shift right 5 units, down 2 units. Also 3
times steeper. **11.** Shift left 1 unit. **13.** Shift
right 2 units, up 3 units. Also half as steep.
15. V(2, 3); $x = 2$ **17.** V(5, 2); $x = 5$

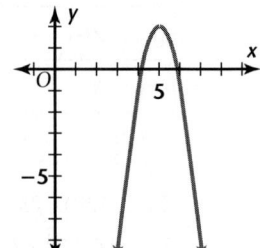

19. V(−5, 7); $x = −5$

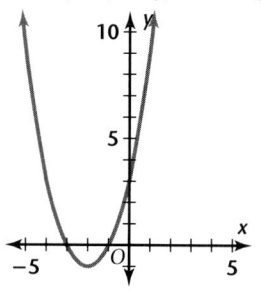

21. 10 **23.** $1, -7$ **25.** $4, -6$ **27.** $1, -4$
29. zero at $x = -9$; axis of sym. $x = -9$; min $= 0$ **31.** zeros at 2 and 3; axis of sym. $x = 2.5$; min $= -0.25$ **33.** 5 seconds
35. $x = 5$ **37.** 75 m **39.** $\begin{bmatrix} 4 & 2 \\ 11 & 9 \end{bmatrix}$
41. $\begin{bmatrix} 3 & 15 \\ 3 & -8 \end{bmatrix}$ **43.** $(x + 7)^2$

Lesson 12.2

9.

t	$-14t^2 + 300$
6	-204
5	-50
4	76
3	174
2	244
1	286
0	300
-1	286
-2	244
-3	174
-4	76
-5	-50

crosses the x-axis at about 4.63 and -4.63.
11. 12 **13.** 6 **15.** 9.49 **17.** 4.36 **19.** ± 13
21. ± 25 **23.** ± 4.90 **25.** ± 7.35 **27.** $\pm \dfrac{7}{11}$
29. ± 0.2 **31.** ± 6.71 **33.** ± 7.94 **35.** 2, 8
37. $4.45, -0.45$ **39.** $4.41, 1.59$ **41.** $14, -10$
43. $4.32, -2.32$ **45.** $-2.13, -9.87$
47. $V(-2, -1)$; $x = -2$; zeros $-3, -1$

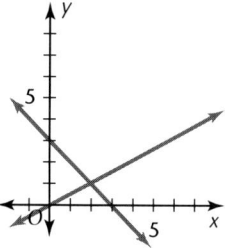

49. $(-4, -4)$ **51.** $-2, -6$

53.

55. ≈ 2.68 m **57.** ≈ 6.12 seconds
59. $\begin{bmatrix} 62 & 70 \\ 20 & 28 \end{bmatrix}$ **61.** b^2c^3 **63.** $5x^2 + 2x + 7$
65. $(a + b)(a - b)$; $(a^2 + b^2)(a + b)(a - b)$; $(a^4 + b^4)(a^2 + b^2)(a + b)(a - b)$; 7; 64 is 2^6, so $a^{64} - b^{64}$ has $6 + 1$ factors. **67.** 0

Lesson 12.3

7. $x^2 + 14x + 49$ **9.** $x^2 + 8x + 16$ **11.** $x^2 + 6x + 9$ **13.** $x^2 + 12x + 36$ **15.** $x^2 + 7x + \dfrac{49}{4}$
17. $x^2 + 15x + \dfrac{225}{4}$ **19.** $x^2 + 16x + 64$
21. $x^2 - 9x + \dfrac{81}{4}$ **23.** $y = (x + 4)^2 - 16$
25. $y = (x - 5)^2 - 25$ **27.** $y = (x - 8)^2 - 64$
29. $y = (x + 5)^2 - 25$ **31.** minimum at -25
33. -4 **35.** -1 **37.** -5 **39.** $y = (x - 1)^2 - 4$; $V(1, -4)$ **41.** $y = \left(x - \dfrac{1}{2}\right)^2 + 1\dfrac{3}{4}$; $V(0.5, 1.75)$ **43.** $y = \left(x + \dfrac{1}{6}\right)^2 - 3\dfrac{1}{36}$; $V\left(-\dfrac{1}{6}, -3\dfrac{1}{36}\right)$ **45.** $V(6, 576)$ **47.** 6 seconds
49. 1 second **51.** 4 seconds **53.** $(2, 1)$

55. $\dfrac{1}{2}$ **57.** m^7 **59.** $x^2 + 8x + 16$

Lesson 12.4

9. $4, -2$ **11.** 2 **13.** $1, -5$ **15.** 4, 6 **17.** 3, 5
19. -5 **21.** $5, -3$ **23.** $5, -4$ **25.** $6, -2$
27. $\approx 1.45, \approx -3.45$ **29.** $\approx -2.27, \approx -5.73$

31. 4, 6 **33.** 2, − 6 **35.** ≈ 3.30, ≈ − 0.30
37. 0, 1 **39.** 5, − 1 **41.** 6, − 2
43. (3, 4), (− 1, 4) **45.** (0, 3), (5, 8) **47.** 10 yards by 6 yards **49.** 2 rows, 2 columns
51. 0.4 seconds and 4.6 seconds

53. $\begin{bmatrix} 5 & -2 \\ 3 & 5 \end{bmatrix} \begin{bmatrix} x \\ y \end{bmatrix} = \begin{bmatrix} 11 \\ 19 \end{bmatrix}$

55. mean ≈ 14.3, mode 10.5 **57.** 65,000,000
59. $2(2b^2 + 3b - 18)$ **61.** $8p(p^3 - 2)$

Lesson 12.5

7. $a = 1, b = -4, c = -5$ **9.** $a = 5, b = -2,$
$c = -1$ **11.** $a = 3, b = 0, c = -45$ **13.** 13
15. 12 **17.** 128 **19.** 89 **21.** 0 **23.** 2, − 8
25. 0, 4 **27.** 3, − 3 **29.** 1, − 0.33 **31.** 1.5,
− 1 **33.** − 0.55, − 5.45 **35.** 1.90, − 1.23
37. 2, − 6 **39.** 0.5, − 3 **41.** 4, − 1.5
43. 2.70, − 0.37 **45.** 0.44, − 3.77 **47.** 27 seats in 43 rows **49.** about 0.1035 **51.** 6.7×10^4
53. 9×10^{-7}

55.

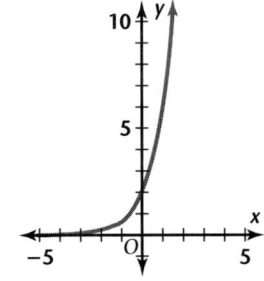

57. $x^2 - 2x - 15$

Lesson 12.6

9. $x < 2$ or $x > 4$

11. $2 \le x \le 7$

13. $1 < x < 6$

15. $3 \le x \le 5$

17. $x \le 5$ or $x \ge 6$

19. $x < -7$ or $x > 2$

21.

23.

25.

27.

29.

31.

33.

35.

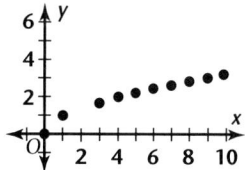

37. between 4 and 16 seconds   **39.** $-a^3b^3c^4$
41. 8×10^{-1} **43.** 7.4×10^{-6} **45.** $(a+3)$
$(2x+b)$

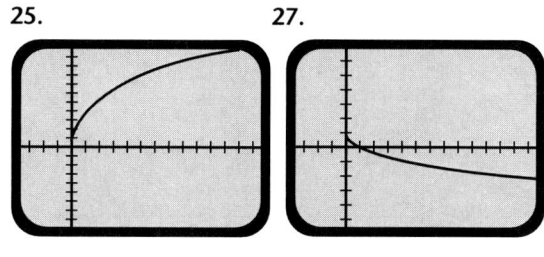

Lesson 13.1

7. 15   **9.** 3.32   **11.** -6.32 **13.** 31.62
15. -0.2 **17.** 15.81 meters   **19.** 5.29 miles
21. 211 miles, approximately   **23.** The square
root of a negative number squared is the
positive value of the number.

25.   **27.**

29.

31. a.   b.

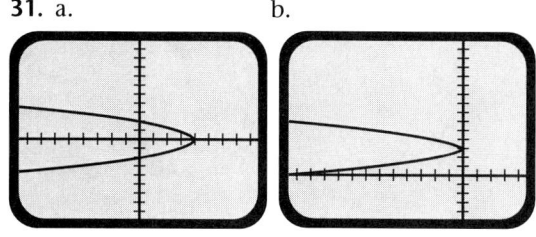

a. (4, 0); domain: $x \le 4$, range: all numbers.
b. (0, 3); domain: $x \le 0$, range: all numbers.
33. $x \ge 0$ **35.** $x \ge -5$ **37.** The graph of the
parent function is shifted horizontally to the
left by 3 units, then shifted vertically down 4
units, and stretched upward by a scale factor of
5.   **39.** $y = \sqrt{x+3}$ **41.** $y = \sqrt{x-2}+6$
43.

45. no   **47.** yes

49.   **51.**

53.   **55.**

Lesson 13.2

7. 7 **9.** 24 **11.** $5\sqrt{3}$ **13.** $18\sqrt{5}$ **15.** false, if $a \neq 0$ or $b \neq 0$; true if $a = 0$ and/or $b = 0$
17. true **19.** 6 **21.** 12 **23.** 2 **25.** $\dfrac{5}{2}$

27. $5\sqrt{3}$ **29.** $\sqrt{7}$ **31.** a^2b^3 **33.** $\dfrac{p^4\sqrt{p}}{q^5}$ **35.** $7\sqrt{5}$
37. $\sqrt{7} + \sqrt{29}$ **39.** $2\sqrt{3}$ **41.** $2 + \sqrt{2}$ **43.** 45
45. $6\sqrt{2}$ **47.** $3\sqrt{5} + 27$ **49.** $6\sqrt{6} + 6\sqrt{3}$
51. $-5 - 2\sqrt{3}$ **53.** $10 + 4\sqrt{6}$
55.

Area of Square	6	12	24	48	96
Length of Side	$\sqrt{6}$	$2\sqrt{3}$	$2\sqrt{6}$	$4\sqrt{3}$	$4\sqrt{6}$

57. (7, 13) **59.** x^3y^4 **61.** $4x^2 - 16x + 16$
63. $18b^2 - 3b - 1$ **65.** $1, -15$

Lesson 13.3

7. 9 **9.** 18 **11.** 20 **13.** 2 **15.** 2, 3 **17.** 1
19. 2 **21.** 2

23.

no solution
25. 0.81 feet **27.** 9 **29.** $v = \dfrac{\sqrt{2Em}}{m}$
31. $3\sqrt{10}, -3\sqrt{10}$ **33.** $3, -3$ **35.** 4
37. $3, -3$ **39.** $9, -9$ **41.** not possible
43. $W = \sqrt{W_l W_r}$ **45.** 12 pounds
47. $\begin{bmatrix} 0.175 & -0.025 \\ 0.05 & -0.15 \end{bmatrix}$

49.
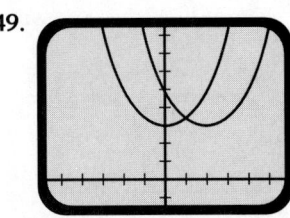
Both are parabolas that open upward with a vertex of (0, 3) and the other has a vertex of (2, 3). They intersect at (1, 4). Second graph is shifted 2 right. **51.** $4(x^2 - 2y^2)(x^2 + 2y^2)$

53. $x < -3$ or $x > 4$

55. $-2 < a < 8$

Lesson 13.4

	Leg	Leg	Hypotenuse
7.	24	45	51
9.	15	8	17
11.	12	21.9	25
13.	30	40	50

15. 10 **17.** 2.45 **19.** \$168.32 **21.** about 7.07 inches
23. **Equilateral Triangles**

Sides	4	6	8	10
Half of a Side	2	3	4	5
Altitude	$2\sqrt{3}$	$3\sqrt{3}$	$4\sqrt{3}$	$5\sqrt{3}$

Sides	12	20	n	34
Half of a Side	6	10	$\dfrac{n}{2}$	17
Altitude	$6\sqrt{3}$	$10\sqrt{3}$	$\dfrac{n}{2}\sqrt{3}$	$17\sqrt{3}$

25. $\sqrt{11}$ **27.** $\dfrac{-1 + \sqrt{127}}{2}, \dfrac{-1 - \sqrt{127}}{2}$; 5.13, -6.13

	T-Shirts Sold	Total (dollars)
29.	1	20.50
31.	3	21.50

33. $w = 20.00 + 0.50t$ **35.** 58 T-shirts
37. perpendicular

Lesson 13.5

7. $(7, 2)$; 5; 7; about 8.60 **9.** 5 **11.** about 6.40

13.

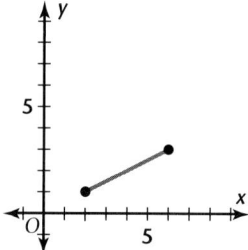

right scalene; the lengths of PQ, QR, and RP are 5, 10, and $5\sqrt{5}$ units, respectively, and $PQ^2 + QR^2 = RP^2$. **15.** Midpoint $(5.5, 5)$
17. Midpoint $(-3, 2)$ **19.** $(-1, 4)$ **21.** $(6, 5)$
23. $M(4, 1)$ **25.** $P(12, 10)$

27.

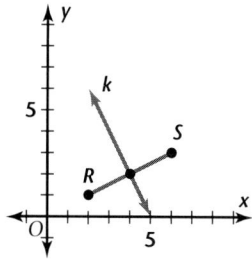

29. slope is $\dfrac{1}{2}$
31. $y = -2x + 10$;

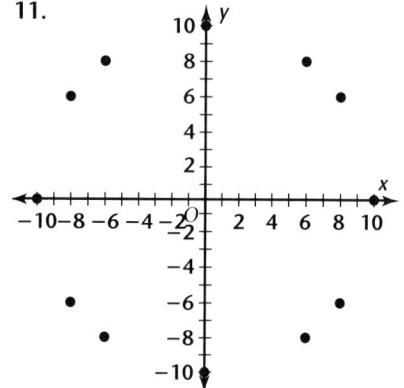

33. a. Let s represent the speed in miles per hour, and d represent the breaking distance at a particular speed in feet. Then $s = 4.26\sqrt{d}$
  b. 1. 42.6 mph
    2. 166.69 feet
  c. According to the table, 36 ft is halfway between 22 and 50 ft. The corresponding

speeds at 20 and 50 ft are 20 and 30 mph, respectively and 25 mph is halfway between the two speed values. Therefore, the passenger was correct. However, by substituting the braking distance into the equation, the speed is 25.56 mph, therefore, the officer was correct. In either case, the driver was incorrect.
35. $(y + 20)(y + 15)$ **37.** $(2, 3)$ **39.** $(1, -4)$
41. $3 + \sqrt{19}, 3 - \sqrt{19}$ **43.** 5

Lesson 13.6

7. $x^2 + y^2 = 49$ **9.** $x^2 + y^2 = r^2$

11.

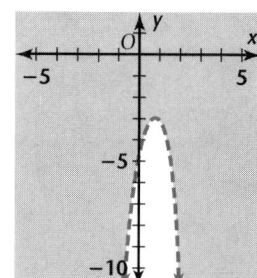

13. $10, -10$
15. $(x - 2)^2 + (y - 3)^2 = 25$
17. $(x - 2)^2 + (y + 3)^2 = 25$
19. $(x - 2)^2 + (y - 3)^2 = 49$; $(x + 2)^2 + (y - 3)^2 = 49$; $(x - 2)^2 + (y + 3)^2 = 49$; $(x + 2)^2 + (y + 3)^2 = 49$ **21.** $(1, 5)$; 3
23. $(2, -4)$; $5\sqrt{2}$ **25–27.** $(3, 4)$; $(7, 2)$; $2\sqrt{5}$ or 4.47 **29.** The length of line segment QR is twice the length of MN. **31.** parallel
33. $\sqrt{x^2 + (y + 4)^2} + \sqrt{x^2 + (y - 4)^2} = 10$
35. -8 **37.** 1.25×10^{-6} **39.** $6b^2 - b - 16$
41. 25

43.

Lesson 13.7

5. 1 **7.** 1.7 **9.** 0.8 **11.** 57.3 **13.** about 60°
15. about 15° **17.** 6.93 meters **19.** 4 centimeters **21.** about 10 inches **23.** 2.55
meters **25.** about 4° **27.** $\frac{14}{3}$ **29.** yes;
$2\frac{1}{2}, -3$ **31.** $x > -4$ and $x < 4$

33. $(7, 7)$ **35.** $\sqrt{13}$

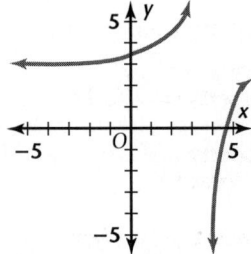

Chapter 14

Lesson 14.1

7. $y = 2$ **9.** $x = 1, x = -5$ **11.** undefined; $\frac{8}{3}$
13. 0.75; 0 **15.** vertical stretch of 2
17. vertical stretch of 0.5, and shift to the left
1.5 units **19.** reflection in the x-axis, vertical
stretch of 2, shift to the right 1 unit and up 4
units
21. $x \neq -2$ **23.** $x \neq 1$

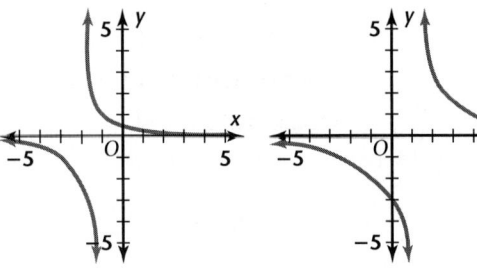

25. $x \neq 4$

27. 3437.5 cubic inches

29.

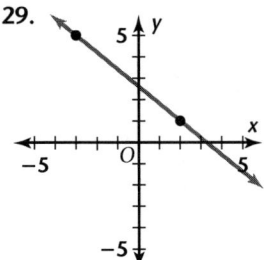

31. 6561 **33.** 4, -5

Lesson 14.2

7. yes **9.** yes **11.** no **13.** no **15.** 3
17. 6 **19.** $\frac{1}{3}$ **21.** 375 rpm **23.** 5 hours
20 minutes **25.** 3.375 hours **27.** $\frac{1}{2}$ the
original length **29.** $5 + 6x$ by $7 + 6x$
31. $(-2, 4)$

Lesson 14.3

7. 5 **9.** 4 **11.** 3, -4 **13.** 0, 2, 3 **15.** 3
17. $r + 3$ **19.** $\frac{a + b}{2(a - b)}, a \neq b$ **21.** $\frac{x - y}{2}$,
$x \neq -y$ **23.** 7, $t \neq -3$ **25.** $\frac{d(3d + 2)}{3d + 1}, d \neq -\frac{1}{3}$
27. $\frac{1}{x + 4}, x \neq 2, -4$ **29.** $\frac{-1}{k + 3}, k \neq 4, -3$
31. $\frac{3}{n - 3}, n \neq 3, 4$ **33.** $\frac{a - b}{a + b}, a \neq -b$
35. 3:8 **37.** $6x^2$ **39.** $29.50
41. $\begin{bmatrix} -9 & -3 \\ 7 & -21 \end{bmatrix}$ **43.** $g(x)$ is obtained from
$f(x)$ by a shift of 2 units to the left and 5 units
down. **45.** $y = -(x - 2)^2 + 3$ **47.** $\frac{-11\sqrt{6}}{3}$

Lesson 14.4

7. $\frac{3}{x + 1}, x \neq -1$ **9.** $\frac{2x - 3}{a - c}, a \neq c$
11. $\frac{-5}{3t}, t \neq 0$ **13.** $\frac{5p - m}{p^2 q}, p^2 q \neq 0$
15. $\frac{x^2 + 3x + 4}{x + 4}, x \neq -4$ **17.** $\frac{ad + bc}{bd}, bd \neq 0$
19. $\frac{x}{x^2 + 4x + 3}, x \neq 0, -1, -2, -3$
21. $\frac{2}{x^2 + x}, x \neq 0, -1$

25. $\dfrac{a^2 + 6a - 1}{a^2 + 4a + 3}$, $a \neq -3, -1$　**27.** $\dfrac{r - r^2}{(r + 5)(r + 1)}$,

$r \neq -1, -5$　**29.** $\dfrac{9y - 8}{y^2 - 4y - 4}$, $y \neq 2$

31. $\dfrac{x - 1}{1 - x}$, $x \neq 1$　**33.** $C = \dfrac{45}{750 + w}$, where w

represents the number of gallons of pure water added.

35. 375 kg　**37.** 20　**39.** $\begin{bmatrix} 4 & -6 \\ 27 & 13 \end{bmatrix}$

41. $\dfrac{1}{16}$　**43.** 2.5×10^{10}

Lesson 14.5

7. $\dfrac{11}{12}$　**9.** $3\dfrac{1}{5}$　**11.** no solution　**13.** 0.5

15. 3　**17.** no solution　**19.** 6　**21.** 3.2

23. $-\dfrac{1}{3}$, -1　**25.** 36 and 20　**27.** 5 meters per

minute　**29.** Tom $\dfrac{x}{5}$, Shelley $\dfrac{x}{7}$, where x

represents the time to paint one car together

31. $\dfrac{10}{7}$

33. (4, 1)

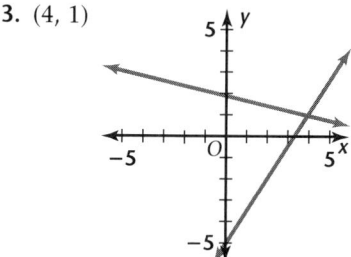

35. $15x^2 - 34x - 16$

Lesson 14.6

7. means: 10, 2; extremes 5, x　**9.** means: 8.5,
2; extremes: 17, x　**11.** means: 5, w; extremes:
3, $2w - 3$　**13.** means: 5, $c - 2$; extremes: 4,
$c + 3$　**15.** 11　**17.** 25　**19.** 3　**21.** 2
23. 2, -6　**25.** 12, -14　**27.** -27　**29.** $-6, 4$
31. 3 cm　**33.** 25 muffins　**35.** 600 miles
37. 20,000　**41.** 2　**43.** $\dfrac{2}{5}$

45. $y^2 - 10y + 25$　**47.** $c^2 + c + \dfrac{1}{4}$

7. Definition of Opposites　**9.** Division
Property of Equality　**11.** Identity for
multiplication　**13.** Commutative Property of
Addition　**15.** Definition of Opposites

17.

$2y + 3y = 5y$	Given
$y(2 + 3) = 5y$	Distributive Property
$y(5) = 5y$	Simplify.
$5y = 5y$	Commutative Property for Multiplication

19.

$(-3a)(-2a) = 6a^2$	Given
$(-3)a \cdot a(-2) = 6a^2$	Commutative Property for Multiplication
$(-3)a^2(-2) = 6a^2$	Definition of a Power
$(-3)(-2)a^2 = 6a^2$	Commutative Property for Multiplication
$6a^2 = 6a^2$	Multiplication of integers

21.

$3(x + 2) = -15$	Given
$3x + 6 = -15$	Distributive Property
$3x + 6 - 6 =$ $-15 - 6$	Subtraction Property of Equality
$3x + 0 = -21$	Property of Opposites
$3x = -21$	Identity for addition
$\dfrac{3x}{3} = \dfrac{-21}{3}$	Division Property of Equality
$x = -7$	Division of integers

23.

$\dfrac{3}{4}m + 8 = -1$	Given
$\dfrac{3}{4}m + 8 - 8 = -1 - 8$	Subtraction Property of Equality
$\dfrac{3}{4}m + 0 = -9$	Property of Opposites
$\left(\dfrac{4}{3}\right)\left(\dfrac{3}{4}m\right) = \left(\dfrac{4}{3}\right)(-9)$	Multiplication Property of Equality
$m = -\dfrac{36}{3}$	Property of Reciprocals
$m = -12$	Division of integers

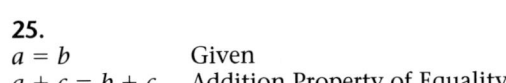

25.

$a = b$	Given
$a + c = b + c$	Addition Property of Equality

27.

$(ax + b) + ay$	Given
$= (ax + ay) + b$	Commutative Property for Addition
$= a(x + y) + b$	Distributive Property

29. If n is odd, then $n = k + 1$, where k is an even number; $n^2 = (k + 1)(k + 1) = k^2 + 2k + 1$ from question 28, if k is even then k^2 is also even. Since $2k$ is also always even, the sum $k^2 + 2k + 1$ must be odd.

31.

$(xy)\left(\dfrac{1}{x}\right) = (yx)\left(\dfrac{1}{x}\right)$	Commutative Property for Multiplication
$= y(x)\left(\dfrac{1}{x}\right)$	Associative Property for Multiplication
$= y(1)$	Property of Reciprocals
$= y$	Identity for Multiplication

33. about 20.3 cm **35.** $y = -\dfrac{3}{2}x + 3$

37. Stretch vertically by a factor 2, reflect through the x-axis, shift 1 unit to the right and 3 units up **39.** $-2, 4$ **41.** $(0, 5)$